T5-BAO-819

COMPUTERS
AN INTRODUCTION TO HARDWARE AND SOFTWARE DESIGN

McGraw-Hill Series in Electrical Engineering

Consulting Editor

Stephen W. Director, *Carnegie-Mellon University*

Circuits and Systems
Communications and Signal Processing
Control Theory
Electronics and Electronic Circuits
Power and Energy
Electromagnetics
Computer Engineering
Introductory
Radar and Antennas
VLSI

Previous Consulting Editors

Ronald N. Bracewell, Colin Cherry, James F. Gibbons, Willis W. Harman, Hubert Heffner, Edward W. Herold, John G. Linvill, Simon Ramo, Ronald A. Rohrer, Anthony E. Siegman, Charles Susskind, Frederick E. Terman, John G. Truxal, Ernst Weber, and John R. Whinnery

Computer Engineering

Consulting Editor

Stephen W. Director, *Carnegie-Mellon University*

Bartee: *Computer Architecture and Logic Design*
Bell and Newell: *Computer Structures: Readings and Examples*
Garland: *Introduction to Microprocessor System Design*
Gault and Pimmel: *Introduction to Microcomputer-Based Digital Systems*
Givone: *Introduction to Switching Circuit Theory*
Givone and Roesser: *Microprocessors/Microcomputers: Introduction*
Hamacher, Vranesic, and Zaky: *Computer Organization*
Hayes: *Computer Organization and Architecture*
Kohavi: *Switching and Finite Automata Theory*
Lawrence-Mauch: *Real-Time Microcomputer System Design: An Introduction*
Levine: *Vision in Man and Machine*
Peatman: *Design of Digital Systems*
Peatman: *Design with Microcontrollers*
Peatman: *Digital Hardware Design*
Ritterman: *Computer Circuit Concepts*
Rosen: *Discrete Mathematics and Its Applications*
Sandige: *Modern Digital Design*
Sze: *VLSI Technology*
Taub: *Digital Circuits and Microprocessors*
Wear, Pinkert, Wear, and Lane: *Computers: An Introduction to Hardware and Software Design*
Wiatrowski and House: *Logic Circuits and Microcomputer Systems*

COMPUTERS

AN INTRODUCTION TO HARDWARE AND SOFTWARE DESIGN

LARRY L. WEAR
JAMES R. PINKERT
LARRY C. WEAR
WILLIAM G. LANE

McGraw-Hill, Inc.

New York St. Louis San Francisco Auckland Bogotá Caracas
Hamburg Lisbon London Madrid Mexico Milan Montreal New Delhi
Paris San Juan São Paulo Singapore Sydney Tokyo Toronto

To Harold G. Doty, a gentleman, an officer, and an engineer.

Larry L. Wear

To all of my wonderful friends at Sierra Nevada.

James R. Pinkert

This book was set in Times Roman.
The editors were Roger L. Howell and David A. Damstra;
the designer was Wanda Siedlecka.
R. R. Donnelley & Sons Company was printer and binder.

COMPUTERS
An Introduction to Hardware and Software Design

2 3 4 5 6 7 8 9 0 DOC DOC 9 0 9 8 7 6 5 4 3 2 1

ISBN 0-07-068674-2

Library of Congress Cataloging-in-Publication Data

Computers: an introduction to hardware and software design / Larry L. Wear ... [et al.].
p. cm. — (McGraw-Hill series in electrical engineering.
Computer engineering)
Includes bibliographical references and index.
ISBN 0-07-068674-2
1. Computer engineering. 2. Software engineering. I. Wear, Larry L. II. Series.
TK7885. C653 1991
004 — dc20 90-29073

About the Authors

Larry L. Wear is a professor of computer engineering at California State University, Chico and is responsible for developing engineering education classes at Tandem Computers. He has been a consultant for several computer companies including Hewlett-Packard and IBM.

Dr. Wear received bachelor's and master's degrees in electrical engineering from the University of Washington and a master's degree in applied mathematics and a Ph.D. in electrical engineering from the University of Santa Clara.

Professor Wear was instrumental in establishing thc curriculum for the computer engineering program at CSU, Chico. His teaching and research areas include microprocessor system design, operating systems, and fault-tolerant system design. He has published articles on microprogramming, system simulation, and the user interface.

James R. Pinkert is a professor of computer engineering at California State University, Chico. He holds bachelor's degrees in electrical engineering and philosophy, master's degrees in accounting and computer science, and a Ph.D. degree in computer science from the University of Wisconsin. Professor Pinkert has been elected to five national honorary societies: Tau Beta Pi (engineering), Upsilon Pi Epsilon (computer science), Beta Alpha Psi (accounting), Beta Gamma Sigma (business), and Sigma Xi (scientific research).

He has authored articles in journals and proceedings, both in business and in computer science. Professor Pinkert coauthored the book *Operating Systems: Concepts, Policies, and Mechanisms* with Larry L. Wear. He is also the recipient of a grant to design and build a computer system devoted to high-speed data encryption with public key algorithms.

Larry C. Wear is an engineer working on the development of computer-aided design tools at Tandem Computers in Cupertino, California. He holds a bachelor's degree from the University of California, Davis. He graduated with honors and has been elected to the Tau Beta Pi (engineering) and Pi Mu Epsilon (mathematics) honorary societies. Larry is currently a graduate student in electrical engineering at Stanford University.

William G. Lane is a professor of computer engineering at California State University, Chico. He founded and chaired the computer science department at Chico and also served as dean of the School of Engineering and Computer Science. He is currently a consultant for the government and private industry.

Dr. Lane received a bachelor's degree from Stanford University, a master's degree from the University of Southern California, and a Ph.D. from the University of California, Davis. All of his degrees are in electrical engineering. He is also a registered professional engineer in the state of California.

Professor Lane's teaching specialities include computer architecture and system development methodologies. He recently received a national award as an outstanding student advisor.

Contents

Part II
AN OVERVIEW OF COMPUTER HARDWARE

Preface

In recent years, the computer industry has seen far-reaching changes in the organizational structures of its companies and the job responsibilities of their employees. Many computer firms have created barriers in their organizational structures between hardware engineers and programmers. Companies that produce only software have grown rapidly in size and number, and employees of these firms have become quite far removed from the hardware design process.

Educational institutions have followed the industry trend toward separation of hardware and software design functions. They have developed computer curricula that stressed *either* hardware *or* software, with little or no crossover. Fostered by both education and industry, the hardware/software gap has grown steadily. With that gap has come corresponding problems.

On too many design projects, neither the hardware nor the software group fully comprehends what the other group is doing. Such a lack of mutual comprehension can lead to problems in both the resulting hardware and software. Furthermore, as Brooks points out so well in his *Mythical Man-Month*, the time spent in communication can become inordinately large.

A second problem in having hardware or software *exclusivists* has surfaced with the rapid advancement of microcomputers. These tiny electronic marvels have presented a new and formidable challenge to designers. Microcomputers have made it possible for companies to put the power of computers into a wide spectrum of consumer products. Splitting into hardware and software groups might have seemed economically feasible for design teams on large computer systems, but when it comes to incorporating a microprocessor into a relatively inexpensive product such as a dishwasher or radio, dedicating separate teams to hardware and software often results in unnecessary duplication of effort.

What is needed is a group of people knowledgeable in both hardware and software. By having such professionals on design teams for large projects, the total number of people working on the project can be minimized. Problems from the lack of mutual comprehension can be reduced and communication delays can be lessened. Small projects can also benefit. One individual or a few general-purpose people can efficiently complete all the design tasks for projects that include microcomputers.

In summary, now more than ever before there is a demand for a group of computer professionals who are familiar with all aspects of computer systems. Our goal in this text is to give such individuals a broad introduction to the varied aspects of computing, including history, theory, hardware, software, tools, interfaces, reliability, and system design. We want to start preparing people to become multitalented computer *generalists.*

Most professional disciplines have developed rigorous standards for educational programs in their fields and have empowered boards to evaluate specific academic programs. When such a board determines that a given program meets established standards, that program is **accredited**. The Accrediting Board for Engineering and Technology (ABET), which establishes the criteria for computer and similarly named engineering programs, has specified the following requirements for these programs:

> The curriculum structure must provide breadth across the field of computer science and engineering, both hardware ad software.... The curriculum must provide a balanced view of hardware, software, hardware-software tradeoffs, and basic modeling techniques used to represent the computing process. Normally this is addressed by an integrated sequence including algorithms, data structures, digital systems, computer organization, interfacing, architecture, software engineering, and operating systems.

We believe that this text provides an excellent introduction to computing that follows these overall program guidelines closely. It provides "breadth across the field of computer science and engineering, both hardware and software." It also provides "a balanced view of hardware, software, hardware-software tradeoffs, and basic modelling techniques used to represent the computing process." Finally, we introduce students to "digital systems, computer organization, interfacing, architecture, software engineering, and operating systems."

This text is primarily directed toward an audience of computer science and engineering students in their freshman or sophomore year. We assume that these students have had one or more courses introducing them to programming in a high-level language. That knowledge of software is necessary to enhance the student's understanding of the material in Part III. No background in hardware analysis or design is assumed or necessary to understand the material in this text.

There are several possible ways to incorporate this book into a course or to use it for personal educational goals. The first is as the text for an introductory course in computer engineering. It can also be used in conjunction with an assembly-language reference manual for an assembly-language programming class. Since this book covers a wide variety of topics, it is also appropriate for a survey class in computing for majors in other engineering or science disciplines. Lastly, the broad coverage with minimal prerequisites makes it an appropriate reference for people who are just interested in a good overall understanding of computers.

Covering such a diversity of areas, without being superficial, has been a monumental task. During early development, this text grew to encyclopedic proportions, sufficient for two or three volumes and two or three semesters. It subsequently went through numerous refinements and class testings and was finally pared down to its current size.

The text is organized into four parts. The first part is intended to provide background information. Part II, An Overview of Computer Hardware, contains detailed descriptions of the hardware components of a computer system. In Part III we look at the flip side of the coin, computer software. The last part draws together the previous parts, covering user interfaces, hardware and software reliability, and system design.

Even with all the reductions and refinements, the book contains 20 chapters. We are definitely *not* implying that all these chapters could be covered completely in a single quarter or even a semester. We thought it would be better to provide a somewhat more comprehensive coverage and allow instructors to tailor material to their individual curricula.

Various versions of this text have been class-tested. We have found that approximately 15 chapters seem to be a reasonable amount for a three-unit, one-semester introductory class in computer engineering. We suggest that such a class include Parts I and IV in their entirety and selections from Parts II and III. Chapter 5, the control unit, and Chapter 11, hardware design tools, are a good sampling from Part II. Chapter 12, a (brief) introduction to operating systems, Chapter 16, software applications, and Chapter 17, software tools, give a good sampling from Part III. Other chapters from these parts can be selected at the discretion of the instructor.

The remaining chapters in Part II are probably ranked as follows in terms of increasing depth and difficulty: Chapter 10 (controllers and devices), Chapter 7 (information transfer), Chapter 6 (memory), Chapter 9 (external device communication), and Chapter 8 (the arithmetic and logic unit). There is no implicit ordering required. A few isolated topics (such as tristate gates) are introduced in one chapter and referenced in others. However, these could be covered separately as necessary.

Similarly, the remaining chapters in Part III are probably ranked as follows in terms of increasing depth and difficulty: Chapter 15 (programming languages), Chapter 13 (components of the operating system), and Chapter 14 (low-level system software and layering). Chapter 13 should be covered before Chapter 14 so that the students can better understand the layering example in Chapter 14.

Chapter 20 is a good "buffer" for the end of the semester. The fraction covered, the depth of coverage, and the amount of class time spent on it are all quite flexible.

The instructor of an assembly-language class has to divide available time between programming and computer architecture. A three-unit, one-semester assembly class might be set up as 2 hours per week lecture (two units) and 2 hours per week programming lab (one unit). This text would be used for the lecture, and an assembly language manual for the lab.

Chapter 3, information representation, would be a good place to start in such a class. Chapters 4 through 10 should all be included, although not necessarily in sequential order. The control unit and the arithmetic unit are most immediately applicable to assembly-language programming, so Chapters 6 and 8 might be covered early.

Assembly-language classes also typically cover subprogram linkage and operating system calls. Therefore Chapters 12, 13, and 14 are probably appropriate. This would give 11 chapters total, which is more than enough when taken with the programming load.

For a survey class, instructors can pick and choose at their discretion. We would encourage complete coverage of Chapters 2, 3, and 4 and Part IV, since we think they are crucial for such a class.

We hope that this book will be both interesting and stimulating. If we succeed in our goals, we will have provided motivation for continued studies in computer hardware and software. Also, we will have provided a good foundation upon which to base those studies.

McGraw-Hill and the authors would like to thank the following reviewers for their many helpful comments and suggestions: James Archer, Texas Technical University; John Bennett, Rice University; Lee Coraor, Pennsylvania State University; Dong Ha, Virginia Polytechnic University; David Johnson, University of Washington; and John Kleitch, Iowa State University.

The authors encourage your comments on this text (and any corrections to it). Please address comments and corrections to Larry L. Wear at E-mail address wear@ecst.csuchico.edu, or send them to the publisher.

LARRY L. WEAR
JAMES R. PINKERT
LARRY C. WEAR
WILLIAM G. LANE

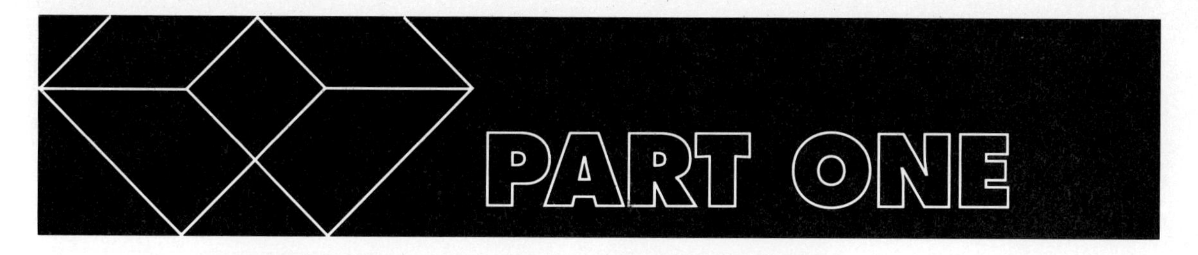

HISTORY AND CONCEPTS

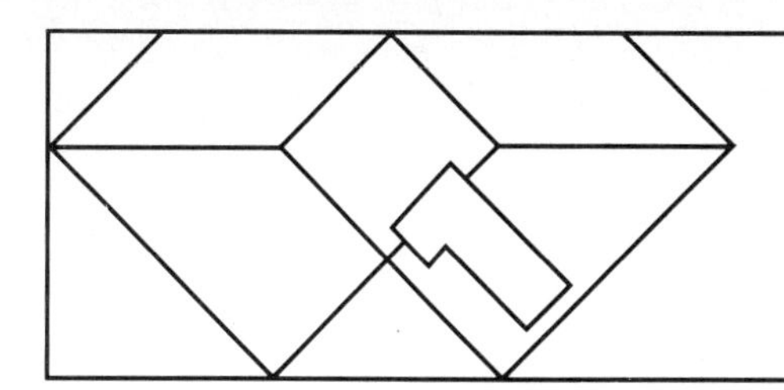

Introduction

Once your friends and relatives discover that you are studying computing, you can almost be guaranteed that you will soon be asked, "What kind of computer should I buy?" How do you answer this question?

There are probably a number of different systems that might be appropriate, but which would be the "best" system for your friend or relative? For example, would it be the system that is the least expensive? Or the one with the most features? Or the easiest to use? Or ... ?

These are important questions, and they are not always easy to answer. One big problem is that different people have different values and needs. Let's take a closer look:

- Least expensive. To many people, least expensive would mean lowest purchase cost. However, if you consider performance or operating expense, the least expensive system might not necessarily result in the lowest total cost. The least expensive systems are frequently the oldest models and may soon be incompatible with many of the new software releases. Investing a little more in the beginning may save a lot later.
- Feature selection. Which system features to include now, and which to buy later, is always a consideration. There is no standard configuration that is appropriate for all computers, and the selection of which ones to include not only affects the price but also the performance. Most people can decide whether to buy a color display, but they might or might not know whether they should buy a larger keyboard, or a mouse, or a hard disk. The size of memory and the kind of operating system to buy are examples of decisions that require more specific technical knowledge.
- Easiest to use. This is an even more difficult choice. Easiest for whom? A system that is easy for a computer professional may be extremely difficult, if not impossible, for a beginner. Conversely, a system that is frustrating for a professional may be "easy" for a novice.

We will find later that none of these questions can really be answered without also knowing the types of programs that are likely to be required, the interests and capabilities of potential users, and many other factors.

A few paragraphs ago, we said that recommending a computer system to a friend is one of the more common requests that you will receive in your professional career. It is also one of the simplest. In the industrial world, the task could just as easily have been to select a system for air traffic control, for control of a nuclear reactor, for forecasting global weather, or for any number of other large applications. (You would not likely be asked to participate in the design of such a system at this point, but it is definitely not out of the question after you graduate.)

Applications that were impossible a few years ago are commonplace now. Large-computer capabilities have migrated to workstations, and supercomputer capabilities are becoming available on systems that sell for a small fraction of the price of last year's supercomputer. The only limitation seems to be our ability to describe new applications and develop systems with sufficient capacity and speed to process the corresponding programs in the time needed.

1.1 THE TOTAL COMPUTING ENVIRONMENT

The development and use of computers is often described as a technological revolution in the way that the development of manufacturing processes was an industrial revolution. Both these revolutions have been the result of major advances, and both have significantly changed the way that large numbers of people work.

Goldstine, in his history of the early years in the development of computers, suggested some conditions that are necessary for progress:

> The development of radical new machines always comes about because some inspired person sees how to adapt a new technology ... and thus to make a major advance in the state of the art. It is usually the convergence of two very different concepts. One is technology ... the other is the recognition of the importance and necessity for the advance. [Goldstine, 1972]

Applying this concept to the computing environment, we can say that these conditions translate to a convergence of the three major technological areas shown as circles in Fig. 1.1:

- The identification of need, characterized by the applications for which the computer is used—the User applications (1)
- The availability of a physical technology that will provide an effective vehicle for the applications—the Computer hardware (2)
- The development of a procedure for using the technology to implement the applications—the Computer software (3)

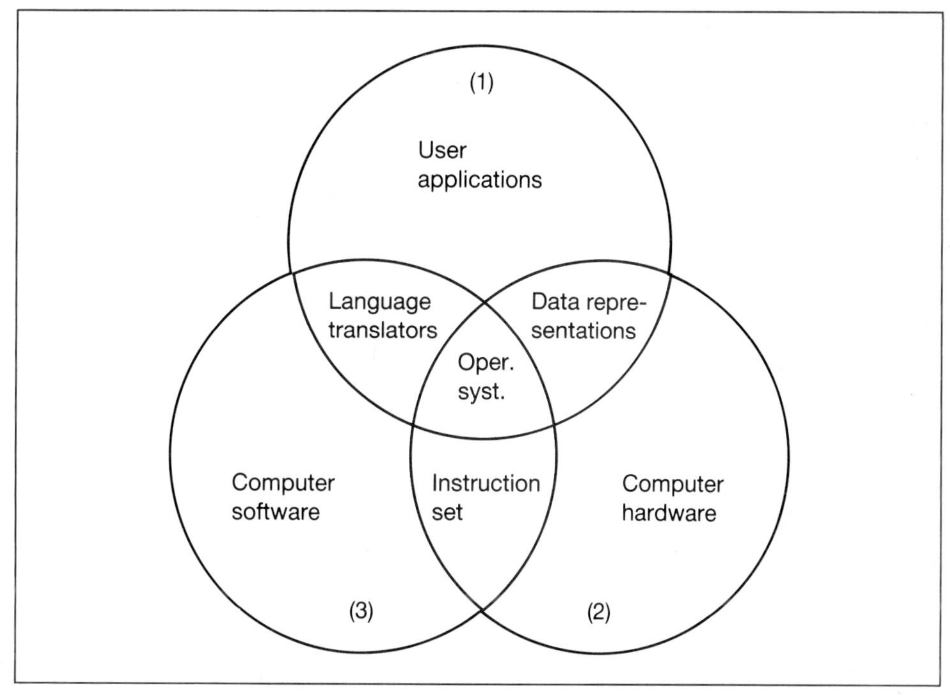

FIGURE 1.1 The convergence of need and two major computing technologies.

The three lens-shaped areas formed by the overlap of each pair of circles in the figure represent the interfaces between the major technological areas and constitute technological areas in their own right:

- The language translators (e.g., for Pascal, C, BASIC, Fortran) provide a way for the programmer to implement the application. They are the interface between the user and the system. The translator for the given language converts the programmer's statements into information that can be understood by the computer.
- The computer's instruction set represents the interface between the software formulation of the application and the computer hardware. The computer uses these instructions to define the sequence of operations it will execute for the given program.
- The data representations form the interface between the application and the computer. They convey various information, such as the types and ranges of numbers that will be used.

The three-sided area that is formed by the overlap of all three circles represents the operating system. The operating system coordinates program interaction, manages the computer's various hardware and software resources, and does the input/output operations.

We will explore all these topics later in the book. But before we do, we need to put a few things into perspective. Computers are so common and new advances are now being announced so often that we tend to forget it was not always so. The entire computer industry is less than half a century old, but it is built on a foundation of technology that required thousands of years to develop.

Computers would have been impossible without a language to describe what we are doing, number systems to support our computations, and mathematics to describe our computational procedures. All these took many, many years to develop, and each provides real examples of the premise that advances in technology often result from the merging of a variety of diverse concepts. Conversely, an advance can be inhibited by the lack of a suitable conceptual foundation or defined need.

1.2 THE EVOLUTION OF LANGUAGE AND MATHEMATICS

Computing systems, like any other advanced technology, must be describable both in terms of how they are to be built and how they are to be used. To do this, we must have:

- A language that is flexible and yet precise enough to allow formulation of the applications to be processed
- A mathematics that will explain the theories upon which these applications are based
- A number system that will support calculation of all the values that will be encountered

Each is a sophisticated technology in its own right. It is appropriate to spend at least a little time to see how they evolved and how the merger of specific key concepts affected their development.[1]

1.2.1 Languages and Alphabets

Early peoples developed oral languages to speak with other members of their societies. As early as 3000 B.C. they drew pictures to describe events and happenings. During the thousand-year period between 3000 and 2000 B.C., the pictures were simplified to represent specific ideas or concepts. These pictures were a large step in

[1] These historical developments are described in greater detail in Encyclopaedia Britannica's *Micropaedia* and *Macropaedia* and in Britannica's *Great Books of the Western World.* Dates for some events are identified in Grun, 1975, and in Ifrah, 1985.

the development of written language, but the approach was self-limiting. Every time a new word or a new concept was developed, a new picture or set of pictures had to be created and learned. If written communication were to flourish, a more efficient method would be needed to represent words and concepts.

An alphabet that represented the unique sounds of a language was the key tool needed. The first true alphabet appeared about 1700 B.C. on the eastern shores of the Mediterranean. We do not know exactly how it evolved, but maybe scholars of the time had seen the limits of picture writing. They had begun to look for some other characteristics of language. Some inspired people must have realized that all spoken words contained distinct, identifiable sounds. They saw that the total set of all such sounds in all the words of the language was small.

The breakthrough was *the use of a separate symbol that had no other meaning for each sound in the language.* This has been called the essential single step in the evolution of writing [Encyclopedia Britannica, 1985]. The symbols became the letters of the alphabet, and the alphabet made it easier to record information. New words could be added to the vocabulary, abstract ideas could be formulated, and it all could be done without increasing the number of "sound symbols" that had to be learned.

The alphabet might have been humanity's first real try at a "standard." Abstract symbols had to be defined and agreed upon for each sound. Then the people had to learn to use them. It probably took many years, and it was probably met with great resistance.

A few different alphabetic forms developed in different geographic areas, but there are far fewer alphabets than languages. It was easier to assign existing letter forms to language sounds instead of creating entirely new forms. The choice of which alphabet to use was also simple. It was imposed by conquering armies, brought by traders, or taught by the church.

The western, or Roman, alphabet is an example of this process. Its evolution followed the sequence shown in Fig. 1.2. Because of the Roman conquest, Latin became the language of scholars, traders, and churchmen. And Latin is based on the Roman alphabet. So naturally, the Roman alphabet was adopted for the written form of languages spoken by conquered peoples.

Speech, however, is more than just a series of spoken words. Meanings are changed by pauses and inflections. Early writings contained no corresponding punctuation and no consistent spacings. Hence, they were difficult to read. An old cartoon provides a classic example. It showed an archaeologist puzzling over the words carved on a post, like the one shown in Fig. 1.3, which was supposedly found outside the Roman Colosseum. The caption read, "It may look like Latin, but it really says 'To Tie Mules To'."

Punctuation began in the fifth century B.C. with two or three vertical dots separating phrases. Critical signs for breathing pauses and word accents were developed about 200 B.C. However, it was not until A.D. 1566 that people began to understand that *clarity of grammar was the real purpose for punctuation.* Another 100 years were needed to define punctuation symbols, to use spaces to separate words, and to use indentation for new paragraphs.

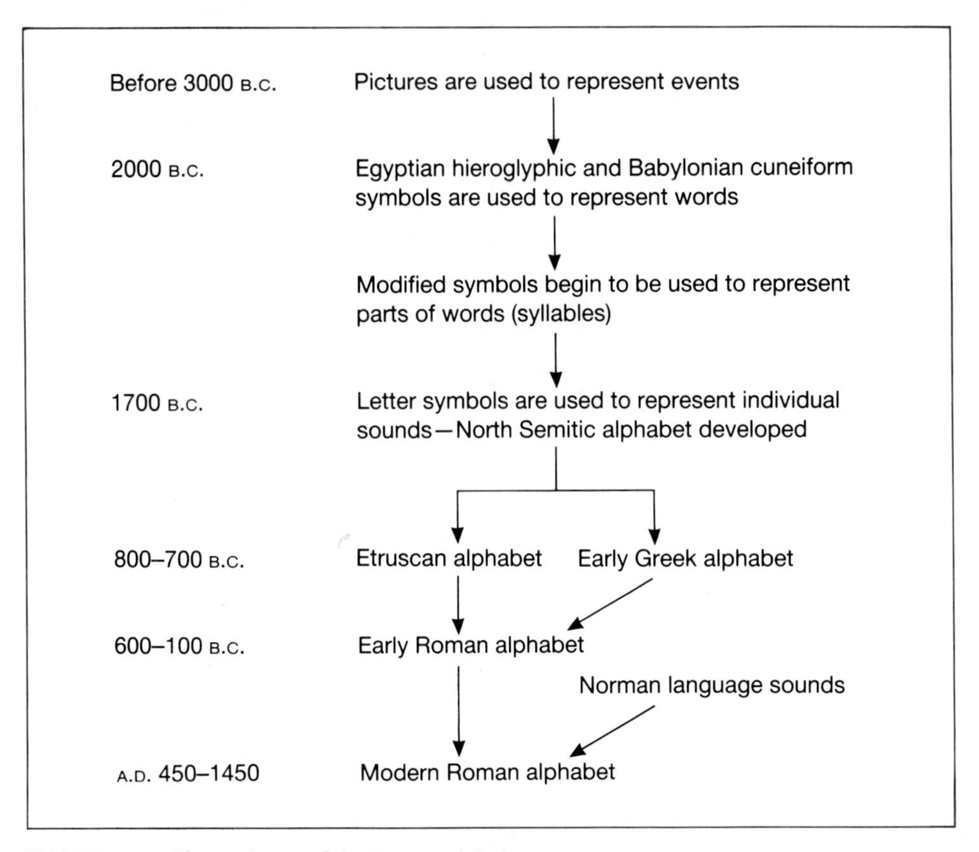

FIGURE 1.2 The evolution of the Roman alphabet.

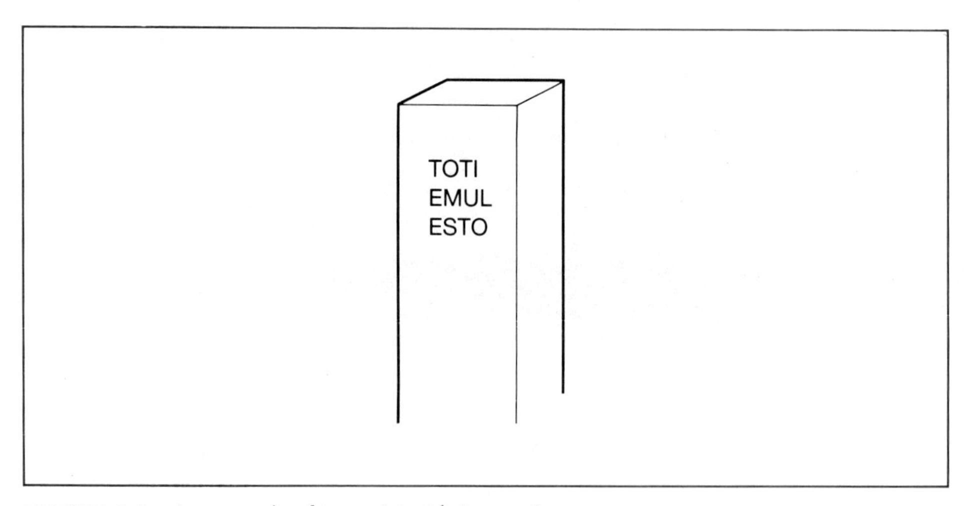

FIGURE 1.3 An example of inconsistent letterspacing.

1.2.2 Numerals and Number Systems

The development of number systems was somewhat similar to the development of the alphabet. Most early societies had some way to express tallies and quantities. Five sheep pictures, for example, could have depicted five sheep. After number symbols were defined, this quantity could also be shown as a five symbol followed by a sheep picture.

Climate and the source of food supply probably had a large influence on the type of number system that was ultimately developed. Early peoples apparently used their fingers, and sometimes their toes, for counting. The duodecimal system (base 20) was common among some people who lived in the tropics. The decimal system (base 10) was developed by people who needed foot covering much of the year. Some bow- and spear-carrying peoples developed a quinary (base 5) system, probably because they normally had only one hand free.

Whatever the circumstance, variations of the decimal system became the most common. Some societies had symbols for specific numbers (such as 1, 5, 10, 50, 100, 500, 1000, ...), and others had symbols for each power of the base of the system (such as 1, 10, 100, 1000, ...). Some included the concept of zero.

Symbols could be combined to represent specific quantities, but the order of their combination followed no consistent rules. None of the early symbolic systems included all the concepts (the *critical set*) necessary for straightforward mathematical computation.

The key here was *the development of a suitable place-value system.* Separate symbols were needed for each possible digit value (including zero), and other symbols were needed to separate the whole part of the number from the fractional part. The digit symbols of the number system are similar to the letters of the alphabet. The point marks and commas we use to separate parts of our numbers are similar to punctuation.

In place-value systems, any digit symbol may appear in any digit position of a number. Furthermore, each digit symbol has a weighted value that is determined by its location. Zeros are used to show no value for that position, and alignment of punctuation ensures alignment of numbers for calculation.

The development and acceptance of our modern Hindu-Arabic number system is shown in Table 1.1. This system represented a major advance for science and mathematics, but its adoption was delayed in Europe. The Roman number system had been adopted, and the power to change rested with the medieval Christian church, not with the governments. Scientific publication was considered to be a challenge to church authority, and the new decimal system had been brought to the west by Moslem scholars.

Leonardo Fibonacci of Pisa grew up in a merchant colony in north Africa and studied mathematics under an Arab master. He is credited with introducing decimal numbers into Europe. His *Liber Abasi*, the first European book on Arabian mathematics, explained the use of the Hindu-Arabic place-value system. The book gave examples of possible commercial applications, but most people continued to use the Roman system. The turning point was Leibniz' generalized treatise on positional

TABLE 1.1 EVOLUTION OF THE DECIMAL NUMBER SYSTEM

3300–2850 B.C.	Egyptian hieroglyphic numerals developed
2700 B.C.	Cuneiform numerals developed
1700 B.C.	Babylonian place-value system developed
250 B.C.	First known appearance of zero, in the Babylonian place-value system
200 B.C.–A.D. 200	Chinese place-value decimal system developed, without zero
A.D. 458	Earliest known use of Sanskrit numerals with an understanding of place values and zero
595	First known written example of the Indian (Hindu) decimal numeration
800	Indian decimal place-value system with zero adopted by the Arabic peoples, becomes the Hindu-Arabic system
1202	Fibonacci introduces Hindu-Arabic decimal system into Europe
1400–1500	Hindu-Arabic system becomes common in Europe
1666	Leibniz publishes a generalized treatise on positional number systems
1679	Leibniz perfects the binary number system

number systems[1] in 1666. Thirteen years later, Leibniz defined the binary system, which later proved to be so important in computer systems.

1.2.3 The Development of Mathematics

The history of mathematics can be separated into the three distinct periods shown in Fig. 1.4. The first of those three periods preceded development of the alphabet. Methods for measurement had been developed and elementary geometric relationships had been defined. "Pythagorean" knotted-rope triangles were being used by Egyptian pyramid builders to lay out right angles in the third millennium B.C. By 1700 B.C., Babylonian astronomers had begun making precise measurements of planetary motion. This early work in construction and astronomy led to the definition of place-value number systems (base 10 or 60). These systems were capable of expressing both fractions and large numbers.

Number systems proved to be the key difference between Babylonian and Egyptian mathematical advancement. The Egyptian (and later Roman) number systems were not well-suited to calculation. Their mathematics never progressed beyond measurements and elementary arithmetic.

The second phase in the evolution of mathematics is known as the Hellenic, or Greek, period. Its foundation was the early work and traditions of the Babylonian astronomers, but the Greek mathematicians had an advantage over the Baby-

[1] The first generalized treatment of positional number systems was done by Thomas Harriot (1560–1621), but his work was never published. Credit is given to Leibniz, who developed it independently.

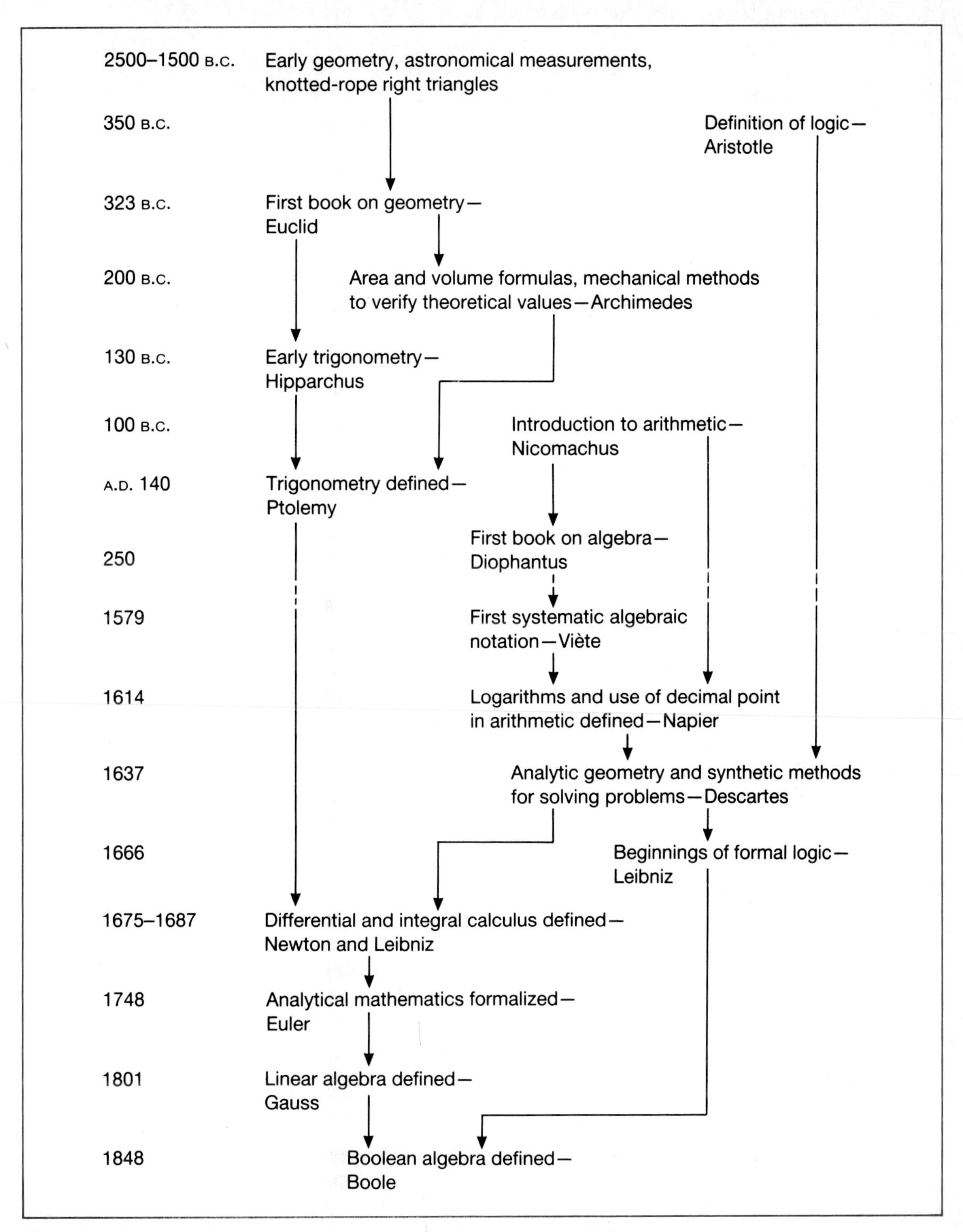

FIGURE 1.4 The evolution of mathematics leading to boolean algebra.

lonians. They not only had a number system that would support their work, they also had an alphabet to record and communicate their results. Algebra, geometry, and trigonometry were all defined during that period.

The third, or modern phase, began about A.D. 1600 after the decimal number system had been developed and introduced into Europe. Scientific observation and measurement had become common, and a systematic method for algebraic notation was in place. The use of letters for unknowns and for general parameters in equations had been introduced. However, mathematics and science were still separate areas of study. The real breakthrough in this phase of mathematics was Galileo's discovery in 1600 that *mathematics could be used to formulate and explain science.*

With this discovery, mathematics had a new purpose and reason for being, and advances came rapidly. Calculus was defined, logic began to be formalized, and analytical methods were developed. By the middle of the nineteenth century, people had developed much of the mathematics that would be needed to support the computer revolution. They had basic arithmetic operations *plus* square root, integration, differentiation, and summation *plus* the constants e and pi. Mathematics had evolved into a design tool.

1.2.4 So What Does It All Mean?

Having just spent several pages reviewing the evolution of language, number systems, and mathematics, we might ask, "What does it all mean?" and "Are there any lessons to be learned?"

We began Sec. 1.1 with a quotation from Goldstine that defined two requirements for a major advance in the state of the art:

- There must be an identified need or problem.
- Technologies for a proper solution must either be available or be easy to develop.

In reviewing the development of language and mathematics, we have suggested that certain key concepts always preceded each major advance.

Sometimes the key is to use a familiar technology in a different way. Alphabets came after people made an important discovery. They could use a short series of abstract symbols for the sounds in a word instead of a picture that represented the word's meaning. Sounds in a language could be limited without limiting the number of words.

The alphabet used a familiar technology (marks on a medium such as sand or rock). However, it used this technology in a radically different way (to represent sounds instead of ideas or things). In doing so, alphabets allowed new words to be added without the need to add new symbols. In addition, they allowed the written language to more closely represent the spoken.

An advance can also be the result of the merger of several existing concepts. The key to the development of the number system was the critical set formed by the merger of:

- A place-value notation
- The assignment of an independent symbol for each digit value
- The use of zero to represent a digit that has no value
- The use of a point mark to separate the parts of a number that are greater and less than 1

Each of these had existed in some combination in other number systems, but until they were used together as a complete set, their power was limited.

Lastly, we saw that advances could be delayed because technology was not available or the problems were not seen as important. For instance, the development of mathematics was interrupted for two long periods. The first gap was from about 1500 to 350 B.C. The alphabet had not yet been developed, and there was no easy way to express more than the simplest of mathematical concepts. The second gap was from about A.D. 300 to 1600. The church and the state saw no need for scientific investigation, and the decimal number system was generally unknown in Europe.

1.3 THE HISTORICAL APPROACH TO COMPUTING SYSTEM DEVELOPMENT

In the previous two sections we looked at a global view of computing. We described the development of three related areas: languages, number systems, and mathematics. We turn now to the development of computers per se, focusing especially on the split between the two aspects of a computer—hardware and software.

1.3.1 The Historical Split between Hardware and Software

During the last four decades we have seen the computer industry grow from economic and social insignificance into a trillion-dollar industry. In one way or another it affects all our lives. This tremendous, rapid growth has led to major changes in the organization of computer firms. It has also led to changes in the duties of employees.

The first computers were primarily built by engineers and scientists as tools to do complex numerical computations quickly and accurately. Often, the people who built the computers were also the ones who programmed them. Such an approach seemed to make sense. The engineers and scientists who designed the hardware were usually familiar with the intended application areas. Therefore, they could also write the software.

Soon, however, the use of computers expanded into other areas such as business data processing. Hardware designers were normally not trained in business, and a separate new profession was created—the computer programmer. Programmers could write applications software for many different areas. They worked not only on very specific tasks but also on improving the usefulness of computers. For instance, they developed more comprehensive and easier-to-use programming tools. Their jobs called for them to *use* the hardware resources of the computers. However, programmers typically did not have the opportunity to learn in detail how the hardware was *designed*.

Simultaneously, the people designing and building the computers were fully occupied trying to develop faster and more powerful hardware. This hardware was needed to support the demands of the new users and the new applications. Hardware designers began to have less and less knowledge about the application programs written to run on the computers they were creating.

Most computer companies reinforced this split between the hardware designer and the programmers by adopting organizational structures that included separate hardware and software groups. A typical organization chart for such a firm is shown in Fig. 1.5. As you can see from the chart, people designing the software and hardware (such as Joe and Sally) were separated by many layers of management.

Why did the companies do this? One answer is that computer hardware and software had become larger and more complex. Splitting hardware and software development appeared to be the only choice. It just seemed unreasonable to expect everyone to learn enough about both hardware and software to work on all aspects of the system.

The separation widened in the 1960s and 1970s with the growth of the new computer software industry. Companies that neither designed nor built hardware found ready markets for software that made effective use of other manufacturers' hardware.

The gulf between hardware and software, established by industry, was fostered by our educational system. Colleges and universities tried to reflect the perceived needs of industry. People interested in software studied computer science, while those interested in hardware studied electrical engineering. Some computer science students took a course or two in electrical engineering. Some electrical engineering students took a course or two in computer programming. However, the crossover was minimal.

1.3.2 Pitfalls with the Split Approach

A major drawback of separating the hardware and software designers is that neither group then understands fully what the other group is doing. This lack of mutual comprehension can often lead to two problems. The hardware group designs computer systems that do not really match the needs of the users. The software group designs programs that don't take full advantage of the hardware.

Such a mismatch can manifest itself in several ways. The hardware designers might spend extra effort creating special machine instructions without ever realizing

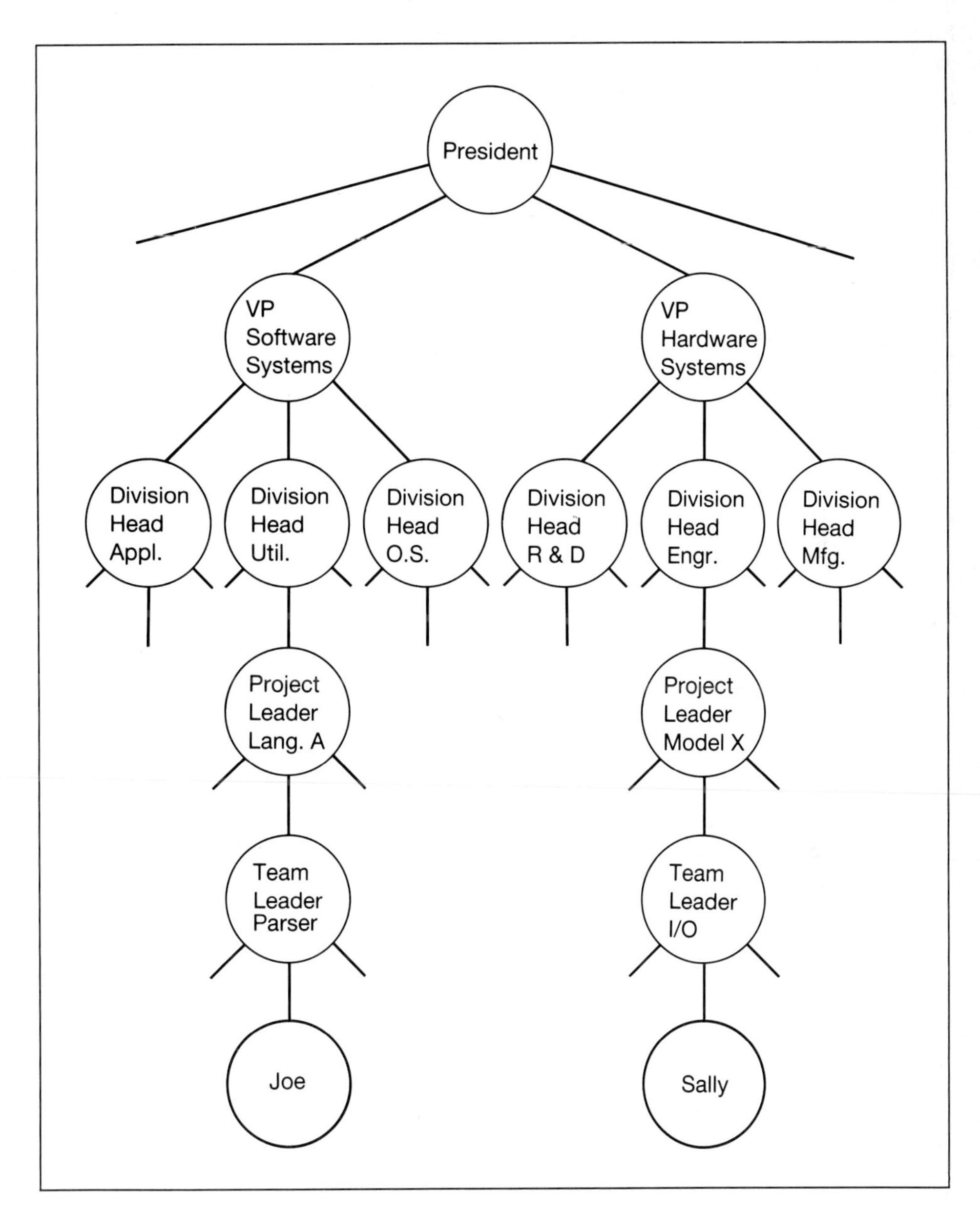

FIGURE 1.5 A sample organization chart for a computer firm.

that the application does not need them. Or, these designers might omit or limit some needed feature in the system. For example, a computer might be needed for process control, but the hardware might not have any convenient way to be connected to the system it is supposed to control.

Sometimes, the hardware designers had no choice in the matter. They had to work within the constraints of those who specified the systems. They had no direct way of determining exactly what was needed.

The software designers were often faced with a different problem. As machines grew more complex, programmers had difficulty deciding how to make effective use of all the hardware features. For example, on some machines there are several different ways to do operations such as data moves. Suppose the programmer does not understand *exactly* how the hardware executes every option for such a move. He or she may make a choice that is inefficient for a particular application. Without knowing the details of the hardware, it is difficult to make the best decision.

A case study often cited to show how software designers have not fully utilized hardware features is described in "The Gibson Mix" [Gibson, 1970]. In his study, Gibson found that about one-quarter of the instruction set accounted for over three-quarters of the instructions executed. This study suggests two possibilities. Either the software designers did not make good use of the instructions available or the hardware designers included instructions that were not necessary. A failure to communicate resulted in a design that was not optimal. What was needed was a design team that understood both the hardware and software aspects.

FIGURE 1.6 Communication paths within project teams of various sizes.

Number of people	Number of paths	Paths
2	1	
3	3	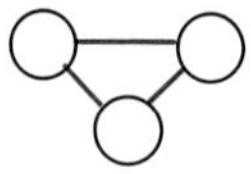
4	6	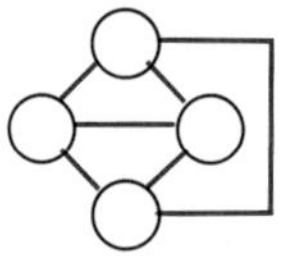
5	10	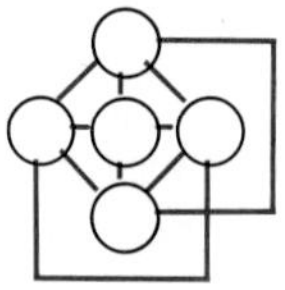

There is yet another problem associated with splitting projects into hardware and software design groups. More time and effort must be expended to guarantee that people who need to share information do so.

In his book *The Mythical Man-Month,* Brooks describes many problems in developing large computer systems [Brooks, 1982]. He states that one of the main problems with large design teams is the increased number of communication paths that must exist. For example, if only four people are assigned to a project, then there will be six possible communication paths, as shown in Fig. 1.6. When the people involved increase to five, the number of paths increases to 10. A problem at the end of the chapter asks you to calculate the number of paths for a project with n people.

When the number of participants on a project can be reduced, the time spent just in interpersonal communications can be significantly decreased. By having team members who can design both hardware and software, the total number of people working on the project can often be reduced.

1.4 COMPUTERS IN CONSUMER PRODUCTS

Another major factor in the need for professionals skilled in both hardware and software is the microcomputer. In 1972, Intel Corporation introduced the 8008 microprocessor; and in 1974, the 8080. These devices were the beginning of a revolution in the electronics industry. They made it possible for product designers to consider putting the power of computers into inexpensive consumer products.

The trend toward putting computers into consumer products has accelerated in the past few years. Consider the partial list shown in Table 1.2. Each of these products typically contains at least one microcomputer, and some contain several.

Microcomputers presented a new and formidable challenge to designers. On conventional computer design teams, it was often reasonable to split into hardware and software groups because of the size of the projects. You might find hundreds and sometimes thousands of people involved in the design of these large systems. However, dedicating separate teams to design the hardware and software for the microprocessor in a digital bathroom scale, for example, could be overkill.

There were two obvious groups from which to choose these small-system designers: the engineers and the programmers. Unfortunately, both groups had difficulty dealing with the total problem. The programmers typically knew little or nothing about digital hardware design, and engineers were usually not familiar with software development. What is really needed is a professional trained in both hard-

TABLE 1.2 SOME OF THE MANY CONSUMER PRODUCTS CONTAINING COMPUTERS

Radios	Cameras	Washing machines
Video recorders	Automobiles	Watches
Refrigerators	Televisions	Scales
Exercise bikes	Microwave ovens	Stereos

ware and software design techniques. As you read the case study in the next section, think about what type of designer you would select for the project.

1.5 CASE STUDY—A VIDEO RECORDER

Consider the sophisticated programming and control systems in home videocassette recorders, or VCRs. They provide a good example of the type of system on which a person with combined hardware and software training might work. A block diagram of the major components of a VCR is shown in Fig. 1.7.

In our discussion of the VCR we will concentrate on the parts of the system that interact with the computer. We will ignore the actual video recording mechanisms. The computer system in this example has a processor, memory, data input devices, data output devices, and output control devices.

The computer in the VCR controls the following functions of the recorder:

- Channel selection
- Record start time
- Record stop time
- Output display

It receives inputs from the buttons on the control panel or from a remote control unit that is probably controlled by its own computer.

FIGURE 1.7 The components of a computer-controlled video recorder.

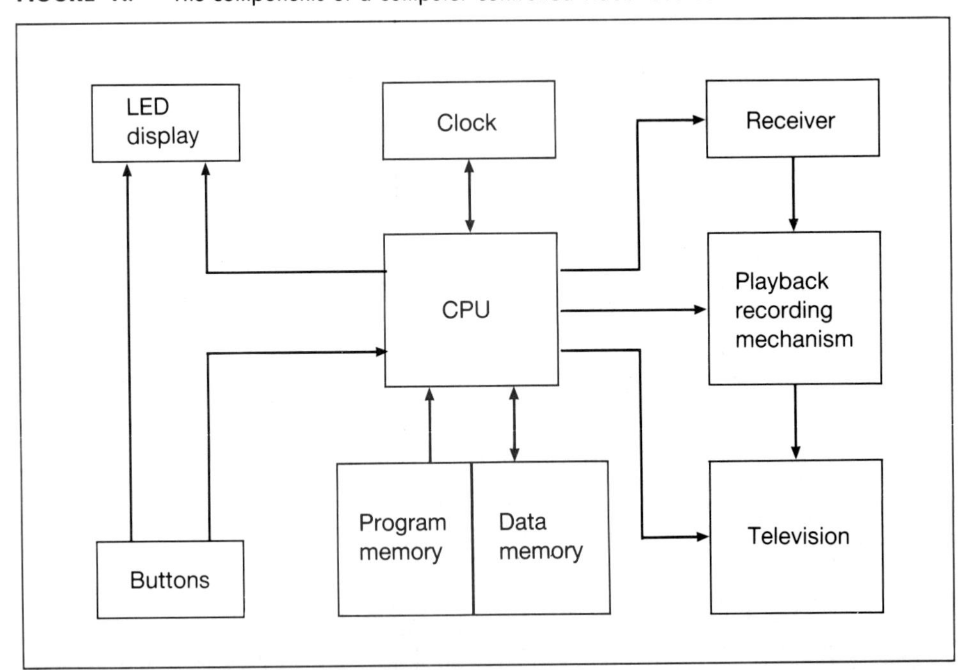

The memory for the computer is divided into two parts: program memory and data memory. The program memory contains the instructions that control the operations of the computer. The data memory could include such things as preprogrammed starting time, length of recording, and channel.

The input devices on the VCR are simple switches or buttons. The program must sense when the buttons are pressed and take the right actions. The outputs from the system appear in several different places: panel lights, the television screen, and relays controlling the recorder.

The programs that control the operation of a computer system can be split into three levels. For our VCR system, the lowest level of software reads data from buttons and sends data to the output devices. The second level is responsible for controlling and monitoring the resources of the system. It does such things as checking the clock to see when the recorder should be turned on and off.

The top level of software is the interface between the user and the system. On a VCR, this interface is the method by which the user programs the VCR. This top level of software gives the system its personality. Its evolution shows how designers have tried to create more user friendly systems. (Some people like to refer to earlier systems as "user abusive.")

Setting up a sequence of instructions for the VCR, which is done by pressing buttons, has been greatly simplified. On the first models the control panel typically had a great many switches and only one small display on which to show information. After the user pressed a switch, the display would show a set of numbers that controlled one of the operations of the VCR. (For instance, a certain set of numbers determined the time at which recording was to begin.) Pressing a different switch would display a different set of information. Programming a VCR was often a confusing process and one very prone to errors.

Newer models have improved software and more hardware that let the computer write messages onto the television screen. The person who programmed the software for this computer unit designed it to send information to the user's screen. This information can guide the user through the VCR's operation.

The designer(s) of the computer system for this type of VCR must have a broad range of talents. The functions that must be performed include:

- Selecting a wide range of components, from the computer to the switches and display
- Designing the circuits that allow the computer to work with the other parts of the system
- Writing the programs that control the operation of the VCR
- Understanding and designing the user interface to provide good human/machine communication
- Testing the final product to verify its correct operation

From this we can see that designing a system such as a VCR calls for a wide variety of skills. Throughout the remainder of this book we will examine more closely how to build complete computer-based systems.

1.6 AN INTEGRATED APPROACH—COMBINING HARDWARE AND SOFTWARE DESIGN

The main point is that while hardware and software specialists will continue to be needed, there is an emerging demand for computing professionals who are familiar with all aspects of computer system design. The VCR design described in the last section is a simple, but typical, job assignment that would not be uncommon for a recent graduate working for a consumer products manufacturer. However, it is just one example. There are many, many more.

Computing professionals are now found in fields such as medicine, space exploration, communications, industrial process control, robotics, and artificial intelligence. They occupy positions involving process and system analysis, hardware and software design, system test and integration, and application development, to name a few. In many cases, they are working on products or systems that must compete in the marketplace.

It is no longer adequate to have systems that meet minimum requirements. Performance, accuracy, reliability, and ease of use are just some of the factors that can spell the difference between success and failure in the marketplace. This, in turn, is forcing a reconsideration of how computer systems are developed. Hardware design, software development, and user application definition can no longer be isolated tasks.

These are also some of the principal reasons for the development of this book. What we are trying to do here is to provide you with an integrated introduction to the concepts upon which good design decisions are based.

1.7 PREVIEW OF THE REMAINDER OF THE BOOK

This text has been organized into four parts. The first part is intended to provide you with some background information. This first chapter has given you an idea of why professionals educated in both hardware and software are needed. Chapter 2 presents some rules of mathematics and logic that enable us to correctly design computer systems. Chapter 3 describes how information is represented and stored within a computer. Chapter 4 introduces the structure, or architecture, of the computer and provides a brief look ahead at the various hardware and software components that make up the system.

Part II, An Overview of Computer Hardware, is just that. The chapters in this part contain detailed descriptions of the hardware components of a computer system. The first five chapters, 5 through 9, describe individual parts of what is usually called the *central processing unit*. Chapter 5 focuses on instructions and how they control the operations of a computer. Chapter 6 discusses memory and information storage. Chapter 7 explains how information is transferred between parts of the system, and Chap. 8 explains the operations of the component that does calculations, the arithmetic and logic unit. The final chapter on the central processing unit describes how external devices communicate with the central processor.

In Chap. 10, we give an overview of the external devices that can be attached to a computer and describe their characteristics. Lastly, Chap. 11 contains an explanation of the hardware development process, including hardware development tools.

In Part III, we look at the other side of the coin, the computer software. Chapters 12 to 16 explain what types of programs form the different levels of software on the system. After introducing operating systems in Chap. 12, we begin with the components of the operating system, Chap. 13, and progress through the layers of the system software, Chap. 14. This is followed in Chap. 15 by a discussion of the languages we use to write our programs. Chapter 16 gives an overview of the outer software layer, application programs. We finish Part III with a look at some tools to aid the software development process in Chap. 17.

Part IV merges the information and concepts presented in the other three parts. The hardware and software aspects of user interfaces are described in Chap. 18, and techniques for reliable system design are given in Chap. 19. Chapter 20 describes the analysis and design of a system requiring both hardware and software components.

We hope that you will find this text interesting and stimulating. If we succeed in our goals, we will have motivated you to continue your studies in computer hardware and software. We will have also supplied you with a good foundation upon which to base those studies.

PROBLEMS

1.1 What are the requirements for a major technological advance?

1.2 What are the major elements that make up the computer environment?

1.3 Why are there fewer *alphabets* than there are *languages*?

1.4 Why are alphabets, human languages, and number systems important to computing?

1.5 What was the key concept that allowed the development of modern languages?

1.6 Consider the words *two*, *too*, and *to*. How do these sounds relate to the key concept of having a separate symbol for each sound in the language?

1.7 What is the primary purpose of punctuation?

1.8 The most common way to identify the beginning or end of a word is with a space. Punctuation symbols can also indicate the beginning or the end of a word. Complete the following table of punctuation symbols that can identify the beginning and end of a word.

Punctuation symbol	Before word	After word end
Space	Yes	Yes
Question mark	No	Yes

1.9 Is the shape or form of the letters of our language really important? Why (or why not)?

1.10 What punctuation characters have you seen used in programming languages?

1.11 Why is punctuation important in a language? What would happen if our language did not have this capability?

1.12 Give an example that shows why punctuation is sometimes necessary for proper understanding of a sentence.

1.13 Why was the Hindu-Arabic decimal number system able to support scientific computation when other number systems with similar concepts could not?

1.14 What were the key concepts in the development of our number system?

1.15 Why is zero important in a number system?

1.16 What is meant by the term *place-value number system*?

1.17 Why did the Egyptian and early Roman civilizations fail to make significant advances in mathematics?

1.18 Discuss some of the problems that might complicate arithmetic operations with Roman numerals.

1.19 The Roman number system uses letter symbols to indicate magnitudes that are either added or subtracted to determine a final value:

I = 1	C = 100
V = 5	D = 500
X = 10	M = 1000
L = 50	etc.

The decision whether to add or subtract a digit value is determined by the symbol's location with respect to other digits in the number. If a lower-valued digit is to the right, it is added; if it is to the left, it is subtracted, as shown in the following example:

$$\begin{aligned} \text{MCMLXXXIX} &= \text{M} \quad \underline{\text{CM}} \quad \underline{\text{LXXX}} \quad \underline{\text{IX}} \\ &= 1000 + (1000 - 100) + (50 + 10 + 10 + 10) + (10 - 1) \\ &= 1989 \end{aligned}$$

Determine the decimal value of the following:

a. MCLXII
b. XIX
c. XVIII
d. CXIV

1.20 Develop an "add" table for Roman numerals (through the value 9). How does the complexity of this table compare to that of the "add" table for the Hindu-Arabic system?

1.21 Develop a "subtraction" table for Roman numerals (through the value 5).

1.22 Why did the computer programming profession evolve?

1.23 Why did some companies create organizations with separate hardware and software design groups?

1.24 What kinds of problems can arise from separating hardware and software groups?

1.25 What advantages are there to separating the hardware and software design functions within a company?

1.26 Name several other computer-based consumer products that are not listed in Table 1.2.

1.27 Figure 1.6 shows the number of communication paths for project teams of two, three, four, and five people. Develop an equation for the number of communication paths in a project team of n people.

1.28 What were some of the "needs" that fostered the development of computers?

1.29 List some examples in other professions where there has been an emphasis on specialization analogous to the hardware/software split in computing.

1.30 What are some professions that are unlikely to be affected by computers?

1.31 What type of designer would you pick for the VCR development described in Sec. 1.5? Why?

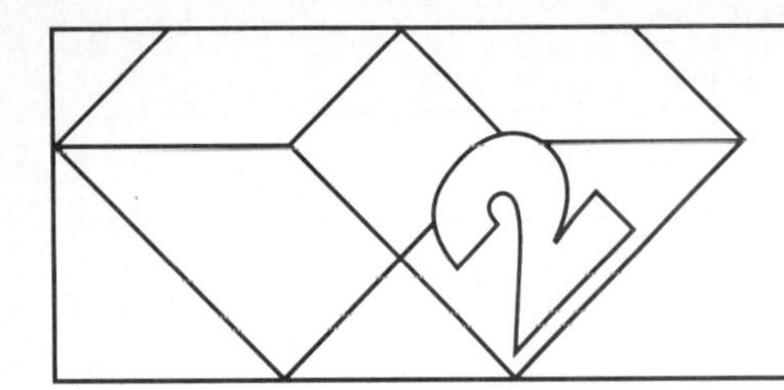

Introduction to Sets, Boolean Algebra, and Logic Circuit Elements

In Chap. 1, we ended our review of the evolution of mathematics with Boole's development of an algebra of logic in 1847. His purpose at the time was to show that logic should be viewed from a mathematical rather than a philosophical standpoint. His algebra is now known as **boolean algebra**, and it has had a significant impact on the subsequent development of mathematics. For the computer designer, the most important use of boolean algebra was discovered by Shannon in 1938. He found that it could be the basis for analysis of relay control circuits, which are conceptually similar to the two-state logic circuits that make up today's computers.

Boole's algebra is based on his reasoning that objects and ideas can be represented by a two-valued system that:

- Involves collecting the objects or ideas into groups or sets according to their properties
- Allows the sets to be analyzed according to the presence or absence of a defined property

The purpose of this chapter is to give you a background in set theory and boolean algebra and to introduce you to the logic circuit elements that make up a computer. We will use boolean algebra to analyze and simplify several example logic functions. After that, we will expand our simplification procedures to include Karnaugh maps and contrast that with the algebraic approach. Then we will take an in-depth look at the operation of a logical storage element, the *RS* flip-flop. We will finish the chapter with a simple design example. Let's begin with a discussion of how objects and ideas can be grouped into sets and treated according to their properties.

2.1 INTRODUCTION TO SET THEORY

Set theory is a branch of mathematics that provides a formal method for dealing with objects that have been separated into well-defined collections which can be treated as a whole. We include it here because it provides an excellent lead-in to both boolean algebra and (in the next chapter) information representation in a computer.

2.1.1 Sets and Subsets

Everything—dogs, cats, people, integers, concepts, computer languages, and so on—has one or more distinguishing features that either set it off from everything else or identify it as being a member of a collection. We call these distinguishing features **characteristics**, or **attributes**, and we refer to the things they are associated with as **entities**. An entity is something that exists as a particular and distinct unit, such as a person, a company, a flag, or a signal. Attributes allow us to classify entities in various ways and to specify criteria upon which similarity can be determined. A signal, for example, could have the attribute on or off.

This leads us to the notion of sets. Since sets are the starting point, we cannot strictly define them; they are the "givens." Intuitively, we can think of a **set** as a collection of entities. The entities in the collection are called the **elements** of the set. We can say that a certain element **is a member of** (or **is in**, or **belongs to**) a set, and we use the symbol $\in$ to denote such membership. Correspondingly, we can say that another element **is not a member of** (or **is not in**, or **does not belong to**) a set, and we use the symbol $\notin$ to denote such nonmembership.

The collection of the authors of this book is a set that we might call *Authors* and write

$$Authors = \{W.\ G.\ Lane,\ J.\ R.\ Pinkert,\ L.\ C.\ Wear,\ L.\ L.\ Wear\}$$

A sample membership statement for *Authors* might be *W. G. Lane* $\in$ *Authors*. *Darth Vader* $\notin$ *Authors* would be a statement of nonmembership.

We need to stress that repetitions and orders are not distinguishable for sets. Therefore, {*J. R. Pinkert*, *W. G. Lane*, *L. C. Wear*, *L. L. Wear*}, {*J. R. Pinkert*, *L. C. Wear*, *L. L. Wear*, *W. G. Lane*}, and {*L. C. Wear*, *L. L. Wear*, *J. R. Pinkert*, *W. G. Lane*} are all the same set as *Authors*.

In the previous example, we defined a set by listing its elements. Another way to define a set is to give **defining properties** that specify the criteria for membership; for instance, we could say that *Authors* is the set of people who coauthored this book.

When determining the elements of a set S according to defining properties P, one must consider the **universal set** U from which the elements of S are taken. Figure 2.1 shows various universal sets that could be used when discussing the set *Authors*. Observe that if we use *Parents* as the universal set, then *Authors* changes, since two of the authors are not parents. Phrased in another way, the set of people who

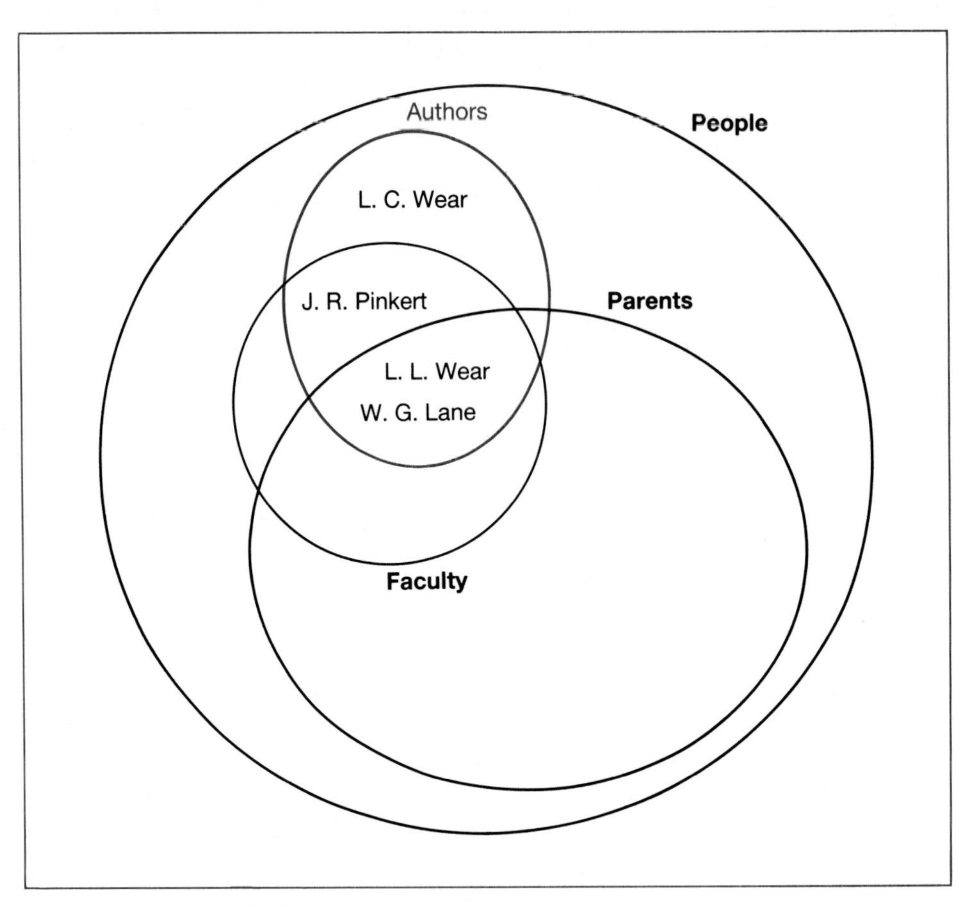

FIGURE 2.1 The set *Authors* as a subset of various universal sets.

coauthored this book is different from the set of parents who are among the coauthors of this book.

If the universal set is understood from the way it is used, then we specify a set S that has the defining properties P by writing

$$S = \{x\text{: } x \text{ has } P\}$$

This is read, "The set S is the set of all x such that x has all the properties P." If the universal set U must be specified, then we write

$$S = \{x \in U\text{: } x \text{ has the properties } P\}$$

and say, "The set S is the set of every x that is an element of U such that x has all the properties P." For example, if we are discussing integer numbers, we could write

$$S = \{x\text{: } 0 < x < 100\}$$

to specify positive integers less than 100. On the other hand, if we did not know that the universal set was integer numbers, we could write

$$S = \{x \in \mathbf{I}: 0 < x < 100\}$$

where **I** is defined as the set of all integers.

We say that a set is **finite** if and only if it has a finite number of elements; otherwise, it is **infinite**. The number of elements in a finite set S is called the **cardinality** of the set and is denoted by $\#(S)$ or $|S|$. A set with one element is called a **singleton** set, and we can define many different singleton sets. A set with no elements is called an **empty set**, or a **null set**. There is only one null set, and it is specified by $\{\ \}$ or $\varnothing$.

Two sets S and T are **equal** (denoted $S = T$) if and only if every element of S is also an element of T and every element of T is also an element of S. S is a **subset** of T (denoted $S \subseteq T$) if and only if every element of S is also an element of T. S is a **proper subset** of T (denoted $S \subset T$) if and only if S is not the empty set, S is a subset of T, and there is at least one element in T that is not a member of S.

It is also interesting to consider the set of all possible subsets of a set S, called the **power set** of S [denoted by $\mathscr{P}(S)$]. The power set of the set *Authors* is

$$\begin{aligned}
\mathscr{P}(Authors) = \{ & \\
& \{\ \}, \\
& \{W.\ G.\ Lane\}, \\
& \{J.\ R.\ Pinkert\}, \\
& \{L.\ C.\ Wear\}, \\
& \{L.\ L.\ Wear\}, \\
& \{W.\ G.\ Lane,\ J.\ R.\ Pinkert\}, \\
& \{W.\ G.\ Lane,\ L.\ C.\ Wear\}, \\
& \{W.\ G.\ Lane,\ L.\ L.\ Wear\}, \\
& \{J.\ R.\ Pinkert,\ L.\ C.\ Wear\}, \\
& \{J.\ R.\ Pinkert,\ L.\ L.\ Wear\}, \\
& \{L.\ C.\ Wear,\ L.\ L.\ Wear\} \\
& \{W.\ G.\ Lane,\ J.\ R.\ Pinkert,\ L.\ C.\ Wear\} \\
& \{W.\ G.\ Lane,\ J.\ R.\ Pinkert,\ L.\ L.\ Wear\} \\
& \{W.\ G.\ Lane,\ L.\ C.\ Wear,\ L.\ L.\ Wear\} \\
& \{J.\ R.\ Pinkert,\ L.\ C.\ Wear,\ L.\ L.\ Wear\} \\
& \{W.\ G.\ Lane,\ J.\ R.\ Pinkert,\ L.\ C.\ Wear,\ L.\ L.\ Wear\} \\
& \}
\end{aligned}$$

WGL	JRP	LCW	LLW
0	0	0	0
1	0	0	0
0	1	0	0
0	0	1	0
0	0	0	1
1	1	0	0
1	0	1	0
1	0	0	1
0	1	1	0
0	1	0	1
0	0	1	1
1	1	1	0
1	1	0	1
1	0	1	1
0	1	1	1
1	1	1	1

FIGURE 2.2 The power set of *Authors* shown in binary-encoded form.

The name *power set* can be associated with the idea that to construct all the subsets of S we look at each element in S and say "include it" or "don't include it." Another way is to represent the "includes" as 1s and the "don't includes" as 0s. These 1s and 0s are the digits, or bits, of the binary number system. They allow us to list the power set in the shortened tabular form shown in Fig. 2.2.

Each line of the figure is the coded equivalent of its particular subset in the power set. The elements of the subset are determined by the location of the 1s in the code, such that 1011 corresponds to {*W. G. Lane, L. C. Wear, L. L. Wear*}. Or said differently, 1011 applied to the power set {*W. G. Lane, J. R. Pinkert, L. C. Wear, L. L. Wear*} results in the subset {*W. G. Lane, L. C. Wear, L. L. Wear*}.

We are also concerned with the size of the power set. Consider the diagram of Fig. 2.3. It shows the power sets that result from sets of one, two, three, and four elements. The size progression is 2, 4, 8, 16, ..., which is 2^n, where n is the number of elements in the set. By inference then, we can see that the cardinality of any power set (S) is

$$\#(\mathscr{P}(S)) = 2^{\#(S)}$$

This relationship is important because it allows us to define two associated equations that will be needed when we discuss coding in Chap. 3:

$$m = \text{number of possible coded combinations in a binary word of length } n$$
$$= 2^n$$

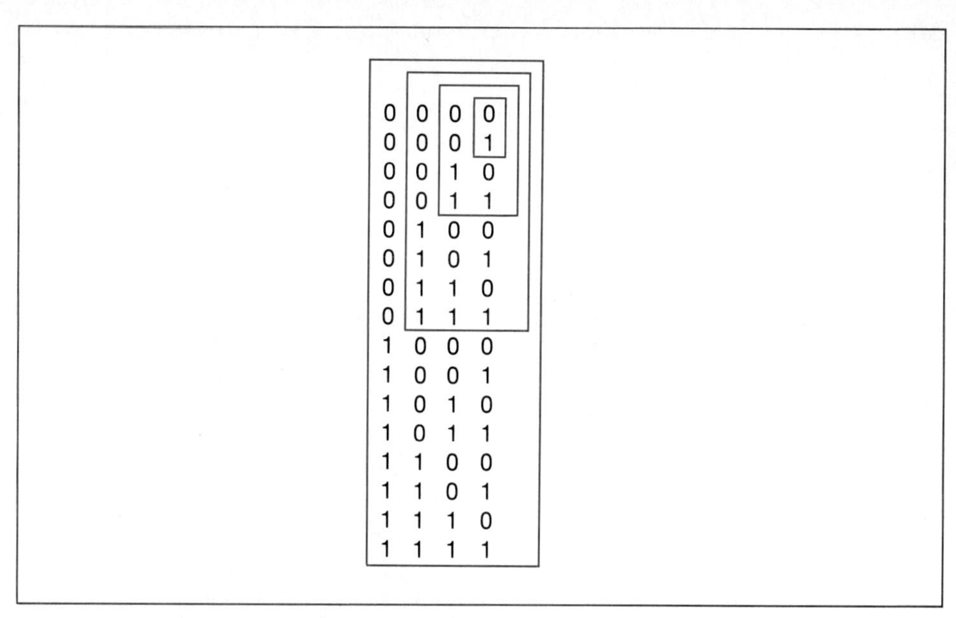

FIGURE 2.3 The power sets for one-, two-, three-, and four-element sets.

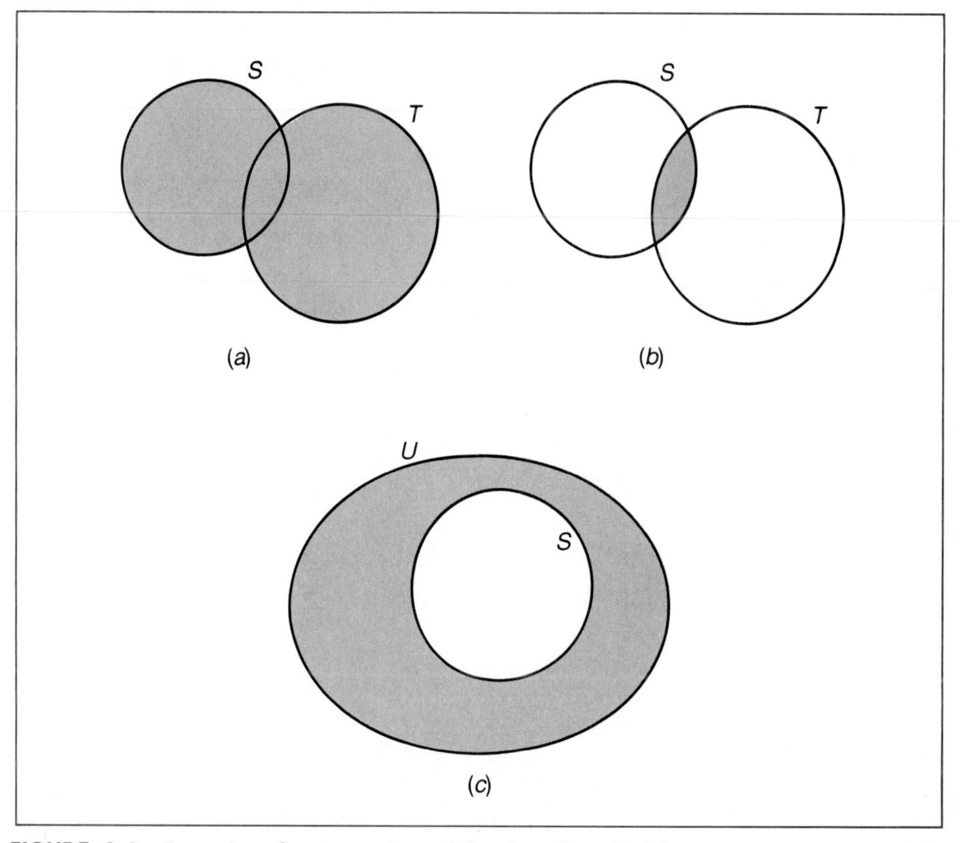

FIGURE 2.4 Examples of set operators: (*a*) union, $S \cup T$; (*b*) intersection, $S \cap T$; (*c*) complement, S'.

and, conversely,

$$n = \text{minimum length of code needed to represent a set of } m \text{ elements}$$
$$= \lceil \log_2 m \rceil$$

where the brackets ⌈ ⌉ mean the integer value equal to or immediately greater than the value computed.

2.1.2 Operations on Sets

There are three important set operators that are used often. The **union** of two sets S and T (denoted $S \cup T$) is the set of all x such that x is either an element of S or an element of T. The **intersection** of two sets S and T (denoted $S \cap T$) is the set of all x such that x is both an element of S and an element of T. The **complement** of a set S relative to some universal set U (denoted by S') is the set of all x such that x is an element of U but is not an element of S. Each of these operators is shown diagrammatically in Fig. 2.4.

2.2 SET ALGEBRAS

In the previous section, we introduced the set of all possible subsets of a set S, called the power set $\mathscr{P}(S)$. This power set, together with the operations of set union, set intersection, and set complementation, can be used to define a mathematical structure called a **set algebra**.

For example, consider set $S = \{a, b, c\}$ with $\mathscr{P}(\{a, b, c\}) = \{\{\ \}, \{a\}, \{b\}, \{c\}, \{a, b\}, \{a, c\}, \{b, c\}, \{a, b, c\}\}$. The definitions of the three operations for this set algebra are given in Tables 2.1 to 2.3.

We can also view the elements of $\mathscr{P}(\{a, b, c\})$ diagrammatically. Such a diagram, organized on the basis of subsets, is shown in Fig. 2.5. Union, intersection, and complementation can be defined in terms of graphical operations on such a diagram. Exact rules for these operations are left as an exercise for the student.

TABLE 2.1

Union	{ }	{a}	{b}	{c}	{a, b}	{a, c}	{b, c}	{a, b, c}
{ }	{ }	{a}	{b}	{c}	{a, b}	{a, c}	{b, c}	{a, b, c}
{a}	{a}	{a}	{a, b}	{a, c}	{a, b}	{a, c}	{a, b, c}	{a, b, c}
{b}	{b}	{a, b}	{b}	{b, c}	{a, b}	{a, b, c}	{b, c}	{a, b, c}
{c}	{c}	{a, c}	{b, c}	{c}	{a, b, c}	{a, c}	{b, c}	{a, b, c}
{a, b}	{a, b}	{a, b}	{a, b}	{a, b, c}	{a, b}	{a, b, c}	{a, b, c}	{a, b, c}
{a, c}	{a, c}	{a, c}	{a, b, c}	{a, c}	{a, b, c}	{a, c}	{a, b, c}	{a, b, c}
{b, c}	{b, c}	{a, b, c}	{b, c}	{b, c}	{a, b, c}	{a, b, c}	{b, c}	{a, b, c}
{a, b, c}	{a, b, c}	{a, b, c}	{a, b, c}	{a, b, c}	{a, b, c}	{a, b, c}	{a, b, c}	{a, b, c}

TABLE 2.2 SET INTERSECTION FOR $\mathscr{P}(\{a, b, c\})$

Intersection	{ }	{a}	{b}	{c}	{a, b}	{a, c}	{b, c}	{a, b, c}
{ }	{ }	{ }	{ }	{ }	{ }	{ }	{ }	{ }
{a}	{ }	{a}	{ }	{ }	{a}	{a}	{ }	{a}
{b}	{ }	{ }	{b}	{ }	{b}	{ }	{b}	{b}
{c}	{ }	{ }	{ }	{c}	{ }	{c}	{c}	{c}
{a, b}	{ }	{a}	{b}	{ }	{a, b}	{a}	{b}	{a, b}
{a, c}	{ }	{a}	{ }	{c}	{a}	{a, c}	{c}	{a, c}
{b, c}	{ }	{ }	{b}	{c}	{b}	{c}	{b, c}	{b, c}
{a, b, c}	{ }	{a}	{b}	{c}	{a, b}	{a, c}	{b, c}	{a, b, c}

TABLE 2.3 SET COMPLEMENT FOR $\mathscr{P}(\{a, b, c\})$

	{ }	{a}	{b}	{c}	{a, b}	{a, c}	{b, c}	{a, b, c}
Complement	{a, b, c}	{b, c}	{a, c}	{a, b}	{c}	{b}	{a}	{ }

It should be emphasized that the *structure* of the tables and the diagram is what is important. We could use different names for the elements of set $\{a, b, c\}$, or we could uniquely name each element of $\mathscr{P}(\{a, b, c\})$. The structure of our set algebra would not change.

Phrased in another way, there is only *one unique set algebra for a set with three elements.* Let's call it $\mathscr{A}_3$. That algebra has $2^3 = 8$ elements, the number of subsets

FIGURE 2.5 A diagram of $\mathscr{P}(\{a, b, c\})$ organized on the basis of subsets.

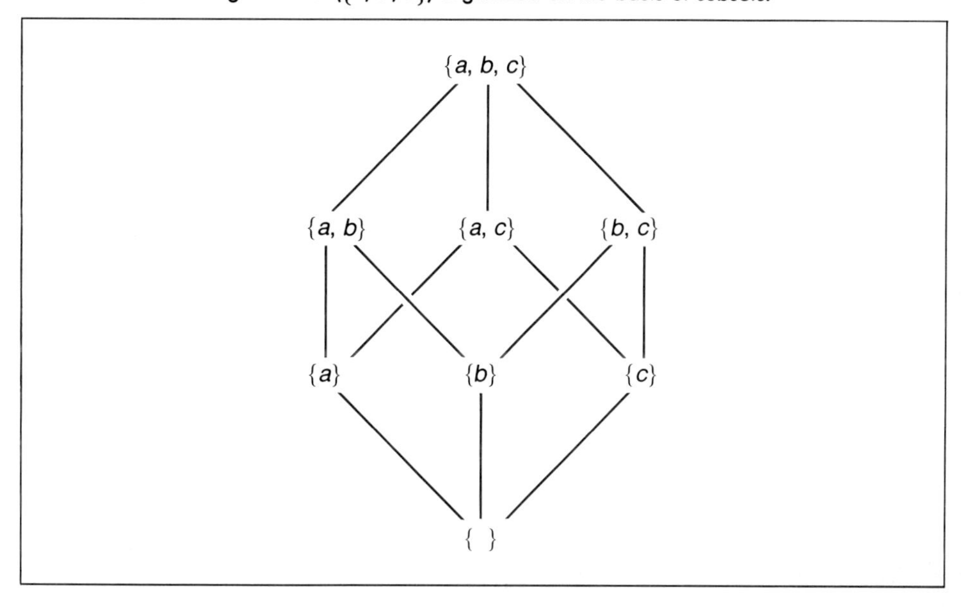

TABLE 2.4 SET UNION FOR $\mathscr{A}_2$

Union	0	P	P'	1
0	0	P	P'	1
P	P	P	1	1
P'	P'	1	P'	1
1	1	1	1	1

TABLE 2.5 SET INTERSECTION FOR $\mathscr{A}_2$

Intersection	0	P	P'	1
0	0	0	0	0
P	0	P	0	P
P'	0	0	P'	P'
1	0	P	P'	1

TABLE 2.6 SET COMPLEMENT FOR $\mathscr{A}_2$

	0	P	P'	1
Complement	1	P'	P	0

FIGURE 2.6 A diagram of $\mathscr{A}_2$.

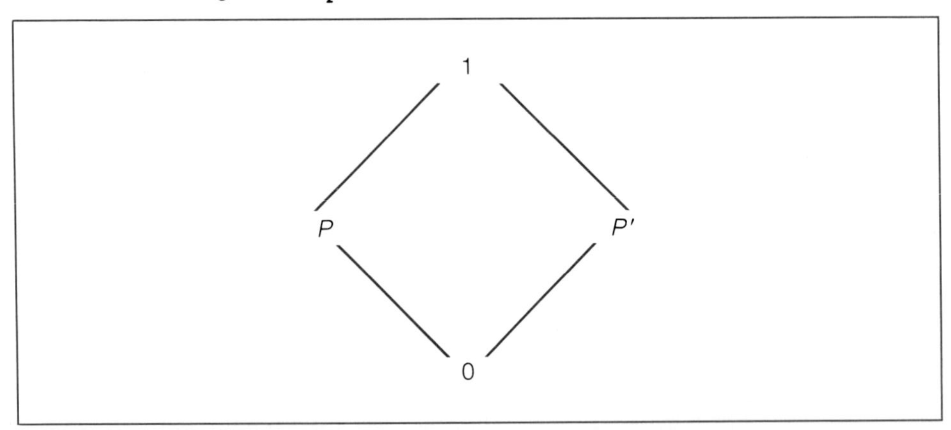

in the power set of a set with three elements. The structure of the tables and the subset diagram for $\mathcal{A}_3$ are as we have shown above.

Similarly, there is one unique set algebra $\mathcal{A}_n$ for every other integer n. Consider Tables 2.4 to 2.6 and Fig. 2.6 for $\mathcal{A}_2$. We have used a different naming convention for the elements of $\mathcal{A}_2$.

2.3 THE GENERAL CONCEPT OF BOOLEAN ALGEBRAS

Now let's apply our discussion of power sets from Sec. 2.1 to the set algebras described in the previous section. We said that the names we give to elements of a given algebra $\mathcal{A}_n$ are not important—only its structure is. Therefore, we could use bit encoding for the names, similar to what we did in Fig. 2.2. With this approach, our structures for $\mathcal{A}_2$ then become as shown in Tables 2.7 to 2.9 and Fig. 2.7.

TABLE 2.7 SET UNION FOR $\mathcal{P}(\{a, b\})$ USING BIT ENCODINGS

Union	00	01	10	11
00	00	01	10	11
01	01	01	11	11
10	10	11	10	11
11	11	11	11	11

TABLE 2.8 SET INTERSECTION FOR $\mathcal{P}(\{a, b\})$ USING BIT ENCODINGS

Intersection	00	01	10	11
00	00	00	00	00
01	00	01	00	01
10	00	00	10	10
11	00	01	10	11

TABLE 2.9 SET COMPLEMENT FOR $\mathcal{P}(\{a, b\})$ USING BIT ENCODINGS

	00	01	10	11
Complement	11	10	01	00

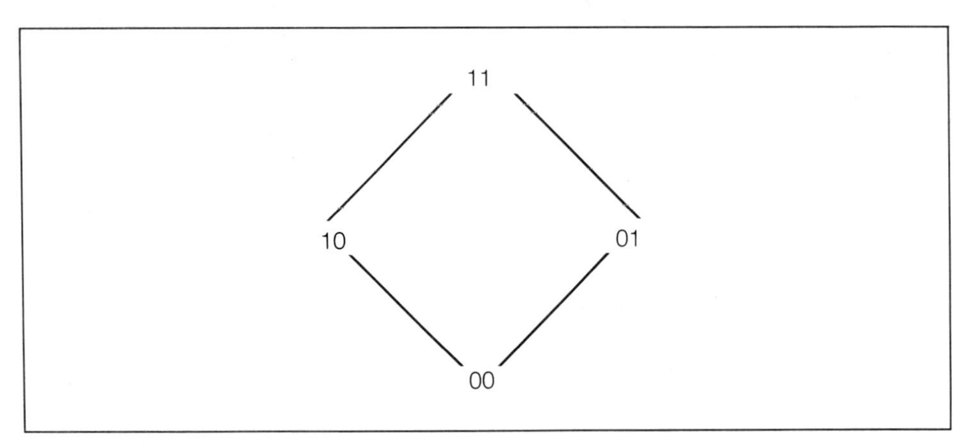

FIGURE 2.7 A diagram of $\mathscr{P}(\{a, b\})$ organized on the basis of subsets and using bit encodings.

The operations of union, intersection, and complementation for set algebra $\mathscr{A}_n$ can be defined in terms of bit encodings as follows:

- $R = S \cup T$: the jth bit in the encoding of R is 1 if and only if *either* the jth bit in the encoding of S is 1 *or* the jth bit in the encoding of T is 1.
- $R = S \cap T$: the jth bit in the encoding of R is 1 if and only if *both* the jth bit in the encoding of S is 1 *and* the jth bit in the encoding of T is 1.
- $R = S'$: the jth bit in the encoding of R is 1 if and only if the jth bit in the encoding of S is 0. Conversely, the jth bit in the encoding of R is 0 if and only if the jth bit in the encoding of S is 1.

These alternate definitions of union, intersection, and complement are called **OR**, **AND**, and **INVERSION** and symbolized by $+$, $\cdot$, and $'$. A set algebra $\mathscr{A}_n$ with operations OR, AND, and INVERSION is called a **boolean algebra** $\mathscr{B}_n$.

The form of boolean algebra we will be discussing in the rest of this chapter is sometimes called a **switching algebra** because of its use to analyze circuits that "switch" between two logical states or voltage levels. It is an important mathematical tool because it allows the logical design of a digital circuit module to be completed before the actual electronic design is even begun. *Boolean functions are not dependent on either the electrical characteristics of the circuits or the analysis procedures that will be used to implement the actual circuit design.* All that is required is your understanding that there will be:

- Two voltage levels corresponding to the two logic states defined by the algebra
- A delay when an electric signal at one point in a circuit is transmitted to another point

2.3.1 Switching Algebra

Let's begin by assuming that there exists a set of elements S consisting of variables and two constants, $S = \{a, b, c, \ldots, 0, 1\}$. The elements of S are subject to an is-equivalent-to relation (denoted by an $=$ symbol) that satisfies the following properties:

- Reflexive: $x = x$ for all x in S.
- Symmetric: If $x = y$, then $y = x$ for all x, y in S.
- Transitive: If $x = y$ and $y = z$, then $x = z$ for all x, y, z in S.
- Substitutive: If $x = y$, then substituting x for y (or y for x) in any boolean expression will result in an equivalent relation.

The algebra includes the two operators $+$ and $\cdot$, corresponding to the set union and set intersection, such that:

If x and y are in S, then $x + y$ is also in S.
If x and y are in S, then $x \cdot y$ is also in S.

For all x, y in S, these operators are commutative:

$$x + y = y + x$$

$$x \cdot y = y \cdot x$$

and distributive:

$$x + (y \cdot z) = (x + y) \cdot (x + z)$$

$$x \cdot (y + z) = (x \cdot y) + (x \cdot z)$$

Evaluation of boolean expressions follows an order of precedence where $\cdot$ operations are done before $+$ operations and where expressions within parentheses are evaluated before nonparenthesized expressions.

The algebra includes the two constants such that for every x in S,

$$x + 0 = x \qquad \text{and} \qquad x \cdot 0 = 0$$

$$x + 1 = 1 \qquad \text{and} \qquad x \cdot 1 = x$$

where the $+$ and $\cdot$ operations on 0 and 1 satisfy the relations shown in Fig. 2.8.

For every element x in S, there exists a complement x' such that

$$x \cdot x' = 0 \qquad \text{and} \qquad x + x' = 1$$

The complement of 1 is 0, and the complement of 0 is 1.

+	0	1
0	0	1
1	1	1

•	0	1
0	0	0
1	0	1

FIGURE 2.8 The boolean + and • operations.

Up to this point, except for the relation $x + 1 = x$, you might think that we are just talking about ordinary algebra. However, there are significant differences:

- Boolean variables are not like variables in ordinary algebra. They can have only two possible values, either 0 or 1.
- Boolean operations are not like operations in ordinary algebra:

  The + operation signifies the logical **OR** rather than add, and the expression $F = x + y$ means that the value of F will be 1 *if and only if either x or y is 1.*

  The $\cdot$ operation is the logical **AND** rather than multiply, and the expression $F = x \cdot y$ means that the value of F will be 1 *if and only if both x and y are 1.*

  The $'$ is the logical complement, or **NOT**, and the expression $F = x'$ means that the value of F will be the inverse of the value of x.

There are several additional basic relations that may appear a little strange when they are compared to ordinary algebra but which are very helpful when manipulating boolean expressions:

$$x \cdot x = x$$

$$x + x = x$$

$$x + 1 = 1$$

$$x + x' \cdot y = x + y$$

$$x' \cdot y' = (x + y)'$$

$$x' + y' = (x \cdot y)'$$

The last two of these relations are known as **De Morgan's theorem** for finding the complement of boolean expressions. They can be applied to any expression, regardless of the number of literals.

2.3.2 The Application of Boolean Algebra to Logic Circuits

It is now time to consider using the relations in the implementation of some simple logic functions. The circuit symbols for the three basic boolean operations OR,

x, y → $x + y$ $\qquad$ x, y → $x \cdot y$ $\qquad$ x → x'

$x \backslash y$	0	1
0	0	1
1	1	1

(*a*)

$x \backslash y$	0	1
0	0	0
1	0	1

(*b*)

x	0	1
x'	1	0

(*c*)

FIGURE 2.9 Symbols and truth tables for logic circuits: (*a*) OR; (*b*) AND; (*c*) NOT.

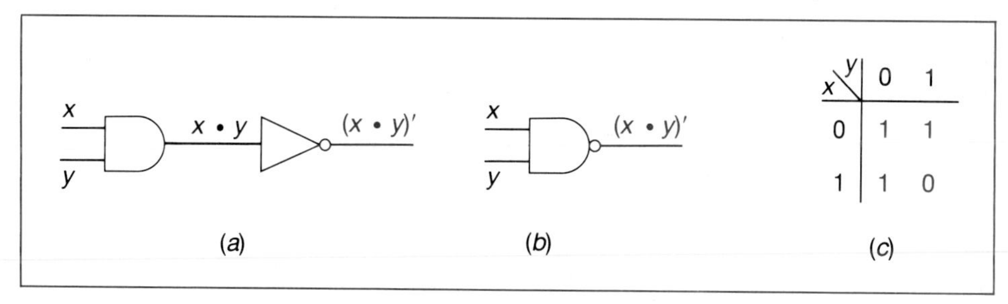

FIGURE 2.10 NAND function: (*a*) equivalent circuit; (*b*) circuit symbol; (*c*) truth table.

AND, and NOT are shown in Fig. 2.9. The actual circuits for the AND and the OR are known as **gates** because they control the passage of signals in their portion of the circuit. The NOT is called an **inverter** because it "inverts," or "complements," the signal. The shape[1] of the symbols can be likened to a stylized arrowhead. The signal flow is in the direction of the arrow. The tables below each symbol are known as **truth tables** because they define when the output of each function is true (equal to 1).

If an inverter is connected in series with an AND or an OR, the function is known as **NOT AND** or **NOT OR**. If the inverter is included as part of the actual AND or OR circuit, the gate is called a **NAND** or a **NOR**. These symbols are shown in Figs. 2.10 and 2.11. The small circle at the output of each gate indicates the complement function.

[1] The *IEEE Standard Graphic Symbols for Logic Functions* allows the use of either distinctive-shape symbols (like those shown in Figs. 2.9 to 2.11) or nondistinctive-shape symbols (labeled rectangles) for the depiction of logic circuit elements.

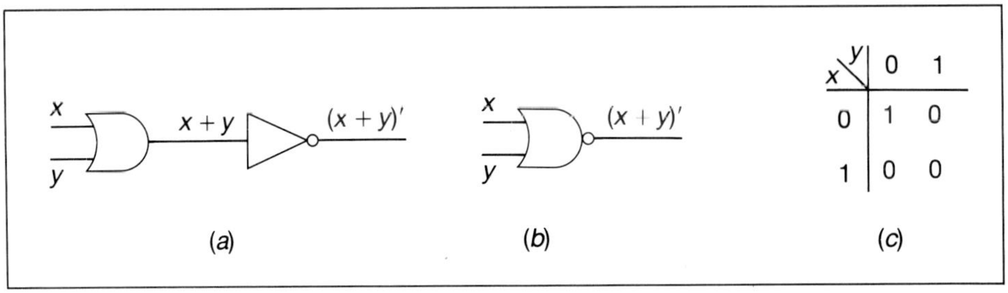

x \ y	0	1
0	1	0
1	0	0

FIGURE 2.11 NOR function: (*a*) equivalent circuit; (*b*) circuit symbol; (*c*) truth table.

For our first example, assume that you have been asked to reduce the following expression to its simplest form:

$$F_1 = x \cdot y + x \cdot y' + x' \cdot y$$

This is very much like the assignments you were given in regular algebra. The first thing we do is scan the terms in the expression, looking for instances of the same variable or its complement. We see that the first two terms both contain an x. This part of the expression can be rewritten as $x \cdot (y + y')$ through use of the distributive rule. Then, since $y + y' = 1$, this part of the expression reduces to $x \cdot 1 = x$.

We also see that the first and third terms both contain y. The same procedure would reduce this part of the expression to y. Then since $x \cdot y = x \cdot y + x \cdot y$, we can expand the expression and reduce it as outlined.

$$\begin{aligned} F_1 &= x \cdot y + x \cdot y' + x' \cdot y \\ &= x \cdot y + x \cdot y' + x' \cdot y + x \cdot y \\ &= x \cdot (y + y') + y \cdot (x + x') \\ &= x \cdot 1 + y \cdot 1 \\ &= x + y \end{aligned}$$

Figure 2.12 shows the circuit representations for both the function and its simplified equivalent. It also provides a vivid example of the purpose of minimization: to reduce the number and complexity of the circuits needed to implement a logical function.

You may have observed that it is easy to confuse the $\cdot$ terms with the + terms in the expressions. In ordinary algebra, when all expression variables are represented by single letters, the multiply symbol $\cdot$ is left out and the term xy is interpreted as meaning $x \cdot y$. We can use the same approach for the boolean $\cdot$ symbol and interpret a boolean term xy as meaning x AND y. The AND operation between a variable and a constant will continue to be shown as $x \cdot 1$ to avoid confusion with a subscripted variable.

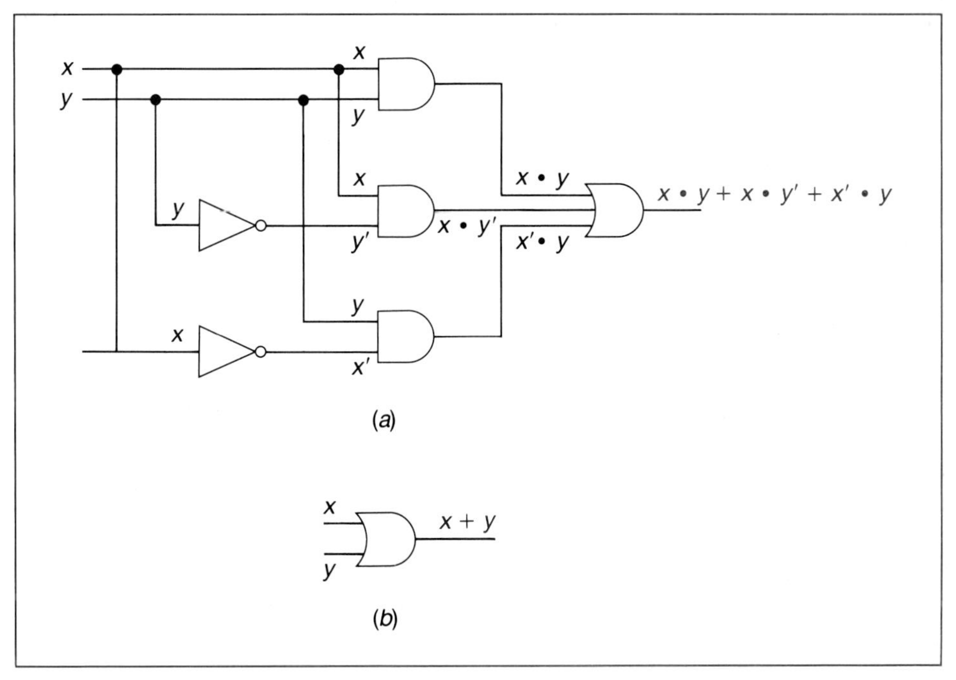

FIGURE 2.12 Circuit representations of F_1 (*a*) as originally specified and (*b*) after simplification.

Let's use this approach in a second example. Assume that you have been given the function

$$F_2 = (x + y)(xy')'$$

Scanning the expression shows that the term xy' is itself complemented. Applying De Morgan's theorem, this becomes $x' + y$. The remaining steps in the simplification are fairly obvious and they are left as an exercise for you to identify what is done in each.

$$= (x + y)(x' + y)$$

$$= xx' + xy + x'y + yy$$

$$= 0 + xy + x'y + y$$

$$= y(x + x' + 1)$$

$$= y \cdot 1$$

$$= y$$

2.3.3 Minterms and Maxterms

The two examples just covered should give you a flavor of algebraic minimization. We now need to make some more definitions. A **product term** in an expression is a series of boolean variables that are related by the $\cdot$ operator. An expression that contains a series of product terms related by + operators is known as a **sum-of-products** expression. F_1 is an example of a sum-of-products expression.

Our second example contains terms that contain variables related by the + operator. These are known as **sum terms**. An expression that contains a series of sum terms related by $\cdot$ operators is known as a **product-of-sums** expression.

To see how these are used, assume that we have been given a function of three variables

$$F_3(x, y, z) = xy + xy'z + x'y'z + x'y$$

Let's expand it so that each product term contains all three variables:

$$F_3 = xy(z + z') + xy'z + x'y'z + x'y(z + z')$$

$$= xyz + xyz' + xy'z + x'y'z + x'yz + x'yz'$$

This expression is now in the **standard-sum-of-products**, or **canonical-sum**, form. Each of the product terms contains as many variables as there are variables in the function. Product terms that are in this form are known as **minterms**.

Sum terms that contain as many variables as there are variables in the function are known as **maxterms**. Functions made up from maxterms are said to be in the **standard-product-of-sums**, or **canonical-product**, form.

Minterms and maxterms are members of the power set of the function variables we saw in Fig. 2.3. Minterms are formed by complementing any variable where the literal value is 0 and by leaving the variable uncomplemented if the literal value is 1, as shown in Fig. 2.13. Maxterms are formed the opposite way: by complementing variables with literal values of 1 and by leaving variables with literal values of 0 uncomplemented.

The minterm/maxterm table is based on the definition of the boolean constants 0 and 1. Recall that

$$x + 0 = x \qquad \text{and} \qquad x \cdot 1 = x$$

which by substitution becomes

$$x + 0 = x \cdot 1$$

This relationship is known as the **principal of duality**. It means that any boolean switching expression remains true if all the operations ($\cdot$ and +) and all the literal values (0 and 1) are each interchanged, so long as all variables are left in place. It

	x y z	Minterm	Maxterm
0	0 0 0	$x'y'z'$	$x + y + z$
1	0 0 1	$x'y'z$	$x + y + z'$
2	0 1 0	$x'yz'$	$x + y' + z$
3	0 1 1	$x'yz$	$x + y' + z'$
4	1 0 0	$xy'z'$	$x' + y + z$
5	1 0 1	$xy'z$	$x' + y + z'$
6	1 1 0	xyz'	$x' + y' + z$
7	1 1 1	xyz	$x' + y' + z'$

FIGURE 2.13 Minterms and maxterms for three functions.

provides us with an easy way to convert our expressions back and forth between the sum form and the product form without going through any intermediate steps.

Minterms and maxterms give us a shorthand notation for expressing a function. Each term can also be identified by either its binary or its decimal number.

$$\begin{aligned} F_3(x, y, z) &= (111) + (110) + (101) + (001) + (011) + (010) \\ &= \Sigma(7, 6, 5, 1, 3, 2) \\ &= \Sigma(1, 2, 3, 5, 6, 7) \end{aligned}$$

We know that *our sum-of-products expression will be a 1 whenever any of the minterms is a 1.* However, we can also express our function as the complement of the sum of the other minterms in the power set.

$$F_3(x, y, z) = (\Sigma(0, 4))'$$

Then, if we substitute the equivalent minterms and apply De Morgan's theorem, our function becomes

$$\begin{aligned} F_3(x, y, z) &= (x'y'z' + xy'z')' \\ &= (x + y + z)(x' + y + z) \\ &= \Pi(0, 4) \end{aligned}$$

This is the *product form of the expression,* and it *will be 0 when any of the maxterms are 0.*

We now have our example function in a simpler form, but it still may not be in a minimum expression. Minimum means that the function contains *the minimum number of terms and the minimum number of literals.* Let's try reducing our example further.

$$
\begin{aligned}
F_3(x, y, z) &= (x + y + z)(x' + y + z) \\
&= xx' + xy + xz + x'y + yy + yz + x'z + yz + zz \\
&= 0 + xy + xz + x'y + y + yz + x'z + yz + z \\
&= y(x + x' + 1 + z) + z(x + x' + y + 1) \\
&= y \cdot 1 + z \cdot 1 \\
&= y + z
\end{aligned}
$$

That was fairly easy. However, in actual practice we often have functions with many more variables and it can be very cumbersome to keep track of all the terms. We need another way.

2.3.4 Karnaugh Maps

A simpler representation of the power set is the **Karnaugh map**, shown in Fig. 2.14. Each square represents the minterm indicated by the binary code of the row and column. We have shown the variable ranges for the rows and columns by name to

FIGURE 2.14 Karnaugh map of the three-variable power set with rows and columns identified by their respective bit vectors.

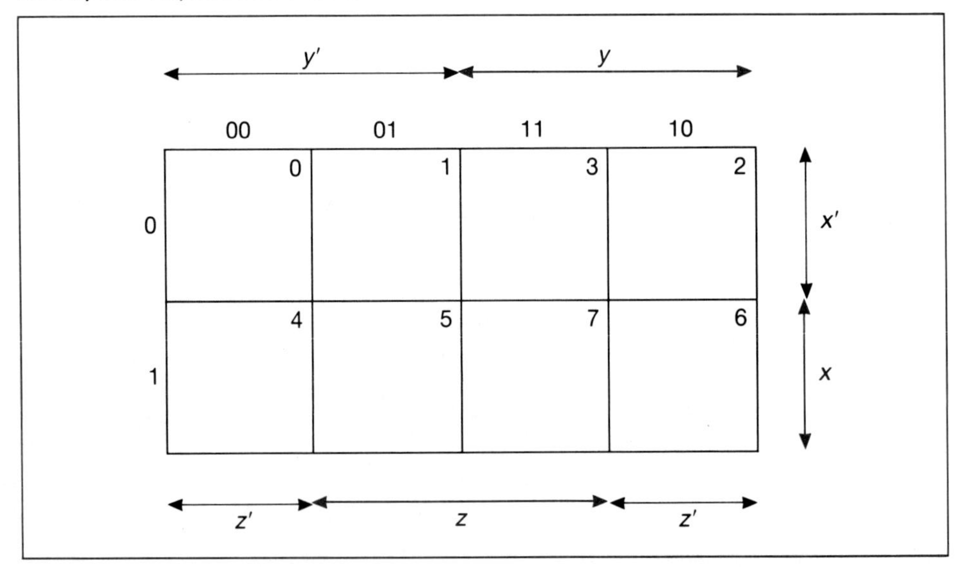

emphasize the construction technique. They are normally left off because they can be derived from the binary codes for the respective rows and columns.

The arrangement of the squares in a Karnaugh map is important. It is chosen so that *the codes for each adjacent pair of squares differ from each other in only one bit position.* We can use our last example to see why.

A function is represented by placing a 1 in the squares corresponding to each of the function's minterms, as shown in Fig. 2.15. The remaining squares are either filled with 0s or left blank. The adjacent 1 terms are then grouped into the largest possible squares or rectangles. The process is complete when all 1s have been included in at least one area, as shown in Fig. 2.16.

Each large area represents a reduced term in the simplified function; the larger the area, the simpler the term. Consider the area indicated by the bold square in Fig. 2.16. It contains the minterms 1, 3, 5, and 7, which is the subfunction

FIGURE 2.15 A Karnaugh map showing the minterms of example F_3 (*a*) with minterms specifically labeled and (*b*) as normally indicated.

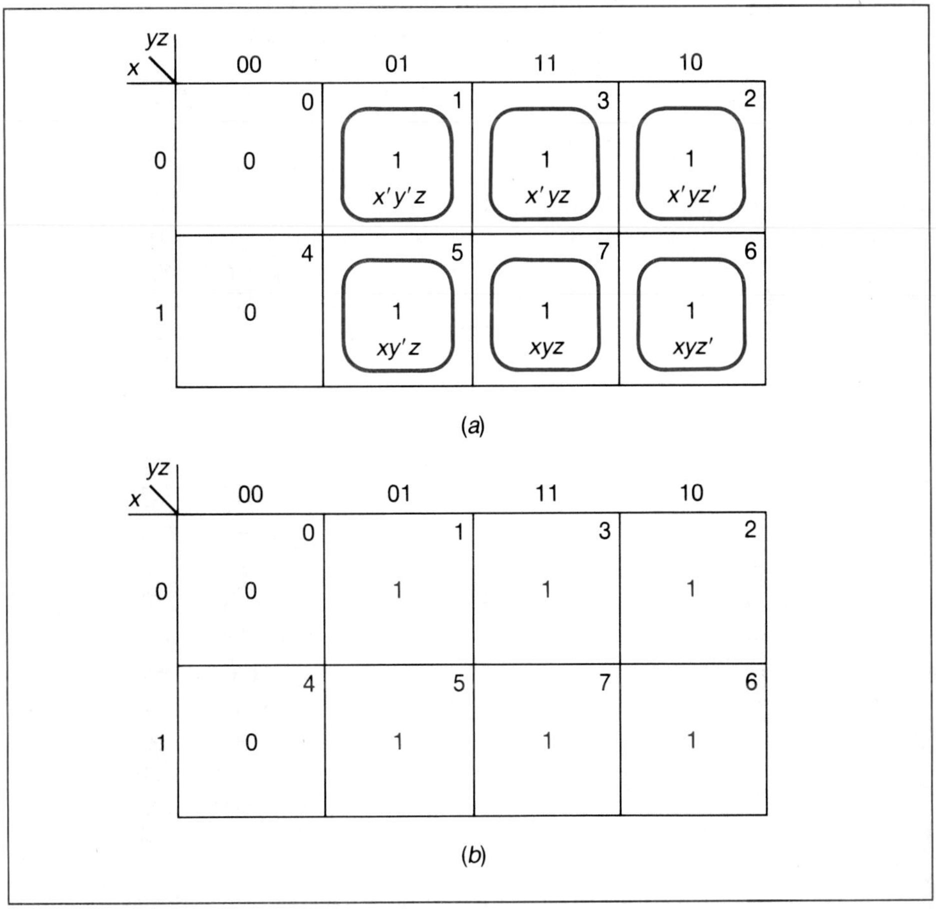

$x \backslash yz$	00	01	11	10
0	0 (cell 0)	1 $x'y'z$ (cell 1)	1 $x'yz$ (cell 3)	1 $x'yz'$ (cell 2)
1	0 (cell 4)	1 $xy'z$ (cell 5)	1 xyz (cell 7)	1 xyz' (cell 6)

(*a*)

$x \backslash yz$	00	01	11	10
0	0 (cell 0)	1 (cell 1)	1 (cell 3)	1 (cell 2)
1	0 (cell 4)	1 (cell 5)	1 (cell 7)	1 (cell 6)

(*b*)

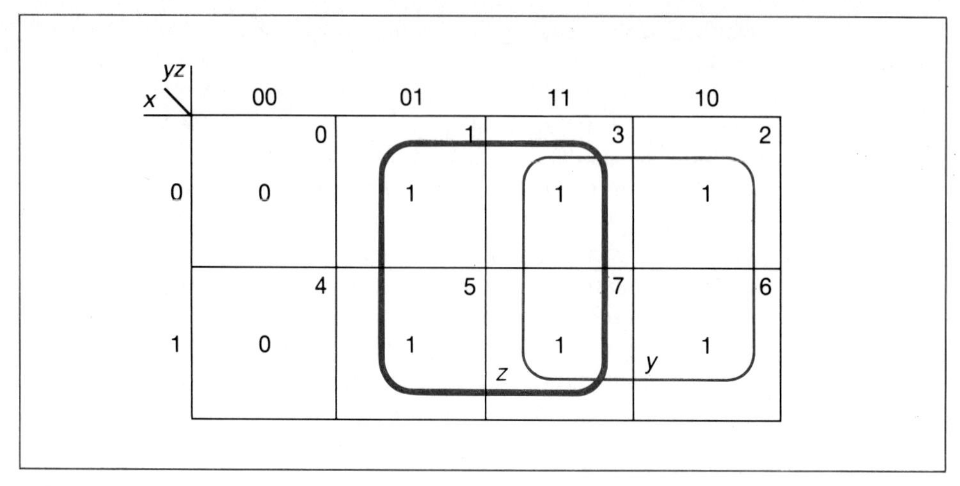

FIGURE 2.16 A Karnaugh map showing 1 terms grouped for simplification: $F_3 = y + z$.

$x'y'z + x'yz + xy'z + xyz$. This can be factored into $z(x'y' + x'y + xy' + xy)$. Then since $(x'y' + x'y + xy' + xy)$ is equal to 1, the reduced term is $z \cdot 1$, which is z.

A general rule of thumb is that one variable can be eliminated each time the size of the area is doubled. A single square of an n-variable function requires a product term of n variables, two adjacent squares require an $(n - 1)$-variable term, four adjacent squares require an $(n - 2)$-variable term, and so on. This is an important concept because it also provides us a way to map our original function (as shown in Fig. 2.17) without first expanding it into the standard-product form.

By now, you may be thinking that this is a simple approach, but a note of caution is in order. Karnaugh maps are not really two-dimensional diagrams. You can think of a three-variable Karnaugh map as an unrolled surface of a cylinder.

FIGURE 2.17 A Karnaugh map of example F_3 with the original product terms.

x \ yz	00	01	11	10
0	0 (cell 0)	1 (cell 1) $x'y'z$	1 (cell 3) $x'y$	1 (cell 2)
1	0 (cell 4)	1 (cell 5) $xy'z$	1 (cell 7) xy	1 (cell 6)

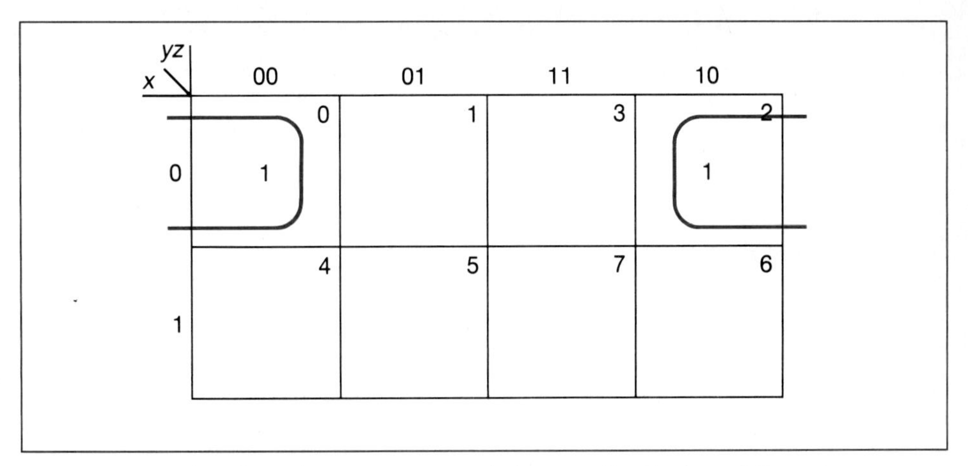

FIGURE 2.18 An edge-adjacent area example (representing the term $x'z'$).

The 0-2 pair is logically adjacent, as shown in Fig. 2.18. A similar, but more complex, situation holds for the four-variable map diagramed in Fig. 2.19. The top and bottom edges are also adjacent. A problem at the end of the chapter asks you to show all edge-adjacent areas for a four-variable map.

FIGURE 2.19 A four-variable Karnaugh map.

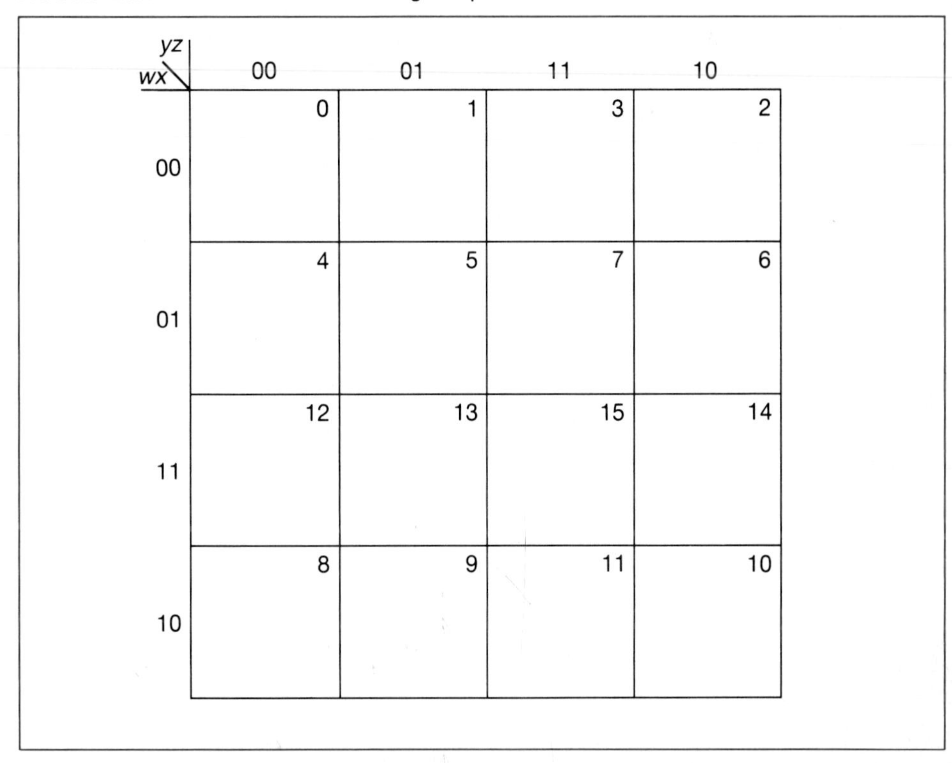

This ends our initial discussion of boolean algebra. However, before we move on to the next section, we need to point out that it is a subject you will frequently encounter throughout your curriculum. You need to develop a proficiency, and we urge you to do a number of the problems that have been included at the end of the chapter.

2.4 COMBINATIONAL AND SEQUENTIAL CIRCUITS

To be precise, the circuits we developed in the previous section should be called *combinational circuits.* In a **combinational** circuit, the output of the circuit at any point in time depends only upon the algebraic combination of its inputs at that time. Put in a different way, the circuit has *no memory.*

There is another important class of circuit, called a **sequential** circuit. The output at any point in time depends upon both the inputs to the circuit at that time *and* upon past events. A sequential circuit *has memory.* Figure 2.20 illustrates the conceptual difference between circuits with and without memory.

Before looking at implementations of sequential circuits, we need to add one stipulation to our discussion. We will limit ourselves to circuits whose elements are *synchronized* by the addition of a *clock* signal. The clock signal will oscillate between 0 and 1, as shown in Fig. 2.21. The percentage of time that the clock output stays at 1 is called the **duty cycle** of the clock signal. The number of times per second that it goes to 1 is called the **clock frequency**. You might have seen the specifications for a personal computer which say that the CPU has a 10-MHz clock; this means the clock will go to 1 at a frequency of 10 million times per second. A typical width for such a clock signal might be around 30 ns, with a +5 V signal representing a 1 and a 0 V signal representing a 0.

Given that stipulation, we can now turn to the development of a logic circuit that has memory, using NOR gates for the design. Let's begin with the circuit in Fig.

FIGURE 2.20 Logic circuits without and with memory: (*a*) a combinational circuit; (*b*) a sequential circuit.

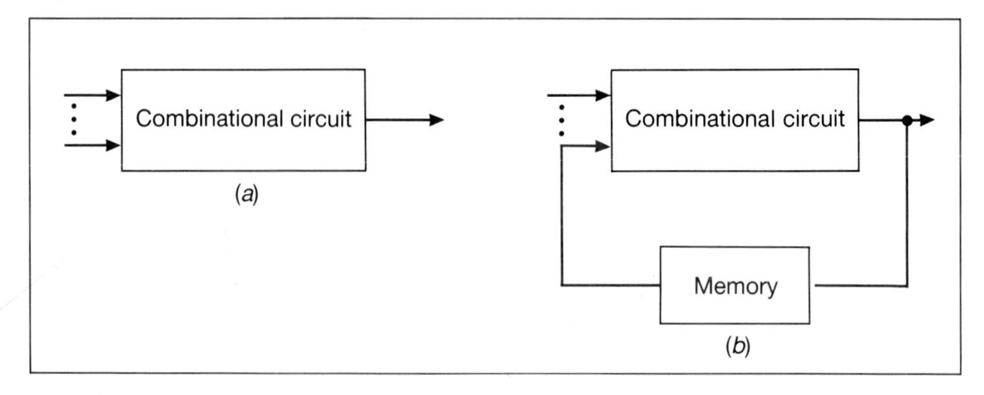

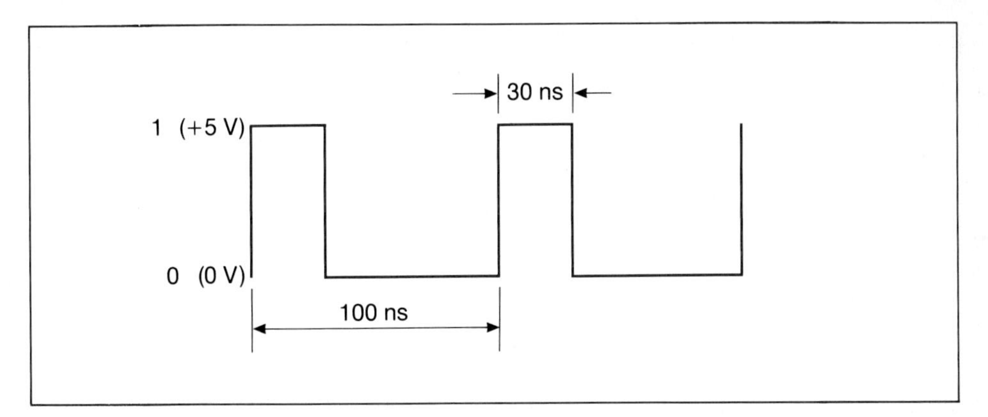

FIGURE 2.21 An illustrative 10-MHz clock signal with a 30 percent duty cycle.

2.22. Assume that at some point in time the logic states at the indicated circuit points are given by the bit encoding

R,	S,	U,	V,	Clock,	Q,	Q'
?,	?,	0,	0,	0,	1,	0

where the ? indicates that the R and S signals need not be known. This is because if Clock is 0, the outputs of both AND gates are automatically 0. Then since Q' is 0, both inputs to the top NOR gate will be 0 and its output will go to 1. This coincides with our statement that Q is 1. Also since Q is 1, the output of the bottom NOR gate will be 0. Again, this coincides with our statement about Q'. The circuit is **stable** or, equivalently, is in a **stable state**, because once it is put into this configuration, it will stay there. Figure 2.23 shows this circuit with the steady-state values of various circuit points identified.

FIGURE 2.22 A clocked circuit with a memory.

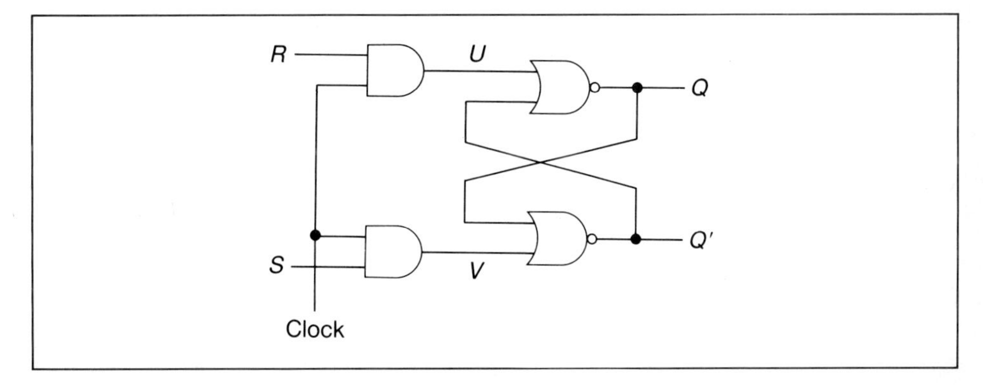

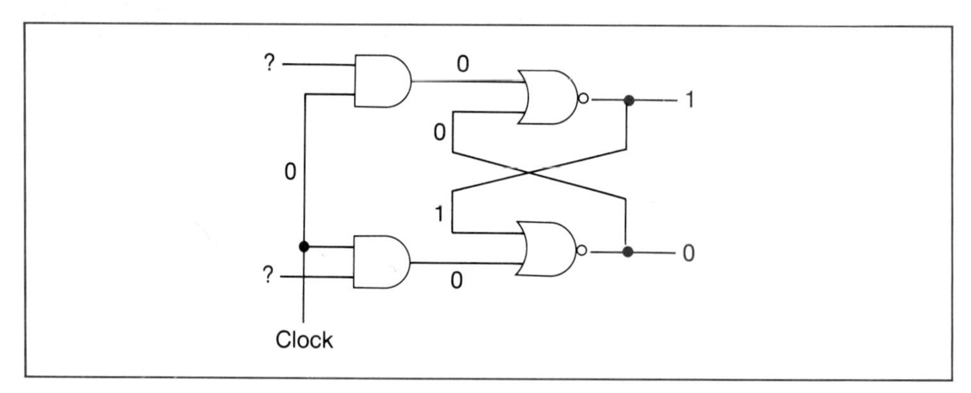

FIGURE 2.23 A stable state of the clocked circuit with memory, Fig. 2.22.

Now suppose that Clock goes to 1 while R is 1 and S is 0. In an idealized circuit, this would instantly change the output of the top AND gate. In an actual circuit, however, it takes a finite amount of time for the change to happen. The time for an individual element to change is called its **switching time**. The time for the change to make its way through the whole circuit is called the **propagation delay**. Because of the switching time, Q will temporarily still be 1, which will hold Q' at 0. The first thing that will happen in this circuit is that the output of the top AND gate will change to a 1 (since both its inputs are 1). That will make the output of the top NOR gate go to 0, which in turn will cause Q' to change to 1. The result is shown in Fig. 2.24. This is also a stable state because of the symmetry of the circuit. We have gone from one stable state to a different stable state. Our circuit has *switched* states.

The sequence of changes and their associated logic values for each point in the circuit can be seen in the timing diagram of Fig. 2.25. Note that both Q and Q' are 0 for a short period prior to reaching the final stable state. A similar condition exists when the Clock and S inputs cause the state of the circuit to be changed back. It also shows the reason why we made our synchronization stipulation at the begin-

FIGURE 2.24 The other stable state of the clocked circuit with memory.

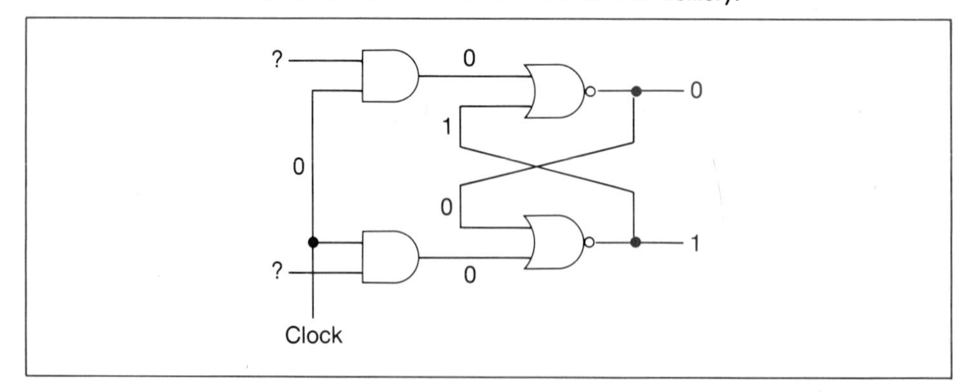

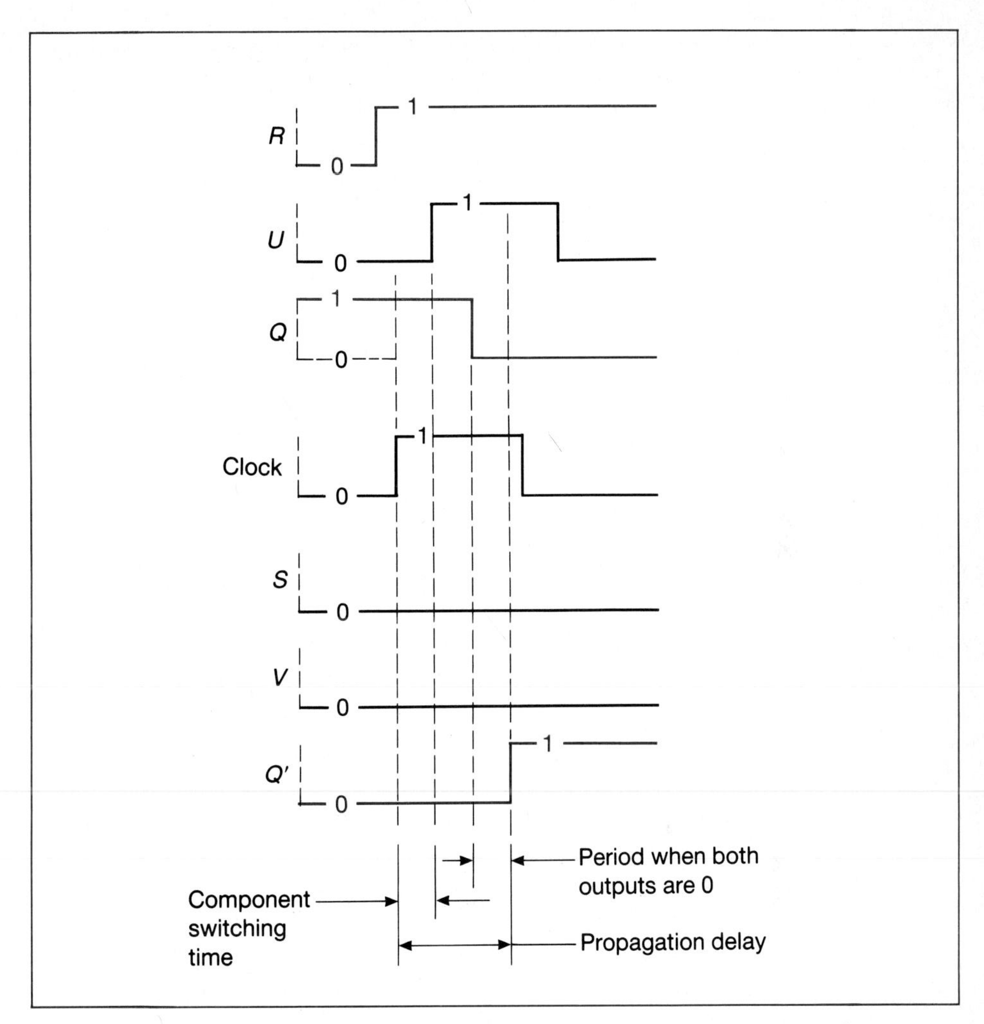

FIGURE 2.25 The timing sequence for the circuit for Fig. 2.22 during a change of state.

ning of this discussion. *Digital circuits often pass through a transition state that could result in an incorrect signal if the output were "read" at the wrong time.*

The operation of this circuit is important, and we want to be sure you understand it. Look at the timing diagram of Fig. 2.25 again. Notice that the Clock signal in our example remains 1 until after Q' has switched to 1. This keeps a 1 input to the upper NOR gate and ensures that Q will remain 0.

This type of memory device is called a **flip-flop**, because the outputs "flip" and "flop" between two stable states. More precisely, this device is called an ***RS* flip-flop**. The reason is that if the *reset* input R is 1 when the Clock goes to 1, then the outputs will go to (or stay in) a stable state with Q being 0 and Q' being 1. If the *set* input S is 1 when the Clock goes to 1, then the outputs will go to (or stay in) a stable state with Q being 1 and Q' being 0. To put it in another way, you could think of S as *setting* Q to 1 or of R as *resetting* Q to 0.

What happens if sometime later, with the flip-flop in the reset state, Clock again goes to 1 while R is 1? The outputs of the flip-flop will stay the same with $Q = 0$ and $Q' = 1$, as shown in Fig. 2.26. If Clock goes to 1 while S is 1, the flip-flop will switch states; the outputs of the flip-flop will be $Q = 1$ and $Q' = 0$. This implies that the circuit is a *sequential* circuit. Its actions at the next clock pulse will depend on both the state of the circuit and the inputs at the time. In other words, this circuit has memory. It remembers (or remains in) the state that it was in after the last clock pulse.

So far so good. We now have a memory device, but we need to examine the timing sequence shown in Fig. 2.25 again. Suppose that the width of the clock pulse had been much narrower. The Clock would go back to 0 before Q' has switched to 1. We would then have the condition shown in the timing diagram of Fig. 2.27. Since the upper NOR gate returns to 0 before Q' becomes 1, Q is allowed to switch back to 1. This in turn causes Q' to return to 0, but the pulse from when Q' was 1 is propagating through the upper NOR gate. This will cause Q to go to 0 for the pulse duration, which starts the cycle all over again. The circuit is now unstable, with its outputs oscillating back and forth between 0 and 1. It will usually stop because there are slight differences in the switching time of the individual components. Unfortunately, we cannot be sure which state it will finally settle into, or how long it will take to do it.

FIGURE 2.26 Flip-flop states during repeated set and reset inputs.

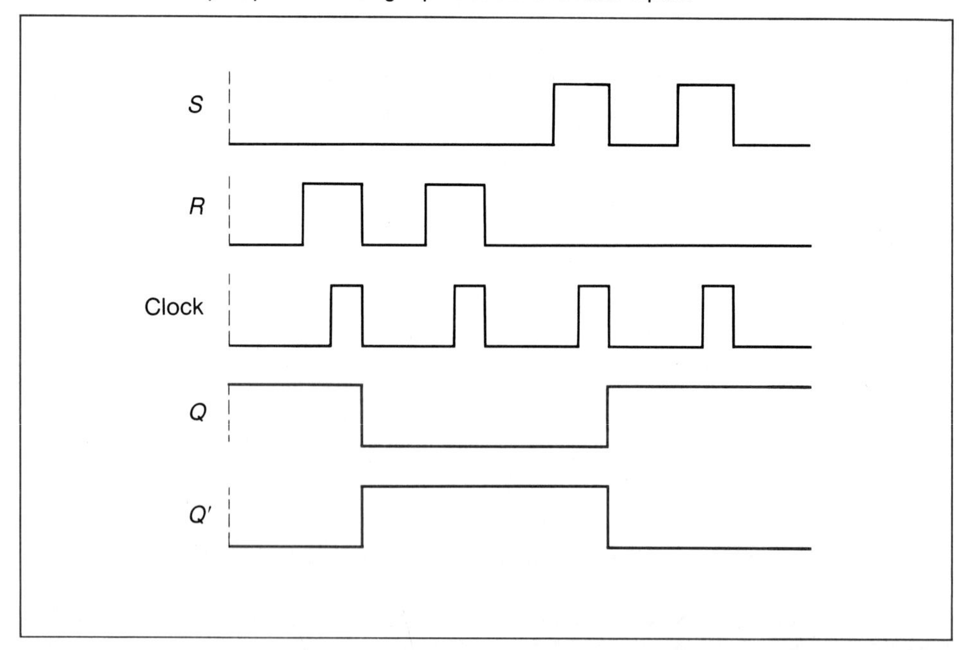

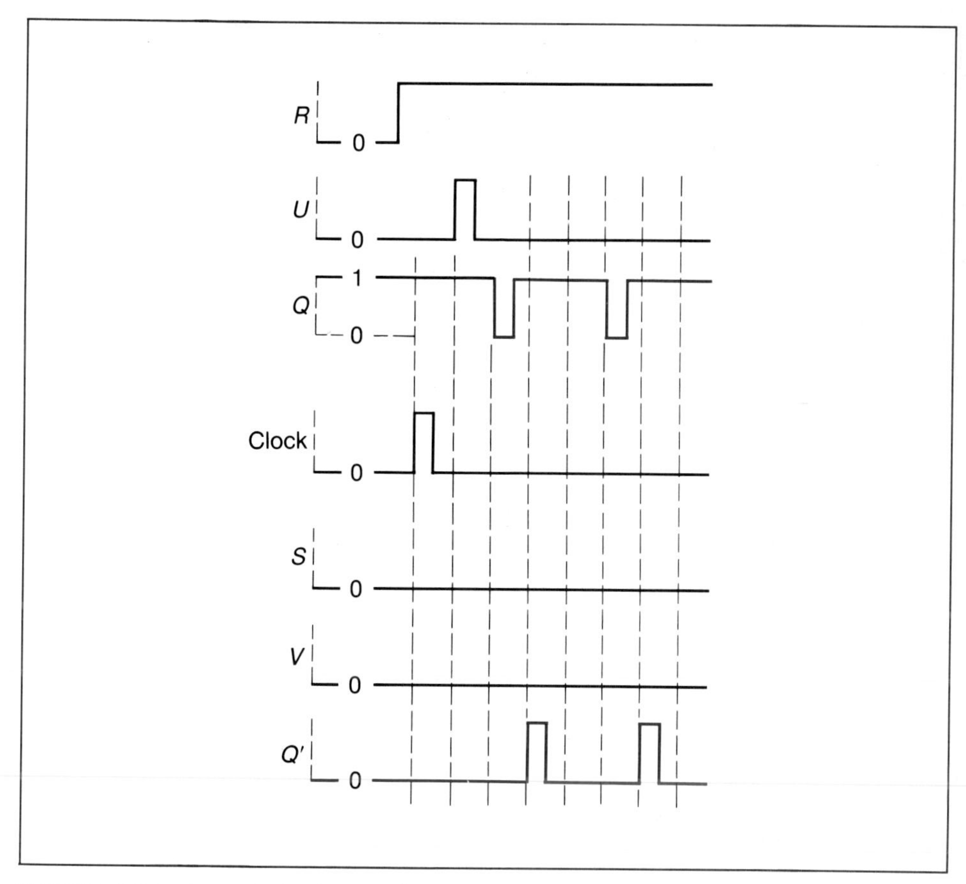

FIGURE 2.27 An unstable change-of-state timing sequence for the circuit of Fig. 2.22, due to a too-narrow clock pulse.

The same condition will result if either the *R* or the *S* signal is allowed to return to 0 too soon. However, we need to note that this is not a problem with the logic design. It is a problem with circuit design. The circuit designer must ensure that the input signals remain on until the circuit has switched and settled into its new state.

A different problem will arise if the clock pulse is too wide. Many sequential circuits feed the change in the *Q* or *Q*′ outputs of the flip-flop back to its *R* or *S* inputs through the combinational circuit. Consider the simple case shown in Fig. 2.28. This is a circuit where the value of the flip-flop at the next clock pulse is designed to be the complement of the value at the current clock pulse. Assume *Q* is 0 and *Q*′ is 1 when *T* and the Clock go to 1. The lower AND gate output now goes to 1 because all its inputs are 1. This makes the output of the lower NOR gate a 0. The flip-flop then changes state, with *Q* going to 1 because *Q*′ is now 0.

That's what we wanted, so things appear to be fine. However, what would happen if the Clock signal stays on too long? Let's assume that *T* and the Clock are

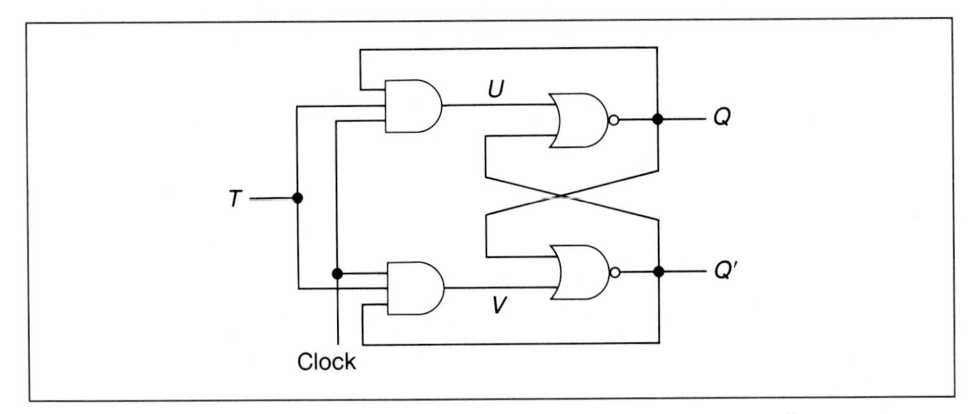

FIGURE 2.28 A proposed circuit that changes state when a 1 is applied to the *T* input.

both still 1 after the change of state. This makes the output of the top AND gate 1, and the flip-flop changes state again, back to where it was before.

We could try to time the flip-flop inputs precisely so that the second (unwanted) change would not happen, but such an approach would be very difficult when we are dealing with millions of gates in a computer. An alternate approach is to isolate the input of the flip-flop from the output through use of a circuit like the one shown in Fig. 2.29. It is called a **master/slave** version of an *RS* flip-flop. The reason is that the output of the flip-flop no longer comes directly from the output of the first set of NOR gates (the "master"). Instead, it comes from the output of the second set ("the slave"). Let's see how this affects the circuit.

FIGURE 2.29 A master/slave *RS* flip-flop.

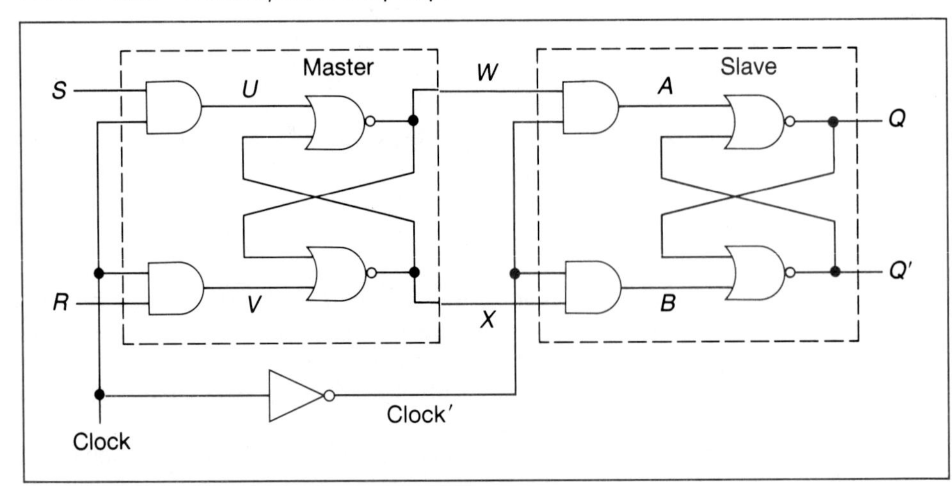

Assume at some point in time we have the flip-flop external states

R,	S,	Clock,	Q,	Q'
0,	0,	0,	0,	1

This means that the internal bit encodings will be

	R,	S,	Clock,	U,	V,	W,	X
Master:	0,	0,	0,	0,	0,	1,	0

	W,	X,	Clock′,	A,	B,	Q,	Q'
Slave:	1,	0,	1,	1,	0,	0,	1

Now suppose that Clock and S both go to 1; this means that Clock′, in the slave, will change to 0 and the slave will be locked in its current state. The master will change state, and the bit encodings will become

	R,	S,	Clock,	U,	V,	W,	X
Master:	0,	1,	1,	1,	0,	0,	1

	W,	X,	Clock′,	A,	B,	Q,	Q'
Slave:	0,	1,	0,	0,	0,	0,	1

This is a stable state, so the circuit will stay there. This implementation has removed the precise timing requirements for our clock; if the clock stays 1 long enough for the master to change, it doesn't matter when the clock output drops to 0.

What happens when the clock does switch to 0? The slave will take on the values of the master. Thus we will go to

$$Q = 1$$
$$Q' = 0$$

The timing for this change also need not be very precise, since the clock is now 0 and fluctuations in the inputs to the master will have no effect. The timing diagram for these sequences is given in Fig. 2.30.

The time needed to change both sides of the flip-flop in the master/slave plus the maximum time for the electricity to travel between connected flip-flops are the primary determinants of the maximum clock rate of a given computer. This is why manufacturers try to make circuit elements smaller, try to make them switch faster, and try to pack them more tightly on circuit boards.

Although there are other ways besides master/slave organizations to avoid the restrictive clock requirements, we will not discuss them here. We assume the problem has been recognized and solved, and we can now use RS flip-flops to

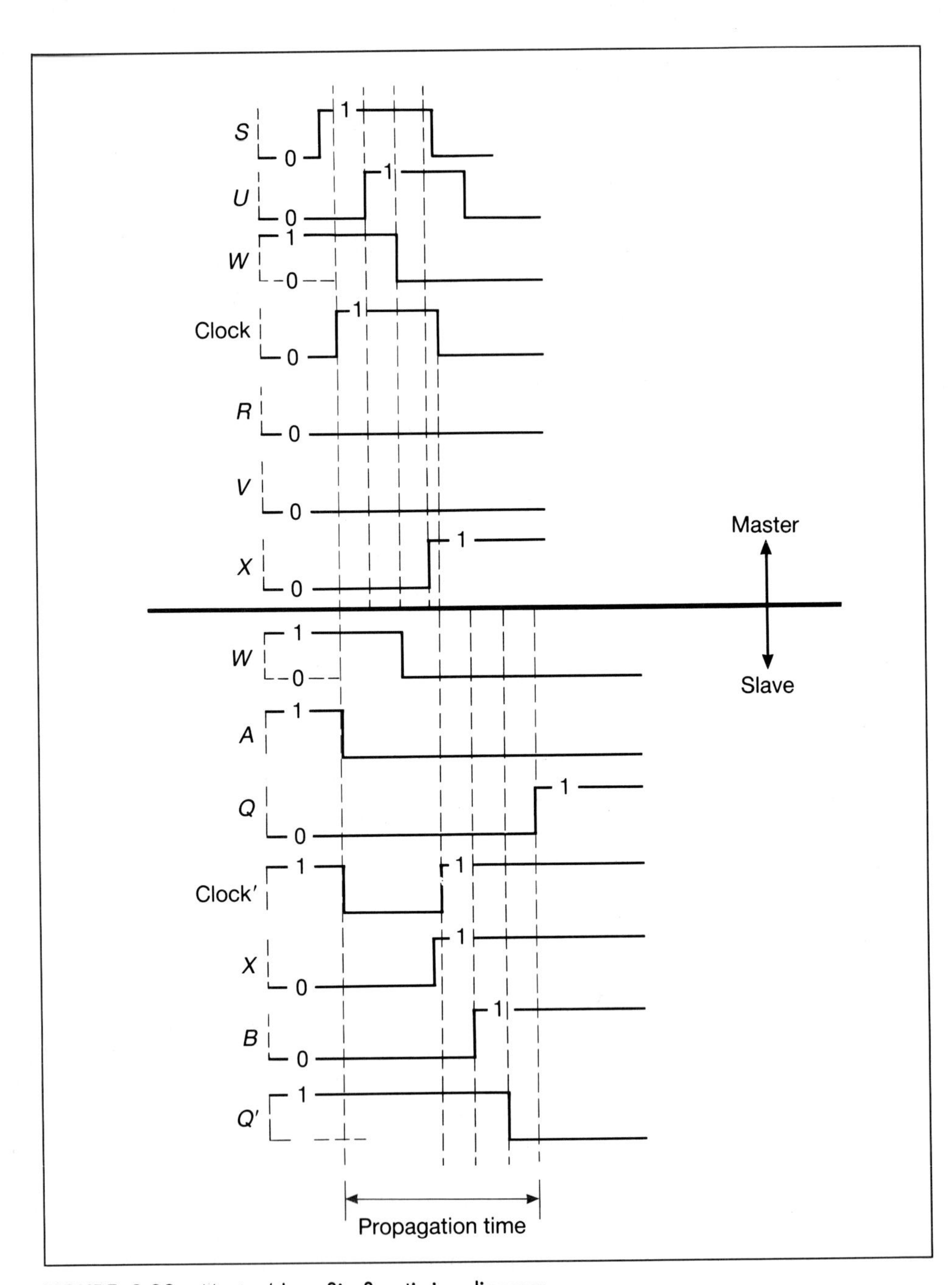

FIGURE 2.30 Master/slave flip-flop timing diagram.

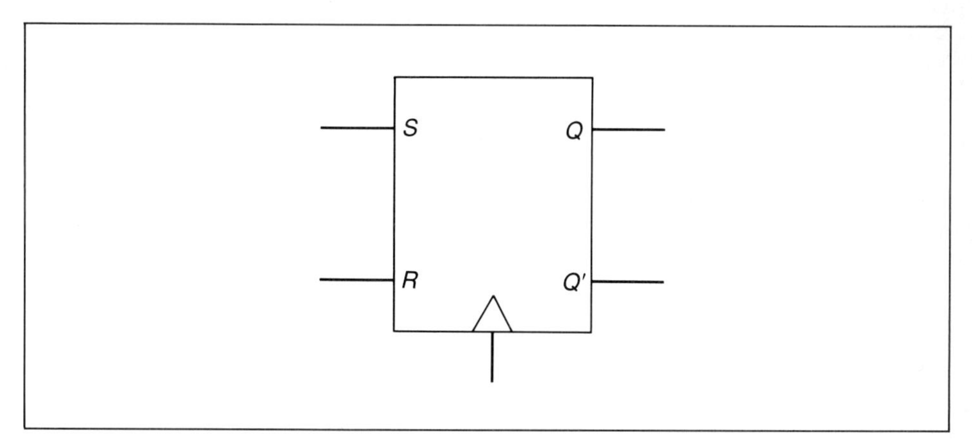

FIGURE 2.31 Circuit symbol for a clocked *RS* flip-flop.

design a circuit. The symbol we will use for this device is shown in Fig. 2.31. (The small triangle at the bottom of the rectangle represents the clock input.)

A table of the various possible states for the *RS* flip-flop is shown in Fig. 2.32. The table is called a **state transition table** because it defines how the output of the flip-flop changes depending on the logical combination of the input and current state. It defines the *transitions between different states.* Not all input combinations are shown because the input pair $R = S = 1$ is not allowed for the *RS* flip-flop.

FIGURE 2.32 State transition table for the master/slave *RS* flip-flop.

Inputs			Current state		Next state	
R	*S*	Clock	*Q*	*Q'*	Q	Q'
0	0	0	0	1	0	1
0	0	0	1	0	1	0
0	0	1	0	1	0	1
0	0	1	1	0	1	0
0	1	0	0	1	0	1
0	1	0	1	0	1	0
0	1	1	0	1	1	0
0	1	1	1	0	1	0
1	0	0	0	1	0	1
1	0	0	1	0	1	0
1	0	1	0	1	0	1
1	0	1	1	0	0	1

The Q output is considered to be the normal output, and Q' is the complement output. When we say that a flip-flop is in the 1 state, we mean that the Q output is 1 (and Q' is 0). Conversely, when a flip-flop is in the 0 state, Q is 0 and Q' is 1.

2.5 A DESIGN EXAMPLE

As a simple example of sequential circuit design, let's suppose that we want to input a random sequence of 0s and 1s and look for the pattern 010. If this pattern appears, we will turn on an indicator light. Otherwise, the light will stay off.

A pictorial description of our machine is shown in Fig. 2.33. In this diagram, the circles represent the various **states** of formation of our pattern. For example, 01 shows that we have seen a 0 followed by a 1. The lines of the diagram represent allowable **transitions** between states. The labels on these lines show the required input for which the transitions happen. The arrows show the direction of the transition. (For example, if you were in state 0 and received a 1, you would go to state 01; a 0 would cause you to stay in state 0.) The circle with an unlabeled arrow going into it is the beginning state of the machine and is usually called the **start state.** The state shown as two concentric circles (labeled 010) is the **final state**, or finish state.

Can you see from this diagram why we need a sequential circuit (sometimes called a sequential state "machine") rather than a combinational circuit for our machine? The effect of inputting a certain bit will be different, depending on what has happened before. Therefore we need a circuit with *memory* to keep track of associated previous events. The situation is similar in a computer; the effect of an instruction can be different depending on what values were stored in memory.

Figure 2.33 is a **finite state machine**, or **finite state automaton.** As explained above, the circles represent the allowable states of the system. *Finite* refers to the fact that the machine has only a finite number of such states. Viewed at a very simplistic level, all current computers, even the largest and most complex, are finite state automata.

FIGURE 2.33 State transition diagram of an example machine operation.

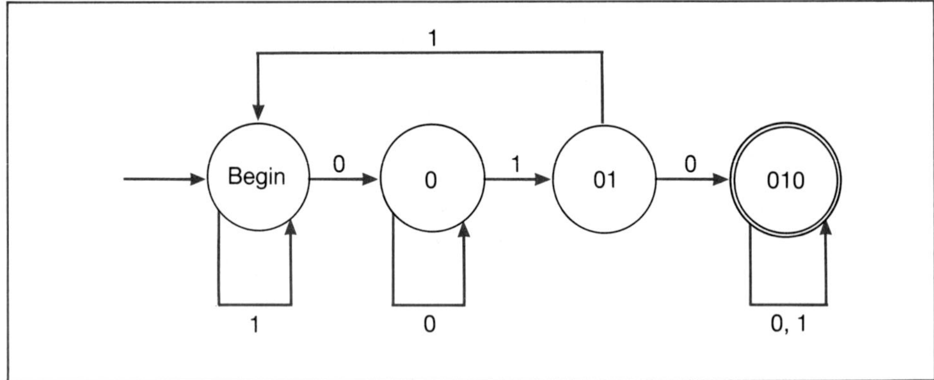

The complete finite state diagram for an entire computer would be *very* large. A machine with 1 million bytes of 8-bit memory elements would have $2^{8,000,000}$ different states of the memory itself. Therefore we usually apply finite-state-machine design techniques only at the individual subcomponents of the machine, such as the arithmetic/logic unit, not to the entire machine at the same time.

Our little machine has only four states, which is an easily manageable number. Each of these four states will be represented by memory elements; for this example, we'll use the *RS* flip-flops we described earlier. How should we establish a correspondence between states and memory elements? This task is a common one, called the **state assignment problem**. It merits a great deal of attention in an actual design, because the manner in which the correspondence is chosen can affect the complexity of the final circuit. However, we will not be concerned with circuit minimization in this example. We just want you to see the general idea.

There are four states, so we will need two flip-flops ($2^2 = 4$ states). Let's call these flip-flops FF_1 and FF_2. We will establish the following state assignments:

- Begin will be represented by FF_1 and FF_2 both $= 0$.
- 0 will be represented by $FF_1 = 0$ and $FF_2 = 1$.
- 01 will be represented by $FF_1 = 1$ and $FF_2 = 0$.
- 010 will be represented by FF_1 and FF_2 both $= 1$.

We can develop the state transition table shown in Fig. 2.34 from Fig. 2.33 and our state assignments. This table tells us what values must be applied to the set and reset terminals of the flip-flops in order to get the transitions specified by the diagram.

FIGURE 2.34 State transition table for our sample machine.

State reference	Current state		Input	Next state		Required sets and resets			
	FF_1	FF_2		FF_1	FF_2	R_1	S_1	R_2	S_2
Begin	0	0	0	0	1	×	0	0	1
	0	0	1	0	0	×	0	×	0
0	0	1	0	0	1	×	0	0	×
	0	1	1	1	0	0	1	1	0
01	1	0	0	1	1	0	×	0	1
	1	0	1	0	0	1	0	×	0
010	1	1	0	1	1	0	×	0	×
	1	1	1	1	1	0	×	0	×

Notice that some entries for the set and reset values are specified as ×. This means that we "don't care" what the input is for that corresponding terminal. Why? Because in certain situations we can get the desired result by specifying a value on one terminal, and the value on the other terminal is irrelevant.

For instance, in row one we see that FF_1 must stay at a 0 output. This means we must have a 0 on its set terminal—a 1 would cause the output to change to 1. It doesn't matter what the value on the reset terminal is. A 0 will leave the flip-flop in its current state, and a 1 will cause it to reset. Both result in a 0 output, which is what we want.

Don't care cases often occur in computer hardware design. For example, we might use 2 bits to represent whether a comparison resulted in greater than, less than, or equal. There are three possible results, but 2 bits can represent four situations. Thus we don't expect one of the possible bit configurations to ever occur, and we could specify a don't care for that situation.

We can now develop the boolean expressions for each of the set and reset conditions, based on the input to the machine and the current state of the flip-flops. We will do one for an example to show the value of including don't care conditions in your simplifications. Figure 2.35 shows all the set conditions for FF_2. Notice the difference between including and not including the don't cares.

The four expressions for the flip-flop set and reset terminals are

$$S_1 = FF_2 \cdot \text{input}$$

$$R_1 = FF_2' \cdot \text{input}$$

$$S_2 = \text{input}'$$

$$R_2 = FF_1' \cdot \text{input}$$

FIGURE 2.35 Karnaugh map for S_2 including don't care conditions.

Input \ FF_1 FF_2	00	01	11	10
0	1	×	×	1
1			×	

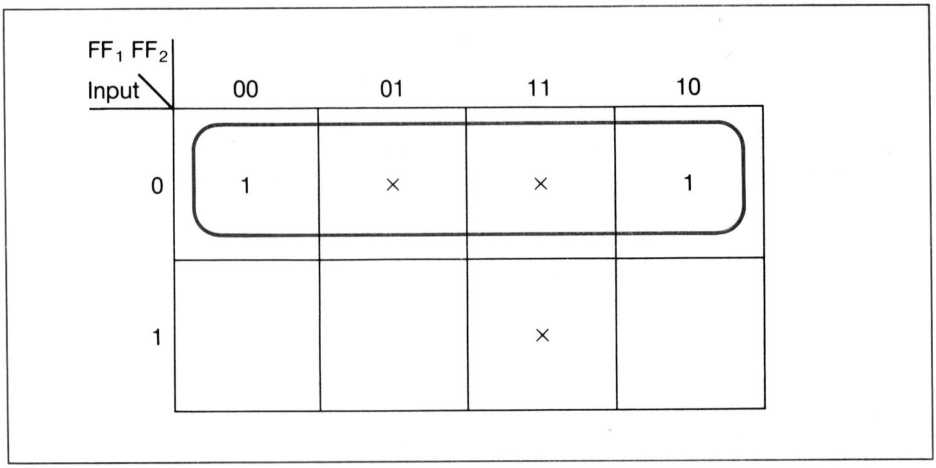

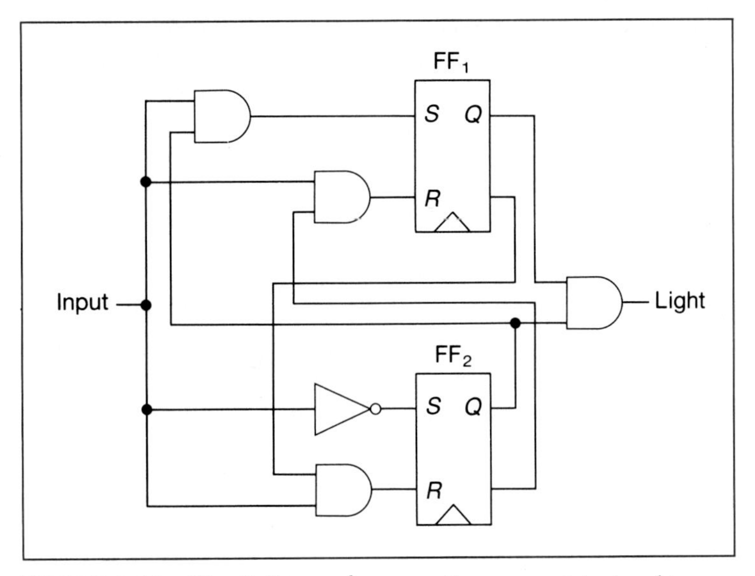

FIGURE 2.36 Circuit diagram for our pattern-recognizing machine.

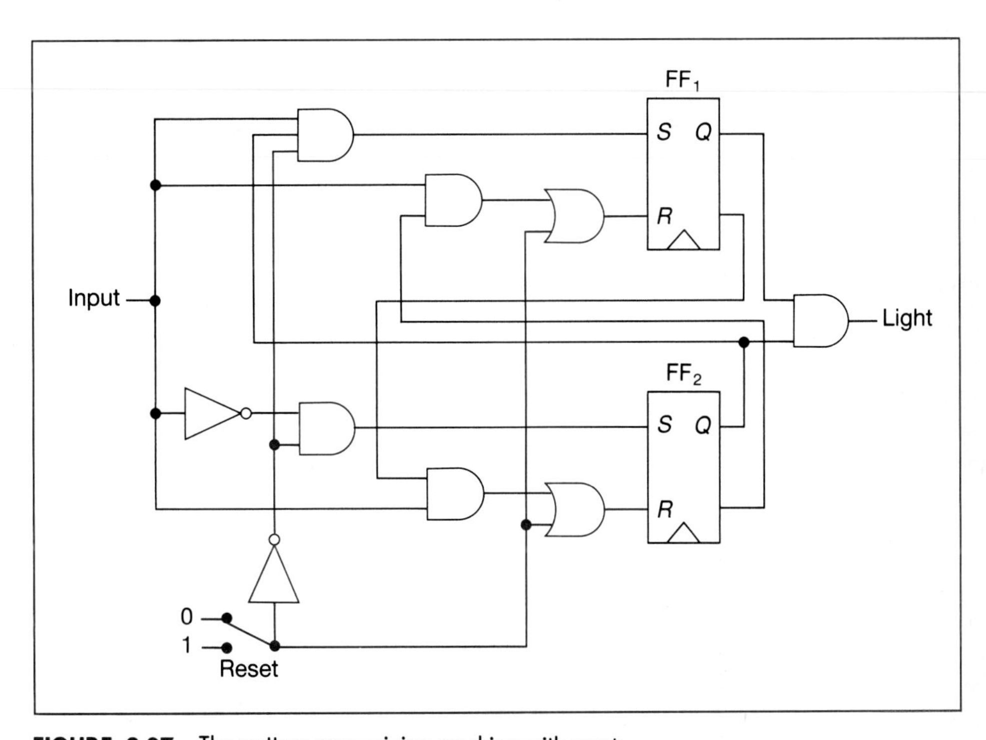

FIGURE 2.37 The pattern-recognizing machine with reset.

The circuit for these expressions is shown in Fig. 2.36. As shown, it will output a 1 to Light when the chosen pattern has been detected. Note that we need both combinational elements (the AND and INVERTER gates) and sequential elements (the flip-flops) in this circuit.

To make the circuit a little more practical, we should add one feature. As it is, a user cannot reset the circuit back to the original starting position, with both flip-flops 0. To add such a *reset* option, we incorporate a switch that can force a 0 on the set terminals and a 1 on the reset terminals, which resets both flip-flops. This is shown in the corresponding circuit of Fig. 2.37.

This ends our design example. You have now seen the basic ideas of combinational and sequential circuits and how logic elements can be used to implement such circuits. We must stress that AND, OR, NOT, and *RS* flip-flops are not the only devices used to design circuits; we have selected them only to provide you with an introduction to this area.

2.6 SUMMARY

In this chapter we have looked at selected concepts from discrete mathematics, then progressed to boolean algebras. Having this groundwork, we introduced logical circuit elements and considered the design of simple combinational and sequential circuits. Although the discussions have been brief, and we have skipped many details, you should have the fundamental ideas of discrete mathematics and computer circuits. You will learn later that there are additional types of circuit elements and still other procedures for simplifying and designing logic circuits. Before then, we need to consider how information is represented in the computer.

PROBLEMS

2.1 Describe the members of your immediate family as a set, using the equation form

$$\text{(Your family name)} = \{\ldots\}$$

2.2 Diagram the members of your immediate family set as elements of various universal sets, such as men, women, parents. (See Fig. 2.1.)

2.3 What is the cardinality of your immediate family set?

2.4 Using Fig. P2.4, identify the areas that satisfy the following set relationships:

a. $S = \{x \in C: x \text{ has properties } B\}$

b. $S = \{x \in C: x \text{ has properties } A \text{ and } B'\}$

c. $S = \{x \in B: x \text{ has properties } A \text{ or } C\}$

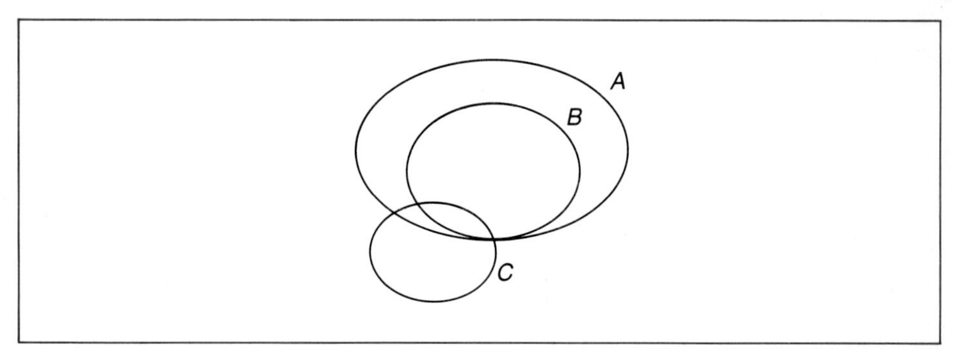

FIGURE P2.4

2.5 Write the set expressions for the cross-hatched areas in Fig. P2.5.

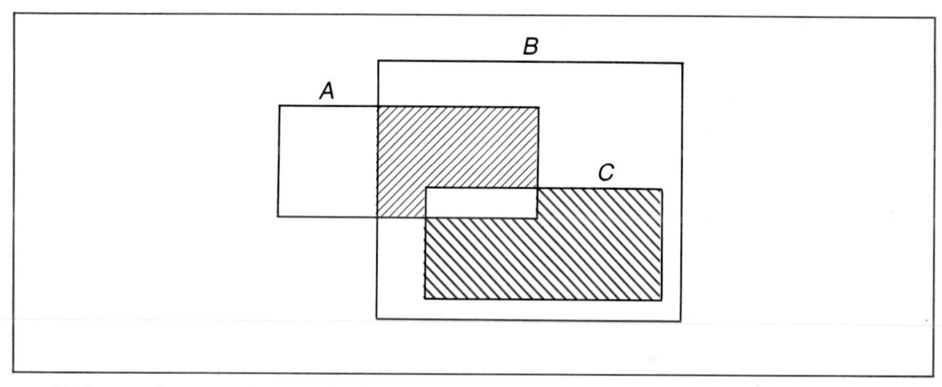

FIGURE P2.5

2.6 The three-variable Venn diagram of Fig. P2.6 contains eight distinct areas that represent all the possible AND combinations of the variables A, B, and C and their complements.

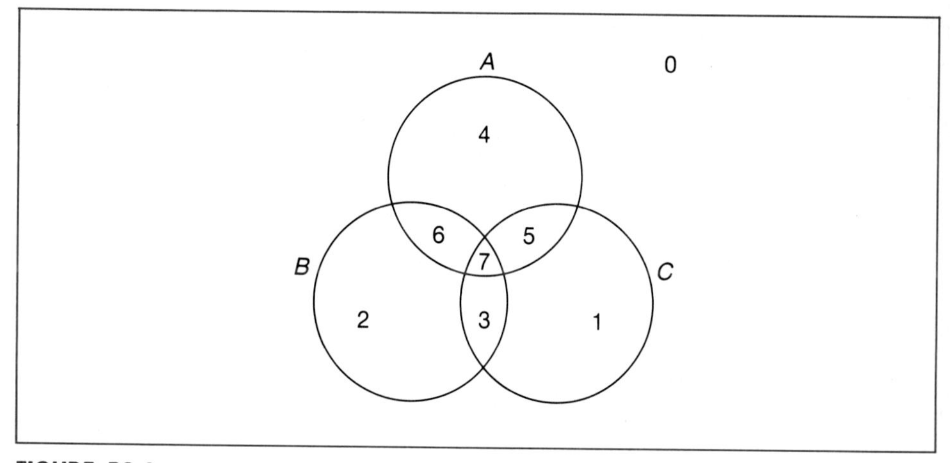

FIGURE P2.6

Prepare a table of these eight possible AND combinations, and specify which area in the figure corresponds to each entry in your table.

2.7 Identify the area or areas in Fig. P2.6 that correspond to each of the following boolean expressions:

a. $A + B + C$
b. $A' + B' + C'$
c. $A \cdot (B + C)$
d. $B + A \cdot C$
e. $A + B$
f. $A \cdot B'$
g. $A' + B + C$

2.8 Find a boolean expression for the function $F(A, B, C)$ defined in the following table:

A	B	C	F
0	0	0	1
0	0	1	1
0	1	0	1
0	1	1	0
1	0	0	0
1	0	1	1
1	1	0	0
1	1	1	0

2.9 Simplify the following boolean expressions algebraically and verify your work by comparing areas of both the original expression and your result on a Venn diagram like that shown in Fig. P2.6.

a. $A' \cdot B + A \cdot B + A \cdot B \cdot C + A \cdot C$
b. $(A + B)' + (A + C)' + (B + C)'$
c. $B \cdot (A + C') + A \cdot B + A \cdot (B + C)$
d. $A \cdot B \cdot C + A \cdot B \cdot C' + A \cdot B' \cdot C + A \cdot B' \cdot C'$

2.10 One way to evaluate a digital circuit implementing a given function is to count the number of logic gates and the number of inputs to those gates. Diagram the original and simplified expressions in Prob. 2.9 and compare the number of logic gates and inputs required for each case.

2.11 Expand the expressions of Prob. 2.9 into their equivalent minterms.

2.12 Write the following functions of three variables as boolean expressions:

a. $\Sigma(1, 3, 4, 6, 7)$
b. $\Sigma(0, 1, 3, 4)$
c. $\Sigma(0, 2, 4, 6)$
d. $\Pi(1, 3, 4, 6, 7)$
e. $\Pi(0, 1, 3, 4)$
f. $\Pi(0, 2, 4, 6)$

2.13 Use Karnaugh maps to simplify the functions of Prob. 2.12.

2.14 Draw a clock signal with a 10 percent duty cycle.

2.15 What is the advantage of a master/slave flip-flop over an ordinary clocked *RS* flip-flop?

2.16 Repeat Prob. 2.11 with the following don't care conditions:

a. Don't care (2, 5)
b. Don't care (2, 7)
c. Don't care (1, 5, 7)
d. Don't care (2, 5)
e. Don't care (2, 7)
f. Don't care (1, 5, 7)

2.17 How must the logic expressions of the example given in Sec. 2.5 be modified to include a reset option?

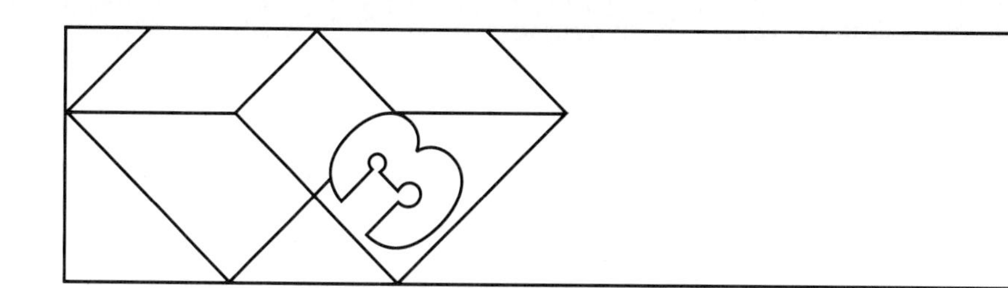

Information Representation in a Computer

In Chap. 1, we learned that the development of mathematics and its subsequent use to formulate and explain science depended on the existence of:

- A number system that would support numeric calculation
- An alphabet and language that would allow us to explain and record what we were doing

A similar situation exists with computing, but computers cannot directly use either our alphabet or our number system. In the last chapter, we saw that digital circuits can only recognize two states, equivalent to the binary 0 and 1. This limits our options, but it is not an insurmountable problem. The common solution is to code each of the symbols in our alphabet as strings of binary digits (bits) and to adopt the binary number system for all computations.

It sounds simple enough, but the details of how it's done need a little more explanation. Our alphabet and number system are both very familiar objects that we use every day. We do not have to think before we use them. It's automatic, and that can be a problem. We have to understand how we actually represent and use our information before these "automatic" capabilities can be translated to a computer.

This chapter will concentrate on information representation, both numbers and text. We will begin by expanding our definition of sets to include sequences, alphabets, and strings. Next we will spend a little time looking at how alphabetic information is stored in a computer. This will be followed by an in-depth discussion of number representation and methods for converting between number systems. We will end with a short discussion on errors in the representation of numeric data.

3.1 SEQUENCES

Sets, as we described them in the last chapter, are an important concept in the theory of computing. You will probably see them many times in your studies. However, in certain applications you will need a structure with two characteristics that are different from sets:

- The order of the elements is important.
- The elements need not be distinct.

Data points in space are an example. Suppose you need to calculate several points on a curve like that illustrated in Fig. 3.1. Each point must be identified as a pair of numbers x_i and y_i, where x_i and y_i are the respective values along the axes of a graph. You could not use sets for this representation, because in a set:

- *Order cannot be specified* (e.g., a point $\{6, 5\}$ would be interpreted as being equal to $\{5, 6\}$).
- *Replication is ignored* (e.g., the point $\{7, 7\}$ would be reduced to $\{7\}$).

Therefore, we will use a different representation called an ordered pair. An **ordered pair** is the structure $\langle x, y\rangle$, where x is the first component and y is the second component of the pair. The ordered pair $\langle 6, 5\rangle$ is not the same as $\langle 5, 6\rangle$, $\langle 7, 7\rangle$ is a valid ordered pair, and $\langle 7\rangle$ is not a pair at all.

The term *ordered pair* was used because it is fairly common; more formally, we call these structures **ordered 2-tuples**. It is easy to generalize the concept to ordered tuples of arbitrary length. Points in three-dimensional space would be represented with ordered 3-tuples. An ordered n-tuple is a structure $\langle a_1, a_2, \ldots, a_n\rangle$, where a_i is the ith component of the n-tuple.

FIGURE 3.1 Example of data requiring ordered representation.

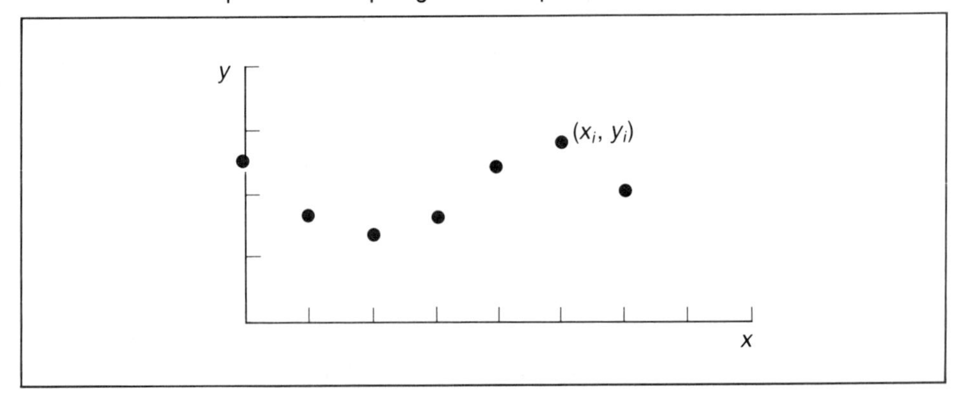

Before we go any further, we need to point out that spatial coordinates are not the only form of n-tuples. We normally specify that each a_i is an element of a certain corresponding set A_i. For example, consider the students in our class. We might define several sets

$$A_1 = \{1, 2, 3, 4, 5\}$$

$$A_2 = \{\mathit{male}, \mathit{female}\}$$

$$A_3 = \{\mathit{freshman}, \mathit{sophomore}, \mathit{junior}, \mathit{senior}\}$$

$$A_4 = \{\mathit{major}, \mathit{nonmajor}\}$$

Then we could use 4-tuples $\langle a_1, a_2, a_3, a_4 \rangle$ to describe the top five students in the class; a_1 gives the student's ranking, a_2 gives his or her sex, a_3 indicates class level, and a_4 says if the student is a major or nonmajor.

Given the sets from which the elements are drawn, we can describe all the possible ordered n-tuples that can be generated. The **cartesian product** of sets A_1, A_2, ..., A_n is the set $\{\langle a_1, a_2, \ldots, a_n \rangle : a_i \in A_i\}$. Any other set of ordered n-tuples using the same A_i will then be a subset of this cartesian product. For the example described in the previous paragraph, there would be $80 = 5 \times 2 \times 4 \times 2$ 4-tuples in the cartesian product.

The term *sequence* is also used to describe n-tuples. Usually a **sequence** is defined as an ordered n-tuple, where the n is unspecified. Thus $\langle 1 \rangle$, $\langle 1, 2 \rangle$, and $\langle 1, 2, 4 \rangle$ are sequences.

3.2 SYMBOLS, ALPHABETS, AND STRINGS

Although general forms of sequences are very important in their own right, the primary concern here is the representation of information. Therefore, we will restrict the remainder of our discussion of sequences to a special form that we will use often for our representations—the string.

When developing a new representation, we first decide on all the elements that will be included. These elements are called the **symbols**, and the set of symbols is called an **alphabet**. The symbols of an alphabet are often arranged in some order that is fixed by custom and that establishes a relative "value" for each of the symbols. The Roman alphabet $\{a, b, c, \ldots\}$ and the Hindu-Arabic decimal digit set $\{0, 1, 2, \ldots\}$ are examples of alphabets with this form of ordering. The punctuation set {?, ", !, ...} is not ordered.

Several alphabets can also be combined to make a larger alphabet. A familiar example is the **character set** we see on computer and typewriter keyboards.

A **string** over an alphabet is a finite sequence of symbols from that alphabet. Unlike the sequences in our previous description, these strings are not written with braces and commas. We simply write them as adjacent symbols. Thus *cat* and 2001 are strings over the Roman and decimal digit alphabets, respectively. Also note that we speak of strings over a single set, the alphabet. With n-tuples, we allowed each element of the tuple to be drawn from a different set.

Strings can be connected to form new strings. If S and T are strings over an alphabet A, then the **concatenation** of S and T, written $S \sqcup T$, is simply the string S followed by the string T. For example, if

$$S = rtwo$$

$$T = dtwo$$

are strings over our Roman alphabet, then

$$S \sqcup T = rtwodtwo$$

Since a string is a sequence, the length of a string is simply the number of elements in the sequence. Similarly, the jth element of a string S, usually denoted $S[j]$, is the jth element of the sequence. For string

$$W = dobedobedo$$

the length of W is 10 and

$$W[1] = W[5] = W[9] = d$$

A string can contain no symbols at all. In this case, it is called the **empty string**. It is denoted by the symbol λ when it is being spoken of by itself. However, when it is concatenated to another string, it vanishes. For example,

$$\lambda \sqcup a = a \sqcup \lambda = \lambda \sqcup \lambda \sqcup a = a$$

Now that you have been given a definition and description of strings, perhaps you will appreciate that everything stored in a computer (both data and instructions) can be regarded as strings. All manipulations done by the computer are then operations on strings. Input to and output from a computer are strings. Often research done in the theory of computing is expressed in terms of operations on strings.

Several computing languages such as GPM, Snobol, and LISP were developed especially for string manipulation, and many other languages have procedures for string manipulation. Examples are such things as PREFIX, SUFFIX, CONCAT, and SUBSTR.

3.3 ALPHABETIC REPRESENTATION IN A COMPUTER

On first look, the task of developing an encoded representation for the symbols that make up the character set for a computer would appear to be relatively easy. The binary representation of an arbitrary set is a straightforward and precise task. All we have to do is define a unique code for each symbol. However, it is not quite that easy. Before we start, we need to agree upon:

- Which specific symbols (alphabetic, numeric, punctuation, scientific, mathematical, ...) are to be included
- How these symbols should be ordered
- Whether we want to represent each symbol with the same number of bits

Let's take the last concern first. We could decide to represent each symbol by the same number of bits, or we could choose to make some codes shorter than others. It has been done both ways, and it is a decision that can affect the design complexity of the system. Perhaps a review of history will help.

Binary codes were used to represent information long before there were computers. Consider the Morse code, which was developed in the 1830s for the electric telegraph. It used patterns of one to six dots and dashes to represent the symbols in the character set (see Fig. 3.2). The reason was telegrapher efficiency.

Information was transmitted by momentarily closing a switch (the telegraph key) to create short or long signal pulses (dots or dashes) in a circuit between distant locations. Each dot or dash required a separate manual operation. However, Morse designed his code to save telegrapher time. Shorter codes were used to represent the most commonly used letters; longer codes were assigned to less commonly used numbers and punctuation symbols. A short pause indicated the end of a letter code, and a longer pause signified the end of a word.

The Morse code proved to be very effective for human telegraphers but not for teletypewriters and teleprinters. Therefore, early versions of these machines used the 5-bit Baudot code, which was developed in 1874. The reason was decode complexity. Decoding variable-length codes requires a sequential machine similar to our final design example in Chap. 2, except that it has many more states and is more complex.

The state diagram for a variable-length decoder is a "tree" structure like that shown for the Morse code decoder in Fig. 3.3. A Morse decoder requires three types of transitions—for dot, dash, and pause. A dot or dash causes the machine to go to the next state in the code sequence. A pause returns the machine to the beginning and causes an output of the symbol associated with the current code state. This is indicated on the diagram by a curved transition marked p/ (for pause) followed by the appropriate symbol. A problem at the end of the chapter asks you to consider what would be needed to detect the end of a word.

Fixed-length codes, on the other hand, represent each symbol with the same number of bits. This can sometimes require more space than would be needed for variable-length codes. Recall our discussion of power sets in Chap. 2. We showed that the minimum number of bits needed to represent a set of m elements is

$$n = \lceil \log_2 m \rceil$$

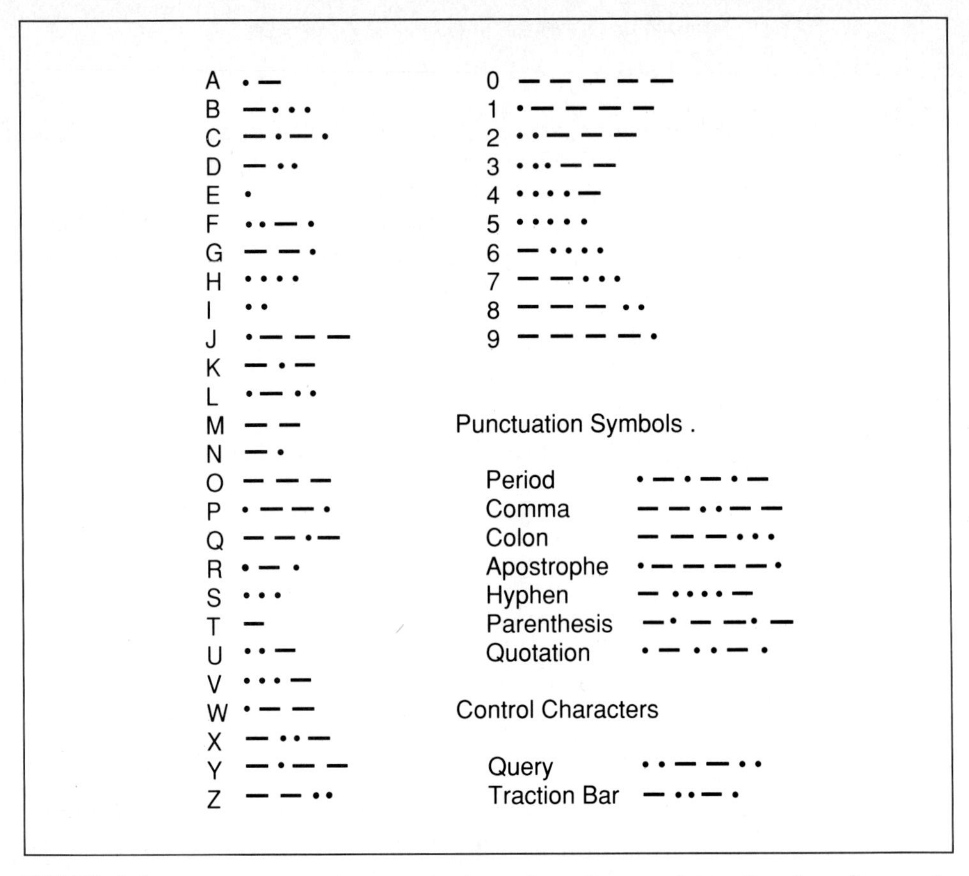

FIGURE 3.2 The international Morse code—adopted at an international conference in 1851.

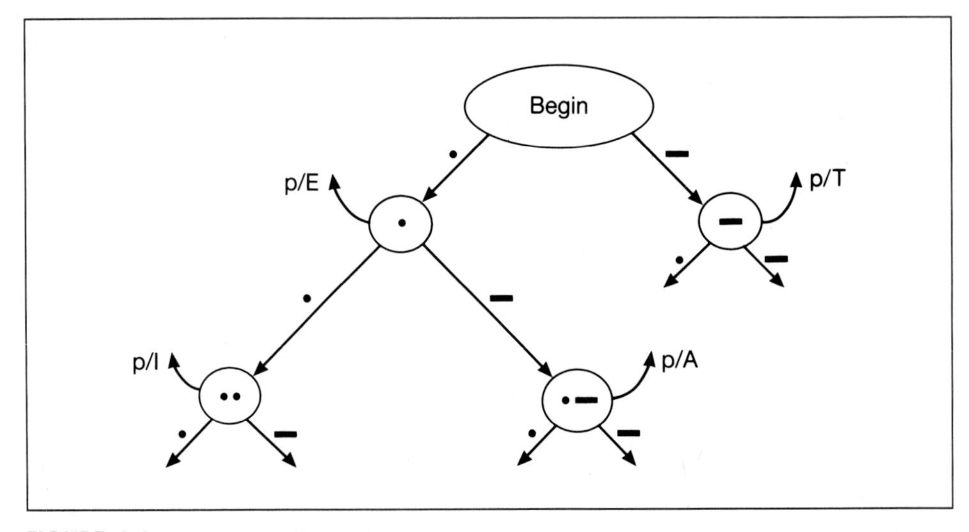

FIGURE 3.3 Part of the state diagram for a Morse code decoder.

The Morse code varies from 1 to 6 bits, with an average code length of 4. Fixed-length coding of the same 45-character set would require 6 bits for each symbol, and 19 possible code combinations would go unused.

Even though fixed-length coding requires more code space than variable-length coding, it has become the preferred approach for internal computer codes. Fixed-length decoding is faster. All that is required is a simple two-level combinational circuit like the one shown in Fig. 3.4.

We need to pause for a moment to emphasize a point made earlier in this section. We have just discussed two conceptually different approaches for encoding symbols of an alphabet. The first approach required a shorter average code length but resulted in a more complex and slower decode design. The second required a longer code length but allowed us to use a simpler and faster decode design. Together, these two approaches provide an example of a **space/time design tradeoff**. Increasing the code space allows us to simplify the decode circuitry and save processing time. It is an important concept that you will encounter many times in your career.

Once we know how many bits will be needed, we can begin to assign each symbol with its own unique code. This requires consideration of how the symbols in the set are to be organized. Sorting procedures typically involve comparing key sections of the records to determine which is numerically larger or smaller. This merely requires our character set to be encoded with binary values assigned in the order we wish the symbols to appear in our expanded alphabet.

So far so good. We have decided that our character set will have fixed-length coding and that the symbols will be in alphabetical and numerical order. This still

FIGURE 3.4 A 3-bit fixed-length code decoder.

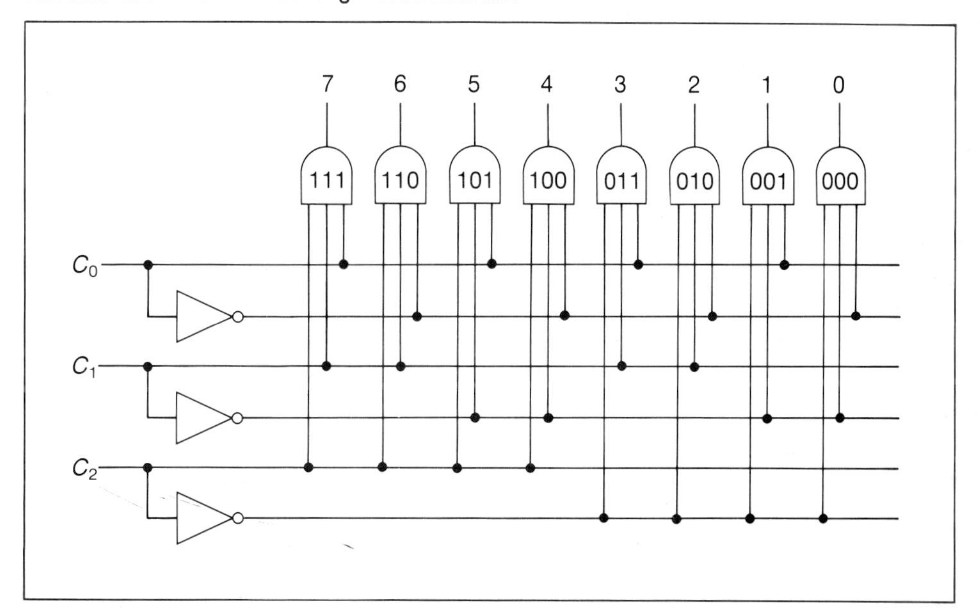

leaves us considerable latitude for the actual assignment of code patterns. We have not decided which punctuation marks or control characters to include, nor where to put them in the symbol order. However, you have a better idea of the problem.

In early computers, there were probably as many variations to code assignment as there were manufacturers. Character code lengths varied from 5 to 8 bits, and assignments of codes with the same length frequently did not match. It was very difficult to exchange information between computers.

Finally, the users demanded that something be done. Two dominant coding schemes emerged:

- ASCII (American Standard Code for Information Interchange)
- EBCDIC (Extended Binary-Coded Decimal Interchange Code)

The ASCII code had been developed by the American National Standards Institute (ANSI) as part of the industry standardization program. EBCDIC had been developed by IBM for its 360 series computers.

Unfortunately, the two codes were not compatible and information transfer was still difficult. The logjam was broken by official action of the U.S. government. An executive order was issued by the President requiring any machine acquired under government contract to have ASCII code compatibility.

The ASCII character set is an accepted national standard today. However, we need to clarify a point. *The standard only requires that ASCII be used for information transfer between computers.* It is still quite common to find computers that use EBCDIC as the internal code and ASCII for external communication.

Which is the better code? EBCDIC's 8 bits allow up to 256 possible symbols to be represented. ASCII was designed for 7 bits, and it is limited to 128 possible combinations. That is not necessarily bad, if the additional combinations are not needed. However, the 7-bit code length is incompatible with several standard numeric formats that are based on multiples of 8 bits. Because of this, internal ASCII is implemented as an 8-bit code, where the eighth bit is a 0. Both ASCII and EBCDIC characters are processed in *8-bit code groups* known as **bytes**.

FIGURE 3.5 A typical computer information storage format.

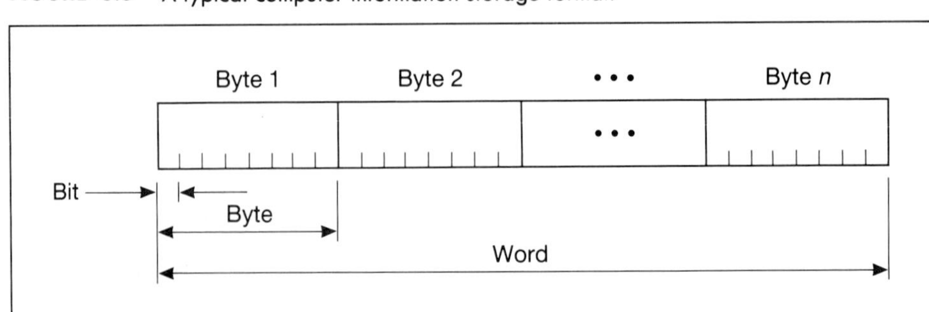

This defines the internal codes for our character set. Let's look at how they are actually stored. Computer memories are designed to store a fixed number of bits in each memory location. The group of bits in each location is known as a **memory word**. The specific number of bits is a tradeoff between several factors that we have yet to discuss. For the moment, let's consider just the relationship between alphabetic symbols and word length.

We obviously do not want a situation where the first part of a character is in one word and the second part is in another (it might be akin to splitting a w into two v's, or an x into)(). Therefore, we can state that *the length of a memory word should be an integer multiple of bytes*, as shown in Fig. 3.5.

3.4 NUMERIC REPRESENTATION IN A COMPUTER

Most computers use only one internal representation for alphabetic information. However, this is not the case with numeric representation. We often must deal with numbers that represent a wide range of values, from the very small to the very large, and we use several ways to represent any given value. Consider the number 1824.771. It could have also been written as 1.824771×10^3, or it might have been rounded to 1825. We typically use all these approaches when we represent numbers outside a computer, and we need to allow for the same capability within the computer as well.

3.4.1 Unsigned Integers

Let's begin our discussion of numeric representation with a familiar number set—unsigned decimal integers. These numbers use the set of symbols

$$\{0, 1, 2, 3, 4, 5, 6, 7, 8, 9\}$$

in an unambiguous format that can be specified as

$$D = d_{n-1} \cdots d_1 d_0 \qquad \text{where } D = 0 \text{ or } d_{n-1} \neq 0$$

(The "where" clause disallows leading zeros, which prevents multiple representations for the same quantity.)

The **semantics**, or meaning, of our representation will be that D represents a quantity

$$Q = d_{n-1}(10^{n-1}) + \cdots + d_1(10^1) + d_0(10^0)$$
$$= \sum_{i=1}^{\text{length } D} d_{i-1}(10^{i-1})$$

This is a formal definition of the **positional number system** we mentioned in Chap. 1. The semantics depends on the positions of the symbols in the string. Furthermore, since the number being raised to the power is 10 (or, equivalently, since there are 10 symbols in the decimal alphabet), the **base**, or **radix**, of this representation is 10.

Because positional number systems are so important, let's look at this concept again. We know that the position of a digit in a number implies the weighting factor of that digit. The weighting factor is the base of the number system raised to the power indicated by the position. In the decimal number 1834, the 1 has a weighting factor of 10 cubed, the 8 has a weighting factor of 10 squared, the 3 has a weighting factor of 10, and the 4 has a weighting factor of 1. Stated in another way,

$$1834 = 1 \cdot 10^3 + 8 \cdot 10^2 + 3 \cdot 10^1 + 4 \cdot 10^0$$

There is nothing special about using a base of 10, and positional number systems are definitely *not restricted* to base 10. In computing, the most common bases are 2, 8, and 16. We know that the base 2 is called binary. The base 8 is called **octal**, and the base 16 is **hexadecimal**.

To illustrate the general concept of bases further, consider base 7 and base 11 systems. The digit set for base 7 includes seven symbols:

$$\{0, 1, 2, 3, 4, 5, 6\}$$

The form of numbers written in the system becomes

$$S = s_{n-1} \cdots s_1 s_0 \qquad \text{where } S = 0 \text{ or } s_{n-1} \neq 0$$

and our summation is

$$Q = \sum_{i=1}^{\text{length } S} s_{i-1}(7^{i-1})$$

Base 11 is a little tougher, since when we try to write our digit set we run out of the standard numeric symbols. Another symbol is needed to represent the decimal quantity 10. We could invent a new symbol or we might use a letter of the alphabet, like T (for 10). Our symbol set is then

$$\{0, 1, 2, 3, 4, 5, 6, 7, 8, 9, T\}$$

and our summation becomes

$$Q = \sum_{i=1}^{\text{length } S} s_{i-1}(11^{i-1})$$

Remember that for any s_i equal to symbol T we must use the decimal quantity 10 in our summation.

Most of the time the base of our representation will be clear from its **context**, or how it is used. If there is any possible confusion, the base is specified as a subscript after the number, such as 12345_{10}. Numbers can use almost any base in their representation. Let's look at what this means for computers.

For the base 2, the alphabet is

$$\{0, 1\}$$

Unsigned integers in this representation will be strings of 0s and 1s, such as

$$B = 10111001$$

and the "value" of a string S will be

$$Q = \sum_{i=1}^{\text{length } B} s_{i-1}(2^{i-1})$$

The value of 10111001_2 would be

$$\begin{aligned} Q &= 1 \cdot 2^7 + 0 \cdot 2^6 + 1 \cdot 2^5 + 1 \cdot 2^4 + 1 \cdot 2^3 + 0 \cdot 2^2 + 0 \cdot 2^1 + 1 \cdot 2^0 \\ &= 128 + 32 + 16 + 8 + 1 \\ &= 185_{10} \end{aligned}$$

Again, since the semantics of this representation depend on the position of the symbols in the string, this is also a positional number system.

Besides the natural correspondence between binary numbers and electronic states, another distinct advantage of binary representations can be seen by considering the complexity of circuits to do arithmetic operations. The addition and multiplication tables for decimal numbers contain 100 entries each; the tables for binary numbers contain only 4, as shown in Fig. 3.6.

FIGURE 3.6 Binary addition and multiplication tables.

x \ y	0	1
0	0	1
1	1	10

Addition
$x + y$

x \ y	0	1
0	0	0
1	0	1

Multiplication
$x \cdot y$

There is a disadvantage of the binary system for human-to-computer communication when an exact representation is required. Binary strings are often just too long and intractable for easy human use, and we will learn later that not all decimal numbers have exact binary equivalents! Because of this, computer professionals normally rely on external notations that are equivalent to binary but which are somewhat less cumbersome. The two most common are octal and hexadecimal.

The octal digit set is {0, 1, 2, 3, 4, 5, 6, 7}. How do octal numbers relate to binary numbers, and why treat them differently than any other arbitrary base? Each octal digit corresponds to three binary digits; to convert a binary number to octal, we simply split up the binary number into 3-bit groups and write the corresponding octal digits:

$$\underset{7}{\underline{111}}\ \underset{3}{\underline{011}}\ \underset{2}{\underline{010}}\ \underset{5}{\underline{101}}\ \underset{1}{\underline{001}}{}_2 = 73251_8$$

This is a representation that is more concise and understandable for people, and it still has a direct correspondence to internal computer representations.

A similar situation holds with hexadecimal except we take the binary bits in groups of four. This gives an even more concise form, but it requires a little adjustment on our part. The hexadecimal system needs 16 symbols, so we must enlarge our decimal alphabet. The normal procedure is to take the first six letters of our Roman alphabet, as shown in Table 3.1. As with octal, it is easy to group binary

TABLE 3.1 BINARY, DECIMAL, OCTAL, AND HEXADECIMAL REPRESENTATIONS

Binary	Decimal	Octal	Hexadecimal
0	0	0	0
1	1	1	1
10	2	2	2
11	3	3	3
100	4	4	4
101	5	5	5
110	6	6	6
111	7	7	7
1000	8	10	8
1001	9	11	9
1010	10	12	A
1011	11	13	B
1100	12	14	C
1101	13	15	D
1110	14	16	E
1111	15	17	F

numbers and write the corresponding hexadecimal number:

$$\underset{F}{\underline{1111}}\ \underset{6}{\underline{0110}}\ \underset{A}{\underline{1010}}\ \underset{9}{\underline{1001}}_2 = \text{F6A9}_{16}$$

3.4.2 Number Conversions

We have seen how unsigned integers can be represented using different bases. The next problem is converting a number in one base to another. Normally, you will be working with decimal, binary, octal, and hexadecimal number systems. However, we will take a general approach that can be applied to any combination of bases.

Often, you will want to convert between decimal and some other base, so let's look at that first. We have already described going from a nondecimal representation to decimal; to convert string

$$S = s_{n-1} \cdots s_1 s_0$$

in base B, you simply expand the summation

$$Q = \sum_{i=1}^{\text{length } S} s_{i-1}(B^{i-1})$$

using normal decimal arithmetic. Going the other direction, from decimal to a nondecimal representation, is a little more involved.

To understand what must be done to convert from a base B to decimal, we need to remember that our decimal number

$$D = r_{m-1}10^{m-1} + \cdots + r_1 10^1 + r_0 10^0$$

can also be written as a polynomial of any desired base:

$$D = s_{n-1}B^{n-1} + \cdots + s_1 B^1 + s_0 B^0$$

This is equivalent to saying that

$$D = r_{m-1} \cdots r_1 r_0 \qquad \text{or} \qquad S = s_{n-1} \cdots s_1 s_0$$

We are trying to find the s_i, so if we divide the second polynomial form of D by our desired base B, we get a quotient

$$D' = s_{n-1}B^{n-2} + \cdots + s_2 B^1 + s_1 B^0$$

plus a remainder s_0. This remainder is the first digit we wanted! Now repeat the process, dividing D' by B to get

$$D'' = s_{n-1}B^{n-3} + \cdots + s_3 B^1 + s_2 B^0$$

plus a remainder s_1. Continuing to divide by B until the quotient is reduced to zero will give us the entire string S. Figure 3.7 shows an example of converting the decimal number 11000 to the octal number 25370.

We can check our answer by expanding the result.

$$2 \cdot 8^4 + 5 \cdot 8^3 + 3 \cdot 8^2 + 7 \cdot 8^1 + 0 \cdot 8^0$$

$$2 \cdot 4096 + 5 \cdot 512 + 3 \cdot 64 + 7 \cdot 8 + 0 \cdot 1$$

$$8192 + 2560 + 192 + 56 + 0$$

$$11000$$

Conversion from one nondecimal base to another nondecimal base is usually done by first converting to decimal and then converting from decimal to the desired base. This circuitous route is technically unnecessary, but to go directly, you would have to be able to do arithmetic in nondecimal bases. Dividing 40_7 by 12_7 using base 7 arithmetic (to convert from base 7 to base 9) gives a quotient of 3 and a remainder of 1, so $40_7 = 31_9$. That wasn't too hard, but how about converting 23423265236_7 to hexadecimal?

Fortunately, we normally will not have to face conversions between arbitrary number bases in our computer applications. Usually, we need to deal with only decimal and binary or hexadecimal numbers.

FIGURE 3.7 Conversion of an example number from decimal to octal.

```
                0 + rem 2 = s4
            8 | 2 + rem 5 = s3
           8 | 21 + rem 3 = s2
          8 | 171 + rem 7 = s1
         8 | 1375 + rem 0 = s0
START --> 8 | 11000
```

$11000_{10} = 25370_8$

3.4.3 Signed Integers

So far, we have discussed only unsigned integers. We now need to consider how to represent the sign of the number. There are two common ways to do this: sign-magnitude and complement representations.

Sign-magnitude is the representation we humans are taught in grade school. We first expand our decimal alphabet to include the two "sign" symbols (+ and −), giving {+, −, 0, 1, 2, 3, 4, 5, 6, 7, 8, 9}. Then, to represent a signed number, we simply attach a + or − sign to the string representing the number's magnitude. Hence, the name **sign-magnitude**. (Usually the + sign is optional.) In computers we can use the same idea, although we usually do not use an expanded binary alphabet. We simply take the binary string representing magnitude and add another binary digit to represent the sign. Usually, a 0 represents + and a 1 represents −. Signs can be at either end of a number. We normally see numbers written with the sign on the left, but in certain applications it might be on the right.

Complement representations use the idea that a negative number can be viewed as a corresponding positive number minus a function of the base. There are different types of complement representations; we will discuss one in which a power of the base is used directly. This type is called *2's complement* for binary numbers and *10's complement* for decimal numbers; generally, it is known as a **radix**, or **true, complement.**

To illustrate the technique, suppose we want to represent decimal numbers in the range [−4999, +5000]. A positive number P will be represented in our usual way; a negative number N will be represented by the string S corresponding to the term $(10000 + N)$. Therefore −2348 would be represented by $(10000 + (-2348)) = 7652$.

Be cautious! String S still represents a number, but not in the usual way; 7652 does not represent seven thousand six hundred and fifty-two; it represents $7652 - 10000 = -2348$.

In essence, then, we have two separate representations, one for positive numbers and one for negative. How can you interpret some arbitrary string in this language, since all numbers appear to be positive? If the string is in the range [0, 5000], then that string represents a positive number. If it is in the range [5001, 9999] it represents a negative number, and to find its actual value you must subtract 10000.

This whole scenario might seem overly complex until you start designing electronic circuits to do arithmetic operations. For example, consider how we add two sign-magnitude numbers. If the signs are opposite, we must determine which number has the smaller magnitude and subtract that magnitude from the larger magnitude. The sign of the result is the same as the sign of the number with the larger magnitude. If both numbers are negative, ... There are many rules, though we might not consciously think about them. In computers, these rules usually mean added complexity, added cost, and reduced speed.

With radix complement representations, we essentially add two numbers simply by adding them; the signs take care of themselves. The only little hitch is that we must ignore the carry out of the highest-order digit. To add −2348 (that is, 7652 in complement representation) to 3621, we add 7652 to 3621 and get 11273. Since we are using four-digit information units, 11273 is 1273 with a carry of 1 out of the high-order digit. Ignoring the carry, we get our answer, 1273.

We aren't really ignoring the carry. The 7652 has an implicit -10000 attached to it; $-2348 = 7652 - 10000$. When we did the add, we were really adding

$$\begin{aligned} 3621 + (7652 - 10000) &= (3621 + 7652) - 10000 \\ &= (11273) - 10000 \\ &= (10000 + 1273) - 10000 \\ &= 1273 \end{aligned}$$

The high-order carry is equivalent to a $+10000$, which cancels out the -10000 term.

Most current computers use some form of complement representation for integer numbers. The binary equivalent of the version we just explained is called a 2's complement representation. The number $-26_{10} = -11010_2$ in a machine with 16 binary bits per integer, using 2's complement representation, would be

$$\begin{aligned} 2^{16} + (-11010_2) &= 10000000000000000_2 - 11010_2 \\ &= 1111111111100110_2 \end{aligned}$$

Note that we must know the size of the representation in order to specify the complement form; otherwise, we don't know how many leading digits there are.

Another common version of complements for binary numbers is called 1's complement, or **diminished radix complement**. Instead of using $2^k + N$ to represent negative number N, we use $(2^k - 1) + N$. A problem at the end of the chapter asks you to consider advantages and disadvantages of this version.

In later chapters, we will discuss the arithmetic unit and the operation subtract. However, we can note here that in many computers "subtract" is implemented as an "add complement."

3.4.4 Floating-Point Numbers

We have discussed unsigned and signed integers. In these numbers, the radix point (the decimal point in decimal numbers) is always understood to be immediately to the right of the least significant digit. The next step in our progression will be to consider numbers with fractional components. In them, we assume that the radix point can appear at an arbitrary place in the string of digits; that is, it can "float." Hence, these numbers are also called **floating-point numbers**.

The standard way of representing floating-point numbers can be equated to a "scientific notation" that some of you might have already used in physics and chemistry. For a number S in base B, we write the number in the form

$$S = 0.s_{-1}s_{-2} \cdots s_{-n} \times B^E$$

where s_{-1} is a nonzero digit. Next, we decide how many digits to use in the representation of the fractional part, and either round S or "pad" it with trailing zeros to get this form. The digits and exponent might have changed during this conversion; for example, rounding 0.9999×10^4 to three digits would give 0.100×10^5. Therefore, let's rewrite the number as

$$S = 0.t_{-1}t_{-2} \cdots t_{-m} \times B^F$$

Lastly, we decide how many digits to use in the representation of the exponent and add leading zeros if necessary; assuming

$$F = f_{k-1} \cdots f_1 f_0$$

let's define a new form of the exponent as

$$F = g_{h-1} \cdots g_1 g_0$$

Then we can write our floating-point number as a two-part string

$$\begin{aligned} S &= t_{-1}t_{-2} \cdots t_{-m} \sqcup g_{h-1} \cdots g_1 g_0 \\ &= t_{-1}t_{-2} \cdots t_{-m} g_{h-1} \cdots g_1 g_0 \end{aligned}$$

where the base B is understood from the context.

For instance, consider the decimal number 12345.6789. Suppose that we want an eight-digit fraction and a two-digit exponent in our representation. The steps would then be as follows:

1. Transform number to fractional form: $S = 0.123456789 \times 10^5$
2. Adjust the fraction length to eight digits: $S = 0.12345679 \times 10^5$
3. Determine exponent: $F = 5$
4. Adjust exponent to two digits: $F = 05$
5. The representation then equals: $S = 1234567905$

For the number 99999.9999, the steps would be:

1. $S = 0.999999999 \times 10^5$
2. $S = 0.10000000 \times 10^6$
3. $F = 6$
4. $F = 06$
5. Representation = 1000000006

The representation of the fraction (12345679 or 10000000 in our examples) is typically called the **mantissa**. The representation of the exponent (05 or 06 above) is called the **characteristic**, or **exponent**.

So far, we have carefully avoided negative fractions and exponents in our floating-point numbers. We have discussed complement form for integer representation. There are two components to our floating-point representation, and it might seem logical to use an individual complement for the fraction and another for the exponent. However, the industry has defined a standard representation for floating-point numbers. The mantissa is represented in sign-magnitude form, and the exponent is represented using a technique known as biasing.

When dealing with floating-point numbers, we are often interested in the relative difference in the sizes of the exponents, not on their signs and magnitudes. To add 0.12345679×10^3 and $0.24335465 \times 10^{-2}$, it is necessary to shift one number by five places. However, the same would hold if the exponents were 12 and 7 or 37 and 32.

Instead of signed exponents, we use a **bias**, or an offset, to store exponents. For a base B and an exponent representation of h digits, we allow an exponent range of $[-L, U]$, where $L + U = B^h - 1$. Exponent E is then represented by biased exponent $E + L$. This representation may sound a little strange, but it is really no different than the familiar temperature scales shown in Fig. 3.8. All we do is shift the zero point to some positive value in the range (typically the midpoint) and then represent any negative exponent in the lower half of the scale.

For decimal numbers with a two-digit exponent, we would probably choose an exponent range $[-49, 50]$, which would give us a bias of 49. Exponents of -27 and $+35$ would become $(-27 + 49) = 22$ and $(35 + 49) = 84$.

FIGURE 3.8 An example of a different biased scale.

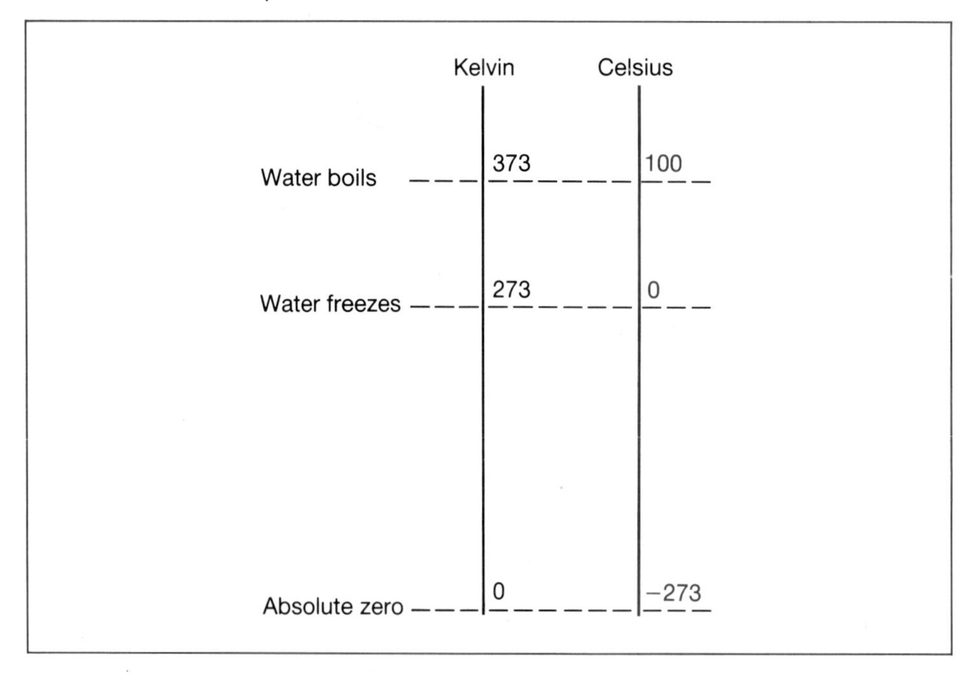

The transformation process described earlier also needs to be changed. For a number S in base B, we write

$$S = \pm 0.s_{-1}s_{-2} \cdots s_{-n} \times B^{\pm E}$$

where s_{-1} is a nonzero digit. Next we either round S or pad it with trailing zeros to get the necessary number of digits for the mantissa. Since this could change the mantissa and possibly the exponent if rounding or padding were required, let's write

$$S = \pm 0.t_{-1}t_{-2} \cdots t_{-m} \times B^{\pm F}$$

Assuming our actual exponent is

$$F = \pm f_{k-1} \cdots f_1 f_0$$

we can write the biased exponent as

$$G = F + \text{bias} = g_{h-1} \cdots g_1 g_0$$

The string representing the magnitude of our floating-point number is then

$$Q = \pm t_{-1}t_{-2} \cdots t_{-m} g_{k-1} \cdots g_1 g_0$$

For example, consider the decimal number -0.000123456789. Suppose that we want an eight-digit fraction and a two-digit exponent in our representation, with an exponent range of $[-49, 50]$. The exponent bias would be 49 and the steps would then be as follows:

1. $S = -0.123456789 \times 10^{-3}$
2. $S = -0.12345679 \times 10^{-3}$
3. $F = -3$
4. $G = 46$
5. Final representation $= -0.1234567946$

As a second example, consider the decimal number -0.000999999999. The steps would now be as follows:

1. $S = -0.999999999 \times 10^{-3}$
2. $S = -0.10000000 \times 10^{-2}$
3. $F = -2$
4. $G = 47$
5. Final representation $= -1000000047$

With these examples, you should have a good general understanding of what floating-point numbers are. We will discuss them more in Chap. 8 when we take up the arithmetic unit.

3.5 REPRESENTATIONAL ERROR

Machine computation often deals with mathematical models of actual situations that exist in the physical world. The values that are used in these models can be either computed theoretically or obtained by making measurements of physical parameters. In both cases, accuracy is a concern. Our systems must account for at least three possible types of errors that relate to the representation of numerical values:

- **Measurement errors** that happen when the method or device used to get the value is not completely accurate
- **Scaling errors** that happen when the number system cannot exactly represent each of the values
- **Truncation** and **rounding errors** that happen when the word length is not long enough to store the needed number of digits

Measurement errors are dependent on the methods and types of instruments used to get the values, so we will not discuss them further here.

3.5.1 Scaling Errors

Scaling errors are generally under the control of the user or the programmer who chooses how the variables in the model should be represented. Assume for the moment that we have decided to represent all our values as integers. If an actual number falls between two integer points on the scale, the best we can do is to represent that value by the closest integer. The representation then contains an error whose magnitude is dependent on its position on the scale with respect to the closest integer, as shown by the error curve of Fig. 3.9. The maximum error in a closest-value representation is one-half the length of the space between representable points on the number scale. In the integer number system, the maximum error is a constant over the entire representational range allowed by the word length. However, there are absolute limits on the size of numbers that can be represented with a fixed word length. The representational error beyond these limits grows without bound, as shown in Fig. 3.10.

The relationship and the general shape of the error curves also apply to floating-point numbers. However, because floating-point numbers include fractional values, the spacing or interval between representable points can be different. The actual spacing is dependent on the number of digits in the mantissa and the interval

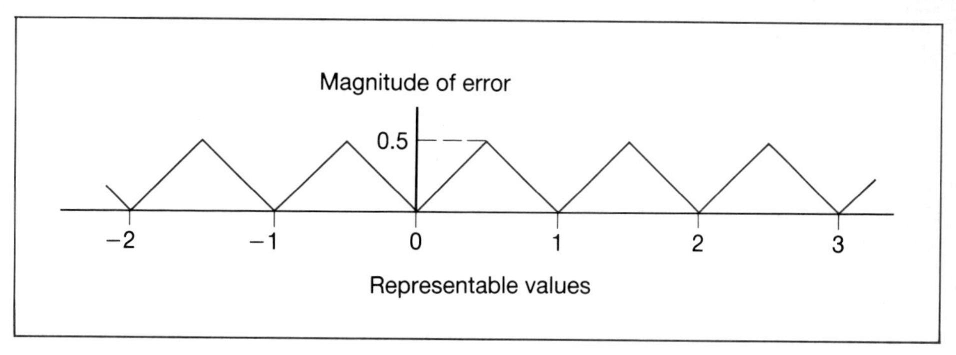

FIGURE 3.9 Representational errors in small integer numbers.

defined by the exponent, according to the equation

$$\text{Repr. error} = \frac{\text{length of scale interval for exponent value}}{\text{no. of possible combinations in mantissa}}$$

$$= \frac{\text{radix}^{\text{exponent}} - \text{radix}^{\text{exponent}-1}}{\text{radix}^{\text{no. of digits in mantissa}}}$$

Floating-point numbers have a different interval for each different exponent value (see Fig. 3.11), each with a separate set of constantly spaced points similar to those we saw in Fig. 3.10 for integer numbers. The spacing between representable points and the magnitude of the maximum error is the radix times larger than the error in the previous interval for each successively larger exponent (that is, 2 × for binary, 10 × for decimal, etc.).

The scaling factor of the radix^{exponent} in floating-point representations can make the least significant digit of the mantissa much larger or much smaller than 1. Normally, we assume floating-point numbers are "more accurate" than integers, but when the word lengths of integer and floating-point numbers are the same, *integer*

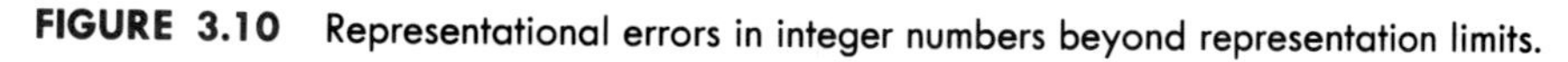
FIGURE 3.10 Representational errors in integer numbers beyond representation limits.

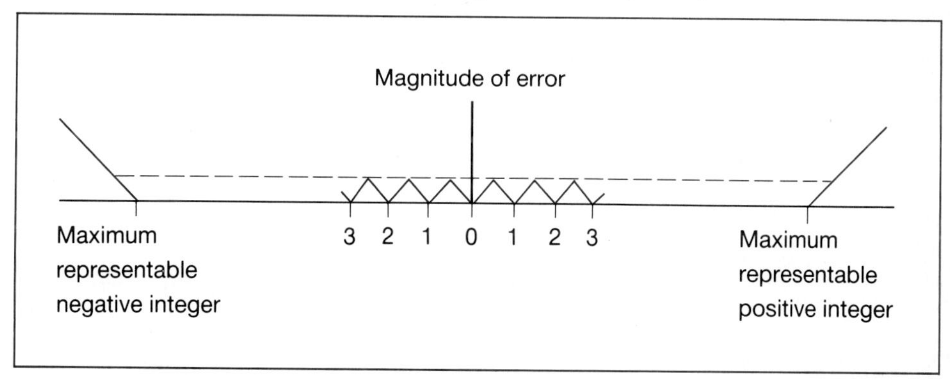

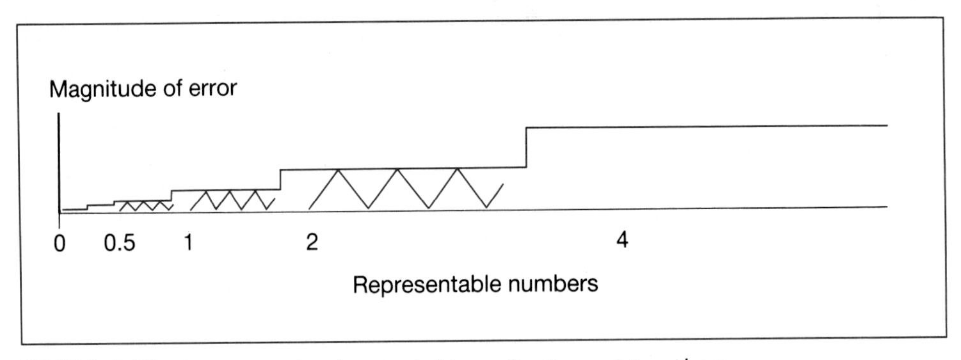

FIGURE 3.11 Representational errors in binary floating-point numbers.

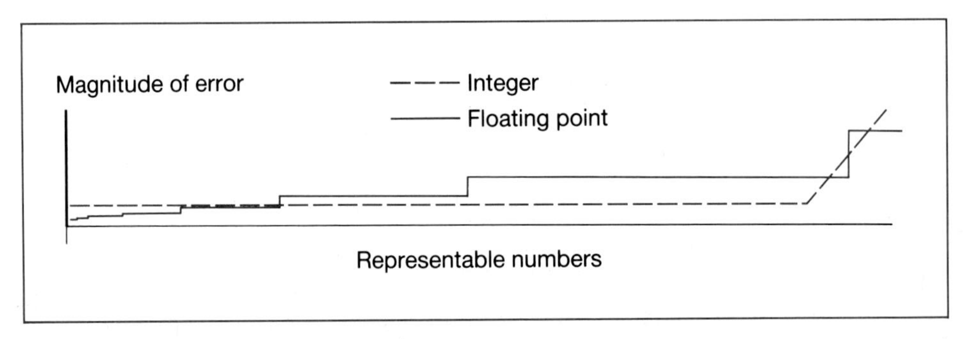

FIGURE 3.12 Maximum integer and floating-point representational error, assuming equal sizes for the representations.

representation can be more accurate for certain number values than floating-point. This is true because not all the word is available for the mantissa. Part of the floating-point word is used for the exponent, as illustrated in Fig. 3.12.

3.5.2 Truncation and Rounding Errors

The next type of error we will discuss is probably the most familiar. If you have filed an income tax form, you know that the IRS allows you to round your entries to the nearest dollar. This shortens the length of the reported values by two digits but creates a small error in the individual entries. That error is never more than 50 cents. Some errors are in your favor and some are in the IRS's. Ultimately, they will tend to balance out. We use similar techniques to fit long data values into shorter formats for the computer. Three specific procedures, truncation, half-adjusting, and statistical rounding, have been commonly used.

The simplest is **truncation**, but it introduces the most error. It consists of truncating (cutting off) the word at the stated length and ignoring all lower-order bits or digits. Figure 3.13 shows that because any value between two representable points is

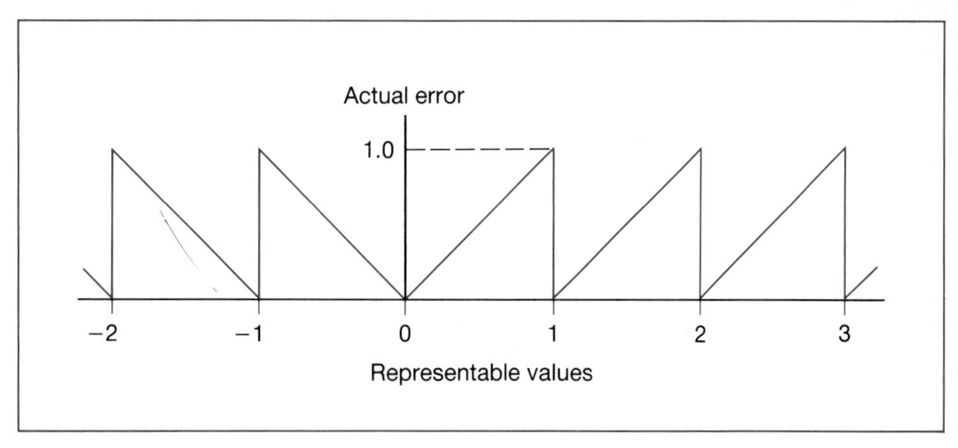

FIGURE 3.13 Truncation error in integer numbers.

always represented as the lower point on the scale, the maximum truncation error is twice the error obtained by choosing the nearest-number approximation.

Half-adjusting is a procedure that is almost the nearest-number representation shown in Fig. 3.9. It is done by first adding one-half to the digits or bits to the right of the desired least significant digit or bit and then truncating the number, as shown in Fig. 3.14. Unfortunately, this creates a biased error when the actual value falls exactly in the middle between two representable points. That value is always represented by the larger representable number.

The most accurate approach is **statistical rounding**, which is recommended in the *IEEE* (Institute for Electrical and Electronic Engineers) *Floating-Point Standard.* The procedure is essentially the same as half-adjusting, except it treats the "exact-middle" case differently. Half the time the value is rounded to the larger representable number, and half the time it is rounded to the lower. It is usually done in a

FIGURE 3.14 Examples of half-adjusting: (*a*) binary; (*b*) decimal.

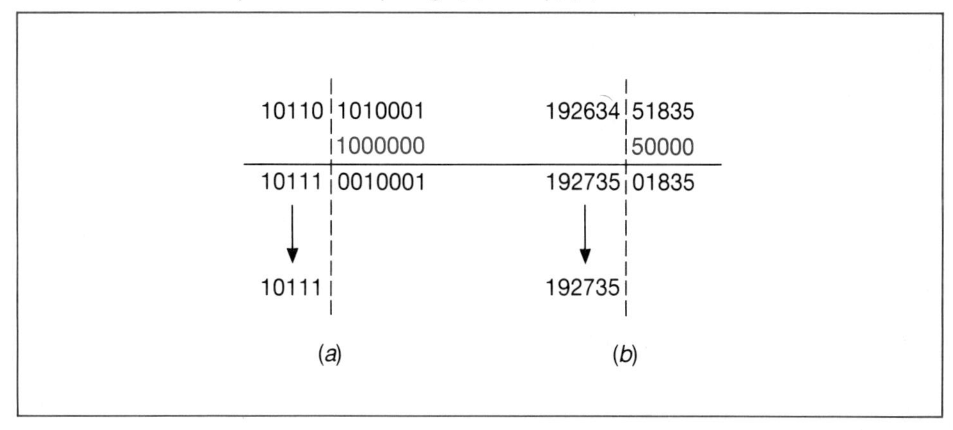

Number to be shortened	Shortening procedure			
	Truncate	Half-adjust	Round to even	Round to odd
1823 ¦ 4999	1823	1823	1823	1823
1823 ¦ 5001	1823	1824	1824	1824
1823 ¦ 5000	1823	1824	1824	1823
1824 ¦ 5000	1824	1825	1824	1825
1010 ¦ 0101	1010	1010	1010	1010
1010 ¦ 1001	1010	1011	1011	1011
1010 ¦ 1000	1010	1011	1010	1011
1011 ¦ 1000	1011	1100	1100	1011

FIGURE 3.15 Comparison of rounding, truncation, and half-adjusting.

manner that results in the desired least significant position being even or odd, as chosen by the programmer.

Examples of truncation, half-adjusting, and statistical rounding are given in Fig. 3.15.

3.6 SUMMARY

We began the chapter with an extension of sets to sequences, alphabets, and strings. This led to a brief discussion of ASCII and EBCDIC character codes, which was followed by a more in-depth discussion of the representation of numeric information.

Before we leave these topics, we need to note that not all computers process numeric data as binary or hexadecimal numbers. Some computers, particularly those designed for business applications, use representations known as packed and unpacked decimal. **Unpacked decimal** representation is the same as the character code (ASCII or EBCDIC) used in the computer. Both these codes represent decimal digits in an 8-4-2-1 code in the lower four bits of the character. The upper four bits are unused, except to identify the character as a number. **Packed decimal** representation strips these upper four bits from each numeric character and "packs" two decimal digits into a single 8-bit byte. Decimal representation avoids the need for decimal-binary and binary-decimal number conversion in computers that contain decimal arithmetic capability. It also eliminates any errors that occur in the conversion process.

We also introduced the concept of integer representation in the binary number system. You can use the binary number system to represent fractional values in the same way that fractional values are represented in the decimal system except the digit (bit) values are

$$\cdots 16 \quad 8 \quad 4 \quad 2 \quad 1 \;\cdot\; \tfrac{1}{2} \quad \tfrac{1}{4} \quad \tfrac{1}{8} \quad \tfrac{1}{16} \cdots$$

↑

Binary point

Bits to the left of the radix point (the binary point in this case) represent the whole-number portion, and the bits to the right represent the fractional portion.

PROBLEMS

3.1 The following sequence of 2-tuples lists the coordinate locations of a sequence of holes to be drilled by a robot:

$$\langle\langle 1, 8\rangle, \langle 11, 6\rangle, \langle 2, 3\rangle, \langle 9, 0\rangle, \langle 6, 11\rangle, \langle 1, 5\rangle, \langle 8, 6\rangle\rangle$$

Assuming that $\langle 0, 0\rangle$ is the home (beginning and ending position) of the robot, graph the location of each hole and then reorder the hole sequence in a way that lessens the travel distance of the robot.

3.2 Would a procedure that sorts hole locations according to their value on one axis be a good way to order the hole sequence of Prob. 3.1? Give examples where it would and would not be.

3.3 Figure 3.3 shows a portion of a state diagram for a Morse code decoder. Show what modifications are necessary to allow detection of the end of a word.

3.4 Suggest another (simpler) way that the end of a word coded in Morse code might be determined.

3.5 Would Morse code be appropriate for use with today's computer keyboard? If not, why not?

3.6 Consider the keyboard of a standard typewriter. What is the minimum size of a fixed-length code that would be required to uniquely represent each of the characters?

3.7 The old Baudot code that was used for teletypewriters was only 5 bits long, yet it allowed more characters than the Morse code. Discuss how this could be done.

3.8 Describe a situation in which fixed-length coding would be preferred over variable-length coding.

3.9 Why is the relationship between byte length and word length important in a computer?

3.10 Convert the following hexadecimal numbers to their binary and decimal equivalents:

a. 19A c. 398

b. EF1 d. 241

3.11 Convert the following decimal numbers to their binary and hexadecimal equivalents:

a. 28 c. 255
b. 256 d. 33

3.12 Represent the following decimal numbers in 16-bit binary 2's complement form:

a. −29 c. 130
b. −64 d. 29

3.13 Convert the following four-digit 10's complement numbers to sign-magnitude decimal integers:

a. 5041 c. 6410
b. 2348 d. 9234

3.14 Name a common biased scale other than biased exponents in a computer.

3.15 Represent the following as floating-point with an eight-digit mantissa and a two-digit exponent biased at 50:

a. 48,299,112,180,000,000 c. 0.000000000256
b. 129.000005 d. 3.891×10^{38}

3.16 What is the maximum representational error (within the representational range) of a number represented in integer binary? In integer decimal? In integer hexadecimal?

3.17 What is the maximum rounding error (within the representational range) of a number represented in integer binary? In integer decimal? In integer hexadecimal?

3.18 What is the maximum truncation error (within the representational range) of a number represented in integer binary? In integer decimal? In integer hexadecimal?

3.19 Why is statistical rounding the preferred technique?

3.20 Write the following as packed decimal:

a. 3612 c. 5
b. 23 d. 133

3.21 How could the sign of a number be represented in packed decimal?

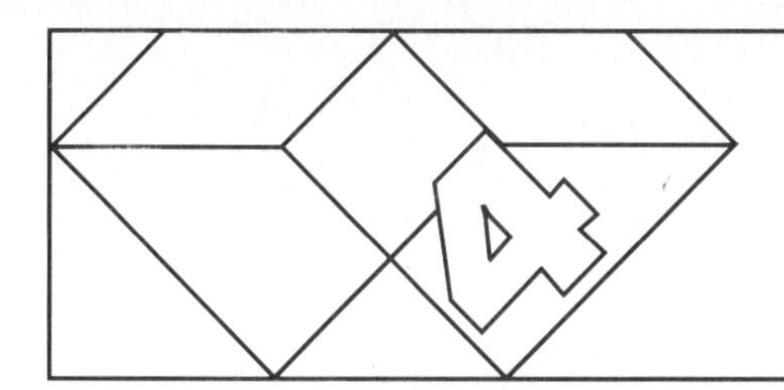

The Components of a Computer System

We have defined computing in the global sense. We have also introduced the mathematics that form the basis for computing systems. It is now time to take a closer look at the computer itself. Let's begin with Fig. 4.1, an amplification of a diagram that was first presented in Chap. 1.

Earlier, we saw that the computing environment can be depicted as an overlapping of three concept rings representing applications, software, and hardware. These rings can also be viewed as system layers that each have their own individual technologies. These layers interact with each other through the interfaces discussed in Chap. 1. We will introduce each of these layers in this chapter, starting with the hardware. But first, to set the stage for that discussion, we will take a brief look at some early computing devices.

4.1 EARLY "COMPUTING" MACHINES

The history of computing hardware is fascinating because it is both very old and very young. The earliest "computing" devices evolved 4000 years ago, yet computers as we think of them are only about 50 years old.

The first calculating tool was a board or table that had a series of parallel grooves which held small rocks or tokens to keep a tally or count. The grooves represented the digits in the number system, and the number of rocks in the groove showed the value of that digit. When the number of tokens in the groove reached the maximum value for a digit, they were removed. Another token was added to the next-higher groove.

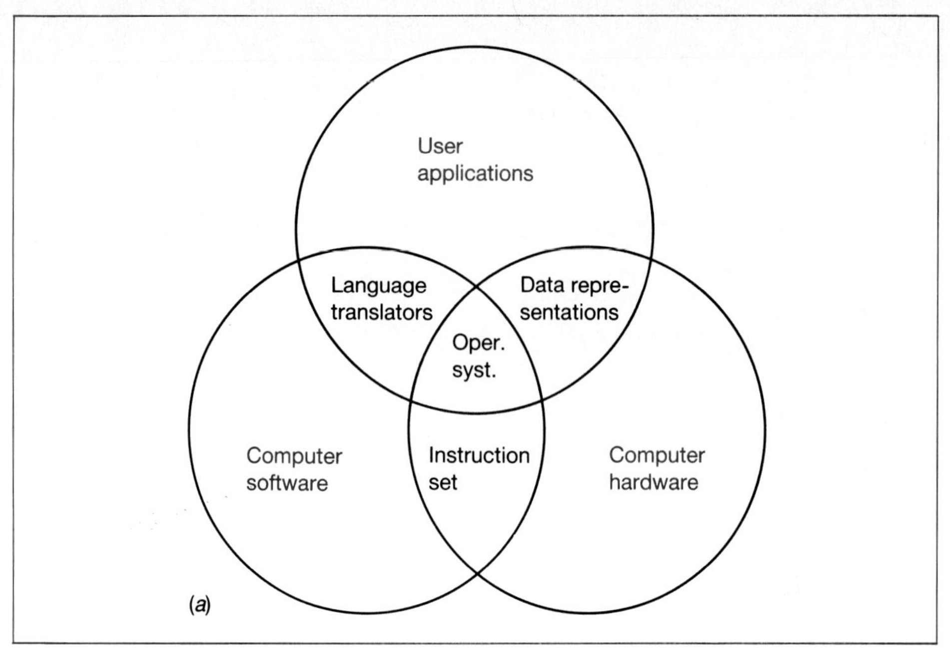

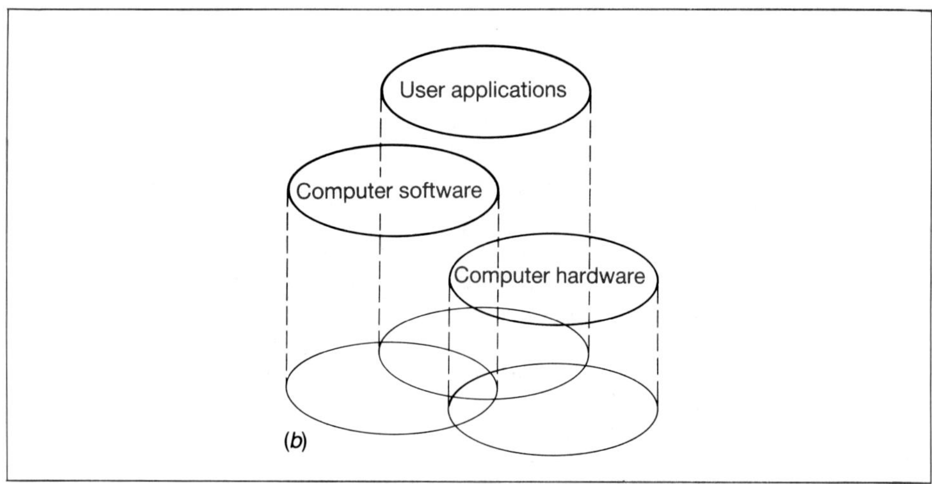

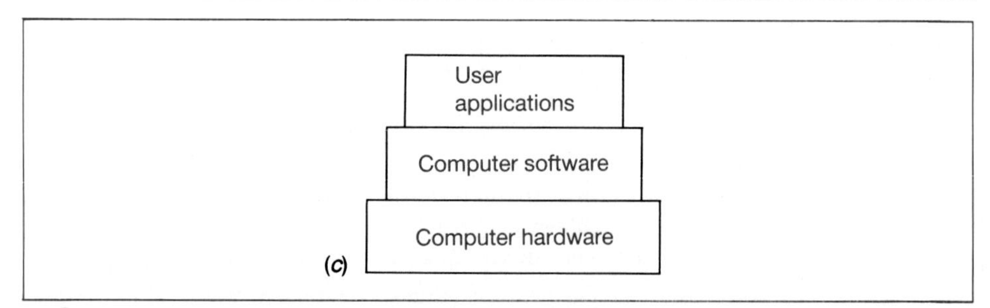

FIGURE 4.1 The convergence and layers of computing system design; (*a*) the computing design environment; (*b*) a second view of the overlapping needs and technologies; (*c*) a layered view of the design environment.

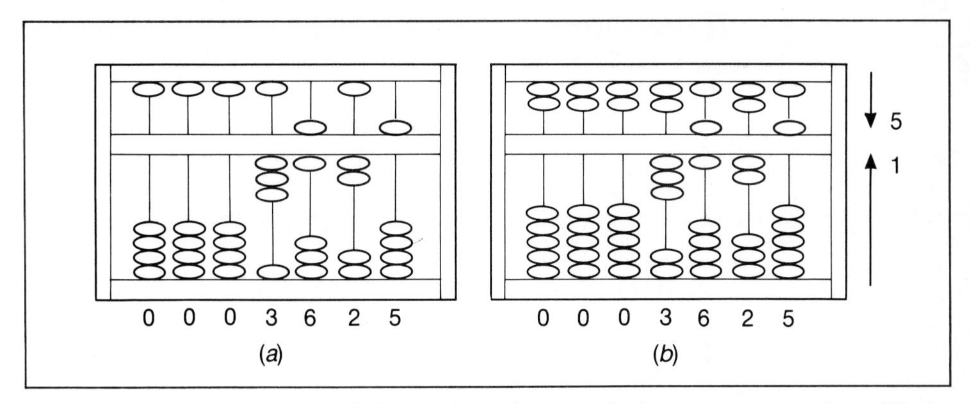

FIGURE 4.2 Two examples of the modern abacus; (*a*) the Japanese *soroban*; (*b*) the Chinese *suan pan*.

Eventually, the counting board was replaced with the abacus. The modern version is a two-section frame device with rods and captive tokens (beads) to represent base 10 digit values, as shown in Fig. 4.2. Each rod represents one place location in the decimal number. The beads in the lower section are each equal to 1, those in the upper section, 5. The values for each digit position are represented by the sum of the beads that have been moved toward the center divider (beads that are next to the outside edges are considered to have been "removed" and carry zero value).

Addition or subtraction is done by increasing or decreasing the number of beads for each digit (moving the indicated number of beads toward or away from the center). Then the digits are adjusted for carries or borrows. Suppose we wanted to add 7 to the current value shown in Fig. 4.2. We could do this in the Chinese *suan pan* by moving the remaining 5 bead plus two 1 beads in the units digit toward the center. The represented value in this digit is now 12, which is interpreted as a digit value of 2 plus a pending carry of 1. To clear the carry, both 5 beads in the units digit are removed (to the top edge) and a 1 bead is moved to the center in the tens digit. The procedure for the Japanese *soroban* is similar, except that carries must be propagated during the initial addition. Two 1 beads are added to and the 5 bead is removed from the units digit, then a 1 bead is added to the tens digit.

Multiplication can be done by resetting the abacus to zero and adding the value of the multiplicand to the abacus the number of times in the units digit. Next, the multiplicand is shifted left one digit and the process repeated. This is done until each digit in the multiplier has been processed. Division is done in a similar fashion by successive subtractions.

The abacus proved to be a very effective, low-cost calculating device that is still being used in many places of the world. Unfortunately, its design imposed a limitation because direct human control is needed for all the calculating procedures. It is what might be termed a "limited potential" design.

The chronology in Fig. 4.3 shows that the first "automated" calculators were not designed until A.D. 1642.[1] These were gear-driven "boxes" that were designed by Pascal to do addition and subtraction. They were later improved by Leibniz to include multiplication and division. Motivation for these efforts was a wish to find a faster, more accurate way to compute tables of logarithms and trigonometric functions. Goldstine [1972] quotes Leibniz: "It is unworthy of excellent men to lose hours like slaves in the labor of calculation which could safely be relegated to any one else if machines were used."

Unfortunately, once these tables were completed, there was no further pressing need for this computational power. No additional significant progress was made until 1823 when Babbage became concerned about errors he had found in the existing logarithmic tables. He developed an extension to the Pascal/Leibniz calculator known as the "difference engine." It was designed to compute sixth-order polynomials and to provide greater accuracy in table computations.

By 1833, Babbage had expanded his designs into an "analytical engine" that contained the basic concepts of a programmable, general-purpose computer. It was to be controlled by a library of calculating procedures stored as hole patterns in groups of small boards or cards. Mechanical pins detected the presence or absence of holes and controlled the operation of the calculator. The punched-card technology had been previously developed (1804) and successfully used by Jacquard to store different warp and weave patterns for his automated loom.

Unfortunately, as with many new ventures, Babbage couldn't obtain enough funding. His machines were never fully completed, probably for several reasons. The most likely appear to be the following:

- There was no commercial market for his machine.
- A needed technology was not available (he could not make gears with the necessary accuracies).
- There was insufficient speed improvement (his machine did not appear to be much faster than the manual methods it was designed to replace).

These are some of the same reasons that projects fail today. For any new product to succeed, there must be a need, an adaptable technology, and a cost/performance improvement over existing products or methods.

Babbage's ideas were valid, but the development of workable digital computers would have to wait for more than a hundred years. His ideas were not put to use until 1937, when Shannon discovered that Boole's algebra could be used as a digital design tool. But even this was not sufficient to stimulate much research in computing. There was still not enough *need*.

The opposite situation led to the development of the punched-card sorter for the 1890 census. The Census Bureau had recognized that it would not be possible to tabulate all the 1890 information before it was time to do the 1900 census unless

[1] The historical developments cited in Fig. 4.3 are described in greater detail in Goldstine, 1972. Dates for some of the specific events are listed in Grun, 1975, and Ifrah, 1985.

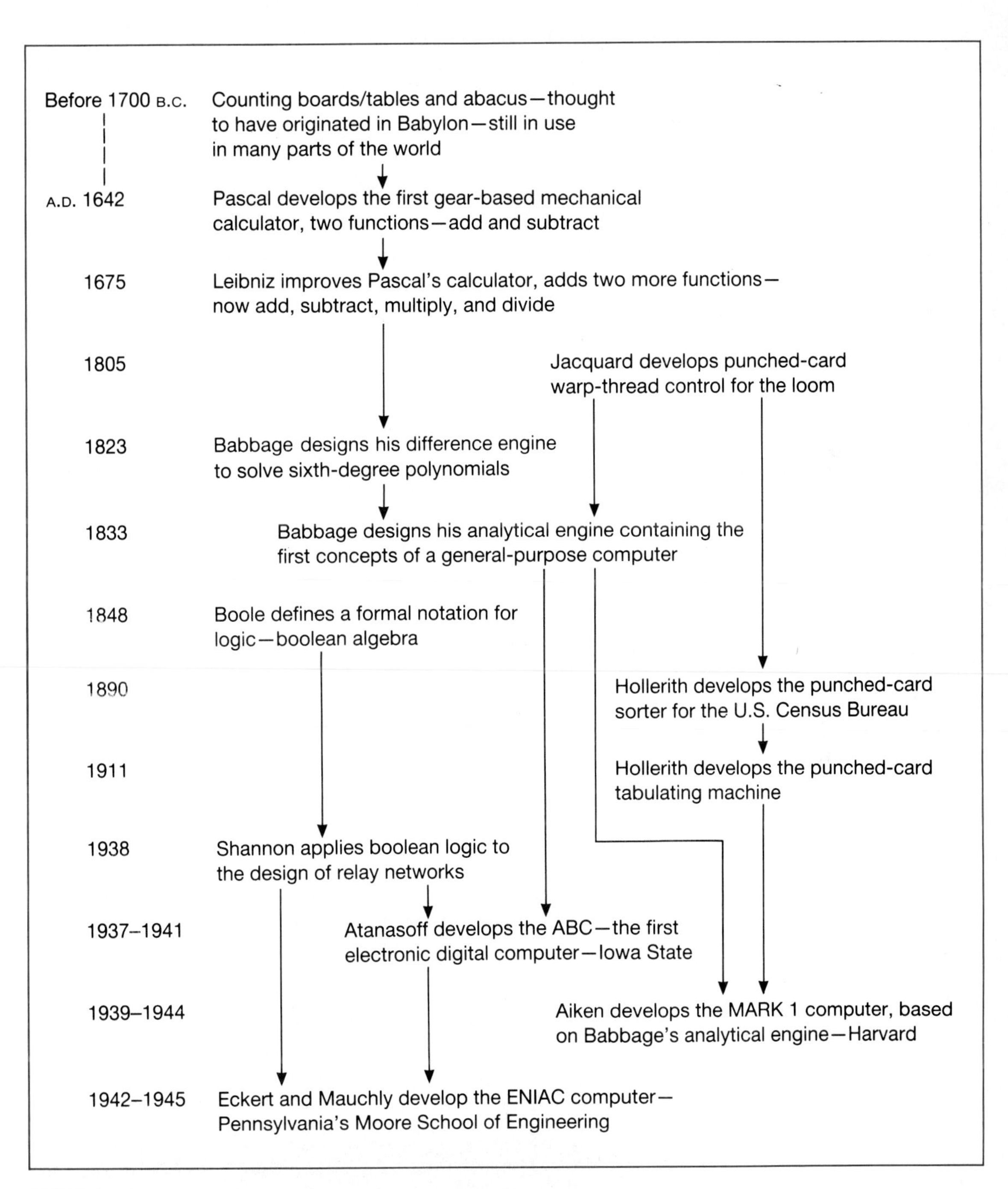

FIGURE 4.3 The evolution of computing devices prior to 1945.

some form of automated procedure could be developed. Herman Hollerith, a bureau employee, was able to adapt Jacquard's punched-card concept into a successful design in time for the 1890 census. He used it again in his later development of the punched-card tabulating machine.[1]

4.2 VON NEUMANN'S PROPOSAL—THE STORED-PROGRAM COMPUTER

By the beginning of World War II, electromechanical tabulating machines were being rather widely used. Ballistic trajectory calculations had intensified the need for *automatic* computation. Several projects for experimental electronic digital computers had been started: the ABC[2] at Iowa State, the Mark I[3] at Harvard, and the ENIAC[4] at Pennsylvania's Moore School of Engineering. These computers were designed with vacuum-tube and relay technology. They all used externally stored programs that were wired on plugboards or punched into paper tape.

During the summer of 1944, John von Neumann, an applied mathematician from the Princeton Institute for Advanced Study, became interested in the ENIAC project. He had been working on mathematical models of nuclear reactions for the Los Alamos Laboratory and needed a way to speed up his computations. But von Neumann brought more than just a desire for rapid computations. He had great insight into problem formulation and was also interested in formal logic. Because of this, he was asked to join the development team for the ENIAC's successor machine. The value of this addition can be seen from Goldstine's comment that

> von Neumann may have been the first person who understood explicitly that a computer essentially performed logical functions, and that the electrical aspects were ancillary.... Von Neumann made a precise and detailed study of the functions and mutual interactions of the various parts of a computer ... rethinking the storage requirements for both numbers and instructions. [Goldstine, 1972]

The result of von Neumann's work was a 1945 draft report[5] defining a logical structure for the internally programmed, general-purpose computing machine shown in Fig. 4.4.

[1] Hollerith founded the Tabulating Machine Company in 1896. It later became International Business Machines.

[2] Atanasoff Berry Computer, proposed by John Atanasoff in 1937–1938 and built with Clifford Berry during 1939–1941.

[3] The first version of the Automatic Sequence Controlled Calculator, proposed by Howard Aiken in 1939, finished in 1944.

[4] Electronic Numerical Integrator and Computer, proposed by John Mauchly in 1942 and built with J. Presper Eckert during 1943–1945.

[5] "Preliminary discussion of the logical design of an electronic computing instrument," by Burks, Goldstine, and von Neumann, prepared for the U.S. Army Ordnance Department and reprinted many times [Bell and Newell, 1971].

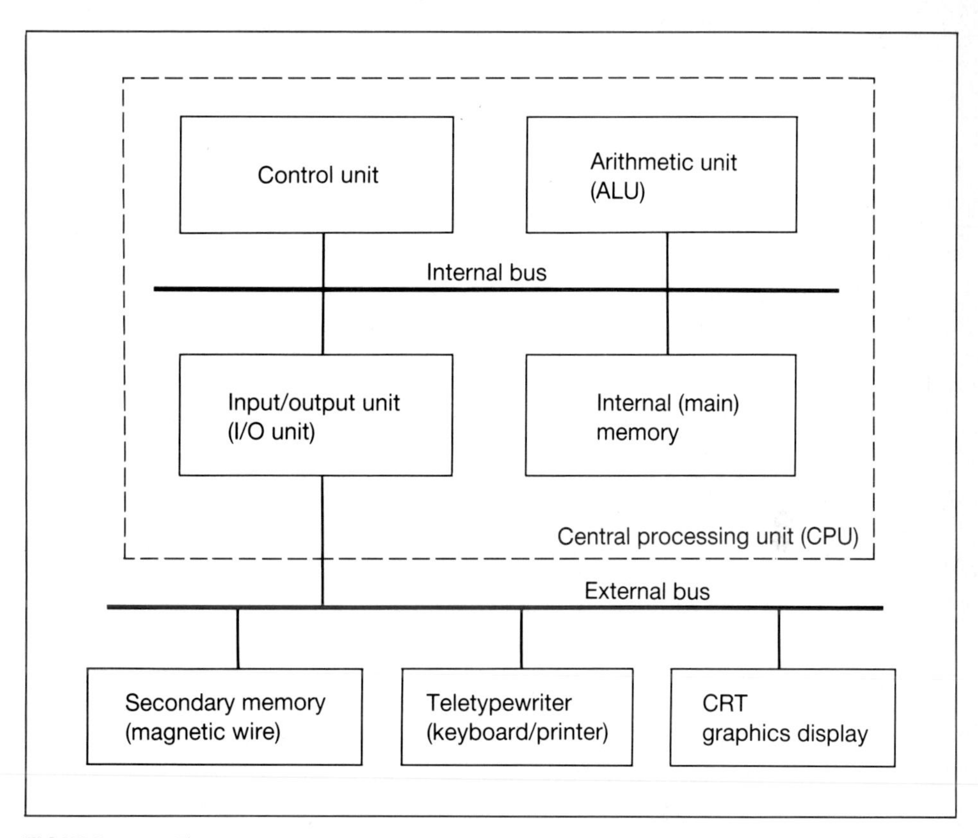

FIGURE 4.4 The von Neumann computer architecture with his suggested peripherals.

Von Neumann's proposal represented a major technological advance. It gave a design framework *that was essentially independent of any particular electronic technology.* His proposal became the basis for most computer design, including the present. It also provides us with a framework to begin our study of modern computing systems. We will look at this architecture in the next section.

4.3 HARDWARE—THE PHYSICAL PART OF THE SYSTEM

The hardware is the part of a computer system that can be seen and touched. From the outside, it looks like a collection of plain boxes or cabinets interconnected with several electric cables. It typically has at least one keyboard and a video display unit for the users and a tape or disk drive to load programs and data sets. Some type of printer is usually included to provide "hard copies" of results.

If you were to open the boxes, you would typically see some smaller boxes, several electronic circuit boards, possibly a fan or two, and some more wires or

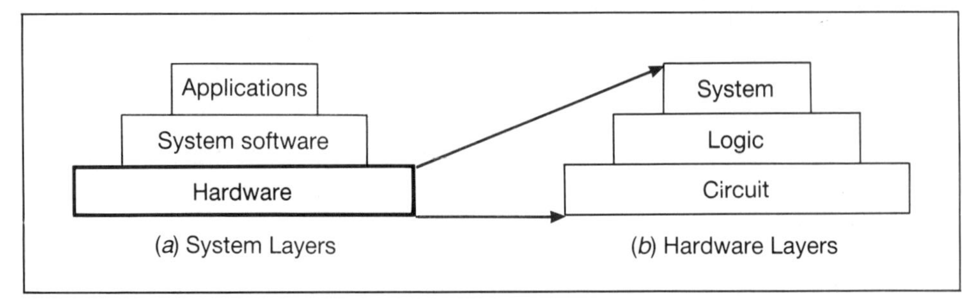

FIGURE 4.5 The hardware layers of a computer system.

cables. Unfortunately, you would not know much more about the system than you did before you opened it. (However, you might be impressed with how much "stuff" is inside the box.) To really understand the hardware, we need to look at how each of the three hardware layers (shown in Fig. 4.5) is defined.

The system level is defined by the instructions that the computer can execute. It is also defined by the size and operations of its modular units and peripherals. The logic level describes the modules by the logical combination of variables and the current state of the logic elements. The circuit layer is defined in terms of the necessary magnetic responses, voltage, or current of the gates and storage elements. These elements constitute the combinational and sequential circuits we saw earlier.

The modern computer system shown in Fig. 4.6 contains functional modules similar to those proposed by von Neumann, except the circuit technology is now completely different. The primary improvements have been in the speed, size, and type of active circuit elements used and in the power they consume. A single vacuum-tube circuit (such as a 1-bit memory element) in a mid-1950s computer needed approximately 6 in^3 of space. It dissipated about 1 W of power and could be switched from 1 to 0 at about 100,000 kHz. Today's technology is based on very large-scale integrated (VLSI) circuits. A $\frac{1}{4} \times \frac{1}{4}$ in memory chip can now store more than 4 million bits while dissipating only about 0.5 W and operating at 50 MHz.

Before we go any further, we need to point out a subtle change in terminology that has occurred over the years. In the early days, the term **central processing unit**, or **CPU**, referred to all the functional modules enclosed within the dashed lines in Fig. 4.4. However, this has now changed. In today's terminology, CPU refers to just those parts of a computer that interpret the instructions and manipulate data (as shown by the dashed lines in Fig. 4.6). Systems can have multiple CPUs that share the same internal memory (sometimes called the storage unit) and internal bus. It is a small, but important difference, and it allows greater flexibility for system definition.

Modern system capacities are also many times greater than those that were envisioned in 1945, and the range of applications has increased manyfold.[1] Today's

[1] One of the first IBM computers was supposedly designed with the assumption that 16 machines would saturate the world market. More than 1300 were built.

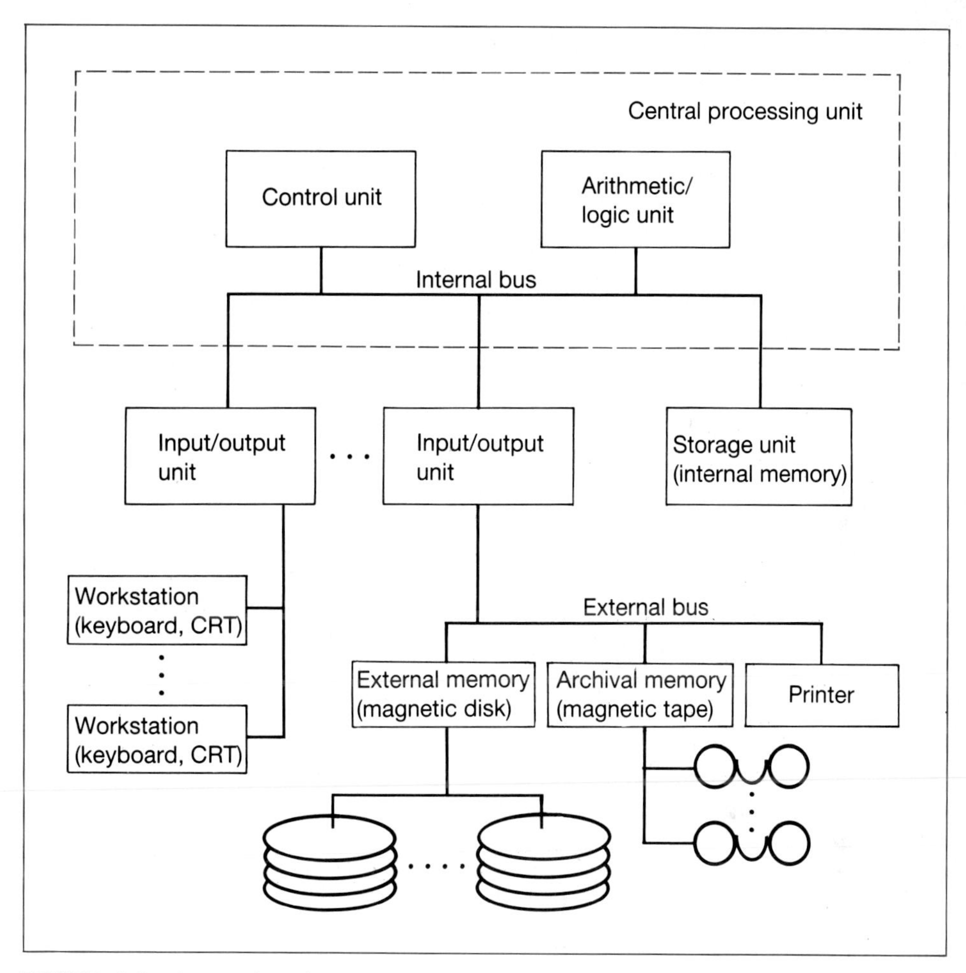

FIGURE 4.6 A typical modern computer system structure.

computers are no longer operated in a strictly sequential manner (that is, where a single program has exclusive use of the system and finishes before the next is loaded). Several tasks can now share one computer, and new hardware provides functions previously considered impractical or impossible.

The remainder of this section provides an introduction to the purpose and function of the major hardware modules. In Part II of the text we will look at hardware design philosophies, the logical considerations of each module, and peripherals. The circuit layer will be covered in later courses on electronics and logic circuit design.

4.3.1 Control Unit—The Central Controller

The control unit is responsible for determining what operations are called for by the program and in what order they should be carried out. When the power is first turned on, the computer does a **bootstrap** operation. This reads an instruction from a predetermined location in memory and transfers it to the control unit for execution. The following instructions are then fetched (read) from memory and executed in the sequence in which they are stored. The computer's program counter provides a way to keep track of the location of the next instruction. Execution order is changed by moving a new instruction location to the program counter before the next fetch is done.

Each instruction is a small, self-contained task description that can be likened to a short imperative sentence. A typical instruction contains an implied subject (the computer), a verb (an operation code) indicating what is to be done, and objects (operands) identifying data values or memory locations. When the instruction is received by the control unit, the operation code activates the proper logic sequence to execute the instruction.

Let's assume we have an arithmetic instruction that requires a memory access to obtain a data value. An example would be to add a value in memory to the current contents of a storage element in the arithmetic unit. The control unit first sends the proper signals to transfer the value from the memory to the arithmetic unit. The arithmetic unit then adds the memory value to the current value in the arithmetic storage element. While this is being done, the control unit increments the program counter so that it will point to the location of the next instruction. When the arithmetic operation is finished, the next instruction is fetched and the cycle begins again.

Modern computers are not generally as simple as the above example, but they do obey the same general rules. Instruction sets and architectures vary from computer to computer. However, once you understand the design concepts behind one system, you should have little trouble understanding another. We will look at two approaches to the design of the control unit in Chap. 5.

4.3.2 Memories and Storage Devices

Memory provides storage space for both the data needed during computations and the instructions that control the program. Large, fast memories allow more space for program and data storage. They can simplify program development and save execution time. The problem is that designers cannot have both large size and high speed without paying a high cost. Therefore, they must often trade off size, speed, and cost.

The typical computer has several kinds of memory, organized in a hierarchical fashion as shown in Fig. 4.7. The organization is based on assigned purpose, location within the system, and speed of access. The three types shown in Fig. 4.7 are considered to be the minimum for a usable computer system:

- **Internal memory** is also called **primary** or **main memory** because it is the storage element that is accessed by the control unit for the programs and data elements currently active in the system.
- **External memory** is also called **secondary memory** because it provides a second, longer-term storage for programs and data needed frequently but not

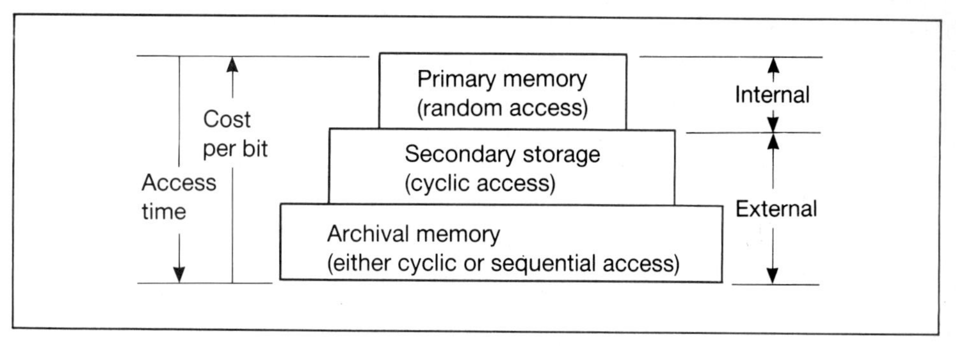

FIGURE 4.7 The hierarchy of computer memory.

currently active. It usually consists of disk units electronically connected to the system. An I/O operation is required to access information stored in secondary memory.

- **Archival memory** is on removable disks or magnetic tapes. It provides extended-term storage for:

  Extra copies of the programs and data stored on the disk. These can be used to back up or restore the system if some programs or data are inadvertently changed or lost.

  Other programs and data not frequently needed. They might be used again sometime later.

Typically only two operations, read and write, are allowed on memory elements. How fast the memories do these operations has a direct bearing on their cost.

Primary memory is the most costly of the levels shown. It is designed so that any location can be accessed in the same length of time (usually called a *memory cycle*). The type of memory used for this level is called **random-access memory**, or **RAM**. The term is descriptive because *RAM exhibits a uniform access time*. Accessing a random sequence of locations can be done in the same time as would be required to access the same length sequence in contiguous locations.

The number of bits that can be accessed in one memory cycle is known as the **memory width** or **memory word length**. On modern computers, memory words are usually an integer number of bytes long, as shown in Fig. 4.8.

Secondary memory is external to the main computer and is usually implemented with magnetic or optical disks. Disks belong to a general group of devices where the data cycles past the access mechanism (read/write heads in a disk) only once each revolution. These devices are called **rotational-**, **direct-**, or **cyclic-access** storage devices. Data on a disk is recorded on concentric areas called **tracks**, as shown in Fig. 4.9.

When a program in the computer needs to access information on the disk, it asks the I/O unit to do the transfer. Time is needed for the system to move the read/write heads to the proper track (if they are not already there). Additional time is required

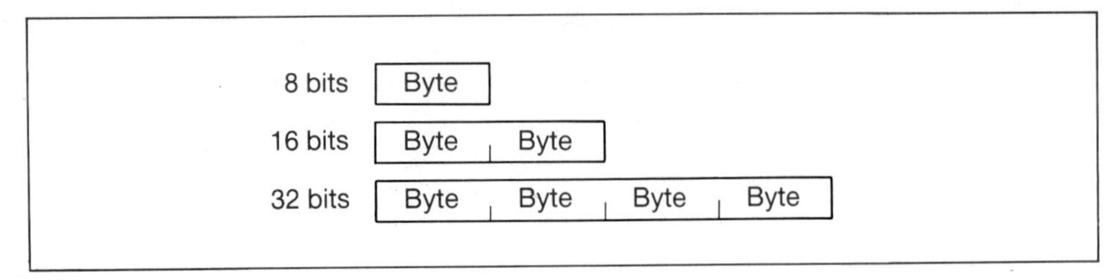

FIGURE 4.8 Some typical primary-memory-word formats.

to rotate the data around to the head. The necessary time is usually much greater than the time for a RAM access. However, the cost per bit of storage is also much less.

The third major memory-access type is characterized by sequentially recorded information on a recording medium, such as magnetic tape. The information is available only once during a read or write operation. Magnetic tape drives are known as **sequential-access** storage devices. They must be reversed or rewound before any specific section of the data can be reread or rewritten. This is like cassette tapes, which must be rewound before they can be played. Magnetic tape saves space that would be used in secondary memory. It also provides a low-cost means of easily restoring data that has been removed from secondary storage.

There are other specific types of memory within the three categories we have just described. However, they all operate with random-, cyclic-, or sequential-access concepts. An in-depth look at internal memories will be given in Chap. 6. Rotational and sequential memories will be discussed in Chap. 10 in Sec. 10.2.3.

FIGURE 4.9 Tracks on a disk memory.

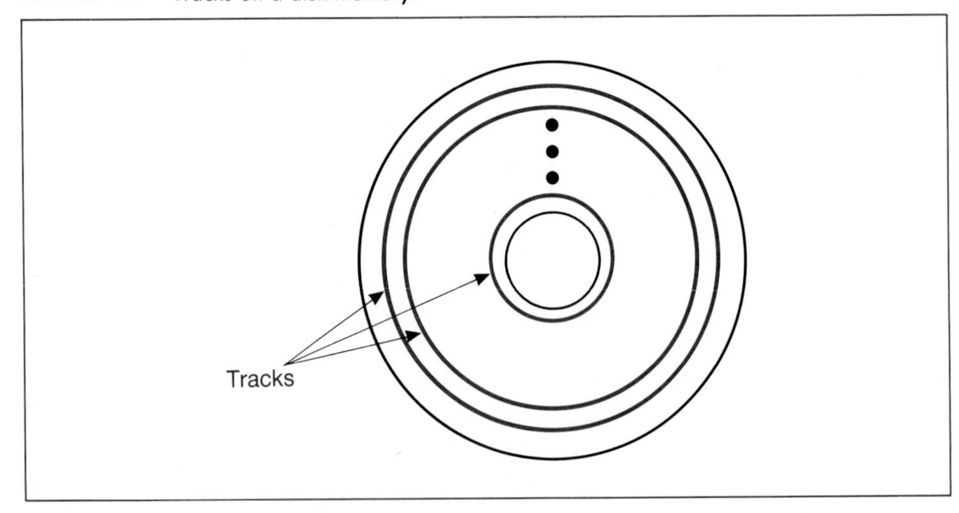

4.3.3 Buses—The System Data Paths

The system data paths, or **buses**, are the physical link between the various modules within the computer. They also link the computer and the outside world. Some buses provide a path for moving data and instructions to and from memory. Others are used for the control signals that ensure proper sequencing of data and instruction movement.

Data may be sent as either a stream of bits on a single line or as groups of bits on several lines. The method used depends on a number of factors and often varies from manufacturer to manufacturer. Each bus transfer must be coordinated or synchronized in some fashion. The receiver must be ready to accept the information when it is sent. The sender must not send it so fast that some information is lost.

The bus configuration is one of the earliest decisions that must be made in a computer design. Everything else connects to the bus or buses. A major question is whether to use an existing bus design or to develop a new one. Use of an existing bus can allow the design of processor modules and peripherals to begin sooner. It can also help ensure compatibility with other systems. In recent years, several buses have been adopted as industry standards. We will discuss the different types of buses and how they operate in Chap. 7.

4.3.4 Arithmetic/Logic Unit—The Processing Element

The **arithmetic/logic unit**, or ALU, is the primary processing element in the computer. Data in the unit can be converted in form or changed in value depending on the operation.

Most ALUs can do the elementary arithmetic operations of addition, subtraction, multiplication, and division on integers. Most can also do the logical operations of AND, OR, and INVERT. Some provide added capability for processing numbers in floating-point and decimal formats. Some advanced units are also capable of doing trigonometric and exponential operations. We will discuss the ALU in Chap. 8.

The ALU is a major factor in determining whether programs written in high-level languages will produce the same results on another computer. The format and word length, plus the procedure used for adjusting fractions to the space available, can allow errors to develop during computations. These errors are studied in an area of computer science and engineering called **numerical analysis**.

4.3.5 I/O Unit—The User Interface

The input/output unit has a twofold purpose. It provides a path between the computer and the user and a link between the computer and secondary memory.

In the extended sense, the I/O subsystem includes keyboards and displays, graphics tablets, printers and plotters, and voice facilities. It includes anything that allows the user access to the applications running on the system. The I/O unit is responsible for:

- Choosing and controlling the peripherals
- Isolating the CPU from I/O devices
- Compensating for speed differences between I/O devices and the computer

The I/O unit is controlled by software that is a part of the operating system.

The I/O system can be one of the major factors in determining the success of a computer system. I/O operations have been known to require as much as a third of the operating system instruction space. They can also use up to two-thirds of the available processor time. The design effort needed for a good I/O system often exceeds that required for the rest of the CPU by manyfold. We will discuss more about this in Chaps. 9 and 10.

4.4 SOFTWARE—USING THE HARDWARE

The upper two layers of the computing system are both implemented in software, as shown in Fig. 4.10. Software furnishes the means for defining the operations and procedures that the hardware will execute. It is the interface between the user and the system.

The lower of these two layers represents the system software. It is responsible for managing the system's resources and for providing services to the application layer. It also isolates the application software from the details of how the hardware works.

The application layer is the software level that gives the computer its "personality." It is the layer most users directly interact with, typically through preprogrammed applications packages. These packages provide specific services or capabilities, such as word processing and graphics design, and are discussed in Chap. 16.

The next two sections give an overview of the various components in the two software layers of a typical computer system.

FIGURE 4.10 The software levels of the computing system.

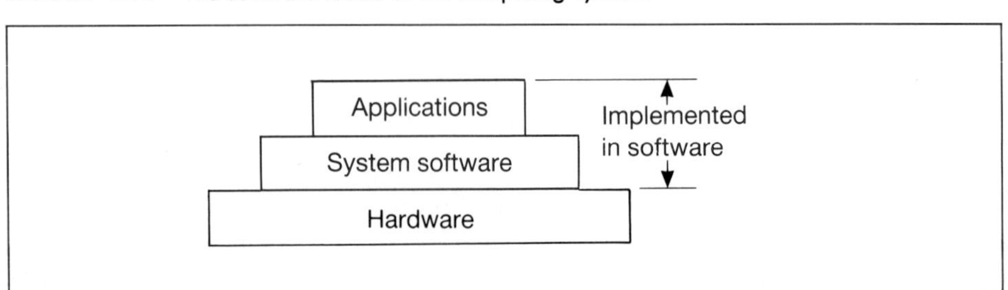

4.5 SYSTEM SOFTWARE

The system software is generally viewed as the interface layer between the applications and the hardware. It manages resources, provides services to the application layer, and isolates the application software from hardware details. The system software usually includes components such as the operating system, language translators, utility programs, and the command interpreter. You see these components when you run programs on an interactive system.

4.5.1 The Operating System

We like to view the operating system from two perspectives: the functions it does and the components it includes. An operating system functions as the following:

- A resource manager, allocating resources like memory, the CPU, printers, disk drives, and other peripherals
- An interface, managing the boundary between the user and the hardware to provide a friendlier environment where users do not have to worry about low-level hardware operations
- A coordinator, providing facilities so that complex activities can be constrained to be done in a predefined order
- A guardian, providing access controls to protect files and to restrict reading/writing/executing data and programs
- A gatekeeper, controlling who can log onto the system and what they can do once they are logged on
- An optimizer, scheduling some users' inputs, others' database accesses, others' computations, others' outputs, etc., to increase system utilization
- An accountant, keeping track of CPU time, memory usage, I/O calls, disk storage, terminal connect time, and so forth
- A server, providing services users frequently require, either implicitly or explicitly, such as file access mechanisms

Those are the general functions of an operating system. We now need to consider its specific components. Many different structures are possible for an operating system, but often it will include the eight layers shown in Fig. 4.11:

- **Kernel**—directly accesses the hardware of the system, controls operations and communications between parts of the system
- **Memory manager**—decides how much memory, and which areas of memory, to give a user
- **I/O system**—services a user program's I/O requests
- **File manager**—manages devices that store multiple information files such as disks and magnetic tapes

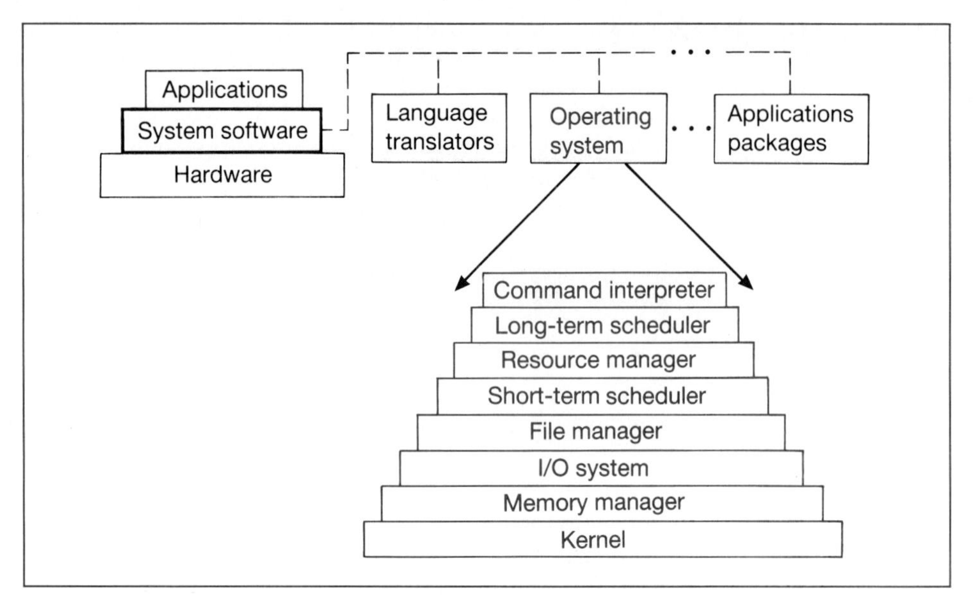

FIGURE 4.11 A layered organization for an operating system.

- **Short-term scheduler**—picks the next process to execute on the CPU
- **Resource manager**—allocates resources (other than memory and the CPU) among competing processes
- **Long-term scheduler**—decides which processes are allowed to enter the system
- **Command interpreter**—interacts directly with the system user, analyzes user commands, and calls on system routines to satisfy these user requests

The kernel is small, specialized, and hardware-dependent, and it will be discussed in Chap. 14. The other parts of the operating system are described in Chap. 13. The operating system and related components will also be covered in other courses you might take, such as Operating Systems or Systems Programming.

4.5.2 Language Translators

Many of you have written programs in a high-level language such as Pascal, C, Fortran, or Ada. Your program statements were written in a code that both you and the language translator could understand. Unfortunately, most computers cannot yet execute this code directly.

The language translator must convert the statements into a language that the computer can use. The language a computer understands is called a **machine language**, and it varies from computer to computer. The software that does the translation from a high-level language to machine language is called, not surprisingly, a **language translator**.

Translator is a generic name that applies to a number of different translation programs. We often hear the terms *compiler* and *assembler*. They are more specific terms. A **compiler** translates from a human-oriented, high-level language such as Pascal to the computer's machine language. An **assembler** translates from a symbolic version of the computer's machine language to the corresponding numeric code that the computer can execute.

For instance, consider a high-level statement that reads a sequence of characters into a variable called "line" from a data file called "data1":

```
read(data1, line);
```

Both you and the compiler know what you mean because both "data1" and "line" have been defined in your program. However, the statement does not yet contain enough information to allow the operating system to execute your request. It might be first translated into a call to an operating systems procedure for controlling input/output operations:

```
IOCS(operation, data1, line, length, errors, status);
```

where the procedure and parameters are as follows:

- IOCS: the name of the input/output control system procedure
- operation: contains the I/O command that IOCS is to do (in this case, a read)
- data1: the logical unit name for the unit on which the operation is to be done
- line: the address of the variable that is to receive the data
- length: contains the number of characters to be read
- errors: tells IOCS whether the user process wants to be responsible for handling I/O errors
- status: a variable that receives the resulting status of the operation (e.g., "successful" or "tried to read past end of file")

This, in turn, might be translated into a sequence of assembler-language statements. These statements transfer all the parameters to a storage area for IOCS, then pass control to IOCS:

```
SAVE read
SAVE data1
SAVE line
SAVE length
SAVE errors
SAVE status
CALL IOCS
```

Lastly, these assembler statements are translated into a series of numeric instructions that might appear as

```
1011000010101011
1011000011001101
        ⋮
1101000010101000
```

where the codes 10110 and 11010 at the beginning of these instructions are the binary codes for the computer operations SAVE and CALL. The rest of the bits in each of the instructions give the address where the parameters and the called procedure are located.

The language that is input to a translator is called the **source** language, and the output of the translator is called the **target** language. We have now taken the source statement

```
read(data1, line);
```

and converted it into the target statements such as

```
1011000010101011
1011000011001101
1101000010101000
```

It is easy to see the human situation that is similar to our computer example. A human *translator* might have to read a source language such as Russian and *translate* it into a target language such as English.

Many computer users interact only with the "language" or interface of an application such as a database or spreadsheet. In your role as a computer professional, you will probably often be involved directly with programming languages such as Pascal, C, Fortran, or Ada. We will therefore cover this topic in much greater detail in Chap. 15.

4.5.3 Loader/Linkage Editor

In the previous section, we described how the compiler produced a binary target code to call an I/O procedure. The compiler has assigned memory locations for all the variables in our data set and for all the instructions in our program. It does this as though we would be the only users in the computer when our program is executed.

When the program is brought into memory, however, both it and the data may not necessarily be placed in the locations assigned by the compiler. Instead, they may have to be placed into another area in memory because others may also be using the system. This is known as **code relocating**. It is done by a **relocating loader** or **linkage editor**. Although this last step is also technically a translation, relocating loaders and linkage editors have traditionally been treated separately from other translators.

4.5.4 Debuggers

You probably have already experienced how difficult and frustrating it is to find and correct, or **debug**, programming errors. Most systems now include interactive source-level debuggers to help you with this task. They probably include some or all of the following features:

- Printing the value of a variable each time it changes
- Indicating where in the program the change took place
- Allowing you to enter new values for specified variables and continuing execution with these new values
- Tracing the flow of execution within a program
- Tracing the flow of execution among the various procedures you have used
- Notifying you when some prespecified condition happens
- Notifying you when an error occurs
- Allowing you to change some code and then reexecute the program

Some of you probably used such a debugger in your introductory programming class.

4.5.5 The Command Interpreter

As we said earlier, the command interpreter is the outer layer on the operating system and is designed to interact directly with the user. It is usually responsible for sending the prompt to the user's screen or displaying the menus or icons on that screen. It analyzes user commands, and it calls the necessary system routines to satisfy the user requests.

Since command interpreters are such an important and visible part of the system, we want to give a brief illustration of three standard formats for command interpreter interfaces. Figure 4.12 shows the following:

- A command line form, where the system (in this case an HP 3000) typed the colon (:) as an operating system prompt and the user typed the rest of the line.
- A menu form, where the system (in this case Microsoft's Fortran Version 2.2 compiler) displayed a menu and the user has moved the cursor to select one option from the menu, shown by the inverse video. (Only the menu portion is shown here; for the full screen, see Chap. 17.)

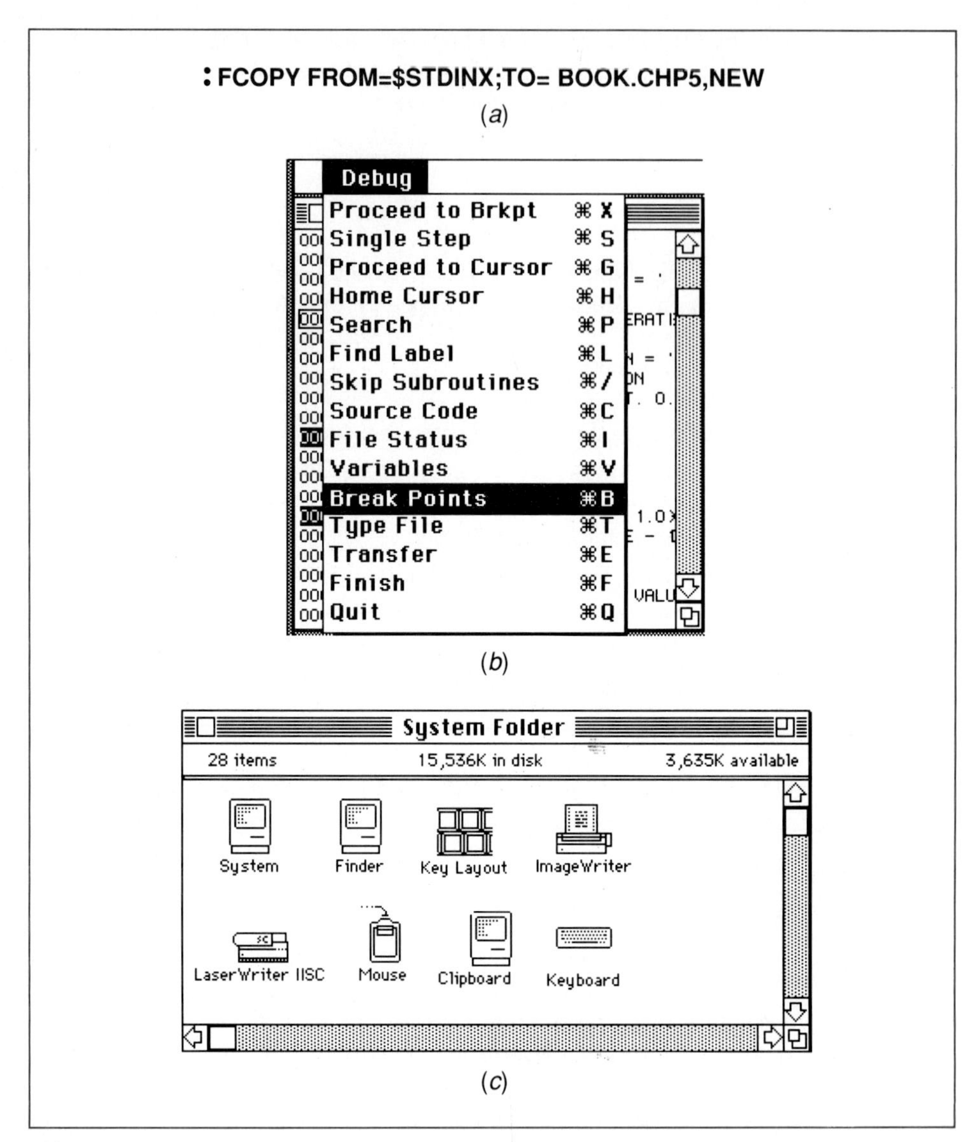

FIGURE 4.12 Three formats for command interpreter interfaces: (*a*) command line; (*b*) menu; (*c*) icons.

- An icon form, where the system (in this case a Macintosh) has displayed a set of icons and the user is moving the cursor to select one of those icons.

Sometimes, people want to take an existing operating system and change the look or "face" it presents to the user. That is, they want a different form of command interpreter. Often they do this by adding a layer above the command interpreter, a layer called a **shell** (see Fig. 4.13).

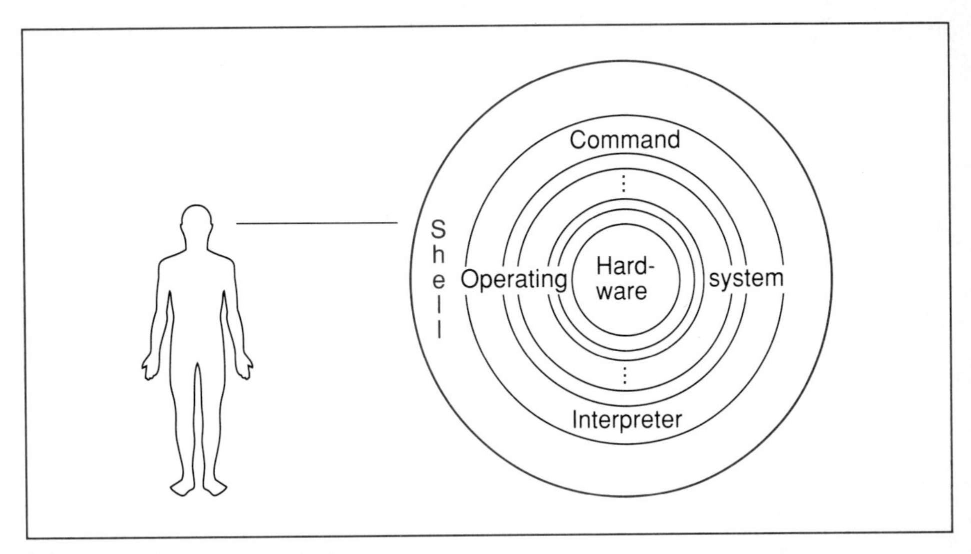

FIGURE 4.13 Adding a shell to an existing operating system.

For instance, menus have become a popular way for users to interact with computers. Several companies have developed menu-based shells that "hide" the details of the existing operating system from the user. Instead of typing commands directly to the operating system's command interpreter, the user picks the desired command from a series of menus. He or she does this by moving a cursor around with a mouse. The menu shell then sends the appropriate command string to the operating system. This process is illustrated graphically in Fig. 4.13. The user "sees" the menus provided by the shell instead of the standard commands of the operating system's command interpreter.

4.5.6 Editor

Editors are programs that allow a user to create and change program and data files interactively. Usually, editors are classified into two types:

- Line editors, which treat text in units called lines
- Screen editors, which treat text in blocks of whatever size fits on a screen

For example, we might want to delete the first bullet entry (the entry identified by the first "■") in the above paragraph. This could be done with a line editor, as shown in Fig. 4.14*a*, or with a screen editor, as shown in Fig. 4.14*b*.

4.5.7 Expanded File Management Utilities

We have mentioned that the operating system has a layer which provides file services to an executing program. System software also includes file management util-

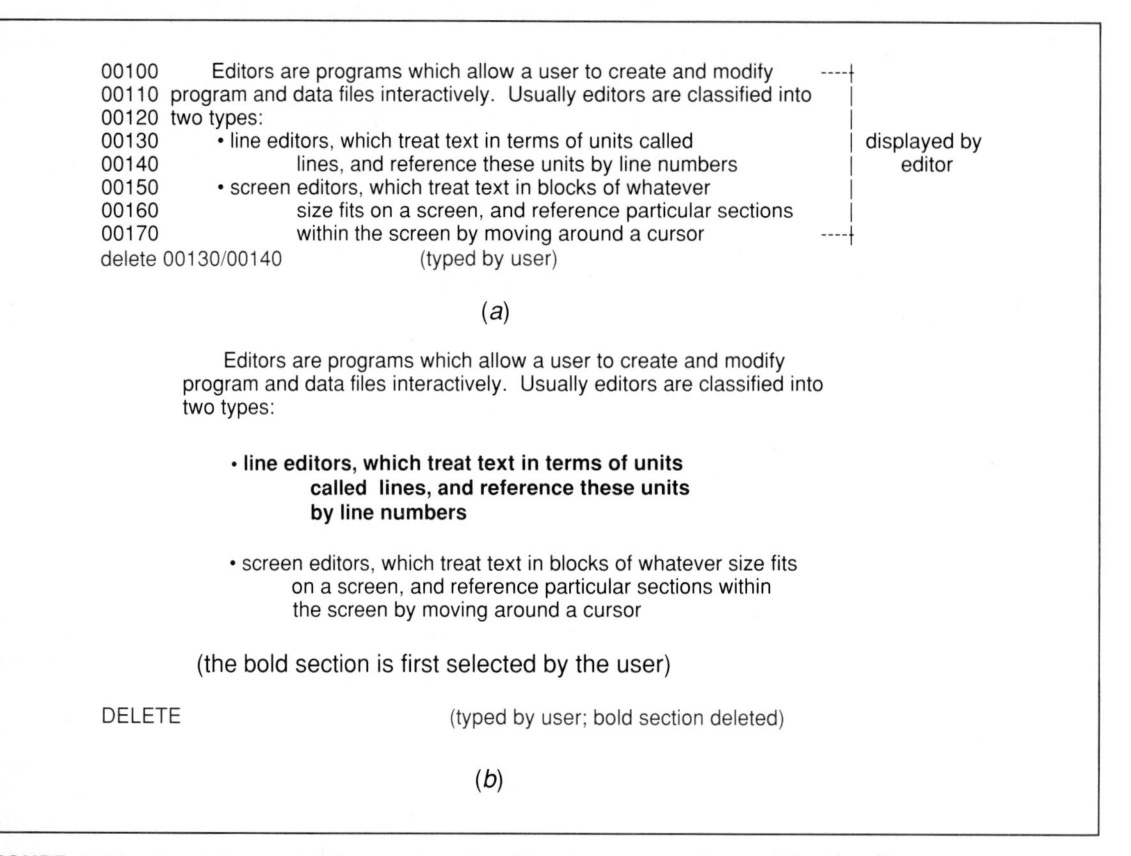
00100 Editors are programs which allow a user to create and modify
00110 program and data files interactively. Usually editors are classified into
00120 two types:
00130 • line editors, which treat text in terms of units called
00140 lines, and reference these units by line numbers
00150 • screen editors, which treat text in blocks of whatever
00160 size fits on a screen, and reference particular sections
00170 within the screen by moving around a cursor
(00100–00170: displayed by editor)
delete 00130/00140 (typed by user)

(*a*)

Editors are programs which allow a user to create and modify program and data files interactively. Usually editors are classified into two types:

• **line editors, which treat text in terms of units called lines, and reference these units by line numbers**

• screen editors, which treat text in blocks of whatever size fits on a screen, and reference particular sections within the screen by moving around a cursor

(the bold section is first selected by the user)

DELETE (typed by user; bold section deleted)

(*b*)

FIGURE 4.14 Example use of different editors for deleting a section of text: (*a*) with a line editor; (*b*) with a screen editor.

ities that can be performed interactively by a user. Some typical ones are:

- Archiving (keeping an orderly sequence of successive versions of a file)
- Protecting (specifying who can access a file, and what each can do)
- Comparing two files; computing a checksum and block count
- Copying, renaming, deleting, or printing a file
- Undeleting a file by trying to retrieve it after it has been deleted
- Packing (obtaining a more condensed form of a file)
- Sorting the contents of a file or searching a file for a string
- Splitting a file into smaller component files

Working on a project involving many files (such as an introductory computer text) would truly help you to appreciate these utilities.

4.5.8 Disk Management Utilities

Operations discussed in the previous section are done with individual files. There are also some operations done with an entire disk. These might include functions such as:

- Determining and reporting the available free space
- Printing the disk directory or a subdirectory
- Making a new subdirectory or removing an existing subdirectory
- Formatting, packing, or reorganizing a disk
- Copying one disk to another disk
- Comparing two disks

Programs which compact the desktop, eliminate fragmentation, and prioritize files on the Macintosh system are excellent examples of disk management utilities.

4.5.9 Security Management

There is an increasing emphasis on computer security, and as a result most systems, even for microcomputers, now include some security features. For example, a user might be able to require passwords for logon, limit accesses rights to files, encrypt sensitive information, and so forth. You can expect to see more security features being built into system software in the future.

4.5.10 Text Output Formatter

Often you do not want to output a text file "as is"; you want to add page numbers, headers, and so forth. Text formatters provide this facility. A few common features are:

- Set margins and adjust text with them (e.g., centered)
- Set line length and page length, line spacing, tab stops, etc.
- Add headers and footers, with or without page numbers, in either Roman or Arabic text
- Insert a specified number of blank lines
- Start a new page
- Print specified text as is, without formatting

These features look similar to some of those provided by a word processor. The difference is that these are normally applied to arbitrary files (such as program output data), rather than to specific word processor files.

4.5.11 Interpreters

Earlier we described compilers and other programs that translate a source language to a target language. Typically, the source language is a high-level language and the target language is the machine language for the computer you are using.

The aim is to execute a program, but sometimes we are more concerned with helping the user than in getting efficient performance. This is especially true if there are errors that must be found and corrected. In this case we might provide an interpreter instead of a compiler. An **interpreter** does not produce target code. Instead, it takes the statements in the source language and does what they specify; i.e., it *interprets* them. This interpretation is done under the interpreter's control, with many error checks. For example, a user might have written the following statement in his or her program:

```
distance := sqrt((x[i] - x[j])^2 + (y[i] - y[j])^2);
```

The interpreter could do the specified operations, while checking such things as

- Have the variables *x*, *y*, *i*, and *j* been declared?
- Have *x* and *y* been declared as arrays?
- Are *i* and *j* the right-type variables for subscripts?
- Are *i* and *j* within the range specified for *x* and *y* subscripts?
- Have the referenced array elements been defined?
- Are *x* and *y* the right type for the function "sqrt"?
- Are the squares and their sum within the range of the specified-type variable?
- Has the variable "distance" been declared?
- Is "distance" the same type as "sqrt," or can a conversion be done?

If any errors happen, the interpreter can tell the user and allow him or her to correct the error interactively.

Compilers could add code to do many of these same checks during execution, but this might greatly slow program execution. Furthermore, some types of errors might automatically transfer to the operating system, without allowing user intervention. Therefore, system software often includes interpreters to provide maximum "user friendliness" and compilers for system efficiency.

4.6 THE APPLICATION LAYER

You have been introduced to the different levels of the application layer in programming courses. You had to analyze applications, define logical procedures to formulate them, and then write and debug source programs to run on the computer. This is shown in Fig. 4.15, and, until recently, it would have been the most common view of the application layer.

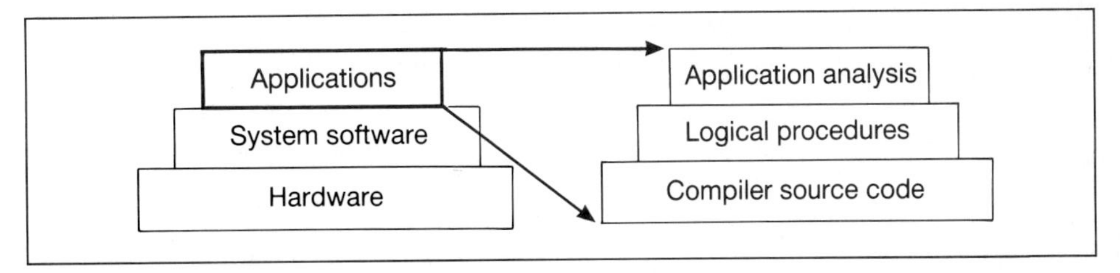

FIGURE 4.15 Levels of the application layer.

The personal computer and computer workstations have brought about major changes in the application layer. Most small-computer users do not want to write their own software. They want a system that is already programmed to do the tasks that they need to have done. This, in turn, led to the creation of an entire new industry dedicated to the development of general-use applications packages, such as:

- Word processors: incorporate many or all of the features of an editor and text formatter, plus additional features such as a choice of fonts (e.g., **Helvetica**, **Rockwell**, Times, or *Bembo*), sizes (e.g., 9 pt, 12 pt, or 18 pt), and formats (e.g., **bold**, *italics*, underline, outline, or *script*). The original manuscript of this book was written using the Microsoft Word word processor on a Macintosh computer.
- Database managers: facilitate information-handling operations, such as data storage and retrieval, data manipulation, and report generation.
- Graphics: give the user various shapes, line types and thicknesses, fill patterns, and other features. The illustrations in this book were originally produced using DeskPaint, DeskDraw, MacDraw, and MacPaint, four graphics editors for the Macintosh computer.
- Spreadsheets: provide accounting, budgets, forecasts, and other financial operations.
- Electronic mail: facilitates transmission of the electronic form of mail.
- Computer-based teaching: allows students to learn via interactive sessions with a computer. These packages can also include automated testing, evaluation, animation, alternate assignment, and automatic problem generation.
- CAD (computer-aided design): have features similar to graphics packages but are specifically designed to assist in making building, electronic, and mechanical drawings. (Some CAD tools are described in Chap. 11.)
- CAM (computer-aided manufacturing): provides facilities for the direct control of manufacturing equipment, such as milling machines and lathes.
- Hardware and software design aids: help the designer develop hardware and software systems. (Some are described in Chaps. 11 and 17.)
- Decision systems: assist management in making decisions when a large number of factors and possible outcomes are present.

- Simulation systems: allow users to build and work with computer models of devices, buildings, service facilities, and many others.
- Sound recognition/production: facilitates recognition of sounds, such as voice input, and generation of sounds, such as synthesized music.

FIGURE 4.16 Software component relationships within a computer system.

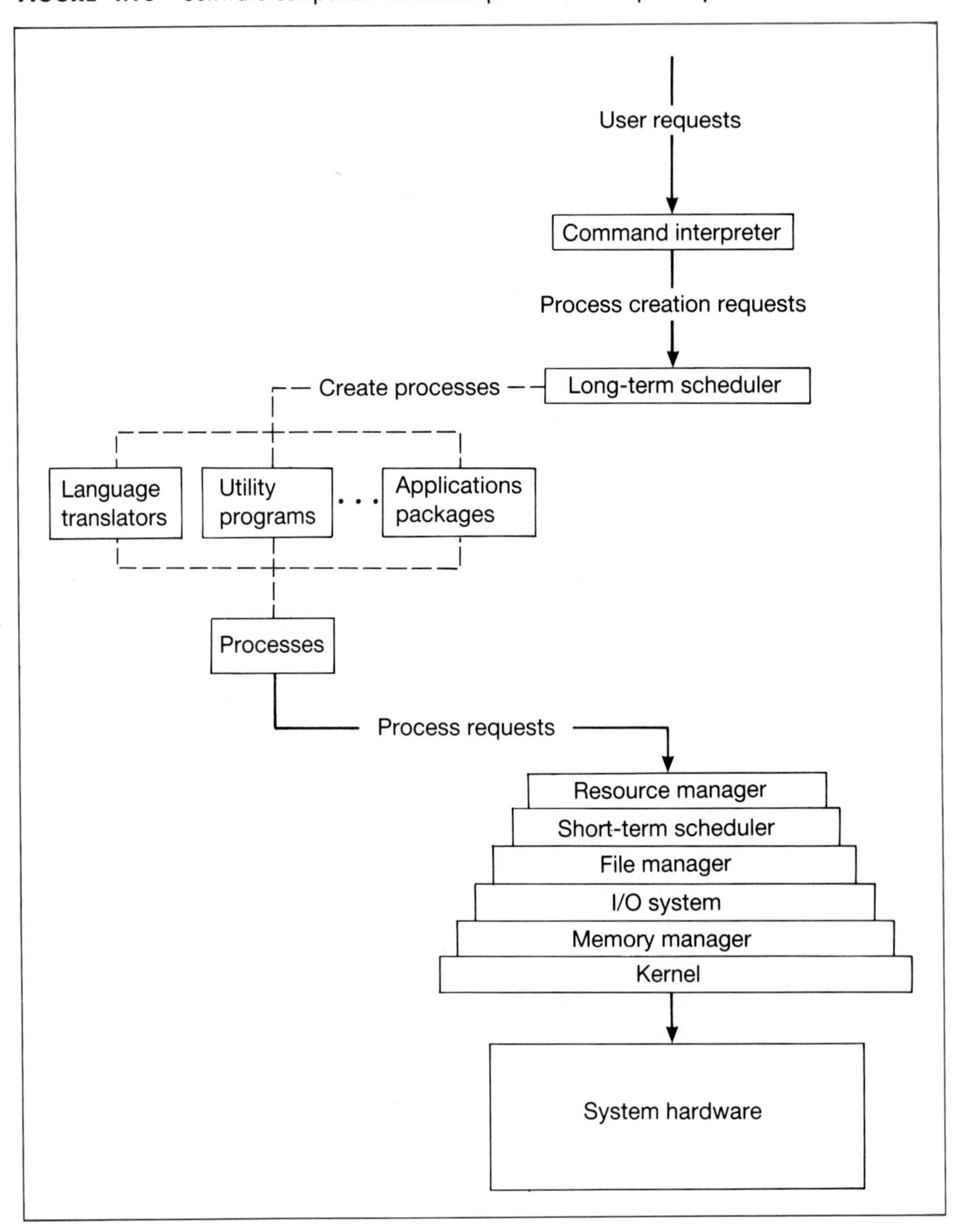

The value of this type of software should not be underestimated. The available applications often determine whether prospective customers will be able to use the system effectively and efficiently. People want to use the computer as a tool. They do not want to spend excessive amounts of time learning peculiar vagaries of a system.

When software is easy to use and simple to learn, we usually say that it is **user friendly**. This is an important, but subjective, measure of the "worth" of the software. It often is a deciding factor in the selection process, and will be discussed in Chap. 18.

4.7 SUMMARY

We have represented each of the components in a computing system as a series of layers, but we need to put this into perspective. The system hardware is a collection of functional modules that can communicate with each other over a system bus. In hardware, the layers are more of a design concept than an actual implementation constraint.

This is not always the case in software. In software, layering is often seen as both design and implementation constraints. For example, the modules of an operating system are often designed as the layers shown in Fig. 4.16. The implementation requires that this layering be enforced by strict communication protocols between each of the layers.

PROBLEMS

4.1 List the operations that would be required to add 28 to 813 on the Chinese *suan pan* and on the Japanese *soroban*. Which do you prefer? Why?

4.2 Why do you think there was a large gap in the development of computing machines after 1700 B.C.? Were these the same reasons there was a gap after Babbage developed his difference engine? Discuss any obvious differences.

4.3 List the operations that would be required to multiply 21 by 34 on an abacus. Which number should you use as the multiplier? Why?

4.4 Why did Babbage's attempt to build a computer fail?

4.5 What was the motivation for Hollerith's card sorter? How did the card sorter impact the development of the tabulating machine?

4.6 What was the major conceptual difference between von Neumann's proposal and previous computer designs?

4.7 An abacus can be thought of as an externally programmed computer. Where is the program stored for an abacus?

4.8 Von Neumann's design was independent of any particular technology. Why was this important?

4.9 How has the usage of the term *central processing unit* changed over time?

4.10 How does the modern computer know which instruction is to be performed next?

4.11 How does the modern computer know which instruction is to be performed when it is first turned on?

4.12 For a computer to which you have access, find out how its bootstrap operation works.

4.13 What are the elements of a computer instruction?

4.14 Where does the program counter point while an instruction is being executed?

4.15 What would you normally expect to find in primary memory? In secondary memory? In archival memory?

4.16 Estimate the volume of physical space it would take to store this book (*a*) in solid-state memory chips, (*b*) on a floppy disk, (*c*) on a hard disk, and (*d*) on a CD.

4.17 Why is primary memory usually much faster than secondary memory?

4.18 Many old computers stored system programs on magnetic tape devices. Why was this method of storing commonly used programs discontinued?

4.19 What are the arithmetic operations that should be included in the ALU?

4.20 What is the purpose of the system bus? Why is its early selection in the design process important?

4.21 Why is an I/O unit needed?

4.22 List 10 computer I/O devices you have seen.

4.23 What are the functions of operating system software?

4.24 Describe the type of command interpreter you used in your beginning programming class.

4.25 What function does a compiler perform?

4.26 Explain the differences between high-level languages and machine languages.

4.27 What are some examples of *source* languages?

4.28 Why is it sometimes necessary to relocate programs when they are loaded into memory?

4.29 Why is the undelete function an important feature of a file system?

4.30 List the operations that a debugger lets a programmer do.

4.31 How can passwords be used with file access?

4.32 What is the difference between an interpreter and a compiler?

4.33 Is there anything that the application programs listed in Sec. 4.6 have in common?

4.34 What is the ultimate limit on the variety of application layer programs that can be developed?

AN OVERVIEW OF COMPUTER HARDWARE

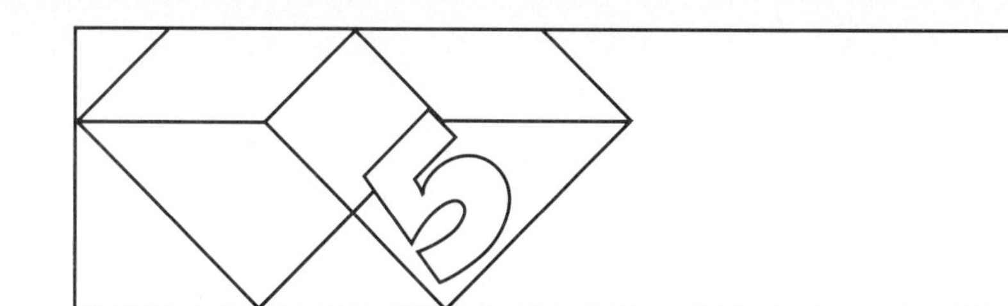

The Control Unit

We learned in Chap. 4 that the control unit is responsible for performing the operations specified by the instructions that make up a program. In the simplest case, it does this one instruction at a time, beginning with the first, then stepping sequentially through the program.

The sequence of operations that the control unit performs for an individual instruction is called the **instruction cycle**. Three steps are required for each instruction processed:

1. **Fetch** the instruction from memory and transfer it to the control unit.
2. **Decode** the instruction to determine both what is to be done and which data values are to be used.
3. **Execute** the instruction by selecting the indicated operation and by controlling any required data movement.

The purpose of this chapter is to see what information the control unit needs and to learn how it uses that information. We will first define a simplified computer upon which to concentrate our discussion of instructions and control concepts. A detailed examination follows, focusing on how to design a control unit for a simple computer.

Lastly, we will examine some extensions to our simple computer and discuss how different designs affect the computer's performance. The discussion begins with an explanation of the kinds of instructions that are typically used by a computer.

5.1 COMPUTER INSTRUCTIONS

We said earlier that computer instructions are small task descriptions that can be likened to short imperative sentences directed to the computer. Each instruction contains several parts:

- An implied subject (the computer)
- A verb (called the **operation code**, or **opcode**) that defines what operation is to be performed
- Objects (called **operands**) that specify which values to use and where to place the result
- Adjectives and adverbs (called **modalities**) that change how the various parts of the instruction are to be interpreted

Each computer instruction is an independent logical command to the computer. This is an important concept, because all the information necessary to do the requested operation must either:

- Be included in the instruction itself
- Or have been previously placed in the registers or memory locations that are referenced by the instruction

The amount of information required varies according to the operation specified. The way the information is provided depends on the instruction itself and on the overall design philosophy of the computer.

Some computers are known as **memory-memory computers**, since all data values come from and go back to memory. During the execution phase, these computers:

1. Fetch operand values from memory
2. Execute the requested operation
3. Return the result to memory

This approach is illustrated in Fig. 5.1*a*.

Figure 5.1*b* diagrams a type of computer that uses a register for intermediate storage of the data values needed in the computation. One instruction moves a data value from memory into the register. Sometime later, another instruction:

1. Moves the second operand from memory
2. Performs the given operation with the value in the register
3. Stores the result back into the register

These computers are known as **memory-register computers**, since one data value comes from memory and the other comes from a register. The result goes back either to the register or to the memory location from which the operand came.

A third class uses a set of registers for intermediate storage of all data values that will be used in the computations. The data must be fetched from memory by a series of instructions before the computation begins. The computation is then performed and the result returned to a register, as shown in Fig. 5.1*c*. These computers are

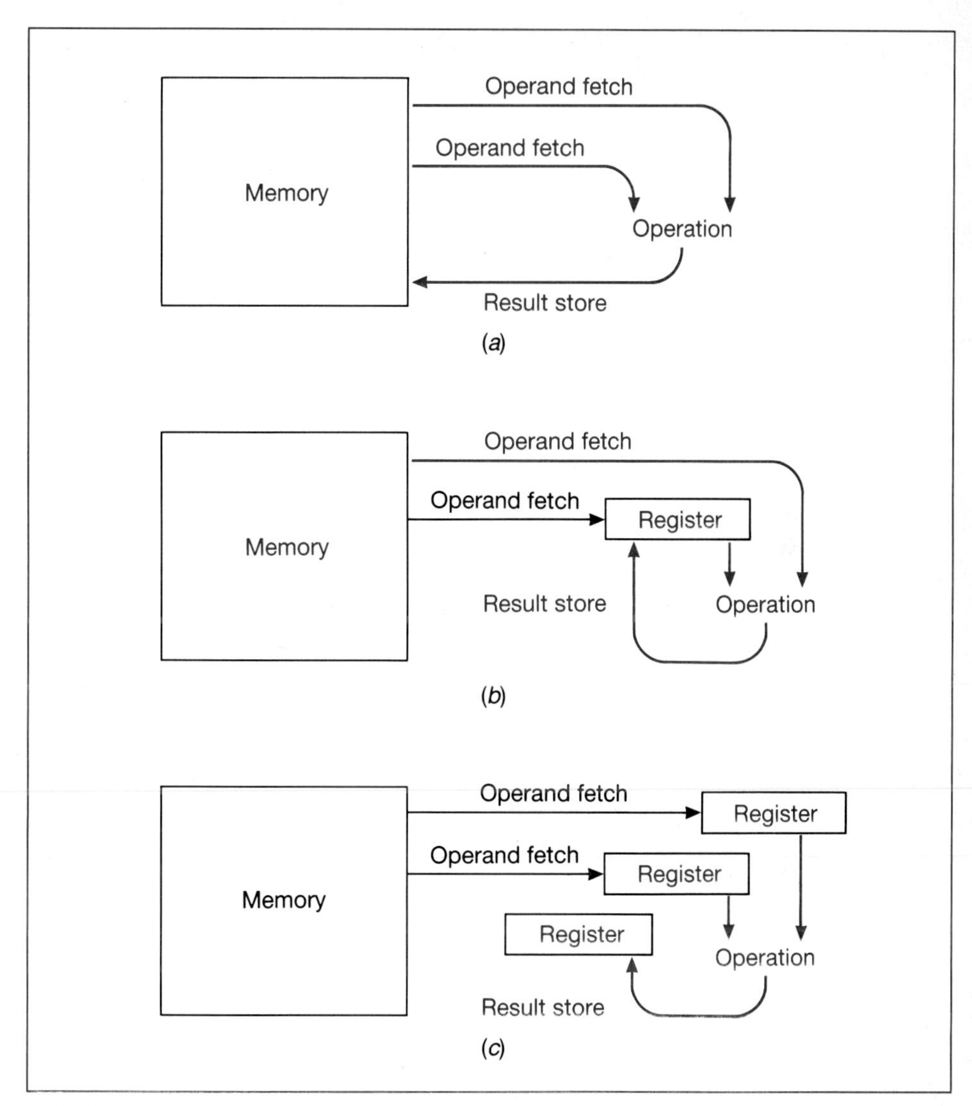

FIGURE 5.1 Alternate design philosophies: (*a*) memory-memory; (*b*) memory-register; (*c*) register-register.

known as **register-register computers** and are sometimes called load/store computers.

The simple computer that we will be using to illustrate an instruction set and the operation of the control unit is shown in Fig. 5.2. This computer uses an arithmetic/logic unit (ALU) with a single accumulator register, so it is a memory-register computer. The ALU also contains a status register composed of 3 bits. These bits are

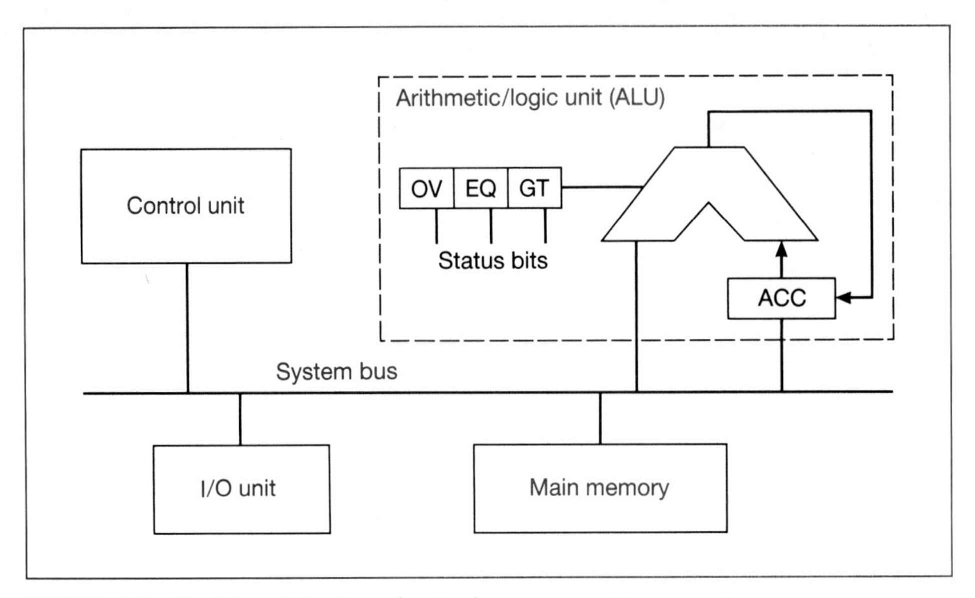

FIGURE 5.2 The internal structure of a simple memory-register computer.

changed by each arithmetic or compare operation:

- EQ is set to 1 if the result of a computation is zero or if a compare is equal.
- GT is set to 1 if the result of an arithmetic computation is positive or if a compare is "greater than."
- OV is set to 1 if the result of a computation is too large to fit (overflows) the register.

Other computers might have additional status bits.

5.1.1 Instruction Formats

A computer instruction is a group of bits, such as the one shown in Fig. 5.3. The organization of this pattern is known as the **computer instruction format**. The opcode is normally the first part on an instruction. The remainder of the instruction identifies the operand used by the instruction.

Instruction formats in most computers are more complex than the one shown in Fig. 5.3. They often contain additional information, some of which we will discuss in later sections, and they are not always the same length. (In most cases, however, they are an integer number of bytes long.)

Computers designed with instruction formats that are always the same length are said to have **fixed-length instructions**. To simplify our discussion, we will assume our computer has a fixed-length instruction format of one memory word, which is 2

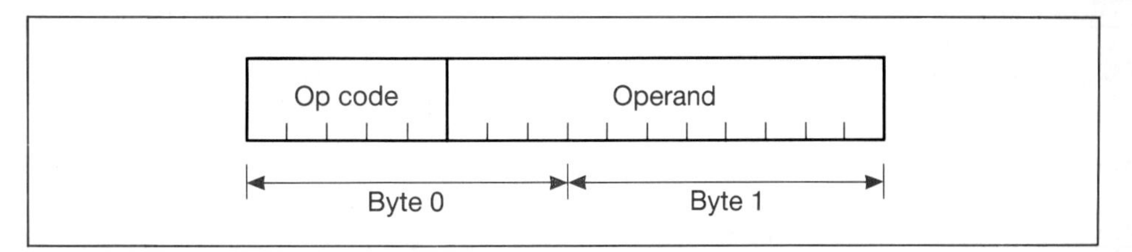

FIGURE 5.3 A simplified machine instruction format.

bytes, or 16 bits, long. We will also assume an instruction can be fetched in one memory cycle.

Computers that can process instructions with different lengths are said to have **variable-length instructions**. This feature can save memory space for the program, but it increases the complexity of the instruction fetch and decode logic. The control unit cannot predict where the next instruction will start until it decodes the current opcode.

5.1.2 Opcodes and Instruction Classes

The opcode portion of the computer instruction defines the operation that will be executed. Since the control unit must be able to distinguish exactly which operation is required, separate opcodes are needed for each different operation. Many general-purpose computers contain an extensive instruction set designed to support a wide range of applications, but some computers are designed to only have a few relatively simple instructions. The latter type, which are described in Sec. 5.5.4, are called reduced instruction set computers. The size of an instruction set typically varies from around 40 to more than 500 instructions. The number depends on which functions are implemented in hardware and which are provided in the system's software library. The system architect chooses the number of instructions to include at system design time, based on:

- The intended applications
- The required performance
- The target selling price
- Compatibility with previous designs

There is no standard instruction set that is included with every computer, nor is there a standard instruction format. But nearly all computers include a common set of instructions that perform operations such as:

- Arithmetic and logic operations that require use of the ALU
- Data move operations that transfer information between functional units within the computer
- Program control operations that change the flow of the program

These groupings are somewhat arbitrary, but they provide a framework to begin our study of instruction types.

To make the instruction readable to humans, opcodes are normally represented by mnemonic abbreviations (such as JMP), and memory locations are given symbolic names. Within the computer, however, both the opcode and the operand are represented by combinations of bits.

The instructions discussed in the following sections are typically found in all computers. We have defined them using the single-operand format that we selected for our example computer. Later in the chapter, we will look at alternate forms of the same instructions to see how the other design philosophies affect the format.

5.1.3 Arithmetic/Logic Group

As the name implies, the arithmetic/logic group includes instructions that execute arithmetic and logic operations. They typically refer to one or more of the registers in the ALU and include operations such as those shown in Table 5.1. (In this table, the operand M means the memory location M, and the accumulator means the contents of the ACC register of Fig. 5.2.)

Most computers have more instructions in the arithmetic/logic instruction group than are defined in Table 5.1. We have included only enough instructions to give you a flavor of the types of operations that are typical to this group and to let us illustrate how the control unit could execute these instructions.

TABLE 5.1 TYPICAL ARITHMETIC/LOGIC GROUP OPERATIONS

Operation	Typical mnemonic	Instruction meaning
Add	ADD M	Add the value in M to the accumulator
Subtract	SUB M	Subtract the value in M from the accumulator
Multiply	MUL M	Multiply the accumulator by the value in M
Divide	DIV M	Divide the accumulator by the value in M
AND	AND M	Logically AND the accumulator with the contents of M
OR	OR M	Logically OR the accumulator with the contents of M
Exclusive OR	XOR M	Exclusively OR the accumulator with the contents of M
Complement	NOT	Logically invert the accumulator
Shift left	SHL	Shift the accumulator 1 bit to the left, move the leftmost bit to the overflow register, insert a 0 on the rightmost bit of the accumulator
Shift right	SHR	Shift the accumulator 1 bit to the right, discard the rightmost bit, insert a 0 in the leftmost bit of the accumulator
Rotate left	RRL	Shift the accumulator 1 bit to the left, but move the leftmost bit around to the rightmost bit position
Rotate right	RRR	Shift the accumulator 1 bit to the right, but move the rightmost bit around to the leftmost bit position

TABLE 5.2 TYPICAL DATA MOVE INSTRUCTIONS

Operation	Typical mnemonic	Instruction meaning
Load	LDA M	Load the accumulator with the value in M
Store	STA M	Store the accumulator in M

5.1.4 Data Move Group

Data move operations are used to transfer data from one location to another within the computer. They vary considerably from computer to computer, depending on whether it is a memory-memory, memory-register, or register-register architecture. Table 5.2 shows two instructions, LDA and STA, that are typical of memory-register computers. Other computers use a more generic move instruction to transfer data to and from memory.

5.1.5 Program Control Group

Program control instructions provide a way for the programmer to halt the program's execution or to change the instruction execution sequence. Typical examples of operations in this group are shown in Table 5.3. (In this table, the operands L and M mean the memory locations L and M.)

TABLE 5.3 TYPICAL PROGRAM CONTROL INSTRUCTIONS

Operation	Typical mnemonic	Instruction meaning
Unconditional jump	JMP L	Execute the instruction at L instead of the next instruction in sequential order
Compare	CMP M	Compare the value in M to the accumulator; set the EQ bit to 1 if the values are equal, set the GT bit to 1 if the value in M is more positive than the accumulator
Conditional jump	JXX L	Transfer control to the instruction at L if the compare operation indicated that the value of M, relative to the accumulator, was: $>$ JGT Jump Greater Than $\geq$ JGE Jump Greater Than or Equal $=$ JEQ Jump Equal $\leq$ JLE Jump Less Than or Equal $<$ JLT Jump Less Than $\neq$ JNE Jump Not Equal Else, execute the next sequential instruction
Jump to subroutine	JSB L	Store the location of the next instruction and then transfer control to the instruction at L
Return jump	RTN	Transfer control back to the instruction at the location saved by the JSB instruction

The first instruction in the table is an unconditional jump. When executed, it transfers control to the specified program location. The unconditional jump is illustrated in Fig. 5.4*a*.

With a conditional jump, control is transferred to the specified location only if the condition shown by the opcode is true. It is implemented here as a pair of instructions. The first compares the given value with the value in the accumulator; the second determines whether the jump should be taken.

Figure 5.4*b* shows how a Compare followed by a Jump Less Than can be used to loop through the instructions starting at location L. The loop is repeated as long as the value in memory location M is less than the value in the accumulator. When the condition is no longer satisfied, control "falls through" the jump instruction. The computer then executes the instruction following the JLT L.

The subroutine jump and the return jump are instructions that have been designed more to help the programmer than to provide computational capability. This pair allows the computer to transfer to a different sequence of instructions (a subroutine) and then to return to the original sequence of instructions when the subroutine is completed. These operations are shown in Fig. 5.5.

The JSB instruction automatically stores the return address where it can be accessed by the RTN instruction. On some computers return addresses are stored in a dedicated part of main memory. On other computers a special register, or set of registers, is used to store the addresses. In a problem at the end of the chapter you are asked to describe how multiple addresses can be stored and retrieved.

In Chap. 9 we will add a group of I/O instructions to our computer that allow it to communicate with external devices. Some computers implement I/O instructions as part of the standard data move instruction group. We have omitted the more advanced instructions that move blocks of data and that perform complex arith-

FIGURE 5.4 Program jump examples: (*a*) unconditional—the jump is always taken; (*b*) conditional—the jump is taken only if the condition is true.

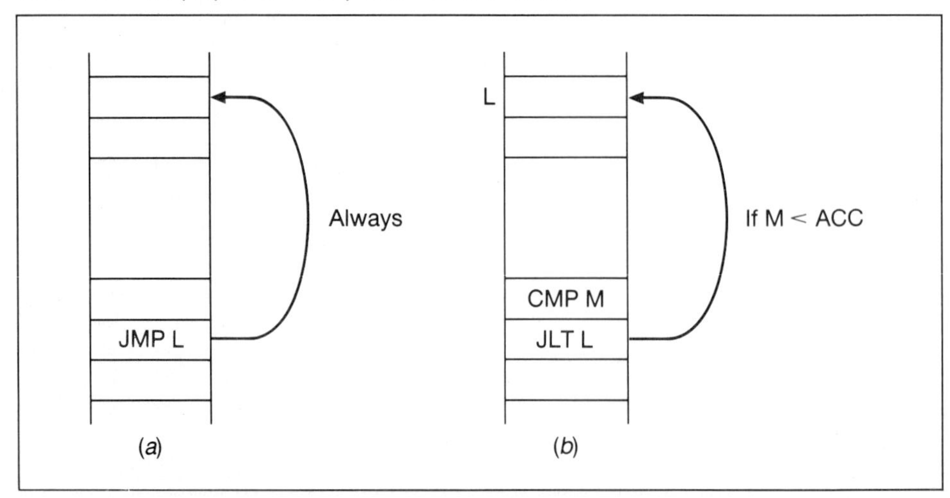

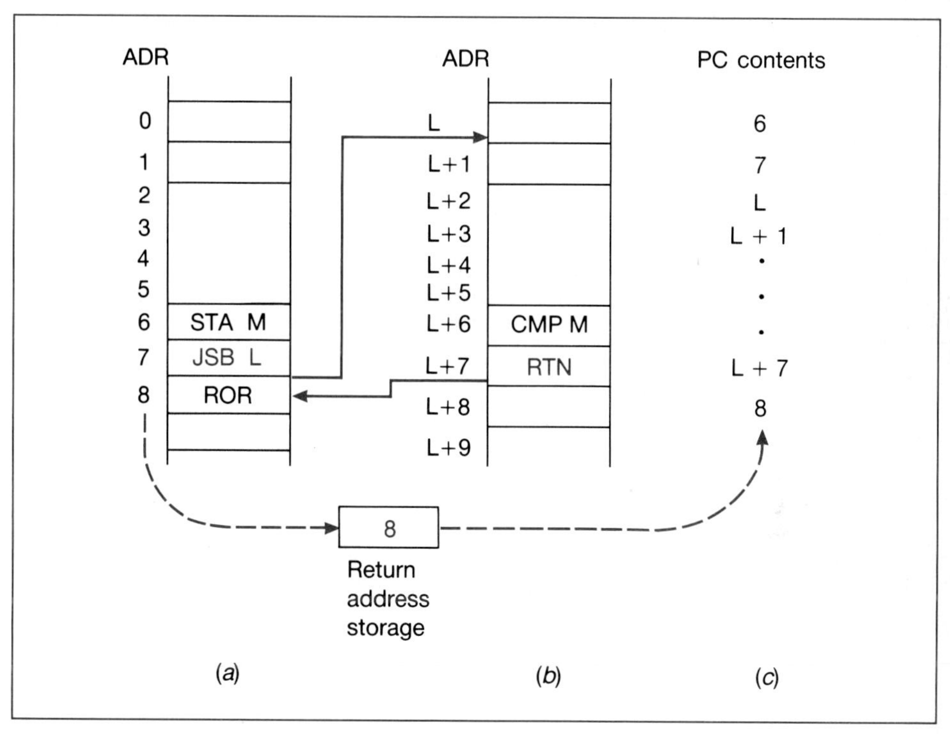

FIGURE 5.5 Subroutine jump with return. (*a*) The subroutine jump causes a jump to memory location L after the contents to the program counter are saved. (*b*) A return jump causes a return to the locations saved by the subroutine jump instruction. (*c*) The program counter contents prior, during, and after subroutine execution.

metic operations. Other computers you study will probably include some of these instructions.

5.1.6 Operand Addressing

The instruction format for our example computer specifies only one operand. This doesn't seem to fit our imperative sentence analogy given at the beginning of Sec. 5.1. We said there that an instruction could include multiple operands.

Consider the Load M instruction, which is actually "Load the value in memory location M into the accumulator." It clearly has both a direct object (the value in memory location M) and a prepositional object (the accumulator). We get around the single-operand limitation of our instruction format by noting that the load operation always involves the accumulator. Therefore, this instruction *implies* that we are referring to the accumulator. The accumulator in this example is known as an **implicit operand**, or an operand that is implied by the opcode. The value of M is *explicitly* identified by the operand part of the instruction. It is called an **explicit operand**.

Explicit operands must be specified by the programmer. So far we have referred to them only as generic memory locations, but explicit operands can take several forms. They can specify the actual value itself, the location of the value, or an address where the location of the value can be found. Furthermore, the operand can also refer to a register. Figure 5.6 shows how addresses from the instruction register are interpreted.

FIGURE 5.6 Instruction operand addressing examples: (*a*) a direct address—the desired information is at memory location 387; (*b*) an immediate operand—the data is contained in the instruction itself; (*c*) an indirect address—memory location 387 contains the address of the desired information.

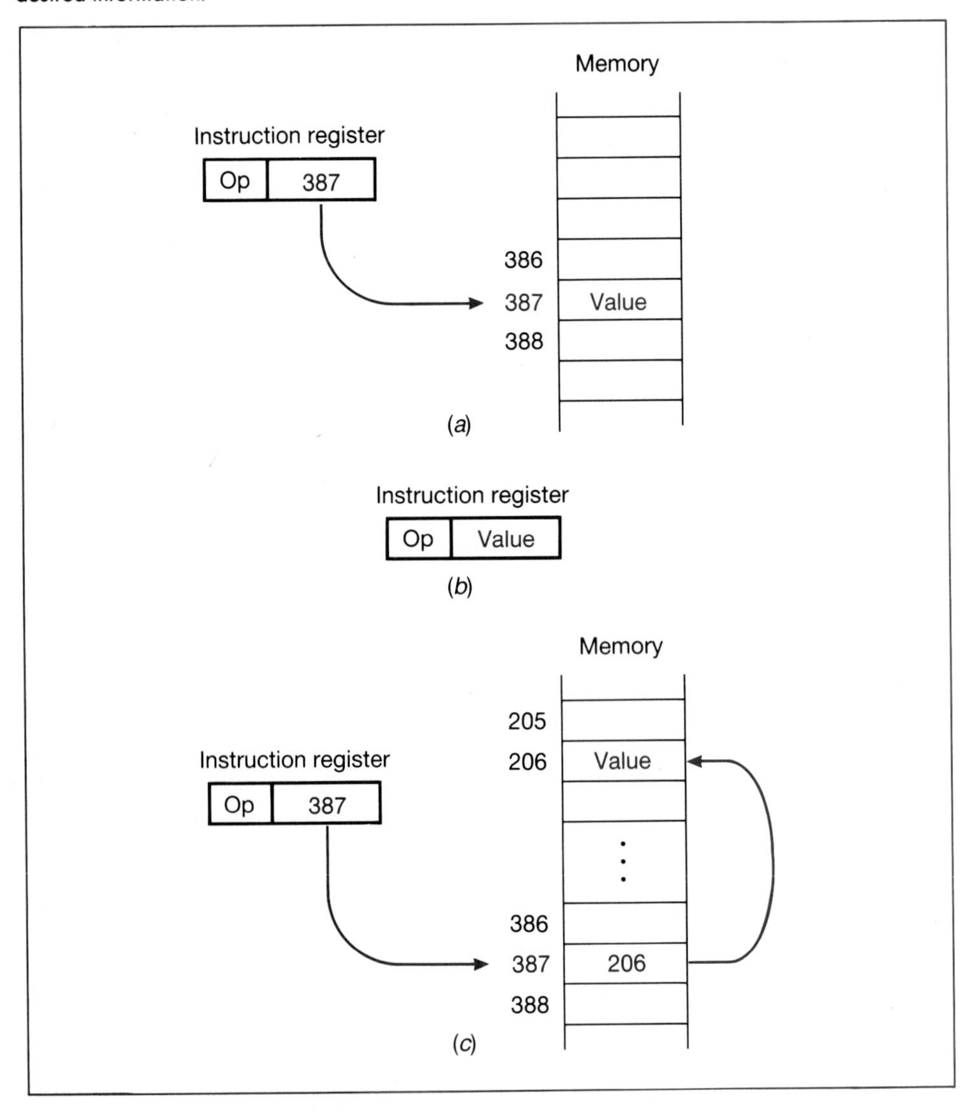

- If the operand is a "variable," say M, we define the operand as the address in main memory where the value of M is stored. This type of explicit operand is known as a **direct address**. It *points directly to the location* where M is stored, as shown in Fig. 5.6*a*.
- If the operand is a "constant," the value can be stored in the operand portion of the instruction and can be *immediately available*. We call this an **immediate operand**. These are useful unless the value has to be changed in the future. Then, a program change would be required each time the value is changed, and this is seldom desirable. The format of an immediate instruction is shown in Fig. 5.6*b*.

FIGURE 5.7 Instruction operand addressing examples: (*a*) program counter relative addressing; (*b*) index-register addressing.

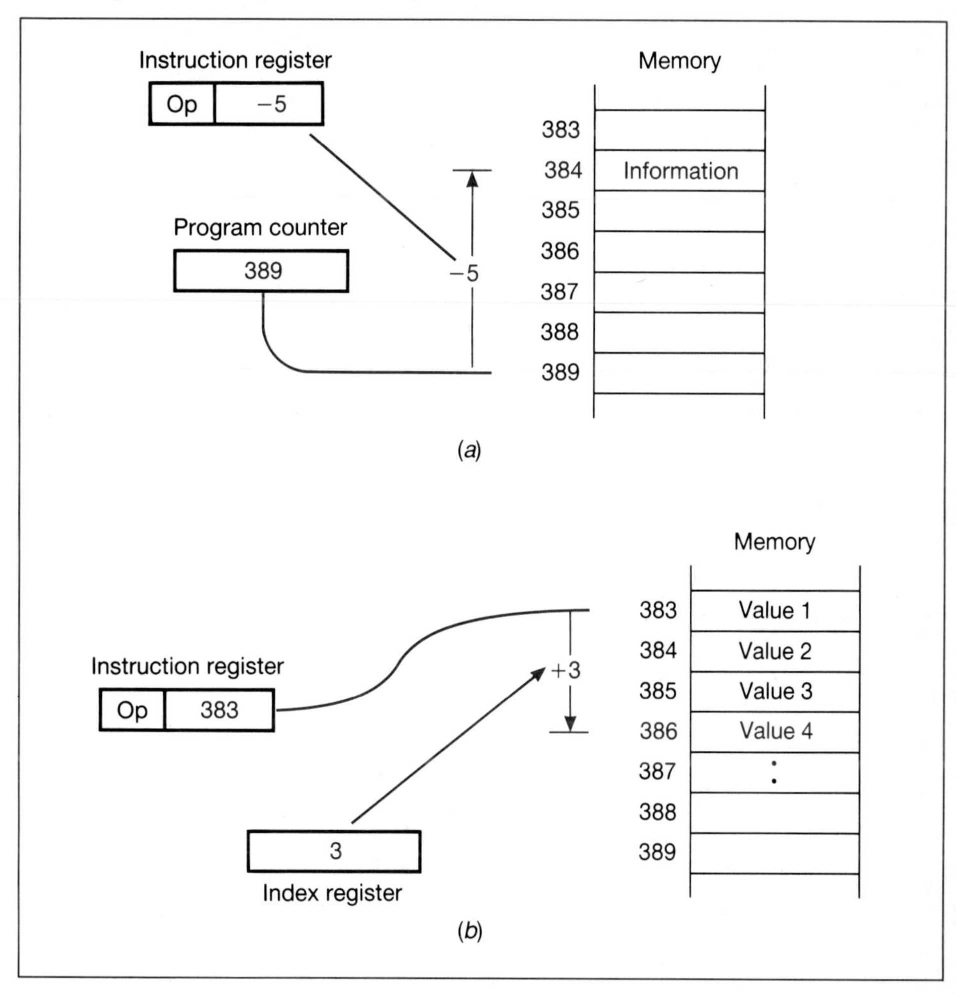

- We might want our operand to point to a location in memory where we could keep the address of the current value being processed. If we did this, we would be *indirectly pointing to the location of the value*, as shown in Fig. 5.6*c*. This is an **indirect address**. It provides programming flexibility, but it also requires a second memory access for each data value referenced. It is useful in applications such as array manipulations.

The way the computer determines whether the operand is direct, indirect, or immediate will be discussed in Sec. 5.5.1, after we have learned a little more about the instruction decode process.

Another common addressing mode is called **relative addressing**. It uses the operand as a relative distance from the current address in the program counter or some other address register, as shown in Fig. 5.7*a*. The contents of the referenced register are not changed by the address computation, and the relative distance is limited by the operand length.

The final addressing mode we will discuss is **indexed addressing**. When this mode is used, the contents of a register, the **index register**, are added to the operand address. Indexed addressing is a convenient way to step or index through the elements of a table or array. This type of addressing is shown in Fig. 5.7*b*.

These are the most common types of operand addressing that you will encounter. For the moment, we will use only direct addressing as we discuss possible control units for our simple computer.

5.2 CONTROLLING THE OPERATIONS OF A COMPUTER

Controlling a computer can be simple and straightforward. Program execution is little more than a sequence of information transfers between functional units within the computer. The control unit's task is to see that the correct information is delivered to the appropriate functional unit at the proper time.

5.2.1 The Internal Structure of a Simple Computer

We begin our discussion of the control unit by defining the internal structure of a simple computer. The structure in Fig. 5.8 is similar to the structure shown in Fig. 5.2, except for the additional detail to see how information transfers are accomplished.

The small rectangular boxes between the major functional units and the system bus are bus/module interface registers. They serve as temporary storage locations for the information being transferred to or from the functional units. Each of the bits in the registers is connected to a separate line in the bus. For simplicity, the registers are shown as single units in Fig. 5.8. An expanded view of a register that shows the connections to the system bus is presented in Fig. 5.9.

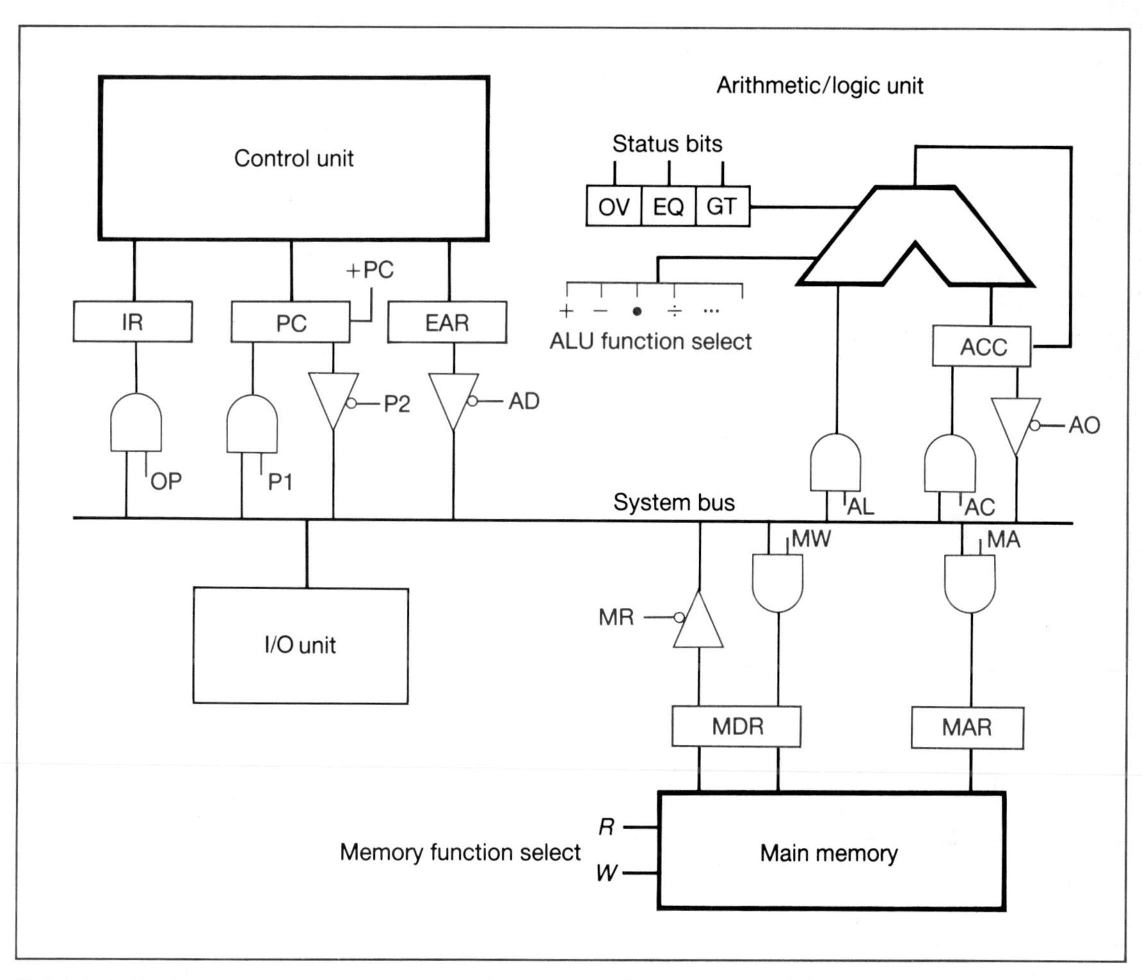

FIGURE 5.8 The internal structure of a simple computer showing bus/module interface registers and control gates.

Selection of which pair of registers to use in the transfer is controlled by Select signals from the control unit. These signals enable logic gates connected between each register and the system bus. Notice that we have used two different gate symbols in the diagram. The reason for this is that a transfer *to* the bus is different from a transfer *from* the bus.

First, consider the case where information is to be transferred *from the bus* to one of the registers. The control gates for each register are all connected to a single source, the bus. Logical ANDs are used to control the transfer of information between the bus and the selected register. The input to any of the register flip-flops is 1 only when both the signal from the bus and Load Register are 1. If either signal is 0, the input to the flip-flop is 0.

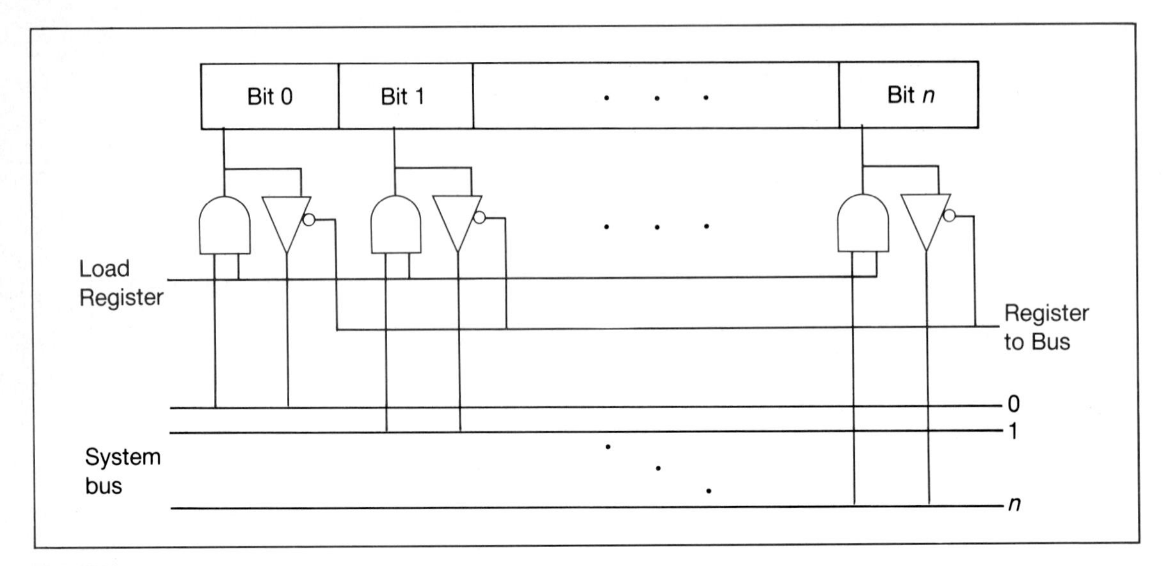

FIGURE 5.9 Connections between a register and the system bus.

Now consider the second case where information is to be transferred from one of the registers *to the bus*. Any of the registers could be a source, and this creates a problem. We learned in Chap. 2 that the output from a conventional AND gate is a logical 0 except when all inputs are 1. This means that we can have a situation where a flip-flop in the selected register is outputting a 1 while another register is outputting 0. This, unfortunately, could cause an electric overload that would destroy some of the circuits.

The gates that we use to control the transfer of information from a register to the bus are called **three-state**, or **tristate**, devices. Tristate gates have three output states: logical 1, logical 0, and a high impedance. Their function is to "connect" the gate to the bus when the enable signal is 0 and to "disconnect" the gate when the enable signal is 1. This high-impedance state essentially disconnects the output of the gate from the circuit when the gate is not enabled. Figure 5.10 shows the truth table, circuit symbol, and operating states for a two-input tristate gate.

Suppose there were a value in the memory data register, MDR, that needed to be transferred to the accumulator. The MR and AC signals would simultaneously select the MR and AC gates to establish the path between the MDR and the ACC, as shown in the functional diagram of Fig. 5.11. Since no other control signals would be enabled during this time, other tristate gates would be effectively removed from the circuit. The input gates not selected would block the bus signals from reaching any of the other registers.

5.2.2 The Bus/Module Interface Registers

Now that we have seen examples of bus transfers in a computer and have an idea of how information is transferred between functional units, we need to look closer at each of the registers in Figs. 5.8 and 5.11.

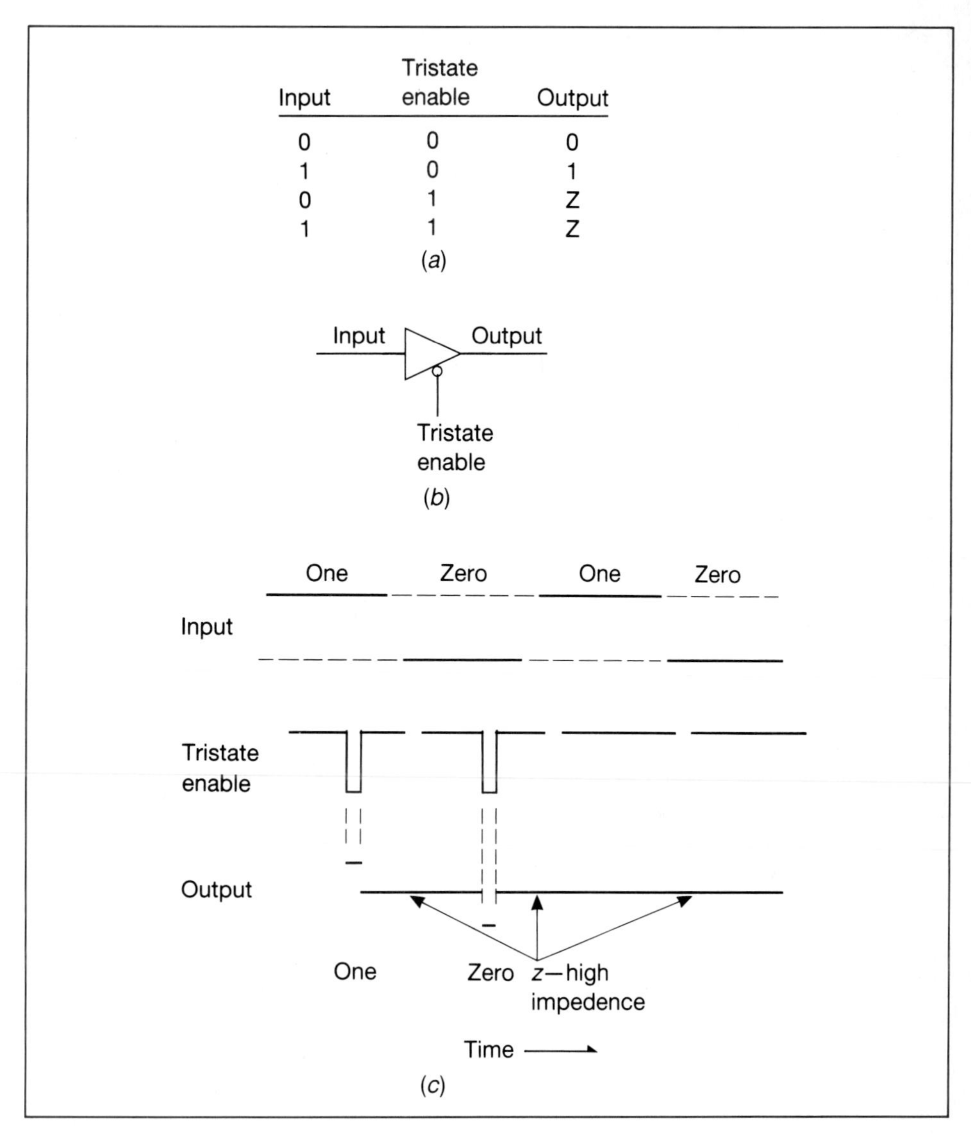

FIGURE 5.10 A two-input tristate gate: (*a*) truth table; (*b*) tristate gate circuit symbol; (*c*) bus transfers using tristate gates.

- **PC** is the **program counter**. It keeps track of the location of the next instruction to be fetched. The PC is incremented at the end of each fetch cycle.
- **IR** is the **instruction register**. It provides a place to store the instruction while it is being decoded and executed.
- **EAR** is the **effective address register**. The control unit translates the operand part of the instruction into its equivalent memory address. It then stores this memory address into the EAR.

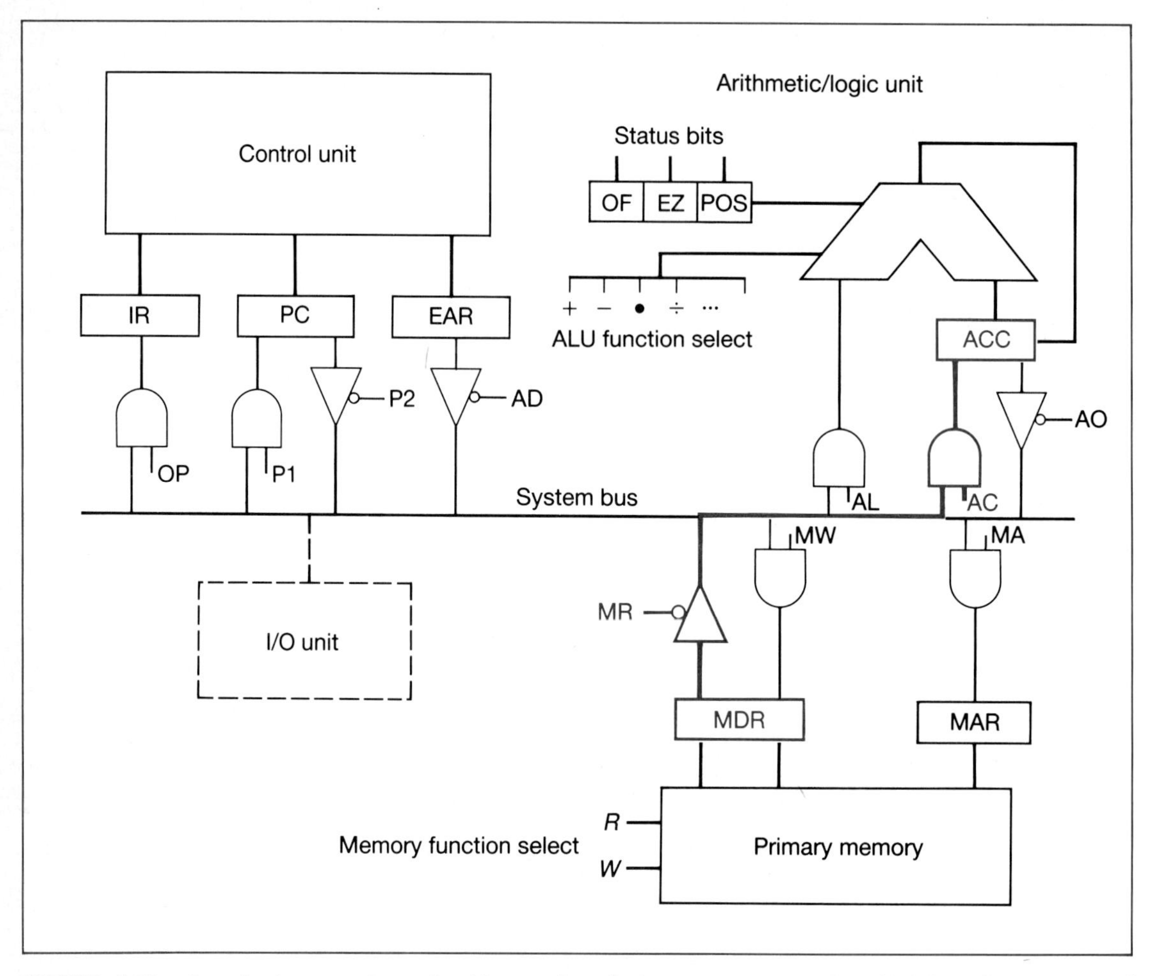

FIGURE 5.11 Sample data transfer path with nonselected tristate output gates in a high-impedance state and nonselected input gates blocking signal transfer to their associated registers.

- **MAR** and **MDR** are the **memory address register** and the **memory data register**. The address of the memory location being accessed is placed in the MAR at the beginning of the memory access cycle. The MDR provides temporary storage for the information going to or coming from the location in memory given by the MAR.
- **ACC** is the **accumulator**. It is a general-purpose register that provides temporary storage of data, usually data that is being used in computations.

Modern computers have more registers than are described here, but these are enough for illustrative purposes.

5.3 THE INSTRUCTION CYCLE—FETCH, DECODE, AND EXECUTE

The instruction cycle is probably best illustrated by example. Suppose we wish to evaluate the expression

$$Z \leftarrow X + Y$$

where X, Y = names of variables whose values are already in memory
Z = location in memory reserved for the result
Since our computer allows us to define only one explicit operand at a time, we will need three instructions to evaluate this expression.

1. *Load* the value of X into the accumulator.
2. *Add* the value of Y to the accumulator.
3. *Store* the contents of the accumulator (the result) into Z.

We begin by fetching the first instruction, assuming that its memory location is in the PC. The fetch sequence requires three operations and uses the following registers and gates:

1. The contents of the PC (the location of the first instruction) are sent over the system bus to the MAR (in Fig. 5.11, P2 and MA are enabled).
2. The contents of the memory location specified by the MAR are copied into the MDR (R is selected). The PC is simultaneously incremented (+PC in Fig. 5.8 is enabled).
3. The contents of the MDR (the instruction) are sent over the system bus to the IR (MR and OP are enabled).

The first instruction is now in the instruction register, and we are ready to decode it. By **decoding** the instruction we mean that the bits in the IR are examined by the hardware and the specified operations are performed. The way instructions are decoded depends on how the control unit is implemented. If the control unit is microprogrammed (see Sec. 5.4.1), decoding is done by converting the opcode into the starting address of a microprogram. If the control unit is hardwired (see Sec. 5.4.2), decoding is done by combinational logic attached to the IR. In either case, the decoding of each instruction enables signals to be sent from the control unit to the other functional units.

Now that the first instruction has been fetched and decoded, the control unit issues the signals necessary to execute the instruction. For the Load X instruction, execution means the following sequence of events takes place:

1. The address of the operand X is transferred into the EAR.
2. The contents of the EAR are placed on the address bus (AD and MA are enabled).

3. The contents of the memory location specified by the MAR are copied into the MDR (R is selected).
4. The contents of the MDR (the value of X) are sent over the system bus to the ACC (MR and AC are enabled).

The Load X instruction is now complete, and we are ready to fetch, decode, and execute the Add Y. Since the PC was incremented while the first instruction was being copied from memory into the MDR, it now points to the second instruction. This instruction is fetched using the same sequence described above. The execute cycle for Add Y is similar to the load, except this time the value is added to the ACC.

1. The address of the operand Y is transferred into the EAR.
2. The contents of the EAR are placed on the address bus (AD and MA are enabled).
3. The contents of the memory location specified by the MAR are copied into the MDR (R is selected).
4. The contents of the MDR (the value of Y) and the contents of the ACC (the value of X) are added. The result is placed back into the ACC (MR and AL are enabled, + is selected).

This completes the Add, and we are ready to write our result back into memory. The Store Z instruction is fetched and decoded as before. For this instruction, the information transfer between the ACC and the MDR happen before the memory function (write) select.

1. The address of the operand Z is transferred into the EAR.
2. The contents of the EAR are placed on the address bus (AD and MA are enabled).
3. The contents of the ACC (the new value of Z) are copied into the MDR (AO and MW are enabled).
4. The contents of the MDR are written into the memory at the location specified by the MAR (W is selected).

The simple program segment that evaluates the expression

$$Z \leftarrow X + Y$$

is now complete. Each instruction was fetched, decoded, and then executed. Upon completion of this segment, the PC is pointing to the next instruction.

5.4 IMPLEMENTATION OF A CONTROL UNIT

The previous section showed the sequences that make up the instruction cycle. These sequences are similar to small-computer programs that are repeated over and over again, with variations for selecting different registers and control signals. This implies that the control unit may be nothing more than a small computer within a computer, and this is sometimes the case. All we have to do is build a small computer, or its equivalent, and we will have our control unit. The next two sections discuss two different ways that this can be done.

5.4.1 Microprogrammed Control

The first type of control implementation we will consider is based on the concept of a computer within the computer. It was suggested by Wilkes in 1953. He proposed that control gates could be selected by bits in a "control word" read from a small separate memory in the control unit [Wilkes and Stringer, 1953].

Wilkes called these control words **microorders** and his control unit, a **microcontrol unit**. Microorders are also called **microinstructions**. A group of microinstructions is called a **microroutine**, or a **microprogram**. Referring to the example in the previous section, one microinstruction is required for each step in the fetch and execute sequences. Since the fetch is the same for all instructions, it is implemented as one microprogram. In a microprogrammed control unit, instruction decoding is accomplished by translating the opcode of the instruction into the address of the microprogram that executes the instruction. The microprograms execute different sequences for each opcode.

Implementation of a control unit based on microprogrammed control is shown in Fig. 5.12. This implementation has a separate memory called the **control store** for storing the microprograms. There are also registers and control gates to control the execution of the microprograms. The control store is similar to main memory except that it is usually implemented with memory that is 5 to 10 times faster than main memory. Its registers are similar to main memory's MAR and MDR:

- The **CSAR** is the address register for the control store. It is used much like the PC in that it points to the next instruction to be fetched.
- The **CSBR** is a buffer register, sometimes called the **microinstruction register,** where the control word (the microinstruction) is stored temporarily. During this time, the bits in the word are controlling and selecting gates in the computer.

Figure 5.13 shows a possible control word format for our control memory. It is very different from the format of the machine instruction we saw previously. There are no opcodes or operands per se. Separate bits are provided for each control gate and function select signal. Each microinstruction controls the computer for one microcycle much the same way a machine instruction controls the main computer for one instruction cycle. The decoding of an instruction is therefore done by having the bits in the selected microinstruction control the computer's operation.

A second difference between microinstructions and machine instructions is the way that jumps are implemented. Jumps are required much more often at the microprogram level because the individual programs are very short. To keep from having

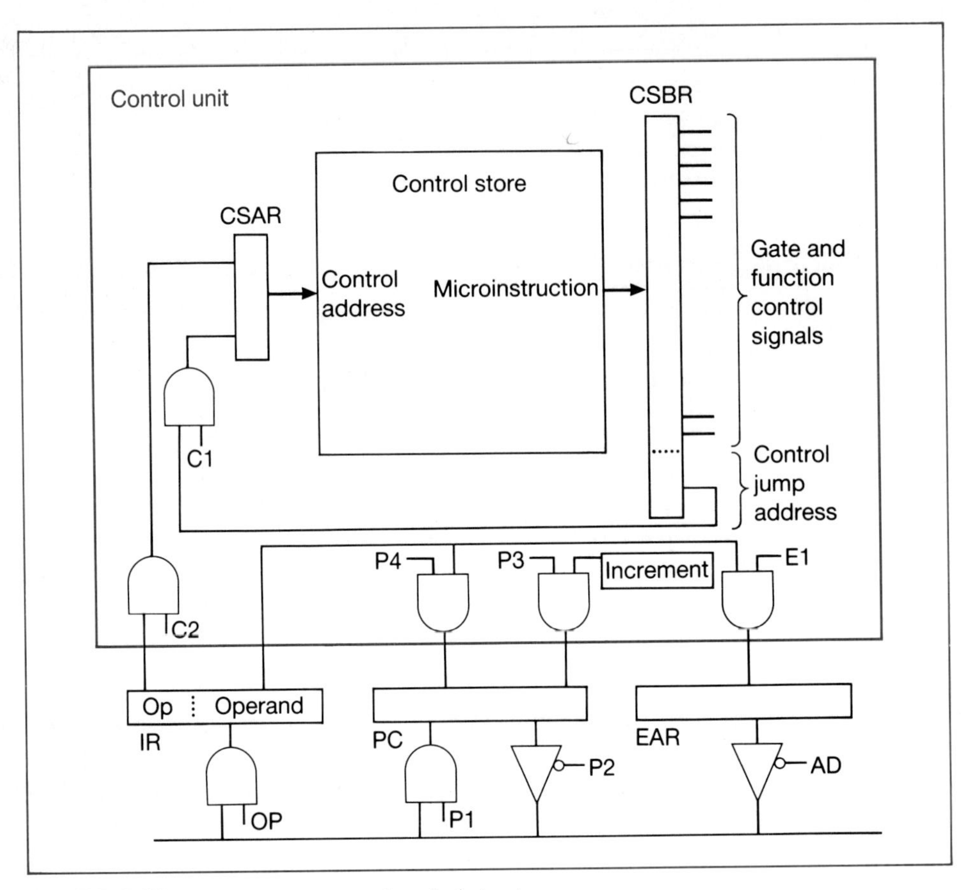

FIGURE 5.12 A microprogrammed control structure.

to add a separate microinstruction each time a jump is needed, we include room for a jump address within the microinstruction. It makes the control word longer, but it also has important advantages:

- We save a microinstruction fetch each time a control store jump is required. We don't have to access a separate microinstruction for the jump. We also save the space that the separate microinstruction would require.

FIGURE 5.13 An example of a microinstruction format.

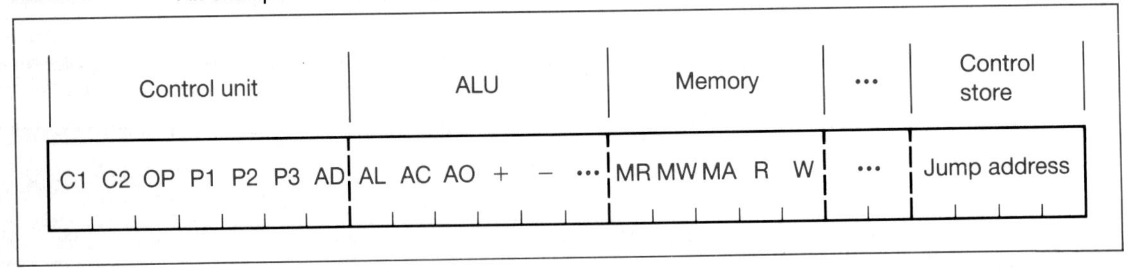

TABLE 5.4 OPCODE AND CONTROL STORE ADDRESS CORRELATION

Operation	Opcode	Control store location	Microprogram
...	...	...	...
Load	12	12	Load
Store	13	13	Store
Add	14	14	Add
Subtract	15	15	Subtract
...	...	...	...

- We can place the first microinstruction of every execute routine at the control store address corresponding to the binary value of the opcode (see Table 5.4). The control store jump address in the first microinstruction of each microprogram points to the rest of the microroutine. This is located elsewhere in the control store. Such an approach saves having to design special circuitry to translate the opcode into an equivalent control store address.

All this can be seen by the example microprogram sequences of Fig. 5.14 and the diagram of a microprogram-controlled computer shown in Fig. 5.15. Assume an LDA M instruction has just been fetched from memory and is in the IR. The opcode we assigned for a load instruction is 12. The C2 gate in the control unit has

FIGURE 5.14 Microprogram for LDA M followed by instruction fetch.

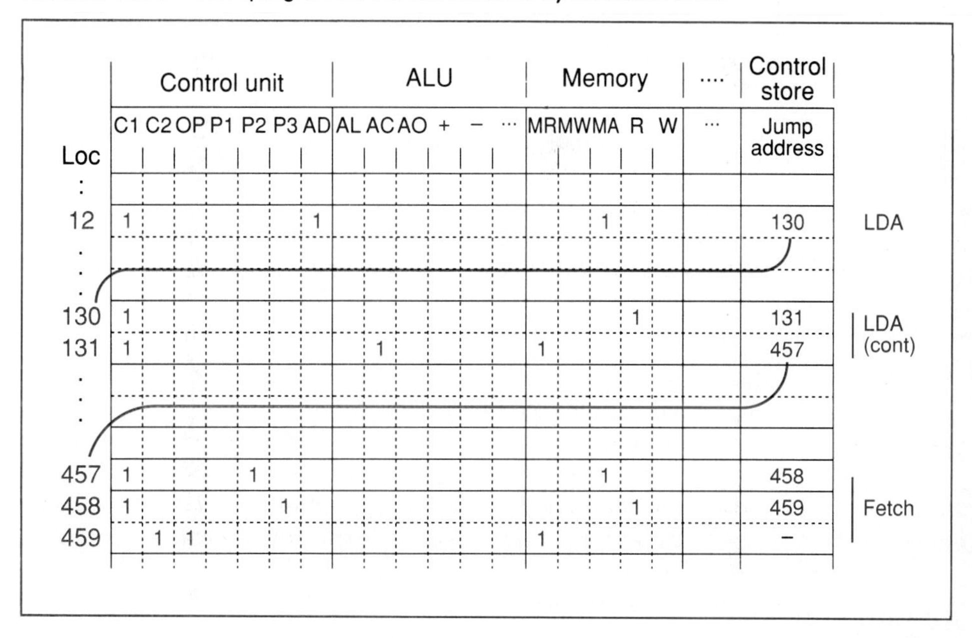

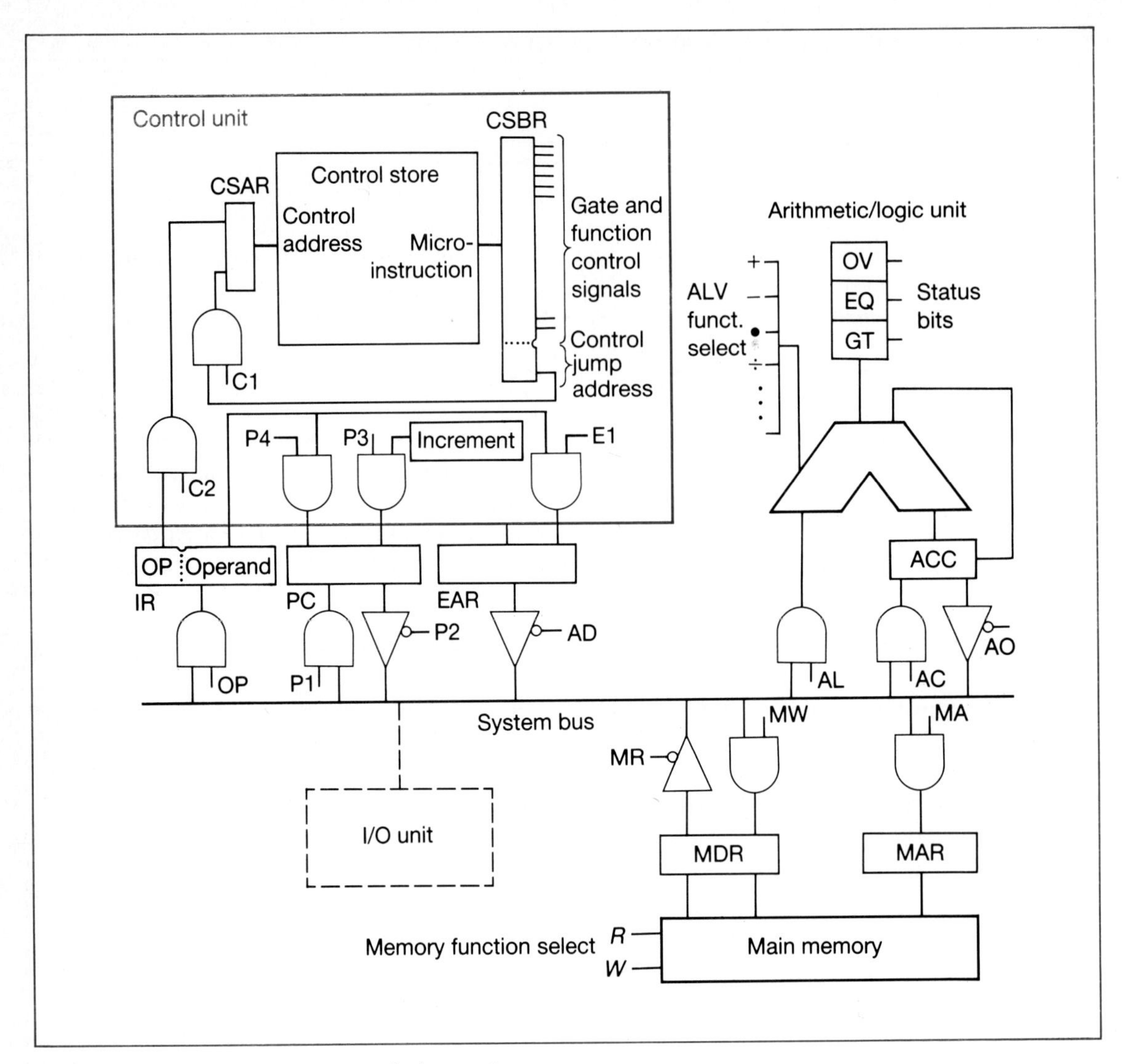

FIGURE 5.15 Microprogram controlled computer.

been activated, the opcode 12 has been copied into the CSAR, and the operand in the IR has been loaded into the EAR.

The microinstruction at control store location 12 is now copied into the CSBR. Logical 1s in the AD and MA bits enable the AD and MA gates to copy the M address in the EAR into the MAR. The logical 1 in the C1 bit enables gate C1, which copies the control store address of the next microinstruction (130 in this example) into the CSAR. The control bits in the CSBR remain active until the contents of the CSBR are replaced.

The second microinstruction is now transferred from the control store into the CSBR. The R bit in the instruction causes the contents of the memory location M to be copied into the MDR. The control store address of the next microinstruction is simultaneously loaded into the CSAR.

The last microinstruction in the routine is transferred to the CSBR, and the location of the first microinstruction in the Fetch microroutine is transferred to the CSAR. The Fetch microprogram, located at 457, 458, and 459, is now executed to load the next machine instruction into the IR. The only new element in this routine is in the last microinstruction, which does not need a control store jump address. The C2 bit causes the new opcode in the IR to be transferred to the CSAR to identify the next microinstruction.

The LDA M instruction was the first of three that were needed to evaluate the example expression ($Z \leftarrow X + Y$) we detailed in Sec. 5.3. Microprograms for the add and store instructions are given in Fig. 5.16.

Not all routines for the execute phase are the same length. Instructions in the program control group illustrate this. Figure 5.17 shows that an unconditional jump can be microprogrammed in only one microinstruction. (The operand address is copied into the PC.) The conditional jump either copies the operand address into the PC or does nothing, depending on the status bits.

However, the control unit cannot process a conditional jump as it is currently designed. We need to add a way to override the control store address that is in the

FIGURE 5.16 Microprogram routines for LDA, STA, ADD, and JMP.

	Control unit							ALU						Memory					....	Control store	
Loc	C1	C2	OP	P1	P2	P3	AD	AL	AC	AO	+	−	···	MR	MW	MA	R	W	···	Jump address	
:																					
12	1						1									1				130	LDA
13	1						1									1				132	STA
14	1						1									1				134	ADD
:																					
130	1																1			131	LDA (cont)
131	1								1					1						457	
132	1									1					1					133	STA (cont)
133	1																	1			
134	1																1			135	ADD (cont)
135	1							1			1			1						457	
:																					
457	1				1											1				458	Fetch
458	1					1											1			459	
459		1	1											1						–	

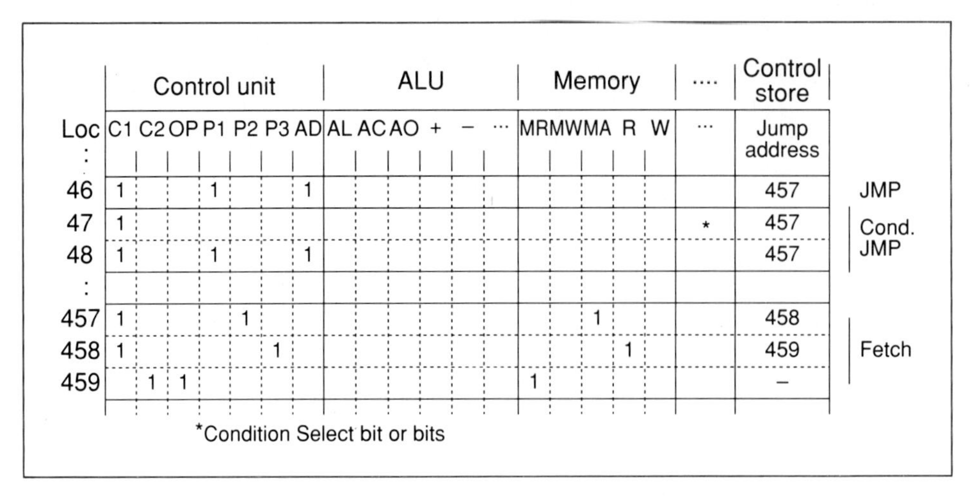

Control unit
ALU
Memory
....
Control store
Loc C1 C2 OP P1 P2 P3 AD AL AC AO + − ··· MR MW MA R W ··· Jump address
46 1 1 1 457 JMP
47 1 * 457
48 1 1 1 457
Cond. JMP
457 1 1 1 458
458 1 1 1 459
459 1 1 1 −
Fetch
*Condition Select bit or bits

FIGURE 5.17 Conditional and unconditional jump microprogram.

microinstruction and provide an alternate location. This can be done by adding a Compare Condition True control signal to the C1 gate and an incrementing capability to the CSAR.

The increment is controlled by a Compare Condition False, as shown in Fig. 5.18. If the specified condition is false, the first microinstruction of the conditional jump will transfer directly to the Fetch microprogram. If the condition is true, the first microinstruction does nothing and the next microinstruction is executed. This microinstruction will transfer the jump address from IR to the PC and then go to the Fetch.

Jump to Subroutine requires saving the PC in memory and then transferring the operand address to the PC. The details of this operation and its associated return jump are left as a problem at the end of the chapter.

5.4.2 Hardwired Control

The primary advantage of microprogrammed control is the flexibility that it provides the system designer. Instructions can be easily added or changed by adding or changing the microprograms. Design errors can also be easily corrected by changing the microprograms. However, microprogrammed control units are generally slower than control units based on hardwired logic.

A **hardwired control unit** uses combinational and sequential circuits, as shown in Fig. 5.19, to control the operation of the computer. It serves exactly the same purpose as the microprogrammed control unit. The **decoder matrix** accepts the opcode as input, and outputs control signals. These signals enable the bus/register selector gates to select the correct operation in the associated functional unit. The difference is that the gate signals are generated by combinational and sequential

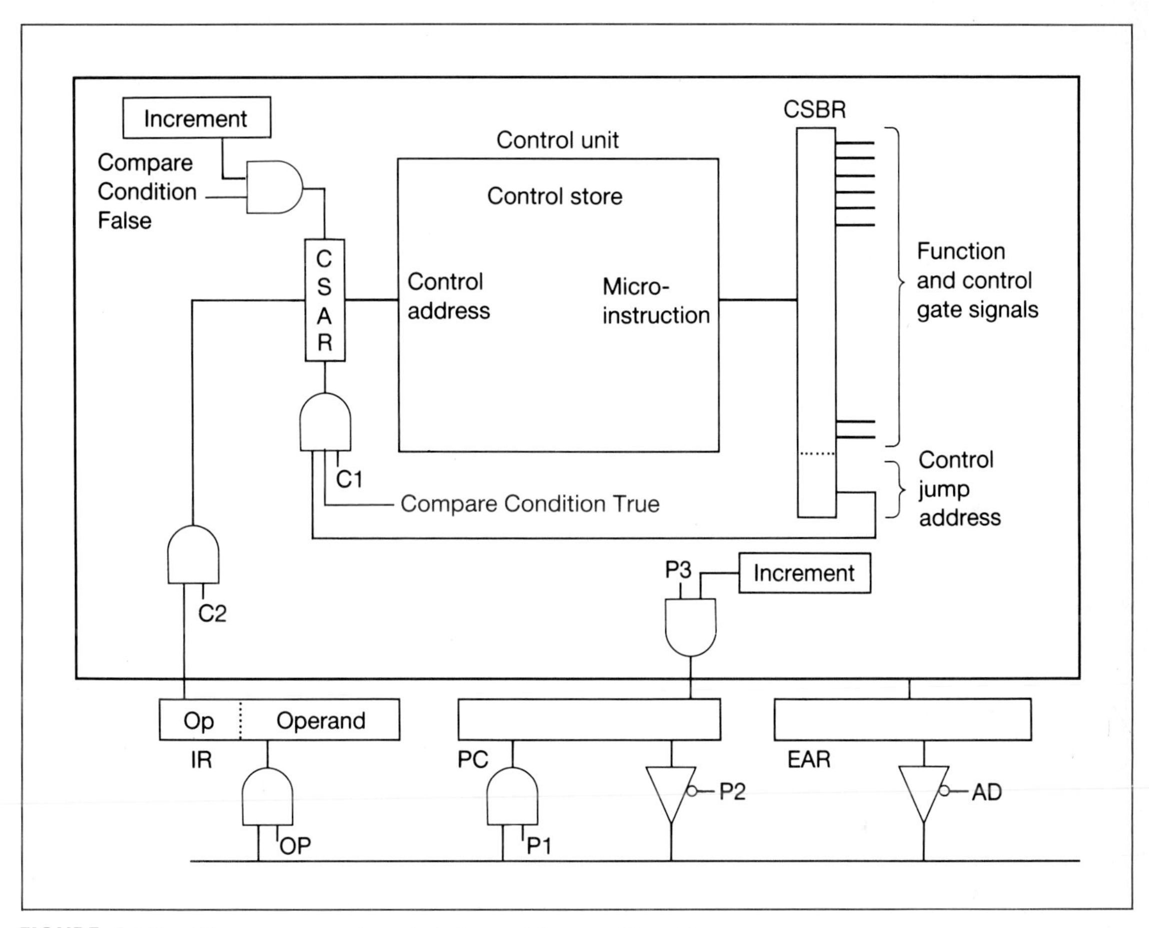

FIGURE 5.18 Microprogrammed control changed for conditional jump.

circuits instead of by bit patterns in a microinstruction. Also, no control address information is needed.

The instruction fetch and most of the execute cycles we have seen so far require a sequence of several steps to activate control gates and transfer information. With hardwired control, a sequence counter is used to generate timing signals. These signals activate combinational logic circuits that provide the gate control and function-select signals.

Consider the MR gate enable conditions that were shown in the code list of Fig. 5.16 as an example. We see that the MR gate is selected (enabled) by three different conditions, the three inputs to the OR gate. If the instruction register contains either an LDA or an ADD opcode, during the third step of the execute cycle the MR signal is activated. It is also selected by the third step of the Fetch routine. These are the logic conditions that are necessary to select the MR gate. They can be implemented by a two-section combinational logic circuit like the one shown in Fig. 5.20.

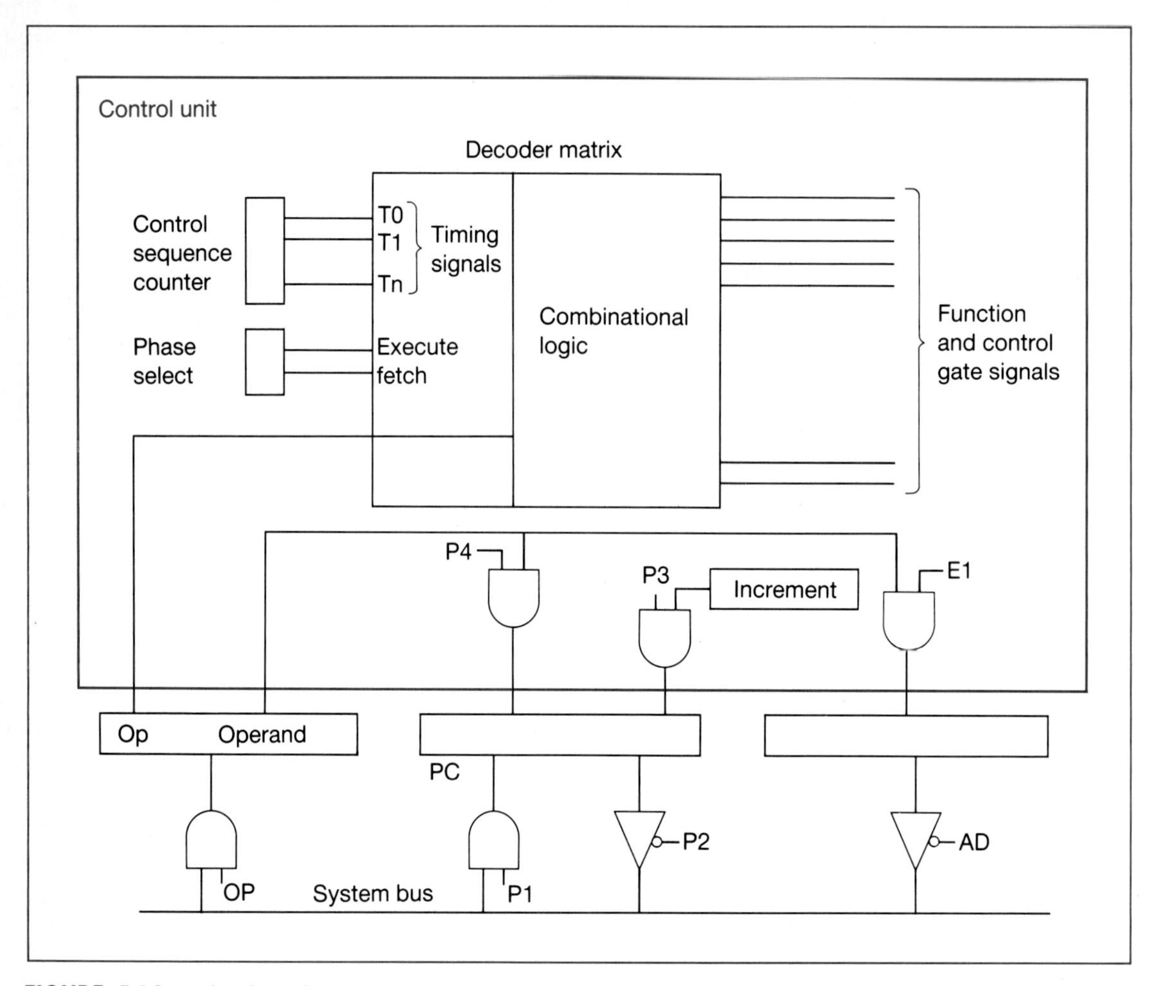

FIGURE 5.19 A hardwired control unit structure.

A hardwired function and control-select decoder matrix in an actual computer is much larger and more complex than our example of Fig. 5.20. However, we can see that even the simple control unit of Fig. 5.19 would require the following:

- A separate decoder circuit for each of the different control gates and function-select signals in the computer.
- A separate decoder circuit for each of the different opcodes in the instruction set.
- A Count Reset is at the end of each fetch and execute phase to reset the control sequence counter so that the next phase will start correctly. (This signal can also be used to switch the phase-select flip-flop between fetch and execute.)

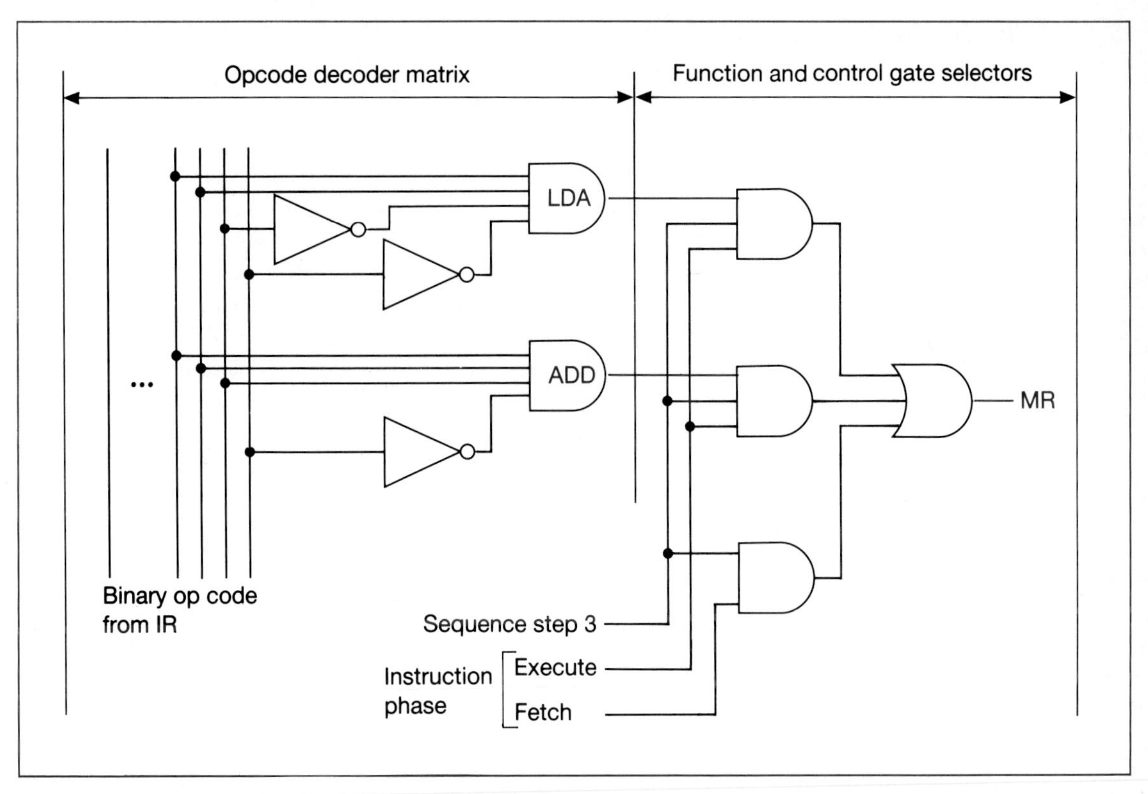

FIGURE 5.20 Example decoder matrix circuit for MR gate-select signal showing the lower 4 bits of the opcode connected: LDA = 12 (... 1100_2) and ADD = 14 (... 1110_2).

The main point is that hardwired control units, although they may contain a lot of circuitry, are conceptually quite simple. They are just combinational and sequential circuits that implement the instruction fetch and execute sequences of the microprogrammed control in Sec. 5.4.1. However, because hardware must be altered, changes to execution sequences in a hardwired computer are more difficult than in a microprogrammed computer. This was one of the reasons microprogrammed control became popular. Hardware changes have become less of a problem in the last few years with the development of new programmable hardware devices.

5.5 EXTENSIONS TO THE SIMPLE COMPUTER

By this time, you should have a reasonably good idea of what an instruction is and how a basic control unit uses it to control the rest of the computer. We now want to consider how different instruction formats and control philosophies can affect the operation and performance of a computer.

5.5.1 Opcode and Operand Modalities

Most instruction formats include additional bits called **modalities** that change how the control unit treats the opcodes and operands. The arithmetic instruction group provides a good example of this. An add instruction, for instance, has to process all the different types of numbers allowed on the computer. (These types include integer and floating-point.) In addition to the opcode, the ALU needs to know the types of the data (numbers) that will be used.

In most computers, the operand itself does not contain any information that identifies its type. There are two common solutions to this dilemma. The first is to provide separate opcodes that imply the type of number that will be used in the operation (such as Add Integer and Add Floating). The second is to include separate **opcode modality bits** that change the meaning of the opcode and identify the types of numbers that will be processed.

Modality bits for addressing can also be used to specify the type of addressing used with each instruction. Different types of operands can be identified in a similar manner. Direct and indirect operands are typically identified by a separate **direct/indirect modality bit**. A format that was used in some minicomputers is shown in Fig. 5.21. If the direct/indirect bit is 0, the operand value that is fetched from memory is treated as a direct address. If it is 1, the operand is treated as an indirect address.

5.5.2 Variable-Length Instructions

We stated in Sec. 5.1.1 that many computers have been designed to allow instructions to have different lengths. This is because some operations do not require operands, while others may require several. Having variable-length instructions allows some instructions to be short and saves some memory space, but it requires control units that are more complex. Also, these control units are more difficult to design and test than ones for fixed-length instructions.

Figure 5.22 illustrates one of the problems with variable-length instructions. The control unit must fetch three words for instructions 1 and 5, two words for instruction 3, and one word for instructions 2 and 4. The instruction register in the computer will also have to be able to hold instructions of different lengths.

FIGURE 5.21 Common minicomputer instruction format with an operand address modality bit.

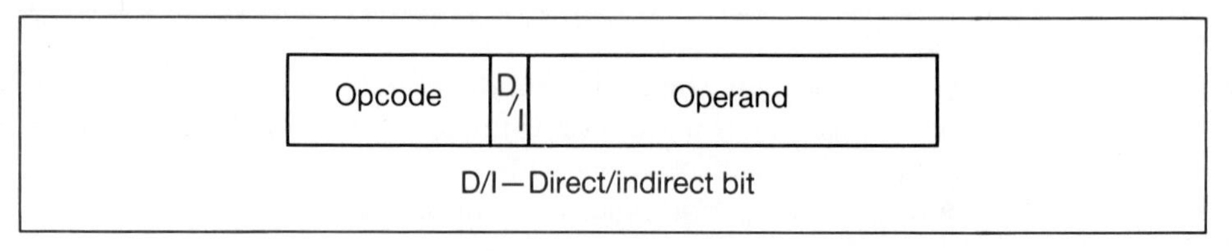

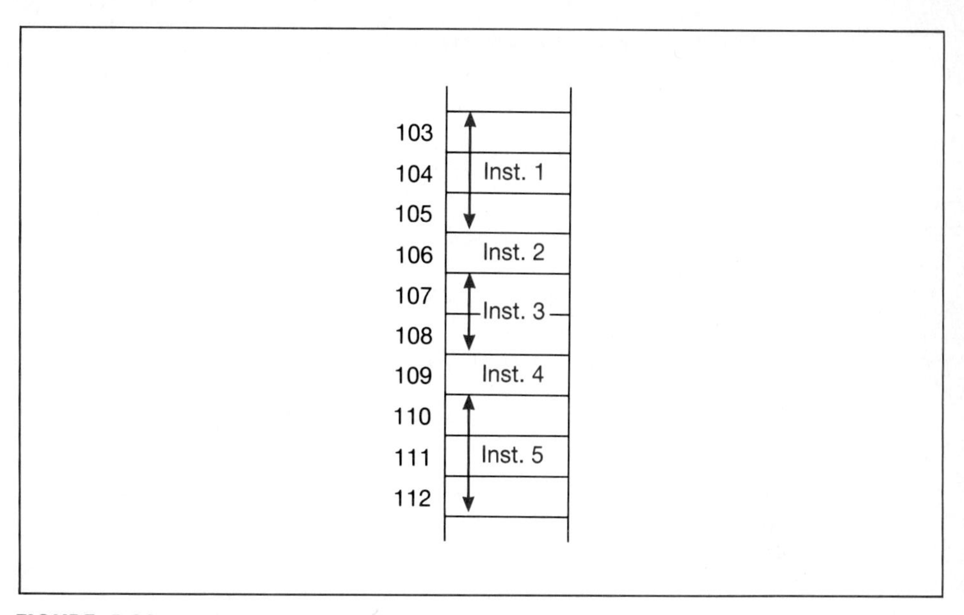

FIGURE 5.22 Portion of memory containing several instructions of different lengths.

5.5.3 Multiple-Operand Formats

In the earlier sections, we limited our discussion of the control unit to single-operand formats. This constraint made it necessary for us to use three instructions to process our example expression

$$Z \leftarrow X + Y$$

Suppose instead that our instruction format allowed three operands that could be interpreted as

Add the value of X to the value of Y, and store the result in Z

Such an instruction is said to have two **source operands** (X and Y) and one **destination operand** (Z). This is an important concept because, as Table 5.5 shows, all the basic arithmetic and logic operations require two source values plus a place to store the result.

Multiple operands can also be applied to other groups of instructions. Table 5.6 shows Load/Store and Move instructions with both two and three operands. A variation of the Move instruction that is often included in all three types of architectures is the Block, or Extended, Move. It differs from the previously defined moves in that it transfers a block of successive data elements instead of a single value. Two operands are required to define the first "to" and "from" memory locations for the source and destination arrays. A third operand is needed to show the number of data elements to be moved.

TABLE 5.5 TYPICAL THREE-OPERAND ARITHMETIC/LOGIC INSTRUCTIONS

Operation	Typical mnemonic	Instruction meaning
Add	ADD X, Y, Z	$Z \leftarrow X + Y$
Subtract	SUB X, Y, Z	$Z \leftarrow X - Y$
Multiply	MUL X, Y, Z	$Z \leftarrow X \cdot Y$
Divide	DIV X, Y, Z	$Z \leftarrow X \div Y$
AND	AND X, Y, Z	$Z \leftarrow (X)$ AND (Y)
OR	OR X, Y, Z	$Z \leftarrow (X)$ OR (Y)
XOR	XOR X, Y, Z	$Z \leftarrow (X)$ XOR (Y)

TABLE 5.6 TYPICAL MULTIPLE-OPERAND DATA MOVE INSTRUCTIONS

Operation	Typical mnemonic	Instruction meaning
Load	LDR R, M	Load register R with the value in M
Store	STR R, M	Store the contents of register R in M
Move	MOV M1, M2	Move the value in M1 to memory location M2
Block move	BMV N, M1, M2	Move N data elements contained in memory locations M1, (M1 + 1), (M1 + 2), ..., (M1 + N) to memory locations M2, (M2 + 1), (M2 + 2), ..., (M2 + N)

The conditional jump is another example where multiple operands can be used. The single-operand computer required that the Jump instruction be preceded by either an arithmetic instruction or a Compare. Table 5.7 shows that with multiple operands, both the Compare and Jump and the logical Test and Jump can be single instructions.

Increased performance is the primary motivation for using multiple-operand instructions. These instructions still require an access for each of the data values, but fewer instructions have to be fetched to accomplish a function. However, these instructions also require more bits to make room for all the operands. This can be implemented at design time by changing the IR, memory and bus word lengths, and the fetch, decode, and execute sequences in the control unit.

Two possible formats for multiple-operand instructions are shown in Fig. 5.23. If the computer has a number of general-purpose registers in the ALU, the individual operands can represent either register pointers or memory addresses. Figure 5.23*a* includes modality bits with each operand to show if it is register- or memory-reference (R/M). Modality bits also show if the addressing is direct or indirect (D/I). Figure 5.23*b* shows an example of a two-operand format that separates register operands from memory operands. This instruction format allows either operand to

TABLE 5.7 TYPICAL MULTIPLE-OPERAND CONDITIONAL JUMP INSTRUCTIONS

Typical mnemonic	Instruction meaning	
Compare and jump group	Compare the value in X to the value in Y, and jump to the instruction at L if the condition specified is true; Else execute the instruction following the jump:	
JGT X, Y, L	If $X > Y$	Jump to L
JGE X, Y, L	If $X \geq Y$	Jump to L
JEQ X, Y, L	If $X = Y$	Jump to L
JLE X, Y, L	If $X \leq Y$	Jump to L
JLT X, Y, L	If $X < Y$	Jump to L
JNE X, Y, L	If $X \neq Y$	Jump to L
Test and jump group	Test the logical value in X, and execute the instruction at L if the value is as shown; Else execute the instruction following the jump:	
JPT X, L	If X = true	Jump to L
JPF X, L	If X = false	Jump to L

be a source or destination and simplifies the decode process. The memory and register operands are always in the same location in the format.

5.5.4 Alternate Control Structures—RISC versus CISC

Up to now, we have concentrated on a few instructions and on how a simplified control unit operates. We have also seen that the control unit could be designed to

FIGURE 5.23 Two possible multiple-operand instruction formats for a register-register computer: (*a*) arithmetic/logic and block move instructions; (*b*) load/store operations.

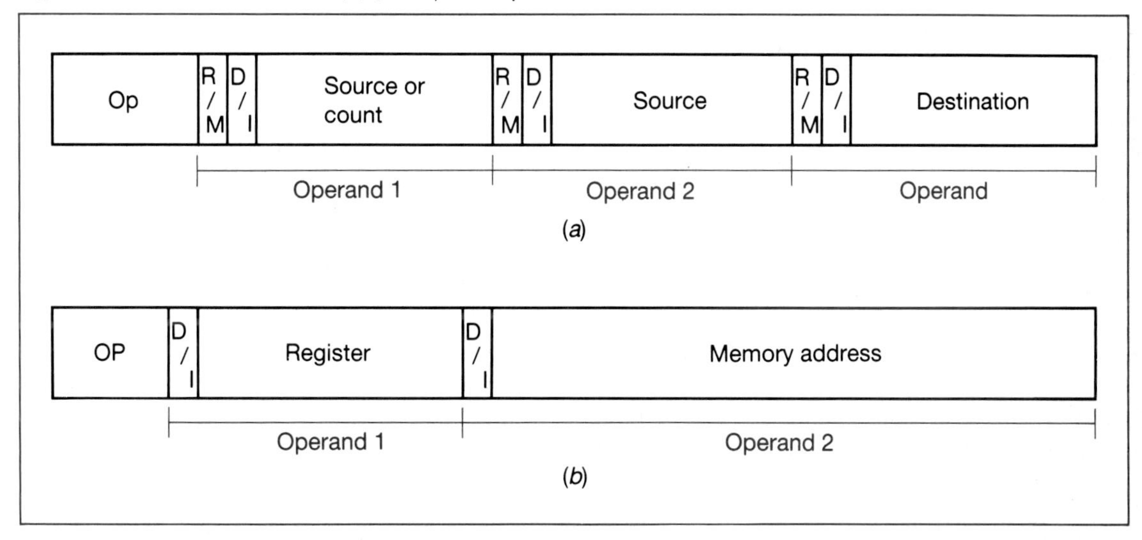

recognize various opcode and operand modalities. We have covered both fixed- and variable-length instruction.

The types of computer structures that we have been discussing generally belong to a computer class called **complex instruction set computers**, or **CISCs**. From a control standpoint, they are characterized by:

- Large instruction sets with both simple and complex instructions
- Instruction formats with several different types of opcode and operand modalities
- Microprogrammed control units

Each of these characteristics has added to increased complexity in the control unit.

Consider the format modalities that were shown in Fig. 5.23. The three **R/M** modality bits alone can result in eight different combinations of operands, as shown in Fig. 5.24. We know that each of these operands can also be either a direct, an indirect, or an immediate address. As you can see, the number of possible combinations becomes large very fast.

This complexity problem bothered some systems designers at Stanford, U.C. Berkeley, and IBM during the late 1970s and early 1980s [Radin, 1983; Hennessy,

FIGURE 5.24 Format combinations and lengths for three-operand variable-length instructions.

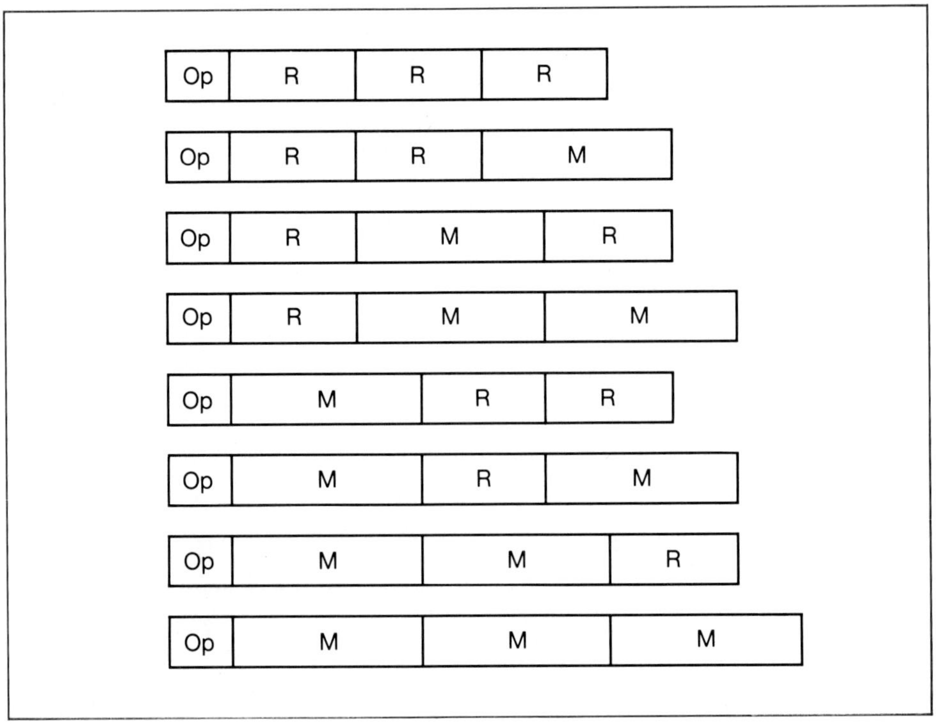

Op	R	R	R
Op	R	R	M
Op	R	M	R
Op	R	M	M
Op	M	R	R
Op	M	R	M
Op	M	M	R
Op	M	M	M

1984; Patterson, 1985]. They found that many instructions and modality variations were seldom or never used. They proposed a new structure class now known as the **reduced instruction set computer**, or **RISC**, that incorporates several of the concepts discussed in this section.

The first of these is the size and complexity of the instruction set, which the designers reasoned would directly affect the size and complexity of the control unit. They chose a register-register architecture with a small number of simple two- and three-operand instructions, like the ones in Tables 5.5 and 5.6. The instruction formats are all fixed-length, one word long, as shown in Fig. 5.25. The addressing modes that are typically provided are the direct and indexed modes that we saw in Figs. 5.6*a* and 5.7*b*.

The second concept is related to performance. In a typical CISC, the control sequence follows a pattern like that shown in Fig. 5.26*a*. The time required for each of the fetch, decode, and execute phases can vary depending on the specific opcode and the operand modalities.

The fixed-partition, fixed-length format in RISC instructions greatly simplifies the decode process. Register-reference operands, for example, can be used directly without translation. There is never more than one memory-reference operand to translate in any instruction.

The RISC design teams chose to make the time periods for each phase of the fetch-decode-execute cycle equal. This allowed them to implement a three-stage hardwired control unit that can decode one instruction while it is executing the previous one and fetching the next. This sequence is shown in Fig. 5.26*b*. When the cycles are overlapped, the time for each cycle must be the same. This means that all the cycles will have to be as long as the longest cycle required for either the fetch, decode, or execute. We would typically waste a little time in the decode cycle, for instance, to make all the cycles equal in length. Memory-access time is often the limiting factor in the performance of a RISC.

FIGURE 5.25 Typical RISC instruction formats: (*a*) arithmetic/logic operations; (*b*) load/store operations.

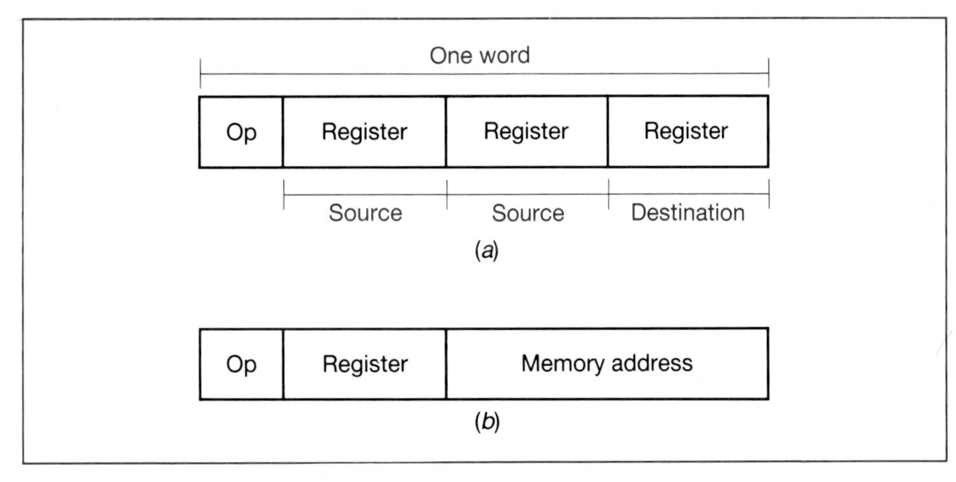

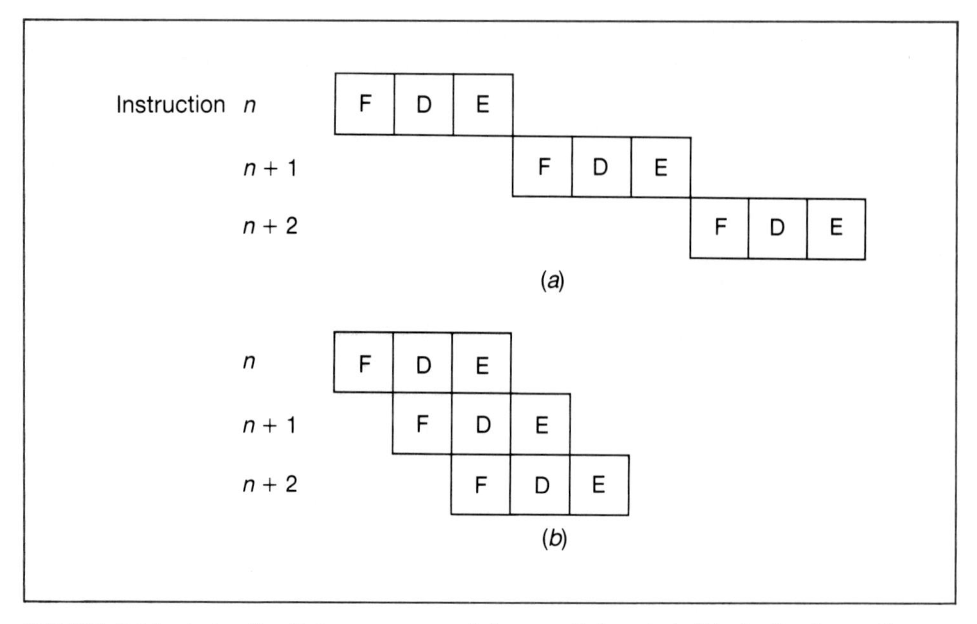

FIGURE 5.26 Instruction timing sequences: (*a*) sequential control; (*b*) pipelined control.

A block diagram for an overlapped, or **pipelined**, control unit is shown in Fig. 5.27. Conceptually, it works like a three-station production line. The individual stages are similar to the hardwired control that we saw in Fig. 5.18. The first stage controls the fetch sequence and transfers the instruction from memory to the first-stage IR (IR_f). Control signals for the fetch phase are provided by the fetch control buffer (FCB). At the end of the time period, the instruction is passed to the second-stage IR (IR_d). The instruction is decoded and the information needed for the execute process is temporarily stored in the decode control buffer (DCB). At the end

FIGURE 5.27 A pipelined control unit.

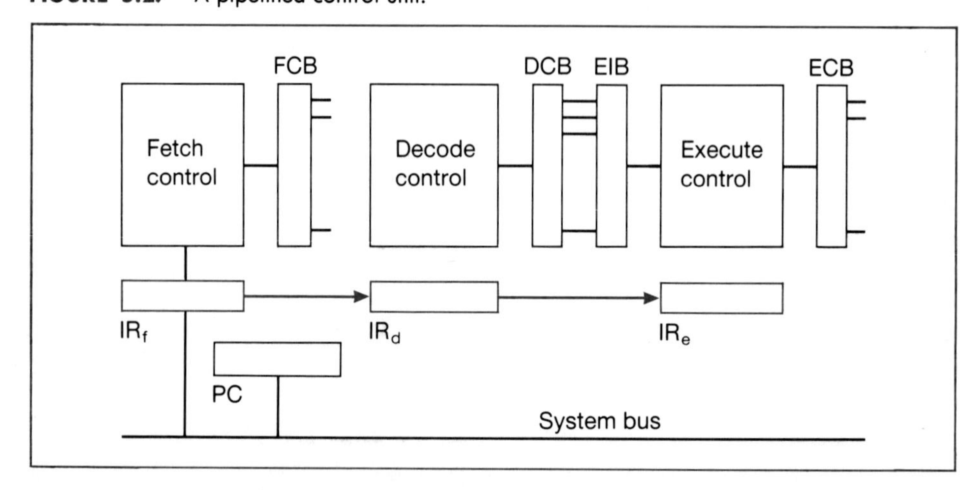

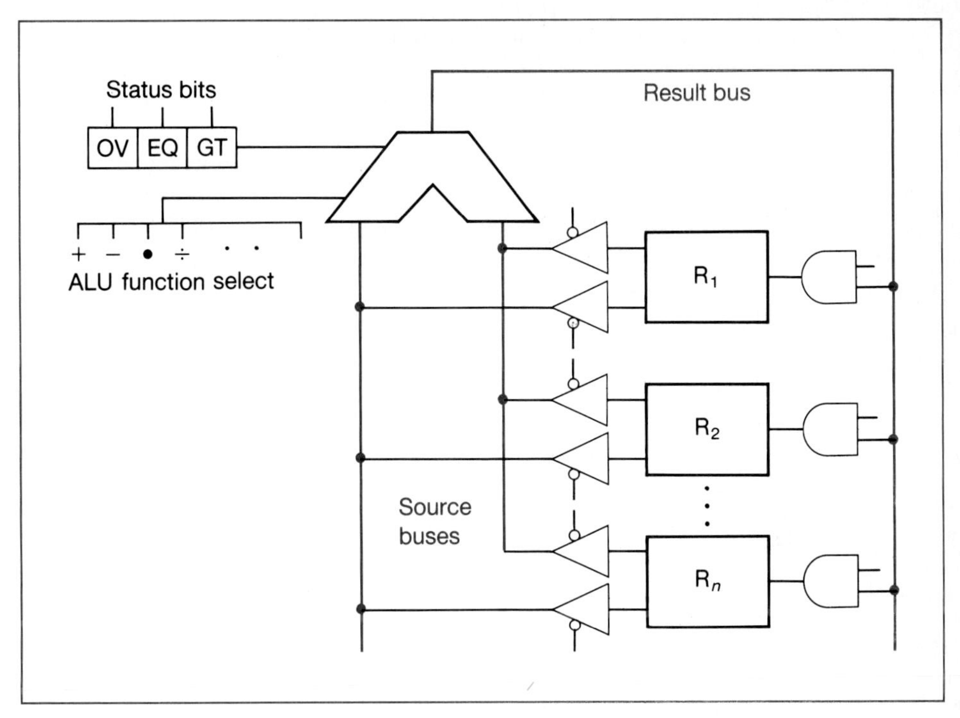

FIGURE 5.28 A three-bus multiregister ALU.

of the second time period, the information in the DCB and IR_d is passed to the execution input buffer and IR (EIB and IR_e). Execution control signals are provided by the execution control buffer (ECB).

Once the pipeline is full, we could need as many as four simultaneous storage accesses. There could be one for an instruction fetch and three for the operands (two sources and the result). Since all of these must happen in the same time period, we need to provide additional data paths to allow simultaneous data movement. This can be done by using an ALU similar to the one diagramed in Fig. 5.28. The registers provide intermediate storage for data that will be used in the computations, and the three internal buses provide the data paths for the two sources and the result. This leaves the system bus free for instruction fetches.

Lastly, before we leave this section, we need to note that a pipelined control decoder has some problems with delayed opcode availability. A jump, for example, would not be identified until the second stage of the pipeline. By then the next instruction in the program sequence would already have been fetched. A simple way to solve this is to **flush** the pipe, that is, remove the instructions, after each jump.

5.6 SUMMARY

At the beginning of this chapter, we said that the purpose of the control unit was to interpret the instructions in the program and then use that information to control the computer. Each instruction is fetched, decoded, and executed in sequence as a complete and separate logical entity. Instructions must contain all the control information that is needed to execute the specified operation. We saw that there can be considerable variation in the formats and capabilities and that the format can have a large influence on how the control unit is implemented.

Performance of a computer is affected by the number of operands and modalities that are included in each instruction. It is also affected by the amount of overlap between the fetch, decode, and execute phases of successive instructions. This discussion of performance should have given you an understanding of the relationships between the instruction set, the control unit structure, and the rest of the computer architecture.

A word of caution is also in order. In most computers, the control unit will treat any information transferred to the IR as an instruction, even if that is not the case. Programmers must take care to ensure that the PC always points to the correct program location and not to data values or to uninitialized locations that are not even in use.

PROBLEMS

5.1 What steps make up an instruction cycle?

5.2 Describe any limitations associated with *immediate operands* discussed in Sec. 5.1.6.

5.3 Give an example of an imperative sentence directed to a computer. Explain the parts of the sentence.

5.4 Where are all operands stored in a memory-memory computer?

5.5 Given the instruction format in Fig. 5.3, how many instructions can the computer have and how much memory can it access?

5.6 What are the disadvantages of using variable-length instructions?

5.7 Most computers contain similar types of instructions. What are the common instruction types?

5.8 Using the instructions listed in Sec. 5.1, write a program loop that would decrement the accumulator until it reaches zero.

5.9 Using the instructions listed in Sec. 5.1, write a program loop that would increment the accumulator until it reaches the value of a specified memory location.

5.10 Write a program that will compute the value of the expression

$$Z \leftarrow (A \cdot X + B)/Y$$

5.11 Describe two additional move instructions that might be added to Table 5.2.

5.12 Describe two additional shift/rotate instructions that might be added to Table 5.1.

5.13 Describe two additional program control instructions that might be added to Table 5.3.

5.14 Some computers do not have an unconditional jump. How is it possible to implement the effect of an unconditional jump on such a computer?

5.15 In Sec. 5.1 we described the subroutine jump and return instructions. We stated that the return address could be saved in a special area of memory. Show how return addresses could be stacked into memory for easy access by the return instruction.

5.16 How do implicit and explicit operands differ?

5.17 Give a noncomputer example of indirect addressing.

5.18 Write a simple program segment to show how an index register can be used to calculate the sum of a series of memory locations.

5.19 Repeat Prob. 5.18 using indirect addressing rather than an index register.

5.20 Describe two addressing modes not included in Sec. 5.1.6.

5.21 Draw the symbol for a tristate two-input AND gate. Define the truth table for the gate.

5.22 Why are tristate drivers used to drive the system bus?

5.23 What part of the CPU generates select signals for the computer?

5.24 Explain the sequence of events that occurs when an OR instruction is executed.

5.25 Explain the sequence of events that occurs when an RTN instruction is executed.

5.26 Microprogrammed computers are sometimes described as having "a computer within a computer." What is the justification for this description?

5.27 Discuss how you might determine the length of a microinstruction for a computer.

5.28 Why is it necessary to have a *microinstruction register* in addition to the *instruction register*?

5.29 Why are microprogrammed computers generally slower than hardwired computers?

5.30 Why is it necessary to have microprograms stored in very fast access memory?

5.31 Write microprograms to implement the logic instructions of Table 5.1.

5.32 The microinstructions for some computers do not include a jump address field. How are microprogram jumps implemented on such machines?

5.33 Since an ALU can perform only one operation during any cycle, it seems inefficient to have a whole group of bits in the microinstruction when we know that at most 1 bit will be 1 and all the rest will be zero. Can you think of a way to reduce the number of bits required to control the ALU's operation?

5.34 Can the technique from Prob. 5.33 be applied to other fields in the microinstruction? Explain your answer.

5.35 Is there any disadvantage in reducing the bits in the microinstruction (see Prob. 5.33)?

5.36 Design a decode circuit similar to Fig. 5.20 for the MW gate. Would the same design procedure be used for the C1 gate? Explain its operation.

5.37 In a microprogrammed computer, what performs the function of the decoder (in a hardwired computer)?

5.38 Why are hardwired computers more difficult to modify than microprogrammed computers?

5.39 What additional modality bits might be used in a machine instruction (see Sec. 5.5.1)?

5.40 What are the advantages of having variable-length instructions?

5.41 How could the shift instruction be modified to use multiple operands?

5.42 Why do computers that use multiple-operand instruction often have variable-length instructions?

5.43 Certain types of instructions reduce the benefits of a pipelined computer. What are they and why do they reduce performance?

5.44 How can the problem of confusing instructions and data be eliminated?

5.45 Why are the fetch, decode, and execute times identical on a RISC?

5.46 Why does the execute part of the instruction cycle usually take longer on a CISC than on a RISC?

5.47 Why does a fixed-length instruction format simplify the decoding of the instruction?

5.48 Most RISC designs have a large number of registers. Why?

5.49 What is the purpose of the flush operation in a pipelined computer?

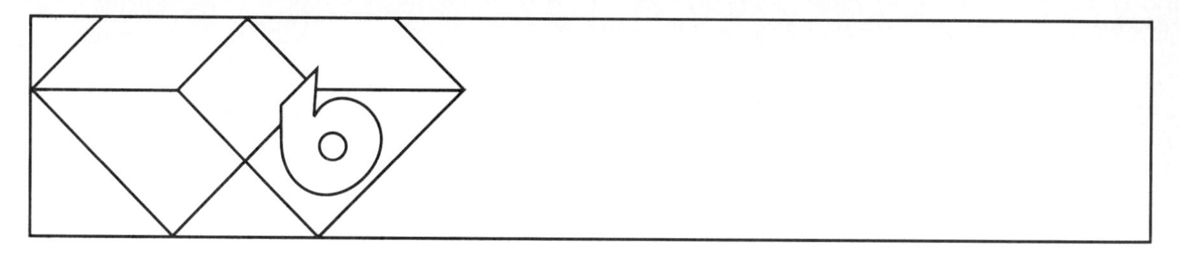

Memories and Storage Devices

We have discussed memories and information storage in previous chapters, but we have not yet described what types of memory devices are used in computer systems. Memory devices can take many different forms and can come in a wide range of capacities. In Chap. 4 we showed a block of memory attached to the system bus. This is usually called *main memory* and is part of the internal memory of the system. Nowadays, it is generally implemented in solid-state electronic circuits. Memory that is accessed through the I/O system is called *external memory*. It is implemented with a variety of devices that include electronic circuits, magnetic media, or optical media.

The purpose of this chapter is to describe some of the different memory types that are currently used. We will also discuss some of the factors that are involved in the intelligent use of memory. We will begin with an expansion of the memory hierarchy from Chap. 4.

6.1 THE REASON FOR A MEMORY HIERARCHY

If we could, we would like our computers to have the fastest memory available, to have a storage capacity large enough to support any application we might envision, and to be low in cost. Unfortunately, large, fast memories are expensive, so design tradeoffs usually must be made.

In fast electronic memory, the speeds are limited by the time it takes for a signal to propagate through the connected combinational logic circuits and to change the state of the storage elements. In slower memory units, which require some form of physical movement, speeds are further limited by the mechanical characteristics of the devices. Memory speeds have historically had an inverse nonlinear relationship to cost, as shown in Fig. 6.1.

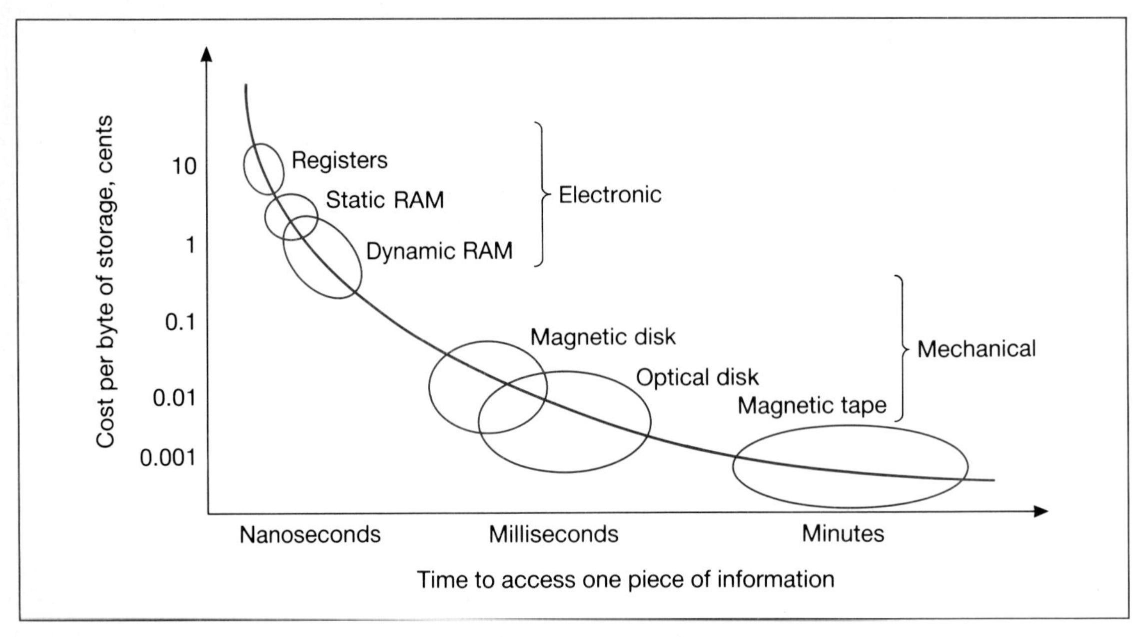

FIGURE 6.1 The cost/access time relationship in different types of memory.

Great advances in memory capacity, speed, and cost have been made over the years, and there will be more in the future. However, the shape of the curve and the relative performance of the different technologies have remained essentially unchanged. Memory speeds that were the state of the art a few years ago are considered relatively slow today, and today's state of the art will be commonplace in the future. The only certainty is that memory speeds and capacities will continue to increase and relative costs per byte of storage will continue to decrease.

There are vast differences in speed and price among the various memory devices. We try to organize our system memories in ways that optimize processing speed and minimize cost. To do this we equip systems with several types of memories, each serving a different purpose. Registers are used to store information within the CPU. Their speed is needed to keep the processor running as fast as possible. Instructions and data that are part of an executing program need to be accessed quickly (within nanoseconds or microseconds). This ensures that the CPU will not have to spend time waiting for information from memory. Files, such as programs that we are developing or letters we are writing, need to be readily accessible, but we can usually wait a while (milliseconds or seconds) to view and change them. Backup copies of programs we developed last year may be needed at some time, but we are usually willing to wait a long time (minutes or hours) to access them.

The classes of data described in the previous paragraph are primarily stored in two areas within the system. Internal memory, which is within the CPU or attached directly to the system bus, is the fastest and is reserved for the programs that are currently executing or ready to execute. External memory, which is accessed

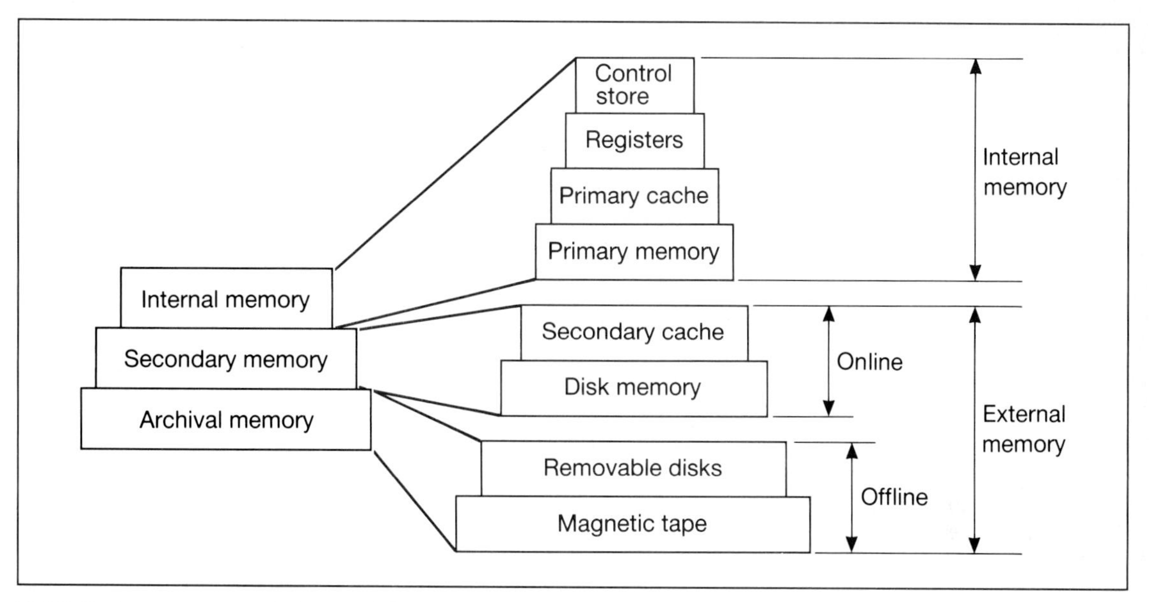

FIGURE 6.2 An expanded view of the memory hierarchy.

through the I/O system, is slower and much larger. External memory is most often used to store files of data we are currently using and to store backup (archival) data.

Figure 6.2 shows that internal memory can occupy up to four layers, depending on how the system is designed. External memory is often split into two classes based on access speed. Programs and data that are regularly used by the system are typically stored **online**. With today's technology this usually means the data is on fast disk devices.

Backup copies of important programs or data as well as programs and data that will not be needed soon are usually stored **offline**. On most systems this is usually some type of removable media such as magnetic tape or disk cartridges.

6.2 THE LOGICAL VERSUS PHYSICAL VIEW OF STORAGE

The way memory and data storage are viewed depends on our perspective. As programmers you have used data elements such as integers, floating-point numbers, and strings. You reference these elements from your high-level programming language as though each were exactly one memory location. From our discussion in the previous chapter, we saw that when the processor requests data from memory, it receives a fixed number of bits equal to the width of the data bus. This implies that there are two ways to view memory. One of them is determined by the hardware, and the other is defined by the application program. To distinguish these two views, we define the following terms:

- A **physical memory element** is the number of bits that can be retrieved in one storage access.
- **Logical memory elements** contain the number of bits necessary to represent distinct pieces of information. (The number can vary from one type of information to another.)

The hardware in the CPU is often designed around the size of the data elements or words in main memory. Registers such as the accumulator, instruction register, and memory data register hold the same number of bits as the memory elements. This enables the CPU to process data efficiently as it comes from memory. On some systems the number of bits in a memory word is equal to the number of bits in an integer.

The logical view of memory is created for the user by the software that runs on the system. Language translators define how many bits will be used to represent each of the various data types supported by the language. By changing the software running on a system, we can change the logical view of memory. The physical view, however, can only be changed by modifying the hardware. In the remainder of this chapter we will focus on the physical view of memory.

6.3 INTERNAL MEMORY

As Fig. 6.2 shows, internal memory can include four components. Internal memory is used to store the instructions and data that are needed by the executing program. Speed of access is critical for this type of memory.

In Fig. 6.3 we have redrawn the block diagram of the computer with the four components of internal memory highlighted. Three of them, the control store, the registers, and the main memory, should be familiar since they were discussed in Chap. 5. We have also added a second bus and a cache memory between the main memory and the "processing" modules (the control unit and the ALU). We will see later in the chapter that cache memory can be viewed as a fast substitute for main memory. Each of these "memories" is represented as a separate layer in the hierarchy on the right in Fig. 6.2. It must be pointed out that not all systems will have all four types of internal memory. Many systems have been built with only registers and main memory for internal memory.

6.3.1 Main Memory

In Chap. 5, we saw that main memory was accessed by loading an address into the memory address register and selecting the desired memory function:

- A read causes the contents of the memory word at the location indicated by the contents MAR to be copied into the MDR.

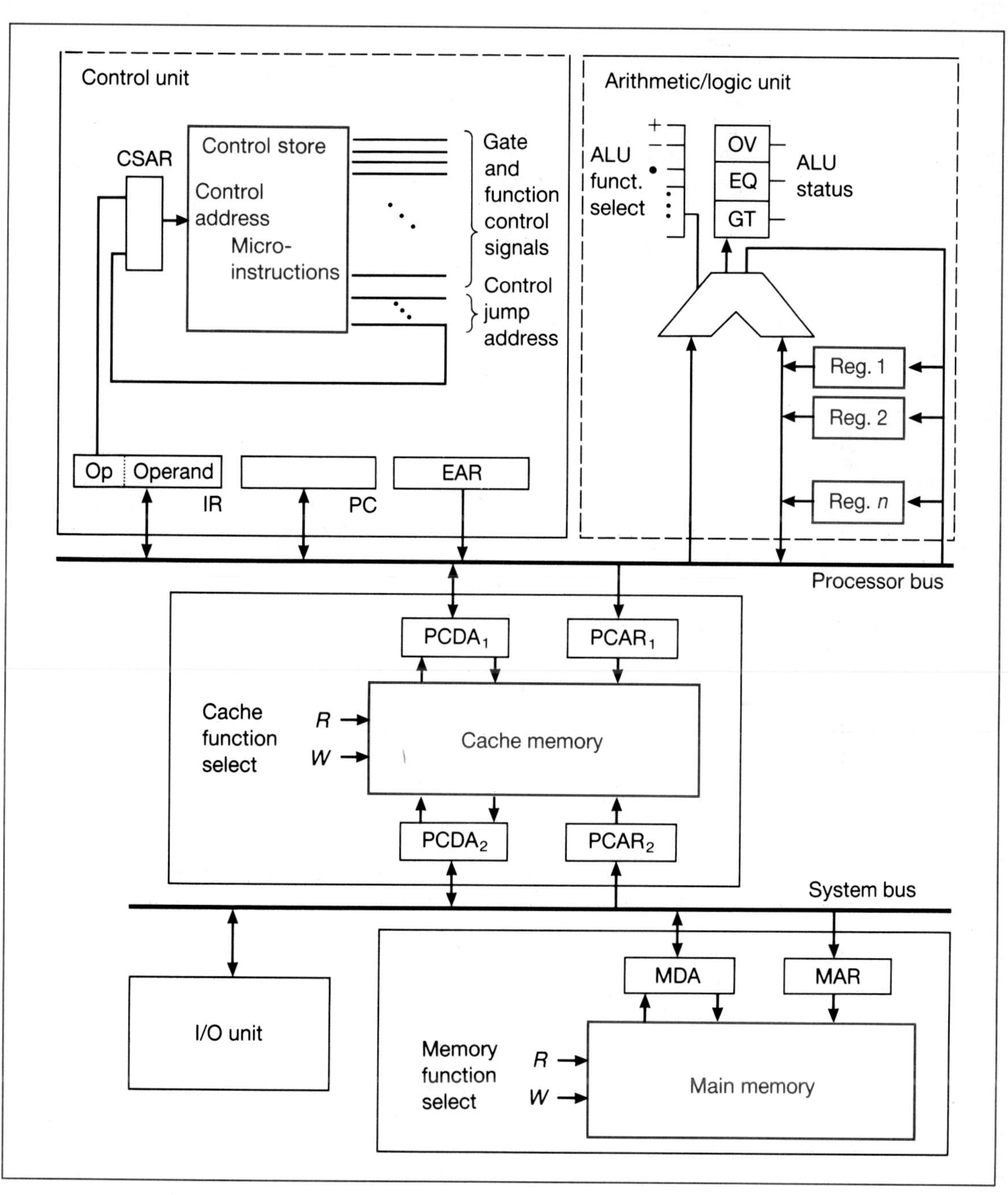

FIGURE 6.3 An internal CPU structure with four levels of memory.

- Conversely, a write causes the contents of the MDR to be copied into the memory word at the location indicated by the MAR.

We also said that main memory is implemented using random-access devices, which means that the memory access time is the same for all locations. This may not be true if we implement part of main memory as read-only and the rest as read-write, but it is an adequate approximation for this discussion. It also allows us to define a simplified set of design constraints for the individual memory cells in particular and for the main memory in general:

- Each memory element must be individually selectable by an address value in the MAR.
- The selected memory element must respond to an input (Write) or an output (Read) command.

As we have seen, a main memory read is done by a sequence of steps that starts by transferring the desired address to the MAR. The memory unit then copies the contents of the designated memory element into the MDR. Finally, the desired information from the MDR is transferred to the destination register. The first thing that memory must do after the address is in the MDR is to translate it into a select signal that will enable the desired memory element. The translation can be done with a circuit like the one shown in Fig. 6.4. The circuit translates three address lines into eight select lines.

FIGURE 6.4 A 3 × 8 address decoder.

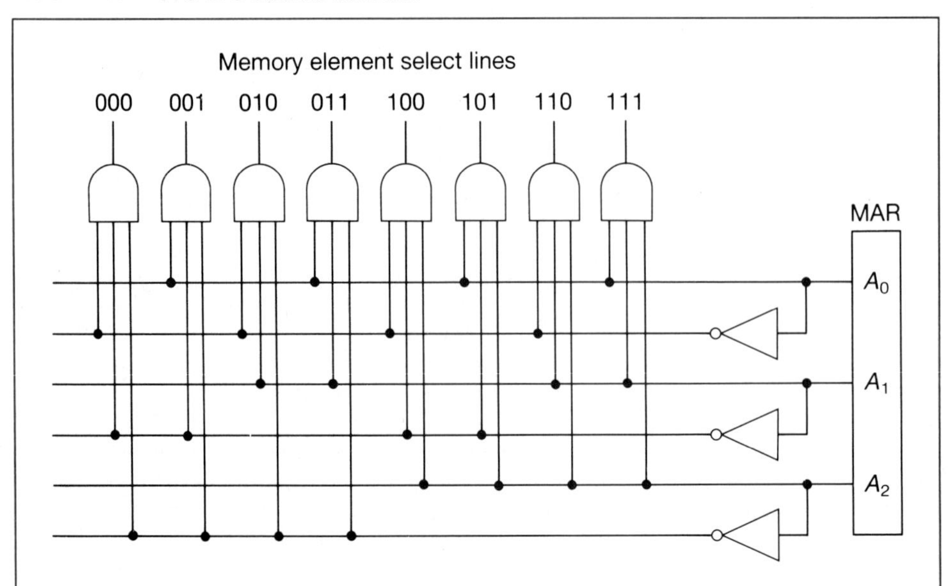

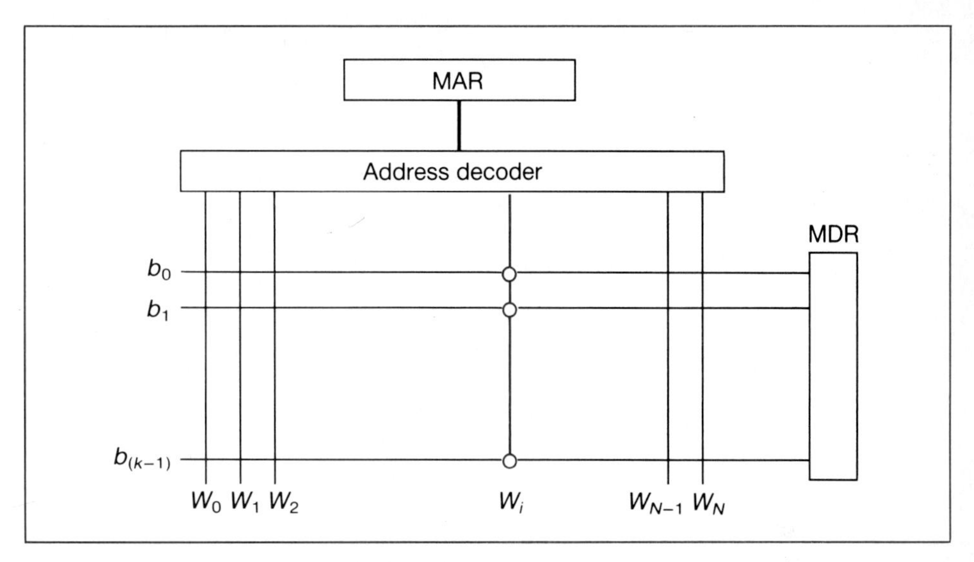

FIGURE 6.5 A one-dimensional addressing structure.

There are two ways that the output of an address decoder like the one in Fig. 6.4 can be used to select the desired memory element. The method used depends on the way the individual memory devices are organized. Figure 6.5 shows a memory organization that is commonly used in ROM chips. The address in the MAR enables one of the (vertical) select lines, which in turn enables each of the bit positions in the selected memory location. In this case, the lines from the address decoder are often called **word select lines**.

Another common organization used in large RAM chips splits the MAR into row and column select lines. The individual memory elements are arranged in a two-dimensional array of rows and columns, as shown in Fig. 6.6. The organization shown in Fig. 6.6 is actually an array, or "plane," of single-bit **memory cells**. To build a complete memory, we will need as many **bit planes** as there are bits in the computer's word.

Figure 6.7 shows how the bits of word $W_{m,n}$ are located at the intersection of row m and column n in each of the corresponding bit planes. All the memory cells in each bit plane are connected to the corresponding bit position in the memory data register by a single data line. This line is used for both read and write operations. The address decoders and the row and column address select lines are shared by all bit planes.

The individual memory cells can be implemented with either a flip-flop or a transistor circuit that uses a capacitor as the storage element. If the storage element is a flip-flop, the memory is known as a **static RAM**, or **SRAM**, because a flip-flop retains its state until it is changed. Access time of a fast SRAM is typically 5 to 50 ns. Because of its speed, it is often used for cache memories.

If the storage element is a single transistor attached to a capacitor, the memory is known as a **dynamic RAM**, or **DRAM**, because the charge on each capacitor must

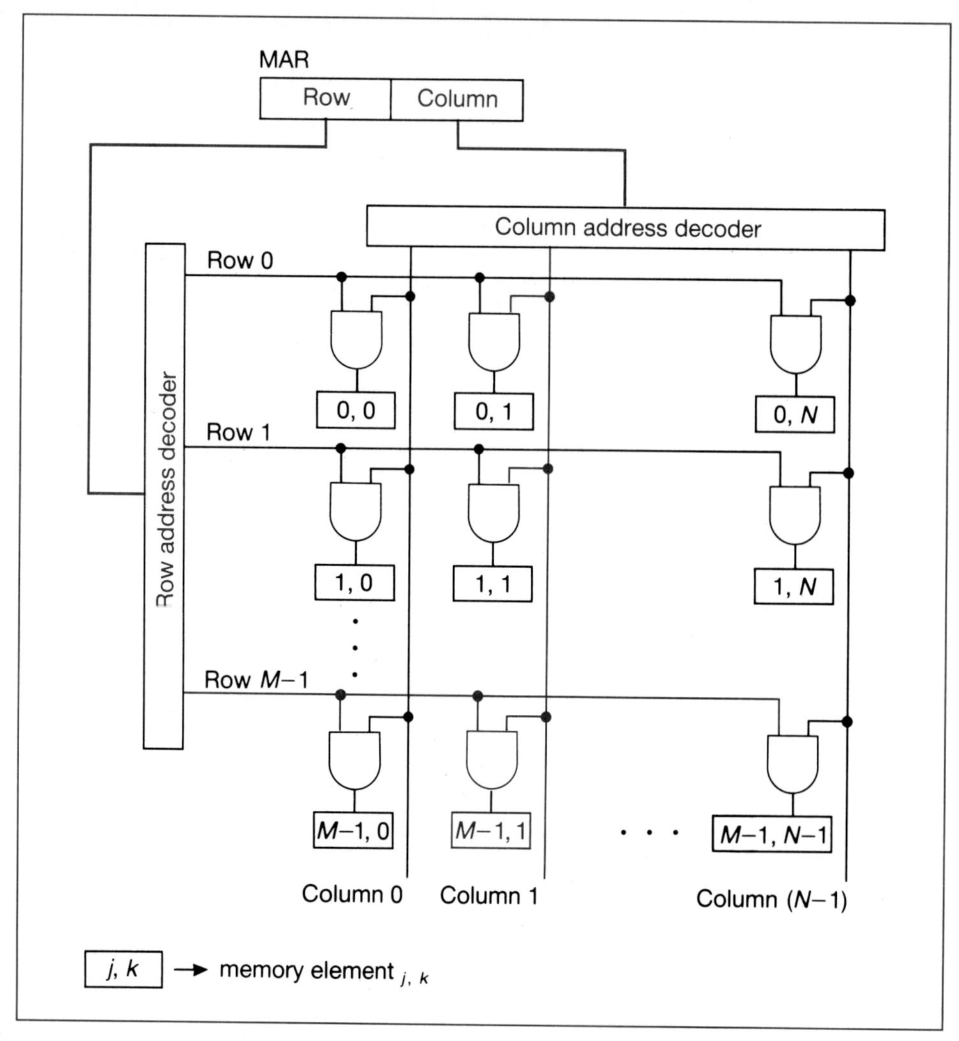

FIGURE 6.6 Two-dimensional row and column memory selection.

be periodically restored or refreshed by special circuitry to maintain the data. The access time for DRAMs is typically 10 to 50 times longer than that of SRAMs. Even though they are slower, DRAMs are currently the most common choice for main memory because they are less expensive and usually store more bits per chip than SRAM devices. The circuit diagrams for both SRAM and DRAM elements are shown in Fig. 6.8. As Fig. 6.8*b* shows, a DRAM cell (1 bit) contains only one transistor and one capacitor. On the other hand, each gate that is used to implement the SRAM cell contains many individual transistors.

Most computers are designed to accommodate a fairly wide range of sizes for several of the levels in the memory hierarchy. The size range of main memory in the

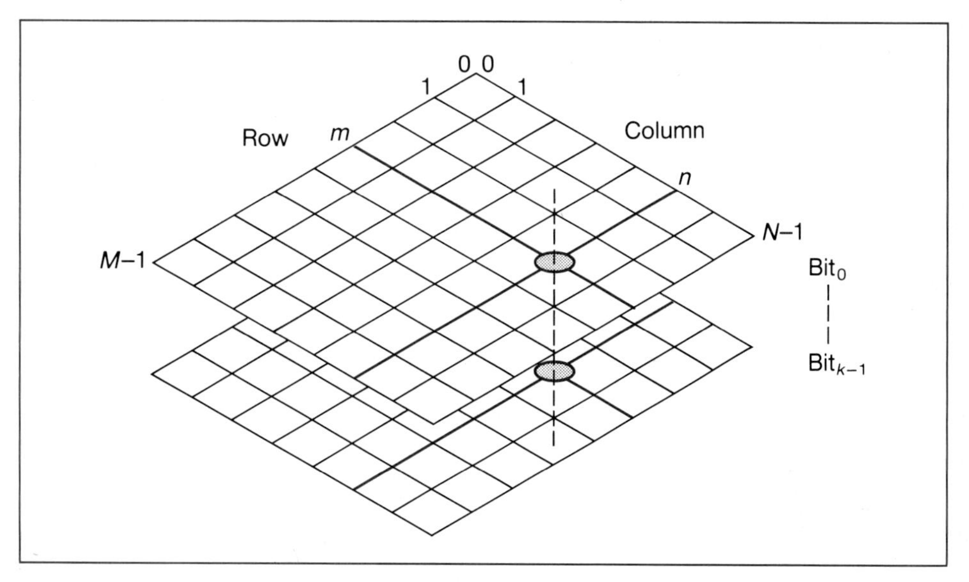

FIGURE 6.7 A bit-plane depiction of three-dimensional memory.

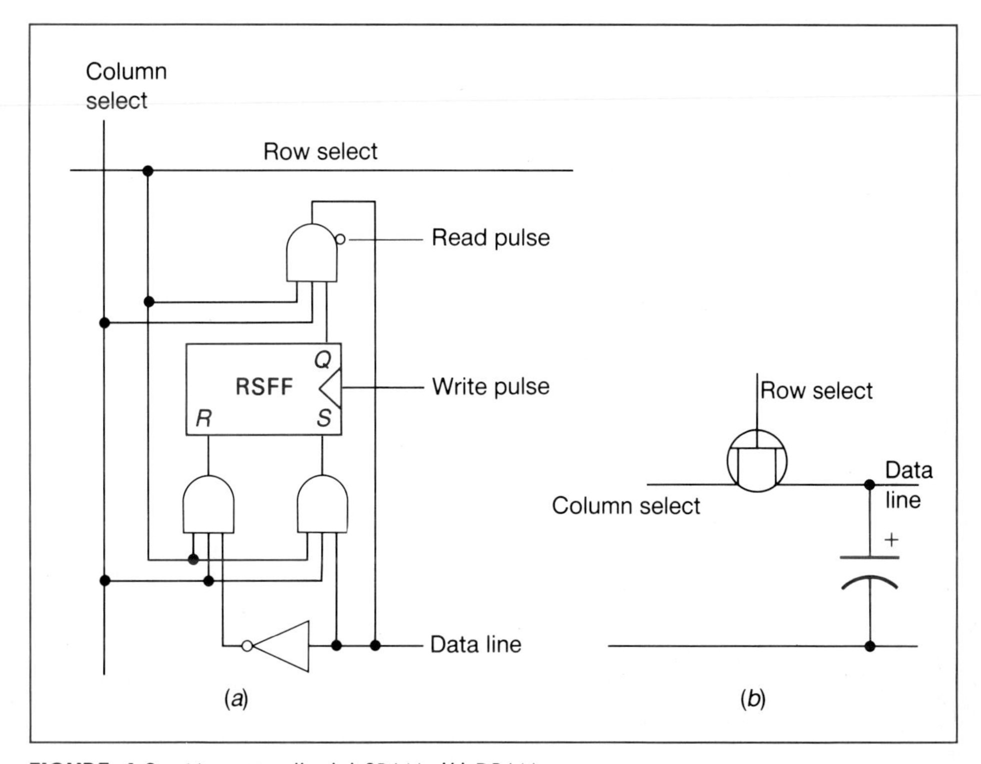

FIGURE 6.8 Memory cells: (*a*) SRAM; (*b*) DRAM.

Apple Macintosh computer that was used to prepare this chapter, for example, is 1 to 32 megabytes. Supercomputers, such as a Cray X-MP, are being designed with main memory capacities up to 4 gigabytes. Providing a range of memory sizes lowers the initial cost of a minimum system, and this is often important to customers. It also provides room for growth in later versions of the operating system and new larger applications.

6.3.2 Cache Memory

Computer programs and their associated data sets often exhibit a characteristic known as **locality**. This is a condition where a group of instructions (such as loops) or a subset of data elements (such as a table) are located close to each other and are not scattered randomly throughout memory. We can use locality to improve performance by placing small blocks of instructions and data into cache memory.

The **cache memory** is a small, fast memory that is located between the main memory and the ALU and control units. It temporarily stores instructions and data during processing and provides the following:

- A faster access to data than is possible with main memory
- A means to allow data and instruction access during times when the main memory is busy with I/O operations

Care must be taken to ensure that the information in cache and in main memory is consistent.

Figures 6.3 and 6.9 show the location of the cache with respect to the other parts of the system. The bus that connects the cache to the control unit and ALU is called the **processor bus** or **local bus**. When the control unit wants to read data from memory, an address and read request are sent on the processor bus to the cache. If the desired data is in the cache, we say that a **cache hit** occurred. When this happens, the cache can respond quickly, and the processor can continue without having to wait for the relatively slow main memory to respond.

Since the cache is much smaller than main memory, sometimes the instructions or data requested by the processor will not be in cache. If this happens, we say there was a **cache miss**, but misses occur less than 10 percent of the time in well-designed systems. When this happens, the control circuitry in the cache passes the address and read request from the processor bus to the system bus. The main memory will then respond by reading the desired location. A question at the end of the chapter asks you to suggest a way to improve this performance when a miss occurs.

Unfortunately, because the main memory is slower than cache, the processor will be forced to wait or pause while main memory responds. When the processor is forced to pause, we say it is placed in a **wait state**. The timing associated with data transfers from cache and main memory is shown in Fig. 6.10.

The use of cache is controlled by the system hardware and software. From the user's view, programs are executed as if no cache were present. We say that the use

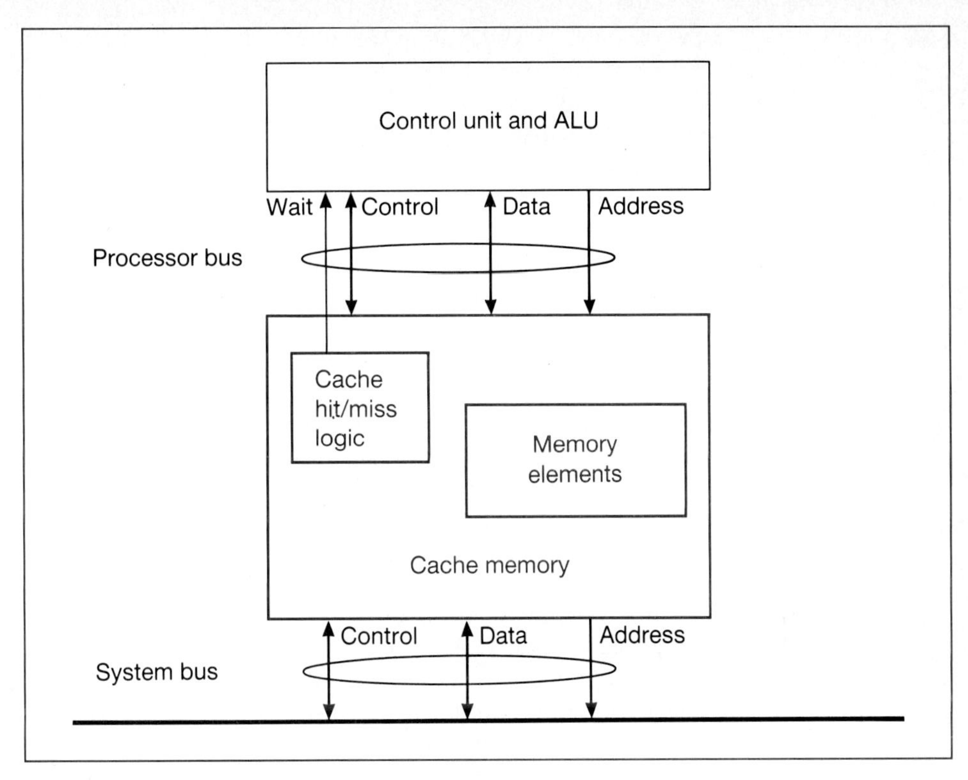

FIGURE 6.9 Cache memory and the processor bus in a system.

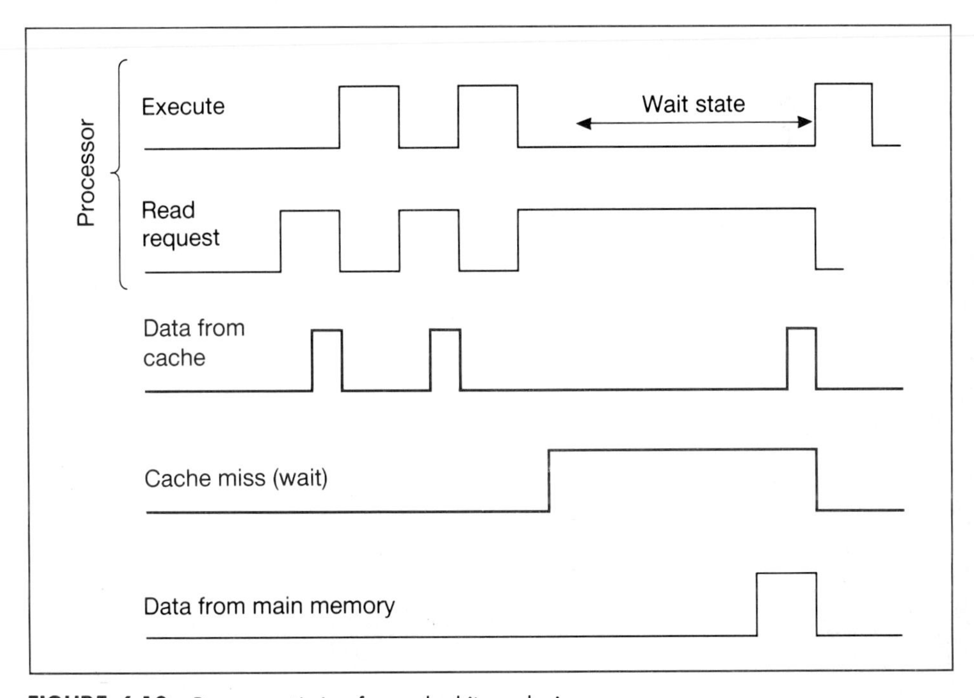

FIGURE 6.10 Processor timing for cache hits and misses.

of cache is **transparent** to the user. There are several techniques that hardware designers use to determine if the data requested from memory is in the cache. The easiest to understand is the direct-address method, which we will now describe.

In **direct-addressed cache**, the memory address from the processor is split into two parts. The lower bits are used to directly address a word in cache. There is a problem though. Many words in main memory have the same lower-bit addresses as our current memory address. By looking at the lower bit alone, we cannot be sure whether the word in cache is the one requested. The only way we can know for sure is to compare the upper bits from the MAR with the upper bits that were used to specify the address of the information currently stored in cache.

To do this quickly, cache is augmented with a one-column **cache tag** where the upper bits for the main memory address associated with each word can be stored. This configuration is shown in Fig. 6.11. When a word is read from cache, the tag is compared with the upper bits of the memory address register. If there is a match, the data requested is in the word. If the tags and address bits do not match, a cache miss signal is generated, and a cache load request is issued to transfer the proper word from main memory. After it has been copied into cache, the tag for the new word is updated and the requested access is completed.

Reading information from cache is straightforward, but the method of writing to cache requires a design decision. If a word in cache is changed, the contents of cache

FIGURE 6.11 Cache memory with tag to identify words.

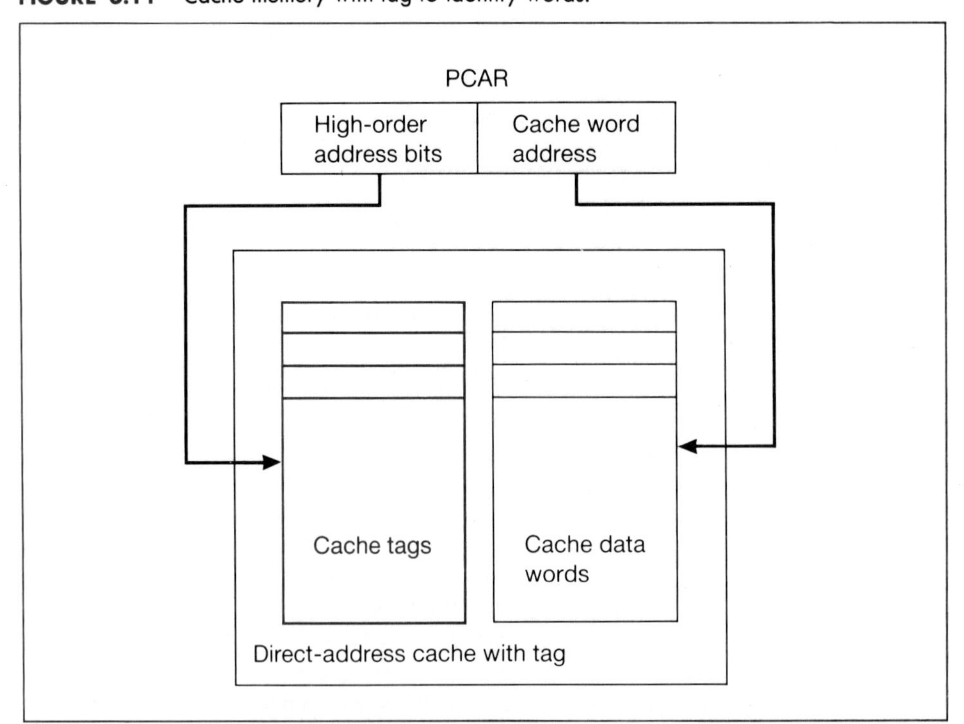

are no longer "consistent" with the corresponding information in main memory. This can cause problems if the main memory is not updated before the information in cache is changed. There are two common ways to approach the problem.

The first is to change the word contents in main memory each time the contents of cache are changed. This approach is known as **write-through**. It is a simple but often inefficient solution. Because main memory is several times slower than cache, processing the next instruction may be delayed until the main memory write has been completed. Also, there is the possibility that the same word could be changed several times before it was needed by other programs. In such a case, the write-throughs would have been wasted.

An alternate approach is to group words of cache together in a block and to write the entire block back to main memory only when the block is needed for other instructions or data and some word in the block has been changed. This procedure is known as **write-back**. It requires us to add a "change bit" to each element of the cache. This bit is set to 1 when any word in the block is changed. The change bit is checked during each cache access at the same time the tag is compared. When a cache miss occurs, the information currently in cache must be written back to main memory before the new block is loaded. The write-back approach saves time in programs with high degrees of locality. It can slow the system if the programs contain large numbers of long jumps and scattered data that cause frequent cache misses.

Not all computers have cache memories, and some that do make them an extra-cost option. The typical cache is often three orders of magnitude smaller than main memory and is an order of magnitude faster. In commercially available computers, cache operations are somewhat more complicated than we have indicated here.

The performance increase in the system is never as much as the speed difference between cache and main memory would lead you to believe, because extra time is required whenever a cache miss occurs. The actual performance improvement depends on the programs being executed.

There is one other concept that we need to cover before we leave this section. The cache memory we showed in Fig. 6.3 contained two sets of address and data registers, one between the cache and the processor bus and the other between the cache and the system bus. Each set constitutes a **port**, or entry point, to the memory. The cache we have shown is an example of a **two-port** memory module. However, *the existence of two ports does not mean that both can have access to the memory at the same time.* Both ports can accept requests, but only one port at a time can access memory. The other port will be forced to wait until the first access is complete.

The simple cache described in this section has some drawbacks. First, a cache tag is required for each word of data. This means that a lot of memory bits are needed just for tags. A way around this is to use one tag for several words of data. A second problem is that many main memory locations have the same cache locations. This implies that the cache will have to be switched often. One solution to this problem is to have two or more cache locations for one memory location. Descriptions of these advanced techniques will be left for later courses.

6.3.3 Registers

The register layer in the computer memory hierarchy refers to the accumulator or general registers that are used with the ALU. They are usually constructed from flip-flops (one per bit), and, like cache, they are faster than main memory. Although access time for registers may be no faster than cache memory, their operation is different. We said that loading cache is under control of the operating system and is transparent to the programmer. Loading data into registers such as the accumulator, on the other hand, is the responsibility of the programmer or compiler.

An advantage to using registers for data storage is that there is no such thing as a "miss" when accessing registers. Because of this, their *average* access time is faster than cache memory. Some registers may also include specific functional capabilities that are more related to the arithmetic and logical operations than to main or cache memory. Therefore, we will defer any further discussion of general registers until Chap. 8.

6.3.4 Control Store

The control store is shown at the top level of the internal memory hierarchy in Fig. 6.2 because it is the memory unit that would be accessed first in the instruction fetch-decode-execute sequence. It is usually implemented in high-speed SRAM or ROM. The control store is not used in machines with hardwired control units and is not accessible to the average user. Because it cannot be used to store user programs and data, we will not discuss it further.

6.4 EXTERNAL MEMORY

The external memory or storage layers provide the most options and the widest choice of storage capacities available on the system. Many different external storage devices have been developed over the years, and some of them are no longer being manufactured. Based on the differences in the speed and method of access, we can divide external storage into three classes:

- Rotational devices that are "disk-oriented," where the medium upon which the information is stored repeatedly "cycles" past a read or write station
- Direct-access devices that are similar in operation to main memory except that they have slower access times
- Sequential devices that are "tape-oriented," where the medium upon which the information is stored must be rewound or repositioned before the information can be accessed a second time

These classes of external storage are illustrated in Fig. 6.12.

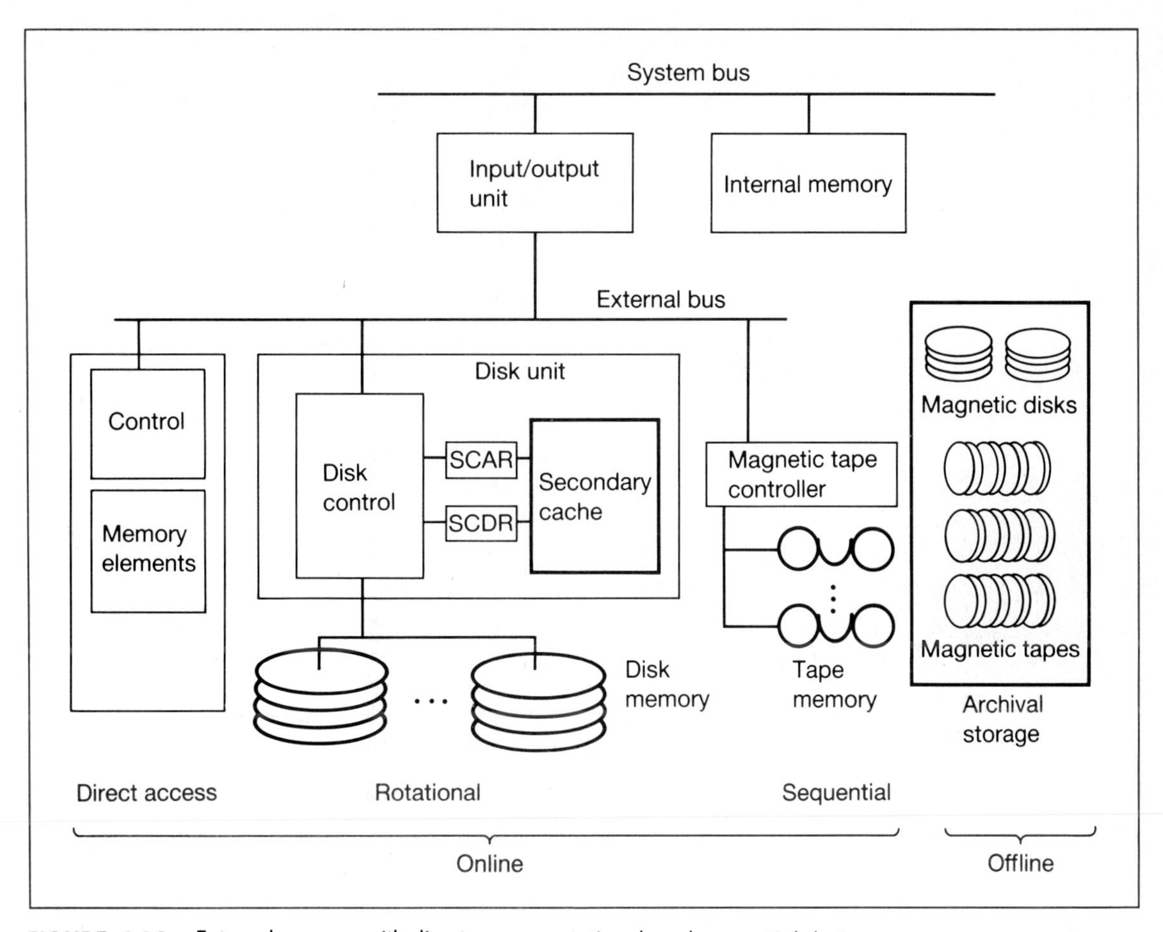

FIGURE 6.12 External memory with direct-access, rotational, and sequential devices.

6.4.1 Online Storage with Rotational Memories

On rotational memory devices the information is stored in concentric rings, or "tracks." The data is formatted in short blocks, or "sectors," and is stored on the surface of the device. The details of how the information is stored and retrieved are discussed in more detail in Chap. 10. For our purposes here, the important characteristic about rotational memories is that they provide much cheaper storage media than either cache or main memory. The bad news about these devices is that they are much slower than cache and main memory.

Access time to disk memory is much longer than the access time for main memory because an I/O operation requiring several steps is required:

1. The program requests a disk access from the I/O unit.
2. The read/write head is positioned over the correct track.

3. The system waits until the correct sector rotates under the head. (The system can do something else while waiting in steps 2 and 3.)
4. The system reads or writes data from or to the disk.

The time to position the head over the track is known as the **seek time**. It is typically on the order of tens of milliseconds. The time to rotate the correct sector under the head is called the **latency time**. In a hard disk, it is typically between 8 and 17 ms; in a floppy disk, it is on the order of hundreds of milliseconds. The **access time** is the sum of the time it takes the system to respond to a disk request plus the seek time (if a seek is required), the latency time, and the read/write time.

The long access time for secondary memory can cause an otherwise fast processor to slow down significantly. One common way to reduce this problem is to take advantage of the locality characteristic of programs and data and install a secondary cache memory. The **secondary cache memory** is a high-speed buffer between the relatively slow disk and the fast CPU. The concept and design are the same as with main memory and its cache.

Secondary cache is a typical option with larger-capacity hard disks and with optical disks. It improves the performance of the system in two ways:

- Disk requests do not always have to wait for mechanical access because the information may already be available in cache.
- If the information is in cache, the transfer rate is a function of electronic circuit speed rather than mechanical rotational speed.

Secondary cache is typically implemented in two ways:

- As a small auxiliary RAM within the disk unit
- By dedicating a portion of main memory as buffer storage

The first approach is common in large, high-performance disk units. The second has been used on many personal computers and has been given the name **RAM disk**. Although it does improve access time to disk data, it reduces the amount of main memory space available for programs.

Before leaving this section, we must mention that much of the world refers to disks as "direct-access storage devices," or DASDs, since this is the term used by IBM. We will be more restrictive in our use of the term and apply it only to devices that do not have mechanical delays.

6.4.2 Online Storage with Direct-Access Memories

The slow access time of disk devices has always caused problems for users who were interested in very high performance. The problem has always been that there is seldom a good alternative for disks. However, there have been some products that have been used to replace disks in particular applications.

When semiconductor memories started replacing magnetic core memories, the manufacturers of core memories had a big problem. Nearly everyone wanted to switch to the higher-performance semiconductor memories. What some core memory manufacturers did to create a new demand for their products was to repackage them into units they called **bulk core storage**. Interfaces were built for the bulk core devices so that they could be attached to the I/O unit rather than to the system bus. These devices were designed to respond exactly like a disk except that they were much faster. They had access times in the order of a few microseconds rather than milliseconds. Although bulk core memories were popular on some large systems for a while, their high cost eventually caused their demise.

About 15 years ago semiconductor scientists developed two new memory technologies that many people thought would eventually replace disk memories. One of these technologies, **magnetic bubble memories**, stored data in a magnetic domain within a chip. The other technology, **charge-coupled devices**, stored data as electronic charges within a chip. The bubbles were electronically rotated past read/write heads and operated much like a disk. Because no mechanical movement was required, bubble memories were much faster than rotating disks. Unfortunately, these devices did not turn out to be as economical as their proponents had hoped. Now they are only used in special applications. Charge-coupled devices suffered the same fate and are now just a footnote in computer history.

Because of the need for faster and faster external memories, designers are still looking for replacements for disks. One method currently being examined is called **wafer-scale integration**. When dynamic or static RAM devices are made, hundreds are built at a time on one semiconductor wafer. After the processing of the wafer is completed, it is sliced up into the small chips that are then placed into the individual RAM chips. The developers who are pushing wafer-scale integration think that they can build cheaper, larger-capacity memories by placing an entire wafer into one memory device. Wafers are being produced that store tens of megabytes of data. They have access times of a few microseconds. If these devices can be built in quantity at low prices, they may replace disks in many applications.

6.4.3 Offline Storage with Sequential Memories

The method of recording on magnetic tape is essentially the same as on a magnetic disk except that the head assembly usually reads or writes several tracks in parallel. A magnetic tape contains a series of variable-length records separated by gaps where nothing is recorded. Addressing of information is by record number on the tape.

Read or write operations can be either in record mode, where the tape stops after each record, or in continuous mode, where the tape moves continuously until the transfer is complete. In record mode, the access time is the sum of the start time, the read or write time, and the stop time. In continuous mode, it is the sum of one start time, the read or write times, the gap times, and one stop time. On older systems, information was usually written using the "start-stop" mode because programs tended to generate data at random times. Newer tape systems are often used only to archive data from disks. In this mode of operation large blocks of disk data may be written without ever requiring that the tape stop to wait for more data from a program.

Magnetic tape drives are available for either reel-to-reel or cartridge recording. They provide a convenient method either for archiving or for exchanging programs and data between systems. Tapes are easily transported and can be stored in locations away from the computer. Tapes provide the cheapest form of storage available on most systems, because the medium is very inexpensive and the same drive mechanism can be used to access many reels of tape. With tape storage, the amount of data that can be stored offline is unlimited. The main disadvantage of tape storage is its extremely slow access time. To find a specific record on a tape may take several minutes. This makes tape impractical for online storage.

6.5 SUMMARY

We have presented several different views of memory in this chapter and have looked at several different types of storage. You should now have a good idea of the concepts underlying the hierarchy and operation of memory. The key is to understand the physical characteristics of each different type of memory and to match these with the requirements of the programs and data that must be stored. Table 6.1 summarizes the capacity, speed, and type of device used to implement the different classes of memory in a system.

Computer systems of the future are likely to include a variety of disk options, and some systems will probably also continue to use magnetic tape. Hard disks currently provide the best performance for active data and programs. Optical disks have the greatest capacity for large data files and archival storage. Floppy disks and magnetic tape provide the most economical methods for program transfer and distribution. All these will be discussed in Chap. 10.

The emphasis in this chapter has been on the physical devices that are used to store instructions and data. In Chap. 13 we will look at how the operating system creates a logical view of memory for executing programs.

TABLE 6.1 CHARACTERISTICS OF DEVICES IN THE MEMORY HIERARCHY OF A SYSTEM

Memory type	Number	Access time	Devices
Control storage	1K–32K words	3–100 ns	SRAM, ROM
Registers	8–256 words	3–10 ns	SRAM
Cache	8 kilobytes–1 megabyte	3–100 ns	SRAM
Main memory	128 kilobytes–4 gigabytes	20–200 ns	DRAM
Secondary cache	32 kilobytes–1 megabyte	20–200 ns	DRAM
Online storage	1 megabyte–100 gigabytes	10–65 ms	Magnetic and optical disks
Offline storage	20 megabytes or more	seconds–hours	Magnetic tape, removable disks

PROBLEMS

6.1 Calculate the cost per bit for registers, static RAM, and dynamic RAM for currently available chips.

6.2 What is the cost per bit for floppy-disk and hard-disk data storage?

6.3 For a computer that has a 16-bit memory word, what logical memory elements would require more than one memory access to retrieve or store their data?

6.4 Do the costs you found in Probs. 6.1 and 6.2 agree with the data in Fig. 6.1?

6.5 Why is the control store separate from main memory?

6.6 On some computers the size of physical memory elements in main memory is a multiple of the physical memory elements of cache. Can you suggest a reason for this?

6.7 Explain why the terms one-dimensional and two-dimensional addressing are applied to the memory-addressing schemes in Figs. 6.5 and 6.6.

6.8 A 3×8 address decoder is shown in Fig. 6.4. Diagram the circuit that would be required for one of the select lines in a 6×64 decoder.

6.9 Why do SRAMs and DRAMs have address decoders built into the chips?

6.10 A typical 1-megabit DRAM chip accepts 10 address lines and Column Access and Row Access signals. This requires that the address be sent to the chip as two parts. Why is this done rather than just having 20 address lines?

6.11 What is the essential difference between static and dynamic memory?

6.12 How does the separation of address information described in Prob. 6.10 affect performance?

6.13 Why do memory chips tend to increase in capacity by a factor of 4 with each new generation of chip?

6.14 In the early 1970s a board containing 1024 sixteen-bit words cost about $1000. How does that compare to today's prices?

6.15 Suggest a way to improve access time to main memory when a cache miss occurs (see Sec. 6.3.2).

6.16 The direct-address cache described in Sec. 6.3.2 has one tag word for each data word. This increases the total number of bits needed to store each data word significantly. Suggest a way to reduce the bits needed for tags.

6.17 Programs with many GOTO (unconditional branch) statements are often called "spaghetti code." How would this type of code affect locality?

6.18 Why do some computers have separate caches for instructions and data?

6.19 The following two program segments each add the elements of a two-dimensional array. One may run 10 times faster than the other. Why?

```
for i := 1 to 100 do
        for j := 1 to 100 do
                sum := sum + array (i, j)

for i := 1 to 100 do
        for j := 1 to 100 do
                sum := sum + array (j, i)
```

6.20 Why might it be useful to have main memory two-ported?

6.21 A cache memory can increase the speed of the machine, but it also increases the complexity of a memory access because the required information may not always be in the cache. Prepare a flowchart of the memory-access process. (*Hint*: Remember that each cache access must check the cache directory to determine whether the access will be a hit or a miss, and assume that the required information is available in primary memory.)

6.22 How do write-through and write-back cache affect performance?

6.23 What are some additional advantages of disk memory over cache and main memory?

6.24 What characteristic of a disk drive determines the worst-case latency time?

6.25 Some disks have been designed with one read/write head for each track of data. How does this affect performance?

6.26 Core memory is an example of a nonvolatile storage media. What does this mean?

6.27 Why are magnetic storage devices inherently nonvolatile?

6.28 Why is it necessary to have a hierarchy of memory devices in a computer system?

6.29 What determines the size of logical data types?

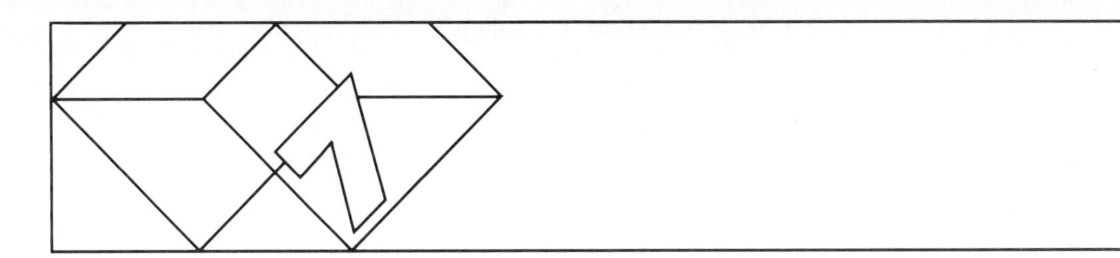

Buses—Information Flow inside a Computer

Buses are the physical links between the functional units within the computer and between the computer and the outside world. They provide the pathways for moving commands, addresses, and data from the control unit to all the other units in the computer.

To some, a bus is nothing more than a set of lines or wires that carry signals throughout the computer; that is an oversimplification. Since all the instruction, data, and control signals in the computer are transmitted over the buses, the design of the buses has a significant effect on the computer's performance. Bus characteristics such as speed, number of bits, and rules for transmitting information play a critical part in the design of every computer system.

In this chapter we expand the concept of information flow in a computer and look at some common approaches to bus design. We begin with another look at the simple computer that was used in our discussions of the control unit.

7.1 INFORMATION TRANSFERS ON A BUS

Early computers were designed with system buses that were similar to the one used in the simple computer presented in Chap. 5 (and repeated here in Fig. 7.1). Signals from the control unit define both the direction of information flow and the specific registers to be used. The system clock synchronizes the actual information transfers between the sending register, the bus, and the receiving register. In the figure, the bus is shown as a heavy line with single control gates to each module's interface register. The heavy line is used to emphasize that the bus contains several individual data lines.

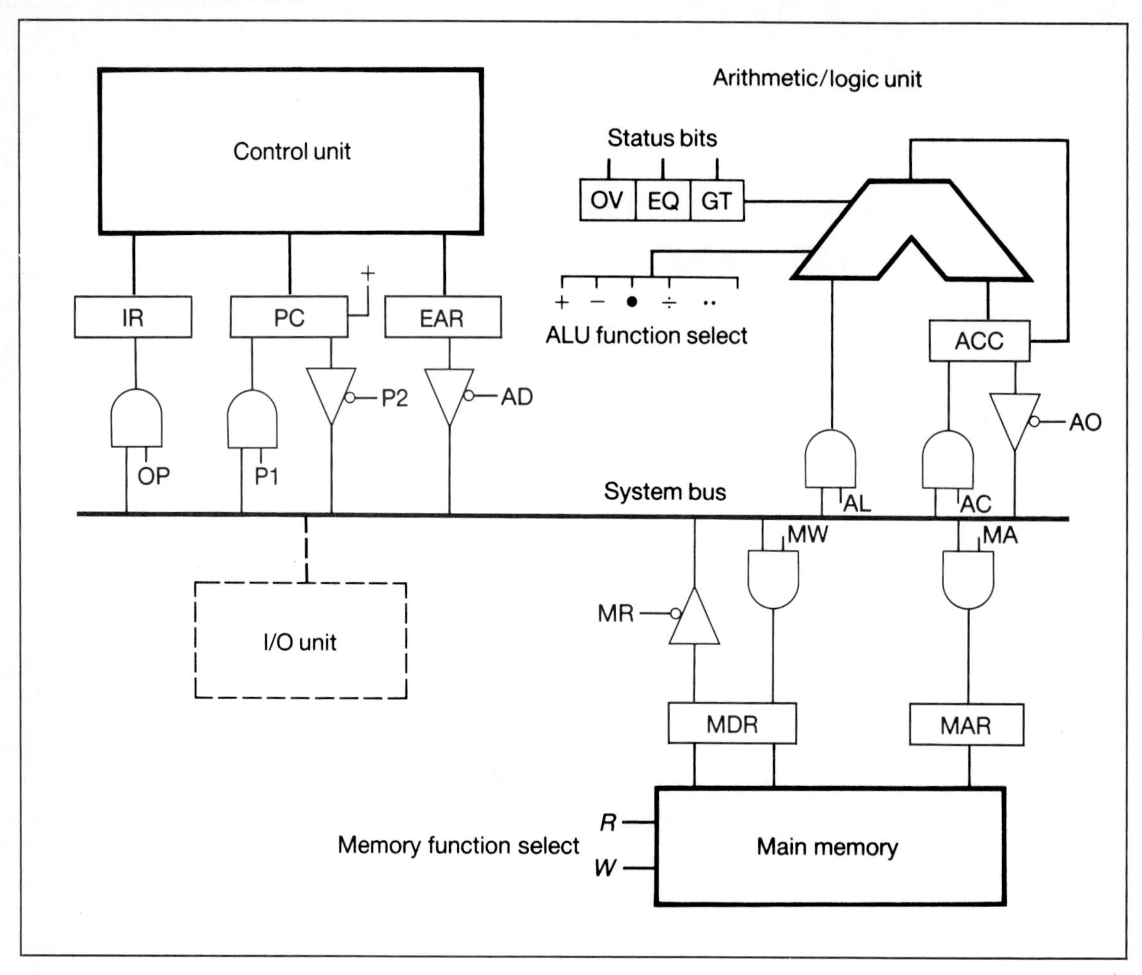

FIGURE 7.1 The internal structure of a simple computer showing the bus/module interface registers and control gates.

In Fig. 7.2 we have expanded the part of Fig. 7.1 that contains the instruction register (IR), the memory data register (MDR), and the system bus. The figure shows one possible implementation of the minimum timing and control circuitry needed to transfer data from the MDR to the IR.

The timing for the data transfer over the bus is shown in Fig. 7.3. The steps needed to complete a data transfer from the MDR to the IR are:

1. Select the MDR for output (Output MDR).
2. Place data onto the bus (Write Bus).
3. Select the IR for input (Input IR).
4. Read the data from the bus (Read Bus).

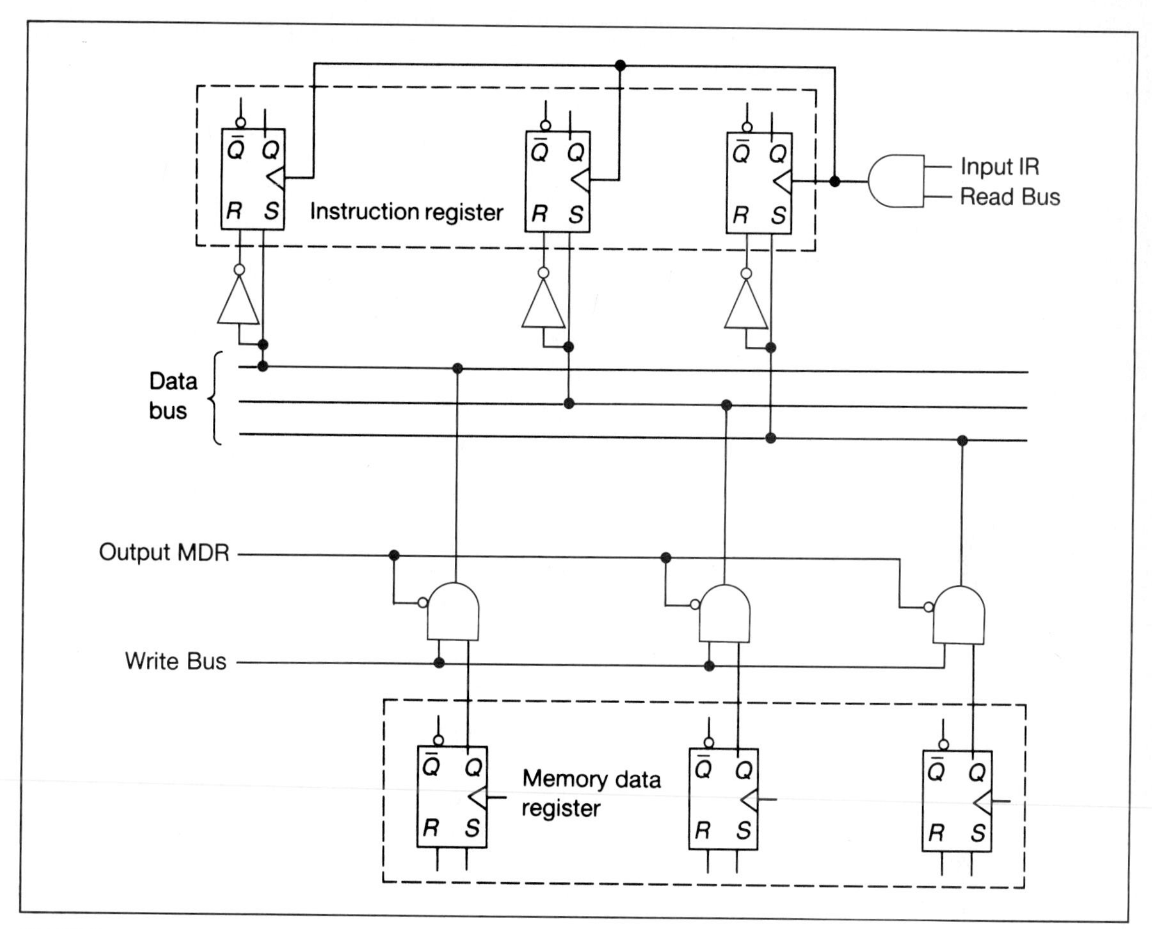

FIGURE 7.2 Logic diagram for a register-to-bus-to-register data transfer.

Since several different registers can be attached to a bus, tristate devices described earlier are used to transmit the register contents onto the bus. To show that the bus is tristated, the Valid Data on Bus signal in the figure starts out as an indeterminate value, neither a 1 nor a 0. The lines then go to either 1 or 0 depending on the signals in the individual bits of the registers. The signal IR Changes State shows that the register contents change at the end of the Read Bus signal.

As described to this point, the system bus is quite simple. It contains a set of lines over which information can be passed from one register to another. The control unit coordinates all transfers and generates all needed timing and control signals.

There are some potential problems with this simple bus. First, it may be rather inefficient for some operations. Consider a write to memory, for example. To execute a memory write, the control unit must:

1. Transfer the memory address into the MAR.

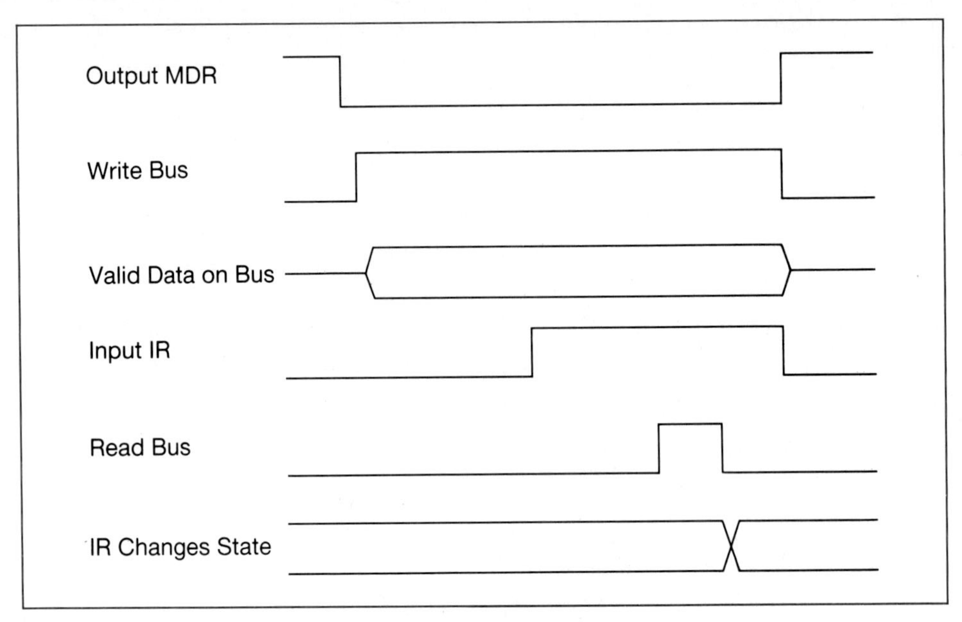

FIGURE 7.3 Timing diagram for a register-to-bus-to-register data transfer.

2. Transfer data from another register to the MDR.
3. Request a memory write.

These operations would call for two sequences like the ones shown in Fig. 7.3. If we are trying to design a fast computer, this would not be acceptable.

A second problem with the simple bus is that all control and timing signals must come from the control unit. This may not seem like a problem, but it is when there is another unit, such as the I/O unit, that also needs to initiate a transfer. In the next two sections, we will look at changes to the simple bus that alleviate these problems.

7.2 A SYSTEM BUS WITH DATA, ADDRESS, AND CONTROL LINES

Our first improvement to the simple bus will be the addition of lines that transmit address information, as shown in Fig. 7.4. The effect of the additional address lines can be observed by looking at a write request from the control unit to memory. Let's assume we want to write the contents of the accumulator into main memory at the location specified by the effective address register (EAR). The following steps are needed to complete this transfer:

1. Select EAR for output to address bus (Output EAR to Address Bus) and select ACC for output to data bus (Output ACC to Data Bus).

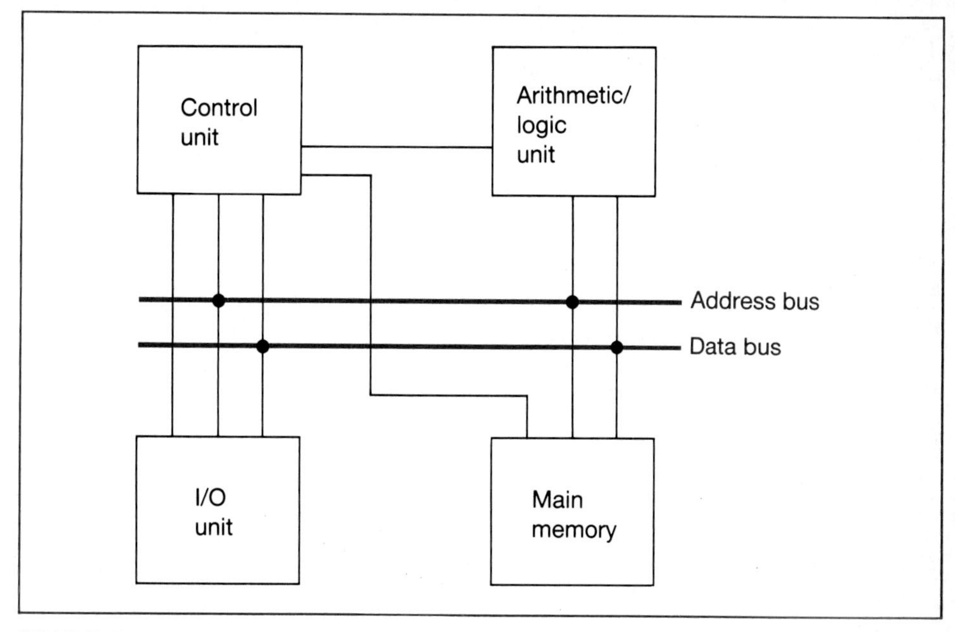

FIGURE 7.4 Bus with data and address lines.

2. Place address onto address bus (Write Address Bus) and place data onto data bus (Write Data Bus).
3. Select MAR for input from address bus (Input MAR) and select MDR for input from data bus (Input MDR).
4. Read address from address bus (Read Address Bus) and read data from data bus (Read Data Bus).
5. Write the contents of the MDR to memory at the location specified by the MAR (Write Memory).

Figure 7.5 gives the timing for these signals. As the figure shows, several operations are done simultaneously (in parallel) on the two buses. This bus operates faster than a single-bus configuration because it has lines for both data and address.

Before moving on to the bus control problem noted above, there is another improvement we should make to the bus. If we look at Figs. 7.3 and 7.5, we see that every transfer needs Read Bus and Write Bus signals. Since these signals must be attached to several different registers, it makes sense to put them in the bus just as we did the data and address. This gives us a complete system bus containing data, address, and control lines. Figure 7.6 shows a complete bus configuration with data, address, and control lines necessary for a simple computer.

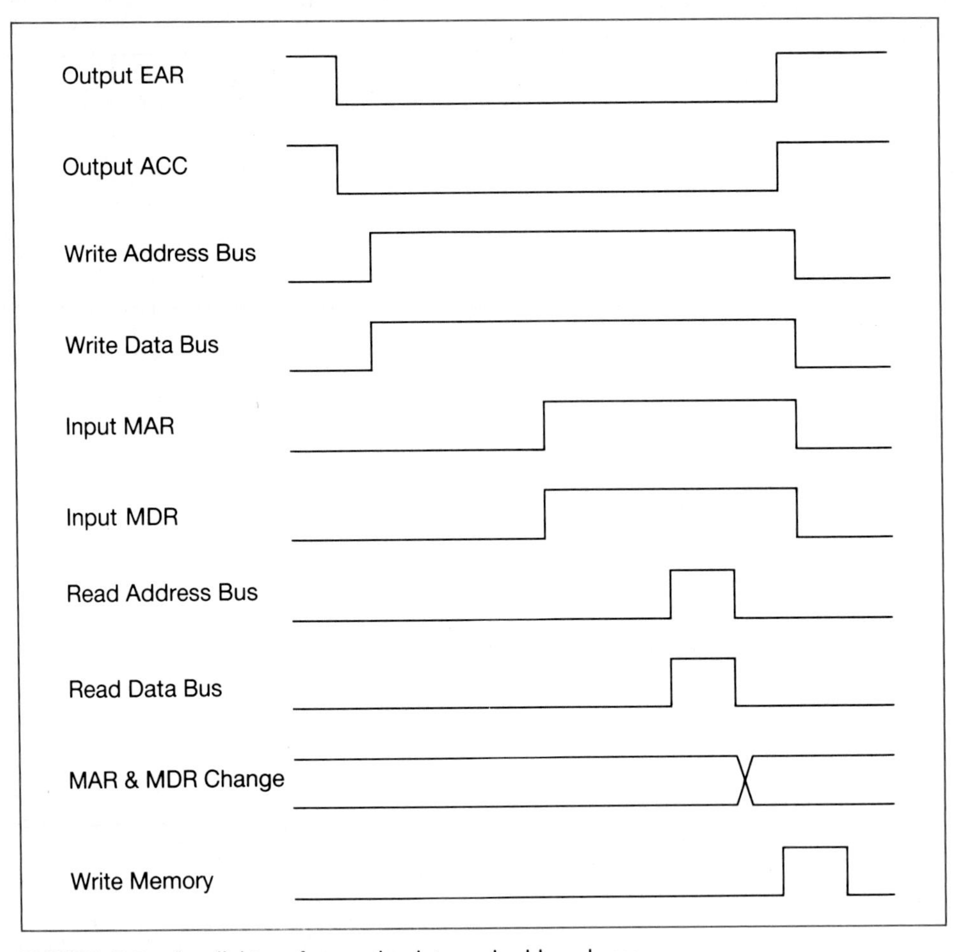

FIGURE 7.5 Parallel transfers on the data and address buses.

7.3 CONTROLLING ACCESS TO THE SYSTEM BUS

In computers such as the simple one we have been using for our illustrations, all signals that coordinate the transfer of data emanate from the control unit. Having all the control logic reside in one unit is sometimes called **centralized control**. Centralized control works well when there is only one unit in the system that needs to make transfers over the system bus. However, it presents a problem if we want to have more than one unit responsible for transfers on the bus.

Suppose, for instance, we wanted to add a second CPU to the simple system and give it the ability to access memory the same way the first CPU does. How would the second CPU get control of the bus, and how would one CPU be prevented from intercepting data that was destined for the other CPU? In fact, we do not even need

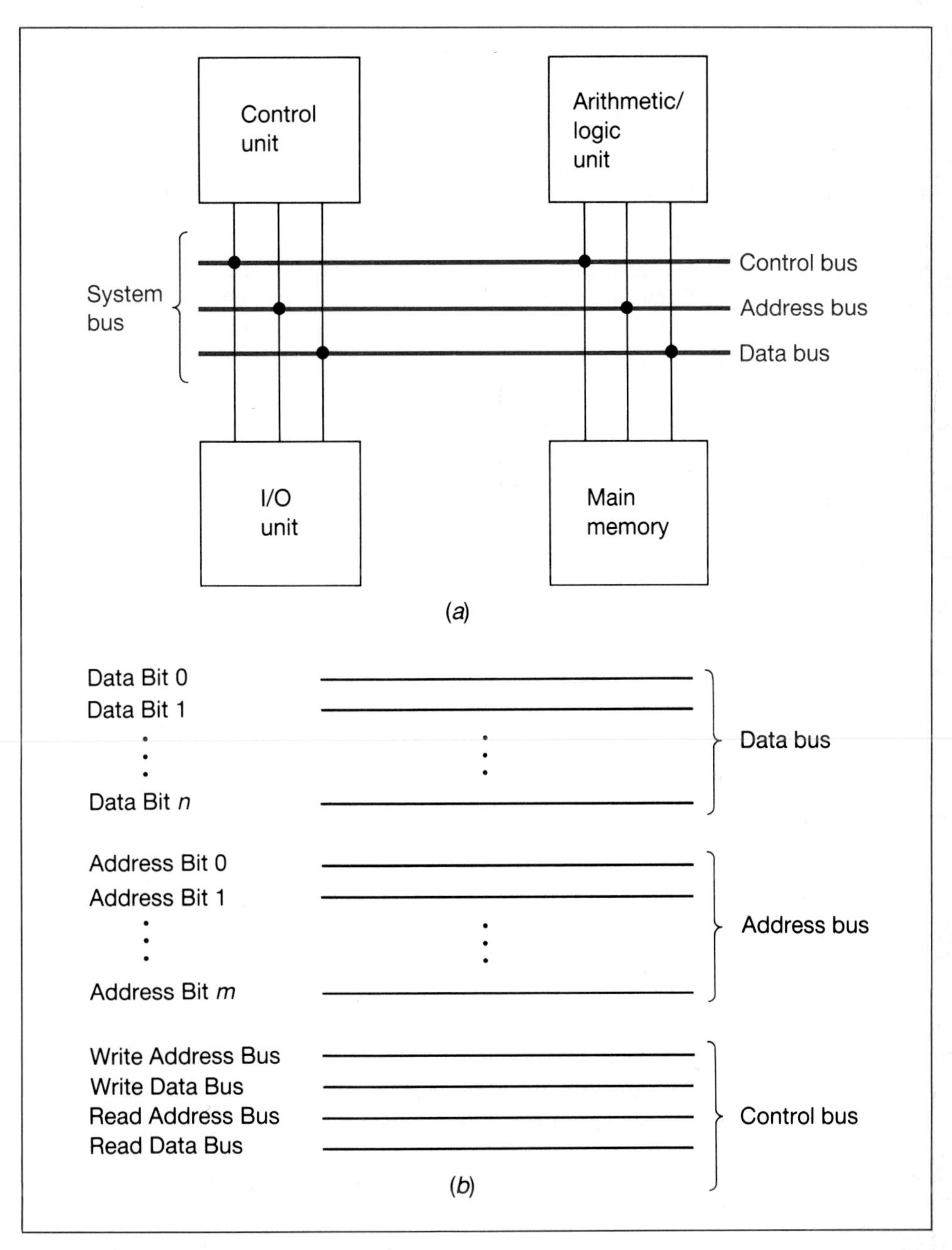

FIGURE 7.6 System bus: (*a*) computer with data, address, and control buses; (*b*) signal lines in the system bus.

to add a second CPU to run into this problem. The I/O units on many computers require the ability to control the system bus just as though they were CPUs. In this section we will examine ways messages are passed between senders and receivers and how control of buses can be shared.

7.3.1 The Requirements for Successful Data Movement

Data movement in a computer is similar in many ways to phone calls between people. First, both the caller and the person receiving the call must have access to a phone. We can think of the phone system as being equivalent to the bus in a computer. If you have ever needed to make a call and found your roommate using the phone, you have experienced one of the problems associated with a shared bus: only one user at a time may access the bus. Assuming you wait until the phone is available, you must know the phone number of the person to whom you wish to speak. The phone number thus corresponds to an address in the computer.

When you dial a number, you are signaling someone that you have some information or data for him or her. If the person you are calling is ready to accept information, he or she will pick up the phone and might say, "This is Bob (or Mary) Smith." It is not really necessary for the person answering the phone to give his or her name or say hello, but it is what we expect to hear. At that point you usually start talking (transferring data). When you have sent your message, you wait for Bob or Mary to tell you that he or she has received the message. When you know that the message has been received, you can hang up and relinquish control of the phone. The elements needed to complete a phone call are:

- Access to a phone
- The number of the person you are calling
- A way to know that the person you are calling is ready to accept a message
- The contents of the message
- A signal that the message has been accepted
- A way to relinquish control of the phone

We have already discussed how the phone system, phone number, and dialing the number correspond to the data bus, address bus, and control bus. Now we need to add one new element to our discussion of message transfers. In the description of the phone call, we said that we *expect* the person answering the phone to give his or her name or say hello. Stated another way, the protocol when answering the phone is to say something that will let the person who initiated the call know you have answered and are listening. A **protocol** is the set of rules which ensure that both the sender and receiver give the correct responses to the signals they receive and that they do so at the correct times.

Protocols are critical components in any message-passing system or computer bus design. In part, they determine the efficiency and ease of use of the system. In the next section, we look at how protocols are used to coordinate the sharing of a bus.

7.3.2 Shared Control of the System Bus

When two or more units attached to a bus want to take control of the bus, some element within the system must decide which one will be given control. The element that makes the decision is called the **bus arbiter**. The unit that is given control is designated the **bus master**, and it retains control of the bus until another unit is given control. It should be noted that some buses do not require arbitration; they are called **self-arbitrating**.

A common approach to the solution of the bus arbitration problem is diagramed in Fig. 7.7*a*. Each of the units in the figure can be a CPU or an I/O unit, and each

FIGURE 7.7 A daisy-chain approach to bus arbitration: (*a*) functional diagram of arbitration logic; (*b*) timing sequence for request, busy, and grant signals.

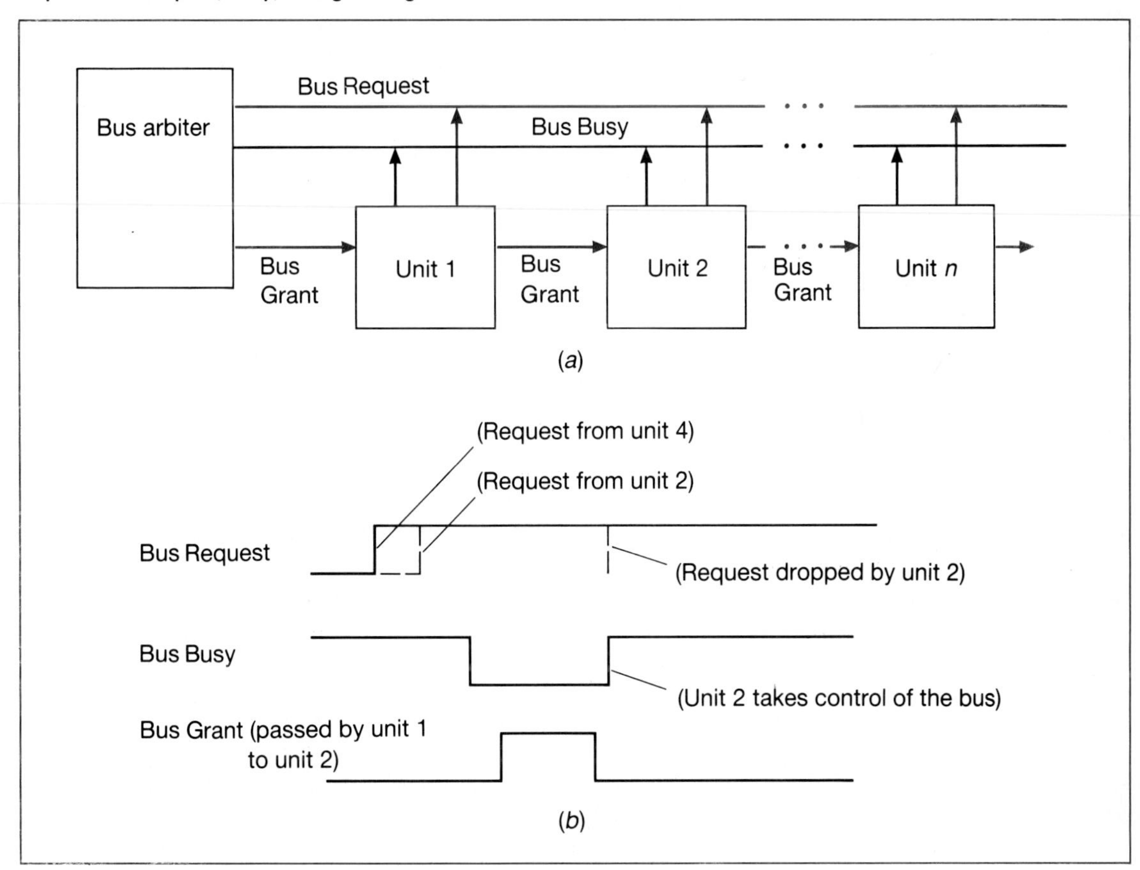

can become a bus master and take control of the system bus. The bus arbiter is responsible for deciding when control should be passed from one unit to another.

The protocol used to acquire the bus is as follows:

1. A unit needing the bus puts a 1 on the Bus Request line. (The circuits from each unit are designed so that more than one unit can place a 1 on the request line. This is sometimes called a **wired OR**.)
2. The bus arbiter monitors the Bus Busy line, and after it goes to 0, indicating that the bus is no longer busy, the arbiter checks the Bus Request line. If the request line is 1, the arbiter sends a Bus Grant signal.
3. If the first unit attached to the Bus Grant line does not want the bus, it passes the Bus Grant signal to the next unit in line. This process, which is called **daisy-chaining**, continues until the Bus Grant signal reaches a unit that requested the bus.
4. When a unit wanting the bus receives the Bus Grant signal, it stops the Bus Grant signal from going to the next unit and places a 1 on the Bus Busy line to let the arbiter know the bus is in use.
5. The unit then uses the bus to transfer information. After the transfer is complete, the unit puts a 0 on the Bus Busy line, and the arbiter starts the cycle over again.

A timing sequence for the signals described above is shown in Fig. 7.7*b*.

Assume that we have five units attached to the bus and that unit 3 currently has control. (It is holding the Bus Busy line at 1.) Unit 4 and then unit 2 both issue requests for the bus by placing 1s on the Bus Request line. (The request from unit 2 is masked because the Bus Request line is already 1.) Unit 3 completes its transfer and lowers the Bus Busy line to 0. The arbiter issues a new Bus Grant signal, which is passed by unit 1 and captured by unit 2. Unit 2 signals Bus Busy by placing a 1 on that line and removes its bus request. (The Bus Request line continues high because unit 4's request is still pending.) Unit 2 holds the bus until its transfer is complete. It then returns control to the system by returning the Bus Busy line to 0.

One problem with daisy-chain arbitration is that the probability of a unit receiving a Bus Grant signal from the arbiter is dependent on where the unit is located on the bus. The unit closest to the arbiter has the highest priority. If the designer wants to change the priority of a particular unit, he or she must change that module's location on the bus, and that is at best inconvenient and sometimes impossible.

Another problem with the daisy-chain approach is that it may be too slow. Each time a unit passes the Bus Grant signal to the next unit on the line, a time delay is added to the Bus Grant signal. This can reduce the overall transfer rate on the bus.

One solution to both problems is to use several Bus Request and Bus Grant lines. The timing problem is solved because the Bus Grant signal is sent directly to the requesting unit instead of passing from unit to unit. The priority problem can be solved by changing the order in which the Bus Request lines are examined by the arbiter rather than by physically changing the order of the units. (A problem at the end of the chapter asks you to define a "fair" way to scan the Bus Request lines.)

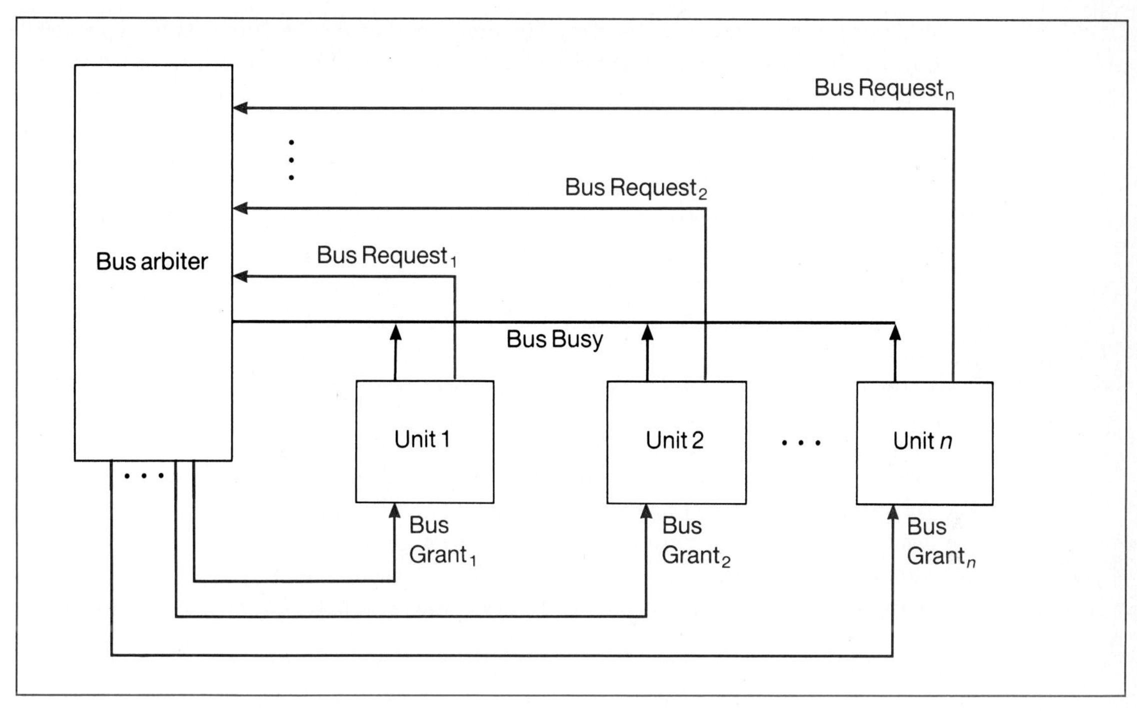

FIGURE 7.8 An independent request/grant line approach to bus arbitration.

The cost of this improvement is an increase in control lines needed in the system bus. This technique is diagramed in Fig. 7.8.

A compromise approach is a technique called **polling**. When a Bus Request is received, the arbiter scans the units in some specified order and "polls" each unit to

FIGURE 7.9 The polling approach to bus arbitration.

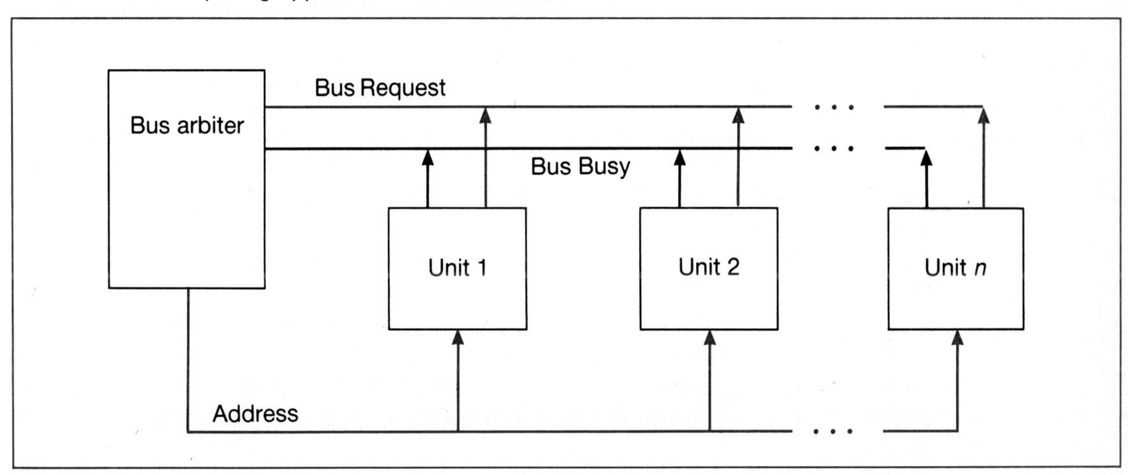

find out if it has requested the bus. The first one that responds positively (by activating the Bus Busy line) is given the bus. Polling calls for the addition of fewer lines than the independent request/grant approach, but it still allows priorities to be changed dynamically. However, since the units must be sequentially polled, there is a service time delay proportional to the number of units that must be polled. A diagram of this approach is shown in Fig. 7.9.

Now that we have found some ways to share buses, we will turn our attention to alternative organizations of the buses.

7.4 ALTERNATIVE STRUCTURES FOR THE SYSTEM BUS

Although we have not stated it explicitly, we have already discussed two different structures for the system bus. In Sec. 7.2 we discussed a bus that has separate lines for data, address, and control information. The lines of these three segments of the system bus are all in parallel with each other. Because of this, we call the organization a **parallel bus**.

In Sec. 7.1 we described a bus that used the same lines to transfer address information and data. When the same lines are used at different times for both *address* and *data*, we say that the bus is an **A/D multiplexed bus**.

The final structure we will examine is an extension of the multiplexed bus. We saw earlier that the parallel bus used one physical line for each signal on the bus. The multiplexed bus we first described used the same physical lines for both address and data. The number of physical lines on the multiplexed bus is reduced because both address and data shared the same lines. The ultimate extension of the sharing concept happens when we reduce the number of signal lines to one. A bus that has only one signal line cannot transmit more than 1 bit at a time. This means that several time periods are needed to transmit an entire address or data element. We call this organization of a bus a **serial bus**.

Figure 7.10 shows how data transfers happen on the three bus organizations. In the remaining parts of this section, we will look at the advantages and disadvantages of each of the bus organizations.

7.4.1 Parallel-Bus Organization

The parallel bus is probably the simplest organization to understand. Each data, address, and control signal has its own private line; there is no sharing. Because of this, the parallel bus requires the most physical space and is also the most expensive. It is expensive not just because of the additional wires needed to build it but because each line in the bus needs circuitry to transmit and receive the signals.

However, the parallel bus is the fastest of the three organizations. As Fig. 7.10*a* shows, the transfer of data, address, and control information requires only one time period. In short, the parallel bus is the fastest, largest, and most expensive of the three organizations.

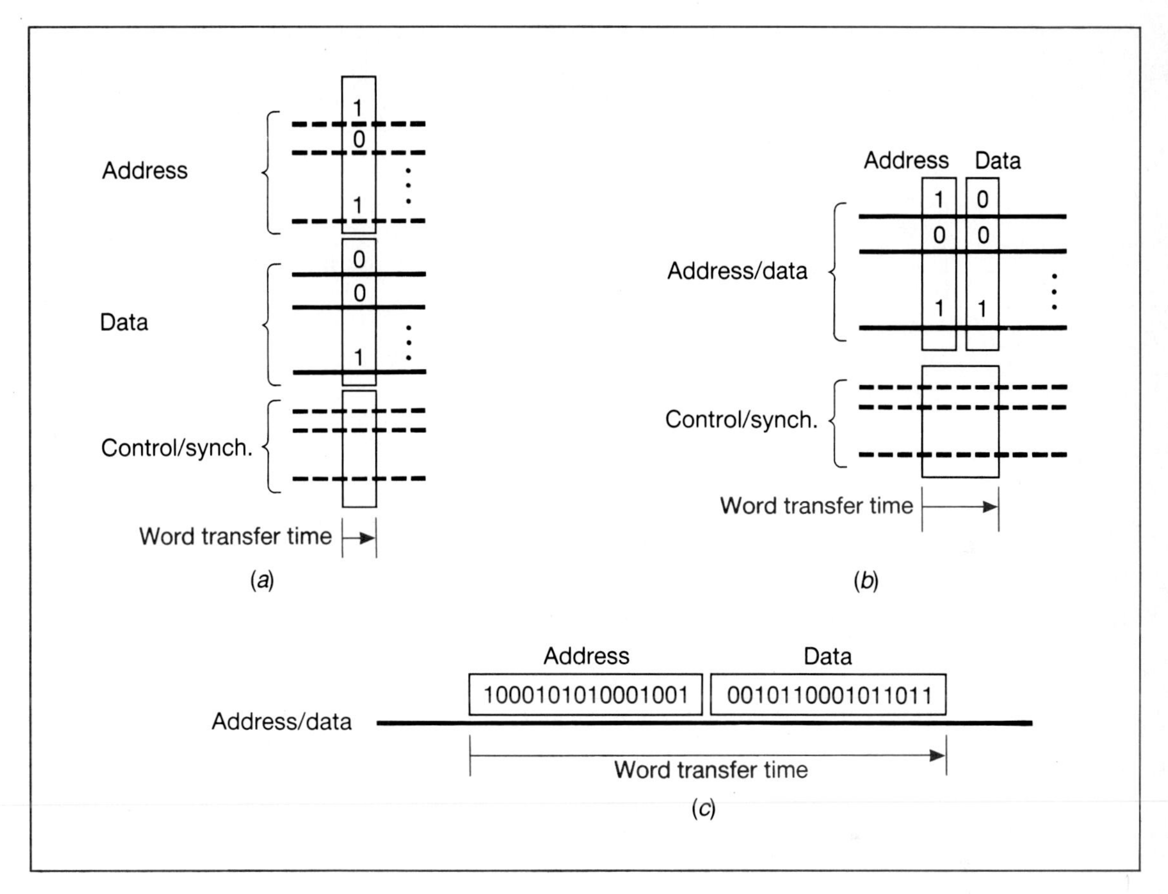

FIGURE 7.10 Alternative bus organizations: (*a*) parallel bus; (*b*) A/D multiplexed bus; (*c*) serial bus.

7.4.2 A/D Multiplexed-Bus Organization

Looking at Fig. 7.10*b*, we can see one obvious disadvantage of an A/D multiplexed bus; more time is needed to complete a transfer. In systems that require very high data transfer rates, this can be a severe limitation.

The advantages of the multiplexed bus are reduced cost of transmitter and receiver circuitry and reduced space needed for the bus lines. One application where the second of these advantages has been important is in microprocessor designs.

As microprocessors grew in complexity and their address and data lines increased, the manufacturers encountered a "pinout" limit. The most common configuration for microprocessors at the time was a 40-pin, **dual in-line package**, or **DIP**. For microprocessors with only 8 data bits and 16 address bits, the 40-pin limit was not a significant problem. However, when processors with 16 data bits and 20 or more address bits were designed, the 40-pin limit presented severe problems.

Motorola chose to solve the problem by changing to a 64-pin DIP for its 68000 microprocessor. Intel decided to stay with the 40-pin DIP for their 8086 microprocessor. This called for the data and address pins on the 8086 to be multiplexed. But, because the Intel 8086 used only 40 pins, it needed less space on a printed circuit board. This was an advantage in some space-sensitive applications. In fairness, we must point out that designs using the Intel microprocessors often required additional integrated circuits that made the overall space savings minimal. Also, the newer microprocessors from both companies now use packages that have far more than 40 or 64 pins.

In summary, we can say that multiplexed buses have the potential to save cost and space when compared to the parallel-bus organization. Their major disadvantage is that they yield lower data transfer rates.

7.4.3 Serial-Bus Organization

The serial bus is the most economical in terms of both circuitry and space. Only one wire and one transmitter and receiver are needed to transmit data, addresses, and control information. (Another wire is also needed for the electrical ground reference.) As Fig. 7.10*c* shows, however, a great deal of time may be needed to complete a transfer.

Because of their slow transfer rates, most computers do not use serial buses to connect the control unit, main memory, ALU, and I/O units. However, because of their low cost, they are often used to connect external or peripheral devices to the computer. In summary we can conclude that serial buses are the slowest and least expensive of the three organizations.

7.5 ERROR DETECTION ON BUSES

Whenever we send information from one place to another, there is the possibility that something will happen to one or more of the data, address, or control bits. Since reliability is becoming more important, many computers are now being designed with error detection logic added to the system bus.

The method most often used is to add parity information to the data, address, and control signals. **Parity** is calculated by counting all the 1s in the signals on the bus and adding one additional bit so that the number of 1s on the bus is odd or even. The "counting" of 1s is easily implemented using XOR circuits. One of the problems at the end of the chapter asks you to design the circuitry that generates parity for a simple bus. In Fig. 7.11 we have reproduced the parallel-bus organization shown in Fig. 7.6 and added a single parity bit in (*a*) and multiple parity bits in (*b*).

By adding a single parity bit with the needed parity generation and checking circuitry, we can detect a change in any single bit. By adding separate parity lines for each of the segments of the bus, we can detect a single error in each of the three

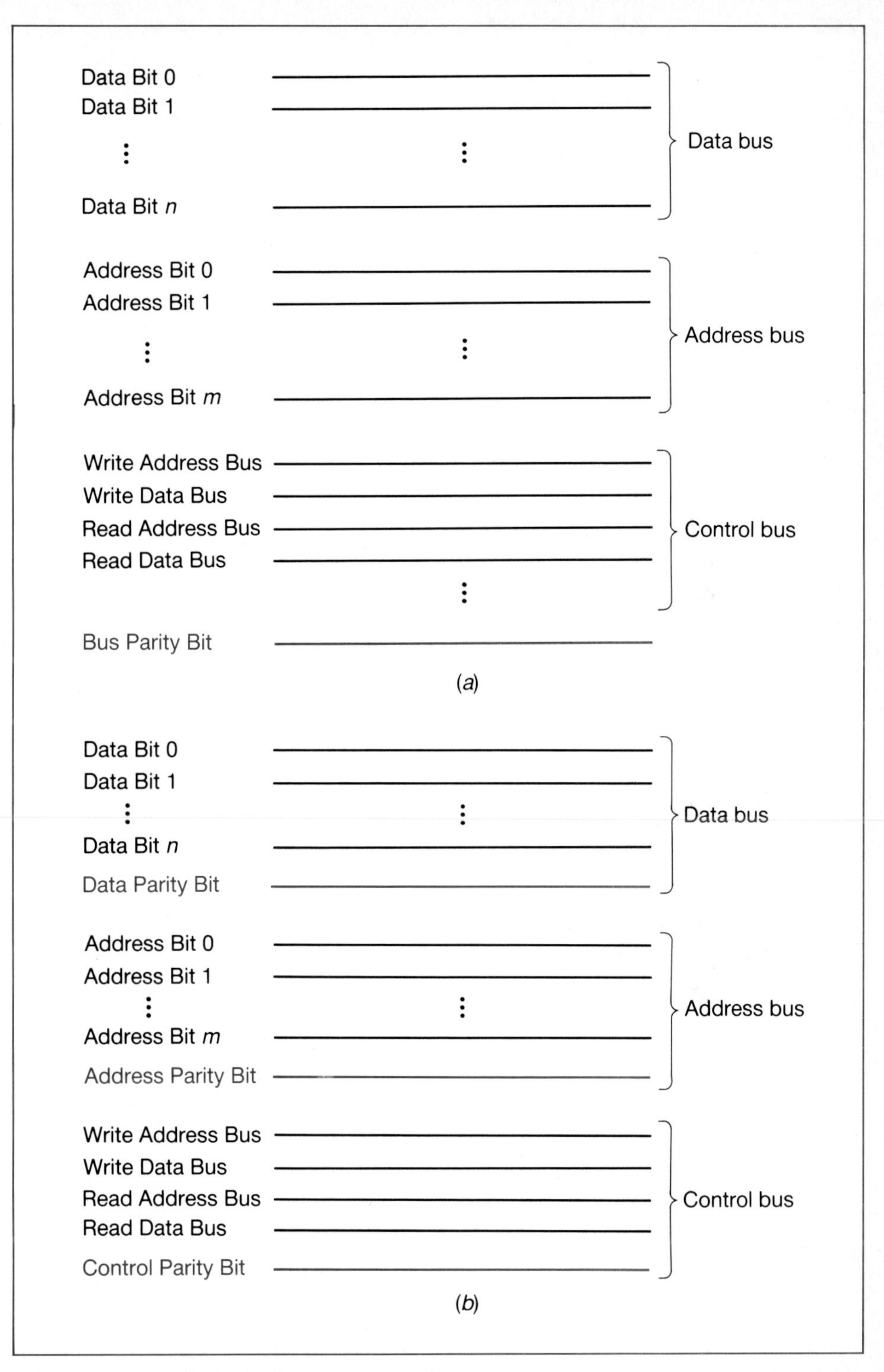

FIGURE 7.11 System buses with parity added: (*a*) 1 bit for entire bus; (*b*) 1 bit each for data, address, and control segments of the bus.

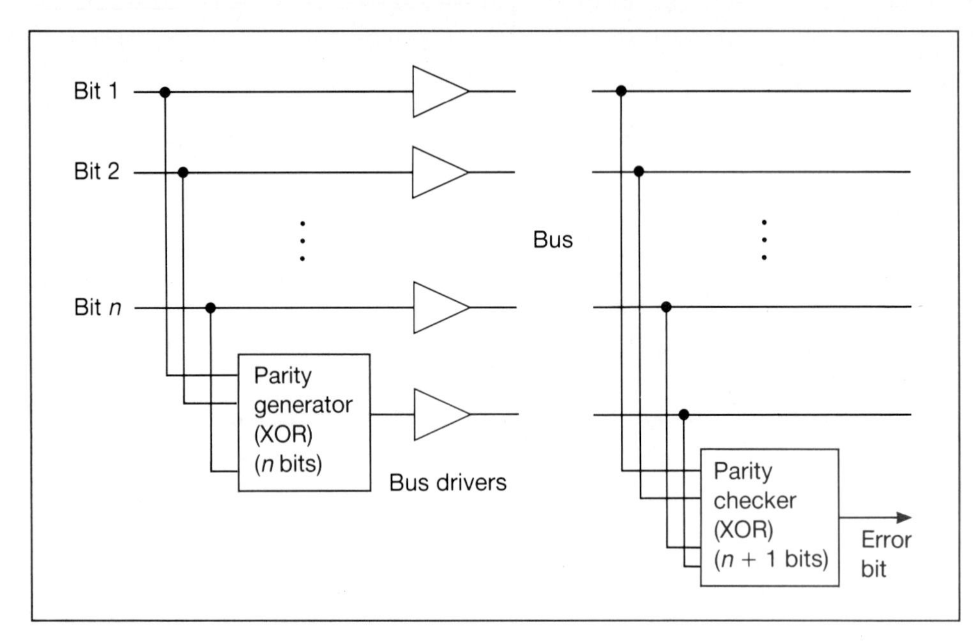

FIGURE 7.12 Circuitry to generate and check parity on a bus.

parts of the bus. A block diagram showing the required circuitry to generate and check parity is shown in Fig. 7.12. Both the parity generator and check are built with XOR circuits. A problem at the end of the chapter asks you to show how timing for bus transfers is changed when parity is generated and checked with each bus transfer.

It is also possible to add a parity line to a serial bus, but here the incremental cost is quite high. To add parity to a serial bus requires that we double the lines (from one to two) and double the transmitter and receiver circuits. Because of the significant cost increases, parity is seldom used on serial buses. This does not mean that no checking is done on serial buses, however.

The technique most often used with serial buses is to add extra bits that can be used to check the parity of the data bits being transmitted. Adding parity increases the length of time needed to transmit data on the bus. Figure 7.13 shows the trans-

FIGURE 7.13 Parity bits on a serial bus.

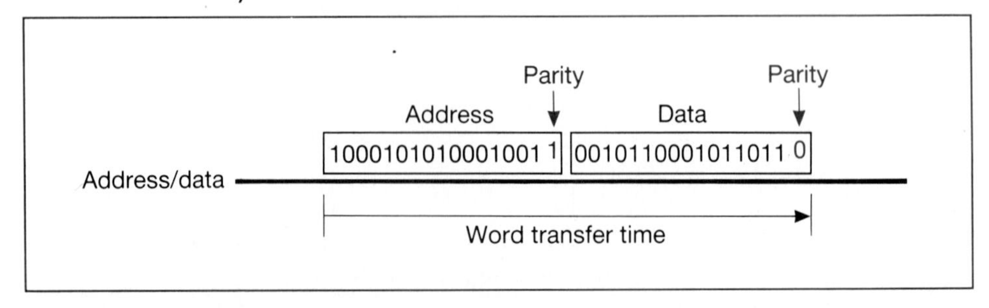

mission from Fig. 7.10*c* with parity bits added to both the address and data "words."

Parity can be a very effective tool for detecting errors in transmission over buses and in other areas of the computer, but there are costs associated with adding parity:

- One or more additional transmission lines are needed.
- Circuitry is needed to generate and check parity.
- A single parity bit cannot be used to detect two errors.
- Generating and checking parity takes time and therefore reduces performance.

Despite these disadvantages, more and more manufacturers are using parity to check data transfers within their computers.

7.6 SUMMARY

By this point, you should have a good understanding of the concepts underlying the design of computer buses. We have seen that the bus design establishes electrical characteristics and the protocol used to send and receive information on the bus. Because the bus affects all the units within the computer, the type of bus to be used must be one of the earliest decisions in the design process.

The bus design impacts the following:

- Address space of main memory
- The size of memory words
- The speed of the processor
- The number of units or node that can be attached to the system

The key for a successful bus design is to recognize the potential impacts and to provide enough capacity to handle the combined requirements of all the connected modules.

PROBLEMS

7.1 Where and how are the signals in Fig. 7.1 (such as MR, OP, and PL) generated?

7.2 Why are there separate Read Bus and Write Bus signals in Fig. 7.2?

7.3 Why do we have simple AND gates providing inputs to registers in Fig. 7.1 rather than tristate gates?

7.4 Draw a timing diagram like the one in Fig. 7.5 that shows the sequence of signals needed to read the next instruction from memory.

7.5 What advantages does the simple bus shown in Fig. 7.1 have over the one in Fig. 7.4?

7.6 Why is the data bus usually referred to as a bidirectional bus?

7.7 Assume a 16-bit microprocessor system uses a bus like the one in Fig. 7.6. How many lines would you expect to find in the bus?

7.8 In Sec. 7.3.1 we described a protocol that could be used to send messages on the phone. What is another noncomputer application that requires a protocol to control interactions? Describe the protocol.

7.9 On a particular 16-bit computer, the data portion of the bus has 18 lines and the address portion has 17 lines. For what purpose are the extra lines used?

7.10 What is the purpose of a protocol?

7.11 It is not always easy to define a protocol that guarantees successful interaction. List some possible problems that would prevent the phone protocol from delivering a correct message.

7.12 How would you modify the phone protocol if you had two or more phones in the house connected to the same line?

7.13 How would you modify the system in Fig. 7.7*a* if you did not want to use a wired OR?

7.14 Describe how a phone arbiter could work in a college dorm where two phones were available for 100 possible users.

7.15 Define a "fair" way to scan the Bus Request lines in Fig. 7.8.

7.16 Figures 7.7 to 7.9 show three different approaches to bus arbitration. Evaluate and discuss the relative advantages and disadvantages of each.

7.17 How would the Bus Busy signal in Fig. 7.9 be generated within each unit?

7.18 List some application where the slow speed of a serial bus would not be a significant disadvantage.

7.19 Give two noncomputer examples where multiplexing is used.

7.20 What type of bus organization would you expect to find on a Cray supercomputer? Why?

7.21 What is the major advantage of the serial bus?

7.22 The diagram in Fig. 7.12 contains a block defined as a parity generator. Develop a logic diagram of this unit for a 16-bit bus segment.

7.23 Does adding parity generation and check circuits have any effect on bus transfer rates for parallel buses? Explain your answer.

7.24 Section 7.6 lists four factors that are impacted by bus design. Discuss how and why the bus design affects each of these.

7.25 How much does parity reduce performance on a serial bus for (*a*) 8-bit and (*b*) 16-bit transmission?

7.26 How is the timing of bus transfer signals affected by the addition of parity generation and checking circuits to a parallel bus?

7.27 What happens when a parity error is detected after a transmission?

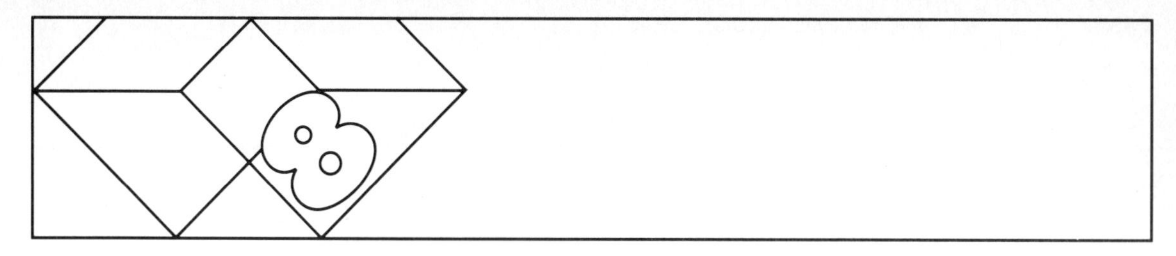

The Arithmetic/Logic Unit

As we said earlier, the arithmetic/logic unit, or ALU, is the part of the computer that executes arithmetic and logic operations. Arithmetic instructions typically reference registers in the ALU and include operations such as those shown in Table 8.1. (An earlier version of these instructions was previously detailed in Chap. 5 on the control unit.)

In our computer, the ALU also "takes part in" conditional jump instructions. Before a conditional branch instruction is executed, the programmer must use the compare instruction to set the status bits in the ALU. A different approach is to incorporate the compare directly into the jump instruction. A typical set of conditions is shown in Table 8.2.

Lastly, the ALU is used for some data move operations, which transfer data from one location to another within the computer. These instructions vary from computer to computer, depending on whether the computer is a memory-memory, memory-register, or register-register computer. They are included here because for our computer they affect the accumulator, which we include as part of the ALU. Table 8.3 shows three instructions typical to memory-register computers. As before, the operand M means memory location M.

Most computers have more instructions in each of these groups than are defined in the tables above. We have included only a sufficient number to give you a flavor of the types of operations in each group. Commercially available computers are also designed to process several different types of numbers (such as integer and floating-point). The ALU must be able to perform different functions, depending on both the specified operation and the type of operands that will be involved.

We will not fully discuss all the various data types in this chapter or the ways the corresponding operations could be implemented in our ALU. For simplicity and clarity, our design examples will assume unsigned binary integers with a limited set of operators.

TABLE 8.1 TYPICAL ARITHMETIC/LOGIC GROUP OPERATIONS

Operation	Typical mnemonic	Instruction meaning
Add	ADD M	Add the contents of M to the accumulator
Subtract	SUB M	Subtract the contents of M from the accumulator
Multiply	MUL M	Multiply the accumulator by the contents of M
Divide	DIV M	Divide the accumulator by the contents of M
AND	AND M	Logically AND the accumulator with the contents of M
OR	OR M	Logically OR the accumulator with the contents of M
XOR	XOR M	Exclusively OR the accumulator with the contents of M
NOT	NOT	Logically invert the accumulator
Shift left	SHL	Shift the accumulator 1 bit to the left, discard the leftmost bit, and insert a 0 on the right
Shift right	SHR	Shift the accumulator 1 bit to the right, discard the rightmost bit, and insert a 0 on the left
Rotate left	RRL	Rotate (shift) the accumulator 1 bit to the left, move the leftmost bit to the rightmost bit position
Rotate right	RRR	Rotate the accumulator 1 bit to the right, move the rightmost bit to the leftmost bit position
Arithmetic shift right	ASR	Shift the accumulator 1 bit to the right, discard the rightmost bit, and duplicate the sign bit in the high-order position (Arithmetic Shift Right)

TABLE 8.2 TYPICAL COMPARE AND CONDITIONAL JUMP INSTRUCTIONS

Operation	Typical mnemonic	Instruction meaning
Compare	CMP M	Compare the value in M to the accumulator; set the EQ bit to 1 if the values are equal, set the GT bit to 1 if the value in M is greater than the accumulator
Jump	JXX L	Execute the instruction at L if the compare operation indicated that the value of M, relative to the accumulator, was: $>$ JGT Jump Greater Than (GT on) $\geq$ JGE Jump Greater Than or Equal (GT or EQ on) $=$ JEQ Jump Equal (EQ on) $\leq$ JLE Jump Less Than or Equal (GT off) $<$ JLT Jump Less Than (GT and EQ off) $\neq$ JNE Jump Not Equal (EQ off) Else, execute the next sequential instruction

TABLE 8.3 TYPICAL DATA MOVE INSTRUCTIONS AFFECTING THE ALU

Operation	Typical mnemonic	Instruction meaning
Load	LDA M	Load the value in M into the accumulator
Swap	SWP M	Exchange the value in M and the value in the accumulator
Clear	CLA	Load 0 into the accumulator

We will first consider the general structure of an ALU. Then we will design logic to implement two operations: addition and comparison. Next, we will look at the accumulator and its associated shifting operations. After that, we will discuss logical operations and integer multiplication. We will finish the chapter with a general discussion of the floating-point operations and the IEEE floating-point standard.

8.1 STRUCTURE OF THE ARITHMETIC/LOGIC UNIT

Before discussing the ALU, we need to examine the structure of arithmetic and logic operations and then state some definitions. We know that the arithmetic operations $+$, $\cdot$, and $\div$ all require two operands. This is also true of the logical $+$ and $\cdot$. The logical "not" requires only one operand, and the arithmetic $-$ can require either one or two depending on whether it signifies negate or subtract.

Arithmetic and logical operations that require only one operand are known as **monadic** operations. Operations that require two operands are called **dyadic** operations. These operations have also been called **unary** (one-operand) and **binary** (two-operand) operations. We will not use the latter terms because of the possible confusion with our other use of "binary" for information representation.

With that in mind, let's look at the structure of the arithmetic unit. A diagram of a simple ALU is shown in Fig. 8.1*a*. It is connected to the system bus via two sets of lines, as shown in Fig. 8.1*b*. The right-hand set of lines is strictly for input into the ALU and usually provides one of the operands for arithmetic or logic operations. The left-hand set of lines is for both input and output and connects the accumulator (ACC) to the bus. The accumulator:

- Contains the second operand in dyadic operations
- Is usually used as the sole operand for monadic operations
- Normally accepts the result of all operations done by the functional unit

It is certainly possible to build an ALU without an accumulator. Temporary storage could be furnished by other registers in the CPU, separate from the ALU. However, there are many computer designs that incorporate accumulators as part of the ALU, and we will use this approach. Also, several operations we described earlier (such as shifts) are normally incorporated into the accumulator itself instead

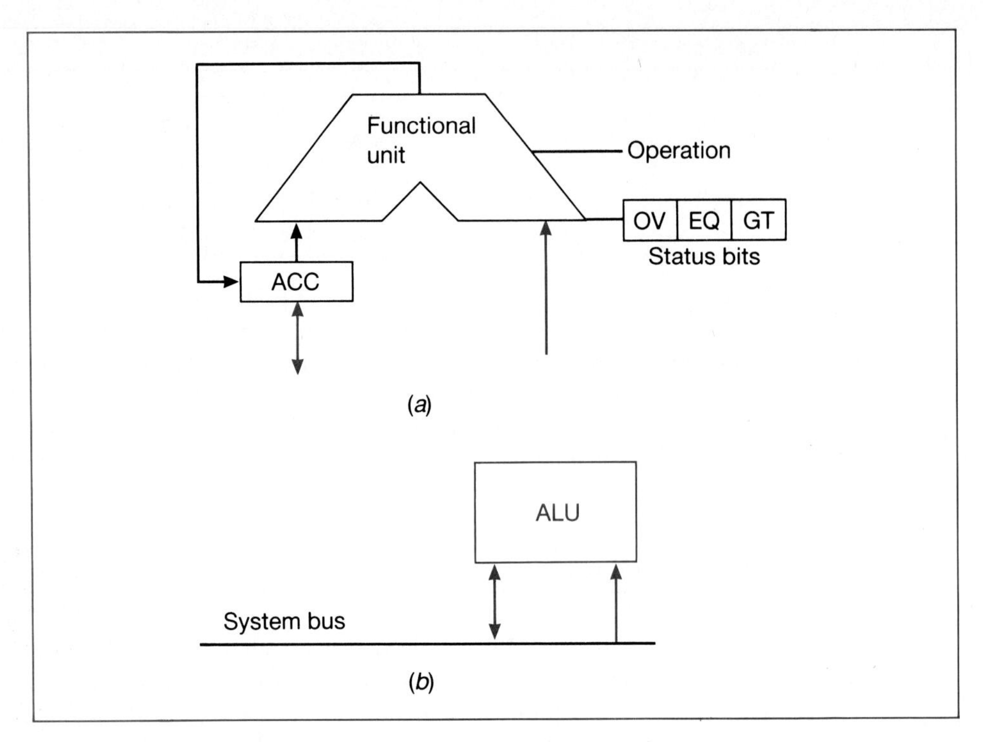

FIGURE 8.1 ALU: (*a*) structure; (*b*) connection to the system bus.

of into the functional unit of the ALU. This reinforces the suggestion that the accumulator be considered an actual part of the ALU.

For our design, then, we need to consider:

- The functional unit of the ALU
- The accumulator
- The ways in which these two units interact

These three aspects will be examined in the next five sections.

8.2 INTEGER ADDITION

We begin with a discussion of how addition hardware in the ALU is built. Instead of using a formal design procedure, let's intuit how this circuit should work. If we were to add 2 bits, say X_i and Y_i, then the sum bit would be 1 if either of the bits, but not both, were 1. Rephrased, this says the sum bit would be 1 if X_i is 1 and Y_i is 0 or if X_i is 0 and Y_i is 1. Therefore, our output should be 1 if either $X_i Y_i' = 1$ or $X_i' Y_i = 1$; in boolean expression form,

$$\text{Sum}_i = X_i Y_i' + X_i' Y_i = X_i \oplus Y_i$$

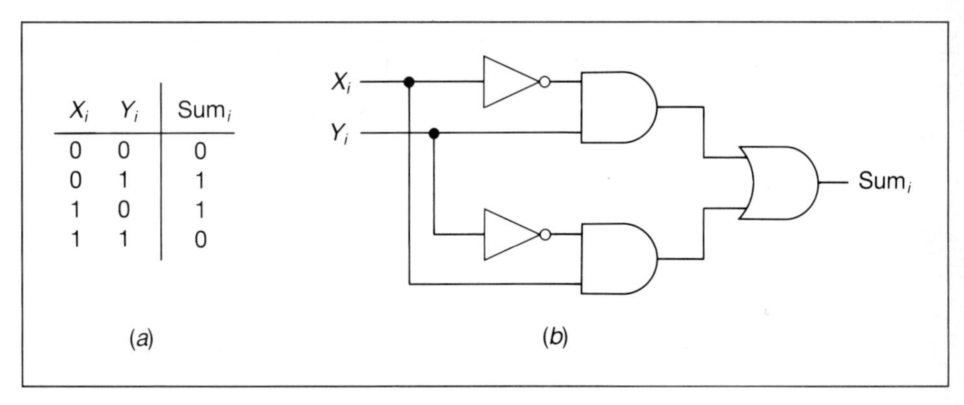

X_i	Y_i	Sum_i
0	0	0
0	1	1
1	0	1
1	1	0

FIGURE 8.2 The addition of 2 bits: (*a*) truth table; (*b*) circuit.

The truth table for this function is shown in Fig. 8.2*a*. One implementation of the circuit (which is called an **exclusive OR**, or **XOR**) is shown in Fig. 8.2*b*. The logic symbol for the XOR operation is ⊕.

This takes care of the sum bit, but there is another important aspect of addition—the carry. We can get a carry from the addition of 2 bits only if both bits to be added are 1, that is, if both $X_i = 1$ and $Y_i = 1$. Therefore, the expression is

$$\text{Carry}_i = X_i\, Y_i$$

Such a circuit is shown in Fig. 8.3.

The two circuits we've just discussed can be combined into a single "box" whose inputs are the 2 bits to be added and whose outputs are the resulting sum bit and carry bit. Such a box is called a **half-adder** and is shown in Fig. 8.4.

FIGURE 8.3 Circuit for carry bit resulting from the addition of 2 bits.

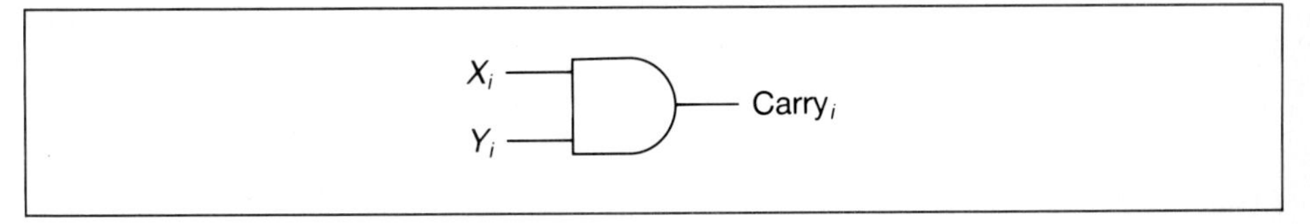

FIGURE 8.4 Half-adder functional diagram.

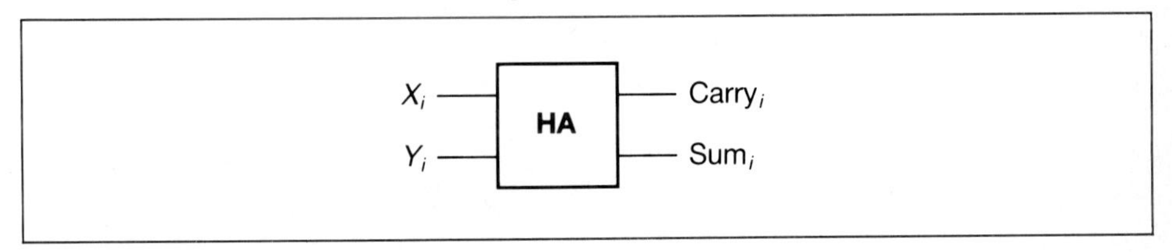

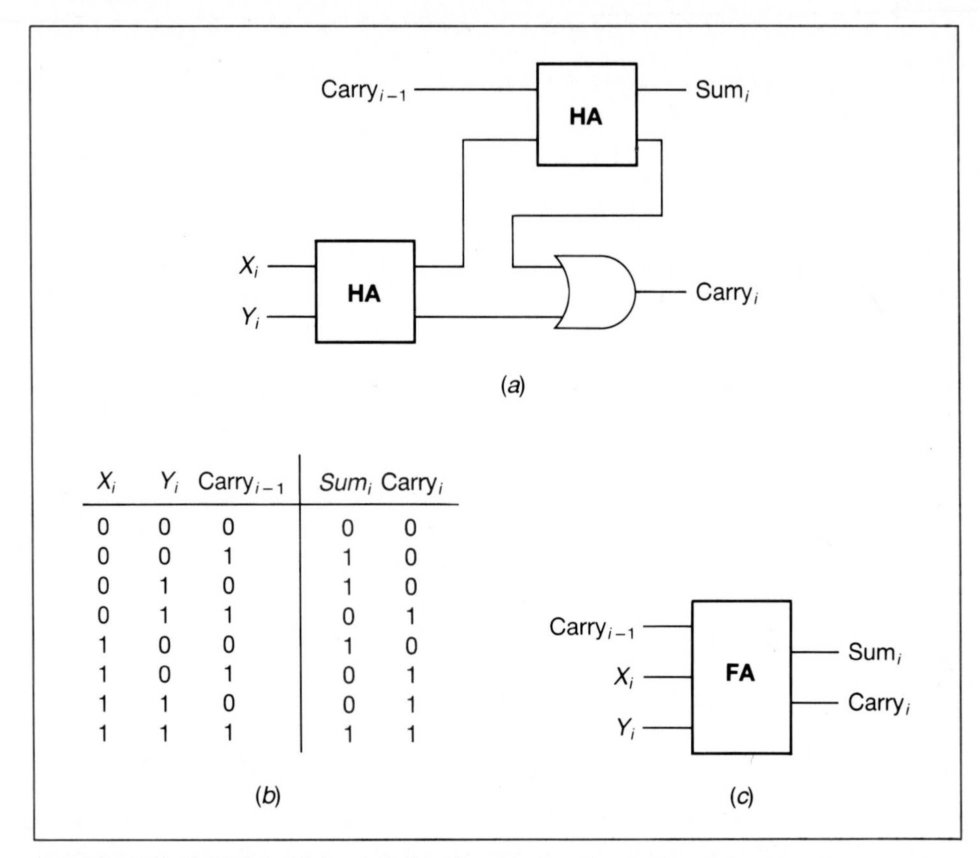

X_i	Y_i	Carry$_{i-1}$	Sum$_i$	Carry$_i$
0	0	0	0	0
0	0	1	1	0
0	1	0	1	0
0	1	1	0	1
1	0	0	1	0
1	0	1	0	1
1	1	0	0	1
1	1	1	1	1

(b)

FIGURE 8.5 Full-adder: (*a*) two half-adders combined to make a full-adder; (*b*) truth table; (*c*) symbolic notation.

The reason the term *half-adder* is used is that we normally are not just adding 2 bits—we must also include a carry bit. Suppose that we are adding two binary numbers. The least significant bit of the sum could be determined by a half-adder, but each of the other higher-order bits would need to be determined by a circuit

FIGURE 8.6 Four-bit adder.

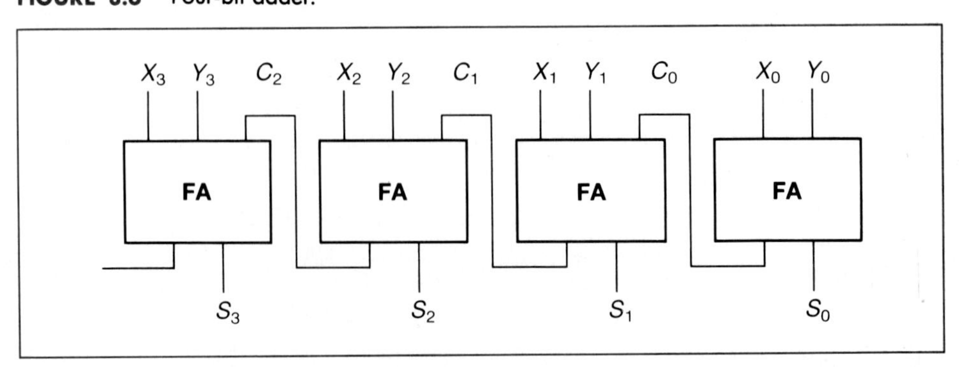

with three inputs: the 2 bits from the numbers themselves and the carry from the previous bit positions.

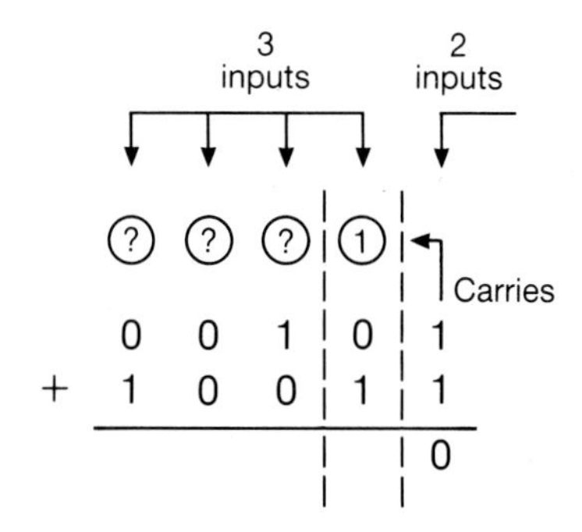

One way to create a three-input adder is to combine two half-adders, as shown in Fig. 8.5*a*. This circuit is called a **full-adder**; its truth table is shown in Fig. 8.5*b* and its notational symbol in Fig. 8.5*c*.

It should be pointed out that the use of two half-adders is not necessarily the fastest way to build a full-adder; however, we think it is the easiest to see and understand. Essentially, we add the 2 bits, then add in the previous carry. This gives us the resulting sum bit. The carry bit will be 1 if we got a carry from adding the 2 bits themselves or from adding their sum to the previous carry.

We now have the tools to build our complete addition circuit. We simply chain together a sequence of full-adders. A 4-bit version of such a circuit is shown in Fig. 8.6.

This completes our discussion of addition circuitry. We turn now to comparison, determining how two unsigned integers are related to each other.

8.3 COMPARISON

One of the operations of the functional unit of the ALU is to set the comparison bits in the status register. In this section, we will see how that can be done.

Testing for equality of 2 bits is straightforward. If they are both 1 or both 0, then the output should be 1. The truth table for this is shown in Fig. 8.7*a* and a circuit to implement it in Fig. 8.7*b*. A symbolic representation for the circuit is shown in Fig. 8.7*c*.

In order for two 4-bit numbers to be equal, all the bits must be equal. A circuit to do this comparison is shown in Fig. 8.8.

Testing for X_i less than Y_i is just as easy as equality. Since we are dealing with binary digits, the only way for X_i to be less than Y_i is if $X_i = 0$ and $Y_i = 1$. This is shown in Fig. 8.9.

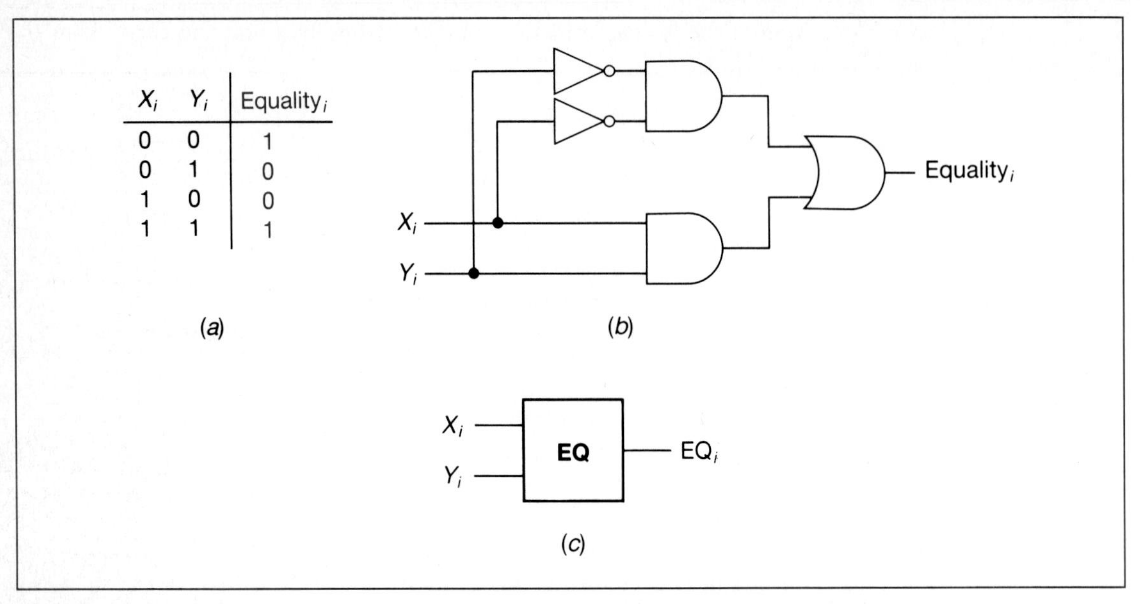

FIGURE 8.7 Checking for the equality of 2 bits: (*a*) truth table; (*b*) corresponding circuit; (*c*) symbolic representation of equality checker.

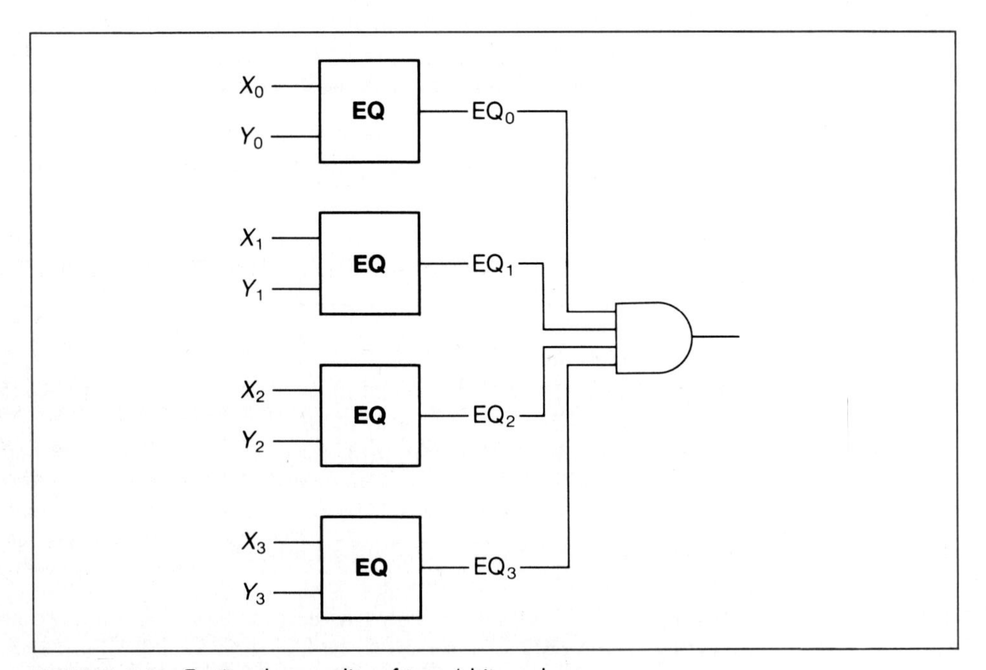

FIGURE 8.8 Testing the equality of two 4-bit numbers.

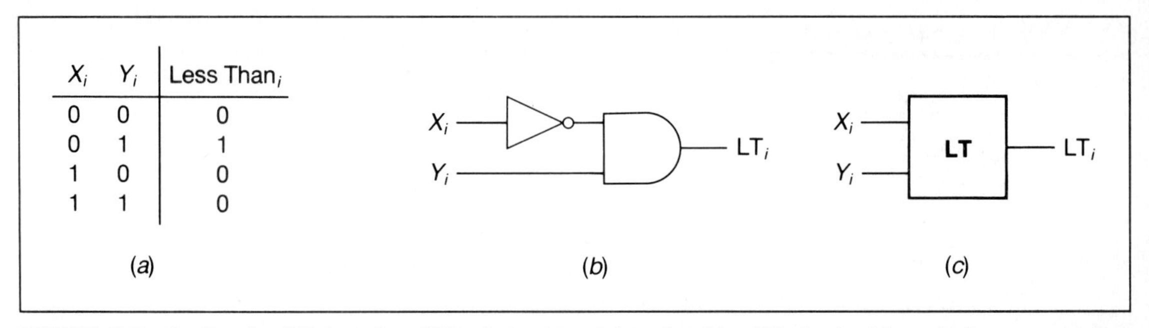

X_i	Y_i	Less Than_i
0	0	0
0	1	1
1	0	0
1	1	0

FIGURE 8.9 Testing for "X_i less than Y_i" relationship: (*a*) truth table; (*b*) circuit; (*c*) symbolic representation.

Testing for "less than" on two positive 4-bit integers is a little more challenging than "equality." We need to start at the high-order bit, then "go down" the bits until we find an X_i less than Y_i. Therefore, X will be less than Y just when:

- X_3 is less than Y_3.
- Or X_3 equals Y_3 and X_2 is less than Y_2.
- Or X_3 equals Y_3, X_2 equals Y_2, and X_1 is less than Y_1.
- Or X_3 equals Y_3, X_2 equals Y_2, X_1 equals Y_1, and X_0 is less than Y_0.

A circuit to implement these conditions is shown in Fig. 8.10.

FIGURE 8.10 Less-than testing on two positive 4-bit integers.

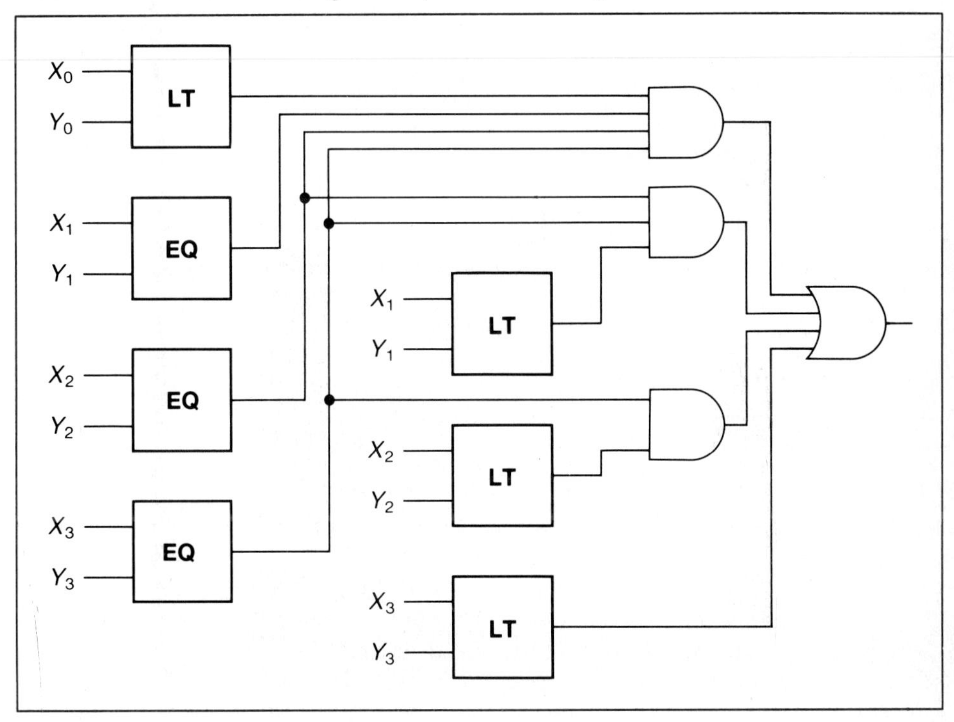

8.4 THE ACCUMULATOR

We must stress that the functional unit of the ALU is a *combinational* circuit—the outputs are produced only while the inputs are applied. To provide temporary storage for operands and results, ALUs usually have data storage registers called accumulators. Besides high-speed temporary storage facilities, these accumulators might also implement some machine operations, such as shifting.

The sample ALU described earlier has one accumulator, which we will now build using the *RS* flip-flop we introduced in Chap. 2. We will not go into great detail for this design, since the essentials were covered in Chap. 6 (the 1-bit storage element) and Chap. 7 (the interface register).

In later designs, we will want an *RS* flip-flop in which the output value is the same as the value that is on the data input line when the clock goes to 1. A possible circuit is shown in Fig. 8.11.

This type of storage element has some uses (for example, in the STR register that we will use in multiplication). Often, however, we do not want to accept a new input value on every clock pulse—we want to do it selectively. Therefore, we will add a Select control line to our basic circuit. This will allow the data on the input line to be accepted only when the control line is a 1. The resulting circuit and its symbol are shown in Fig. 8.12*a* and *b*.

8.4.1 Changing the Value of the Accumulator

Now that we have an accumulator element, we can look at operations on the accumulator. The first two such operations you normally think about are Load and Clear. Load sets the values of the accumulator bits to the values on some specified input line, which in our example could come from the functional unit of the ALU or from the system bus. Clear sets the accumulator to all 0s. (These instructions were introduced in Table 8.3; their opcodes are LDA and CLA.)

FIGURE 8.11 One-bit data storage element.

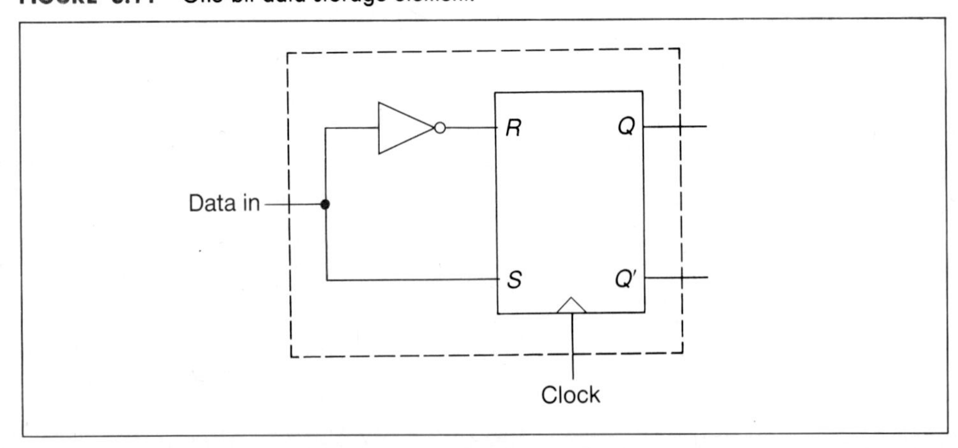

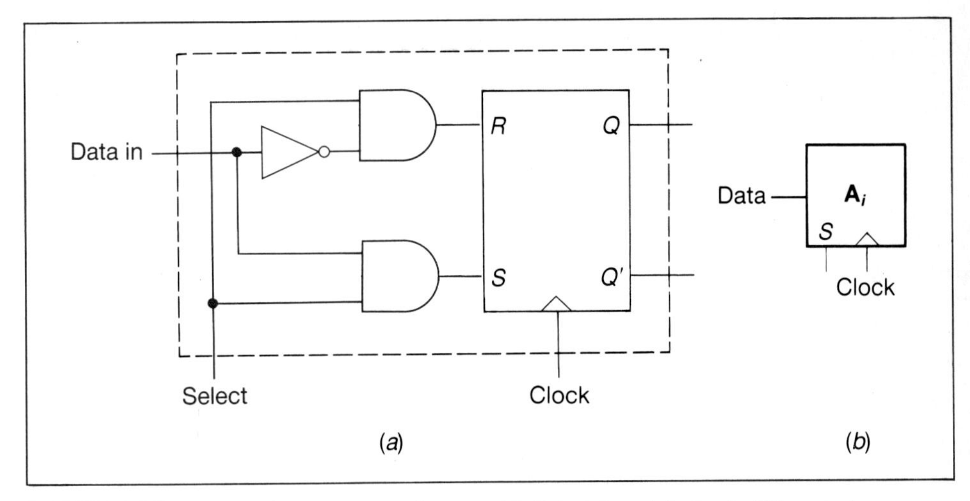

FIGURE 8.12 Accumulator storage element with Select line: (*a*) circuit; (*b*) symbol.

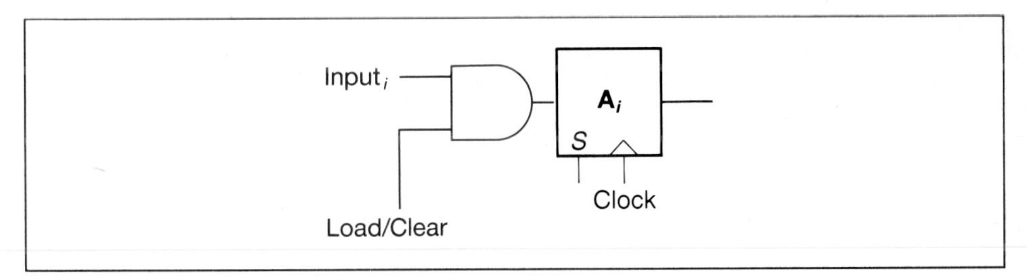

FIGURE 8.13 Logic for Load and Clear operations.

To implement these operations, we recall the use of AND gates as selectors in Chap. 7. We have a control line and an input line going into an AND gate. If the control line is 1, then the output of the AND gate (which is the input to the accumulator storage element) will be the same as the value on the input line; this gives us the "load." If the control line is 0, then the output of the AND gate will be 0; this gives us the "clear." A circuit diagram for this is shown in Fig. 8.13.

We should point out again that the output of the AND gate will be accepted by the accumulator element only if the *S* line is a 1. That *S* line would be activated by the computer's control unit after it has decoded an instruction like LDA or CLA.

On several early systems the Load operation was called Clear and Add. The reason is that the only input connection to the accumulator was from the output of the ALU's functional unit. To load a single value, say Y, the accumulator was first cleared. Then it was set to the sum of its current value (now 0) and Y, which obviously is Y.

ACC ← 0 (clear)

ACC ← ACC + Y (and add)

TABLE 8.4 BIT PATTERNS AND ACTIONS IN FIRST SHIFTER CIRCUIT

Load/Clear	Direct/Shifted	Action
0	Ignored	A_i replaced by 0
1	0	A_i replaced by Input_{i-1}
1	1	A_i replaced by Input_i

Also, some early systems used a different form of clear. They subtracted the value in the accumulator from itself.

8.4.2 Shifting

Besides loading and clearing the accumulator, we frequently want to shift its contents. For example, multiplication can be implemented using repeated shifts and adds. Also, it is sometimes necessary to extract parts of the value in the accumulator (e.g., the first character), and this can be implemented easily with shifting and a mask.

As a first sample implementation, we might replace the Load input to our accumulator bit by a circuit that picks either the directly corresponding bit of the input or the bit immediately to the right. In the latter case, we would be shifting left one position—input bit_{i-1} goes to accumulator bit_i. This circuit is diagramed in Fig. 8.14 and specified in Table 8.4.

To make our accumulator more versatile, we probably want to be able to shift both directions. We can do this by adding a third circuit to replace the Input_{i-1} line; this circuit will allow us to pick either Input_{i-1} or Input_{i+1} and therefore shift either left or right. This circuit is shown in Fig. 8.15 and specified in Table 8.5.

FIGURE 8.14 First version of shifting.

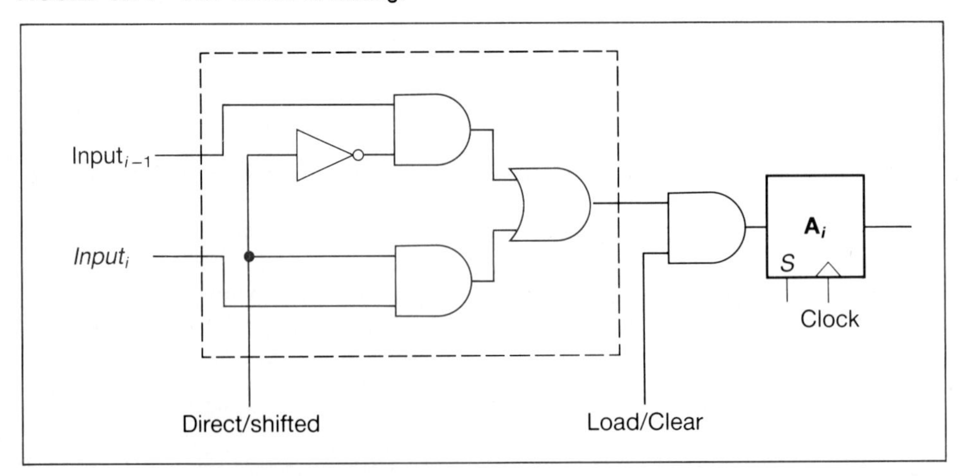

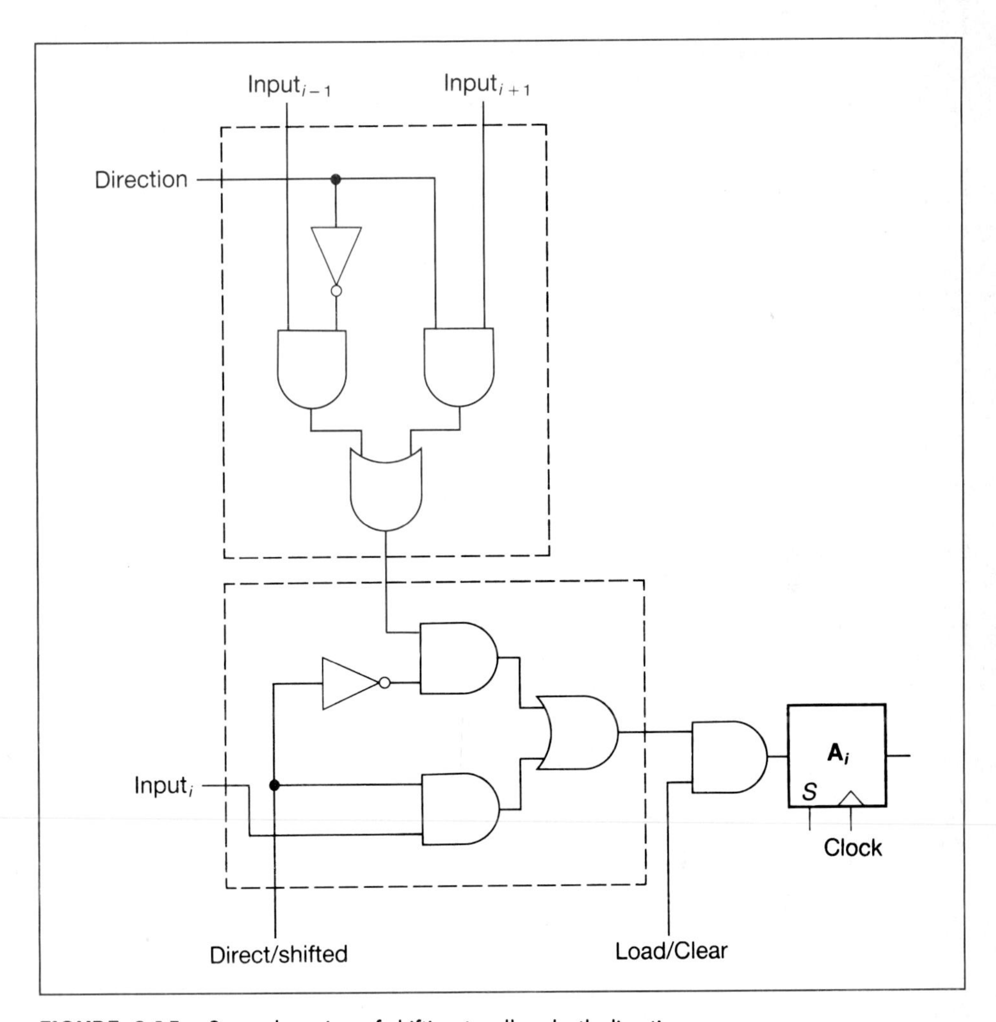

FIGURE 8.15 Second version of shifting to allow both directions.

In the above example, there are 3 bits that specify one of four operations. Each bit corresponds directly to some action. This is called a "horizontal" encoding to control; you might think of it as each control bit aligned with and corresponding to a certain action. It is a viable approach and is used on some systems. The criticism

TABLE 8.5 BIT PATTERNS AND ACTIONS IN SECOND SHIFTER

Load/Clear	Direct/Shifted	Direction	Action
0	Ignored	Ignored	A_i replaced by 0
1	1	Ignored	A_i replaced by $Input_i$
1	0	0	A_i replaced by $Input_{i-1}$
1	0	1	A_i replaced by $Input_{i+1}$

TABLE 8.6 BIT PATTERNS AND ACTIONS IN THIRD SHIFTER CIRCUIT

a	*b*	Meaning	Action
0	0	Clear	A_i replaced by 0
0	1	Right Shift	A_i replaced by Input_{i+1}
1	0	Left Shift	A_i replaced by Input_{i-1}
1	1	Load	A_i replaced by Input_i

of this approach is that we could specify one of four operations using only 2 bits; the other bit might in some senses be considered as "wasted." If you look in Table 8.5, you can see the "waste" as configurations that are "ignored."

A different version of our accumulator circuit, using only two control bits, is shown in Fig. 8.16. In this version, neither of the two control bits by themselves really have any meaning—they must be "decoded" as a pair to decide what is intended. The encoding for the shift circuitry is shown in Table 8.6.

There are two primary types of shift configurations, one that shifts the bit values out of the register and another that shifts the bit values out of the register and back in the other end. Both these configurations are shown in Fig. 8.17.

Shift operations that move the bits out of the register and back in the other end are called **rotate**, or **end-around, shifts**. For a left shift, we place the high-order bit into the low-order position; for a right shift, the low-order bit is placed in the high-order position. All original bit values are still retained in the register and could be restored to their original locations if the operation(s) were reversed.

Shift operations that move the bits out of the end of the register are different in two respects:

- Unless the bits are shifted into another register, they are lost forever.

FIGURE 8.16 More concise logic for load, clear, and left and right shift.

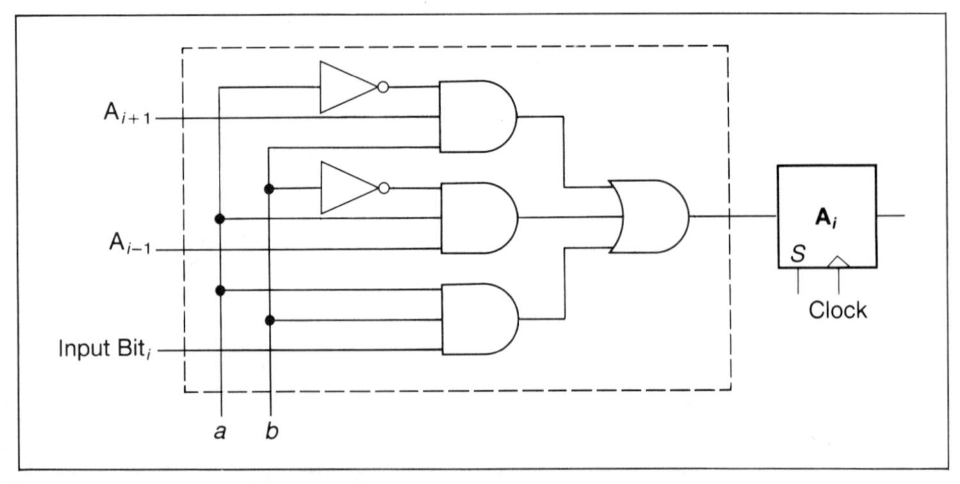

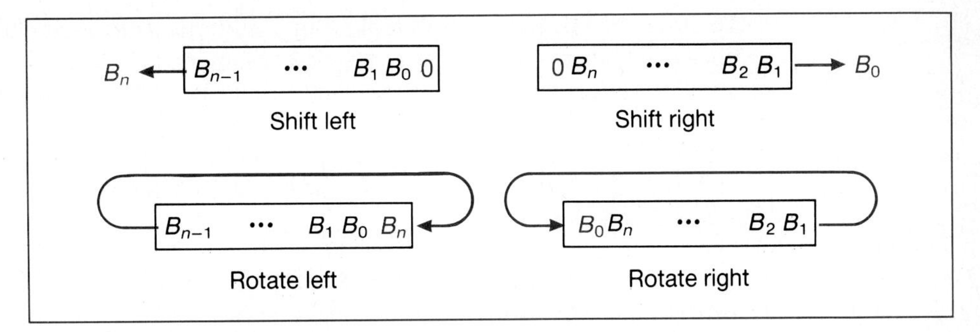

FIGURE 8.17 Graphical descriptions of different shift and rotate operations.

- Additional bits must be shifted into the opposite end of the register to fill the bit positions vacated by the shift operation.

The bit that is inserted into the low-order position in a left shift or into the high-order position in a right shift is usually a 0.

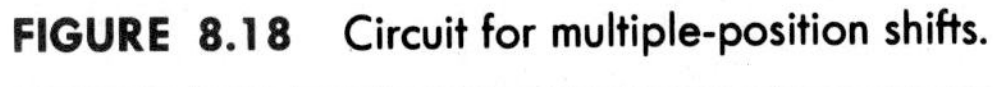

FIGURE 8.18 Circuit for multiple-position shifts.

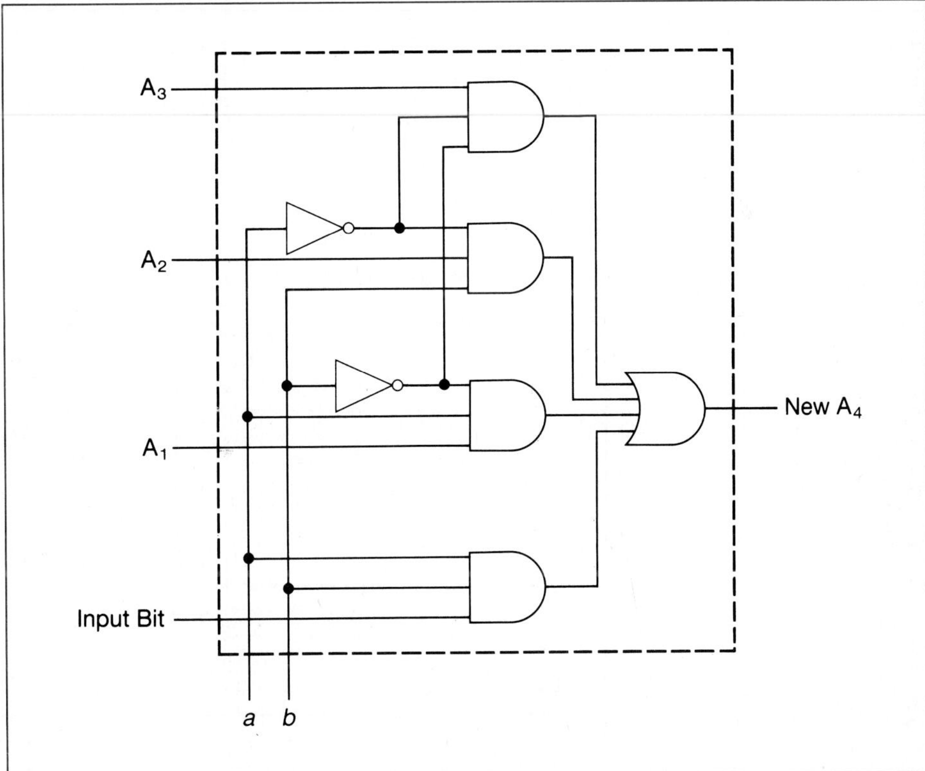

To close out this section, let's look at one enhancement to our shifters. If we use shifting a lot, then it might be too time-consuming to repeatedly shift only one position per instruction. A multiple-shift circuit, allowing left shifting up to four positions per instruction, is shown in Fig. 8.18. We could specify shift options in our instruction by having the unused address field indicate the number of bits to shift.

To further illustrate this circuit, let's use it in a 4-bit configuration for left shifting. This circuit is shown in Fig. 8.19.

FIGURE 8.19 Register with multiple-position-shift capability.

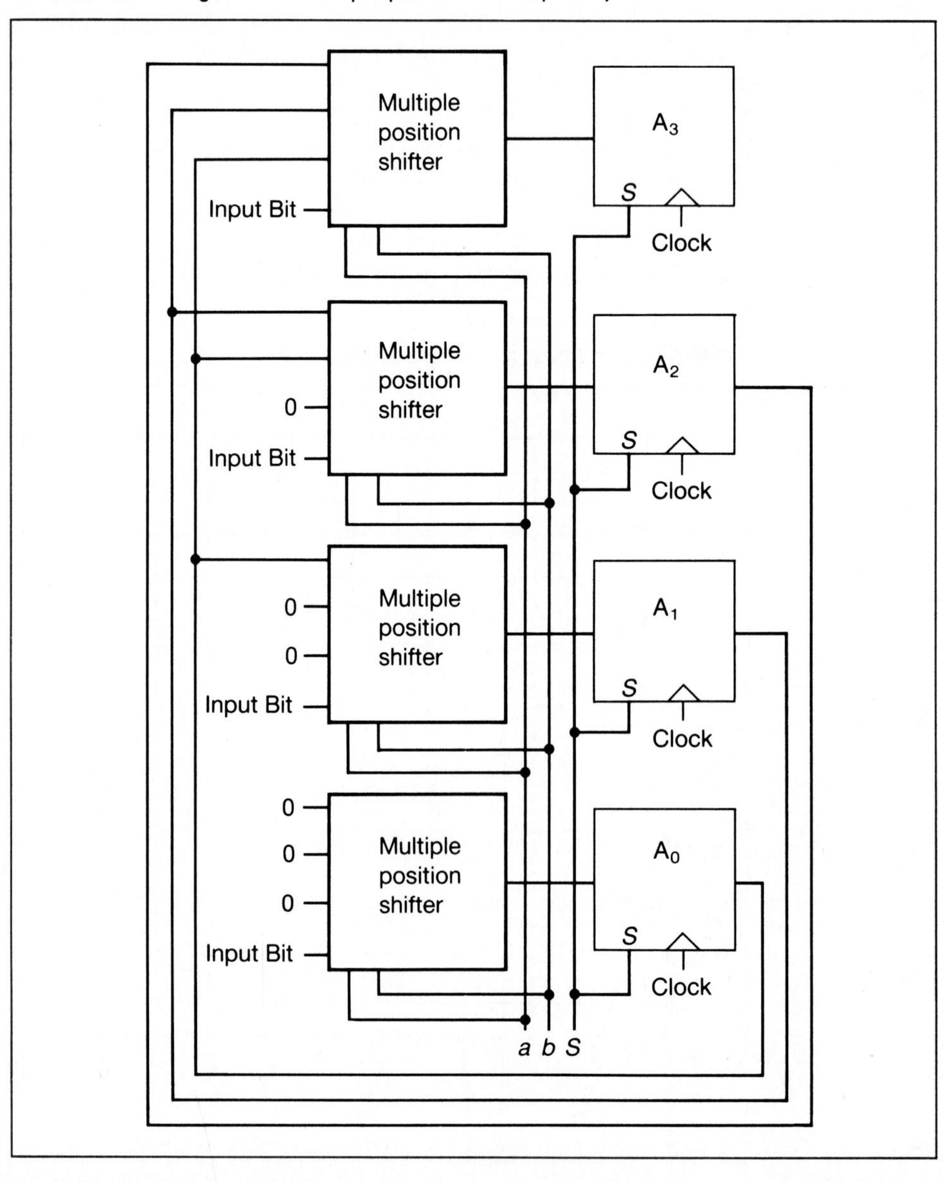

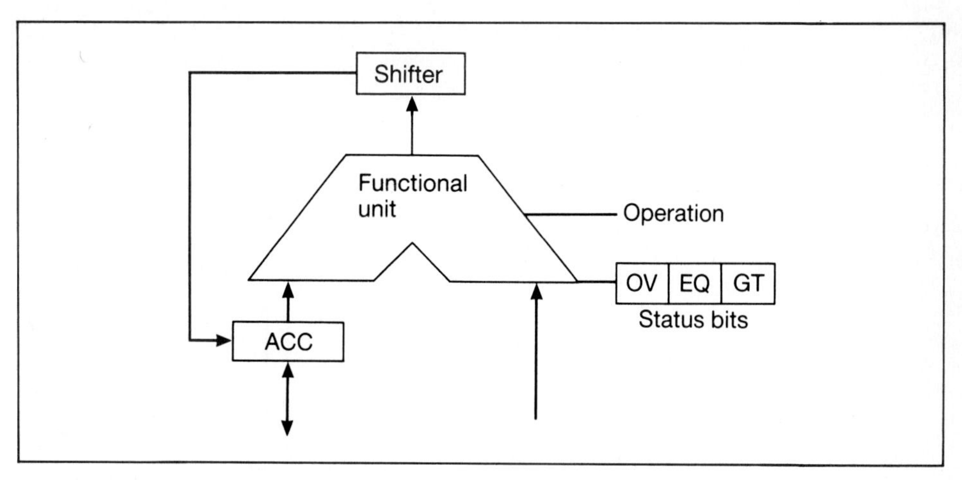

FIGURE 8.20 Incorporating the shifter as a separate unit after the functional unit.

We have been discussing shifting as a part of the accumulator. If we have extensive shifting capabilities, such as those in Fig. 8.18, we might have the shifter as a separate unit. Often, such a separate unit is incorporated on the output of the functional unit, as shown in Fig. 8.20.

8.5 LOGIC OPERATIONS

In Chap. 2, and earlier in this chapter, we covered individual circuits that will perform the basic logic operations AND, OR, NOT, and XOR. We need a way to select and perform the same operations on data elements. This can be easily implemented in our ALU with a circuit like that shown in Fig. 8.21. The X_i and Y_i inputs are connected to the accumulator and system bus in the same manner as the functional unit shown in Fig. 8.20. The output is returned to the accumulator. The logic circuit is replicated for each bit in the data word. Signals on the control lines are used to enable the logic unit and to choose which of the four logic operations is to be performed. Encoding for the function select control is given in Table 8.7.

TABLE 8.7 BIT PATTERNS AND ACTIONS IN LOGIC CIRCUIT

a	*b*	Function
0	0	X_i'
0	1	$X_i \oplus Y_i$
1	0	$X_i \cdot Y_i$
1	1	$X_i + Y_i$

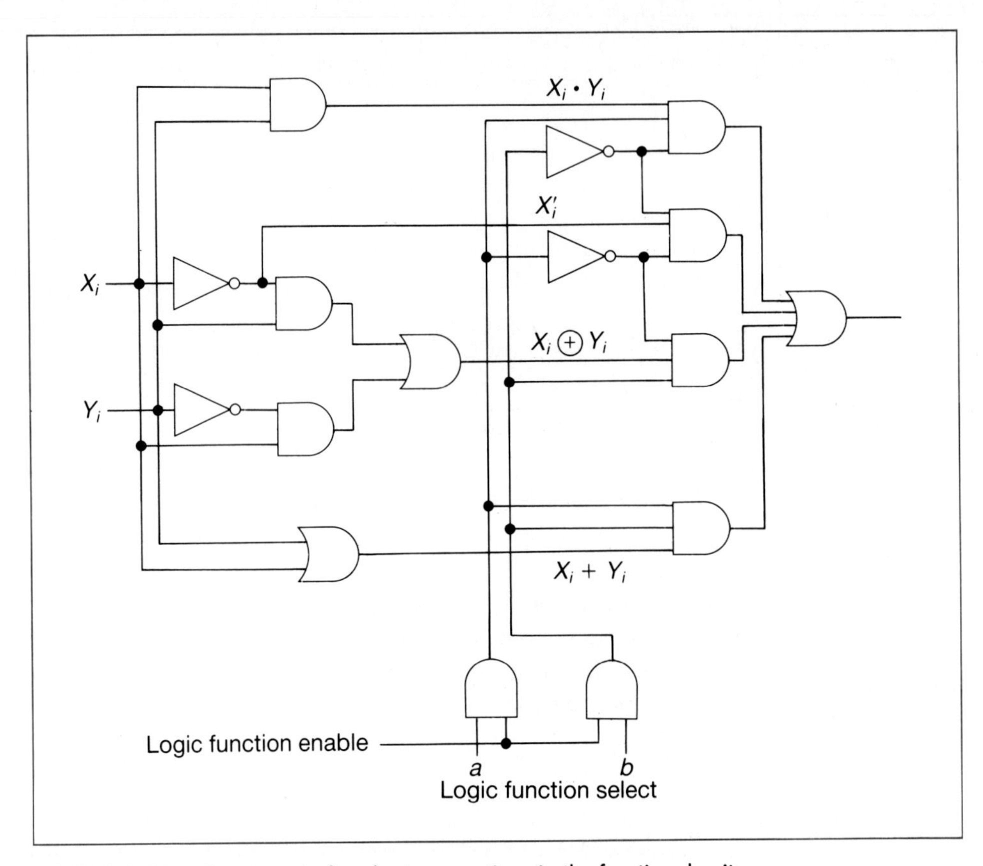

FIGURE 8.21 Circuit to do four logic operations in the functional unit.

8.6 INTEGER MULTIPLICATION

Multiplication can be done in an ALU in much the same way that we would do it on paper. Since the multiplier bits are either 0 or 1, the process is even simpler—if the multiplier bit is 1, we add in the multiplicand; if it is 0, we don't. Another difference is that in an ALU we usually start with the high-order multiplier bit. With this approach, the partial sum is shifted left in the accumulator and partial products are added on the right.

Probably the best way to study this multiply process is with an example. Suppose we have 4-bit integers, $X = X_3 X_2 X_1 X_0$ and $Y = Y_3 Y_2 Y_1 Y_0$, and want to multiply X by Y. The steps are shown in Table 8.8 both for a general case and for the example 0110 times 1101.

Before implementing a circuit for multiplication, we need to do a few preliminaries. This operation is multistep, so we will need a counter to keep track of the steps. Also, the multiplier is shifted and its high-order bit is used to decide whether to "add," so we will build a register for this operation.

TABLE 8.8 BINARY MULTIPLICATION

Initialize: ACC	=	0	0	0	0	0	0	0	0	
If $Y_3 = 1$ then $P_3P_2P_1P_0 = X_3X_2X_1X_0$										$X = 1101$
else $P_3P_2P_1P_0 = 0\,0\,0\,0$										$Y = \underline{0}110$
Add: ACC	=	0	0	0	0	0	0	0	0	00000000
	+					P_3	P_2	P_1	P_0	0000
	=	0	0	0	0	S_3	S_2	S_1	S_0	00000000
Shift: ACC	=	0	0	0	S_3	S_2	S_1	S_0	0	00000000
If $Y_2 = 1$ then $P_3P_2P_1P_0 = X_3X_2X_1X_0$										$X = 1101$
else $P_3P_2P_1P_0 = 0\,0\,0\,0$										$Y = 0\underline{1}10$
Add: ACC	=	0	0	0	S_3	S_2	S_1	S_0	0	00000000
	+					P_3	P_2	P_1	P_0	1101
	=	0	0	T_5	T_4	T_3	T_2	T_1	T_0	00001101
Shift: ACC	=	0	T_5	T_4	T_3	T_2	T_1	T_0	0	00011010
If $Y_1 = 1$ then $P_3P_2P_1P_0 = X_3X_2X_1X_0$										$X = 1101$
else $P_3P_2P_1P_0 = 0\,0\,0\,0$										$Y = 01\underline{1}0$
Add: ACC	=	0	T_5	T_4	T_3	T_2	T_1	T_0	0	00011010
	+					P_3	P_2	P_1	P_0	1101
	=	0	U_6	U_5	U_4	U_3	U_2	U_1	U_0	00100111
Shift: ACC	=	U_6	U_5	U_4	U_3	U_2	U_1	U_0	0	01001110
If $Y_0 = 1$ then $P_3P_2P_1P_0 = X_3X_2X_1X_0$										$X = 1101$
else $P_3P_2P_1P_0 = 0\,0\,0\,0$										$Y = 011\underline{0}$
Add: ACC	=	U_6	U_5	U_4	U_3	U_2	U_1	U_0	0	01001110
	+					P_3	P_2	P_1	P_0	0000
	=	W_7	W_6	W_5	W_4	W_3	W_2	W_1	W_0	01001110
(finally!)										

Table 8.9 shows an annotated form of a 3-bit counting sequence. Observe that the low-order bit of a counter changes every clock pulse. The second bit of the counter changes only when the first bit is a 1, and the high-order bit changes only when the two low-order bits are 1, so its input is an AND of the low-order two.

Before implementing the 3-bit counter, we need to define a circuit that will change state when its input goes to 1 (that is, when specified conditions happen). To do that, we can "cross-connect" the outputs of an *RS* flip-flop and its *R* and *S* inputs. If *Q* is a 1, then we want the circuit to reset, so we AND the input and *Q* and use the result as *R*. Similarly, if *Q'* is a 1, we want the circuit to set, so we AND the input and *Q'* and use the result as *S*. We say "cross-connecting" because usually the *S* input is associated with the *Q* output (since *S* will set *Q* to a 1), and the *R* input with the *Q'* output. With such a setup, we have a circuit that changes state every time the control input goes to 1. It is shown in Fig. 8.22.

We have two choices for a symbol for the circuit in Fig. 8.22. To be precise, it is another type of flip-flop called a "toggle." The reason is that it "toggles" (changes state) whenever its input goes to 1. However, since we are using this circuit to build our counter, we will use the symbol shown in Fig. 8.23.

TABLE 8.9 PATTERNS OF BIT CHANGES IN COUNTING

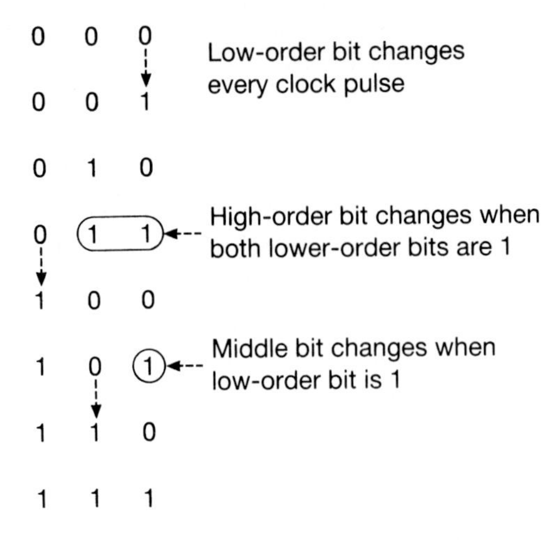

Given the toggle circuit, it is easy to build our counter. Following the earlier discussion, a given element of the counter changes when all elements to the right of it (lower-order bits) are 1. The circuit for three elements is shown in Fig. 8.24.

Our multiplication needs four adds/transfers and three shifts, for a total of seven operations. Hence we cannot use all eight positions of the 3-bit counter. We will therefore change our basic counter circuit to allow loading an initial value.

Our modified counter circuit element is shown in Fig. 8.25*a*. The Count circuit provides the constant 1 input for toggling and counting. We also have the same

FIGURE 8.22 A circuit element that changes state every time its control input goes to 1.

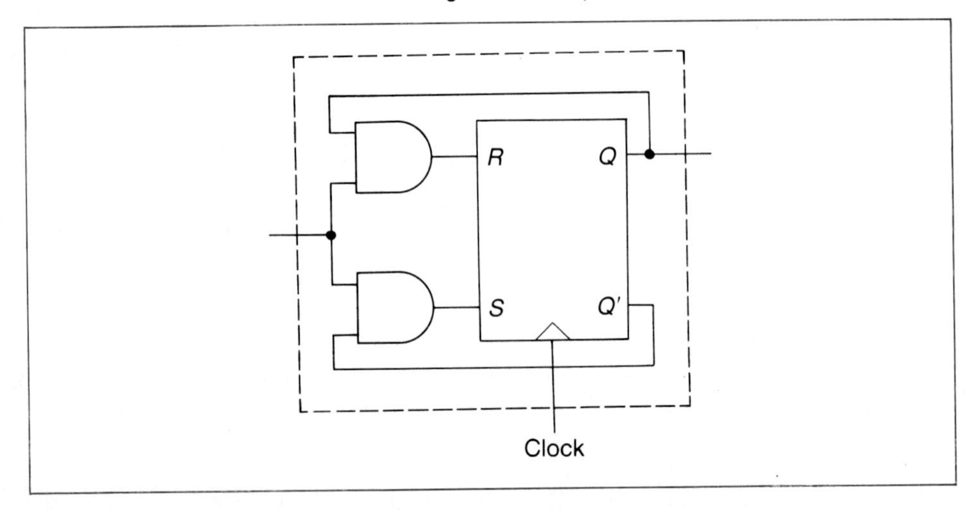

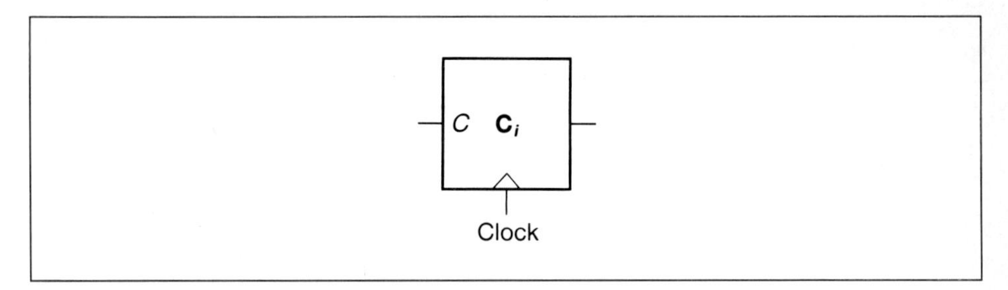

FIGURE 8.23 Symbol for the circuit shown in Fig. 8.22.

Load circuit that we used in our memory element. The symbol for the resulting combination circuit is shown in Fig. 8.25*b*.

Using these elements, we can build a counter circuit with load capabilities. This is shown in Fig. 8.26.

The circuit for the multiplier needs to be able to load that multiplier and shift it left one position at a time. The only output we need is the current high-order position. Design for this circuit is similar to previous ones, so we will not discuss it in detail. It is shown in Fig. 8.27.

Lastly, we need a temporary storage register to hold the multiplicand. All it has to do is hold the multiplicand during the operation, so it can be a set of the memory elements described earlier.

We can now construct our multiplier. A block diagram is shown in Fig. 8.28. To initialize the operation, we:

- Zero the accumulator.
- Move the multiplicand into the STR (temporary storage) register.
- Move the multiplier into the MULT register.
- Enter 001 into the CTR (counter) register.

Then the actual multiplication operation is initiated. The low-order bit of the counter is 1, so there is no shifting. The high-order bit of the MULT register picks

FIGURE 8.24 Three-bit binary counter.

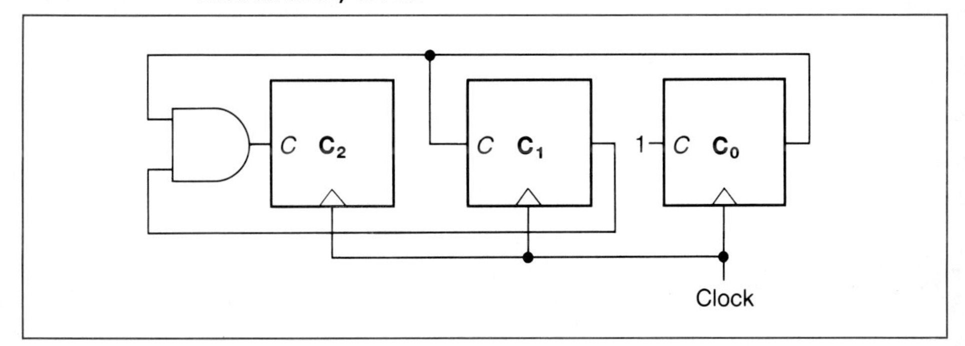

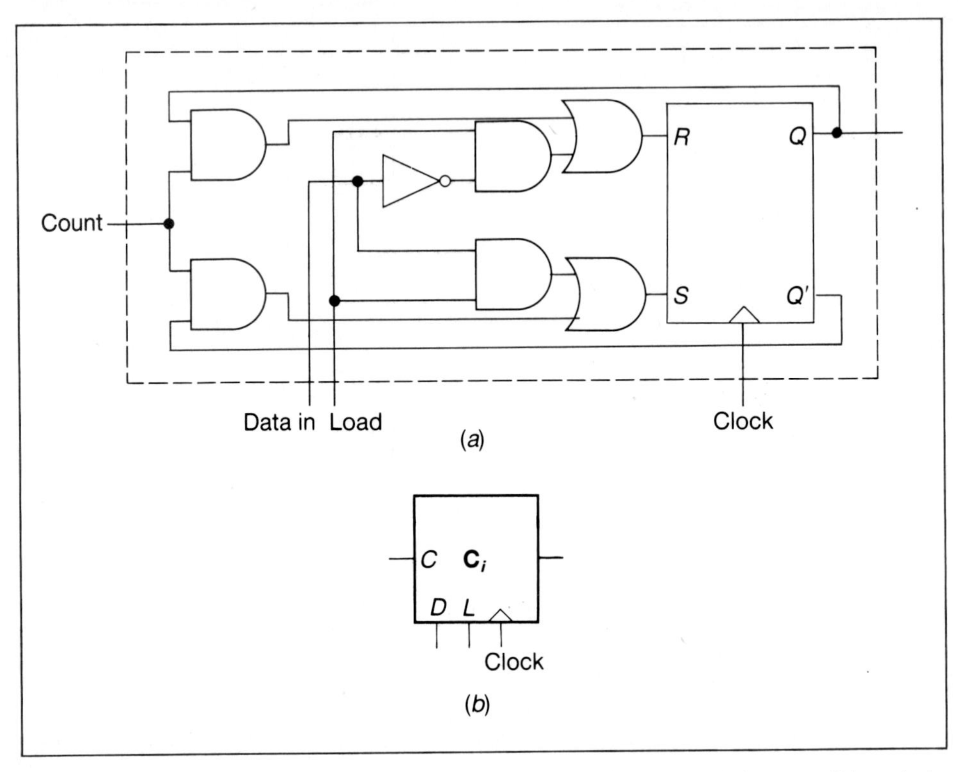

FIGURE 8.25 New counter element to allow loading an initial value: (*a*) circuit; (*b*) symbol.

either Add (bit = 1) or Transfer ACC (bit = 0) in the functional unit. The Load control line of the ACC is also 1, so the output of the add/transfer operation is loaded into the accumulator.

Now the low-order bit of the counter goes to 0. This makes the accumulator shift

FIGURE 8.26 Counter circuit with load capabilities.

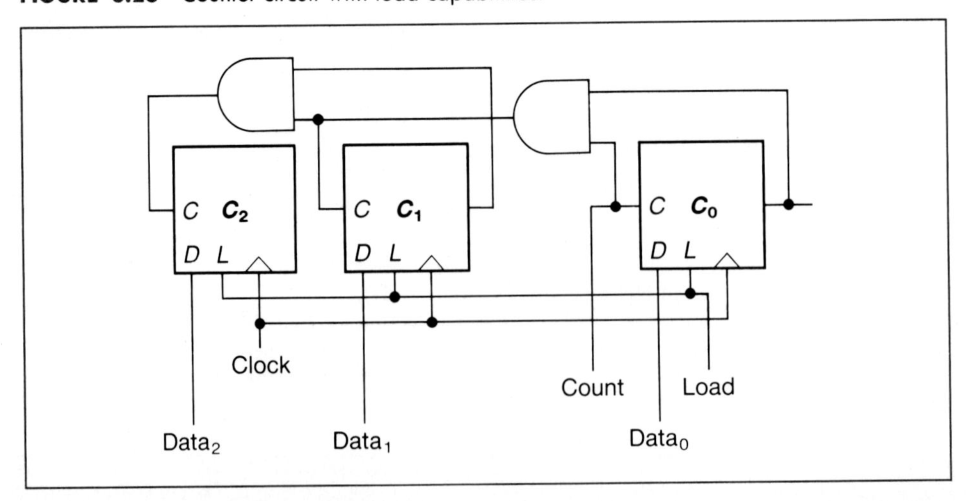

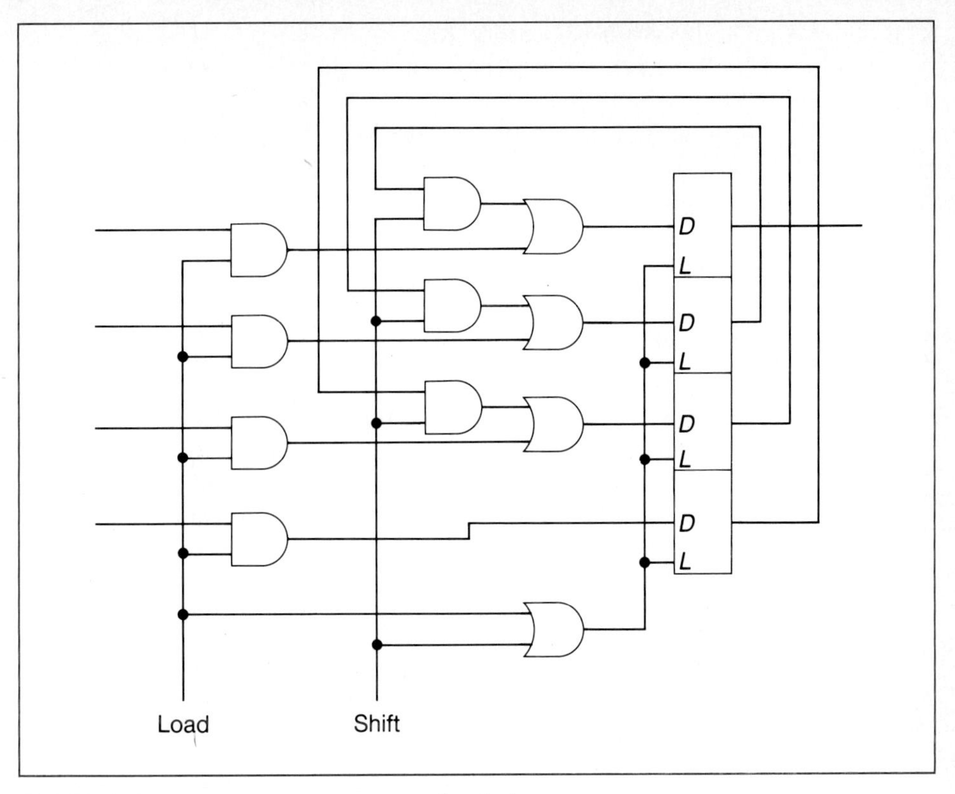

FIGURE 8.27 Circuit for MULT register of multiplier.

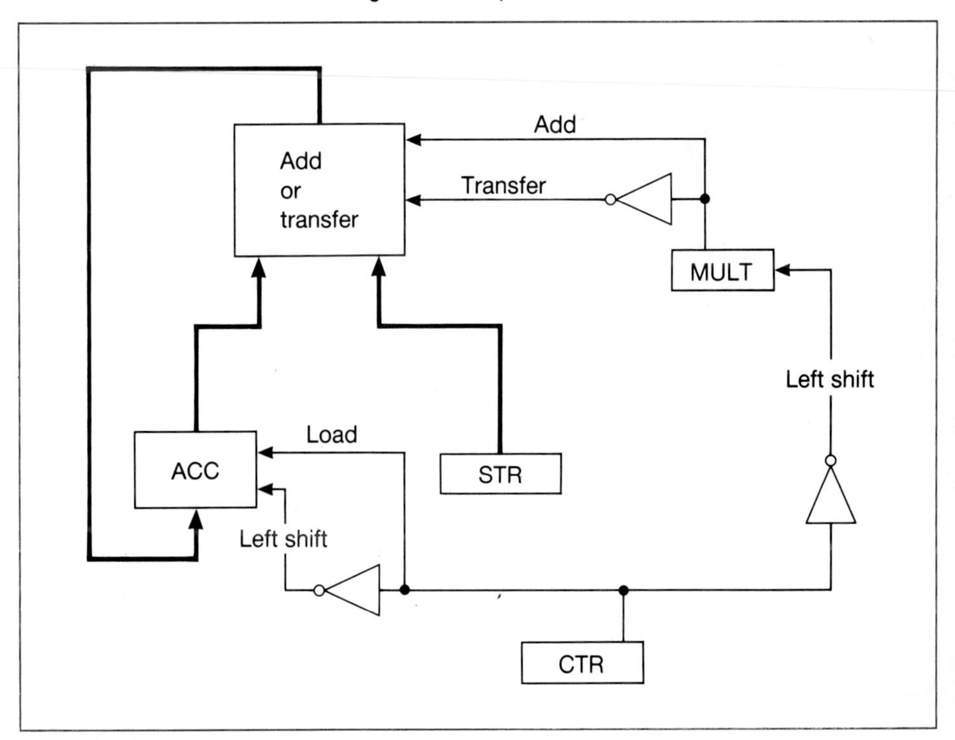

FIGURE 8.28 Multiplier circuit.

one position to the left and the MULT register shift one position to the left. We now have the next multiplier bit in the high-order position and can repeat the sequence. The cycle continues for the rest of the multiplication, as described earlier in our table.

8.7 THE DESIGN OF AN ACTUAL ALU

Before we go on to the next section, let's consider the ALU as we have designed it so far. A block diagram of our unit is shown in Fig. 8.29. It includes clear, load, add, shift, compare, multiply, AND, OR, NOT, and XOR. This is not a complete set of all the operations that would normally be provided. However, it is sufficient to give you a feeling of what is involved.

An actual ALU would use circuits that are optimized for speed within the cost parameters defined for the design. It might also include floating-point operations, which will be covered in the next section. The point to remember is that the concepts are essentially the same as we have been discussing, and they are quite simple.

FIGURE 8.29 Functional diagram of the ALU as designed in this chapter.

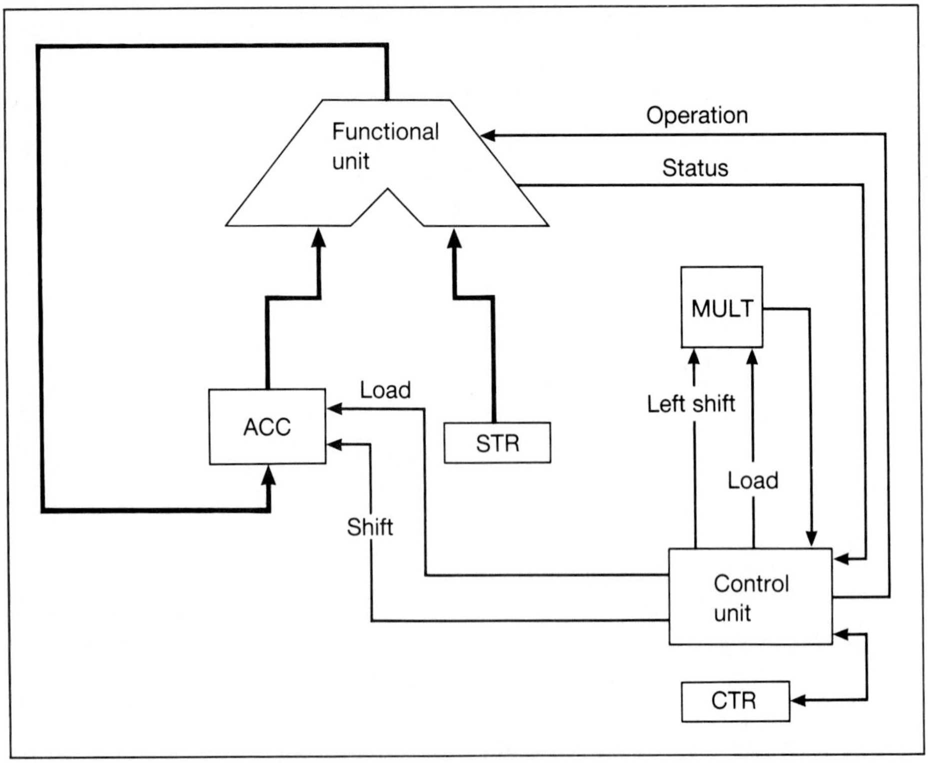

8.8 FLOATING-POINT OPERATIONS

By now, you should have a reasonable idea of how arithmetic operations are performed on integer numbers. We now need to spend a little time discussing these same operations with the other common numeric format—floating point.

In Chap. 3, we learned that floating-point numbers contain both a fractional mantissa M and an exponent E in the form

$$\text{Floating-point value} = \pm M \cdot R^{E+b}$$

where R is the radix, or base, of the number system, b is the exponent bias (typically the center value of its range), and the most significant digit of the mantissa is nonzero. The term for this form is **normalized**. Normalized means that the mantissa has been put in a form that is *normally expected*—where the most significant digit in the mantissa is nonzero. A floating-point number with a zero in the most significant digit in the mantissa must be normalized before it is stored.

8.8.1 Floating-Point Formats

Up to this point in our discussions, we have only alluded to how floating-point numbers are represented in the computer. As you might guess, the actual formats used in the early computers varied considerably. That is not the case today. Two standards have been defined by the Institute of Electrical and Electronic Engineers (IEEE) and agreed upon by the American National Standards Institute (ANSI):

- ANSI/IEEE Std. 754-1985: The IEEE Standard for Binary Floating-Point Arithmetic
- ANSI/IEEE Std. 854-1987: The IEEE Standard for Radix-Independent Floating-Point Arithmetic

They are both similar, and for simplicity, we will limit our specific discussions to one format from the binary standard.

The 754 standard defines the normalized 32-bit single-precision format shown in Fig. 8.30. It is a little different from the floating-point numbers we saw earlier. S is the sign of the mantissa, Exp is a biased exponent, and Fraction is the fractional portion of a mantissa that also includes an "understood" integer 1 to the left of the binary point.

The understood 1 is a form of data compression that extends the effective length of the mantissa by 1 bit. Since we are dealing with a normalized format, the most significant mantissa digit must be nonzero. However, we are also dealing with the binary number system. Its alphabet contains only one nonzero digit, the binary 1. Therefore, the only part of the binary mantissa that is variable is the fractional part. The whole part of the number (the 1.) is "understood." All we have to do is reconstruct the mantissa equation whenever we load the number into the ALU.

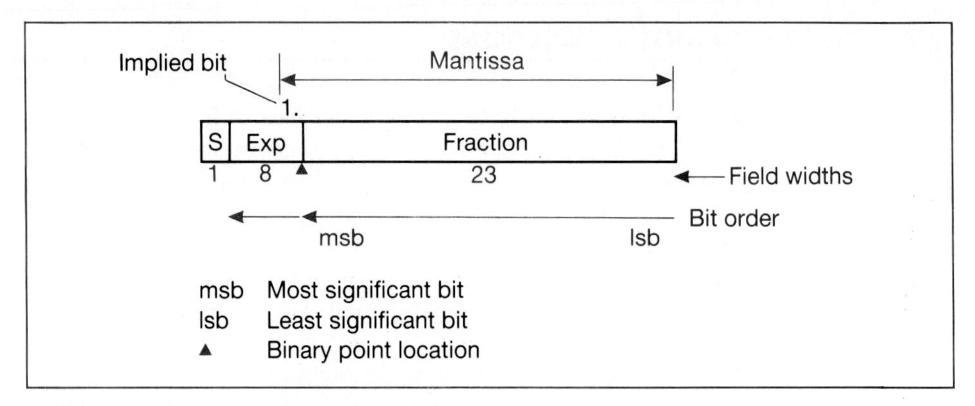

FIGURE 8.30 IEEE single-precision binary floating-point storage format.

8.8.2 Loading Floating-Point Numbers into the ALU

Unfortunately, inclusion of the understood 1 also complicates loading a floating-point number from memory into the ALU. The typical internal ALU floating-point format is longer than the storage formats. Room must be allowed for the longest exponents and mantissas that will occur plus any additional space necessary for developing the result of the arithmetic operations. The minimum recommended length for the ALU format is 80 bits.

The ALU register is reset to zero at the beginning of the Load operation. Then the most significant bit of the mantissa is set to 1 and the fields of the storage format are copied into the ALU register, as shown in Fig. 8.31. Storing a word in the ALU back into memory is a reverse procedure, except that the contents must be rounded and shortened to fit the memory format.

FIGURE 8.31 Loading an IEEE binary floating-point number into the ALU.

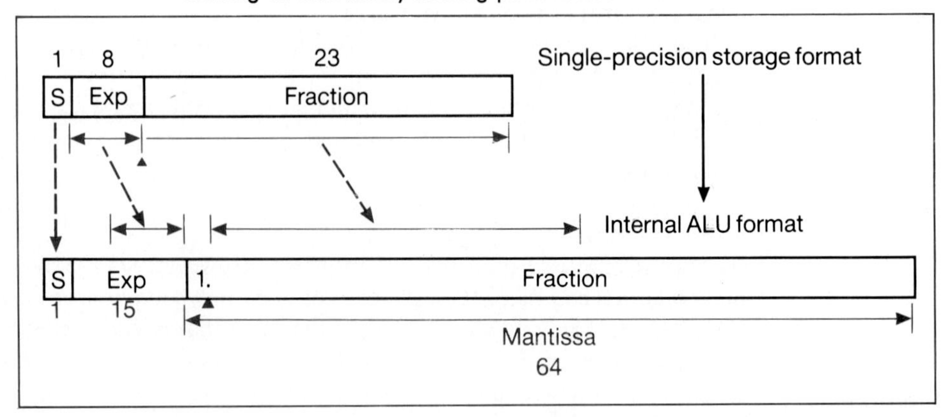

8.8.3 Floating-Point Add and Subtract

Adding or subtracting floating-point numbers is similar to adding or subtracting ordinary numbers with fractional values. The first thing we do is align the radix points. Assume, for example, that we need to add 1.10011 to 111.1001. By habit, we would write

$$\begin{array}{r} 111.1001 \\ +\quad 1.10011 \\ \hline 1001.00101 \end{array}$$

where we automatically offset the smaller number so that the radix points line up.

We do the same thing with floating-point numbers. The radix points are aligned by "denormalizing," or right-shifting, the mantissa of the smaller number until the exponent values are equal, as shown in Fig. 8.32. When both exponents are equal, we can proceed with the add.

There is still a problem because the add operation generated a carry. This, in turn, generated a mantissa **overflow**. There is no room for a carry in the mantissa format. We can think of the mantissa as overflowing its container. The condition is indicated by setting the mantissa overflow bit in the ALU status register to 1. This indicates that the mantissa must be renormalized. It is shifted 1 bit to the right, a 1 is inserted into the leftmost bit position, and the exponent value is increased by 1.

An overflow can occur in either the mantissa or the exponent. The difference is that a mantissa overflow can usually be recovered by renormalization. An exponent

FIGURE 8.32 An example of binary addition with mantissa overflow.

	S	Exponent		Mantissa	Overflow Exp	Overflow Mant
Shift A = 1.10011	0	0 1 1 1 1 1 1 1	1	1 0 0 1 1 0 0 0 0 · · ·	0	0
B = 111.1001	0	1 0 0 0 0 0 0 1	1	1 1 1 0 0 1 0 0 0 · · ·	0	0
Shift A = 01.10011	0	1 0 0 0 0 0 0 0	0	1 1 0 0 1 1 0 0 0 0 · · ·	0	0
B = 111.1001	0	1 0 0 0 0 0 0 1	1	1 1 1 0 0 1 0 0 0 · · ·		
A = 001.10011	0	1 0 0 0 0 0 0 1	0	0 1 1 0 0 1 1 0 0 0 0 · · ·	0	0
+ B = 111.1001	0	1 0 0 0 0 0 0 1	1	1 1 1 0 0 1 0 0 0 · · ·		
Unadjusted sum = 1001.00101	0	1 0 0 0 0 0 0 1	0	0 1 0 0 1 0 1 0 0 0 0 · · ·	0	1
Sum = 1001.00101	0	1 0 0 0 0 0 1 0	1	0 0 1 0 0 1 0 1 0 0 0 0 0 · · ·	0	0

overflow in addition or subtraction indicates that the resultant sum or difference is outside the representable range of the number system, and is thus an error.

A similar condition can also occur when two very small numbers are subtracted and the result is left with leading zeros. When this happens, renormalization left-shifts the mantissa and decreases the exponent. The process can sometimes cause the exponent to go negative and out of range. This is known as **underflow**. You might think of an underflow occurring because the value is below the bottom of the container.

8.8.4 Floating-Point Multiply and Divide

Floating-point multiply and divide rely on the identities

$$X^{\mathrm{Exp1}} \cdot X^{\mathrm{Exp2}} = X^{\mathrm{Exp1}+\mathrm{Exp2}}$$

and

$$X^{\mathrm{Exp1}} \div X^{\mathrm{Exp2}} = X^{\mathrm{Exp1}-\mathrm{Exp2}}$$

To multiply or divide two floating-point numbers, we multiply or divide the mantissas and add or subtract the exponents. This sounds fairly easy, but we need to remember that each exponent is really $E + b$. Therefore, our radix product and quotient equations become

$$R^{E_1+b} \cdot R^{E_2+b} = R^{E_1+b+E_2+b} = R^{E_1+E_2+2b}$$

and

$$R^{E_1+b} \div R^{E_2+b} = R^{(E_1+b)-(E_2+b)} = R^{E_1+b-E_2-b} = R^{E_1-E_2}$$

This means that in addition to the possibility of having to renormalize the mantissa, we must also readjust the exponent. We must either remove the extra bias introduced by the multiply or replace the bias removed by the divide.

8.8.5 The Special Case of 0 and ∞

The floating-point numbers were originally proposed as a concise way to represent a wide range of values. Consider the diagram in Fig. 8.33. There are four ranges and three specific values identified. It might appear that only two can be represented by normalized floating-point numbers. However, the standard has made provision for very large positive and negative numbers and has set aside specific representations for 0 and ∞.

Very large positive and negative numbers are identified by a maximum-value exponent (all 1s) and mantissas that have at least one bit equal to 0. The standard calls these values **not a number**, or **NaN**. The sign of the mantissa indicates whether they are positive or negative. Infinity is represented by a maximum exponent and a maximum mantissa (both all 1s). Again, the sign indicates whether it is positive or negative. Zero is represented by a minimum exponent and a minimum mantissa (all 0s). Zero can also be positive or negative.

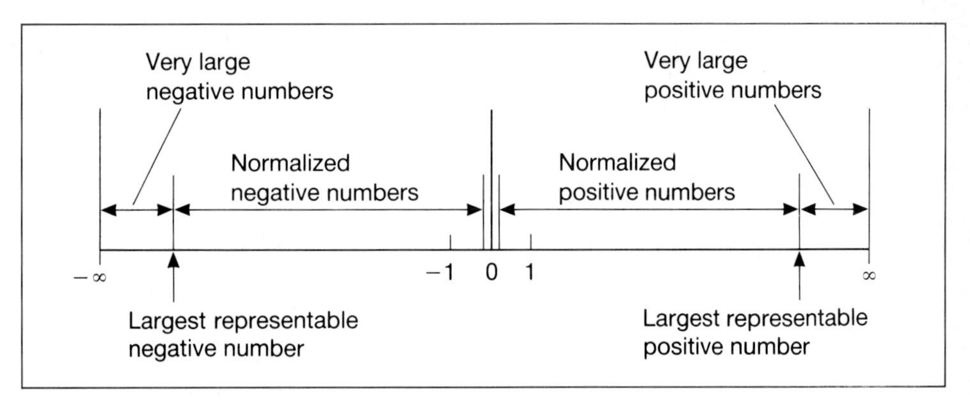

FIGURE 8.33 Representation ranges of floating-point numbers.

Both ∞ and 0 are typically treated as special arithmetic cases. From previous courses in mathematics, we know that

$$
\begin{array}{llll}
x + 0 = x & 0 + x = x & x + \infty = \infty & \infty + x = \infty \\
x - 0 = x & 0 - x = -x & x - \infty = -\infty & \infty - x = \infty \\
x \cdot 0 = 0 & 0 \cdot x = 0 & x \cdot \infty = \infty & \infty \cdot x = \infty \\
x \div 0 = \infty & 0 \div x = 0 & x \div \infty = 0 & \infty \div x = \infty
\end{array}
$$

Therefore, most computers check for these cases before the requested floating-point operation is actually done. If either of the numbers is 0 or ∞, the appropriate value is substituted for the result and a condition flag is set.

8.9 SUMMARY

Our purpose in this chapter was to introduce you to the arithmetic unit. Several arithmetic and logic functions were developed, and we did a rather extensive walk-through of an integer multiply. We also discussed the logic design of the circuits that could be used to provide this functional capability.

This was followed by a discussion of floating-point operations and the IEEE floating-point standard. Here again, we covered only enough to give you the essence of what is involved. The IEEE binary floating-point standard actually defines:

- Floating add, subtract, multiply, divide, compare, and square root
- Conversion between all allowable floating-point formats
- Conversion from binary integer to floating-point formats
- Rounding and conversion of floating-point numbers to integer values
- Conversion between binary and decimal number systems

Most of these operations are conceptually obvious, and you have probably used several of them in your programming classes.

Before we go on to the next chapter, we need to make one more point. Standards were briefly discussed in conjunction with character set representations in Chap. 3. We have just finished discussing floating-point standards in this chapter. It is time to ask what impact standards have on the industry and whether they are necessary.

Some people argue that they could define a better system and that standards stifle creativity. Others argue that standards are required if we are to have computers that will work together. You probably have heard that Fortran, Pascal, and other high-level languages have been designed to make source programs "transportable." In general, they have succeeded. Source programs written in the standard version of these languages typically run without change on a number of different computers. However, there is no guarantee that the same programs running the same data sets will produce the same results.

The problem was that until the IEEE standard was defined, computer manufacturers were free to:

- Adopt any mantissa length, exponent range, and floating-point format
- Use any shortening procedure (rounding, half-adjusting, or truncation) they chose

The result was that floating-point data frequently had to be converted from one format to another before it could be transferred from one computer to another. A second frustration resulted from the fact that it was generally very difficult to determine which computer would produce the most accurate results.

If you were the user, what would be your position regarding the use of standards?

PROBLEMS

8.1 Define the purpose of the ALU in a modern scientific computer. List the operations that need to be implemented to perform scientific computations.

8.2 For a computer you have access to, list the instructions corresponding to those in Tables 8.1 to 8.3. Are there any that don't fit into these tables. Why not?

8.3 Give some advantages/disadvantages of having a separate compare instruction before a conditional jump versus incorporating the compare directly into the jump.

8.4 Is it necessary to have both addition and subtraction functions in the ALU?

8.5 In Sec. 8.1 we described several dyadic operations. List two additional dyadic arithmetic operations.

8.6 Give an example of an operation that requires three operands.

8.7 What is the difference between a half-adder and a puff adder?

8.8 Implement an exclusive OR circuit using only NAND gates.

8.9 Figure 8.5 shows the truth table for a full-adder. Develop the truth table for a full-subtractor.

8.10 What would be the sequence of instructions needed to add five numbers (N1, N2, N3, N4, and N5) in the accumulator of a computer that did not have an LDA instruction? How does the lack of this instruction affect the computer's performance in arithmetic operations?

8.11 Assume the 4-bit adder in Fig. 8.6 is made using the gates as shown in Figs. 8.2 and 8.3. If each gate in the circuit requires 10 ns to change states, how long would it take to perform an addition?

8.12 Figure 8.6 shows the block diagram for a 4-bit adder. Show how this can be extended to provide *n*-bit addition.

8.13 Show that a half-adder is also an exclusive OR.

8.14 Design a circuit for a full-adder that does not use half-adders but determines the sum and carry bits directly.

8.15 Write the boolean expressions for a 4-bit equality checker.

8.16 Discuss why it is possible to implement the functional unit of an ALU as a strictly combinational circuit.

8.17 Design a circuit to test if 2 bits are unequal.

8.18 Use the circuit from Prob. 8.17 to build a circuit to test if two 4-bit numbers are unequal.

8.19 What are the primary uses of the accumulator in the ALU?

8.20 How would the conditions listed for less-than checking have to be changed to allow for greater-than checking?

8.21 What is the difference between the Clear operation in the accumulator and a Reset operation in any other register?

8.22 Discuss how the relational operations in Table 8.2 could be implemented in the ALU.

8.23 Implement the storage element described in Fig. 8.11 with a toggle flip-flop.

8.24 Figure 8.20 shows the shifter as a separate unit from the accumulator. List the operational sequence that would be required to shift a number in the accumulator 2 bits to the left. Would this appear to require more, less, or about the same time that would be required if the shifting were done in the accumulator itself?

8.25 How and where is the Load/Clear signal shown in Fig. 8.13 generated?

8.26 Some computers have shift operations that leave the contents of the most significant bit unchanged. Why is such an operation used?

8.27 For the shift operation described in Prob. 8.26, what happens to the second most significant bit?

8.28 The diagram in Fig. 8.17 shows graphical descriptions of the four basic shifting operations. Some computers have been designed to allow the programmer to select whether a 0 or a 1 is inserted in the low-order bit position during a Shift Left operation. Discuss what would be required to allow this option.

8.29 Why can the Function Select signal in Fig. 8.21 be encoded?

8.30 Figure 8.21 shows a circuit that will accomplish four different logic operations, based on a Function Select input. Show how this circuit could be modified to add NOR and NAND.

8.31 Refer to Table 8.8 and show the steps required to multiply 11101 by 01011.

8.32 How must the design of the ALU be changed to allow multiplication of two numbers whose length is the same as that of the accumulator?

8.33 Compare the time required to add two numbers with the time required to multiply the same two numbers.

8.34 Show how the procedure defined in Table 8.8 would have to be modified to multiply decimal numbers.

8.35 Why is it more difficult to perform arithmetic operations on floating-point numbers than on integer numbers?

8.36 Comment on the alternative structure for the ALU shown in Fig. P8.36.

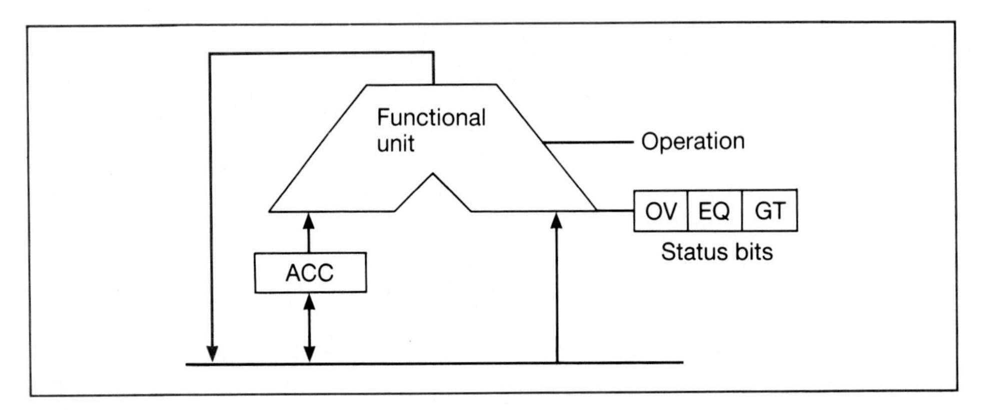

FIGURE P8.36

8.37 Identify any problems you see with the alternative structure for the ALU shown in Fig. P8.37.

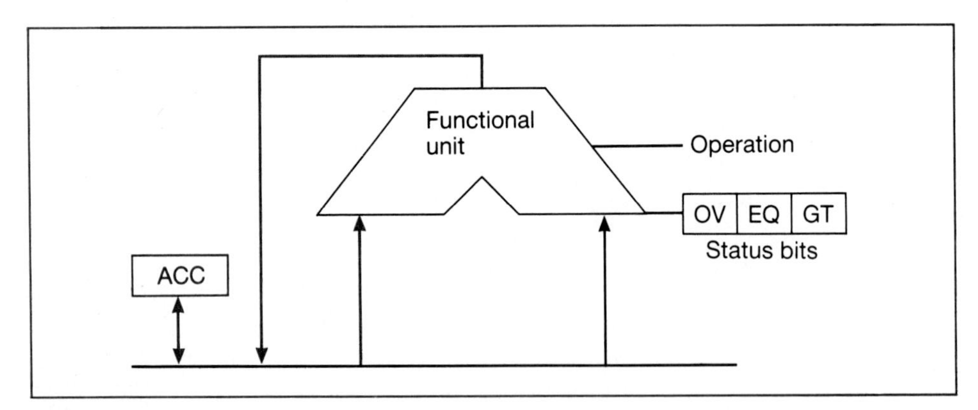

FIGURE P8.37

8.38 Explain how the IEEE floating-point storage format differs from the internal ALU format.

8.39 Explain what is required to store a floating-point number in the ALU back into memory. Include diagrams.

8.40 Give a noncomputer example of a register overflow.

8.41 When should the special-condition checks be done in floating-point operations? Why?

8.42 The chapter summary includes a brief discussion on standards. Do you think that they are good or bad? Why?

8.43 Give four examples of useful standards that are not in the computer field.

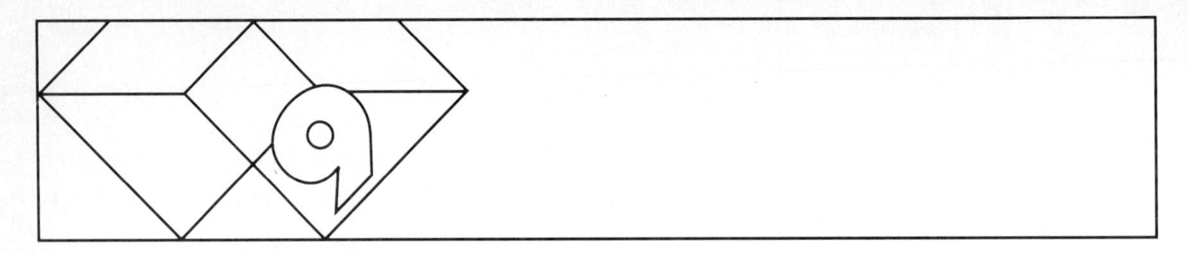

The Input/Output Unit

In the previous chapters of Part II, we described the control, ALU, bus, and memory units of a computer. These units enable the computer to store program instructions and data, retrieve them, execute instructions, and manipulate the data. With these capabilities, the computer can perform useful work. However, there are no provisions for getting programs and data into the computer or results out of the computer. In this chapter we will examine the I/O unit. It gives the computer the ability to communicate with devices external to the computer itself. We call these external devices, such as keyboards, monitors, printers, and disks, **peripherals**.

One possible relation between the control, ALU, bus, memory, I/O unit, and external devices is shown in Fig. 9.1. As the figure indicates, external devices, or peripherals, are connected to the I/O bus by peripheral controllers. The I/O unit transfers information between the I/O bus and the system bus. The peripheral controllers and peripheral devices shown in the figure will be covered in Chap. 10.

Just as the design of the control unit has undergone some significant changes, there have also been major developments in the design of the I/O unit. To show how the I/O unit has evolved, we will describe several different implementations of the I/O unit. Before examining these implementations, though, we will discuss the functions a typical I/O unit performs.

9.1 THE FUNCTIONS OF THE INPUT/OUTPUT UNIT

As stated above, the I/O unit gives the CPU the ability to communicate with peripherals. There are several reasons why the I/O unit is needed. The electric signals needed to transfer information to and from a peripheral may be different from those used on the system bus. In Fig. 9.2*a* we have reproduced transfer control signals from Fig. 7.3. This figure, however, shows only the signals associated with the data bus. It also shows a possible timing diagram for data transfers made on the system

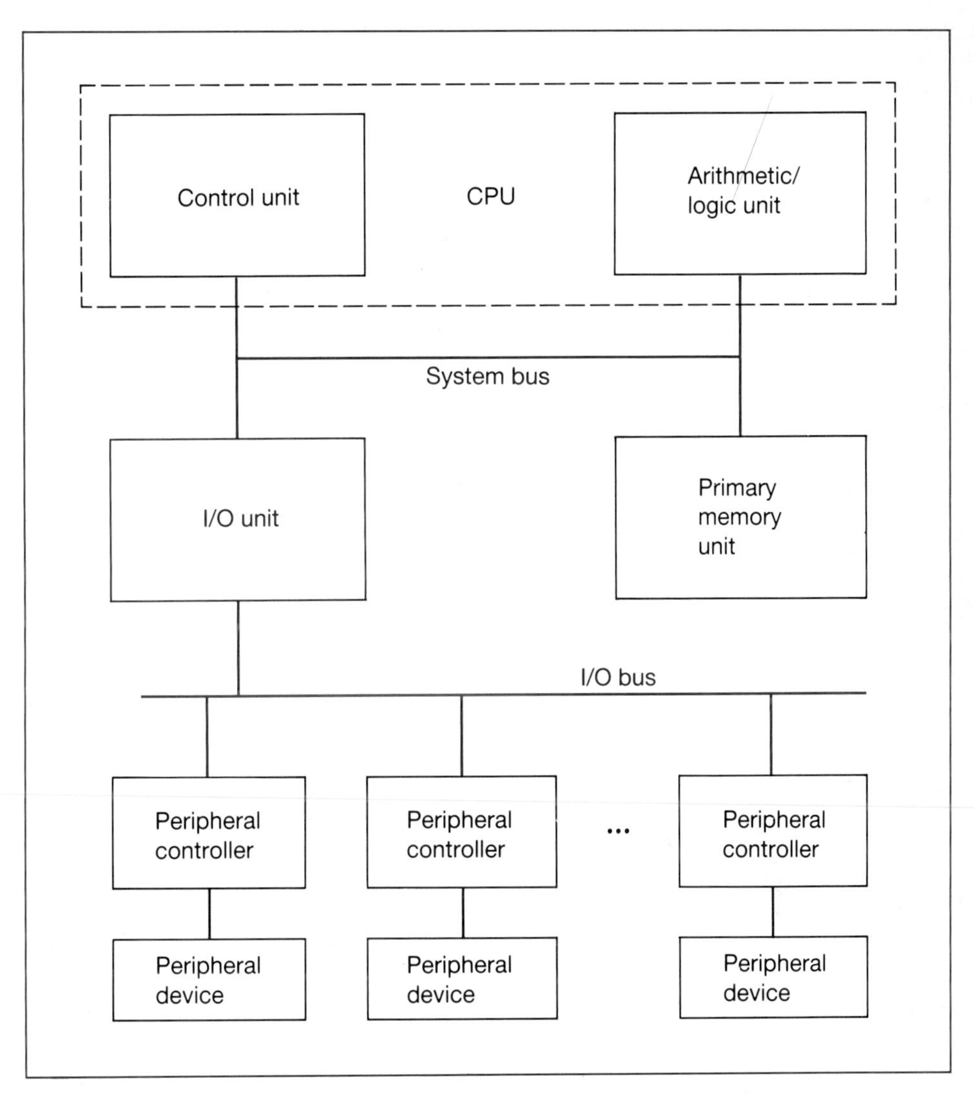

FIGURE 9.1 The I/O unit's relation to the CPU, memory, and external devices.

bus. We see that the time between transfers is 100 ns and that the voltage level of the signals varies between 0 and -2 V. Figure 9.2*b* shows corresponding signals for transfers on the I/O bus. There, the transfer time is 1 μs, and the voltage swings are between 0 and $+4.5$ V.

It is reasonable to ask why we would design a system that used different voltages for logical 1 and 0 on the two buses. The answer is that the two buses serve different purposes. The system bus must be capable of delivering instructions and data from the main memory fast enough to keep the CPU running at full speed. This requires

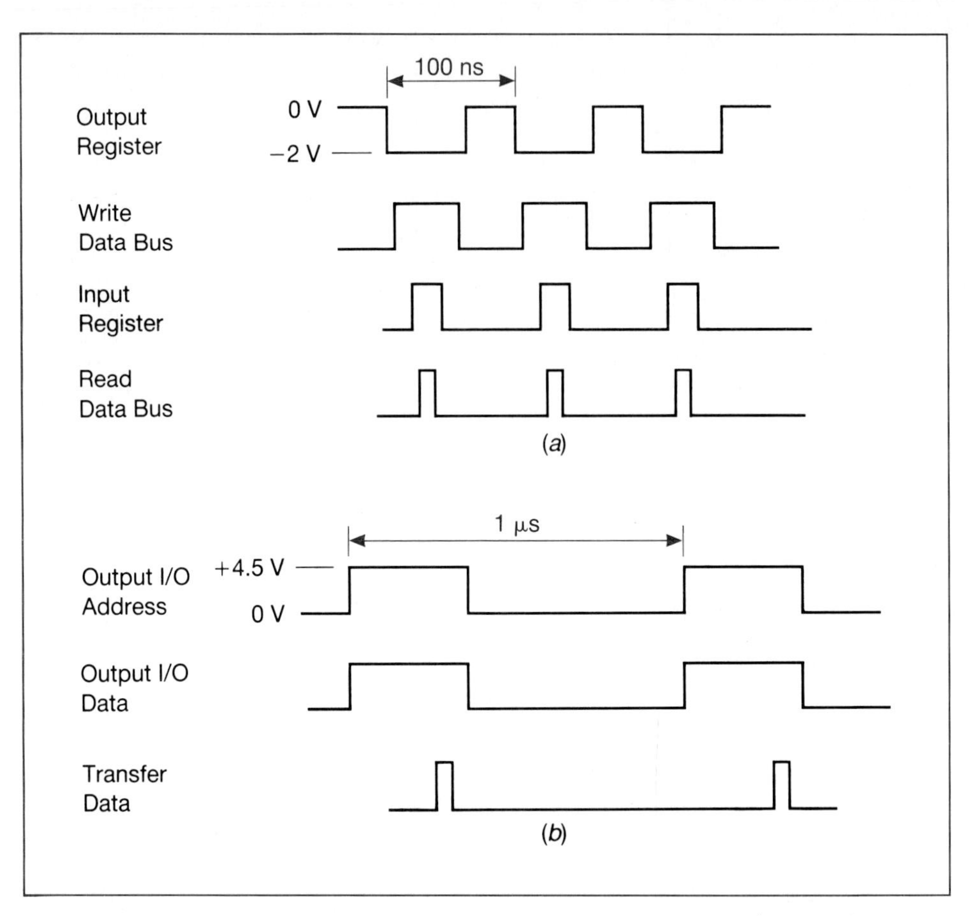

FIGURE 9.2 Timing for data transfers on (*a*) the system bus and (*b*) the I/O bus.

that the bus length be short to reduce the time it takes to send electric signals from one end of the bus to the other. It also requires that the circuits used to send information on the bus have electrical characteristics that facilitate fast transfers.

On the other hand, the I/O bus must be designed to meet different requirements. First, it must be physically longer to allow room to attach many different peripheral controllers. Second, since transfer rates from most peripherals are much slower than memory–ALU–control unit transfer rates, the I/O bus can be built with lower-cost devices. Cost is more critical here because a large system can have dozens of devices attached to the I/O bus. This contrasts with the system bus, which typically has only a few attached units.

Addressing requirements for the I/O bus are also different. The actual number of peripheral devices attached to the computer will always be much smaller than the number of memory locations. Therefore, the number of bits needed to address peripherals will be much smaller than the number needed to address main memory.

Besides a narrower address bus, the data bus will usually have fewer data lines. This is because many peripherals only transfer 1 or 2 bytes at a time. The transfer cycles are also longer since most peripherals are mechanical devices that cannot transfer data as fast as main memory.

Figure 9.3 shows the system and I/O buses in more detail. We can see that the number of lines in corresponding parts of each bus is different. Because the system bus must support high transfer rates to keep the CPU busy, it must have a wide data segment (that is, many data bits). The variations in control buses reflect the fact that there are differences in control and status points between the CPU and peripherals.

The primary functions of the I/O unit are:

- To compensate for speed and timing differences between the CPU and the peripherals
- To adjust voltages from one level to another
- To convert the data lengths between the system and I/O buses

Now that we have seen what the I/O unit does, let's look at some specific implementations of the I/O unit.

FIGURE 9.3 Typical width of the system and I/O buses.

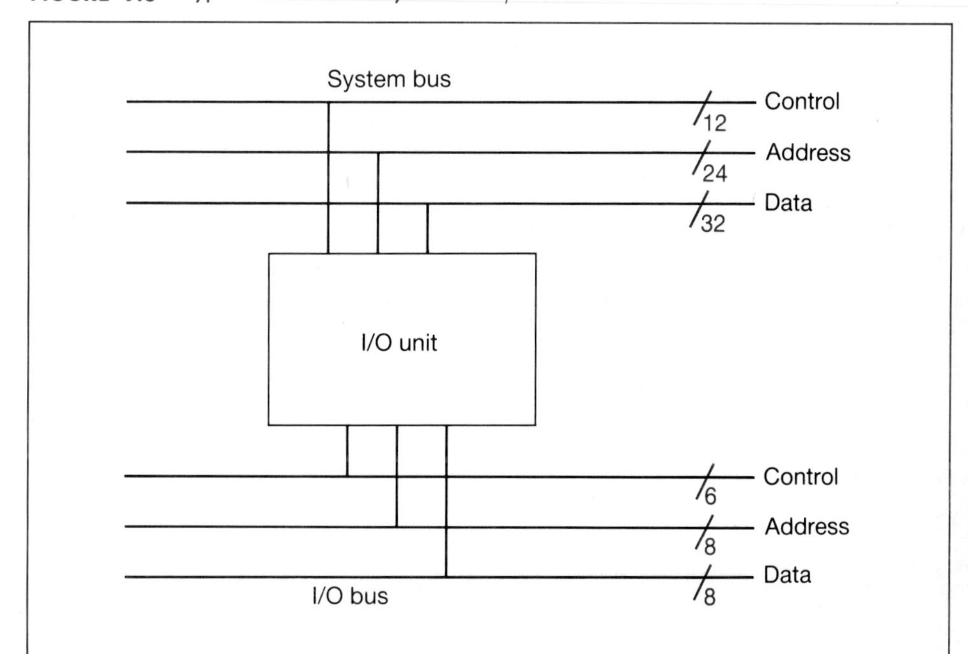

9.2 PROGRAMMED INPUT/OUTPUT—THE SIMPLE APPROACH

The simplest implementation of the I/O unit, which was proposed by von Neumann in his 1945 paper, calls for very little hardware. That original design was incorporated in many systems from early mainframes to current microcomputers. The implementation of the I/O unit shown in Fig. 9.4 is similar to von Neumann's proposal. It contains buffers for data going to and coming from the I/O bus and a simple control unit. Any voltage conversion needed between the buses is done by the AND gates that attach the buffers to the I/O bus.

The I/O unit contains two registers that are loaded from the system data bus: the I/O address and I/O command registers. As you might suspect, the I/O address register holds the address of a peripheral device. The I/O command register is used much like the instruction register in the control unit; it contains the next I/O oper-

FIGURE 9.4 Simple implementation of the I/O unit.

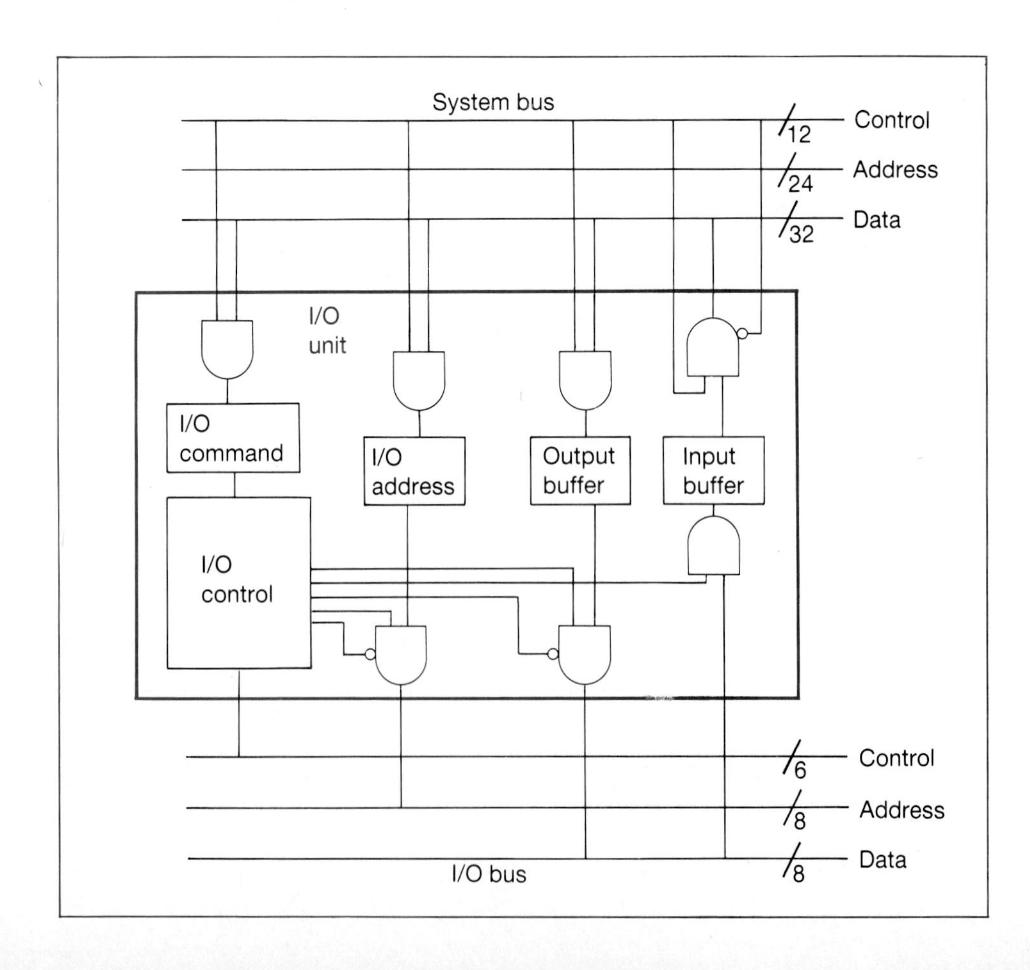

ation to be executed. The block labeled I/O control uses the I/O command register to determine which control signals need to be generated. This control circuitry can be implemented using either microprogrammed control or hardwired control. Since these techniques were described in detail in Chap. 5, we will skip the details here and focus on how the instruction set is affected by the addition of I/O to the system.

Two data move instructions that transfer information between main memory and the accumulator of the computer were defined in Chap. 5:

LDA M Load the accumulator from memory location M
STA M Store the accumulator into memory location M

One possible format of these instructions was shown in Fig. 5.3 and is repeated here as Fig. 9.5*a*. When we add opcodes to the instruction set to allow for I/O operations, something must be done to distinguish between them and the other instructions. One possible way to add extra opcodes for I/O instructions would be to make the length of the opcode field larger. If we did this, it would be necessary either to increase the size of instructions or to reduce the length of the operand field. One of the problems at the end of the chapter asks you to list the disadvantages of lengthening the opcode field. Because of those disadvantages, we will use the method that is shown in Fig. 9.5*b*.

With this method, a single opcode is defined for I/O instructions. The remaining bits in the instruction must be redefined to allow an opcode modality for the type of I/O instruction plus the I/O address. The format we have chosen uses 3 bits for I/O

FIGURE 9.5 Format of (*a*) memory and (*b*) I/O instructions.

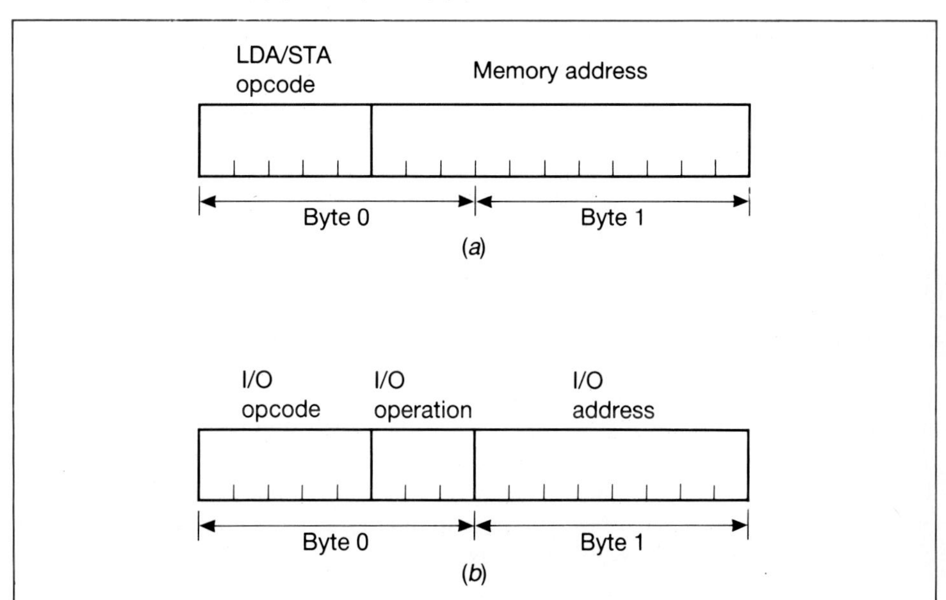

TABLE 9.1 I/O INSTRUCTIONS FOR SIMPLE COMPUTER

Operation	Typical mnemonic	Instruction meaning
Input	LIA D	Input data into accumulator from I/O device D
Output	OTA D	Output data from accumulator to device D
OR	MIA D	OR the data from device D into the accumulator
Start I/O	STR D	Initiate an I/O operation on device D
Stop I/O	STP D	Terminate the current operation on device D
Test I/O	TST D	Test the status of device D and set the EZ bit in the ALU to 1 if the operation requested has been completed
Clear Status	CLR D	Set the status register in device D to 0
No Operation	NOP	Unused I/O opcode

instruction operation (I/O opcode) and 8 bits for I/O address. By decoding these fields we can have 8 different I/O instructions and 256 I/O addresses.

Table 9.1 defines the I/O instructions that we will include on our computer. The first are similar to instructions that were described in Chap. 5. They load data into the accumulator, store data from the accumulator, and logically OR data with the accumulator. The main difference is that they address locations in the I/O address space instead of the memory space. They also require longer execution times.

The next three instructions shown in the table are new and relate only to I/O operations. They are used to start I/O operations, cancel the current I/O operation, and test to see if the last operation has been completed. The clear instruction will reset the status bit in the peripheral controller to 0. The final instruction, the NOP, represents an unused opcode. To illustrate how these instructions are used, we will look at a data transfer operation from a paper tape reader. This is a primitive I/O device that you will probably see only in a museum, but its simplicity makes the example easier to understand.

Figure 9.6 shows that the reader is attached to its controller which in turn is attached to the I/O bus. The controller contains two registers labeled Control and Data. To read a character, it is first necessary to send the device a read command. This is done by addressing the peripheral controller and storing a byte into its control register. The output of the control register will send a signal to the drive motor telling it to start moving the tape.

When the first character has been read, it is transferred into the data register. After transferring data, the reader sets the status bit in the controller to indicate that information is in the data register. The program controlling the read operation will then sense that the status bit is set and read the data into the ALU's accumulator.

We will assume the paper tape reader continues reading until it is given a stop command. With this assumption, we can write an assembly-language program that reads two 8-bit characters from the reader and stores them into a 16-bit memory location. The program and its corresponding flowchart are shown in Fig. 9.7.

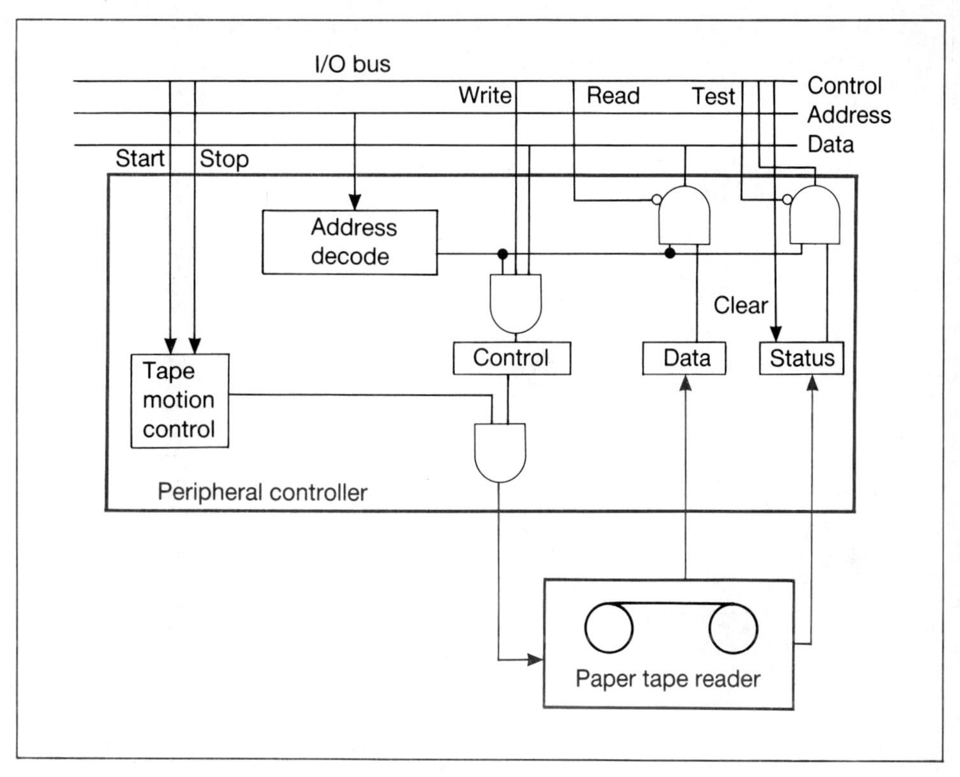

FIGURE 9.6 A peripheral controller and device attached to the I/O bus.

The program begins by sending a Clear command to the status register in the controller. It then loads a Read Forward command into the accumulator and outputs that to the control register in the peripheral controller. Next, the Start command is sent to the controller. This in turn causes an electric signal to be sent to the motor in the paper tape reader.

The program uses a technique known as **wait loop I/O** to determine when a character has been read. This technique requires that the CPU repeatedly test the status of the device to see if the requested operation has been finished. The wait loop is necessary because the peripheral device is much slower than the CPU. The instructions

```
WAIT1   TST   READER   * Set EQ register in ALU to 1 if the read
                       * is complete and to 0 if not complete
        JNE   WAIT1    * Jump to WAIT1 if EQ bit is not 1
```

form the wait loop in this example. When the status of the EQ register indicates that the first read is complete, the program loads the first byte into the accumulator. The

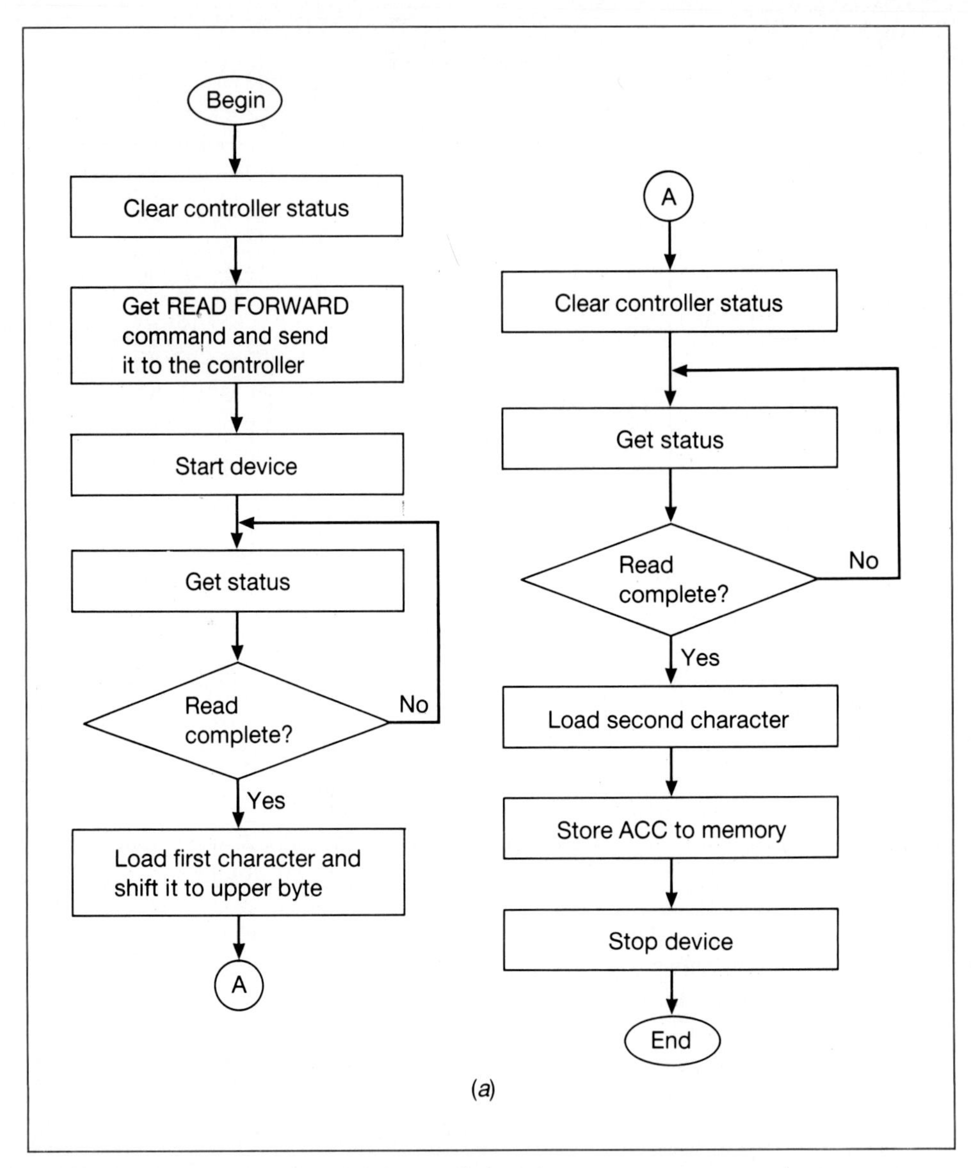

FIGURE 9.7 Program to read 2 bytes of data from a paper tape reader into memory: (*a*) flowchart; (*b*) assembly-language listing.

byte is shifted to the upper half of the accumulator with the multiply instruction. (Most computers have a multiple-bit shift instruction for this operation.) The operations are then repeated to merge a second byte into the accumulator. The 2 bytes are then stored into one word in main memory.

```
Label   Inst  Operand   Comment
START   CLR   READER    * Set the controller status to NOT DONE
        LDA   READCMD   * Load READ FORWARD command from memory
        OTA   READER    * Output the READ FORWARD command to
                        * the command register in the controller
        STR   READER    * Initiate the READ FORWARD command
WAIT1   TST   READER    * Set EQ register in the ALU to 1 if the read
                        * is complete and to 0 if not complete
        JNE   WAIT1     * Jump to WAIT1 if EQ bit is not 1
        LIA   READER    * Load the character from the reader data register
                        * into the lower 8 bits of the accumulator
        MUL   C256      * Shift byte into upper byte of the
                        * accumulator by multiplying it by 256
                        * This was done because the computer has
                        * no multiple bit shift instruction.
        CLR   READER    * Set the status register to NOT DONE
WAIT 2  TST   READER    * Set EQ register in ALU to 1 if the read
                        * is complete and to 0 if not complete
        JNE   WAIT2     * Jump to WAIT2 if EQ bit is not 1
        MIA   READER    * OR character from the data register
                        * into 8 bits of the accumulator
        STA   RWORD     * Store the 2 bytes from the reader
                        * into memory location RWORD
        STP   READER    * Terminate the READ FORWARD command
```

(*b*)

FIGURE 9.7 (*Cont.*)

Many computers can do I/O operations using the wait loop technique. Although it is simple and easy to use, wait loop I/O has a distinct disadvantage. The CPU wastes a lot of time testing and waiting for an I/O operation to complete. This is not a problem when the CPU has nothing else to do; however, in systems that support more than one user at a time, this type of I/O is not acceptable.

There is another major problem with wait loop I/O. Some devices, such as disk drives, transfer data so rapidly that the wait loop program cannot keep up with the device. A question at the end of the chapter asks you to calculate the maximum transfer rate that wait loop I/O can support. In the next two sections we will look at improvements in computer architecture that overcame the problems resulting from wait loop I/O.

9.3 INTERRUPTS AND MULTIPROGRAMMING

There are three problems associated with the method of I/O transfer described in the previous section:

- It wastes CPU time.
- It has a low maximum transfer rate.
- It uses the accumulator for data transfers.

CPU time is wasted because, as we just saw, a lot of time is spent executing instructions in the wait loop. By calculating the maximum transfer rate possible with wait loop I/O, you can easily see that speed using wait loop I/O is very limited.

Use of the accumulator may seem like an unlikely problem, but it also contributes to loss of efficiency. Assume that the accumulator is being used for an arithmetic calculation when an I/O operation is needed to get the next operand. Before the I/O operation can be executed, the contents of the accumulator must be saved, since the accumulator is needed for I/O. After the I/O is completed, the accumulator will have to be restored so that the computation can continue.

An important concept had to evolve before improvements could be made in the methods used to transfer data between main memory and I/O devices. Designers had to find a way to let the CPU know when an I/O operation was complete, but in a way that would not use CPU time. The solution was to adopt a concept that is commonly used in many types of communication. When someone wishes to get your attention, they tap you on the shoulder, call your name, or call you on the phone. Each of these operations interrupts your current activity.

In the computer, an **interrupt** is a signal to the CPU telling it that an event has occurred that calls for CPU intervention. Interrupts are used for many different purposes within a computer, including coordinating I/O transfers. They can signal power or component failure, a special condition within a program, and an exceptional state within the ALU (such as an arithmetic overflow or divide-by-zero status). Just as there are several uses for interrupts, there are several variations in the implementation of interrupts within computers. We will describe only one implementation, and we will limit our discussion to interrupts associated with I/O operations.

The addition of interrupts allows the CPU to start an I/O data transfer without requiring that it go into a wait loop until the transfer is complete. Since the CPU does not go into a wait loop, it can switch execution to another program while the I/O operation for the first program is occurring. The ability to execute two or more programs concurrently is usually called **multiprogramming**, and it will be described in more detail in Part III.

To see how interrupts facilitate multiprogramming, look at the flowchart shown in Fig. 9.8. As the figure indicates, the CPU first starts executing program 1 and continues doing so until that program requests an I/O operation. Since the operating system is responsible for all I/O operations, the program must call the oper-

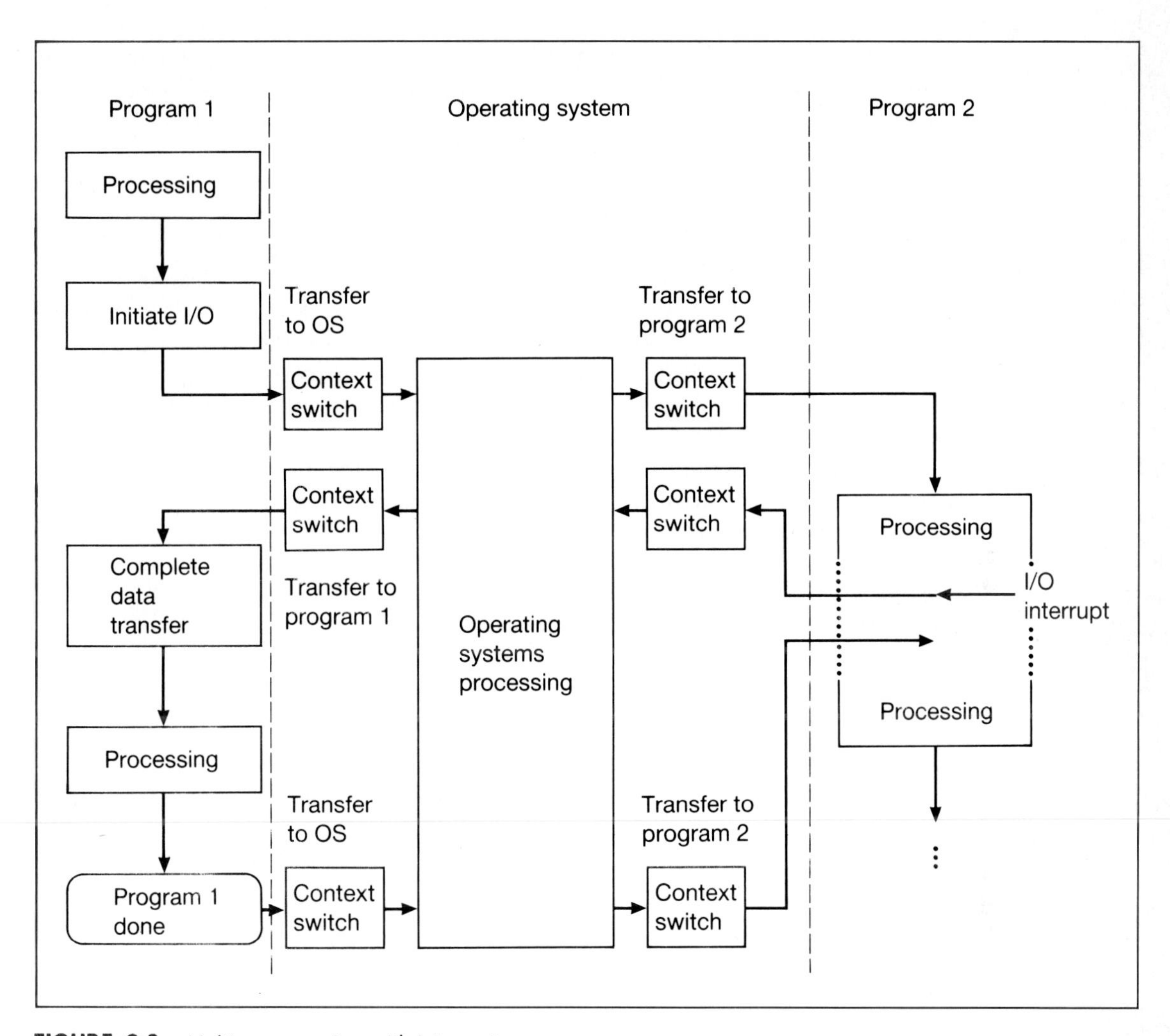

FIGURE 9.8 Multiprogramming with interrupts.

ating system to make the request. In order to switch from the user program to the operating system program, the current CPU status associated with the user program is saved and control is transferred to the operating system. Saving the status of one program and transferring control to another program is called a **context switch**. The program status that is saved before a context switch includes the contents of the:

- Program counter
- Accumulator
- ALU condition register

In machines with multiple accumulators, this can require a significant number of cycles. The part of the operating system responsible for context switching is discussed in Chap. 14.

After the operating system initiates the I/O operation requested by the user program, it looks for another user program to execute while the I/O operation continues. Assuming there is a second user program to execute, another context switch is made to that program and it starts executing. If program 2 calls for no I/O operations, the CPU continues executing it until an I/O Complete interrupt is received from the I/O device that program 1 was using. When the I/O completion interrupt is received and acknowledged, the CPU first saves the status of program 2. Having done that, it then transfers execution back to the operating system. The operating system completes the I/O operation and then transfers control back to the first user program.

The simple example in Fig. 9.8 shows only two programs, but in practice there may be many more. Now that we have seen how interrupts are used, we will turn our attention to the hardware that implements them.

FIGURE 9.9 The I/O unit with interrupt logic.

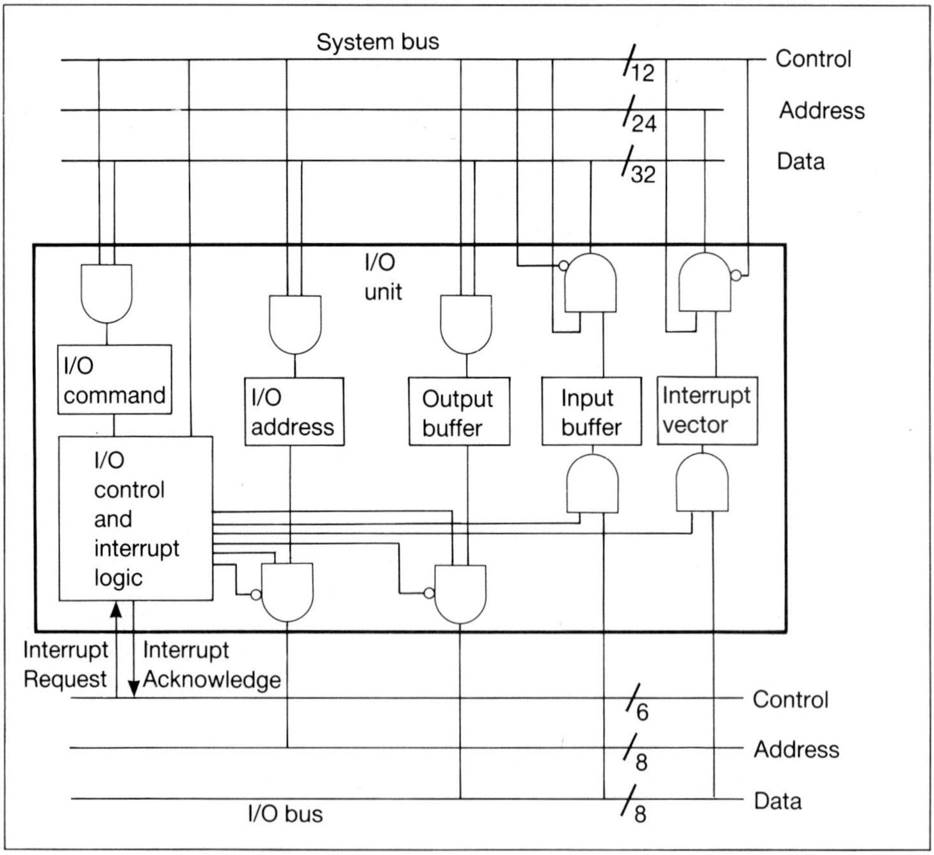

In Fig. 9.9 we have redrawn the I/O unit to show the hardware necessary to implement interrupts. An Interrupt Vector register and interrupt logic have been added to the I/O unit. The **interrupt vector** is a pointer or address that tells the CPU where to find the program that will respond to the interrupt. When a device has completed an I/O operation, it signals its controller, which then initiates the interrupt sequence. The events that follow are:

1. The peripheral device finishes an operation and sends a Completion signal to its controller.
2. The controller sends an Interrupt Request signal to the I/O unit.
3. The I/O unit returns an Interrupt Acknowledge signal to the controller.

FIGURE 9.10 Sequence of events for an I/O completion interrupt.

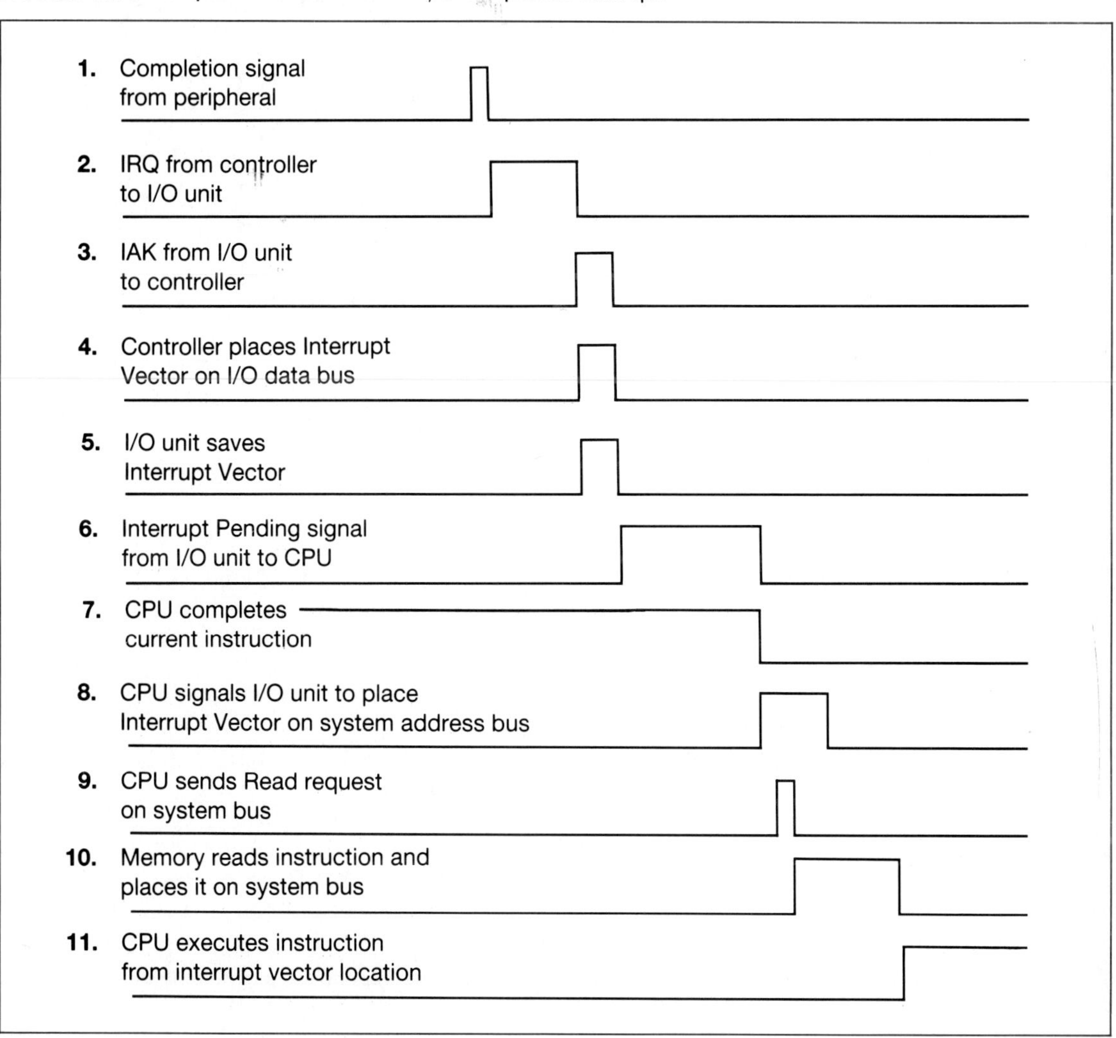
1. Completion signal from peripheral
2. IRQ from controller to I/O unit
3. IAK from I/O unit to controller
4. Controller places Interrupt Vector on I/O data bus
5. I/O unit saves Interrupt Vector
6. Interrupt Pending signal from I/O unit to CPU
7. CPU completes current instruction
8. CPU signals I/O unit to place Interrupt Vector on system address bus
9. CPU sends Read request on system bus
10. Memory reads instruction and places it on system bus
11. CPU executes instruction from interrupt vector location

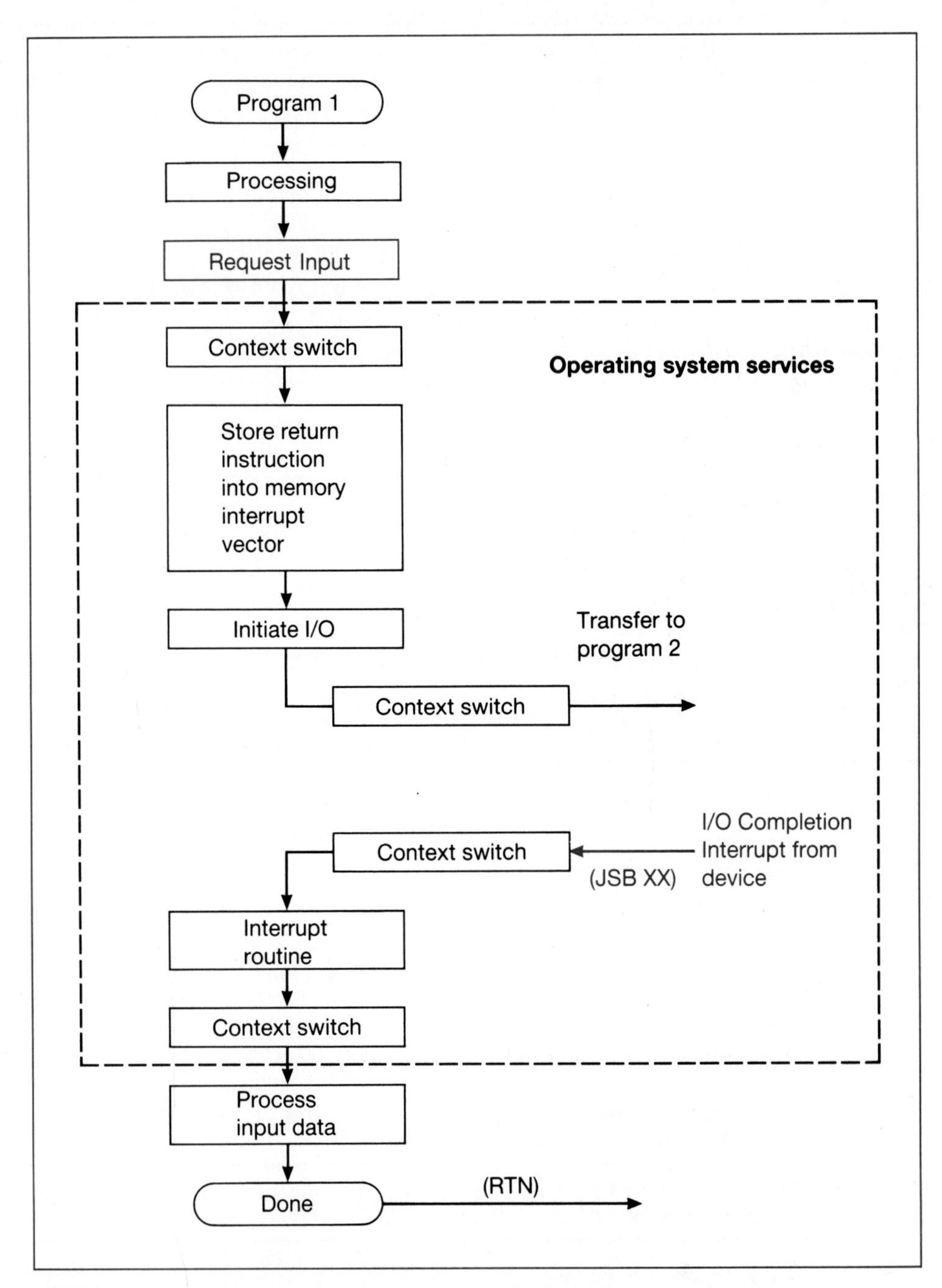

FIGURE 9.11 Preparing for an I/O operation and responding to an interrupt.

4. The controller deposits its interrupt vector onto the I/O bus, and the I/O unit stores this vector in the Interrupt Vector register.
5. The I/O unit saves the interrupt vector.
6. The I/O unit places an Interrupt Pending signal onto the system bus.
7. The CPU completes execution of the instruction in its instruction register.
8. The CPU places an Enable Interrupt Vector signal on the system bus, enabling the tristate AND gates that transfer the interrupt vector onto the system bus.
9. The control unit places a Memory Read signal on the system bus.
10. The memory unit accesses the location specified in the interrupt vector and transfers the contents (an instruction) onto the system bus.
11. The CPU executes the instruction from the location specified by the interrupt vector.

This sequence of events is shown in Fig. 9.10.

A good question to ask now is, "What instruction was fetched and stored into the IR?" If the instruction were a JMP XX, control would be passed to a program at location XX. This would accomplish part of what we wanted, transfer of control to a different program. It would also create a problem. How does the computer return to the interrupted program after the I/O operation has been completed?

We can solve the problem by using a JSB XX instead of a JMP XX instruction. The JSB instruction not only transfers control to a new location, it also saves the address in the program counter so that it can be restored when needed. Figure 9.11 shows the steps that must be performed to prepare for an I/O operation and respond to an interrupt.

As a programmer you might be concerned about the extra steps added to the program to prepare for an I/O operation and to respond to the interrupt. In Fig. 9.11 we have enclosed these steps in a box labeled Operating system services. This is done to show that the steps included are done on behalf of the user by the operating system software. More information about the services provided for the user by the operating system will be given in Chaps. 12 and 13.

Interrupts were a very significant step in freeing the CPU to do useful work while I/O operations are being performed by the system. They do not, however, completely solve the problem. The CPU must still be interrupted each time the device has data to transfer. To reduce the CPU interaction needed during an I/O operation even further, we must add more hardware to the computer.

9.4 DIRECT MEMORY ACCESS BY I/O DEVICES—DMA

The key to reducing the I/O-related work load in the CPU is to add more hardware to the system. This hardware works in parallel with the CPU to transfer data to memory while the CPU is executing user programs. An expanded I/O unit that has additional registers and control logic is shown in Fig. 9.12. This hardware enables the I/O unit to transfer data directly from a peripheral to main memory without

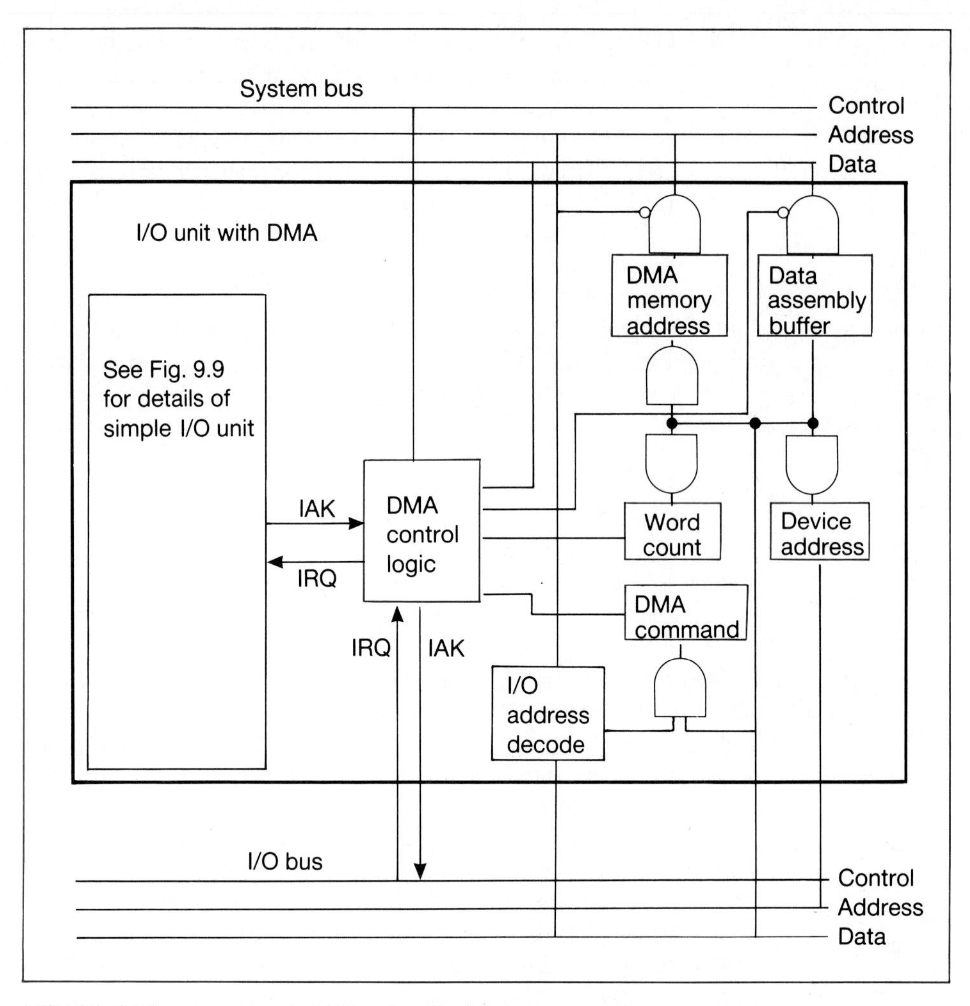

FIGURE 9.12 An expanded I/O unit with direct memory access.

going through the control unit and ALU. Because it provides a direct path to memory for I/O data transfers, it has been given the name **direct memory access**, or **DMA**. The following is a brief description of how the DMA unit operates.

Although the DMA unit has some "intelligence" or decision-making ability in its control logic, it must be told what operations to perform. This is done by having the CPU store information into the DMA's controller. This is called "programming" the DMA. Since the CPU must store information into the DMA controller's registers, these registers must be addressable by the CPU. We will assign I/O addresses to these registers and treat them as though they were peripherals.

Let's assume we want the DMA unit to transfer a block of data from a peripheral into memory and then examine the DMA program for this operation. The program and its corresponding flowchart that set up the DMA are shown in Fig. 9.13.

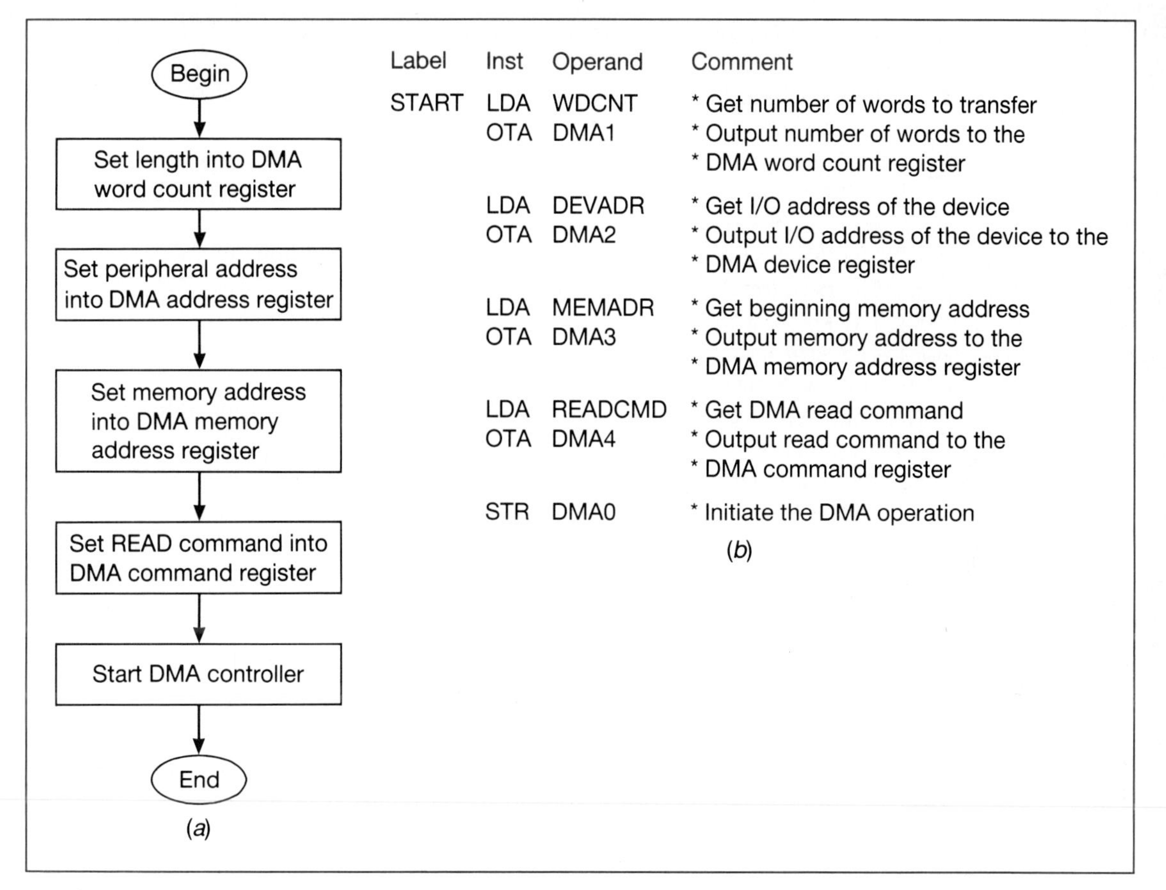
Begin
Set length into DMA word count register
Set peripheral address into DMA address register
Set memory address into DMA memory address register
Set READ command into DMA command register
Start DMA controller
End
(a)
Label Inst Operand Comment
START LDA WDCNT * Get number of words to transfer
OTA DMA1 * Output number of words to the
* DMA word count register
LDA DEVADR * Get I/O address of the device
OTA DMA2 * Output I/O address of the device to the
* DMA device register
LDA MEMADR * Get beginning memory address
OTA DMA3 * Output memory address to the
* DMA memory address register
LDA READCMD * Get DMA read command
OTA DMA4 * Output read command to the
* DMA command register
STR DMA0 * Initiate the DMA operation
(b)

FIGURE 9.13 Program to set up a DMA transfer from a peripheral to main memory: (*a*) flowchart; (*b*) assembly-language listing.

As the program shows, programming the DMA is quite simple. All that is required is to store the necessary information into the registers in the DMA. Once this is complete, the DMA transfer is initiated with the I/O Start command just as though it were a peripheral. Because the simple DMA we have defined performs only data transfers, the CPU must then send commands to the I/O unit that will cause the peripheral device to start the desired operation. The sequence of instructions for this is similar to the sequence used to start the paper tape reader described in Sec. 9.2. After the peripheral has been started, the DMA unit takes over control of the data transfer. It sends all the signals to the system and I/O buses necessary to complete the transfer.

The DMA waits until it receives an interrupt request from the peripheral whose data transfer it is controlling. Starting with the Interrupt Request signal from the peripheral, the following sequence of events happens:

1. The peripheral controller places an Interrupt Request signal on the I/O bus, signifying that it is ready to transfer data.
2. The DMA recognizes that the Interrupt Request is from the device it is controlling and intercepts the Interrupt Request. Interrupts from other peripherals would be passed on to the I/O interrupt logic.
3. The DMA sends an Interrupt Acknowledge to the peripheral.
4. The peripheral places data on the I/O bus.
5. The DMA stores the data into its data assembly register.
6. The DMA takes control of the system bus. (The mechanism to do this is discussed later.)
7. The DMA places the contents of its memory address register onto the system bus along with a Memory Write request.
8. After completion of the Memory Write, the DMA relinquishes control of the system bus.
9. The DMA increments its memory address register and decrements its word count register.
10. If the word count register is zero, the DMA sends an Interrupt Request to the I/O unit.

Each time the device is ready to transfer data to memory, the DMA unit takes over and performs the operations described above. These operations are sometimes called **cycle stealing** because the DMA unit "steals" the system bus and main memory long enough for one bus/memory cycle. The sequence of events described above is shown in Fig. 9.14.

The DMA must share the system bus with the CPU. Therefore, something must be responsible for deciding which will be given control of the bus when both want the bus. The logic that makes this decision is called the bus arbiter and was described in Chap. 7. Figure 9.15 shows that the bus arbiter monitors the bus request lines from the DMA and the CPU. If only one request is received, it is granted. If both request the bus for the same cycle, the arbiter gives control to one of the requesters and forces the other to wait. In most systems, DMA requests are given priority over CPU requests, because some devices may lose data if they are forced to wait. If the CPU is forced to wait, only time is lost, not important data.

When the transfer of a block of words is complete, the DMA issues an Interrupt Request to the I/O unit. This request is processed the way we described in the previous section. The CPU stops executing its current program long enough to acknowledge the interrupt, check status, and send any necessary commands to the device. One of the problems at the end of the chapter asks you to estimate the time the DMA saves the CPU. When calculating this, you must consider the extra work necessary to program the DMA and the cycles stolen by the DMA to do data transfers.

If you reexamine Fig. 9.12 closely, you will see that the DMA shown has a severe deficiency; it can only perform reads from a device to memory. One of the problems

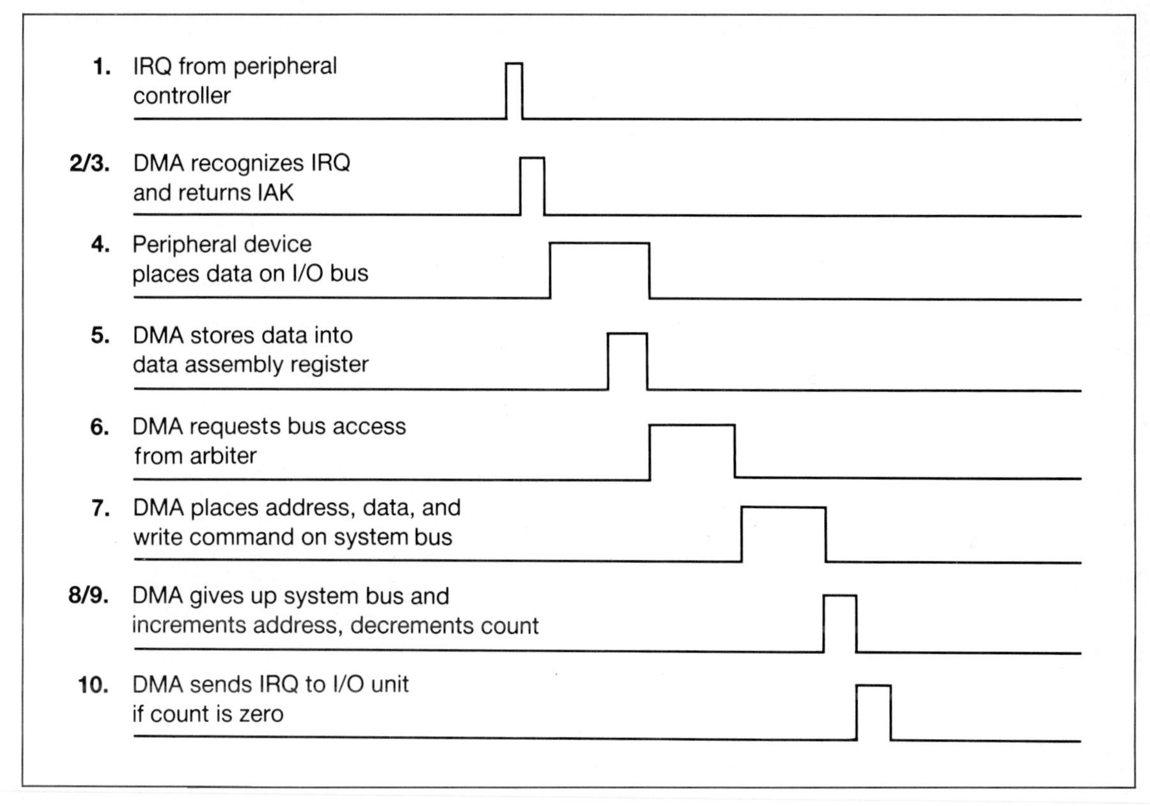

FIGURE 9.14 Sequence of events associated with a DMA transfer.

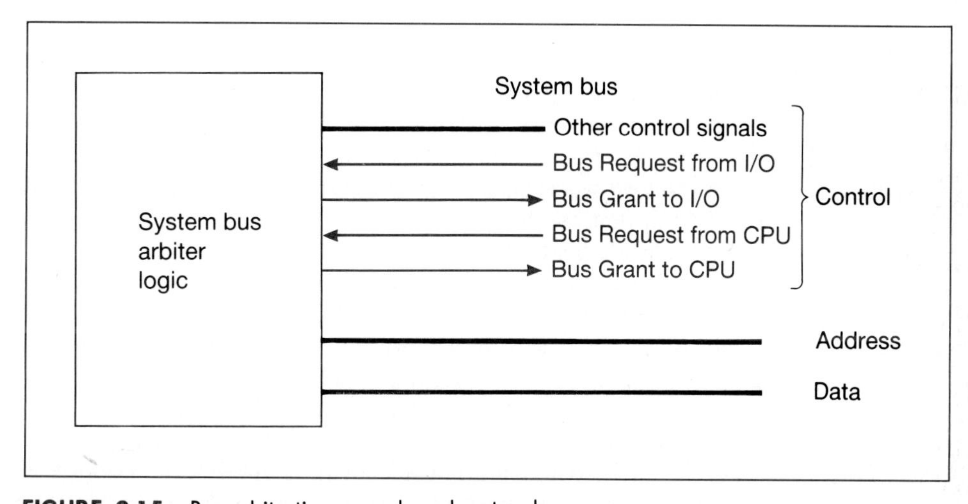

FIGURE 9.15 Bus arbitration on a shared system bus.

at the end of the chapter asks you to modify this diagram to allow for memory writes.

What we have described in this section is a very simple DMA controller. Additional capabilities that have been added to DMAs include:

- Logic to combine multiple I/O bytes into one word whose length is the same as the memory word (**data assembly**)
- The ability to control more than one peripheral at a time
- The ability to start and stop devices
- The ability to store into nonsequential locations in memory (**scatter store**)
- The ability to accept a sequence of transfer programs at one time (**DMA chaining**)

DMA solves the problem of low data transfer rates associated with wait loop I/O. Since it can steal every cycle, it is possible for DMA to transfer data at the maximum rate of the system bus.

Although it was not always credited with being so, the DMA was probably one of the first practical implementations of multiprocessing. By **multiprocessing** we mean that two or more processing units are sharing the work load of the machine. DMA was a significant step in the development of computer architecture.

9.5 I/O CHANNELS AND I/O PROCESSORS

DMA offers significant improvements over simple I/O units with interrupts. Although these improvements were sufficient for most small- and medium-sized systems, there are some deficiencies. First, the DMA must be programmed by the CPU for each transfer that is done. Second, to do multiple I/O operations concurrently, multiple DMAs are required. Some computers solved the second problem by including a DMA for each device on the system. Another disadvantage of this solution is that its cost can be quite high.

Because of the disadvantages of DMA, many computers needed more sophisticated I/O processing units. The device that was used on most IBM 360/370 computers (and on several other computers) is called an I/O channel. An **I/O channel** is an I/O unit that has the power of a simple CPU. By this we mean that it can fetch and execute I/O instructions.

Figure 9.16 shows how the architecture of the computer system is changed when the simple I/O unit is replaced by I/O channels. The system in the figure has two I/O channels, each with peripheral controllers that are in turn attached to one or more peripheral devices. These channels and the CPU share the system bus and require an arbiter like the one described in the previous section.

Some systems have three types of I/O channels: multiplexor channels, selector channels, and block multiplexor channels. **Multiplexor channels** were designed to allow several slow- to medium-speed devices, such as terminals, printers, and card

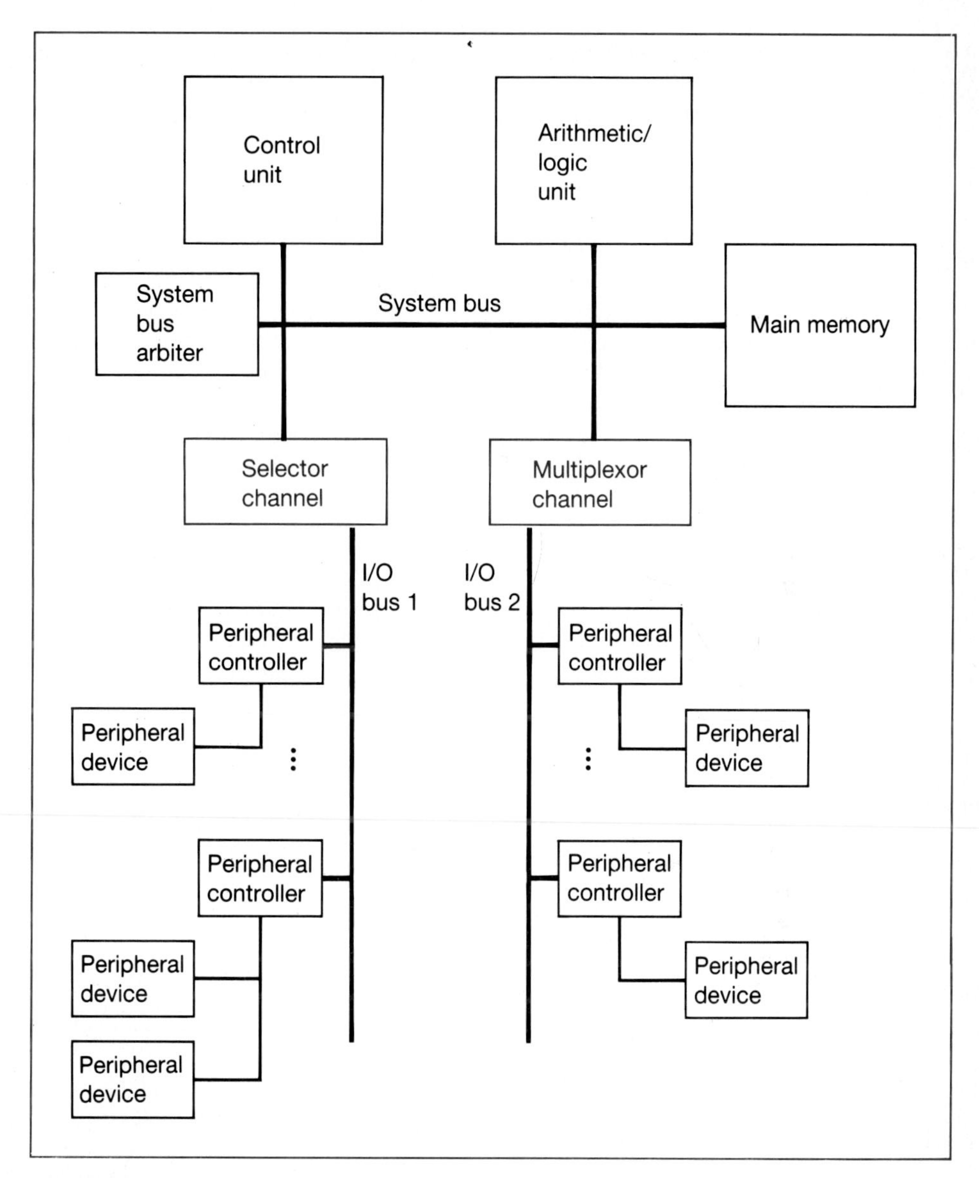

FIGURE 9.16 A system with a CPU and two I/O channels.

readers, to use the system concurrently. You can think of the multiplexor channel as a device that contains several DMAs operating in parallel.

Selector channels were designed for high-speed transfers between a "selected" peripheral and main memory. They are capable of high-speed data transfers and are used to connect high-speed peripherals such as disk drives to the system. A selector channel can only service one peripheral at a time.

Block multiplexor channels have the characteristics of both selector and multiplexor channels. They are capable of transferring blocks of data at high speed from peripherals to main memory. The block multiplexor channel has another advantage over a selector channel. It can start an operation, such as repositioning a disk read/write head, and switch to another operation while the disk head is moving. In this respect, it is like a multiprogramming computer itself.

As you recall, DMAs are programmed by first sending information to each of the control registers and then telling the DMA to start. Programming an I/O channel is more like programming a CPU. A sequence of instructions is first stored in main memory. The I/O channel then fetches and executes these instructions the same way the CPU would. The individual instructions are called **channel command words**, or **CCWs**. A sequence of these CCWs is called a **channel program**.

The CPU is responsible for creating a channel program and then storing it into main memory. The CPU then executes a Start IO instruction that causes the I/O

FIGURE 9.17 A system with both a CPU and an IOP.

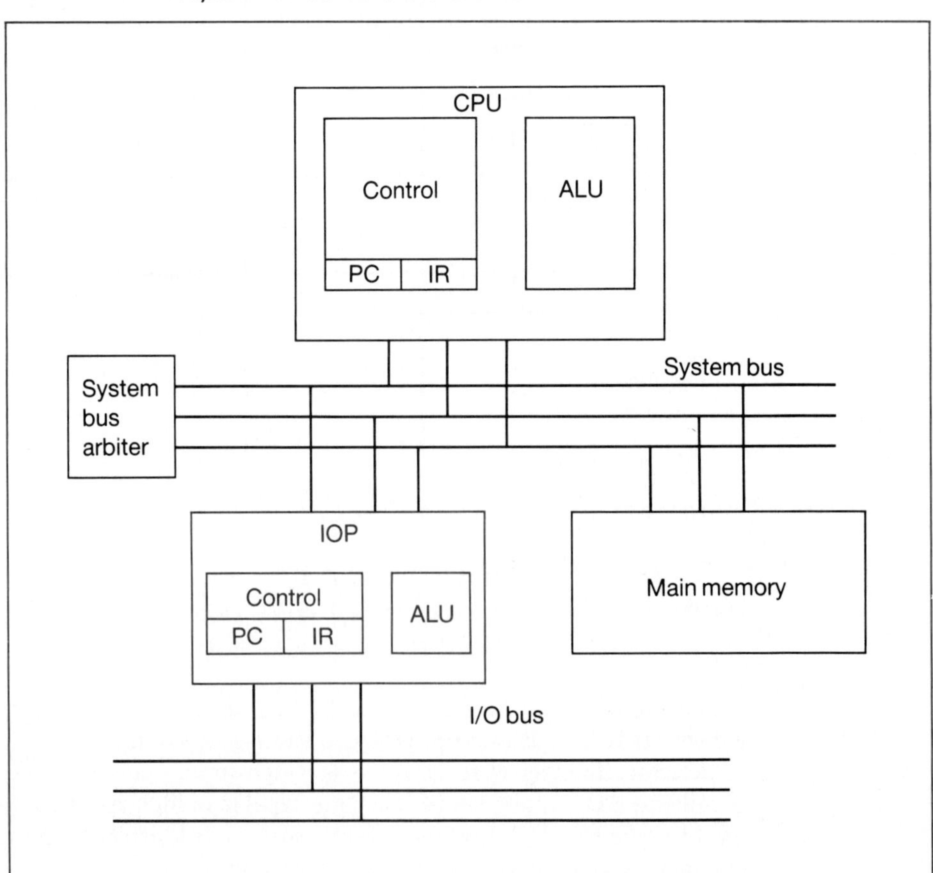

channel to start executing the channel program. I/O channels have most of the characteristics of a CPU but can execute only simple channel programs.

The next logical step in the development of I/O units was to replace the I/O unit with a general-purpose CPU. Since these processors were dedicated to I/O operations, they were called **input/output processors**, or **IOPs**. You can visualize an IOP by considering it to be a second computer sharing the system bus with the CPU. Figure 9.17 shows the essential elements of a system that includes both a CPU and an IOP. You will notice that the IOP has all the features found in a CPU. These include a program counter, instruction register, ALU, and other registers.

There are differences between the capabilities of the two, however. The ALU contained within the IOP is usually much simpler than the one found in the CPU. It is usually unnecessary for the IOP to do calculations using floating-point or double-precision numbers. Therefore, these capabilities are typically not provided in the IOP's ALU. The control portion of the IOP is also simpler than that of the CPU, because the number of different instructions that the IOP needs is less than the number the CPU needs. However, the IOP operates exactly the same as the CPU; it goes through the standard fetch, decode, and execute phases.

Just as with DMA and I/O channels, a bus arbiter is needed to resolve conflicts over access to the system bus. This unit determines whether the CPU or IOP should have the bus for a given bus cycle and forces the one not chosen to wait for the next cycle.

Since IOPs have more capabilities than simple DMA controllers or I/O channels, their programs can be more sophisticated. For example, programs for IOPs can control transfers of blocks of data from multiple peripherals concurrently just as the I/O channel does. Because the IOP has an ALU, it can transform data as it is being transferred from a peripheral to main memory. Consider the case where information stored on a tape is in ASCII format and the program processing the data requires that the data be in EBCDIC. Each time a character is read from the tape, the program in the IOP converts it from ASCII to EBCDIC. It then stores the character into main memory. The data conversion could be done easily by the CPU, but having the IOP do the conversion saves CPU time. Usually, this will improve system performance.

Although this has been a very brief introduction to I/O channels and I/O processors, it has given you a glimpse of the power of such devices.

9.6 A SINGLE-BUS ORGANIZATION FOR A COMPUTER

To this point in the chapter we have assumed that peripheral devices are not attached to the system bus. This was certainly the case for most computers until the early 1970s. At that time Digital Computer Corporation, DEC, introduced a new minicomputer with a very different bus organization. The computer was called the PDP-11, and the bus organization was called the Unibus. The PDP-11 was destined to become one of the most popular computers of its time. One reason for its popularity was its bus organization.

A single common bus gained even more popularity with the arrival of microcomputers. A typical system based on a single-bus organization is shown in Fig. 9.18. In the figure you can see that both main memory and peripheral devices are attached to the common system bus. Let's examine the single-bus organization and investigate its advantages and disadvantages.

If we look at the system from the CPU's point of view, it appears that all devices attached to the bus act like memory. On such a system, the CPU doesn't have to distinguish between a location in memory and an I/O device. **Memory-mapped I/O** is the term that is used to describe the situation where peripherals and memory are treated identically.

When a bus is designed only to provide a transfer path between main memory and the CPU, the implementation can be quite simple. Enough time is allocated to allow the memory to complete its operation after each memory access request. This

FIGURE 9.18 A computer organization based on a single bus.

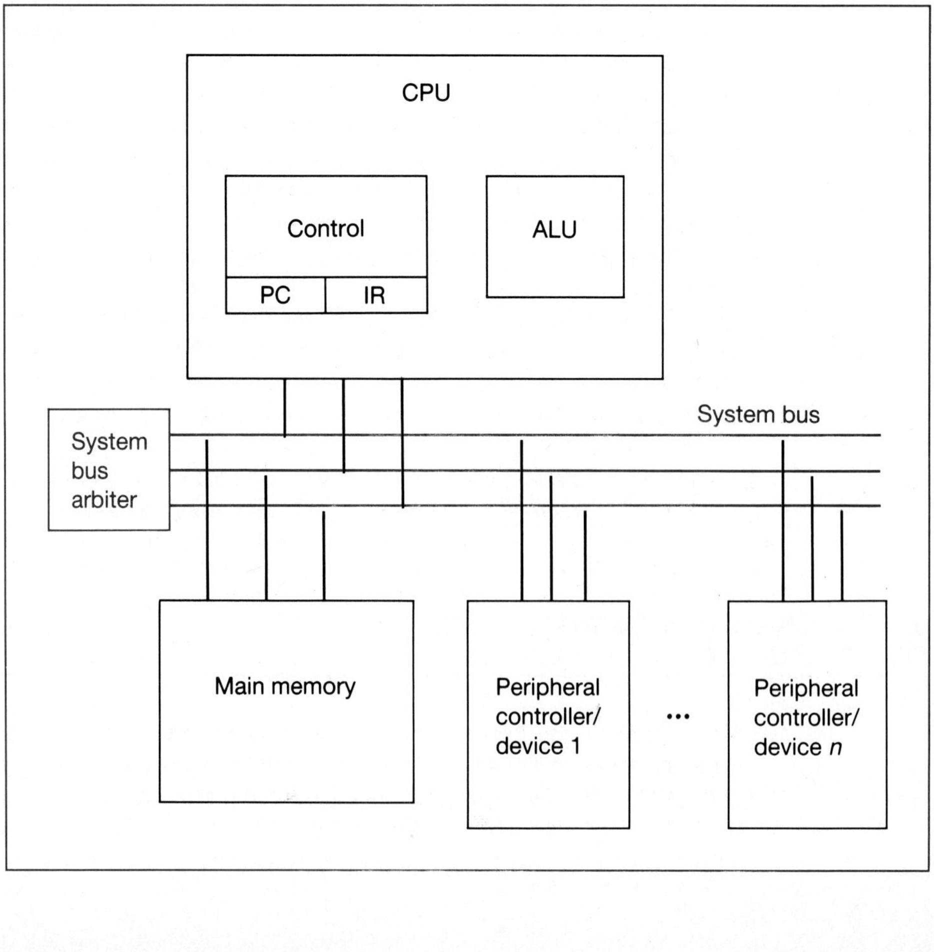

is a very predictable operation once we know the characteristics of the memory elements. When peripheral devices are added to the bus, the design is complicated because not all peripherals respond in the same amount of time.

One way to solve this problem is to add an additional line to the control portion of the bus. Such an addition is shown in Fig. 9.19*a*, where we see that the control bus now includes a Ready line. As Fig. 9.19*b* shows, after the CPU sends out an address and read request, it waits for the Ready line to go to logical 1 before it loads data into its accumulator. When main memory is being accessed, it will probably respond with a Ready signal fast enough so that the CPU will not have to wait. However, when a peripheral is being accessed, very long waits may be required. In the worst case, the wait could be as long as the time a CPU spends in the I/O wait loop described earlier. Interrupts and DMA can be used to eliminate these long waits just as they were in the earlier examples.

There is an advantage to having a Ready line added to the bus and forcing the CPU to check the line before completing each transfer. By using a Ready line, the designer can add memories to the system that have different access times and not have to redesign the bus. This may allow the designer to take advantage of the speed or cost of different memory devices.

In a memory-mapped I/O system the LDA XX instruction used to load the accumulator from memory can also be used to load the accumulator from a peripheral. This implies there is no longer any need for I/O instructions! Not having special I/O instructions is one of the real advantages of this approach over the more conventional multiple-bus approach.

Eliminating an entire subset of instructions from the set of instructions that the CPU must decode and execute should decrease the complexity of the control logic (or microprograms) in the CPU. This in turn decreases the cost of the CPU.

A second advantage is that the single-bus organization simplifies direct transfers of data between any two devices attached to the bus. These transfers take place without the need to pass the data through an intermediate I/O unit. This can sometimes improve performance.

Unfortunately, there are also some disadvantages to the single-bus organization. First, since there is no longer a separation of memory address space and I/O address space, peripherals and memory must share a common address space. Consider the case where the address bus has 16 address bits. Sixteen bits gives an address space of 65,536 unique locations. On a multiple-bus system, this means that main memory size can contain 65,536 words. On the single-bus system, only part of the 65,536 address locations can be assigned to memory, and the remaining addresses must be assigned to peripherals. There is a simple solution to the address space problem—add more lines to the address bus. Increasing the address bits will, however, increase the cost of the system and possibly increase the length of the instructions.

In the multiple-bus organization, the system bus is usually designed to operate as fast as the main memory can respond. This makes transfers over the bus very fast. When the length of the bus must be extended to allow I/O devices to be attached, the speed of the bus must be reduced. The design must also include more complex

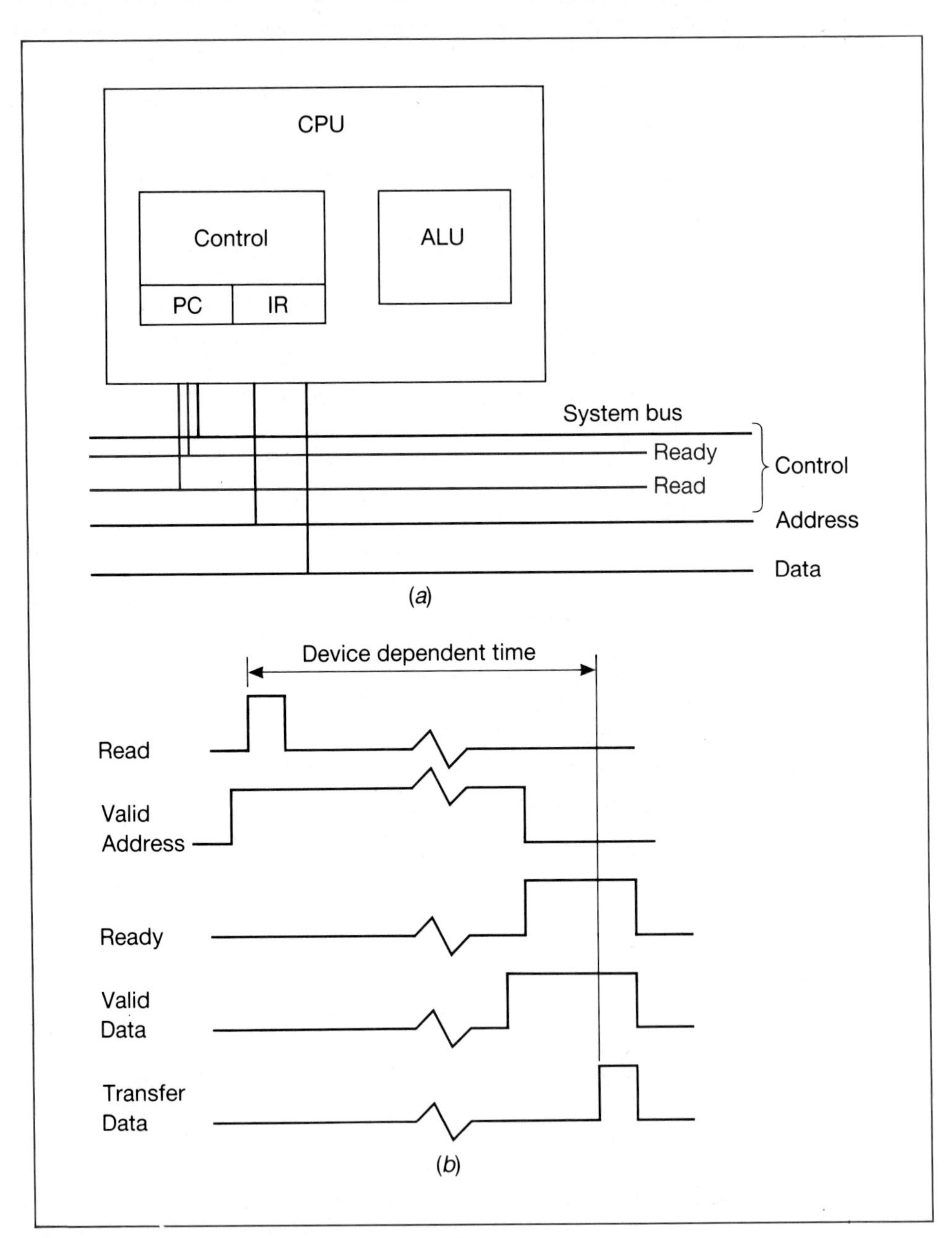

FIGURE 9.19 Synchronization of data transfers using a Ready line: (*a*) system bus organization; (*b*) timing for data transfers.

bus control information. For these reasons, shared common buses are usually slower than the system bus in a multiple-bus organization. Because the single bus is slower, system performance may be reduced. Later versions of the PDP-11 system tried to overcome the speed limitation by adding a separate bus. The added bus only attached high-speed memory to the CPU. This solution obviously defeated some of the advantages of the original Unibus design.

As we have shown above, there are both advantages and disadvantages to the single-bus organization, and there is no clear-cut "winner" between the two organizations. The Intel 808X family of microprocessors, for instance, has been designed to work with either approach.

9.7 SUMMARY

The I/O unit of the computer has undergone several significant changes in the past 30 years. One of the first major advances in the design of the I/O unit was the addition of interrupts to the computer. This feature led to the development of multiprogramming on single CPU systems.

After interrupts were added to computers, the I/O unit progressed through several evolutionary steps from a simple buffer between the system bus and I/O bus into a sophisticated processing unit. In many applications today where large amounts of data must be transferred within a system or where many interrupts must be processed, the performance of the I/O unit is equally important to that of the CPU. We looked at how an alternate bus organization for the computer, the single, or common, bus affected performance. In that section we saw there were both advantages and disadvantages to the single-bus organization. The designer must weigh these before making a decision as to which organization should be used for a particular application.

In the next chapter we will step slightly outside the computer and examine peripheral controllers and peripheral devices themselves in more detail.

PROBLEMS

9.1 In addition to the peripherals mentioned in the introduction to the chapter, list 10 other peripherals that are used on computers.

9.2 Redraw Fig. 9.1 to include cache memories.

9.3 What determines which voltage levels will represent logical 0 and 1?

9.4 We have seen that one of the functions of the I/O unit is to convert the data lengths between the system and the I/O buses. This often involves making a tradeoff between the I/O bus width and the maximum I/O data rate. Considering Figs. 9.2 and 9.3, what is the maximum I/O data rate for this system (with respect to the internal system rate)?

9.5 What would be required to double the maximum I/O rate discussed in Prob. 9.4?

9.6 Would it be reasonable to triple the maximum I/O data rate of Prob. 9.4? Explain.

9.7 What determines the widths of the address and data parts of the I/O bus?

9.8 Section 9.2 stated that there are disadvantages to lengthening the opcode field of the instruction to allow designation of the specific I/O operation to be performed. Explain what these disadvantages are. (*Hint*: Refer to Fig. 9.5 and recall our discussion of opcode modalities in Chap. 7.)

9.9 Why do I/O instructions require longer execution times than instructions that reference memory?

9.10 What do the terms *memory address space* and *I/O address space* mean?

9.11 What type of I/O operations on a computer using wait loop I/O would cause the computer to wait indefinitely?

9.12 How could we eliminate the function of the peripheral controller? Do you think this is ever done?

9.13 Calculate the maximum data transfer rate when using wait loop I/O for a computer with a 100-ns instruction cycle time.

9.14 Figure 9.7 shows the flowchart and the assembly-language program for reading 2 bytes of data from paper tape into memory. Prepare a detailed diagram that shows all required registers and the path of the data between the paper tape reader and the memory.

9.15 The early versions of the Apple II computer used only wait loop I/O and not interrupts. Why do you think they did this?

9.16 Figure 9.7 presumes that the paper tape reader moves the tape in the forward direction. Some paper tape readers were also capable of reading the tape in the reverse direction. What changes would be required to the flowchart and assembly-language listing in Fig. 9.7 to implement read backward for a paper tape reader?

9.17 Estimate the possible percentage savings in CPU cycles that would occur by using interrupts instead of wait loops for a long paper tape read operation (assuming a 100-ns CPU cycle and a 600 byte/s paper tape read rate).

9.18 Give a noncomputer example to illustrate the use of interrupts.

9.19 Estimate the time needed to perform a context switch on a computer with a 100-ns cycle time. What assumptions did you make in determining this number?

9.20 Give a noncomputer example to illustrate context switching.

9.21 Where can return addresses for subroutine jumps be stored in the computer?

9.22 Some computers execute a special Interrupt Jump instruction rather than a Jump Subroutine when an interrupt occurs. What operations do you think the Interrupt Jump performs?

9.23 Why does the interrupt vector come from the I/O controller rather than the CPU itself?

9.24 Referring to Prob. 9.22, what operations do you think an Interrupt Return would perform?

9.25 When we say the DMA has some "intelligence," what do we mean?

9.26 The DMA unit shown in Fig. 9.12 can only read data from devices. Modify it so that writes can be performed also.

9.27 What does the CPU do while the DMA is stealing a cycle?

9.28 If peripherals and DMA were capable of transferring data at the rate of the system bus, what could happen to the CPU?

9.29 If a system had capabilities described in Prob. 9.28, what might you do to the bus arbiter?

9.30 Why did we say DMA was an example of multiprocessing?

9.31 Estimate the CPU time saved by using DMA rather than interrupts to transfer a block of 128 characters from a peripheral to memory in a computer with a 100-ns cycle time.

9.32 If a computer has an IOP, is there any need for the CPU to have its own interrupt system? Explain.

9.33 Why does an IOP need its own ALU?

9.34 Section 9.5 states that one of the functions that is sometimes performed by an I/O processor is the conversion of ASCII data to EBCDIC, and vice versa. Discuss what arithmetic or logical capabilities the IOP must have to perform this task.

9.35 Is there any need for DMA or I/O channels on a system with a single common bus? Explain.

9.36 Write a program segment that transfers two words from a peripheral device to main memory on a system with a single common bus.

9.37 What are the advantages of a single common bus over the system bus I/O bus architecture?

9.38 On the DEC PDP-11 it was possible for each controller to have its own DMA circuitry. How would bus arbitration be implemented on a system like this?

9.39 Why do large mainframe computers usually *not* have a single common bus?

9.40 What is the maximum data transfer rate for a computer with a 100-ns instruction cycle time performing wait loop I/O? Use the program in Fig. 9.7 to estimate the number of instructions required per transfer.

9.41 Is DMA effective when a system only transfers short blocks of data (8 bytes at a time)? Justify your answer.

9.42 Is it possible for a computer to have two CPUs and two IOPs sharing one memory unit? Explain.

Peripheral Controllers and Devices

We saw in the previous chapter that the I/O unit is an integral part of the computer. It is designed by the computer manufacturer to match the system bus. Peripherals, on the other hand, are often designed and built by companies other than the computer manufacturer. In fact, the peripheral manufacturers often don't know which computers will use their products until they are on the market. To further complicate matters, peripheral manufacturers are more interested in making devices that deliver high performance at a low cost than in worrying about how to connect them to different computers.

Since peripheral manufacturers may not know where their product will be used, designing electrical and mechanical interfaces that match all possible host computers is not feasible. In addition, the peripheral manufacturers may not have the expertise or personnel necessary to build a variety of interfaces for their products. Therefore, in many cases an extra hardware unit must be placed between the peripheral device and the I/O unit of the computer. This hardware is often called a **peripheral controller**, **peripheral adapter**, or **peripheral interface**. It provides all the necessary electrical and mechanical conversions needed to connect a peripheral device to the computer.

The following sections cover how several peripheral devices operate and how they are connected to computers. We will also discuss how their performance can be measured and compared. We will then look at some innovative ways to connect peripherals to systems so that they can be shared by many computers.

10.1 PERIPHERAL CONTROLLERS—THE INTERFACES BETWEEN THE COMPUTER AND EXTERNAL DEVICES

As stated above, the peripheral controller provides an *interface* between the computer and a peripheral. In this chapter we take a very narrow view of the term

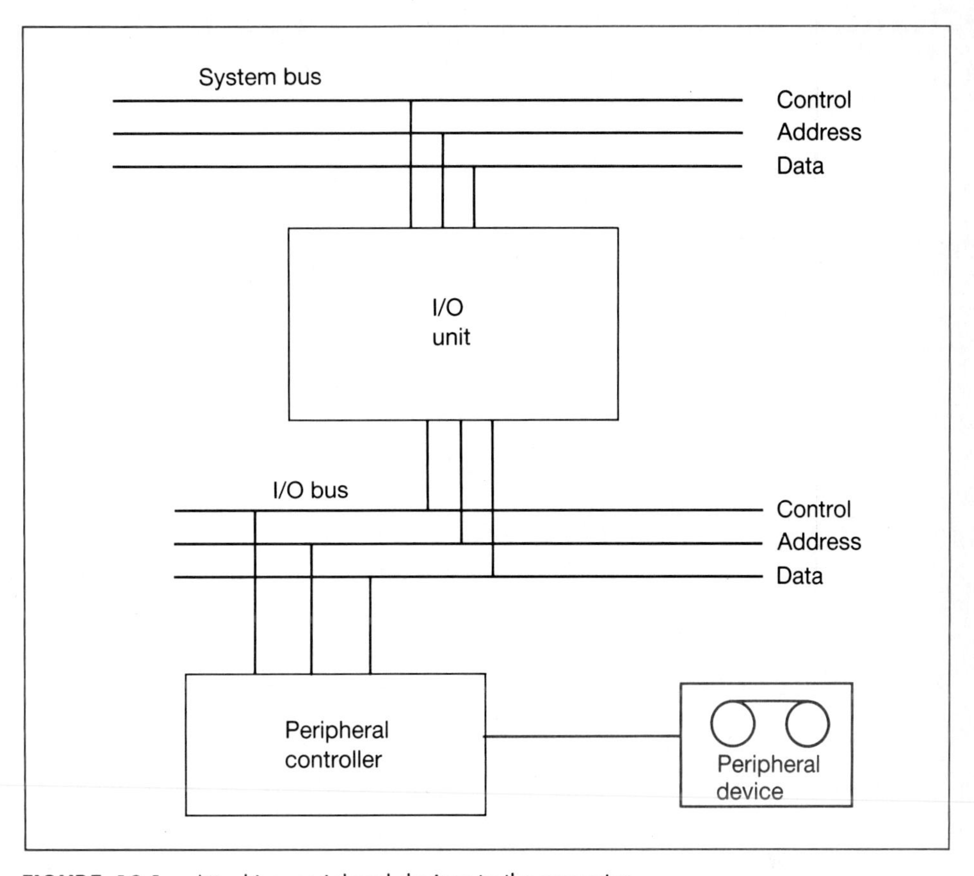

FIGURE 10.1 Attaching peripheral devices to the computer.

interface. We define an **interface** to be the hardware module that facilitates communication between different devices. This relationship between the I/O unit, peripheral controller, and peripheral is shown in Fig. 10.1.

To gain insight into the functions that a controller must provide, we will look at one in some detail. The controller we will examine provides the interface between the computer I/O unit and a dumb terminal. A **dumb terminal** simply sends characters from its keyboard to the computer and displays characters received from the computer on its screen. Figure 10.2 shows the information flow between the computer's I/O bus, the controller, and the terminal.

Most terminals transmit and receive ASCII characters, and from Chap. 3 we recall that the ASCII characters require 7 bits. Therefore, 7 bits of data plus a parity bit are sent each time a character is transmitted or received by the terminal. The two most obvious ways to send the characters are by parallel transmission or by serial transmission. Most terminals have been designed to transmit and receive ASCII characters as serial data. That is what we will assume for this example.

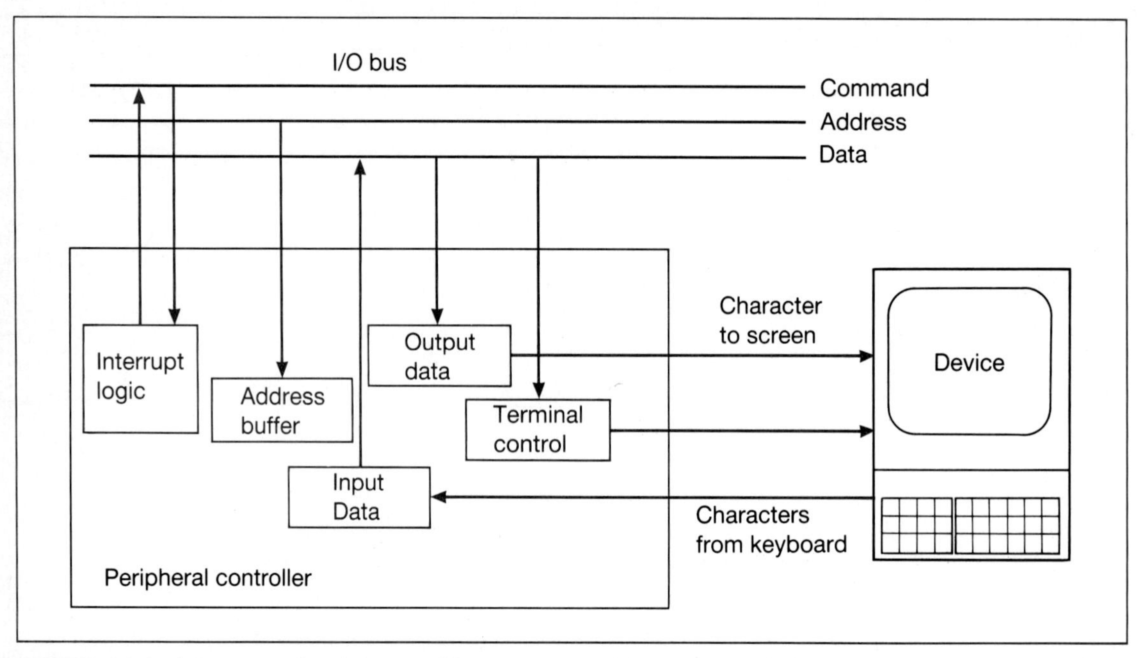

FIGURE 10.2 Information flow between I/O unit, peripheral controller, and terminal.

Our controller must have serial receivers and transmitters to communicate with the terminal. Figure 10.3 shows the voltage levels and timing of the waveform of an ASCII character being transmitted between the terminal and the controller. A logical 1 is represented by a voltage amplitude of +12 V and a logical 0 by a voltage amplitude of −12 V. For the example, we have chosen a transfer rate of 2400 bits/s, which is an industry standard transmission rate. This gives a time period for each bit of 416.67 μs.

From the figure we see that a start bit precedes the data and a stop bit follows the data. The **start bit** defines the beginning of a character, and the **stop bit** defines the end of the character. Since the start and stop bits are different voltage levels, it is

FIGURE 10.3 Voltage waveform for serial data transmitted to and from a terminal.

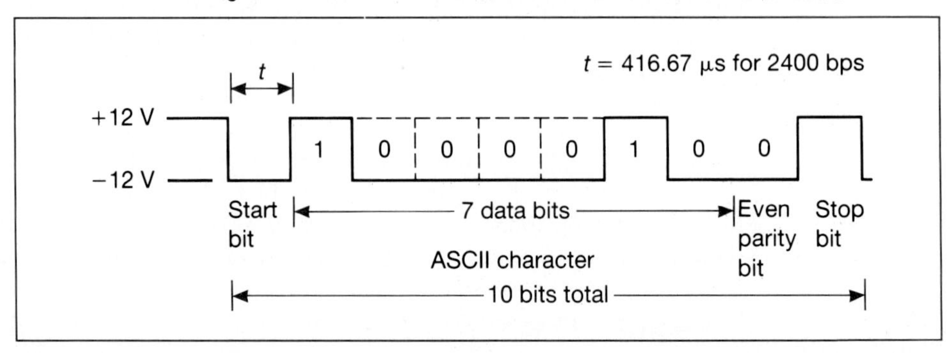

easy to tell where a valid character begins. The data format shown in the figure corresponds to the RS-232 standard. It was developed by the communication industry so that users could connect devices from different manufacturers and have a reasonable assurance that they would work correctly.

The peripheral controller accepts the stream of bits from the terminal and converts it into a form that can be transmitted to the I/O unit on the I/O bus. For this example, assume that the I/O bus is 8 bits wide. This implies that each time the I/O unit requests data from the peripheral controller, 8 bits will be transmitted: 7 data plus parity.

The sequence of events that causes a character to be transmitted from the terminal to the accumulator is described below and shown in Fig. 10.4.

1. The CPU sends an address and read command to the I/O unit. (The specifics of this were described in detail in Chap. 9.)

FIGURE 10.4 Data transmission between the I/O unit, the peripheral controller, and the device.

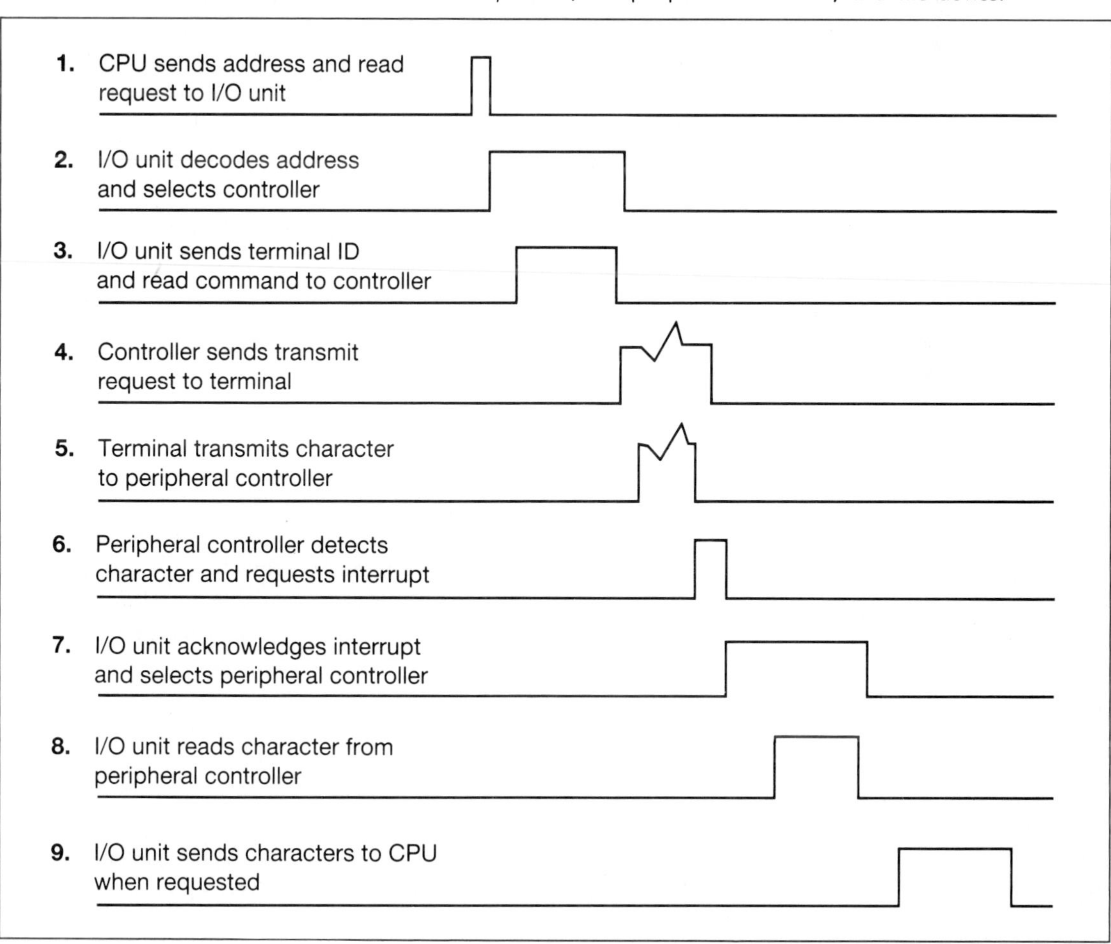
1. CPU sends address and read request to I/O unit
2. I/O unit decodes address and selects controller
3. I/O unit sends terminal ID and read command to controller
4. Controller sends transmit request to terminal
5. Terminal transmits character to peripheral controller
6. Peripheral controller detects character and requests interrupt
7. I/O unit acknowledges interrupt and selects peripheral controller
8. I/O unit reads character from peripheral controller
9. I/O unit sends characters to CPU when requested

2. The I/O unit decodes the address and selects the correct peripheral controller.
3. The I/O unit sends the terminal identification (address) and read command to the peripheral controller.
4. The peripheral controller sends a transmit request to the terminal.
5. After a key on the keyboard has been pressed, the terminal transmits the character to the peripheral controller.
6. The peripheral controller detects that a character has been received and sends an interrupt request to the I/O unit.
7. The I/O unit sends an interrupt acknowledge to the peripheral controller.
8. The I/O unit reads the character from the peripheral controller.
9. The I/O unit sends characters to the CPU when it is ready. (This transfer was also described in Chap. 9.)

In step 3 of the sequence it was stated that the I/O unit sends the terminal identification to the peripheral controller. In general, peripheral controllers can be connected to more than one device. Therefore, a controller must be capable of accepting address information from the I/O unit and converting the information into signals that select one of its peripherals. Figure 10.5 shows a functional diagram of a peripheral controller that includes all the features described above.

The time required for steps 4 and 5 is a function of the speed at which the terminal can transmit data. Such time variations show another function of the controller. It compensates for speed differences between devices and CPUs.

The functions of the peripheral controller we have discussed so far are:

- Converts from one voltage level to another
- Compensates for timing differences
- Supplies device selection and control
- Provides temporary data storage
- Converts from one data format to another

There is one more function that I/O controllers must often provide: mechanical connections. Manufacturers seldom choose mechanical connectors for the I/O buses in their systems that are the same as the connectors provided by peripheral manufacturers. Therefore, the peripheral controller must have a connector on one end that matches the one on the I/O bus and one on the other end that matches the connector coming from the peripheral device.

Now that we have seen what a peripheral controller does, we will describe briefly what a controller is physically. If you have ever opened an Apple or an IBM PC, you have most likely seen an example of an I/O controller. Many of the peripheral controllers for these personal computers are on cards that plug into connectors within the computer. These cards then have connectors to which cables from the devices can be attached. A typical controller on a personal computer is about 25 to 50 in^2 and has 10 to 75 integrated circuits.

On large mainframe computers, peripheral controllers sometimes interface many

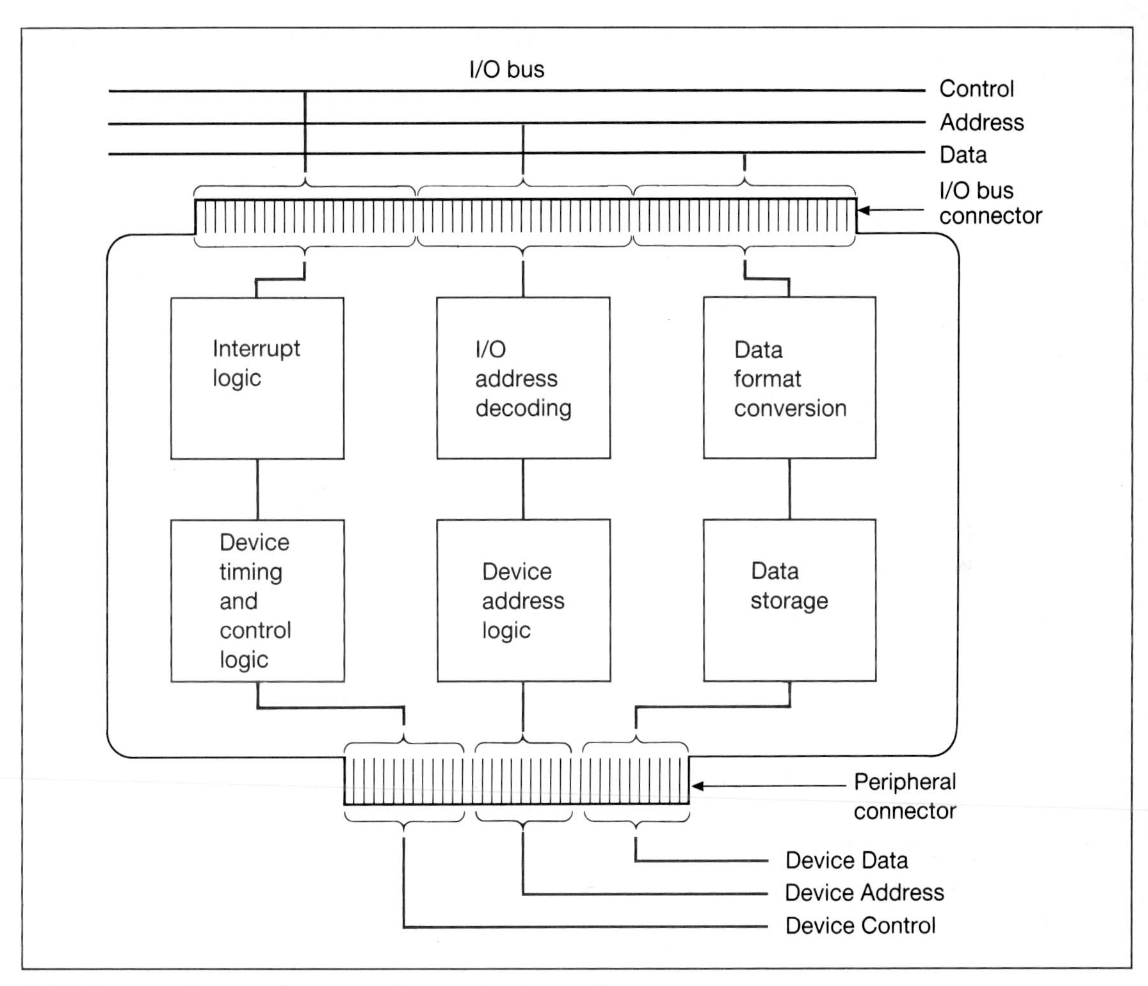

FIGURE 10.5 Functional diagram of a peripheral controller.

devices to the computer and are often quite large. These controllers may require one or more printed circuit boards that are 150 in^2 and have hundreds of integrated circuits. Even though these controllers may have their own power supply and internal processors, they perform the same basic functions that we described above.

The terminal controller we examined in this section is very simple, yet it includes most of the features found in controllers for more complex devices. Now that we have seen how to connect an external device to a computer, we will look at the characteristics of these peripherals themselves.

10.2 CLASSIFICATION OF PERIPHERAL DEVICES

In this section we will look at the types of devices that are typically attached to computers. To organize the discussion, we will group the peripherals into three categories: data input, data output, and data storage. This is not a terribly exact grouping, since some devices overlap the categories, but it will serve our purposes.

10.2.1 Data Input Peripherals

The devices in this category are primarily used to transmit data from the user or the outside environment to the computer. The most familiar element in this category is the keyboard. It allows the user to input single characters by depressing individual keys. Keyboards generate data at very low data rates, since human operators are quite slow compared to computer speeds. Because of the slow rates, most keyboards are interfaced to the computer over a serial data bus. Keyboards have been the standard way to input characters into the computer for many years. However, a number of variations have been designed that can be used by individuals who are physically unable to use the standard keyboard.

Early mainframe computers often did not have user keyboards and terminals directly attached. Instead, the keyboards were connected to devices that punched holes in cards. The punched cards were then placed into a card reader that then transferred data from one card at a time to the computer. On minicomputers, punched paper (or Mylar) tape was used rather than cards because the cost of the tape punches and readers was much less than the cost of card readers and punches.

Figure 10.6 shows the format of data storage for both punched cards and punched tape. As the figure shows, a "character" on a card has 12 bits, whereas one on tape has 8 bits. Cards, tape, and keyboards all have one element in common: they were designed to transmit information in the form of characters.

If users wanted to transfer only characters to the computer, keyboards would be sufficient. However, many of today's applications require more complex data input. Consider programs that manipulate graphic images, for example. They require information such as position, size, and color. We could use sequences of characters to input this information, but this would be a slow, tedious, and error-prone process. Therefore, new devices have been developed for graphic applications. The most common devices used for these applications are:

- Mouse
- Track ball
- Joystick
- Light pen
- Graphic tablet
- Touch screens

All these devices allow the user to provide coordinate information to the computer quickly and easily. Like keyboards, they transmit data at a low rate and therefore are often interfaced to a peripheral controller that transmits and receives serial data.

The **mouse** has become a standard input device for Apple Macintosh users, and it is also popular on many other personal computers and workstations. Simple mech-

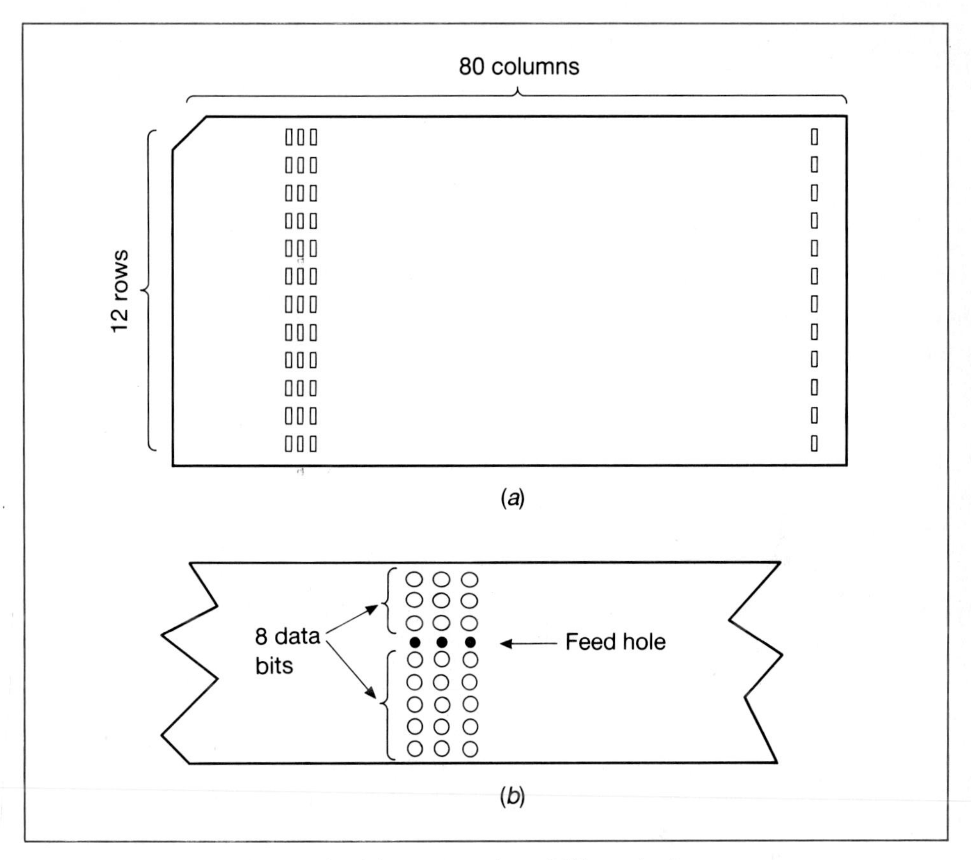

FIGURE 10.6 Data formats for (*a*) paper cards and (*b*) punched tape.

anical mice have a ball enclosed in a small plastic box as shown in Fig. 10.7. When the box is slid over a flat surface, the ball turns, and electronics within the mouse detect this movement.

The controller to which the mouse is attached keeps track of the movement and provides position information that can be displayed on the display screen. Mice also have one or more control buttons that can be pressed to tell the program when the mouse is currently pointing to a desired location. In the mouse world, the term *click on* is used to describe the operations of pointing the mouse cursor at an object on the screen and then pressing and releasing the mouse control button. Mice are also very convenient for selecting items from menus and moving elements from one location to another. Chapter 18 describes these operations in more detail.

A **track ball** is essentially a mouse turned upside down. With this device, the user rotates the ball rather than sliding it across a flat surface. Some users prefer the track ball because the box holding the ball does not have to be moved to generate position information. It stays in one place and requires less desk space.

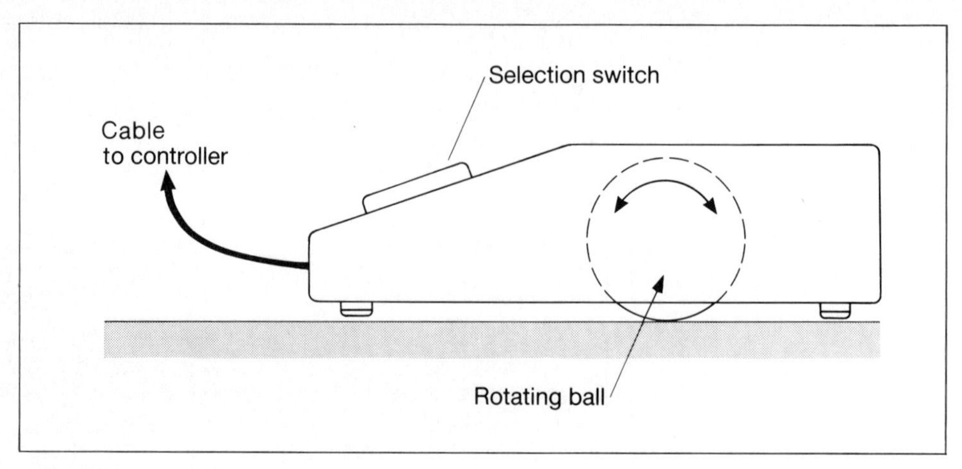

FIGURE 10.7 Diagram of a computer mouse.

Joysticks are devices that were not originally designed for computer applications. They were used to manipulate control surfaces on airplanes long before they were attached to computers. Their function is the same in both cases. Coordinate information is transmitted as the user moves the stick in the X and Y directions. A diagram of a joystick is shown in Fig. 10.8. Joysticks also have one or more buttons that are used to interrupt the computer when a desired location is reached.

Light pens are devices that detect the light emitted by a display. The circuitry driving the display moves an electron beam across the display, causing it to emit light. When circuitry in the controller detects that the light pen is over a spot that is currently being illuminated, the position of the pen is calculated. Since the light pen produces coordinate information, it can be used for the same applications as mice, track balls, or joysticks. The name *light pen* was chosen because the light sensor was placed in a device that looked like a pen with a cable attached. Light pens also have buttons attached so that objects may be selected by clicking on them. A diagram of a light pen is shown in Fig. 10.9.

Several years ago the graphic tablet was the primary graphic input device. The **graphic tablets** themselves were flat surfaces that had sensors around their edges. The sensors could detect the position of a pointer that was touching its surface. In this respect the graphic tablet and light pens are quite similar. One popular use for graphic tablets was to trace lines from a photograph or drawing that was placed on the tablet's surface.

Touch screen devices are conceptually quite similar to graphic tablets. The major difference is that the sensors are placed around the surface of the video display rather than on an external tablet. You use your finger to point to a position on the screen that you want to select. Since a finger makes a large pointer, it is difficult to select small objects using the touch screen. On systems with a touch screen, one must be careful when pointing to objects on the screen. We know of several people who claimed they accidently reformatted their disks by touching the wrong locations on the screen while pointing out some feature on the display!

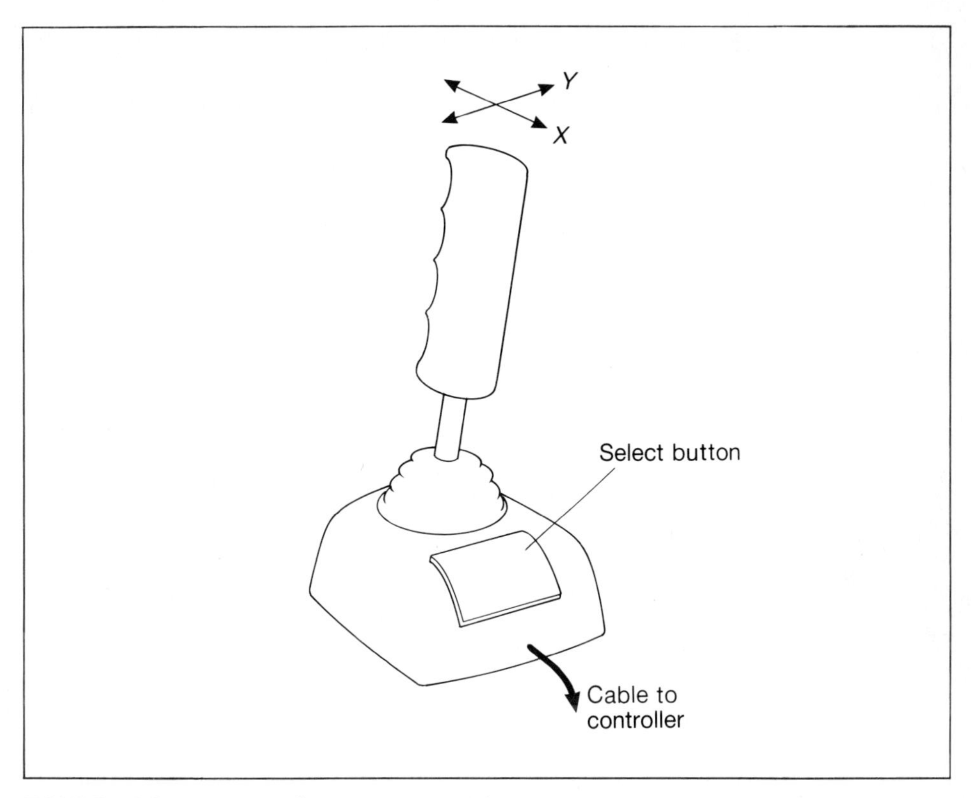

FIGURE 10.8 Diagram of a computer joystick.

The devices described above are useful when information about a single point is needed or when we want to draw lines. They are of little use, however, when we need to transmit an entire page of graphical data to the computer. In the past few years, scanners have provided a solution to this input problem. **Scanners** shine light on the source document and measure the reflected light to determine whether a spot is light or dark. Some of these devices are very fast and generate a large amount of data very quickly. Because of this, they often require an I/O controller that accepts

FIGURE 10.9 Diagram of a light pen.

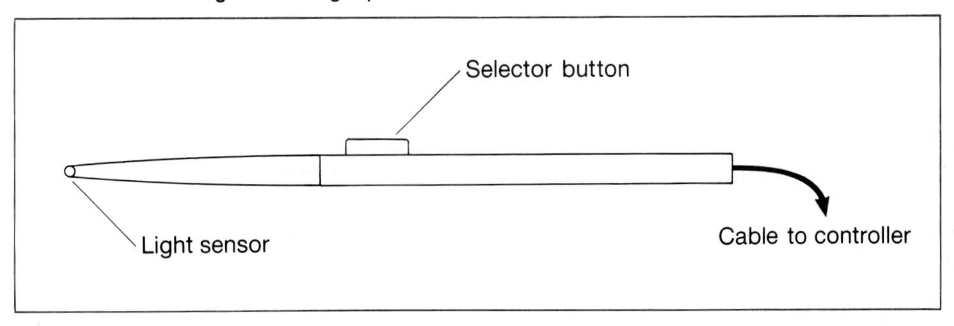

parallel data at very high rates. Some scanners are capable of detecting color in addition to black-and-white images. Because they generate color information in addition to just intensity, they require even higher data transfer rates.

Another input device that has been in the development stage for many years is audio input. One device of this type works very well and is the basis of today's digital audio disks. It simply measures the output of a microphone and converts the measurements into digital values. It then stores the stream of data just as any other digital information would be stored. A second device in this class, which seems to work quite well on *Star Trek*, converts speech into characters. The characters are then transmitted to the computer just like a stream of characters from a keyboard. Voice input has great potential, but because of the drastic variation in individual speech patterns, it has yet to be perfected for general-purpose applications.

There is another class of input device called **transducers** that we will mention briefly. These devices measure a wide variety of physical parameters ranging from air pressure to velocity. They convert the signals to digital values and send them to the computer. (The digital audio described above falls into this class.) Special-purpose computers for medical applications often use this type of device.

10.2.2 Data Output Devices

Just as keyboards are the most common input device, video displays are currently the most common form of output device. Unlike keyboards, however, displays may require very high data rates, and therefore connections via serial controllers may not always be appropriate. This is especially true when complicated graphics are displayed. With video displays, information is "written" to the screen by focusing a beam of electrons onto the desired portions of the screen. The locations where the beam strikes the screen emit light. The patterns of light emission contain the information. For simple character displays, serial transmission is quite adequate. However, for color graphic display terminals, high-speed parallel data transmission may be required for good response times.

The second most common output device is the printer. Printers come in many sizes, shapes, and colors and operate at a wide variety of speeds. Slow printers, such as those found on inexpensive personal computers, are often connected to the computer by a serial link like the one used for terminals. However, many printers are connected to controllers that provide a parallel output. The most common parallel interface on personal computers and some minicomputers is called the *Centronics interface*. The name comes from a brand of printers used on numerous minicomputers. The interface has been adopted by many printer manufacturers.

There are several possible ways to categorize printers, but in this chapter we will limit our discussion to four categories: impact printers, ink-stream printers, laser printers, and thermal printers. **Impact printers** place characters on the paper by pressing an object(s) with the shape of the character against an inked ribbon that then imprints the character's image onto the paper. Typewriters are obvious examples of this technology. Some character imprinters use print balls and print wheels to imprint characters on paper.

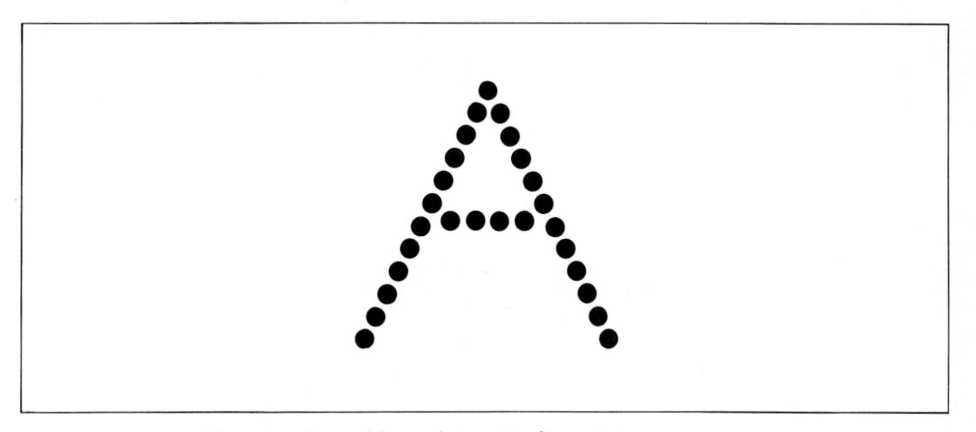

FIGURE 10.10 Character formed by a dot-matrix line printer.

Another type of impact printer in this class forms letters from a series of dots. These printers are typically called **dot-matrix printers**. Figure 10.10 shows a character that has been formed using individual dots. If you look closely at the screen of your video display, you will see that characters on the screen are also formed as patterns of dots.

One disadvantage of some dot-matrix printers is usually quite obvious: the characters are not very pretty. Newer versions of these printers have increased the number of dots used to form a character from about 60 to 300, and this does improve the character quality. These printers are usually described as being "near letter quality," meaning they produce characters nearly as well-formed as those on a typewriter.

One of the major advantages of the dot-matrix approach over the character printing described above is that the dots on the page can be changed to print characters of different sizes and shapes. This ability to change character "fonts" and sizes is very useful when you want to create special effects. There is another closely related advantage. Since dot-matrix printers can place dots at virtually any point on a page, they can be used to print graphic images on paper. When they are equipped with colored ribbons, they can even produce multicolored graphics.

Ink-stream printers, as the name implies, place information onto paper by spraying ink onto it. These devices are actually very similar to video displays. The differences are that paper is used rather than a video screen and a stream of charged ink particles replaces the electron beam. Figure 10.11 shows the elements of this type of printer. Since ink can be produced in different colors, both dot-matrix and ink-stream printers can be used to produce color output.

Ink-stream printers are capable of forming much clearer characters than those created by dot-matrix printers because the dots created by the stream of ink are much smaller than dots formed by a pin pressing on a ribbon. In fact, some ink-squirting printers have dot resolutions that approach the resolution of the laser printers.

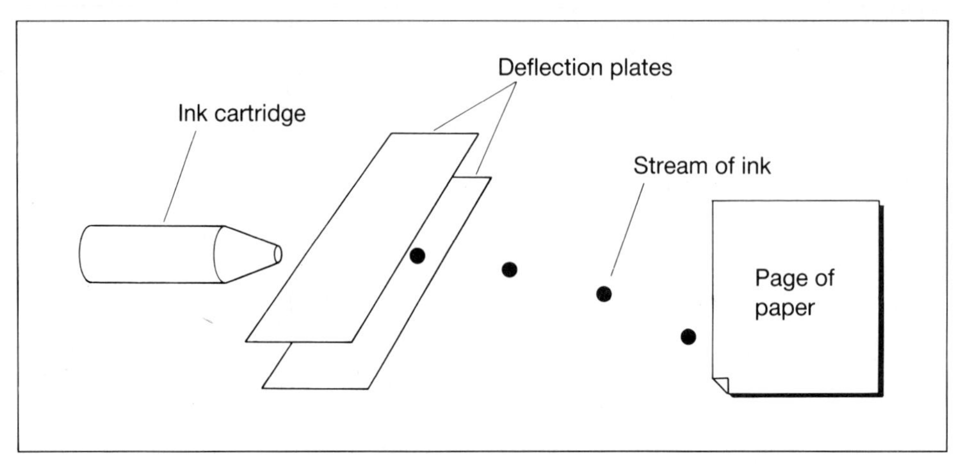

FIGURE 10.11 Elements of ink-squirting printer.

Laser printers use a different mechanism to put dots onto paper. They accomplish this by focusing a laser beam on the surface of a rotating drum at the desired points. The beam causes a small electric charge to be placed on the drum where the beam strikes it. The drum then rotates past a supply of dry ink, called *toner*, which is attracted to the charged dots. A sheet of paper is pressed against the drum, and the toner is transferred to the paper. Then heat is applied to the paper to permanently attach the toner to the page.

Because laser beams can be made to have extremely small diameters, laser printers are capable of forming very high quality characters. They are also capable of printing graphic images. At the present time, however, color laser printers are expensive and not in general use.

The printers described so far all use relatively standard paper and apply ink in some form to the paper. **Thermal printers** use a different technique, however; they burn the desired images onto special heat-sensitive paper. Some printers discolor the paper by heating it, and others melt a material that contains ink. Both types form characters out of dots just as the other printers described above.

Another type of output device that has been in use for many years is the plotter. In fact, the plotter was one of the first devices used to create graphic images on paper. Conceptually, a **plotter** is little more than a mechanically driven pen. The pen is attached to a holder that is usually moved across the paper by a combination of motors and cables. On some plotters the paper is moved along one axis and the pen is moved along the other axis. Figure 10.12 diagrams such a plotter. Plotters are quite good at drawing lines, circles, and other figures but are not very good at creating characters that typically require many short line segments.

As with input devices, many specialized output devices have been designed for specific applications. They span the spectrum from simple light-emitting diodes (LEDs) to image-projection techniques used originally on aircraft and now found on some automobiles.

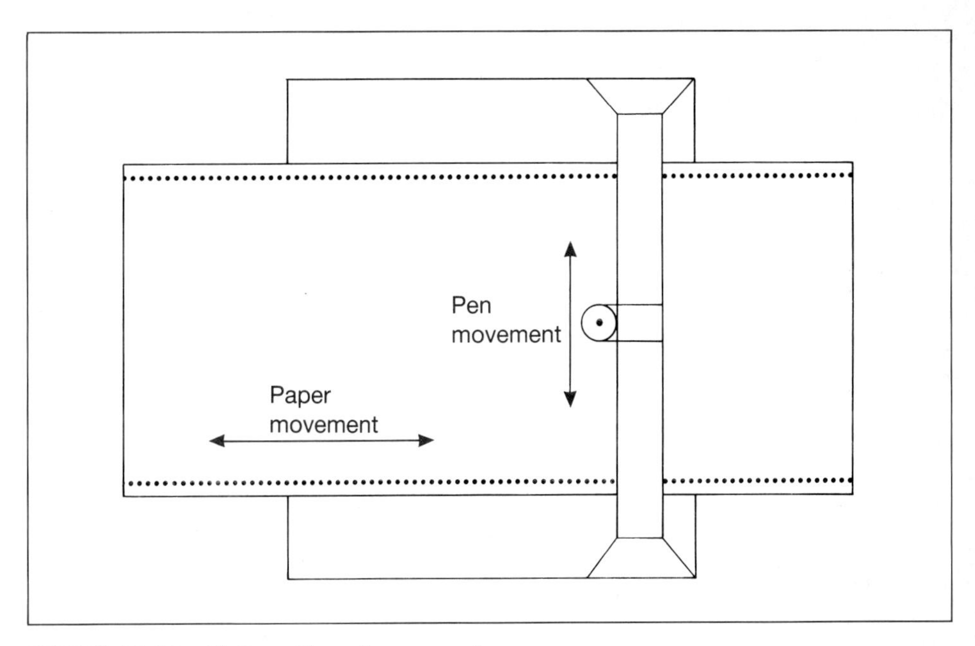

FIGURE 10.12 Plotter with moving pen and paper.

10.2.3 Data Storage Devices

The number and variety of data storage devices that have been used throughout the history of computing is quite remarkable, but many are based on similar principles. In an attempt to organize our discussion, we will separate storage devices into two categories: (1) sequential-access devices and (2) rotational or cyclical access devices.

To illustrate the differences between the two classes, let's look at an example of each. Magnetic tape is a typical sequential-access device, and disks are typical cyclic-access devices. With magnetic tape, you start from the beginning of the tape and search "sequentially" through the records until the desired information is found. This can be a long, time-consuming process. With a disk, however, you can very quickly position the read heads directly over the portion of the rotating medium that contains the desired information. Because of this capability, disks are also known as random access storage devices.

Before going into the operation of magnetic tapes and disks, we will describe how information is stored on magnetic media. First, we need a medium which is coated with a thin film of material that can be magnetized. To store information onto the medium, a device that is capable of inducing a magnetic field is placed next to the medium. The device that induces the field is called a **read/write head**. The field is induced by passing current through the coil of wire attached to the write unit. After the read/write head has moved to a different location on the surface, the magnetic medium "remembers" the direction of the magnetic field that was applied. A sequence of bits is stored onto the medium by alternating the direction of the

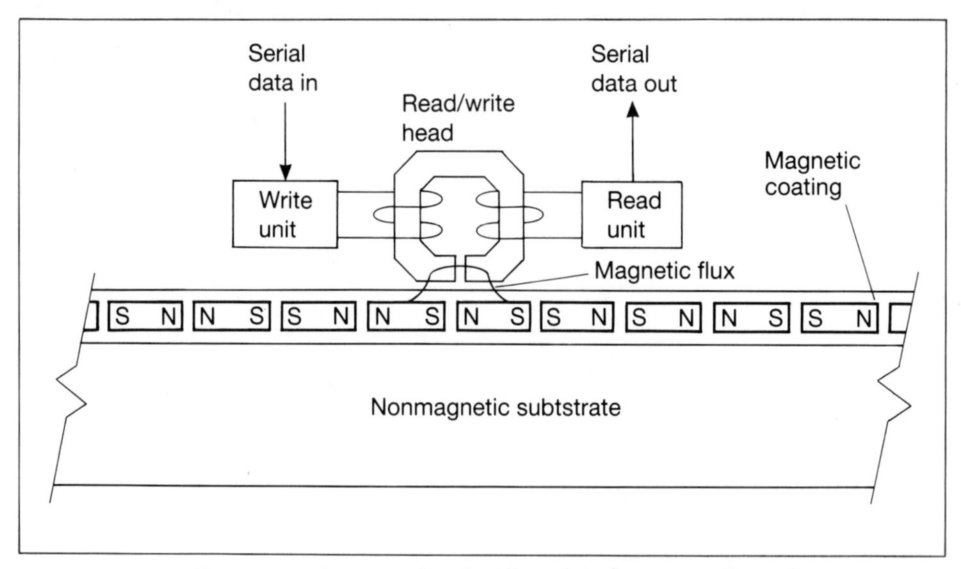

FIGURE 10.13 Elements used to record and retrieve data from magnetic media.

current in the write head, which in turn causes a number of magnetic flux reversals in the medium.

To read information, the medium is again passed by the read/write head, but this time the write circuitry does not force any current through the head. Instead, the field in the medium induces a current in the read head. The induced current waveform contains the information that was originally recorded on the medium. Figure 10.13 shows the components needed to read and write information on magnetic media.

The general mechanism described above applies directly to both magnetic tape and magnetic disk recording. In the case of magnetic tape, the head assembly usually reads or writes several tracks of data in parallel. For $\frac{1}{2}$-in reel-to-reel tape, nine tracks (eight data and one parity) are used; this configuration is shown in Fig. 10.14. As the figure indicates, a parity bit is added to each tape frame. Parity is generated and checked using the techniques described in Chap. 7.

The tape can contain a series of variable-length records separated by gaps where nothing is recorded. Each record contains data bytes with parity followed by two frames of information that are also used to check for errors. These two check characters, the CRC and checksum characters, are used because a single parity bit on each data character will not detect all possible errors. Each bit in the **checksum character** is an even-parity bit for the bits in one track. The **CRC character** detects multiple errors that cannot be found with simple parity bits.

The format of magnetic tapes includes both reel-to-reel and cartridge configurations. They provide a convenient method for both archiving and exchanging programs and data. Tapes are easily transported and can be stored in locations away from the computer.

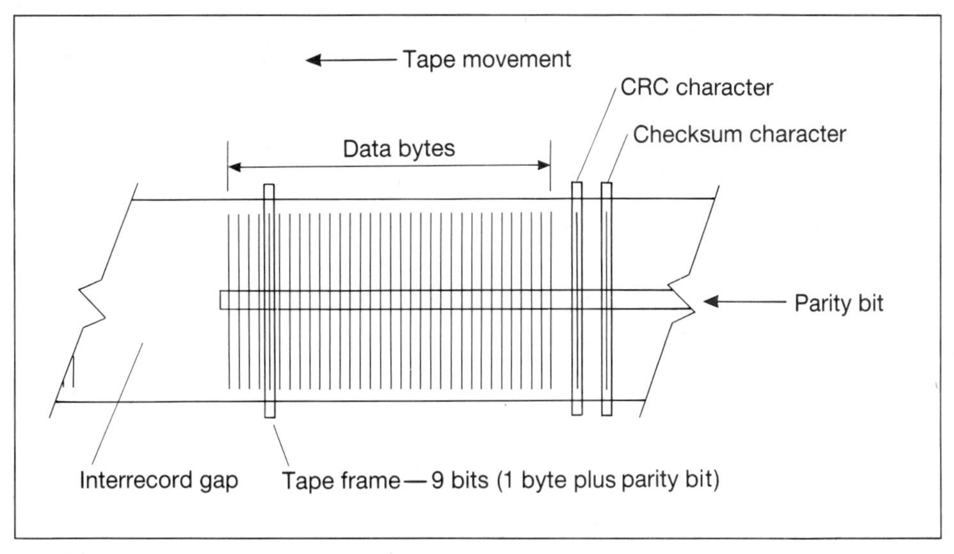

FIGURE 10.14 Nine-track magnetic-tape-recording format.

Magnetic tape as a recording medium for computers has been around since the early 1950s. On early mainframe and minicomputers, it was often the only available form of secondary storage. The tape used on these systems was typically the large reel-to-reel variety. Magnetic tape is still used on today's computers from PCs to mainframes. However, many newer magnetic tape recorders use some type of cartridge rather than separate reels of tape. The cartridges vary from ones that look a lot like an audiocassette to others that resemble videocassettes.

Not all sequential storage devices are based on magnetic tape. Other examples of sequential storage devices used on early systems include:

- Punched cards
- Punched paper tape

Since neither of these devices is used much anymore, we will not give the details of their operation.

Magnetic disks and drums have been in use for over 30 years. Because of continuing technological advances, their designs have kept up with the needs of the industry. They are by far the most commonly used secondary storage devices today. Most computers can be equipped with either soft (floppy) or hard (rigid) disk drives. Both store information in concentric *tracks* in a thin magnetic coating (the recording medium) on the surface of a nonmagnetic platter or "disk." Each platter can be either one-sided, with recording media on one side only, or two-sided, with recording media on both sides. A write operation transforms a serial data stream into current pulses that magnetize a small area of the magnetic coating in the direction indicated by the particular bit (for example, N to S for a 1, S to N for a 0). In a read operation, movement of the data track past the head allows the magnetic fields surrounding the bits to induce current pulses in the read coil. The pulses are then transformed into a serial bit stream by circuitry in the read unit.

Hard disks are available with one or more platters that are either removable or permanently mounted. Figure 10.15 is a conceptual diagram of a six-surface hard disk. The platter rotation creates a thin cushion of air that allows each recording head to "fly" a few microinches above the surface. Read/write heads are mounted on a comblike assembly that moves in and out to position the heads over the desired track during read/write operations. The head assembly is retracted when the disk is shut down.

The data in each track is formatted in short blocks, or sectors, separated by gaps where nothing is recorded. Each sector is preceded by a bit pattern that indicates both the beginning of the sector and the sector number and is followed by error check bits. The length of the data portion typically varies from 256 to 1024 bytes, depending on the particular disk model. The number of sectors per track also varies but is usually between 4 and 32.

FIGURE 10.15 Magnetic disk memory: (*a*) mechanical configuration; (*b*) track and sector configuration; (*c*) track recording format.

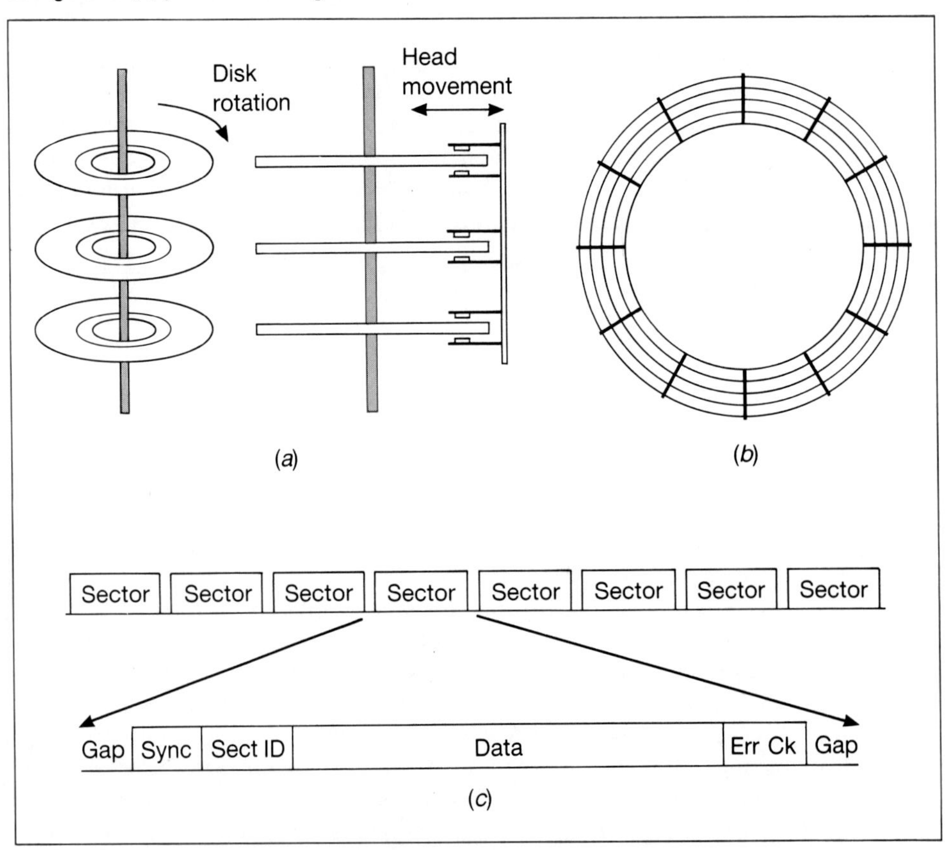

Addressing is conceptually similar to that used for primary memory, but the number of bits retrieved is considerably different. Each disk location is identified by three address components: the track, the surface level, and the sector. The track indicates the head position, the surface level selects which head is to be used, and the sector indicates when the actual read or write operation is to begin.

Although magnetic disks have been the primary devices used for fast secondary storage, there is a new candidate that now shows great potential as a replacement, the optical disk. These devices also use a spinning disk to store the information; however, they use beams of laser light to read the data on the surface. They have a much higher capacity to store information on a given surface area than magnetic disks. When this technology becomes more readily available, it could quite possibly replace the magnetic disk as the primary direct-access storage device.

Because of the very high volumes of data they can hold, some companies are now using optical disks for read-only storage. These devices use exactly the same technology that is used to produce audio compact discs. Apple computer now sells a compact disc player that both plays audio CDs and reads disks that have been recorded with computer-generated information.

The idea of storing high-density digital information on media with a laser beam is actually not very new. Precision Instrument Corporation was selling a device in the late 1960s that was capable of storing information by burning minute spots of metal off strips of Mylar. This was a complicated mechanical device, but it no doubt served as an example for later designs.

All the devices included in this section require relatively high data transfer rates. Therefore, they are all interfaced through peripheral controllers via parallel data buses. There are always exceptions, but this is the general rule.

Describing all the devices that have been used for external storage is beyond the scope of this text. In the next section we will concentrate on some detailed specifications for the devices described above.

10.3 MEASURES OF PERIPHERAL PERFORMANCE AND CAPACITY

Many of the decisions we make about purchases are based in part on the specifications of the products. Car buyers want to know such parameters as miles per gallon of gas, engine horsepower, stopping distance, and price. Stereo buyers want to know output power, distortion, stereo separation, and price, among others. There are similar parameters and specifications that we use to evaluate peripheral performance. In this section we will look at three: speed, resolution, and capacity.

Two important measures that we will not consider in this chapter are reliability and price. Reliability will be discussed in Chap. 20. Prices will not be discussed because they change too fast to be included in a textbook. Besides, price should not be of concern until all the functional requirements have been evaluated.

10.3.1 Speed

There has always been a lust for speed in the computer industry. It does not seem to matter how fast a device is; someone will always want one that is faster. Because of this real or perceived need, nearly every external device we buy for a system will have some published parameter related to speed of operation. For input and output devices, the parameters in which we will be most interested are the access time and data transfer rate. Since access time was discussed in Chap. 5, we will focus here on transfer rate. Data transfer rate is a measure of how fast the device can generate or accept data. We measure transfer rates in units such as bits per second or bytes per second.

For simple input devices, such as keyboards and mice, the actual speed is often limited by the operator's capability and is therefore not as important as other measures of performance. When a data transfer rate is given for a terminal that has a keyboard and video display, the transfer rate is a more important measure of how fast the device can accept data for display purposes. Typical transfer rates for terminals are 1200 to 19,200 bits/s (bps).

Other input devices where transfer rates *are* important are scanners and some devices with analog-to-digital converters. Consider a scanner that can detect 300 individual dots per inch in both the X and Y directions. This gives a dot density of 90,000 dots per square inch. Let's assume that the portion of a sheet of paper that the scanner can read is 10 by 8 in. If each dot on the page is represented by 1 bit, then the total number of bits per page is

$$\text{Square inches/page} = 10 \text{ in} \cdot 8 \text{ in}$$
$$= 80 \text{ in}^2\text{/page}$$
$$\text{bits/page} = \frac{80 \text{ in}^2}{\text{page}} \cdot \frac{90{,}000 \text{ bits}}{\text{in}^2}$$
$$= 7{,}200{,}000 \text{ bits/page}$$

If we were using a serial line capable of 19,200 bits/s, the transfer time for a page would be

$$\text{Seconds/page} = \frac{7{,}200{,}000 \text{ bits/page}}{19{,}200 \text{ bits/s}}$$
$$= 375 \text{ s/page}$$

This calculation points out the need for high-speed, parallel data transfers between scanners and computers.

Let's now turn our attention to the speed of output devices and start with a simple character display screen. First we will assume that the screen contains 25 lines of characters with 80 characters per line. As we described earlier, characters on the screen are created as a combination of dots. We will assume that there is hardware within the terminal that can store characters and convert them into the corresponding combination of dots. Using the transfer rates 1200 and 19,200 bps, we can calculate the transfer times as follows. The total number of characters is

$$\text{Characters/screen} = \frac{80 \text{ characters}}{\text{line}} \cdot \frac{25 \text{ lines}}{\text{screen}}$$
$$= 2000 \text{ characters/screen}$$

We will assume that each character requires 10 bits, taking into account start and stop bits. This gives the following transfer times:

1200 bps transfer rate:

$$\text{Time/screen} = \frac{(2000 \text{ characters/screen}) \cdot (10 \text{ bits/character})}{1200 \text{ bits/s}}$$
$$= 16.67 \text{ s/screen}$$

19,200 bps transfer rate:

$$\text{Time/screen} = \frac{(2000 \text{ characters/screen}) \cdot (10 \text{ bits/character})}{19{,}200 \text{ bits/s}}$$
$$= 1.04 \text{ s/screen}$$

These calculations would also apply to some high-speed printers, since, as we pointed out earlier, they often use serial interfaces to the computer.

High-quality page printers place much greater demands on data transfer rates. In fact, laser printers with some types of graphics output require the same number of bits per page as the scanners described above. For this reason, they are usually connected to the computer through a high-speed parallel interface. We must point out, however, that when printing characters on a page, a laser printer may not require as high a data transfer rate as when creating graphics printouts. This is because most laser printers store information about each character and therefore do not require patterns of dots to create each character.

External storage devices place the greatest demands on computers and controllers with regards to data transfer rates. Let's make a few assumptions about a disk drive so that we can calculate its data transfer rate. We will assume that the disk rotates at 3600 rev/minute and contains 8192 bytes per track. The transfer rate is then

$$\text{Transfer rate} = \frac{8192 \text{ bytes}}{\text{rev}} \cdot \frac{3600 \text{ rev}}{\text{min}}$$
$$= 29{,}491{,}200 \text{ bytes/min}$$
$$= 491{,}520 \text{ bytes/s}$$
$$= 3{,}932{,}160 \text{ bits/s}$$

It should be noted that the bytes per track and revolutions per second used are actually rather modest. In other words, the calculation is quite conservative. Even with these conservative assumptions, the transfer rate is much higher than for any of the other peripheral devices we discussed.

It is probably obvious that a controller with a high-speed parallel data path is needed to meet the requirements for disk data transfers. Many of the devices used for external storage also have transfer rates comparable to those calculated above. They will also require controllers with high-speed parallel interfaces.

10.3.2 Resolution

Resolution is a measure that we often apply to input and output devices. For both the scanners and printers described above, we used device resolution without specifically defining the term. The term we used to describe resolution was dots per inch.

Since both laser printers and scanners use a similar technology, it is not surprising that they have similar resolution. As we stated above, typical resolutions for these devices are 300 dots per inch. Newer laser printers, however, have even better resolution; some are well over 2000 dots per inch.

Resolution is also an important measure of performance on video output devices. If you have used an IBM PC, you may be familiar with terms such as "640 by 480 dots" screen resolution. These are measures of the number of dots on each line of the video display and the number of lines on the screen. Although this resolution is an improvement over what was found on the previous generation of personal computers, it is far below what is used on engineering workstations. Workstations, which use large screens, now display more than 1200 dots per line. They also have a comparably increased number of lines per screen.

For mechanical input devices, such as track balls, mice, and tablets, resolution is measured a little differently. For these devices we are interested in the minimum difference between two points that can be detected by their sensors. With mechanical devices this is typically in a few hundredths of an inch.

10.3.3 Capacity

Just as resolution was an important measure of performance for input and output devices, capacity is an important measure for storage devices. For storage devices we usually measure capacity in kilobytes or megabytes. A few years ago floppy disks had capacities of only a few hundred kilobytes. Today, floppies have capacities in excess of a megabyte.

The capacity for hard disks has changed even more drastically. Personal computers a few years ago typically had hard disks with 5 or 10 megabytes of storage. Now it is not uncommon to find disks on personal computers with 100 to 600 megabytes of storage. Storage on mainframe computers has increased even more impressively. Their storage is now measured in gigabytes.

Magnetic tape capacities have also increased significantly. Ten years ago a 12-in reel of magnetic tape had a maximum capacity of about 45 megabytes. Today tapes only a few inches in diameter have capacities greater than 60 megabytes.

Performance measures such as speed, resolution, and capacity are significant because they affect the design of the controllers that interface devices to the computer. They also significantly influence the price of peripherals. Because of this, it is very important to understand the significance of performance measures before making decisions regarding their use on a computer system.

10.4 SHARING PERIPHERALS BETWEEN COMPUTERS

On most systems the cost of peripheral devices exceeds the cost of the main computer itself. Because of this, users have tried to find ways to share peripherals among systems and thereby reduce the total cost of peripherals on the combined systems. We examine two methods commonly used to share peripherals: switches and networks.

10.4.1 Switching Peripherals between Computers

Probably the easiest method of sharing peripherals is with a switch. The switch transfers the peripheral's data and control lines from one system to another. The simplest form of the switch requires manual operation.

On early personal computers, printers were often considered a luxury. This was especially true when the printer being used was a laser printer. Because of their relatively high cost, many users would purchase one printer and a switch that allowed the printer to be shared between computers. This was such a popular configuration that many companies produce and sell such switches.

Automated versions of the switch allow the computers to switch the peripherals back and forth. These are usually found only on mini- and mainframe computers. Switches of both types are rapidly being replaced by communications networks that allow multiple CPUs to share peripherals and memory.

10.4.2 Sharing Peripherals on Networks

Networks are a very important development in computer system architecture, and their use goes beyond just sharing peripherals. However, in this section we will discuss only how networks facilitate peripheral sharing.

A simple way to look at a network is to think of it as being a path between system buses of several computers. By connecting the computers in this manner, we can easily send information between computers. If the network is also attached to peripherals, then all the computers connected to the network can access any of the attached peripherals.

Networks have some of the same problems that were described in Chap. 7, where we discussed internal CPU buses. Any device connected to the network must have a unique address so that other devices on the network can communicate with it. Also, since several devices share the network, there are often conflicts between devices wanting to transfer data over the network. The network must be capable of resolving conflicts between the attached units.

Networks simplify sharing of devices between computers. However, they also impose some restriction on peripherals that are attached to the network. The peripherals must have sufficient intelligence to communicate with other devices on the network.

10.4.3 Distributed Intelligence

Some devices attached to a network require only enough logic to interpret addresses on the network and to respond to simple read and write requests. There is another important class of device that is often found on networks, servers. A **server** can be thought of as an "intelligent" peripheral. Typical examples of servers are print servers and file servers.

File servers are, in part, external storage devices; they go beyond basic storage devices, however. In addition to simply storing data, they have the ability to organize information into sophisticated file structures. For more information on file systems, see Chap. 13.

Print servers are also more complicated than simple switched printers. A typical print server does not just print information sent from a computer. It can also temporarily store information coming in from several computers and print the information out in a logically correct manner. By doing this, it creates the illusion that each computer is connected directly to a printer. We call this illusion a **virtual printer**. There are many other examples of virtual devices on a computer system.

Distributed intelligence and networking are two topics that often go together. They are also going to be very important topics for research and study in the future. Hopefully this introduction will motivate you to study them in later classes.

10.5 SUMMARY

In this chapter we looked at several peripherals, described their characteristics, and related how they are connected to computers. In the discussions we emphasized the problems caused by electrical and mechanical differences between computers and controllers. In the past few years many companies have addressed this problem. As a result of this work, several standards have been defined that specify how computers and peripherals can be connected. These standards have made it easier for peripheral designers to attach their devices to a wide range of computers.

Peripherals and their controllers are important components of any computer system. Sometimes, however, other parts of the system, such as the CPU and main memory, receive more attention. Everyone wants to design a new processor, but many shy away from the design of peripherals and controllers because they consider such designs less glamorous. In reality, though, designs for peripherals and controllers are often more complex and challenging than the design of the CPU itself.

A computer without peripherals is of little more use than a paperweight. Until we attach peripherals to the computer, none of its power is accessible to the user. Peripherals make computers useful tools.

PROBLEMS

10.1 How would the controller described in Sec. 10.1 be changed if EBCDIC characters were transmitted rather than ASCII?

10.2 Overhead associated with data transfer is proportional to the time when data is *not* being transmitted. What is the percentage of overhead for the type of data transmission shown in Fig. 10.3?

10.3 Figure 10.3 shows the voltage waveform for serial data transfer between the computer and a terminal. Compute t for transmission rates of 1200, 4800, 9600, and 19,200 bits/s. bits/s.

10.4 The peripheral controller and the I/O unit must operate at essentially the same clock frequency to ensure accurate data transfer. Compute both the maximum and minimum clock rates that would be allowed for the 2400-bps waveform shown in Fig. 10.3. Assume a minimum 10-μs overlap on the last bit period.

10.5 What are the primary functions of a typical peripheral controller?

10.6 Compare the functions of a peripheral controller to those of the I/O processor discussed in Chap. 9.

10.7 Some peripheral controllers have their own microprocessor. Why do you think this is done?

10.8 Why does the I/O unit need to send a device address to the peripheral controller?

10.9 List five data input peripherals not described in Sec. 10.2.1.

10.10 List five data output peripherals not described in Sec. 10.2.2.

10.11 What type of information do you think a mouse or track ball sends to the controller?

10.12 Consider the data formats shown in Fig. 10.6. Discuss any additional capabilities that are necessary for a card reader controller but are not required for a paper tape reader controller.

10.13 What advantages does a mouse have over a light pen?

10.14 What advantages does a light pen have over a mouse?

10.15 Why were joysticks used long before anyone ever attached them to computers?

10.16 A medium-resolution video monitor is capable of displaying 640 lines of data on the screen, with each line containing 480 dots. Assume each dot requires eight levels of intensity. How long would it take to draw a new image on the screen if the monitor were attached to the computer via a 9600-bps serial link?

10.17 How would the time calculated in Prob. 10.16 change if each dot were capable of displaying 256 different colors?

10.18 What are the advantages of dot-matrix printers over character printers?

10.19 Why do color dot-matrix printers usually run at about one-third the speed of comparable-quality single-color printers?

10.20 Some character printers must return to the left side of the page before they can begin printing the next line. Others can print in either direction. Both are usually equipped with a "line buffer" that is filled with the contents of the entire line before printing begins. Diagram a possible control sequence for a forward/backward printer controller that can track the current location of the print head and begin printing from the closest end of the next line.

10.21 Why are plotters relatively poor at creating characters?

10.22 Why do disks and tapes require that the media move past the read/write heads in order for data to be read?

10.23 Most disk drives used currently can read or write only one track of information at a time, while tape drives usually read/write multiple tracks. Why don't disks have read/write heads that access multiple tracks?

10.24 If a tape drive record is 65,536 characters long at a density of 6250 bits/in (on each track) and has an interrecord gap of 0.25 in, how much data can be stored on a 1200-ft tape? Assume the read/write head has nine tracks including a parity track.

10.25 The term "head crash" is sometimes heard in the computer room. What does this mean?

10.26 Why don't heads on tape drives experience "crashes"?

10.27 A common disk size is $3\frac{1}{2}$ in in diameter. If the innermost track is recorded 0.75 in from the center and the outermost track is recorded 1.65 in from the center, what is the density of bits on the surface of the disk? Assume the surface contains 150 megabytes of data. What assumption did you make?

10.28 Compare the density of data on a disk surface with that on a tape. See Probs. 10.24 and 10.27.

10.29 What are some applications of the optical disk other than computer data storage?

10.30 How does the cost per megabyte of semiconductor memory compare with that of hard disks and floppy disks?

10.31 What are the access times of semiconductor memory, hard disks, and floppy disks.

10.32 What are some important characteristics of disk drives other than cost, capacity, and access time?

10.33 If the tape drive in Prob. 10.24 moves tape at 25 in/s, how long will it take to transfer the data from the disk described in Prob. 10.27?

10.34 Some color video monitors can display 1200 lines with 2000 dots per line, where each dot can have one of 1024 colors. If we want to write a new image to the screen in 0.25 s, at what rate must data be transferred to the monitor?

10.35 What types of applications require high-resolution video monitors? What types of applications do not require high-resolution monitors?

10.36 How does the bit density of semiconductor chips compare with that of hard-disk surfaces?

10.37 What are some advantages of using a file server to provide disk storage to multiple computers? What are some disadvantages?

10.38 What is another example of a virtual device on a system?

10.39 What is the advantage of distributed intelligence in a system?

10.40 A typical personal computer graphics display has a screen resolution of 640×480 dots and is refreshed 60 times per second. Compute the data rates between the graphics controller and the display, assuming a 16-color display (4 bits/dot). What is the rate if the number of colors is raised to 256?

10.41 Section 10.3.1 discusses the industry's desire for more speed in computer peripherals. Unfortunately, faster peripherals are usually more expensive than slower ones, and we should also consider how fast is fast enough. What criteria would you propose for determining the design speed (and the required data transfer rate) of a keyboard? Of a printer?

10.42 Repeat Prob. 10.41 for (*a*) a graphics tablet and (*b*) a disk.

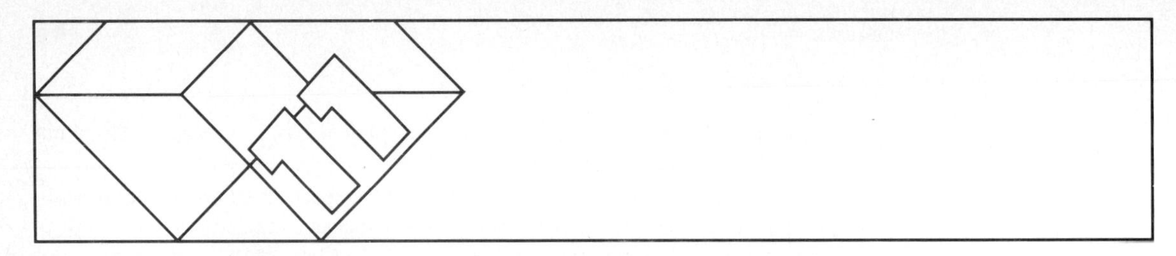

Hardware Design Tools

Early computers were designed by engineers using slide rules, paper, and pencils. This was possible because the designs were relatively simple and no automated tools were available. Today, those early computers are comparatively "trivial" and can be implemented easily on a single silicon chip about the size of a dime. Modern computer designs, however, are so complex that tools which aid in their design are becoming mandatory. For example, on February 15, 1989, a prominent Silicon Valley newspaper ran an article entitled "Intel Shows Off 'Supercomputer on a Chip.'" The article was referring to Intel Corporation's announcement of the 80860, a single-chip RISC-based microprocessor that Intel compares with a Cray I supercomputer. The design of a processor like the Intel 80860 calls for a host of computer-aided design tools. Tomorrow's computers are being designed with the aid of today's computers.

Computer hardware design tools can be divided into two categories: computer-aided design (CAD) tools and hardware debugging tools. CAD tools are used for design entry, design verification, and physical design. The ones we will consider in this chapter are:

- Schematic capture
- Logic simulation
- Printed circuit board layout and routing
- Analog circuit simulation

All the CAD tools we describe can be run on the Apple Macintosh computer. However, large, complex designs may require a lot of CPU time.

Hardware debugging tools aid the engineer in finding errors in the design after the actual computer has been built. The debugging tools we will look at are:

- Multimeters
- Oscilloscopes
- Logic analyzers

The primary example we will use in this chapter is a single-board computer based on an Intel 8088 microprocessor chip. However, we will digress as necessary to give you a better understanding of some of the tools.

11.1 SCHEMATIC CAPTURE

Schematic capture is an interactive computer graphics tool that is used to describe a circuit with the logic symbols we discussed in Sec. 2.3. These logic symbols represent complex physical devices, and they are used because the symbols are much easier for people to understand than the actual devices. Figure 11.1 illustrates this point. It shows (*a*) the logic symbol for an inverter gate, (*b*) a corresponding circuit diagram, and (*c*) how the circuit might be laid out on silicon. Tools are also used to create diagrams like those shown in Fig. 11.1*b* and *c*, but we will focus on creating circuits using just the logic symbols.

FIGURE 11.1 An inverter gate: (*a*) logic symbol; (*b*) CMOS circuit diagram; (*c*) physical layout in CMOS technology. [*Weste, 1985*]

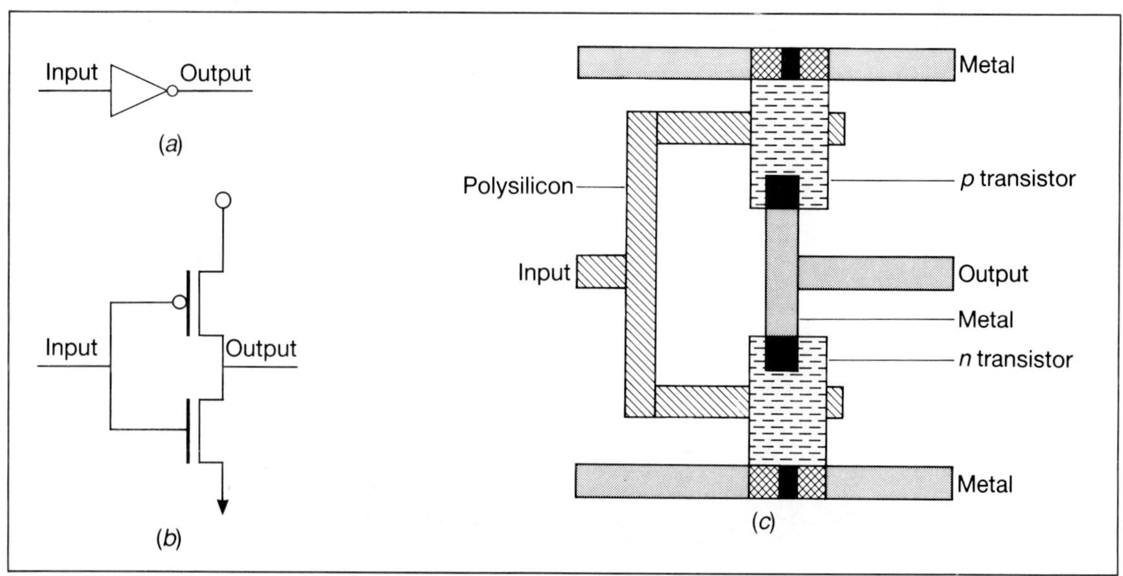

11.1.1 Design Entry with Schematic Capture

In this section, we will discuss how a design is created using a schematic capture tool, DesignWorks [Capilano Computing, 1990]. Most schematic capture tools present the user with a blank schematic page and a list of menus containing the various operations supported by the tool (see Fig. 11.2). A **menu** contains commands that can be used to create and save the schematic along with many other functions. Note that the blank schematic page contains a grid which helps the user place symbols accurately. The menus (located at the top of the page) are File, Edit, Drawing, Options, Window, and Devices.

The first step in entering our design is to select an 8088 microprocessor symbol from the component library. The **component library** contains a list of components that can be chosen and put onto the schematic page. Part of the information about each component is a description of the graphical symbol for the component. The schematic capture tool needs this information to draw a picture of the component on the page. Component libraries also contain a great deal more information about each component, some of which we will discuss in later sections.

The component libraries for the tool we are using are contained in the Devices menu, as shown in Fig. 11.3. Directly under the Devices menu are the Gates, Gates-Inv. Inputs, Generic, I/O, Demo Lib, and 808x LIB submenus. The components in the 808x LIB submenu are 8086 and 8088 microprocessors. Under the Gates submenu is a library that contains gates such as AND-2, AND-3, NAND-2, OR-2,

FIGURE 11.2 A blank schematic capture page.

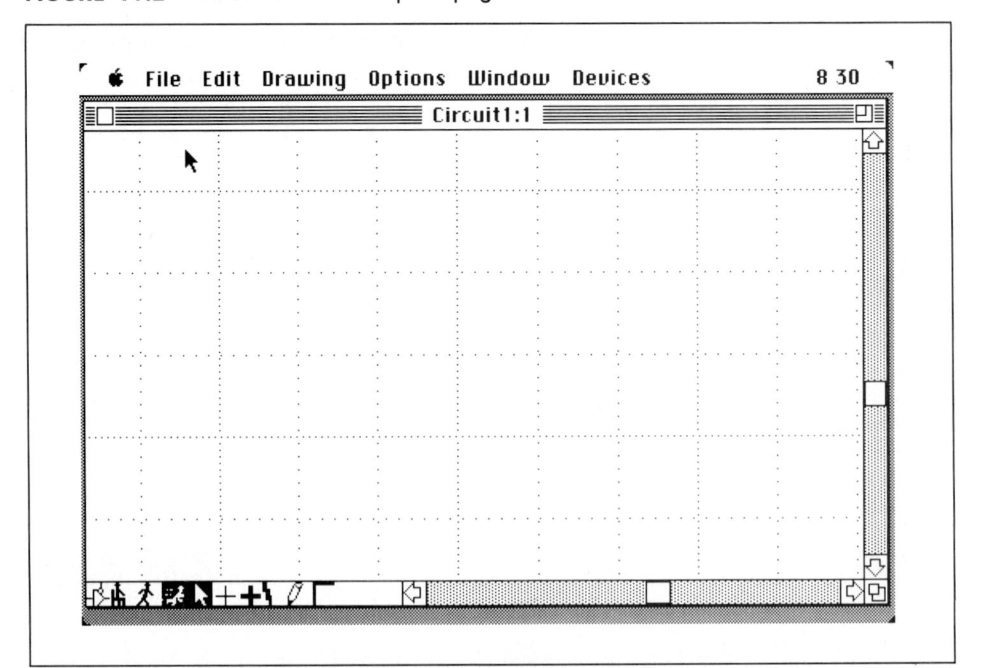

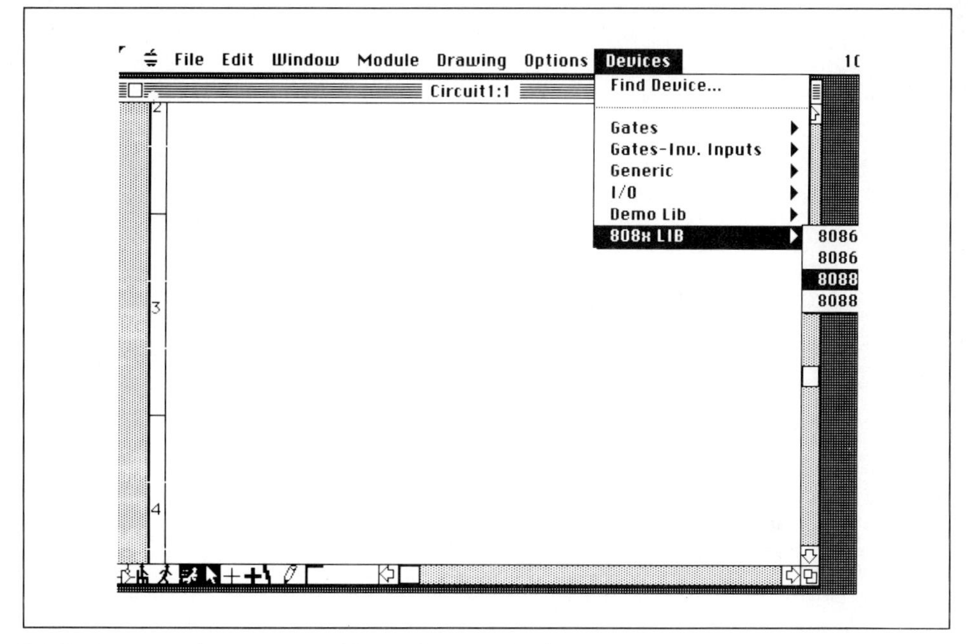

FIGURE 11.3 The component library that is under the Devices menu.

NOR-2. The numbers after each gate signify the number of input pins that each gate has.

It is a simple task to select a part from the component library and place it on the page. A component is chosen from the menu by placing the cursor on the desired name and releasing the mouse button. For our example, the 8088 will be chosen from the 808x LIB submenu under the Devices menu. Once the 8088 is selected, a blinking outline of the chip appears on the screen. The user can reposition this outline before it is permanently drawn. Figure 11.4 shows the schematic page after the 8088 has been selected and placed at the desired position.

Our schematic is not particularly interesting at this point because it has only one component. The 8088 needs external memory to store instructions and data. Therefore, let's select a memory component from the Devices menu, place it on the schematic page, and connect it to the 8088 microprocessor.

Connections between components are created by drawing lines that connect terminal sites on components. Lines in schematic drawings represent wires, which are often called **nets**. A **terminal site** is equivalent to a pin on an integrated circuit or to one end of a discrete component (such as a capacitor or resistor). Terminal sites are represented as short lines protruding from the schematic symbols. Figure 11.5 indicates where the terminal sites are located on a chip, an inverter gate, and a capacitor.

To connect terminal sites, the program must be told that the user wishes to draw a net. This can be done either by selecting a function from a menu or by typing a

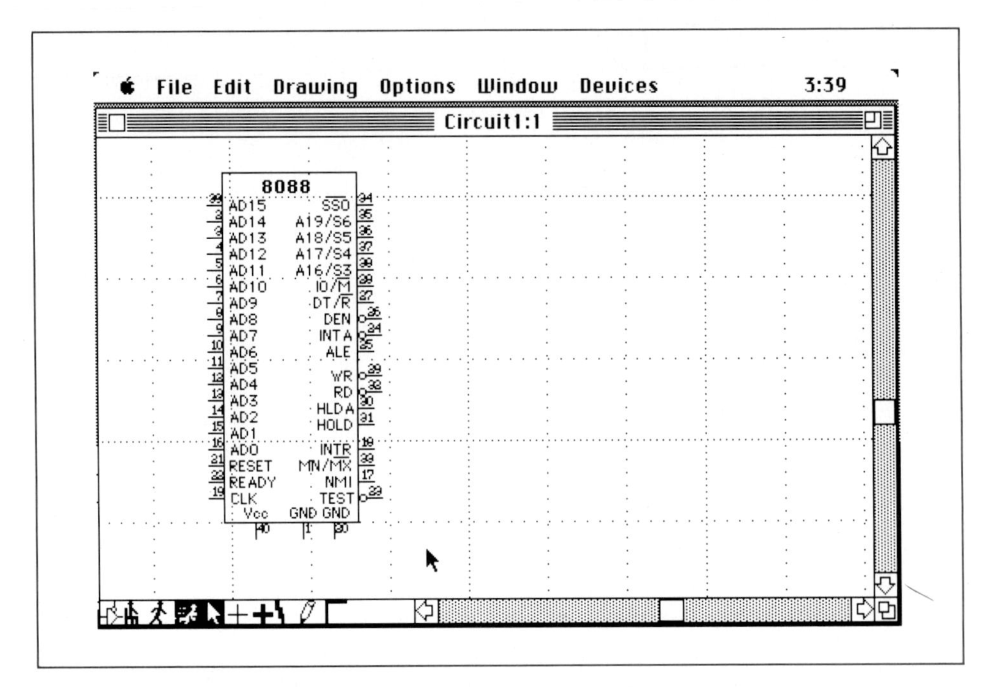

FIGURE 11.4 Intel 8088 microprocessor placed on the schematic page.

command. The program will then display a cursor or pointing device that is typically shaped like a cross or an ×. To draw a net, the user moves the cursor to a terminal site, clicks the mouse button, moves the cursor to another site, then clicks again. The program draws a line connecting the terminals so that the user can see the results. If the user clicks again on the second terminal site, the line is made permanent and the net is complete.

FIGURE 11.5 Terminal sites on a chip, a logic gate, and a capacitor.

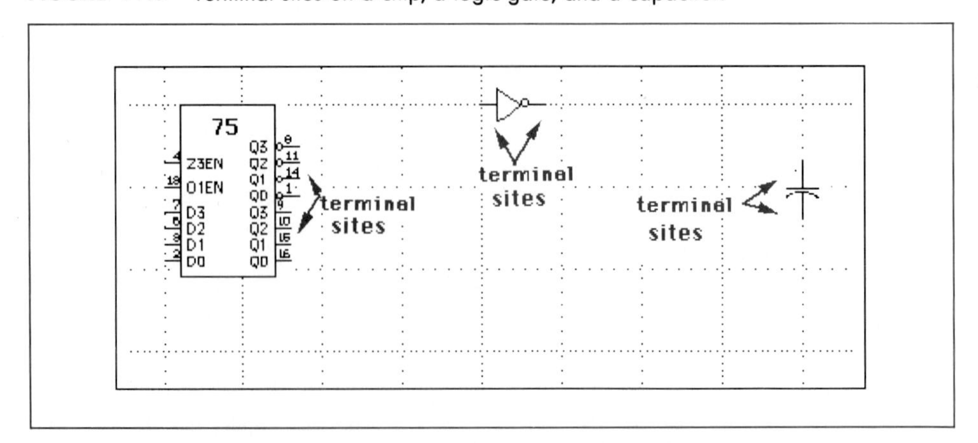

In most schematic capture tools a bus is drawn the same way as a net except that the bus is shown as a wider line than the net. Since a bus may contain many different nets, schematic capture programs require the user to name each net in the bus. If an address bus has 16 nets, the program needs to know that the bus contains the nets ad⟨0⟩, ad⟨1⟩, through ad⟨15⟩. We will not discuss how the nets are named within the bus, since most programs do it differently. Figure 11.6 shows the 8088 microprocessor connected to a 256ROM memory element and a RAM memory element using nets and buses.

Our single-board computer needs several more parts before the design is complete. A clock chip must be included because the 8088 needs an external clock input. The 8284A clock chip generates the system clock that is used to synchronize data transfers between different parts of the computer. The computer also needs a device to communicate with a keyboard and display, and the Intel 8251 chip provides this function. Finally, several NAND gates are needed so that the RAM and 8251 can respond to the CPU requests for reads and writes.

The last important items missing from our schematic are names. Every component, net, and bus must be given a unique name. With these names, components can be identified throughout the design process. Naming parts in the schematic is generally quite easy. The mouse is used to select the component by clicking on it. The designer then types the desired component name, and the schematic tool associates the name with the part.

FIGURE 11.6 Schematic diagram with 8088 and memory elements connected.

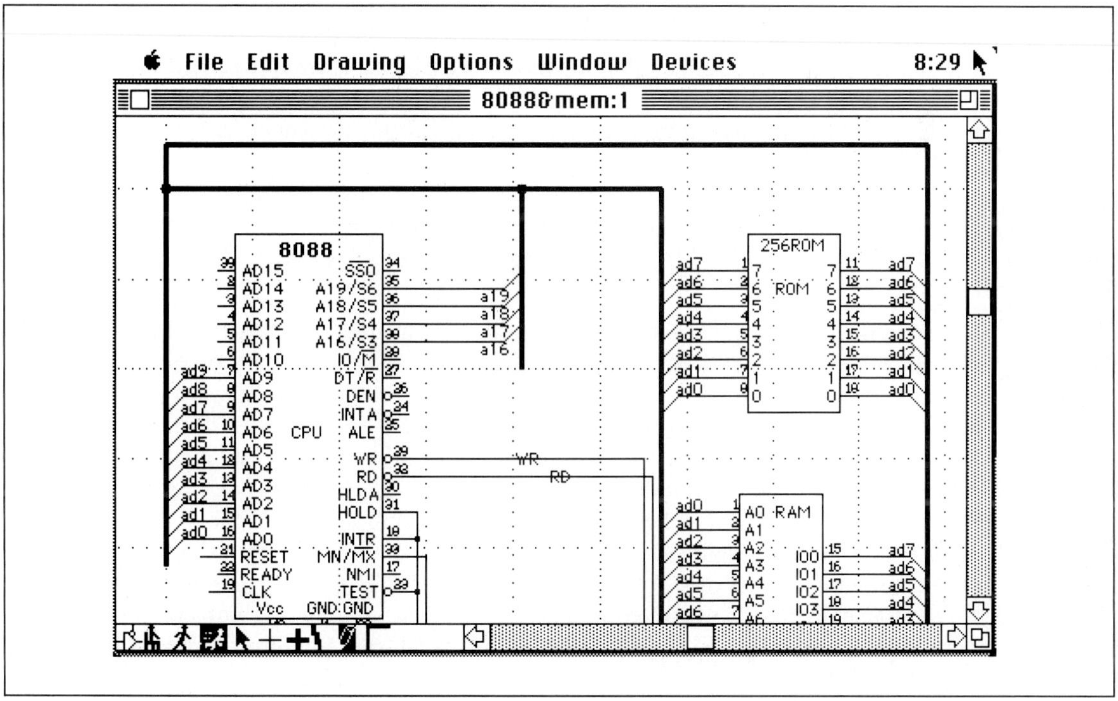

Usually, schematic capture programs will give default names to components as they are placed on the page. The first component might be named COMP0001 and the second component, COMP0002. Similarly, the first net drawn might be named NET0001 and the second, NET0002. Because these names have no significance or meaning, they are not as useful or as easy to remember as names assigned by the designer. Since the 8088 was the first component placed on the page, it was given a default name of COMP0001; a better name for the 8088 is CPU. Likewise, a net that is contained in the address/data bus was named NET0014, and a better name is ad5.

You might ask why ad5 is a good name for a net contained in a data bus. The name ad5 was chosen for bit 5 on the bus because the 8088 microprocessor multiplexes the address and data bus. As we described in Chap. 7, this means that some of the time the bus is used as an *address* bus and other times the bus is used as a *data* bus.

Figure 11.7 shows the completed schematic for the single-board 8088 computer system. The important elements of the schematic diagram are the components, the nets, the buses, and the names. Designing a circuit like the one shown in Fig. 11.7 might be compared to designing an automobile using components such as an engine, the transmission, and the drivetrain. We turn now to designing these individual components.

11.1.2 Chip Design versus System Design

The design example discussed in the previous section was a single-board computer system. The components used to design that system (8088, memory chip, 8251, and so forth) are all commercially available parts. The 8088 and the memory chips are classified as **very large-scale integration**, or **VLSI**, devices.

A VLSI chip has thousands of logic gates. Other classifications are:

- **Small-scale integration (SSI)**: several logic gates
- **Medium-scale integration (MSI)**: 10 to 100 gates
- **Large-scale integration (LSI)**: 100 to 1000 gates [Mano, 1988]

Circuits are also classified by the number of transistors they contain. Intel's 80486 microprocessor and Motorola's 68040 microprocessor each contain more than 1 million transistors. To give you an idea of the relationship between gates and transistors, it takes about 10 transistors to make a logic gate. Although circuits are sometimes classified in terms of the number of transistors they contain, the logic gate is typically the lowest-level circuit with which computer engineers work.

Logic gates have schematic capture tool symbols just as the 8088 and memory VLSI chips do. Figure 11.8 shows the symbols for some common logic gates.

Figure 11.9 shows AND, OR, and INVERTER gates connected together to implement the XOR function. The XOR function shown is implemented with five logic gates, but a microprocessor like the 8088 may require thousands of gates to design. The point is that schematic capture tools can be used to design any of these

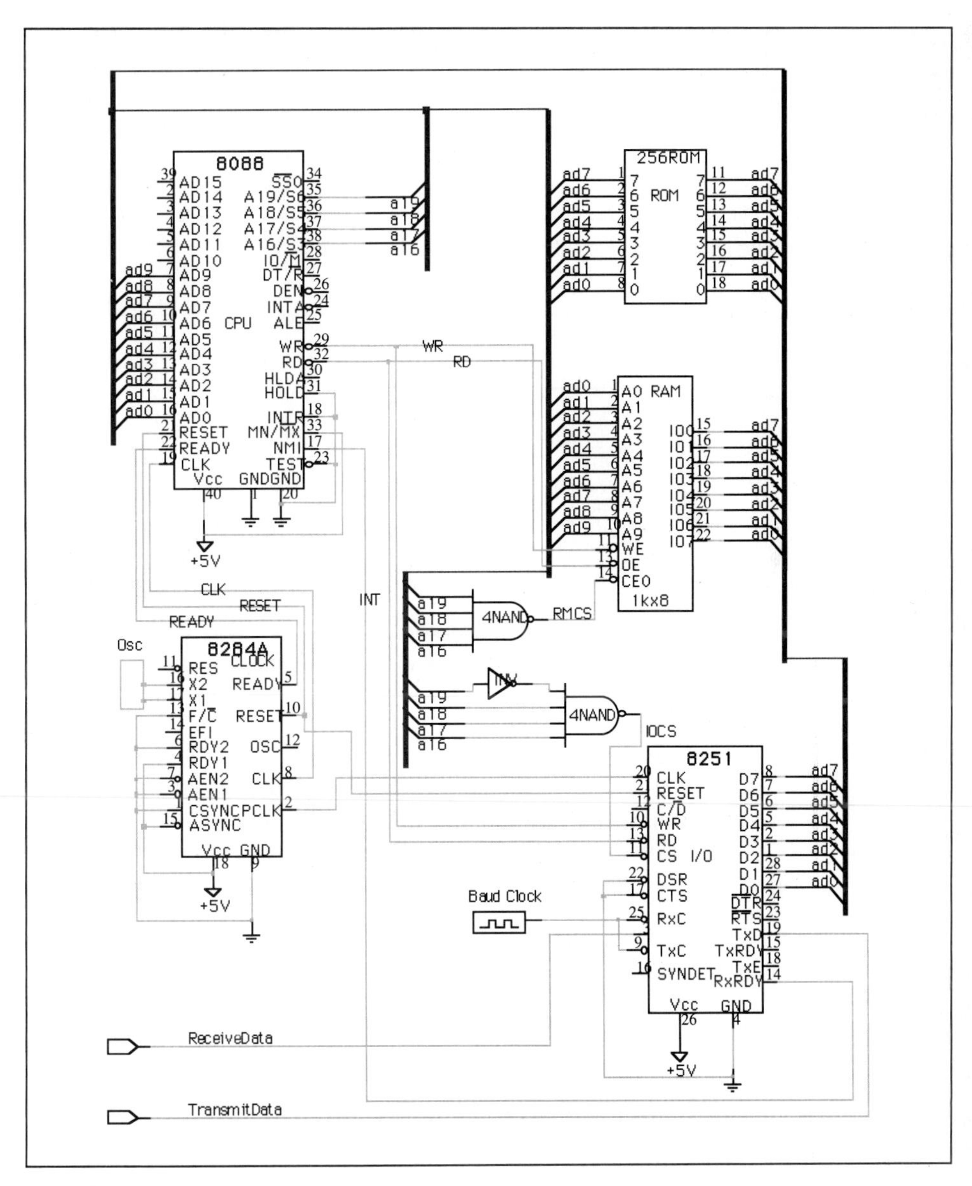

FIGURE 11.7 Schematic diagram of 8088-based single-board computer system.

components, from an XOR function with five gates to an Intel 80486 with tens of thousands of gates.

In the future, designers will be working more with devices the size of MSI and LSI chips rather than with individual gates. An analogy can be made here between software and hardware in this respect. In the past, software was developed in assembly language, and now software is mainly developed in high-level languages

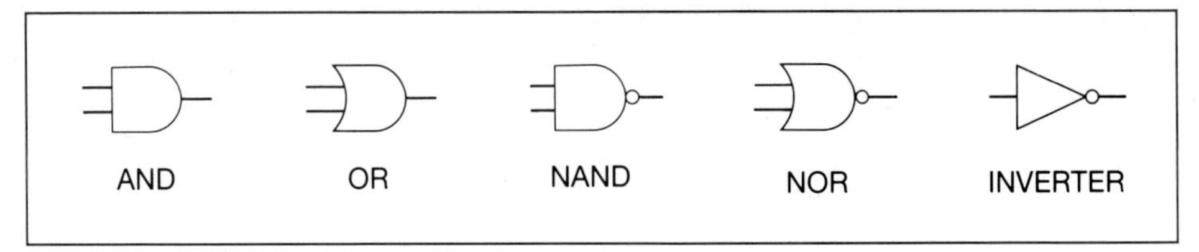

FIGURE 11.8 Symbols for some common logic gates.

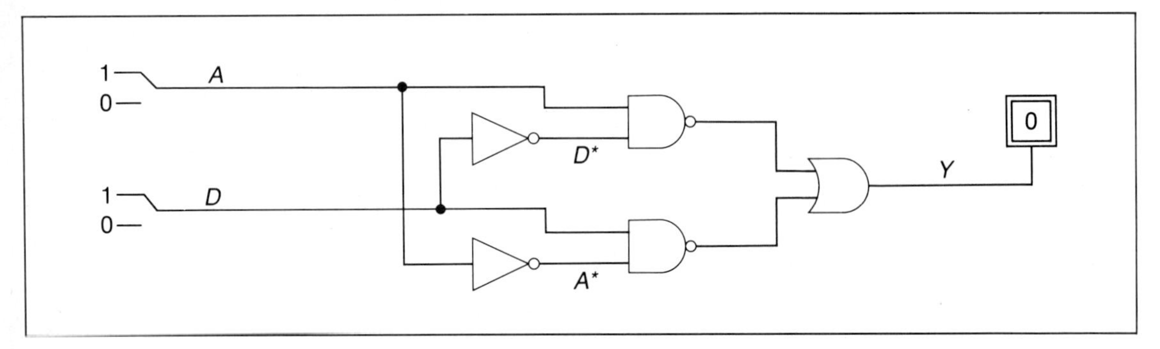

FIGURE 11.9 XOR function implemented with AND, OR, and INVERTER gates.

like Pascal or C. Designing hardware at the gate level corresponds to developing software in assembly language, whereas designing hardware at the MSI or LSI level corresponds to developing software in a high-level language. Regardless of the level of the hardware design, a tool will be useful to "capture" the design data.

11.1.3 Schematic Capture versus Other Formal Design Specification Methods

Three methods are currently used to create formal design specifications:

- Paper and pencil
- Design language
- Schematic capture

The third method, which is most often used, has already been discussed. The paper-and-pencil method of formal design specification is so cumbersome that it can only be used for trivial designs. Imagine trying to keep hundreds of pages of schematics up to date with paper, pencils, erasers, rulers, and gate templates! Design languages were developed prior to schematic capture tools to provide an alternative to designing with the paper-and-pencil method. These languages are similar to programming languages because they have a formal syntax and key words. Using the design language, a designer can specify the components and gates and their interconnections.

```
!
! Build a 2 input NAND gate
!
DEFINE nand2 SUB-BLOCK;
 IN ( a, b );
 OUT ( c );
BEGIN
 g1: $NANDP IN( a, b ),  OUT( c ),  PARAM( DELAY = 1 );
END;
!
! Build an INVERTER gate
!
DEFINE inv SUB-BLOCK;
 IN ( a );
 OUT ( b );
BEGIN
 g1: $NANDP IN( a ), OUT( b ), PARAM( DELAY = 1 );
END;
!
! Build a HALF ADDER
!
DEFINE halfAdder DEVICE;
 IN ( x, y );
 OUT ( sum, carry );
BEGIN
 a: inv IN ( x ), OUT(  x* );
 b: inv IN ( y ), OUT ( y* );
 c: nand2 IN ( x, y* ), OUT ( z1 );
 d: nand2 IN ( x*, y ), OUT ( z2 );
 e: nand2 IN ( z1, z2 ), OUT ( sum );
 f: nand2 IN ( x, y ), OUT ( z3 );
 g: inv IN( z3 ), OUT ( carry );
END;
```

(*a*)

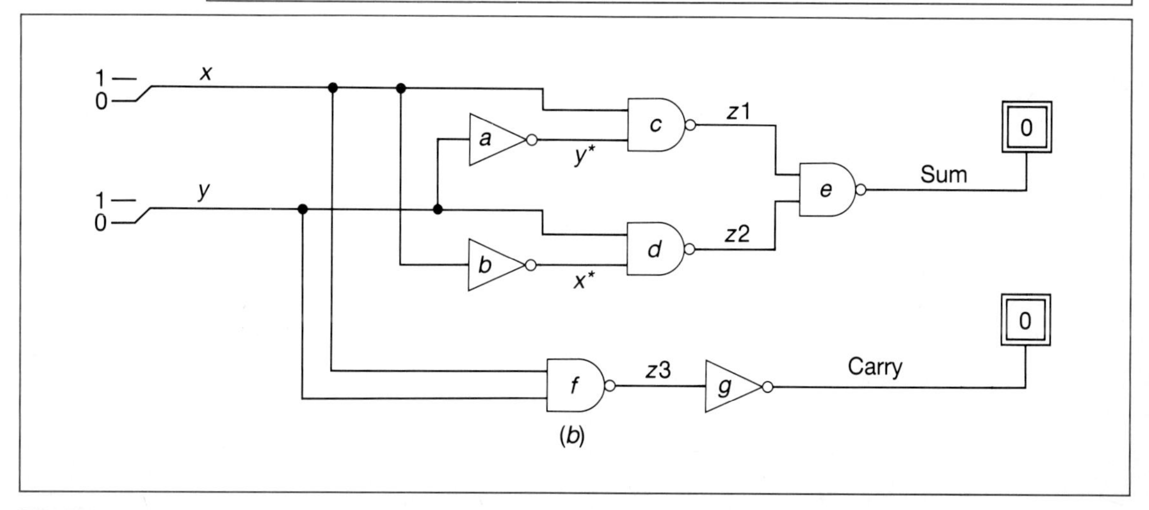

(*b*)

FIGURE 11.10 Half-adder: (*a*) described using Design Automation Language; (*b*) described with DesignWorks schematic capture tool.

The main disadvantage of a design language is that it is difficult to look at words and visualize how the circuit works. Here, a picture is definitely worth a thousand words. Design languages specify all the same information that a schematic capture tool does, but the completed design does not include a symbolic representation. In Fig. 11.10 we compare the formal design specification of a circuit described using Tandem Computer's Design Automation Language (DAL) and the DesignWorks schematic capture tool.

The design description in Fig. 11.10*a* uses a two-input NAND gate as a building block. The "DEFINE nand2 SUB-BLOCK" in the description defines the two-input gate nand2 that is like the standard NAND gate but has a delay of one time unit. The "DEFINE inv SUB-BLOCK" defines the inverter inv. The remainder of the description defines a half-adder. The half-adder has two inputs, x and y, and two outputs, sum and carry. The operation of the half-adder is then described by the series of statements a: through g:. Statement a: says that the signal x is the complement of signal x. The remainder of the statements specify how the outputs sum and carry are generated from the inputs x and y.

Figure 11.10*b* shows exactly the same circuit created using a schematic capture program. The main advantage of schematic capture over design languages is that it provides pictures of the circuits as well as all the other relevant information. An advantage of using a design language is that a design can be created using a simple text editor and terminal with no graphics capabilities. Since each method is useful, computer manufacturers often use both schematic capture and design languages to specify designs. Furthermore, some design-capture programs will generate schematics from a design language description of a circuit.

The purpose of any schematic capture tool is to create a logical description of the design. The final schematic contains the information needed to create a list of components used in the design, a list of nets that connect the components, and names of nets and components. Not surprisingly, the list of components is called a **parts list**, and the list of nets is called a **net list**. We will see in later sections how the parts list and net list are used with the physical board design tools. Schematic capture, or design specification, is the first step in building any new computer. However, creating an accurate schematic or a language description of a design does not guarantee that the circuit will work correctly. The next step in building a useful circuit is logic verification.

11.2 LOGIC VERIFICATION

Logic verification is a technique used to determine if a design will operate correctly. An example will help clarify why logic verification is necessary. Let's suppose you develop the specifications of the plumbing system for a new house. You would like to convince yourself that the plumbing system will operate correctly before the system is installed. What if your piping diagram were incorrect and you had all the outside faucets connected to the hot water line instead of the cold water line? The plumbing system would operate, but it would cause a significant waste of energy.

Furthermore, the grass and shrubs might experience some serious side effects from the hot water.

The system described above is *logically* incorrect because outside faucets should be connected to the cold water line. Computer engineers can make similar mistakes when developing the design of new systems. Figure 11.11 illustrates an incorrectly designed XOR function; both inputs to the lower AND gate are complemented.

Logic verification is used to determine if a specified design will function as it is intended. It is obvious when we look at the truth tables shown in Fig. 11.11*b* that the circuit described in Fig. 11.11*a* does not perform the XOR function. The circuit actually performs the EQUAL function. Unfortunately, for large designs it is not possible to look at the truth table and determine that there is a mistake. Tools are needed to assist the designer in finding logical errors.

Logic verification is generally done with a CAD tool called a **logic simulator**. Logic simulators operate on the principle that for a given set of input signals, a known set of output signals can be expected. Consider the simulation of a simple logical AND function. The operation of an AND gate conforms to the truth table in Fig. 11.12. The AND gate has two input signals and one output signal. One way to test if the gate works is to apply input signals to it and look at the outputs. If, for all combinations of input signals, the gate gives the outputs listed in the truth table, then we would say it works correctly.

Let's discuss simulation in more detail using the incorrectly designed XOR circuit in Fig. 11.11. There are five logic gates shown in that figure: two INVERTER gates,

FIGURE 11.11 An example of a design error: (*a*) incorrect design of the XOR function; (*b*) truth tables for incorrect and correct designs.

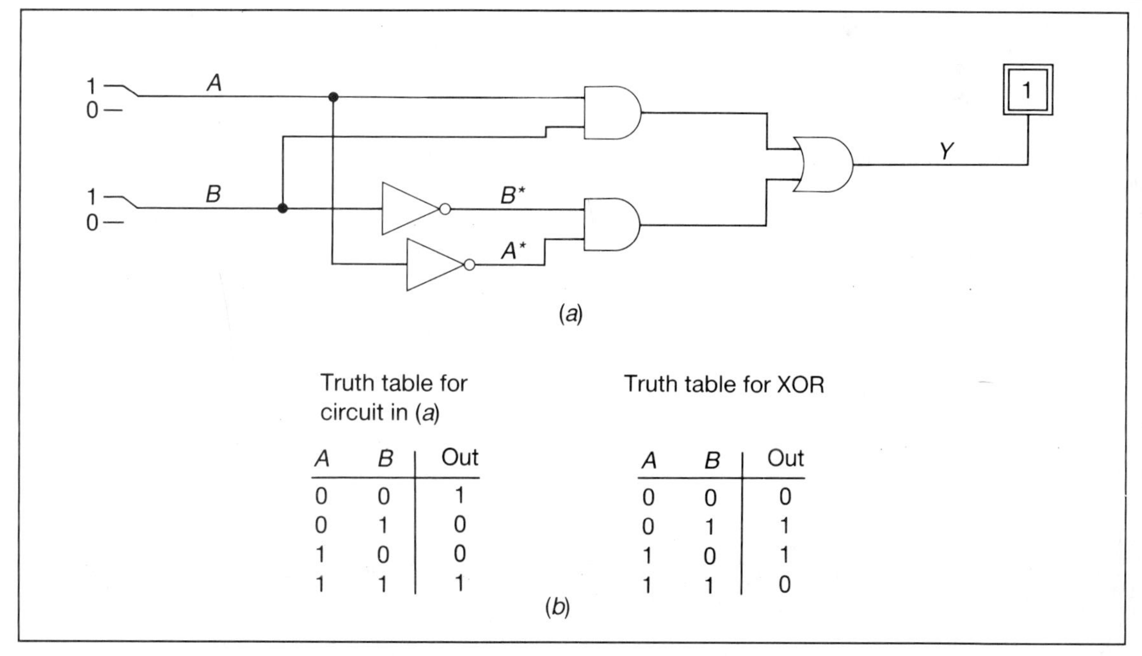

1
0
A
1
0
B
B*
A*
Y
1
(a)
Truth table for circuit in (a)

A	B	Out
0	0	1
0	1	0
1	0	0
1	1	1

Truth table for XOR

A	B	Out
0	0	0
0	1	1
1	0	1
1	1	0

(b)

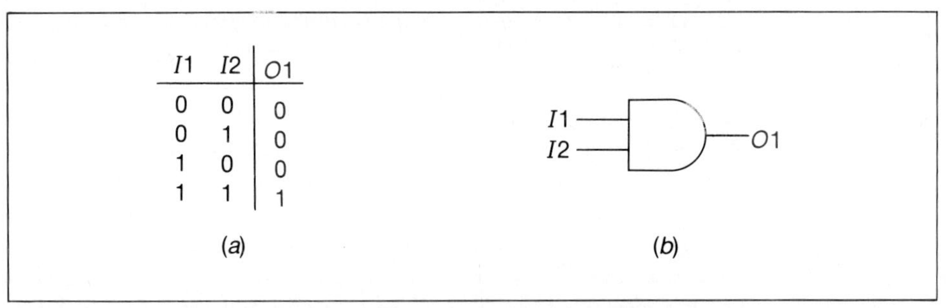

$I1$	$I2$	$O1$
0	0	0
0	1	0
1	0	0
1	1	1

FIGURE 11.12 (*a*) Truth table and (*b*) circuit symbol for AND function.

two AND gates, and one OR gate. Assume we have a tool capable of generating a programming-language description of the circuit shown in Fig. 11.11. A single C language statement that represents the circuit is

$$Y = (A\&\&B)||(!A\&\&!B) \tag{11.1}$$

This programming statement reads, "*Y* equals *A* and *B* or not *A* and not *B*." This statement can be used in a program to determine if the circuit functions as intended.

Figure 11.13 contains a program segment that will test our function and report if it is correct. This program segment shows how an interactive logic simulator could be used to verify a circuit. After each pass through the "while loop," the program asks if the user is finished testing the circuit. If the user is not finished, he or she types "n" and then types in the values from the truth table corresponding to the inputs A and B and the output Y. The program simulates the circuit by executing the statement given in Eq. (11.1). If the simulation yields a value different from the expected value, an error is flagged, the "while loop" finishes, and the user is informed that the circuit does not function correctly.

Some logic simulators can use the data generated by a schematic capture tool to

FIGURE 11.13 Program for testing sample function.

```
while NOT( Error ) AND NOT( Done)
{
  PRINT ("Are you finished? (y/n)");
  READ (Answer);
    IF (Answer != "y") THEN Done = TRUE
  ELSE
           PRINT( "Please type in value for A, B, and Y from the table")
           READ (A, B, Expected_Output);
            Y = ( A && B )  ||  ( !A && !B );
            IF ( Expected_Out != Y ) THEN Error = TRUE;
}
        IF NOT( Error ) THEN PRINT ("The circuit is correct !")
        ELSE PRINT( "The circuit is NOT correct")
```

create the program statements necessary to test the logic in a design. This eliminates the need for the designer to input information about a design twice. Logic simulation is an important and fascinating field, and we have given you a brief introduction to how a simulator can operate.

So far we have a logical description of a single-board computer, and we know that a logic simulator can be used to verify that the logic is correct. Now, we need to determine what the physical computer will look like. The physical characteristics of a computer are designed using physical placement and routing tools.

11.3 PHYSICAL PLACEMENT AND ROUTING TOOLS

Physical placement and routing tools are often combined into one physical design tools package; however, they are really two separate tools. The physical placement tool is used to specify where parts will be located on a printed circuit board and to assign logic gates to physical components. The routing tool is used to design how the nets are placed on the printed circuit board. Nets, remember, are the equivalent of wires on printed circuit boards and are thin traces of metal.

Physical placement and routing tools are used to create a "blueprint" of the printed circuit board. The blueprint of a printed circuit board is similar to the blueprint for a building. The blueprint for a building specifies floor plans for each individual floor; blueprints created using physical design tools specify the planes in the printed circuit board. Before we discuss the physical design tools, it will be useful to learn a little about printed circuit boards and how they are made.

11.3.1 Printed Circuit Boards

A **printed circuit board** is typically a multilayer board that is made from fiberglass, metal sheets, and epoxy cloth. Printed circuit boards are sometimes called **printed wire boards**, because metal traces (wires) are etched or printed onto the board. Figure 11.14 shows a cross section of a printed circuit board that has an electrical ground plane, an electrical power plane, and traces routed on two sides. The ground and power planes are thin copper metal sheets. A piece of epoxy cloth separates the

FIGURE 11.14 Cross section of a typical printed circuit board.

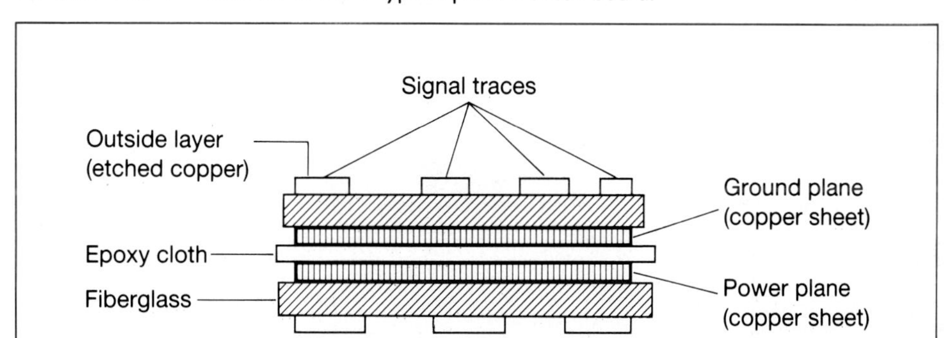

ground plane from the power plane, and fiberglass separates the signal layers from the planes.

Power and ground planes are used for several reasons. Most components have at least one pin connected to power and one pin connected to ground. Power and ground planes are also useful for limiting electrical noise problems that cause signal distortion. Noise is a complicated phenomenon that we will discuss later, but note here that planes help to reduce it.

When designing a board, there are several questions that must be answered:

- How are components mounted on the board?
- How are metal traces etched or printed onto the board?
- How are traces on different levels connected?

There are currently two popular methods for mounting components: (1) through-hole mounting and (2) surface mounting. **Through-hole mounting** is a technique where holes are drilled through the printed circuit board so that the component wires fit in the holes. **Surface mounting** is a technique where components are "glued" onto the printed circuit board using glue and a solder paste.

Figure 11.15 shows views of a through-hole integrated circuit (IC) package and a surface-mount IC package mounted on a two-layer board. Through-hole mounting

FIGURE 11.15 Chip mounting: (*a*) side view of a through-hole IC package; (*b*) end view of surface-mount IC package.

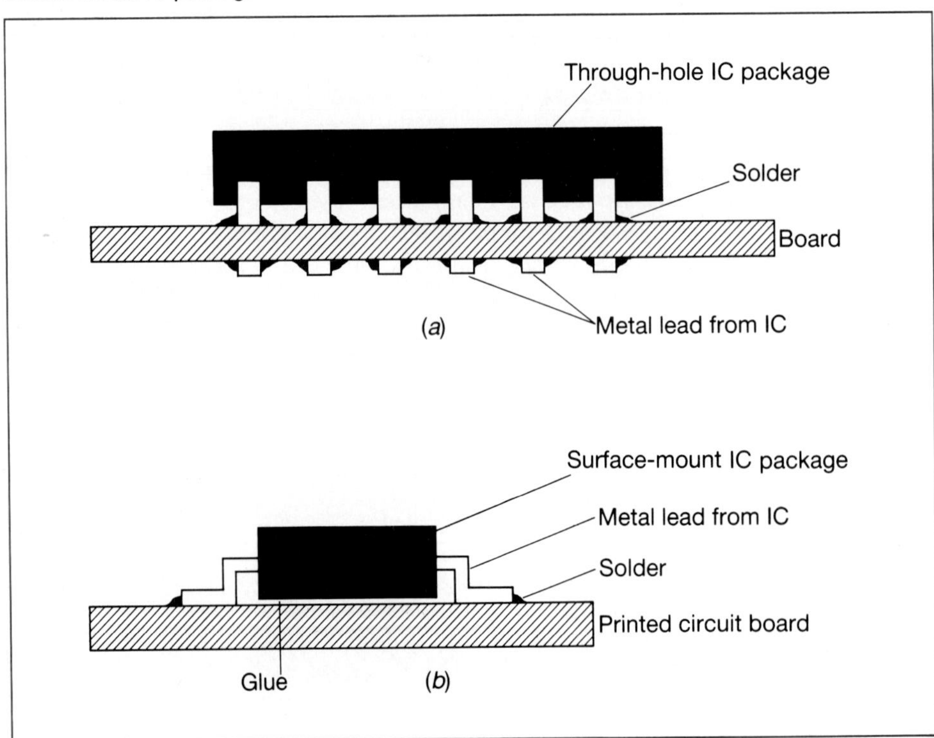

allows the metal leads on an IC to protrude through the board. Components are generally placed on only one side of the printed circuit board when through-hole mounting is used. When the components have all been placed in their correct positions, the board is passed over a solder bath so that the metal leads of the components touch the molten solder. A **solder bath** is essentially a large pool of molten solder. Solder fills the hole around the metal leads of the components. When the solder hardens, it holds the ICs in place and establishes the electrical connections.

Surface-mount technology is growing rapidly, mainly because surface-mount packages are much smaller and more reliable than through-hole-mounted packages. Also, surface-mount packages can be mounted onto both sides of a printed circuit board because they are glued onto the board and not passed over a solder bath. Figure 11.15*b* shows the end view of a surface-mount IC. Notice that the flat portion of the surface-mount package is glued to the board to hold the chip firmly in place and the wires on the ICs are connected to the board with solder.

We have looked at how components are mounted on printed circuit boards, so now let's discuss how the nets or wires are created. Figure 11.16 shows one possible technique for creating traces on a board. First, a fiberglass board is laminated with a metal conducting layer. The conducting layer is cleaned and then coated with a light-sensitive resist layer. The resist layer is exposed to light through a pattern **mask** to imprint the desired connection pattern. The mask for each layer of the board is designed using physical placement and routing tools. The sensitized pattern is then flushed away, leaving exposed areas where the metal is to be removed. The metal conductor not protected by the resistant layer is then etched away with acid. Finally, the resistant material is removed from the conducting layer, and all that remains are the desired metal traces on the fiberglass board.

FIGURE 11.16 Four-step process for creating traces on a circuit board.

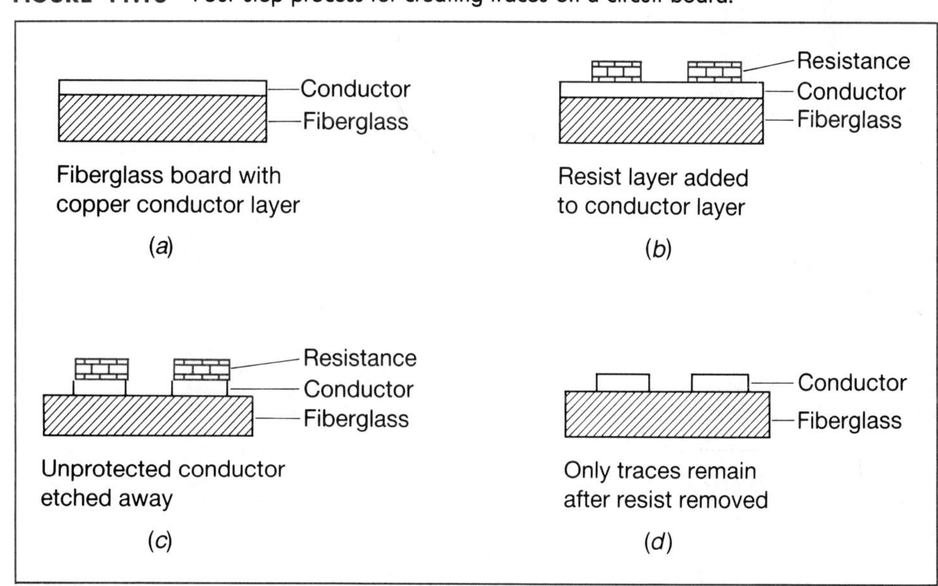

The last topic we will mention regarding printed circuit boards is **feed-through holes**, commonly called vias. A **via** is used to connect metal traces that are on different layers of a printed circuit board. A trace could be made to go from the top of the board, to the edge, down the edge, and onto the bottom layer of the board. It is much easier, however, to drill a hole through the board and directly connect the traces.

In this section we explained how components are mounted on printed circuit boards. We also discussed how traces are put onto the boards and described how traces on different layers of the board are connected. This section was intended to familiarize you with printed circuit boards so that you might better understand the CAD tools used to design them.

11.3.2 Physical Placement Tools

A physical placement tool is similar to a schematic capture tool because both are interactive graphics programs used to create pictures of circuits. There are important differences between schematic capture and physical placement, however. Remember that, with the schematic capture tool, we simply placed components onto schematic pages. The relationship between gates and physical components was not considered when the schematics were created. When designing a board, however, all the components are real devices that must be placed so that they fit onto a printed circuit board. The printed circuit board becomes the "page" used to place physical components.

The component symbol used in physical placement is different from a schematic capture symbol of the same component. Figure 11.17*a* shows a schematic diagram of a 74LS373 part, and Fig. 11.17*b* shows the footprint of the corresponding physi-

FIGURE 11.17 A 74LS373 IC (*a*) schematic and (*b*) footprint.

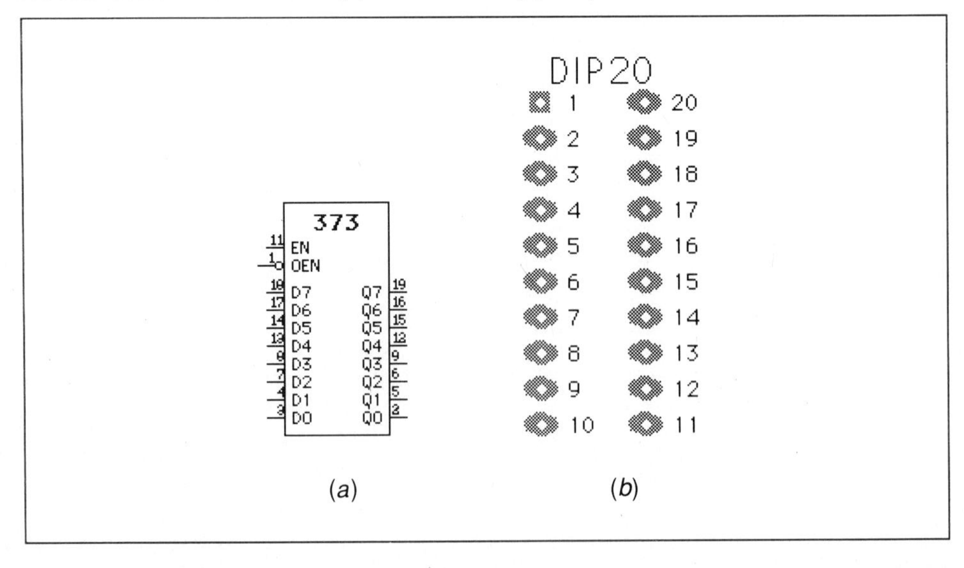

cal layout part. **Footprints** represent holes that will be drilled into the printed circuit boards to hold the ICs for through-hole mounting. For surface mounting, they represent the positions where the component leads will be soldered onto the board.

Each component that is used in the design must be placed onto the printed circuit board. Many components will have the same footprint, but separate footprints must be specified for each part in the design.

Another important function performed by the placement tool is gate assignment. Often, physical devices contain more than one gate or element. Gate assignment is the process of assigning logic gates from the schematic drawings to physical devices. Figure 11.18 shows three devices that contain more than one gate. Gate assignment is necessary so that traces or nets can be connected to the proper pins on each part.

Let's return to our single-board computer example and discuss how a typical physical placement tool is used to describe the printed circuit board layout. The first step in physical placement is to define the characteristics of the printed circuit board. The designer must specify the dimensions of the printed circuit board, the number of layers in the board, through-hole or surface mounting, and other physical attributes. For this discussion we assume the connections will be made using traces on the top or bottom layer of the board. On boards designed for real computers, it is not uncommon to find 10 or more layers of traces.

After specifying the characteristics of the board, the next step in physical layout is to transfer the parts list and net list to the placement tool. Typically, this is done by choosing a menu item or typing a command and then typing in the name of the parts list and the net list. The layout tool will then present the parts list on the screen so that the designer can select parts and place them onto the board.

Physical placement may seem like an easy task, but there is more to it than meets the eye. The location of the parts directly affects the lengths of the interconnections between parts. Two parts that have multiple interconnections should be placed in close proximity so that long traces will not have to be created to connect them. One of the most important criteria for physical placement is the reduction of the total trace length. This, in turn, reduces noise and signal distortion problems.

The layout designer must analyze the schematic diagrams to determine which parts are connected together so that he or she can put these parts in close proximity

FIGURE 11.18 Gates: (*a*) quad two-input AND; (*b*) quad two-input OR; (*c*) HEX inverter.

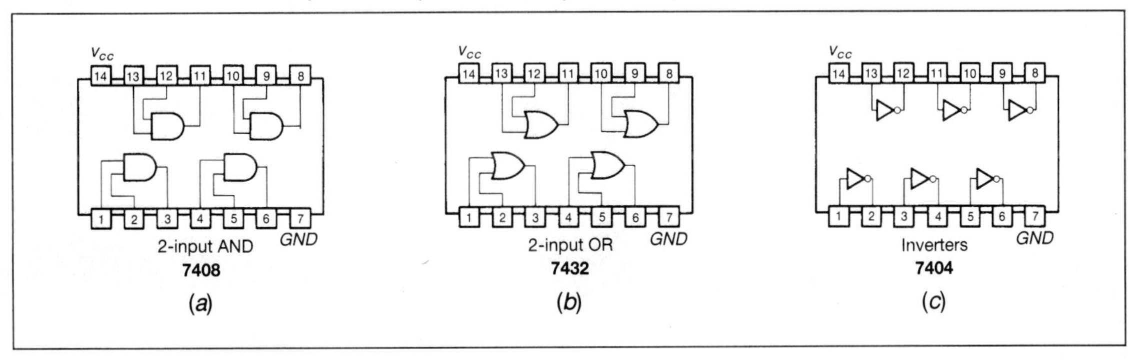

to one another. After the parts have been placed on the "board," gate assignments are made. The designer selects a menu item or types a command to enter the gate assignment mode. A list of gates is displayed on the screen just as the parts list was displayed. The designer then picks a gate and specifies which part of the physical device will be used.

Since it may not be clear to you why gate assignment is necessary, let's refer to our single-board computer design example. In the schematic shown in Fig. 11.7, you will notice that there are three gates, two 4NANDS and one INV. These are the only symbols in the schematic that do not directly correspond with physical parts. In our physical design of the single-board computer we will use a type 7404 (like the one shown in Fig. 11.18*c*) for the inverter and a type 7420 for the 4NAND gates. We will assign the inverter in Fig. 11.7 to the gate that is between pins 1 and 2 on a 7404. The gate assignment of the 4NAND gates is left to the student as an exercise.

After gate assignment has been completed, the next step is to adjust component locations to minimize the logical connection lengths. Often, physical placement tools aid the designer by showing the logical connections between parts. A logical connection is represented as a straight line between two points that are connected. The tool determines how the parts are logically connected by analyzing the net list. The placement tool also calculates the total length of all the connections and reports this number to the designer.

Since the designers can then see how the placement of the parts affects the trace lengths, they can move parts around and minimize the total length of the connections. Figure 11.19 shows what a printed circuit board for the single-board com-

FIGURE 11.19 Placement of 8088-based single-board computer system.

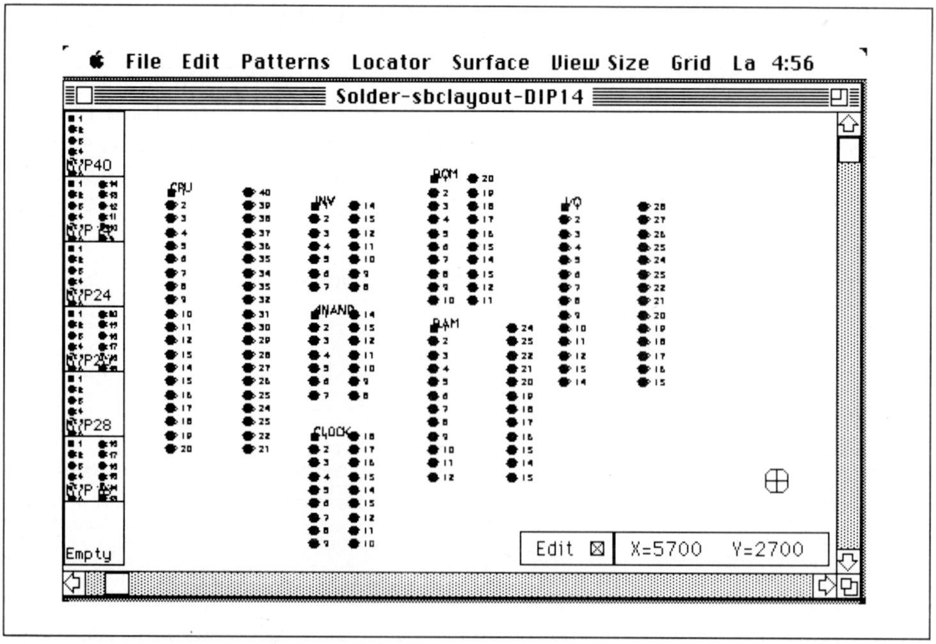

puter might look like after an initial physical placement and gate assignment is completed. We will discuss how the actual physical connections are made in the next section.

The placement shown in Fig. 11.19 is not optimal; it is intended only to show you how the part footprints look when laid out on a printed circuit board. In a real design, the parts would be located much closer to one another. After the designer is satisfied with the layout, physical connections must be made using a routing tool.

11.3.3 Routing Tools

Routing tools, often called **routers**, are used to connect the footprints generated by the placement tool in a manner consistent with the connections specified on the schematic. Routing is the equivalent of a connect-the-dots puzzle, where the dots represent holes for component leads and the lines represent trace connections. However, for real designs, routing is not this simple because there are rules that must be followed.

One important goal in routing a design is to minimize the total length of traces needed to connect the parts. The total trace length of a design is assumed to equal the sum of the lengths of all the individual trace lengths. A good way to evaluate a route's trace length is to compare it to the best possible route, or the optimal route.

The **optimal route** between any two points on a board is a straight line connecting the points. An optimal route may require diagonal lines, however, and diagonal lines are more complicated to work with than horizontal and vertical lines. Figure 11.20 shows the optimal route between two pins and the best Manhattan route between the same two pins. The best **Manhattan route** between two points is the

FIGURE 11.20 Imaginary map of Manhattan illustrates the term *Manhattan route*.

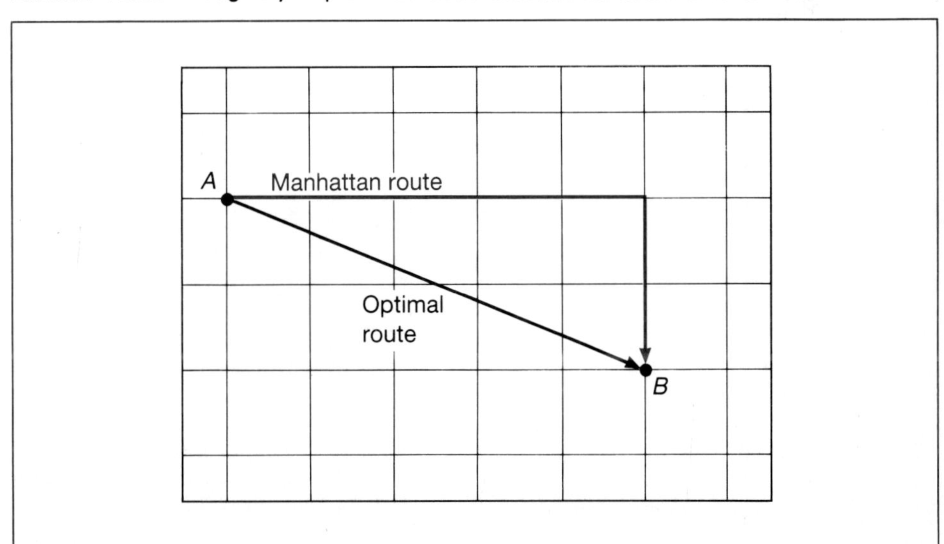

shortest distance, using only horizontal and vertical lines, to connect two points. To get from point *A* in Manhattan to point *B*, one must travel horizontally and vertically with respect to the map shown in Fig. 11.20.

In a complex design, the number of nets that must be routed is on the order of thousands. It is a very difficult and time-consuming task for an engineer to route that many connections. Therefore, the designer will often use a tool called an autorouter to assist in the process. An **autorouter** is a tool that takes the placement information from the layout tool and the net list from the schematic capture tool and routes traces automatically.

Because the areas near large ICs are densely populated with nets, it is unreasonable to assume that all connections in a design can be made using the optimal Manhattan routes. Some nets must be routed around the densely populated areas, and therefore they are not optimal. Since not every net can be perfect, the autorouter is generally given a **derating factor**.

Suppose a derating factor of 1.5 is given to the autorouter. This means that the tool can make a connection between two points that is up to 50 percent longer than the optimal Manhattan route. Consider the following example; if the optimal Manhattan route is 2 cm and the derating factor is 1.5, then any route less than 2×1.5 cm $= 3$ cm is acceptable. A derating factor of 2.0 would specify that the distance can be twice the optimal Manhattan route.

Autorouters can be used to route anywhere from zero to 100 percent of the nets in a design, depending on the complexity of the design. Very complex designs

FIGURE 11.21 Fully routed printed circuit board of 8088-based system.

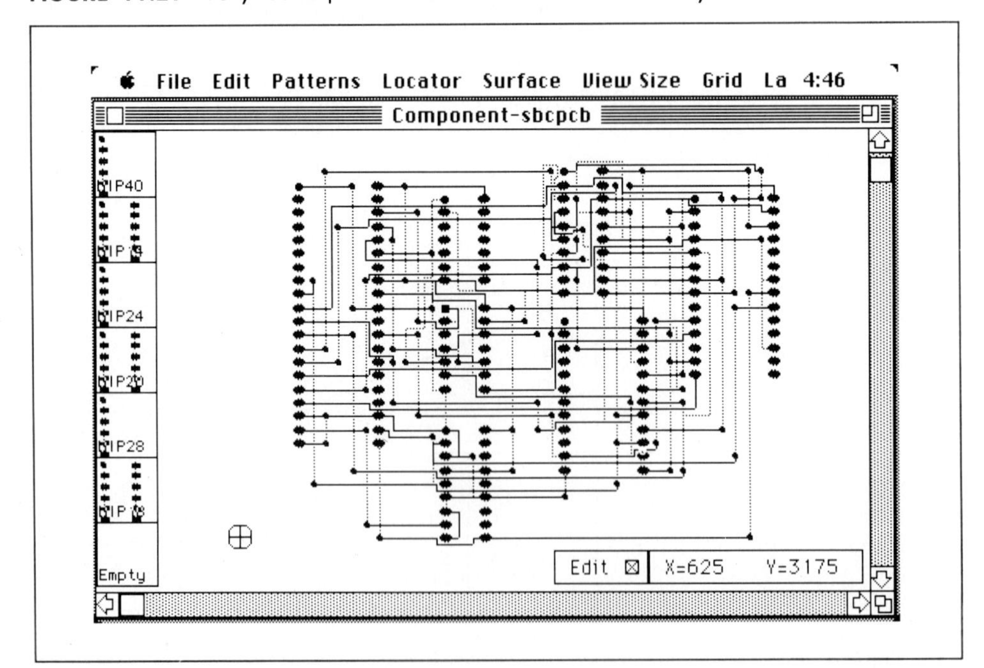

require more hand routing because autorouters cannot find reasonable routes for many of the nets. Figure 11.21 shows the routing for the single-board computer. The schematic for the single-board computer is shown in Fig. 11.7. Figure 11.21 was generated using the autorouter from Douglas Software with no derating factor specified [Douglas, 1990].

Figure 11.21 is a poor design because the components are not placed closely enough together, and therefore the routes are longer than necessary. Despite this, we can still learn from the design. Notice that the horizontal lines are solid and the vertical lines are dotted. All horizontal nets are usually put on one layer of the board, and all vertical nets are put on a different layer. The traces are connected together using vias, which are represented as dots at the points where horizontal and vertical lines intersect. The purpose of creating the design in Fig. 11.21 is so that the mask, necessary to build the printed circuit board, can be made.

We emphasized that one of the main goals of physical design is to minimize trace lengths. For a majority of the traces in a design, this is true. However, timing requirements and circuit characteristics sometimes require that traces be longer than optimal routes. Since there are certain exceptions to this goal of minimizing trace lengths, we felt it was important to note that fact here.

Our next topic is verification of the routed printed circuit board. Once the printed circuit board is laid out and routed, the nets must be simulated to determine if the timing requirements of the design will be met.

11.4 ANALOG CIRCUIT SIMULATION

We will discuss only one analog circuit simulation tool called SPICE. The acronym SPICE stands for *s*imulation *p*rogram with *i*ntegrated *c*ircuit *e*mphasis [Tuinenga, 1988]. SPICE is an analog design tool that is used to determine how signals will propagate along metal traces and through components. There are several practical problems associated with passing information in electric signals that we must mention here.

First, because electric signals travel at a finite speed, they take time to go from one point to another in the circuit. Second, components take a finite amount of time to change states. The delay introduced into a circuit by these factors is called propagation delay and was described in Chap. 2. Propagation delay affects circuit timing, and SPICE can be used to determine the delays before the circuit is built.

The third important problem associated with passing information in electric signals is signal distortion. **Signal distortion** is a phenomenon where some characteristics of a signal are changed as a result of the signal traveling from point *A* to point *B*. Because signal distortion can induce errors, we must analyze our designs in order to reduce distortion. After a brief discussion of how propagation delay and signal distortion can cause a computer to fail, we will discuss how SPICE is used to analyze propagation delays and signal distortion.

A computer is primarily a synchronous digital device. A **synchronous device** is one that is designed around a system clock that keeps time for the circuit. We mentioned earlier that the CPU reads and writes information to the memory unit. In many systems the CPU will request a word of data from memory and expect to get the data from memory on the next clock cycle. Suppose that the clock cycle for a system is 8 MHz. This corresponds to a clock period of 125 ns because

$$\text{Frequency} = \frac{1}{\text{period}}$$

If the CPU requests a piece of data in one clock period and expects to receive the data in the next clock period, then the memory will have to provide the data within 125 ns.

Figure 11.22 shows a timing diagram where the CPU requests data in one clock period and reads the data in the next clock period. Suppose it takes the memory unit 100 ns to retrieve the data and put it onto the data bus. The data word will then have 25 ns to travel on the data bus and reach the CPU. If the propagation delay on the data bus is greater than 25 ns, then the data word will arrive after the CPU reads the data bus. The CPU will read an incorrect data word, and the computer will probably malfunction.

In digital devices, signals carry binary information. A signal in a digital system is either in the 1 state, the 0 state, or the undefined state. Figure 11.23 shows voltage values that correspond to the signal being in one of these three states for one type of

FIGURE 11.22 Timing diagram for CPU reading information from memory.

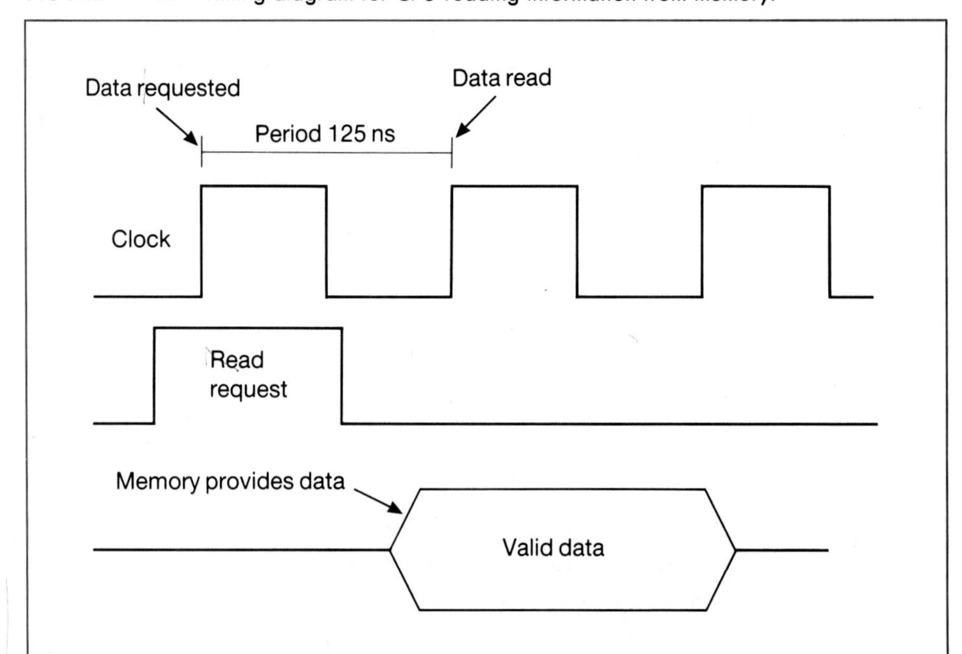

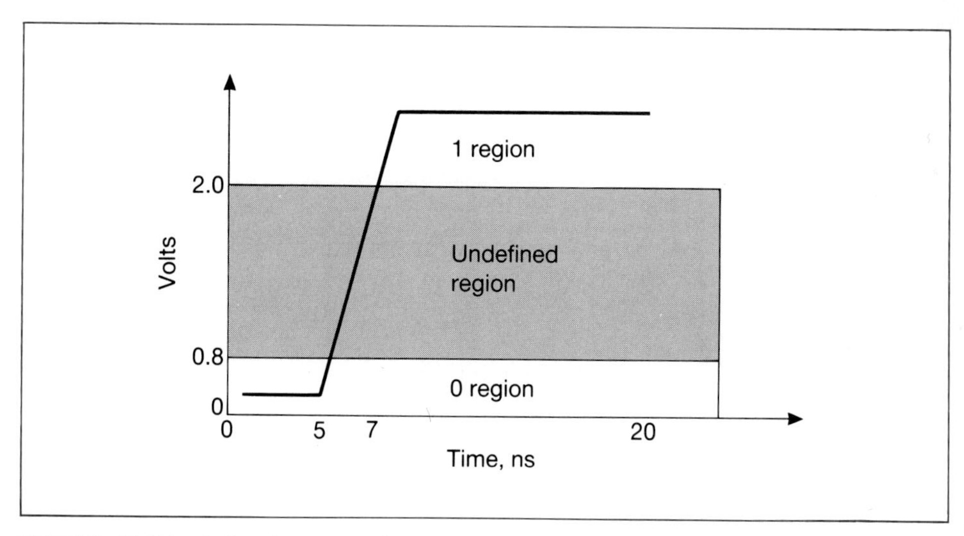

FIGURE 11.23 A signal transition from the 0 state to the 1 state.

IC. If the voltage of the signal is less than 0.8 V, the signal is in the 0 state; if the voltage is between 0.8 and 2.0 V, the signal is in the undefined state; and if the voltage is above 2.0 V, the signal is in the 1 state.

Signals are distorted for many reasons, including trace loading, trace length, and environmental conditions. Suppose that the DEN signal, data enable, from the CPU is connected to five input pins. When the CPU sets the DEN signal to 1, each of the inputs should see a voltage above 2.0 V after the propagation delay. If the data enable trace is overloaded, then some inputs may see a voltage of only 1.9 V because the heavily loaded line caused the signal's amplitude (voltage) to distort. A voltage level of 1.9 V is neither a 1 nor a 0; therefore a device may not pass the CPU the requested information because it did not receive the request. SPICE can calculate signal distortion, given the same information necessary to calculate propagation delay.

There are several important pieces of information that the SPICE program must have in order to correctly calculate delays and distortion: (1) **trace length,** (2) **trace loading**, and (3) **driver type**. The trace length is known because the routing tool connected each net in the design. Input pins, resistors, capacitors, and other elements are considered loads on a trace. The number and type of loads on a trace can be calculated by analyzing the parts list and net list. Finally, SPICE needs to know the characteristics of the device driver that is sending the signal on the trace. Driver information is also found by looking at the parts list and net list.

Given the information described above for each net, SPICE generates diagrams like the one shown in Fig. 11.24, which is a MacSPICE [Deutsch Research, 1990] plot with signal values at the driver and at the receiver or load. The designer can analyze the SPICE diagram and determine if propagation delay or signal distortion is a problem.

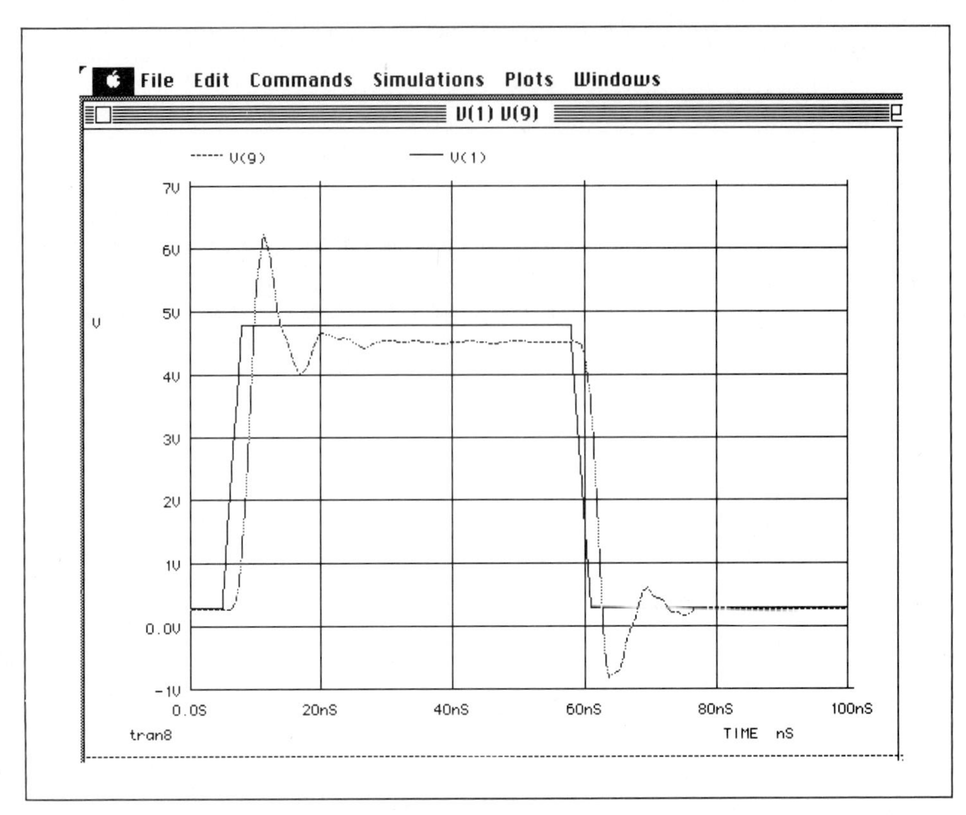

FIGURE 11.24 MacSPICE output showing propagation delay and signal distortion.

SPICE will find errors caused by propagation delay and signal distortion. When errors are found in a design, they must be corrected. Suppose a propagation delay that will cause a timing error is found. The propagation delay can be reduced by shortening the trace with the router. If a trace is overloaded and signal distortion is a problem, the designer may have to go back to schematic capture and redesign the circuit to reduce the loads on a particular net. The design will then have to be rerouted and possibly laid out again.

The final point we would like to make about SPICE is that the tool is only as accurate as the models it uses. SPICE calculates delays for signals based on models that attempt to represent the real traces and loads. It is often difficult to model real-world circuit elements, and the more complete and complicated the models are, the slower SPICE runs. MacSPICE includes a program, Model Maker, that allows you to generate SPICE models directly from product data sheets. This simplifies the use of SPICE and increases its flexibility.

11.5 CAD TOOLS AND THE DESIGN DATABASE

We have introduced a small but important subset of the CAD tools available to today's computer engineers. Figure 11.25 shows a flow diagram of the design process using CAD tools.

Notice that Fig. 11.25 contains the names of each of the CAD tools we have discussed. One feature of this figure that has not been discussed is the **design database**, an entity that contains all the information generated about a design. A design entered using a schematic capture tool will generate a parts list and a net list that are saved in the database. A logic verification tool can then take information from the database to determine which types of parts are included in the design and how the parts are connected. With this information, logic verification is performed on the circuit. Any errors found during logic verification must be corrected using the schematic capture tool.

Physical design tools use the final parts and net lists generated by schematic capture to create a blueprint. The physical placement tool uses the parts list in the database to determine which parts must be included on the printed circuit board. The routing tool uses the net list in the database to determine how the parts are connected so that routing can be done. Finally, SPICE uses information generated by the routing tool that describes the physical characteristics of each net. SPICE

FIGURE 11.25 Flow diagram of CAD tools and their usage.

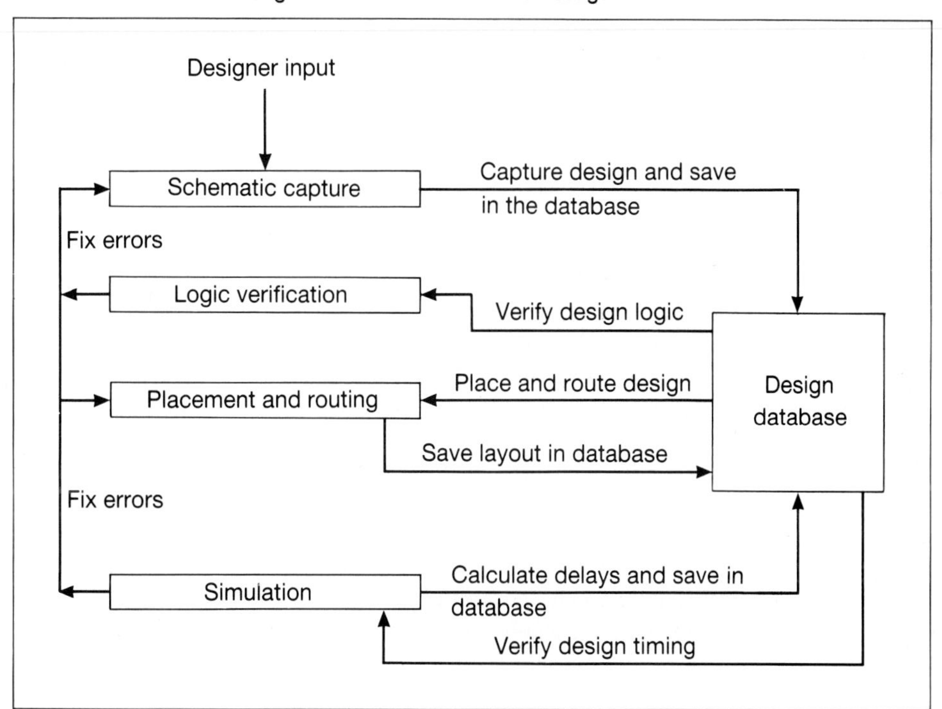

uses the information about nets to determine propagation delays along nets, and this is useful for verifying design timing. When timing errors are discovered with SPICE, the errors are corrected by rerouting nets or changing the logical design in schematic capture.

CAD tools are used to specify how a new computer will be built. The result of their combined operation is a database that contains all the information necessary to build a computer. The database allows different CAD tools to communicate and share design information. The database contains a complete specification of the computer; it describes the computer's function and how it will be built.

11.6 HARDWARE DEBUGGING TOOLS

We have discussed the CAD tools necessary for designing a computer. Unfortunately, using all the tools does not guarantee a perfect board. Because design errors may exist and components fail, the actual computer must be tested to determine if it works correctly. There are several hardware debugging tools available to facilitate finding design errors, or "bugs," and component failures. We will discuss three: (1) multimeters, (2) oscilloscopes, and (3) logic analyzers.

In general, multimeters can measure voltage, current, and resistance. After a printed circuit board has been manufactured, it can be tested using the ohmmeter function; the **ohmmeter** measures resistance. If two points on a printed circuit board are connected, then there will be approximately 0 Ω of resistance between them. The designer can use the ohmmeter to verify that connections exist on the printed circuit board by measuring the resistance between points. A fault, such as a broken trace, will be discovered because the resistance between two unconnected points is very large.

The **voltmeter** measures voltages and is helpful in determining if ICs have the correct voltages. For example, a typical transistor-transistor logic (TTL) IC will not function unless it has +5 V on its power pin(s) and 0 V on its ground pin(s).

An **oscilloscope** is a tool that can also measure voltages and currents like the multimeter, but the oscilloscope provides a video display. The display allows engineers to view signal waveforms, and this feature is extremely useful in debugging hardware. The screen has an *X* and *Y* axis, as shown in Fig. 11.26. The *X* axis represents time, and the *Y* axis represents signal amplitude. Oscilloscopes generally have two or more probes that can be attached to any conducting point in a circuit. Each probe, when attached to a point in the circuit, causes a signal waveform to be displayed on the screen.

We mentioned earlier that an oscilloscope was a more powerful tool than the multimeter. One reason is that the oscilloscope allows individual signal waveforms to be viewed as functions of time. We used SPICE to predict what signal waveforms would look like, and the oscilloscope shows us how the actual signals look. The oscilloscope is useful for determining the propagation delay and signal distortion for any signal. By looking at the actual signals, it is possible to isolate errors. Finding

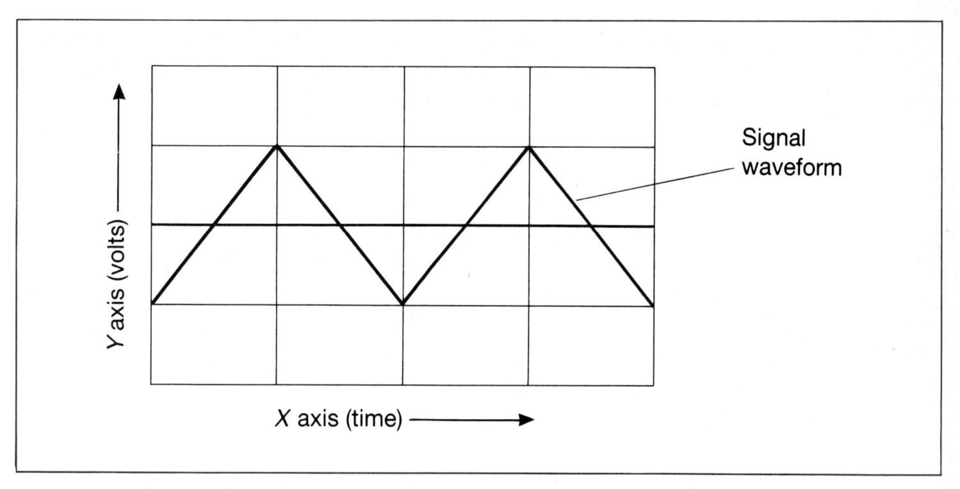

FIGURE 11.26 Oscilloscope screen showing a waveform.

and correcting errors in a new computer is not an easy task, but let's discuss one possible scenario.

If the new computer hardware does not function correctly when it is initially powered up, a good place to start looking for an error is on the clock chip. Assume we designed our clock to generate the waveform shown in Fig. 11.27*a*. Once the probe is attached and the appropriate adjustments are made on the scope, we should see that signal. Suppose, however, we see a signal on the scope like the one shown in Fig. 11.27*b*. This clock signal is different from the signal shown in Fig.

FIGURE 11.27 Clock signals: (*a*) desired; (*b*) actual.

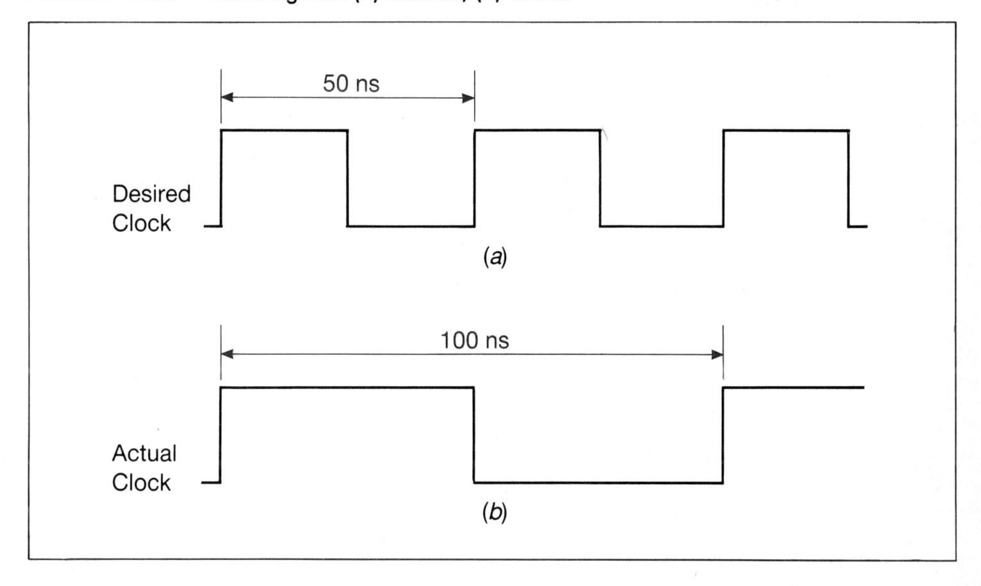

11.27*a* because the signal changes every 100 ns instead of every 50 ns. There are many possible reasons why the clock signal is incorrect. One problem could be that the oscillator driving the clock chip is off by a factor of 2. The important point here is that the oscilloscope allows engineers to view the signals actually produced by the circuit and to compare them with expected results. The next (and last) tool we will discuss is a logic analyzer, which was developed after the oscilloscope.

A **logic analyzer** is essentially a multichannel oscilloscope. Oscilloscopes typically have two to four probes that can be attached to points in a circuit. Logic analyzers often have 50 or more probes that can be attached to a circuit. Logic analyzers not only allow engineers to view signal waveforms like the oscilloscope, they also allow the signals to be grouped and presented in more convenient forms. For example, the system clock is a signal that is important as a single entity. In contrast, a single bit in an address or data bus is not very interesting, although the entire contents of the data bus or address bus is often very helpful.

Logic analyzers allow engineers to view information as either waveforms or binary values. It is useful to view clocks and certain control signals as waveforms, but it is more useful to view the contents of a data bus in binary form. Figure 11.28 shows how a logic analyzer might display the contents of the data bus and address bus. Notice that the contents of the data bus are displayed in binary and the contents of the address bus are displayed in hexadecimal. This type of display capability lets you view the contents of the data and address buses in convenient forms. The information displayed by the logic analyzer is extremely valuable when trying to determine if a computer is functioning correctly.

The logic analyzer is programmable so that it can be instructed to start monitoring signals after a specified event occurs. When the event occurs, the logic analyzer detects the event and starts recording each data transfer that occurs after the event. By looking at the specific state of the computer after a certain event, errors can be found.

For example, let's assume a logic analyzer is set up to monitor the signals connecting the keyboard and the CPU. Also assume that the logic analyzer will begin recording the signals after a key is pressed on the keyboard. A "picture" of the

FIGURE 11.28 Logic analyzer display showing the contents of the data and address buses.

data<7:0>	address<19:0>
XXXXXXXX	FFFF0
11010001	XXXXX
XXXXXXXX	FFFC2
00000000	XXXXX
XXXXXXXX	000A8
10011111	XXXXX

events that occurred after the key was pressed is saved. You can look at the picture and determine if the CPU received the appropriate keystroke from the keyboard. If the CPU did not receive the correct keystroke, then an error occurred somewhere and must be fixed. Errors are not always easy to find, but logic analyzers simplify the process considerably.

11.7 SUMMARY

In this chapter we discussed CAD tools that are used to design computers and verify the correctness of the designs. We saw how schematic capture is used to create a picture of the design, and we discussed an alternative design language method that is used to produce the same information. Both tools produce a parts list and a net list, but schematic capture is more popular because a picture of the design is created as opposed to a textual description. We then saw how logic simulation tools use output from the schematic capture tool to determine if designs were logically correct by simulating the operation of the logic.

Next, we looked at physical design tools, specifically physical layout and routing tools. Physical placement is used to locate each part on a printed circuit board and to do gate assignment. A very important goal of physical placement is to reduce the total logical trace lengths. We then discussed the routing tool that was used to design the connections between parts on a printed circuit board. The end result of doing physical design is a blueprint of a printed circuit board. The blueprint is used to build the masks necessary for actually manufacturing a board. The final CAD tool we discussed was SPICE, which is used to simulate signals traveling on nets. The purpose of simulating nets is to determine if propagation delay and/or signal distortion might cause the computer to fail.

Finally, we looked at three hardware debugging tools: the multimeter, the oscilloscope, and the logic analyzer. We saw how the multimeter could be used to test connections by measuring resistance in ohms and how it could be used to measure constant voltages to determine if chips were properly powered and grounded. Next, we discussed how the oscilloscope was used to look at individual signal waveforms. Lastly, we discussed logic analyzers, which are used to record signal values and display them in a convenient form.

Hardware design tools are becoming increasingly important because modern computers are getting more complicated. As computer engineers, you will no doubt have an opportunity to explore the tools we have presented here and many other more sophisticated ones in the future.

PROBLEMS

11.1 Explain the difference between a logical component and a physical component.

11.2 Explain the differences between a bus and a net.

11.3 What are some advantages that design languages have over schematic capture tools?

11.4 If the total length of a net is 10 in but the net is derated by a factor of 1.3, what is the optimal length of the net?

11.5 How are the two diagrams in Fig. P11.5 related?

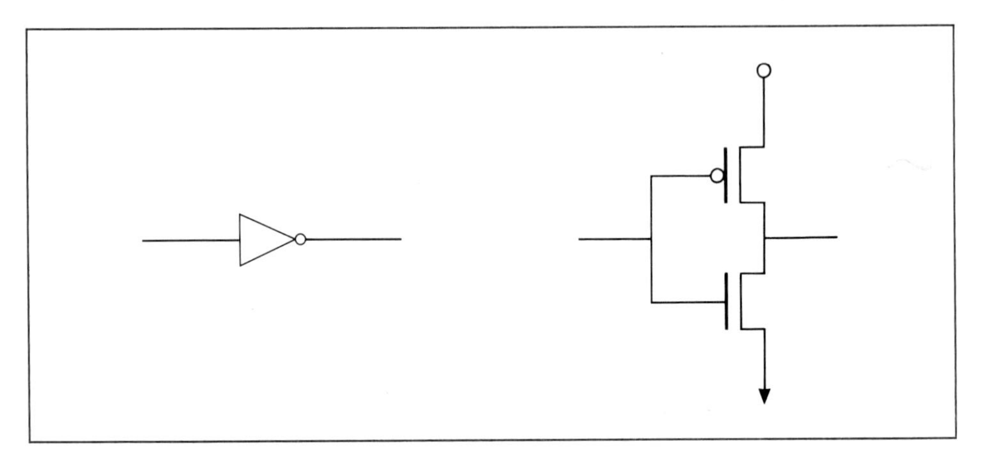

FIGURE P11.5

11.6 What functions do the menus in a typical schematic capture tool let you do?

11.7 Create the XOR function and implement the function (*a*) using only NAND gates and (*b*) using only NOR gates.

11.8 Give a design language description of the circuit implemented in Prob. 11.7*a* and *b*.

11.9 What is the component library in a schematic capture tool?

11.10 Which is the menu and which the submenu in Fig. P11.10? Why do we call it a submenu?

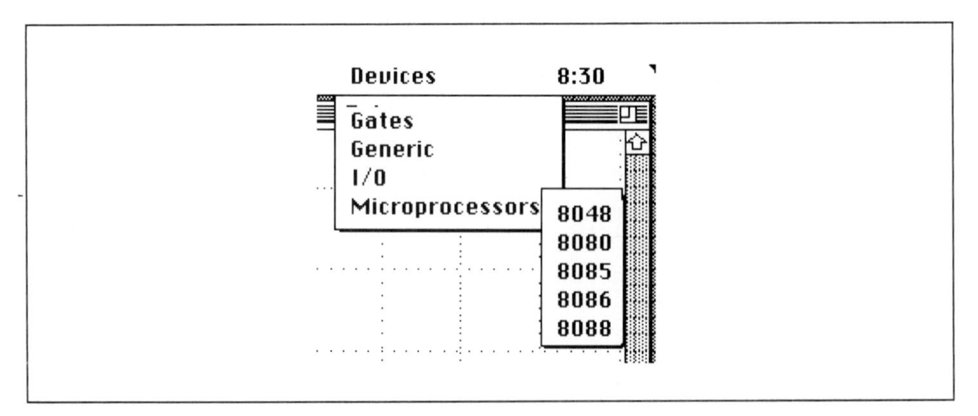

FIGURE P11.10

11.11 What is a terminal site?

11.12 A typical net list is given below. The element number is unique for each part in the design. Notice that some have a "P" in the name; the "P" means partial part because the actual device has several of the same type of parts.

(a) Reconstruct the schematic represented by the net list.

(b) Generate a net list like the one below for the half-adder shown in Fig. 11.10 of the text.

```
Designer: Larry C. Wear
Date: 2-10-90
Version: 1.0
{ Element 1
  Type 74F00
  PIN,1,IN,IN1
  PIN,2,IN,IN2
  PIN,3,OUT,WOW
  PIN,4,IN,DOIT1
  PIN,5,IN,DOIT2
  PIN,6,OUT,DONE
  PIN,7,GND,GND
  PIN,8,IN,ON1
  PIN,9,IN,ON2
  PIN,10,OUT,ITSON
  PIN,11,IN,NOW
  PIN,12,IN,JOE
  PIN,13,OUT,OK
  PIN,14,VCC
}
{ Element 2
  Type 74F74P
  PIN,A,IN,IN3
  PIN,B,OUT,ON1
}
{ Element 3
  Type 74F74P
  PIN,A,IN,IN4
  PIN,B,OUT,ON2
}
{ Element 4
  Type 74F74P
  PIN,A,IN,IN6
  PIN,B,OUT,JOE
}
{Element 5
  Type 74F74P
  PIN,1,IN,CLEAR
  PIN,2,IN,DIN
  PIN,3,IN,CLOCK
  PIN,4,IN,RESET
  PIN,5,OUT,DOIT2
  PIN,6,OUT,NOW
}
```

11.13 What did we mean when we said that the 8088 microprocessor multiplexes the address and data buses?

11.14 Why is it that in most schematic capture tools a bus is drawn the same way as a net except that the bus is shown as a wider line than the net?

11.15 Contrast SSI, MSI, LSI, VLSI.

11.16 What are the three methods currently used to create formal design specifications?

11.17 What are parts lists and net lists?

11.18 What is the purpose of doing logic simulation? What information does the logic simulator need to perform a simulation of a circuit design?

11.19 What is a routing tool?

11.20 Why would it be an advantage to have a logic simulator use the information generated by a schematic capture tool directly to create the program statements necessary to test the logic in a design?

11.21 What is a physical placement tool?

11.22 Describe the typical construction of a printed circuit board.

11.23 Why do printed circuit boards have ground and power planes rather than just running ground and power connections where needed?

11.24 Describe the difference between through-hole mounting and surface-mounting techniques.

11.25 What is a solder bath?

11.26 Describe what is going on in each of the four steps of the circuit board operations shown in Fig. P11.26.

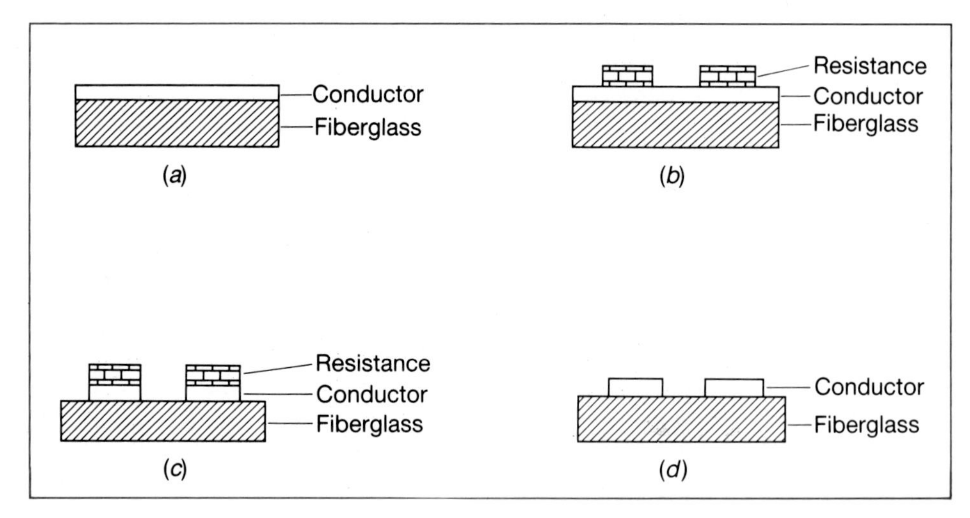

FIGURE P11.26

11.27 What is the footprint of a physical layout part? Why are footprints used?

11.28 Assume that a printed circuit board has four signal layers like those shown in Fig. P11.28.

(a) Explain how a net on signal layer 1 is connected to a net on signal layer 4.
(b) Explain how a net on layer 2 is connected to a net on layer 3.
(c) Are there any other ways to connect the nets on layers 2 and 3?

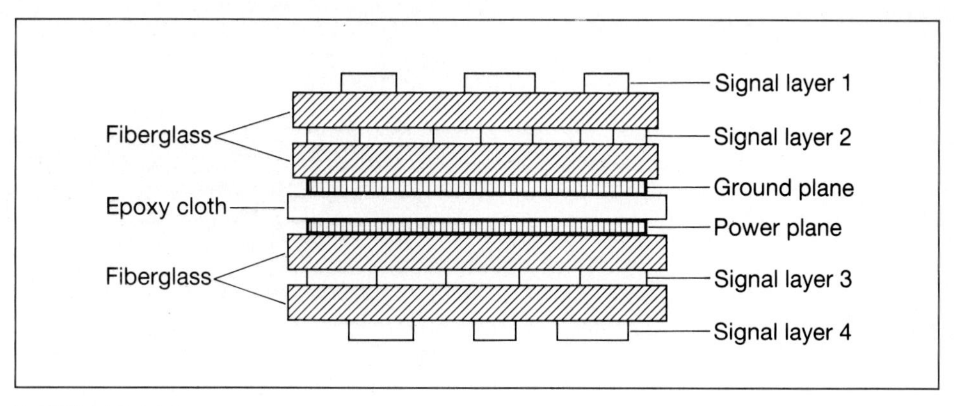

FIGURE P11.28

11.29 What is the first step in physical placement?

11.30 What is the second step in physical placement? Where do we get the necessary information for this step?

11.31 What is gate assignment?

11.32 What do we mean by saying that routing is essentially a connect-the-dots puzzle?

11.33 What are Manhattan and optimal routes?

11.34 What is an autorouter?

11.35 Identify the various elements on the portion of a routed printed circuit board shown in Fig. P11.35.

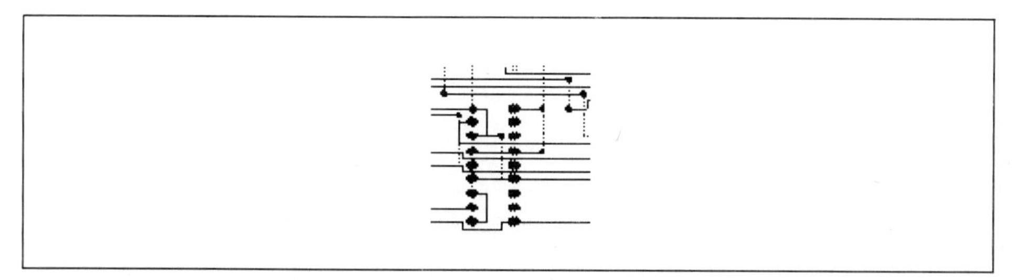

FIGURE P11.35

11.36 Route by hand the schematic of Fig. P11.36. Attach the inputs signal to the IN CONNECTOR and the outputs signals to the OUT CONNECTOR. Use the 16-pin DIP for the 375 and the 14-pin DIP for the 77. Assume that a two-layer printed circuit board will be used and that vertical traces are routed on the top layer and horizontal traces are routed on the bottom layer. Use vias to connect the traces on the top and bottom layers. (Signals on the same layer cannot touch each other unless they are part of the same net.)

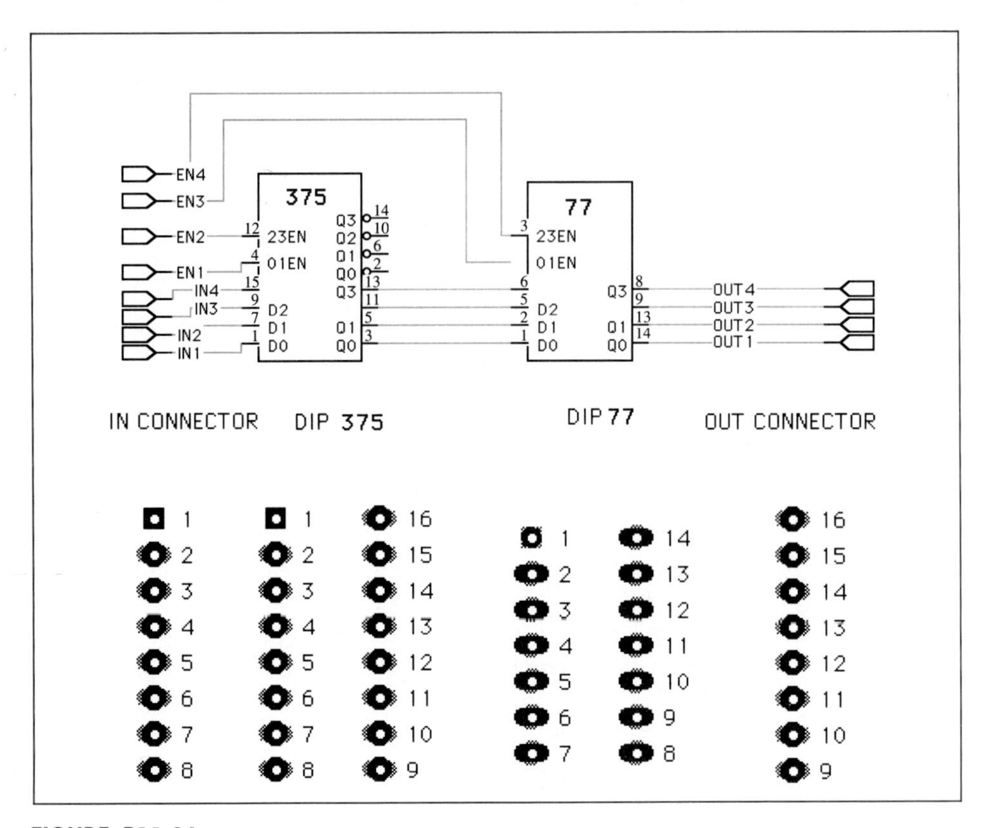

FIGURE P11.36

11.37 What are the goals of routing? Why are they important?

11.38 Comment on the signal shown in Fig. P11.38.

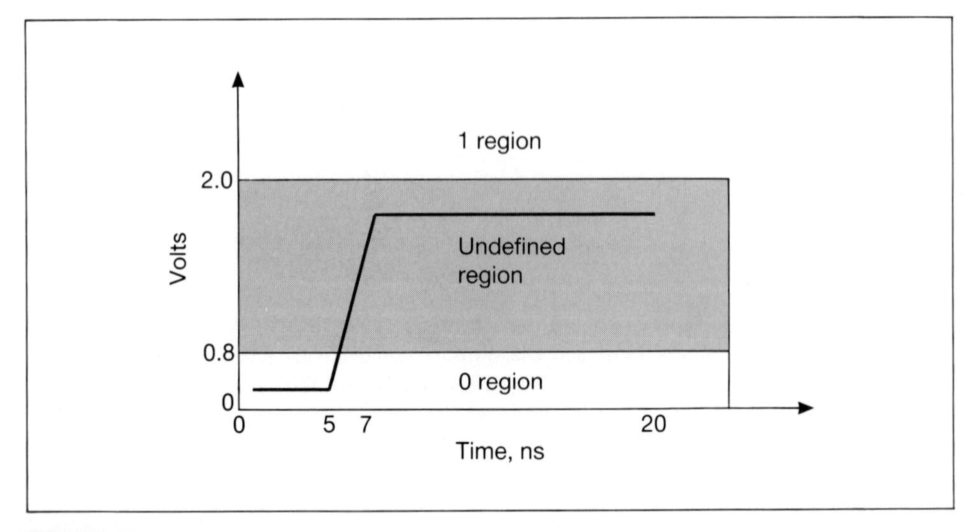

FIGURE P11.38

11.39 What is propagation delay?

11.40 Label the arrows and boxes in the flow diagram of Fig. P11.40 of the design process using CAD tools.

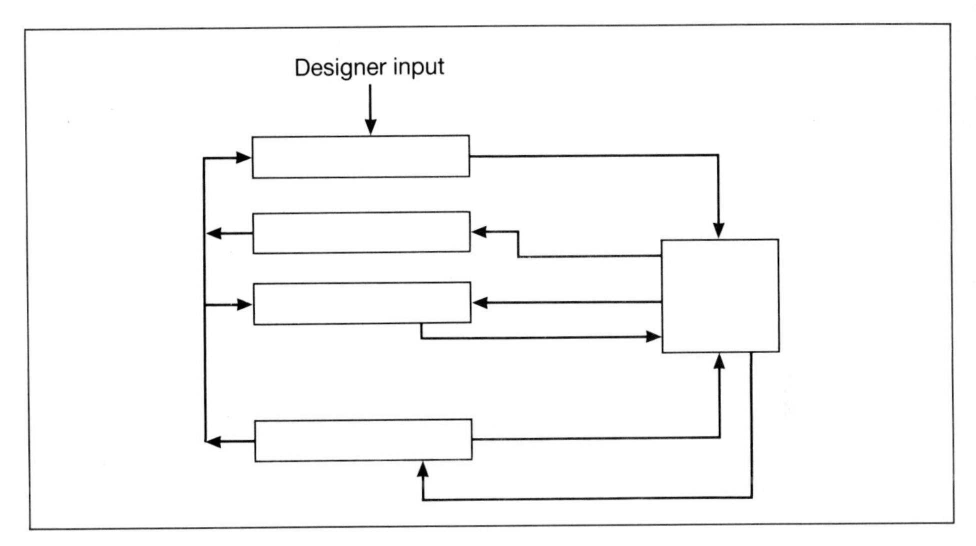

FIGURE P11.40

11.41 Define (*a*) trace length, (*b*) trace loading, and (*c*) driver type.

11.42 What functions does a multimeter perform? How are each of these functions used to debug computer hardware?

11.43 What is an oscilloscope, and what functions does it perform?

11.44 What is a logic analyzer? Why is it often more useful than an oscilloscope?

11.45 How can signal distortion and propagation delay cause circuits to fail?

11.46 Schematic capture, physical placement, SPICE, and many CAD tools share information in a design database. The database can be contained in one file or many files. What problems can occur when tools share information via the design database? (*Hint*: What if one engineer is doing schematic capture and another wants to do physical placement on the same design?)

11.47 Suppose the system clock for a design runs at a speed of 25 MHz. Assume that the CPU requests data from the memory on the rising edge of the clock signal. The CPU expects to read the data from the data bus on the next rising edge of the clock. Assume the memory takes 20 ns to fetch the data. What is the maximum length the data bus can be, assuming that a signal propagates along the bus at 0.5 ft/ns?

AN OVERVIEW OF COMPUTER SOFTWARE

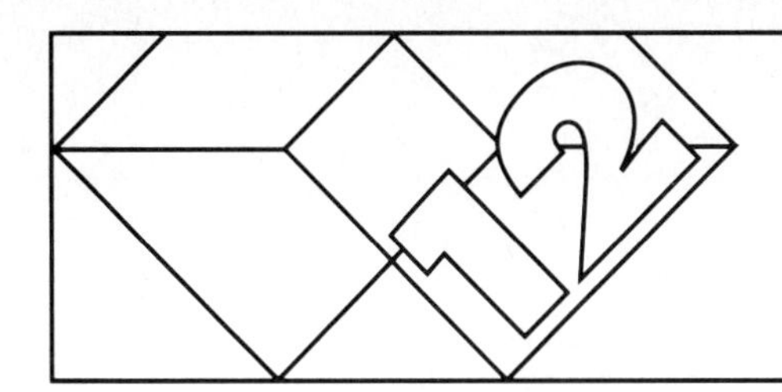

Introduction to Operating Systems

As we described in Chap. 4, the upper two layers of the computing system are both implemented in software, as shown in Fig. 12.1 (a reproduction of Fig. 4.10). This software gives us the means to define the operations for the hardware to execute.

A term for an element that is placed between two layers is an *interface*; we saw examples of hardware interfaces in Chap. 10. In a computer, the system software can be considered as an interface between the user and the hardware. It provides a friendlier environment where users can do things at a "higher level" and not have to worry about low-level hardware operations. It manages resources and provides services to the application layer. The system software usually includes components such as the operating system, language translators, and utility programs. An expanded version of Fig. 12.1, illustrating some of these components, is given in Fig. 12.2.

Our focus in the next two chapters will be the operating system, which is shown further expanded in Fig. 12.3. The kernel is small, specialized, and hardware-dependent. It will be discussed in Chap. 14. The remainder of the operating system is described in Chap. 13.

To help you understand the material in later chapters, there are two preliminary topics we will cover first:

- Computer processes
- A mechanism for computer process synchronization

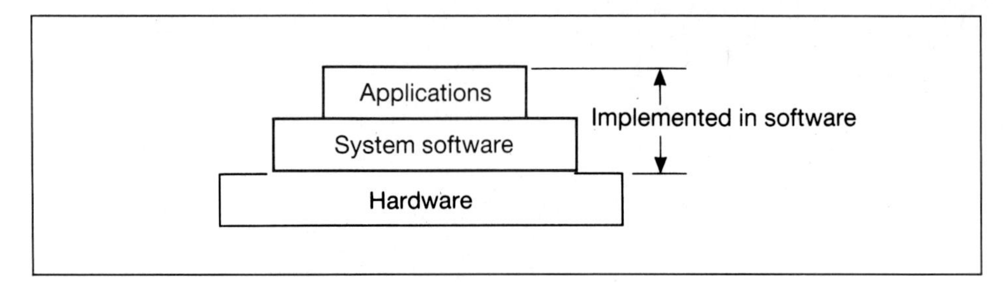

FIGURE 12.1 The software levels of the computing system.

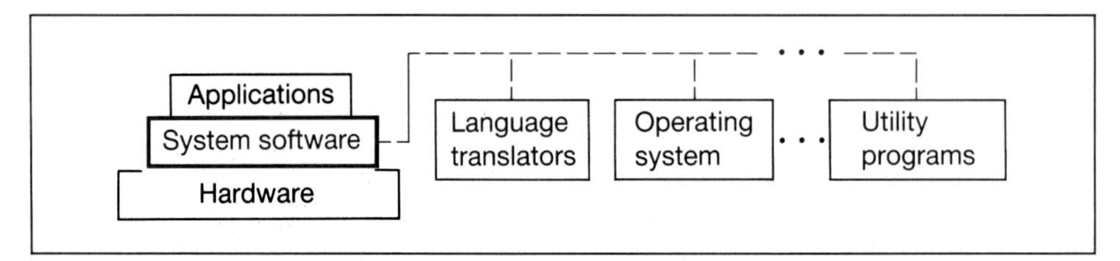

FIGURE 12.2 An expanded version of Fig. 12.1, illustrating some system software components.

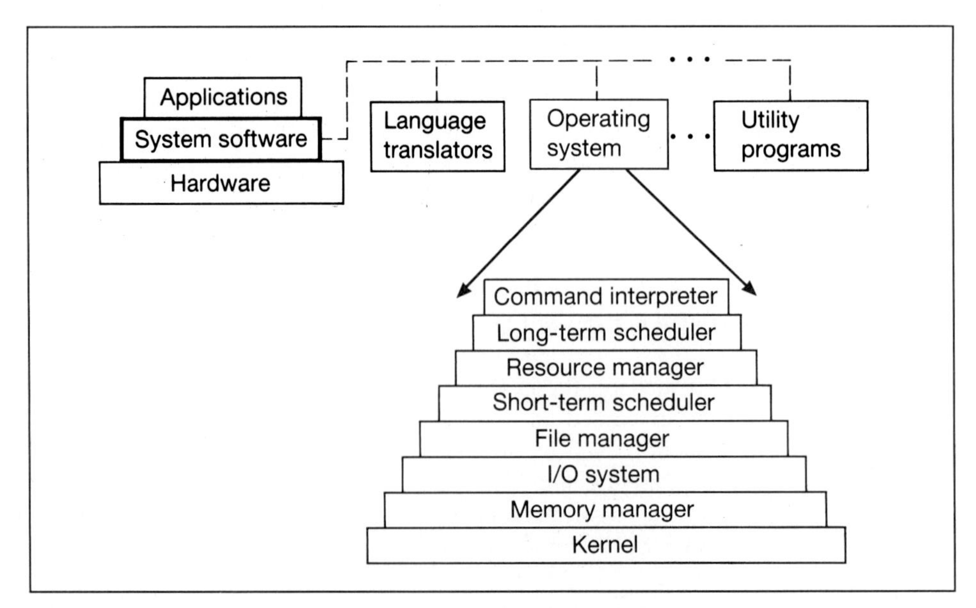

FIGURE 12.3 Another version of Fig. 12.1, showing the components of a typical operating system.

12.1 COMPUTER PROCESSES

In later chapters, you will be told about a process being scheduled, a process being blocked, a process this, and a process that. But what is a process? A **computer process** is the unique execution of a specific program with a specific set of data.

We favor this definition because nowadays many systems keep only one copy of a program in main memory even if two or more users are running it. Each user has a unique execution of that program, using his or her own specific set of data. Therefore, each user has his or her own unique process.

Every process in a computer system is spoken of as being in a certain "state." Figure 12.4 is a simplified diagram of process states and the transitions between them. This figure has *all* the states in which a process can reside and *most* of the transitions between these states. Some lesser-used transitions are not included, since they are unnecessary for your understanding of the material in this chapter. For a complete diagram and a discussion of all the transitions, see Pinkert and Wear [1989].

Every process in the system has an identifier called a **process control block**, or **PCB**. This identifier contains information such as who created the process, a list of the files the process is accessing, and the time that the process has been executing. Saying a process is in a certain state implies that its PCB is on a certain list and has a field whose value is currently an indicator of that state. To move a process to a different state means we hook its PCB onto a different list and change the "state" field in its PCB.

Figure 12.4 contains four ellipses representing the four states in which a process may reside: active, execute, suspend, and blocked.

FIGURE 12.4 A simplified diagram of process states and state transitions.

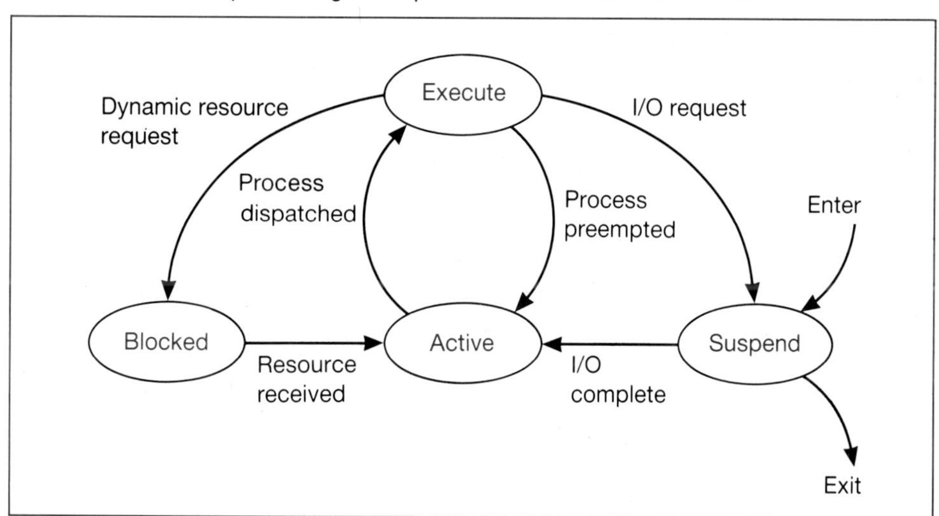

- **Active state**: for processes that have all the resources they need to execute *except* the CPU
- **Execute state**: for processes that currently have a CPU (*several* processes could be in the execute state if the system has *multiple* CPUs)
- **Suspend state**: for processes waiting for completion of an I/O operation on a resource that they already control
- **Blocked state**: primarily for processes that must wait because they have tried to acquire some resource that is not available.

There are two lines in the figure that are connected to the suspend state at one end but then dangle off into space at the other end. The one that points *into* the suspend state represents the path taken by a process when it is created. A newly created process is moved to the suspend state because it must wait for the transfer of its code and data into main memory from secondary storage. The second dangling line points *away* from the suspend state. This line shows the path taken by a process that is being removed from the system. When a process is removed, all the resources it has claimed are returned to the system. The process ceases to exist though the original program file still resides somewhere in secondary storage. As an example, you can think of a process being created when you request an application to be run. The process is removed when you exit the application.

After a process has been created, the most common reason that it is placed into the suspend state is so that it can wait for an I/O operation to finish. Hence, the transition from execute to suspend is labeled I/O request. After the requested I/O operation is finished, the process is moved from the suspend to the active state. Once the process has been moved into the active state, it waits to execute.

There are usually several processes in the system at the same time, and we call the number of such processes the **degree of multiprogramming**. The system will use some algorithm to decide which of these processes will be chosen for execution. Once a process has been chosen, the part of the operating system known as the dispatcher will move that chosen process into the execute state.

Processes end up in the blocked state because they have requested a resource that is not available. This form of request, which is made after the process starts executing, is called a **dynamic resource request**. Normally, after the process has waited in the blocked state for a while, the requested resource will become available, and it will be allocated to the waiting process. That process will then be transferred to the active state.

We have already discussed a process leaving the execute state when it makes an I/O or a resource request. The third line is labeled "Process preempted" and is the path followed when the CPU is *taken away* from the executing process. That is, the process does nothing to relinquish the CPU voluntarily, but some other event forces the departure. The mechanism that starts the action is an *interrupt*, which was described in Chap. 7.

If our system can have two or more processes running at the same time, then we might have to coordinate or synchronize these processes. We turn now to mechanisms for such synchronization.

12.2 PROCESS SYNCHRONIZATION

In discussions of modern operating systems, one frequently hears the term *concurrency*. **Concurrency** means that during some small time interval $[T, T + \Delta t]$ there is more than one process or device running. If Δt is zero, then we have **true concurrency**. If Δt is small, but not zero, then we have **apparent concurrency**. As an example of true concurrency, consider a computer with three CPUs. At time T (or equivalent interval $[T, T]$), there could be three processes running. We could have true concurrency. But, when people are using a timesharing system with one fast CPU, it only *appears* that everyone's process is running at once. What really happens is that during interval $[T, T + \Delta t]$ the CPU switches rapidly among processes. At any instant T, *only one process is running on that CPU*; therefore, this is an example of apparent concurrency. Pictorial descriptions of these two situations are shown in Fig. 12.5.

FIGURE 12.5 Concurrency examples: (*a*) apparent; (*b*) true.

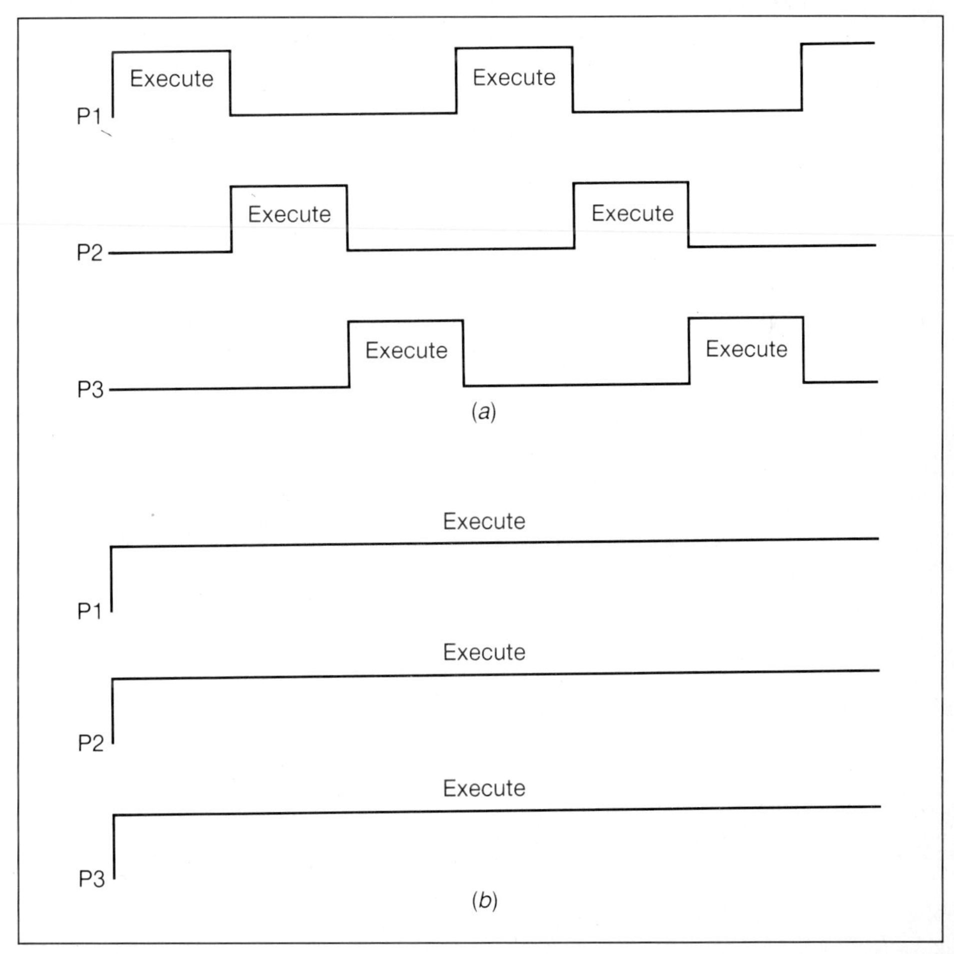

When processes are executing concurrently, it is sometimes necessary to synchronize their activities. To illustrate this, consider a noncomputer example. Suppose that one of the authors has a busy evening planned in which he has many things to do around the house. To be efficient, he tries to do some of them concurrently—for example, printing an assignment, washing some clothes, cleaning the kitchen, and heating some leftovers for dinner. A diagram of his activities is shown in Fig. 12.6.

In this diagram, there are two shapes of figures, squares and ellipses. The squares show happenings for which the author has to wait, such as Leftovers Heating. The ellipses show activities that the author has to do, such as Turn on Microwave.

Now for synchronization. It would be unwise to dry the clothes before they are washed or to rinse the kitchen floor before it has been mopped. Therefore, the graph in Fig. 12.6 is set up to show not only what must be done but any constraints on the order in which they must be done. For example, the edge from Turn on Wash to Mop Utility Room says that he must turn on the wash before mopping the utility room (or else the wet floor will get marks on it). But, Leftovers Heating and Assignment Printing can happen in any relative order, so there is no direct edge between them.

On the diagram, edges show order. What physical happenings could be related to these edges and their orderings? The author might specify the Start oval as listening to Garrison Keillor's radio program, *A Prairie Home Companion*, and Keillor's sign-off as the indicator that this event has finished and the next three events can be started. The light which comes on inside the microwave could tell him that the Turn on Microwave process has been finished. The oven timer buzzer could say that the Leftovers Heating has finished.

To look at synchronization in a little different way, let's go back to Fig. 12.6 and label each node with a single letter. The edges will be labeled with two-letter strings; the first letter of this label will be the event that must be finished before the event designated by the second letter can begin. (For example, R must occur after K, so the edge from K to R is labeled KR.) This labeled version is shown in Fig. 12.7.

We can now develop a chart that specifies the edges, as shown in Fig. 12.8. The left letter (row) is read first, and the top letter (column) is read second. Therefore, edge BE is represented by the second row (labeled B) and the fourth column (labeled E). This chart will be important later for event coordination.

Next, let's make a 3 × 5 card for each of the nodes on our diagram. Each card will contain:

- The node
- "Wait on" edges leading into the node, showing events that must happen before the event on this card can happen
- "Signal" edges leading out of this node, showing events which can now be notified that the event on this card is finished

Some of these cards are shown in Fig. 12.9.

Now the author will put all the cards corresponding to events he does directly (such as Turn on Microwave) into one card file called Jobs. The rest of the cards, for

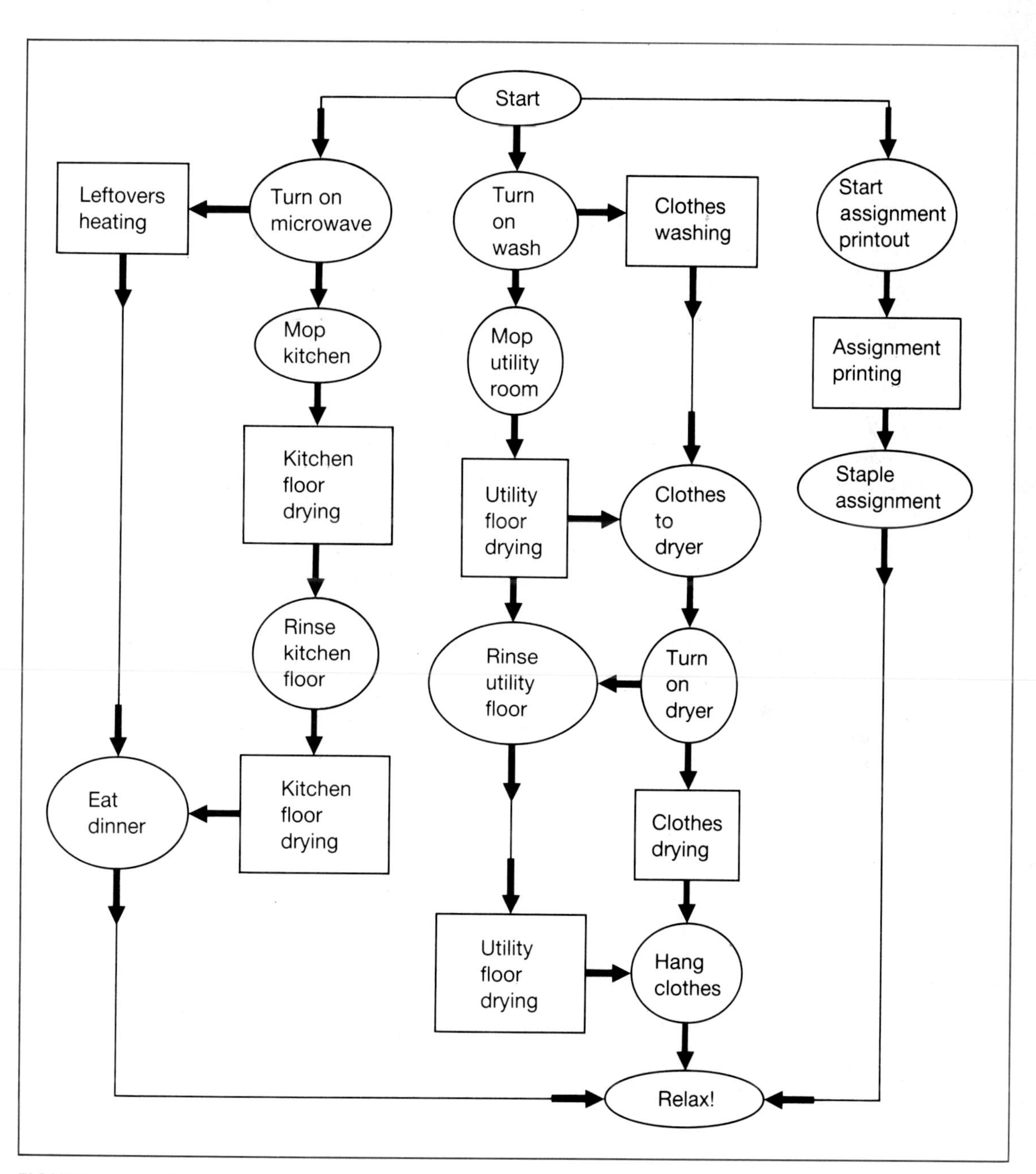

FIGURE 12.6 A sample process flow graph.

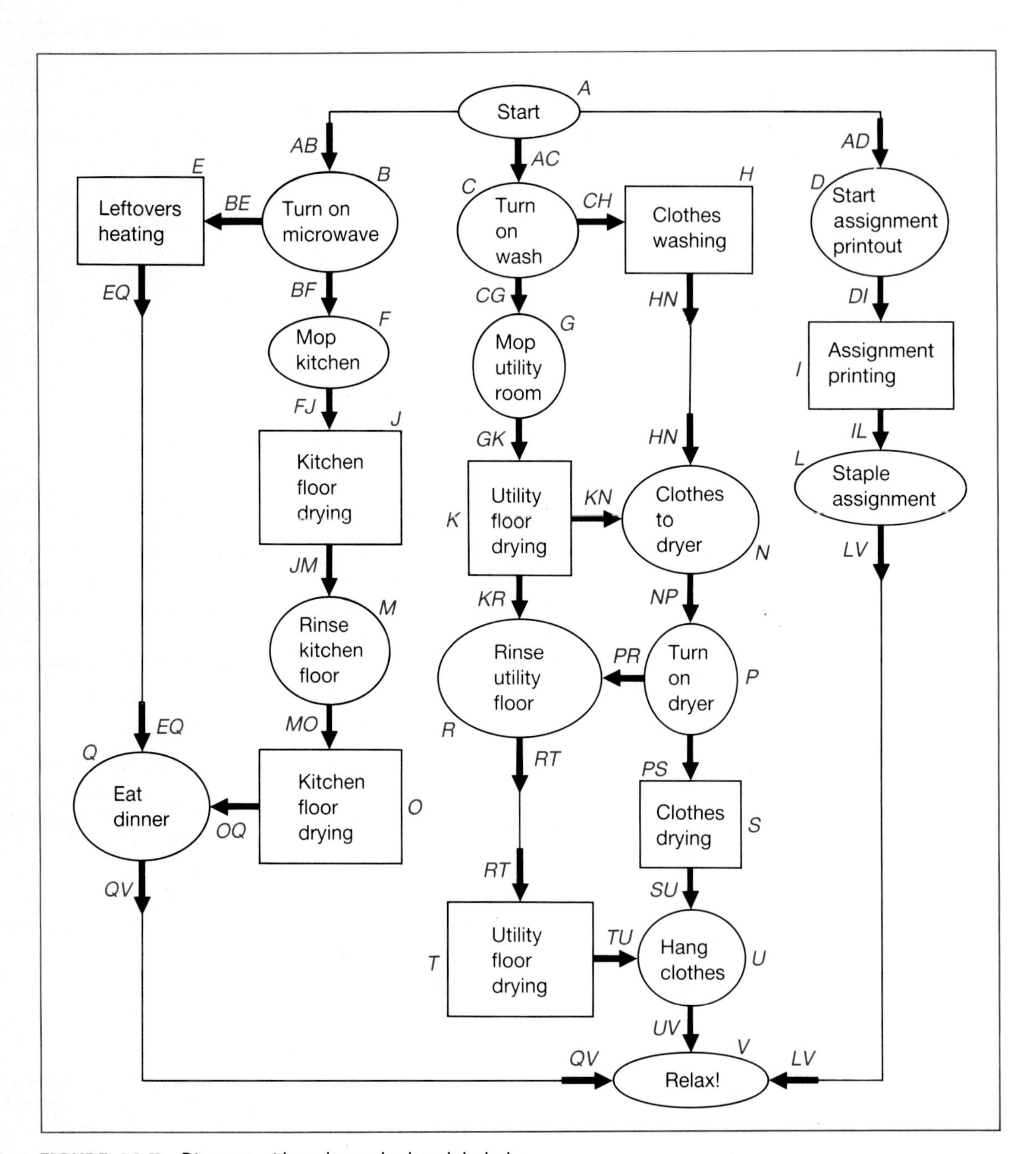

FIGURE 12.7 Diagram with nodes and edges labeled.

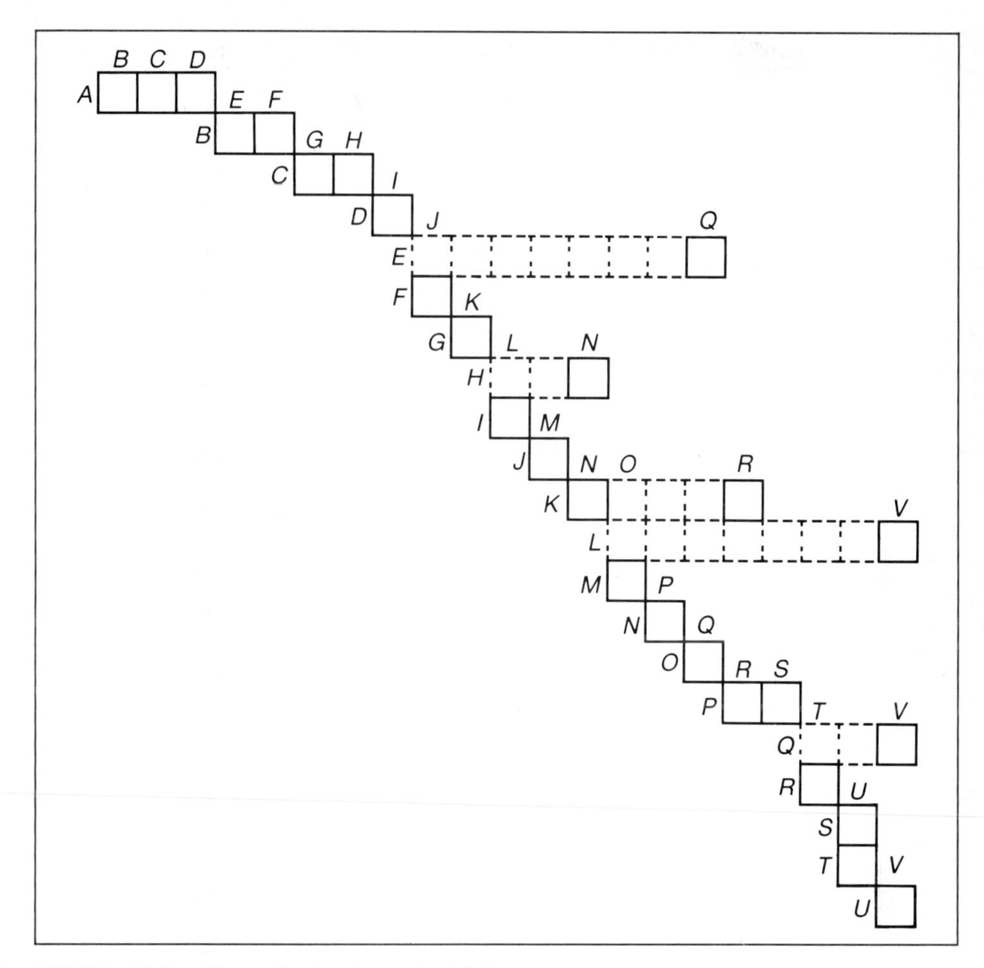

FIGURE 12.8 Chart of edges from Fig. 12.7.

events that don't need direct actions (such as Utility Floor Drying), go into a file called Reference. When the author feels like doing something, he goes through the Jobs file and picks a card. He then looks at the card to see if any other events must have happened before this event can happen (i.e., there are "wait on" specifications). If so, he checks the chart (Fig. 12.8) and sees if the corresponding square has been "signaled" (marked). If it has, he can continue with the event on the card; if not, he puts the card back into his Jobs file.

When he finishes an event, he sees if that event's card shows that any squares on the chart should be "signaled." If so, he marks the proper box. Also, whenever other events finish, such as Utility Floor Drying, he looks in the Reference file for that event's card and marks any indicated squares on the chart.

This approach might seem clumsy, but people do use it (without keeping card files). That's why there are buzzers on ovens and clothes dryers, for instance. Also,

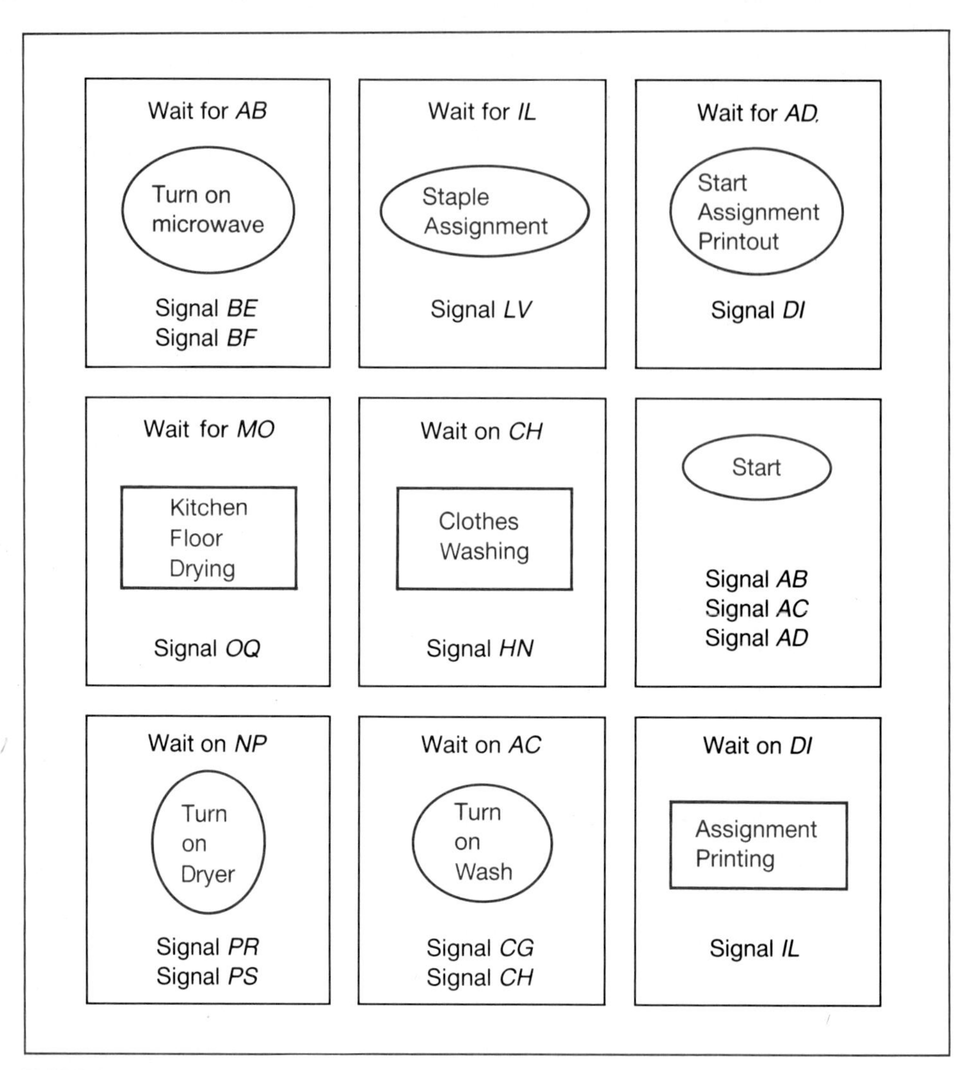

FIGURE 12.9 Some job cards for our example.

this approach gives us an explicit visualization of event synchronization and leads us into computer process synchronization via a technique involving variables called semaphores.

In its most elementary form, a **semaphore** is a protected binary variable whose value can be changed only in specific, predefined ways and only by a special sequence of machine instructions [Dijkstra, 1965]. A user can specify the initial value of a semaphore and then access it via two system calls—Wait and Signal. These two operations are shown in Figs. 12.10 and 12.11.

Note how nicely semaphores fit into our previous noncomputer example of event synchronization. Go back to Fig. 12.7 and assume node *P* is a computer process

```
WAIT(S):   If S = 0
                then put calling process on a waiting list
                for semaphore S
                     {The calling process will be removed
                     from the waiting list at some later
                     time after a signal (S) from another
                     process.}
                else reset the semaphore, S := 0, and allow
                the calling process to continue
```

FIGURE 12.10 The Wait operation for semaphore S.

instead of turning on the clothes dryer. The event sequence for the *synchronized* version of process P could then be

```
WAIT(NP);
{code for process P}
SIGNAL(PR);
SIGNAL(PS);
```

This is very close to the sequence the author wrote on his corresponding Job card. The signals correspond to marks on the chart in Fig. 12.8, and the squares on the chart can be likened to semaphores.

Semaphores are used for many other things, such as process communication, data sharing, and resource allocation. However, we only want to introduce them. You will probably study them more in later courses.

FIGURE 12.11 The Signal operation for semaphore S.

```
SIGNAL(S):   if waiting list for S is not empty
                  then remove a process from the waiting
                  list for semaphore S and put it on
                  the active list
                       { The calling process and the process
                       removed from the waiting list will
                       now be rescheduled for execution by
                       the OS.}
                  else set the semaphore, S := 1, and allow
                  the calling process to continue
```

12.3 SUMMARY

In this chapter, we looked at how the system software fits with the hardware and application software. We also introduced the concept of a process and defined the states in which a process can reside. Next, we saw how concurrent execution of processes can cause problems and how semaphores can be used to solve these problems. We are now ready to look at the other components of a typical operating system (Chap. 13) and the kernel (Chap. 14).

PROBLEMS

12.1 What is the "state" field in the process control block?

12.2 One way to indicate the state field in the PCB is to have a separate bit for each separate process state. What is another way?

12.3 What is the relation between a process and a program?

12.4 Why do you suppose there is a separate active state? Since the CPU is a resource, couldn't a process waiting for that resource be put into the blocked state?

12.5 We said the most common reason a process is placed into the suspend state is so that it can wait for an I/O operation to be finished. What is another reason it might be suspended?

12.6 A common reason for a process to be removed from the system is that the user has signed off. What is another reason?

12.7 Why might a process be "preempted"?

12.8 Construct a process flow graph like Fig. 12.6 for a set of activities you do.

12.9 Explain the difference between *simultaneously* and *concurrently*.

12.10 Referring to Prob. 12.8, develop a chart that specifies the edges of your process flow graph like the one shown in Fig. 12.8.

12.11 What is required to execute programs simultaneously?

12.12 Referring to Prob. 12.8, develop a set of cards for your chart like the ones shown in Fig. 12.9.

12.13 A process that places an object into storage is sometimes called a *producer*, and one that takes the element out is called a *consumer*. Write a program segment using a semaphore to show how to coordinate producer/consumer operations.

12.14 Referring to Prob. 12.13, what must the initial value of the semaphore be?

12.15 How does the semaphore concept apply to a street intersection?

Operating System Components

Many texts try to pin down a precise definition of an **operating system (OS)**. We prefer instead to look at *the functions* that an operating system does and *the components* that make up a typical operating system.

13.1 THE FUNCTIONS OF AN OPERATING SYSTEM

Some people consider operating systems strictly as resource managers. One can easily take this view in situations where the operating system is allocating resources like printers, disk drives, and other peripherals. Furthermore, memory, the CPU, and almost everything else can also be viewed as resources. Therefore, the functions of an operating system can be described in terms of resource management.

We said in the last chapter that in a computer, the operating system is the interface between the user and the hardware. It provides a friendlier environment where users can do things at a "higher level" and not have to worry about low-level hardware operations. A good example of this is giving the user a simple way to read data without being concerned about data rates, device protocols, etc. We illustrated this interface concept in an earlier diagram (Fig. 4.1), which is reproduced in Fig. 13.1.

Although the terms *resource manager* and *interface* are quite descriptive, the operating system is more. For example, it is:

- A coordinator, supplying facilities so that complex activities can be done in a certain order
- A guardian, supplying access controls to protect files and allow restrictions on reading/writing/executing data and programs

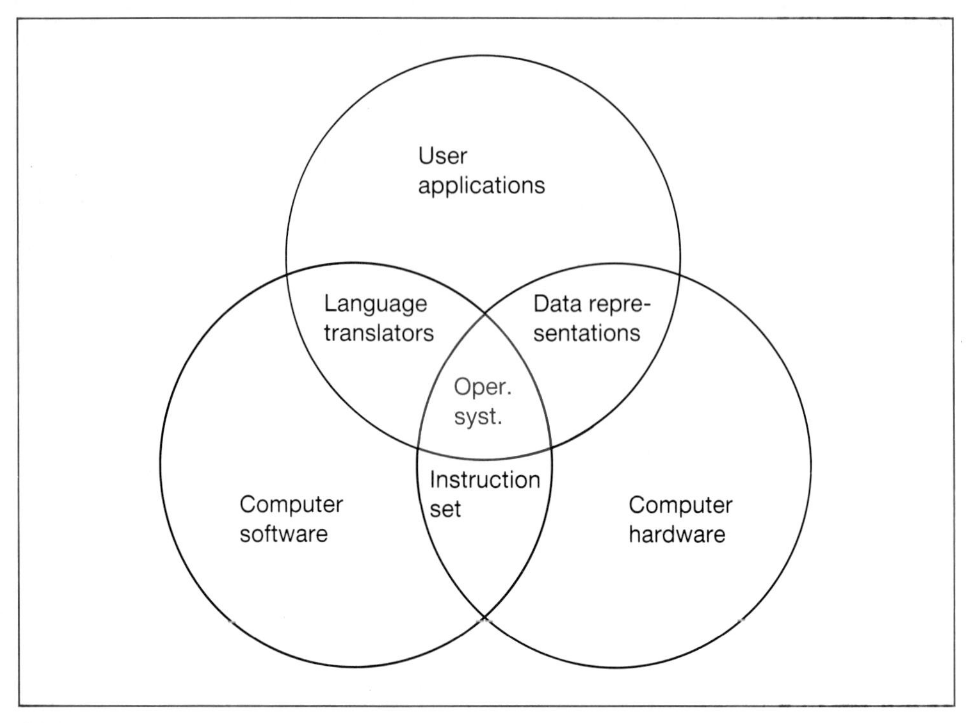

FIGURE 13.1 The operating system as an interface between applications, hardware, and software.

- A gatekeeper, controlling who can log onto the system
- An optimizer, scheduling some users' inputs, others' database accesses, others' computations, others' outputs, etc., in a way to increase system throughput
- An accountant, keeping track of CPU time, memory usage, I/O operations, disk storage, etc.
- A server, supplying services users might need, such as file directory restructuring

These are some of the general functions of an operating system. Let's now look at its specific components in more detail. There are many different ways to structure and design an OS, but most designs will include eight major components. Six of these will be described in this chapter:

- Memory manager: decides how much memory, and which areas of memory, to give a user
- I/O control system: services user program's I/O requests
- File manager: manages devices that store files of information, such as disks and magnetic tapes

- Short-term scheduler: chooses the next process to execute on the CPU
- Resource manager: allocates resources (other than memory and the CPU) among competing processes
- Long-term scheduler: decides which processes are allowed to enter the system

Our descriptions of these components are brief, but they explain the main functions that each provides.

The two components of an operating system that are not covered in this chapter are the kernel and the command interpreter. The kernel, the lowest level in the system, is quite different from the other layers and is treated separately in Chap. 14. The command interpreter is the layer that interacts directly with the user. It sends the prompt to the user's screen, analyzes user requests, and calls system routines to satisfy these requests. The command interpreter is often considered as part of the user interface, which is discussed in Chap. 18.

13.2 MEMORY MANAGEMENT

Probably no other OS topic (with the possible exception of scheduling) has received as much attention as memory management. That attention, and the voluminous discussions in most OS descriptions, is understandable. Memory is an important part of every computer system. In this section we will introduce the topic of memory management.

13.2.1 Memory Management in a Single-User System

On early systems, each user had exclusive use of the computer for the duration of the program's execution. Memory management was nothing more than implicitly allocating all of memory to the next program. "All" included the part of memory containing the OS. This allowed a user's run time error (such as storing data into array elements outside previously defined array boundaries) to "crash" the OS.

Such a crash was best described as an inconvenience. The OS had to be reloaded, but that was usually a simple affair. There were no other processes whose execution could be affected and no online files whose access had to be regulated.

13.2.2 Separating the User and the System Memory

As computer systems evolved, it became necessary to guard the integrity of the OS. There was still only a single process on the system at one time, so other jobs could not be directly affected. However, it was possible for that single job to have an indirect affect on other users or on the system itself. Files, for example, could be stored on online disks, and access was possible without operator intervention. Failure of the OS could leave these files susceptible to accidental or intentional penetration by an unauthorized user.

These and other factors resulted in the development of OS protection hardware. A number of different approaches have been used to protect parts of memory from accidental or intentional change. One example is the use of limit registers. The hardware compared program addresses to the contents of **limit registers** that defined the boundaries of the user's memory space. Attempts to access memory outside this space would be caught, and the OS would be notified that a memory-bounds error had occurred.

13.2.3 Multiple Users

Memory management took on a whole new dimension with the arrival of multiuser systems. Decisions had to be made about how to divide memory, allocate regions, and protect each user's processes.

There have been many approaches to multiuser memory management, too many for us to describe in detail here. We will therefore consider only three: paging, segmentation, and the combination of segmentation using paging.

13.2.4 Paging

With paging, the system divides the memory and each user's process into pieces, in much the same way that the publisher has divided the text space of this book into pages. Each memory piece is known as a **page**. On most systems the size of a page is a permanently fixed quantity.

The OS maintains a list of available pages and assigns them to processes as needed. The pages do not have to be in order so long as the OS knows where they are. A user's logical address is "mapped," or translated, into its physical memory location via a table, appropriately called a **page map table**, or **PMT**. In the simplest (but also probably impractical) version of paging we assume:

- All of a process's pages must be loaded into memory prior to execution.
- At completion, all these pages are written back out to secondary storage (since their contents might have changed).

Let's consider an example. Assume that our arbitrary page size was fixed at 100 memory locations. We would start assigning pages to memory at location 00000, and the pages in memory would then be assigned as shown in Fig. 13.2.

If we assume that a sample process requires five pages and that the system has been running for a while, then the OS might assign scattered memory pages to our process pages as shown in the PMT of Fig. 13.3*a*. During execution, an individual program location (such as location 83 on program page 3) would have to be mapped into its equivalent physical memory location before memory is actually accessed.

Page mapping can be done in two ways. We can either add the displacement ("offset") to the beginning address of the page, as shown in Fig. 13.3*b*, or concate-

Memory page number	Actual memory locations
000	00000–00099
001	00100–00199
002	00200–00299
⋮	⋮
128	12800–12899
⋮	⋮

FIGURE 13.2 An example page/memory mapping for 100-word pages.

Program page number	Assigned memory page number	Memory locations
0	102	10200–10299
1	513	51300–51399
2	631	63100–63199
3	204	20400–20499
4	387	38700–38799

(*a*)

Beginning address of page 3	20400
Offset value	+ 83
Actual memory location	20483

(*b*)

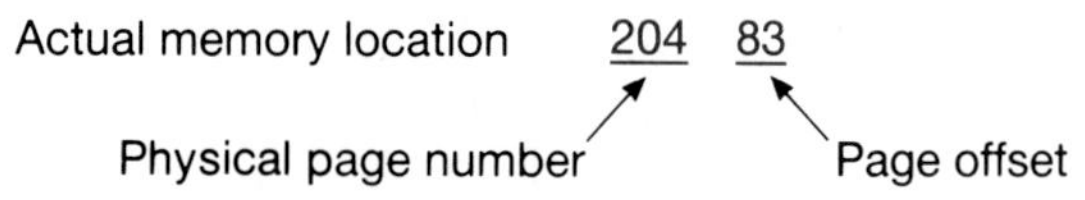

Actual memory location 204 83

Physical page number ↗ ↖ Page offset

(*c*)

FIGURE 13.3 Page address translation: (*a*) a sample page map table; (*b*) translation by addition; (*c*) translation by concatenation.

nate the page number with the page offset, as shown in Fig. 13.3*c*. The end results are the same. Concatenation circuitry is usually faster, but it requires page sizes to be powers of the base.[1]

[1] To see why, assume that the page size is 99 locations rather than 100. Memory page number 204 would then be at location 20196 in memory, and the displacement could not just be concatenated.

When we read a book, we do not read all the pages at once. The same thing is true in a computer program. Not all of a process's pages need to be loaded in main memory prior to execution. In an extreme case, we might have only one page in memory at any given time, with the remainder in secondary storage.

Normally, having just one page of each process available is unreasonable because it causes heavy traffic between the system disk and main memory. However, it is certainly reasonable to keep only a subset of the process's pages in memory and bring in others as they are needed. This approach, usually called **demand paging**, forms the basis of many modern multiuser systems.

13.2.5 Segmentation

Paging is interesting and important, but as we have described it, there are some limitations. In general, they can be traced back to the arbitrary physical division of a process by the operating system rather than a logical division by the software designer. Let's consider two situations where an arbitrary physical division could cause problems.

First, there is the enforcement of access restrictions. Suppose we designed a system for other departments in our company and we want some data areas to have unlimited access, while others are to be restricted. This might not be an easy task with paging because access rights are usually assigned to the page rather than to the word. We have to split up the two data areas and make sure that they are physically arranged to go on pages with the proper access restrictions.

Second, the system may not operate efficiently because we don't take advantage of the user's knowledge about his or her program's structure and performance patterns. In paging, the division is based strictly on physical size, with no consideration of logical usage.

This brings us to the second standard memory management technique, segmentation. The main difference between segmentation and paging is that the user separates and identifies logical components of programs rather than having the system divide them into arbitrary pieces. For example, a user might specify in his or her source program that there are three code segments, INPUT, SORT, and ANALYZE, and two data segments, PERMANENT and NEW_TRANSACTIONS. These, in turn, might be assigned segment ID numbers of 0, 1, 2, 3, and 4 by the compiler. Specific locations within the segment would again be identified as an offset from the beginning of the segment.

Continuing our example, we might have a source-language instruction such as PERMANENT/GROSS_PAY := 0. This would be interpreted to mean "set the variable named GROSS_PAY within the segment named PERMANENT to 0." Assume for the moment that GROSS_PAY is at offset 195 in segment PERMANENT. Its program location would then be segment 3, offset 195.

We still need to translate the program and data locations into their actual physical memory locations at execution time. Here again, we use a table to implement memory mappings. Instead of a PMT with the starting points of pages, we use a

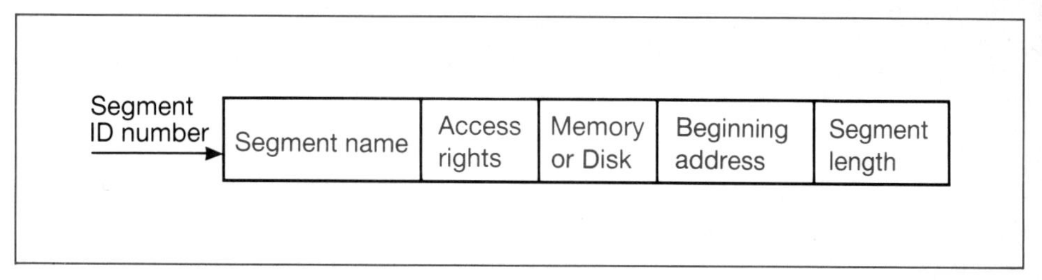

FIGURE 13.4 Typical fields in a segment map table entry.

segment map table, or **SMT**, with entries that indicate the beginning addresses of each segment. However, our address translation is a little different than with paging.

Consider the SMT format shown in Fig. 13.4. The segment ID is similar to the page number. It identifies which segment is being referenced. But segments are different from pages because they are variable, rather than fixed, in length. The *user* defines them.

The segment name identifies the segment table entry associated with the specified segment ID number. The second field is not directly involved in the address translation, but it solves one of our paging problems. It specifies the access rights of the user, such as read-only or execute-only. The third field in Fig. 13.4 is included because, as with paging, we assume that a segment is not loaded until it is needed. This field shows whether the segment is actually in primary memory or whether it is on the disk.

The fourth and fifth fields in Fig. 13.4 provide the information needed to load the segment and to translate the program addresses into their equivalent memory locations. The segment length plus the beginning address defines the segment space in memory and can be used to test for out-of-bounds access requests. The actual address of a program element is computed as shown by the example in Fig. 13.5.

Segmentation is a straightforward approach to memory mapping, but it also has some weaknesses. The primary one is that segments can be quite large. Then, because the *entire* segment must be in primary memory during execution, segmentation tends to require much more memory space at any given time than would be needed with paging. (Recall that with paging we need to have only some of a process's pages in memory at any given time.)

13.2.6 The Best of Both Worlds—Segmentation with Paging

Segments are separate logical entities. In a way, they can be viewed as we viewed programs when we discussed paging. To gain back the advantages of paging, we can apply paging to each of the individual segments.

Before you throw your arms up in dismay, consider the following simple example. It is really not difficult if you follow the mapping step by step. Suppose we again have the instruction PERMANENT/GROSS_PAY := 0. This reference would get translated to a segment number and displacement within the segment, as shown in the top part of Fig. 13.6. However, let's not stop here; let's consider the displace-

Segment number	Beginning address	Segment length
0	10231	752
1	15956	364
2	20400	996
3	00342	355
4	29040	42

(*a*)

Beginning address of segment 3	00342
Offset value	+ 195
Actual memory location	00537

(*b*)

FIGURE 13.5 Segment address translation: (*a*) an abbreviated sample SMT; (*b*) an address translation.

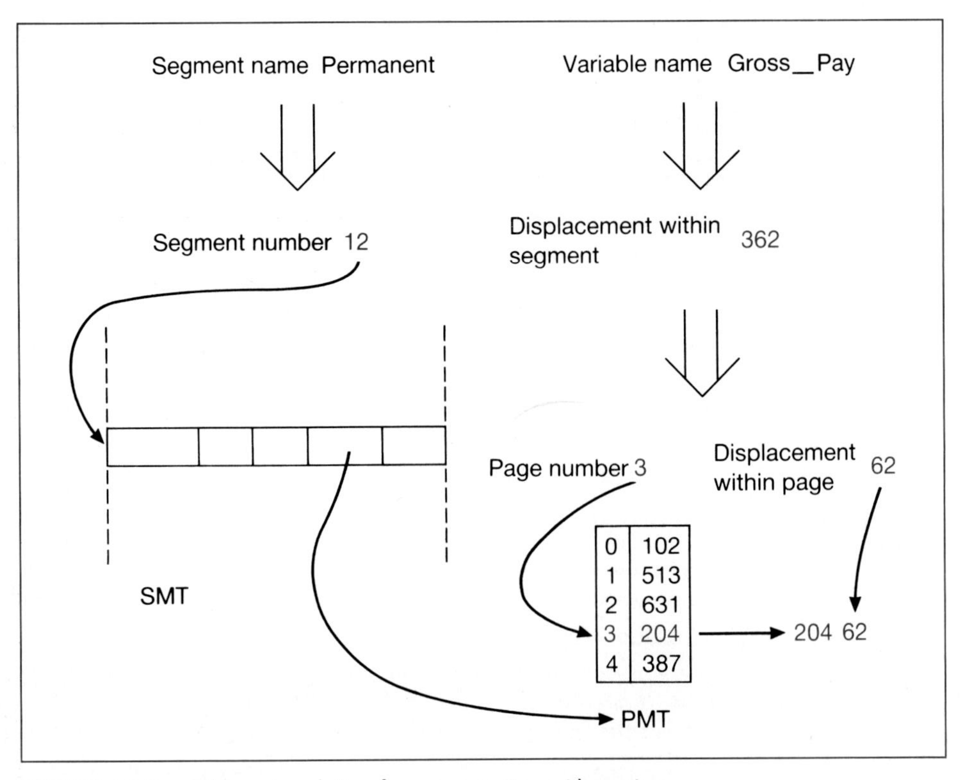

FIGURE 13.6 Address translation for segmentation with paging.

ment within the segment as a page number and a displacement within that page. With a page size of 100, we would have the situation shown in the bottom part of Fig. 13.6.

When this address is referenced, the system goes to the SMT. After checking access rights, it next goes to the field we called location in our previous version of segmentation. However, in this version the location is not the start of the segment in main memory but rather a reference to the PMT for that segment. Once the system has this PMT, it applies the mapping techniques of paging, as we described in Sec. 13.2.4.

One might object that there is a great deal of overhead involved in the mapping process. True. Therefore, systems that use this full segment/page approach usually have much of the mapping done directly by hardware.

13.3 INPUT AND OUTPUT OPERATIONS

One of the primary services supplied for the user by the OS is the ability to access I/O devices in a simple, consistent manner. In this section, we discuss the routines and structures within the OS that give us this capability.

13.3.1 The Path of an I/O Request

To find out how the system helps the user with I/O operations, we will analyze an I/O request. We will look at its beginning in a user process, its path through the OS, and, lastly, its return to the user process. Figure 13.7 shows the steps involved and the operations that cause them to be taken.

Let us start our discussion with a typical input statement from a user program. Consider the Pascal statement

```
readln(input,buffer)
```

This statement provides information necessary to generate code that will request an input operation by the operating system.

What information is in the statement? The keyword "readln" tells the compiler that an input operation is needed. The parameter "input" specifies the peripheral device from which information will be read, input (typically a keyboard or a disk). The parameter "buffer" tells the compiler where the incoming information should be stored into memory. Lastly, the format of the data is implied by the type of the variables into which the information will be stored.

The compiler must convert this source code into an input request that can be processed by the OS. To do the operation, the compiler generates a call to the OS, which on some systems would take the form:

- PUSH BUFFPTR
- PUSH CTRLPTR
- CALL IOCS

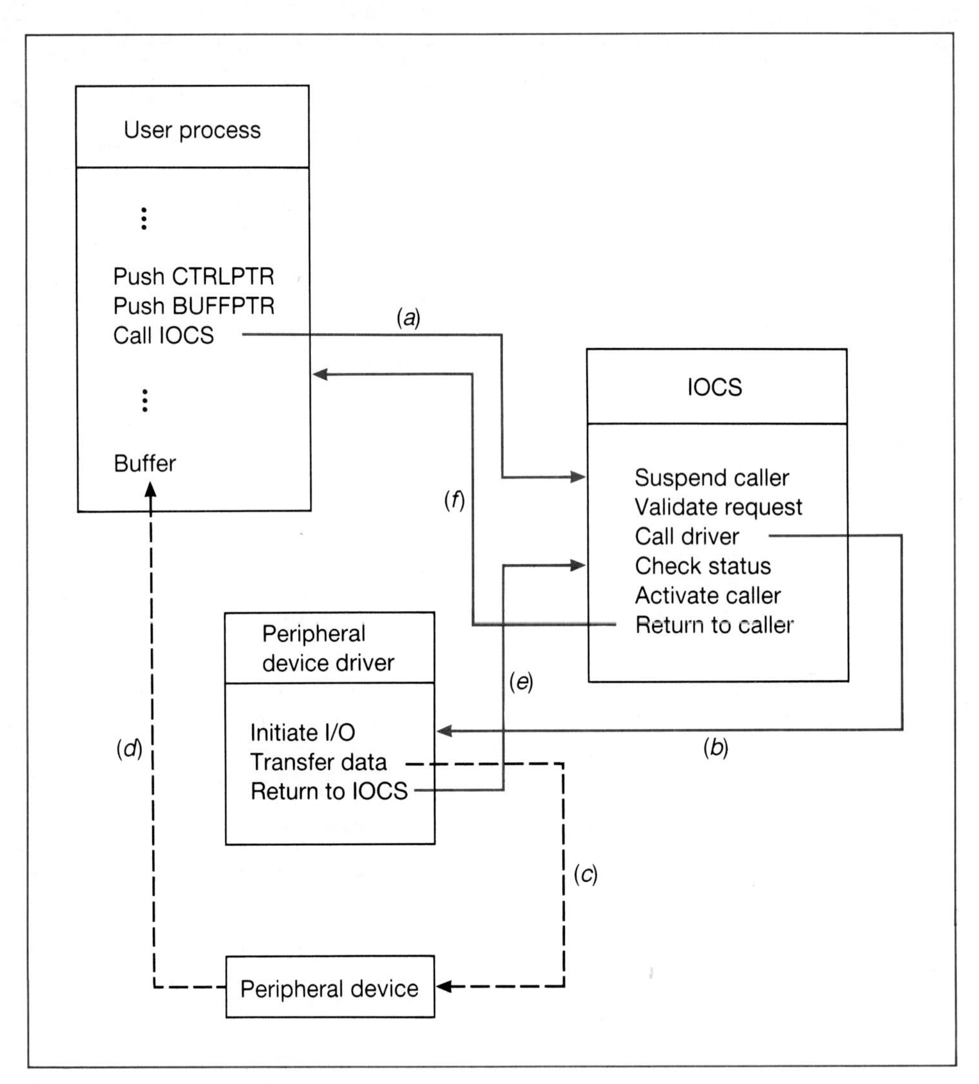

FIGURE 13.7 Steps in an I/O request.

where

- CTRLPTR defines the address of the control information which will tell the I/O system that a read is requested from one of the specified devices.
- BUFFPTR defines the address of the data buffer.
- IOCS is an address within the OS where the routine that processes I/O requests is found.

When the above code is executed, control is passed to the part of the OS that executes input and output operations. This is seen as step *a* in Fig. 13.7.

The **input/output control system, IOCS**, is responsible for processing the I/O requests from the executing process. If the requested device is available, IOCS will transfer control (step *b* in Fig. 13.7) to a routine that can communicate with the device. Common names for the routines that talk directly to a device are **peripheral device drivers** or **I/O handlers**. The control information sent by the driver and the resulting data transfer are shown as steps *c* and *d* in Fig. 13.7.

When the I/O transfer is finished, the driver will return to IOCS. It will pass back a status word showing that the operation has finished (step *e* in Fig. 13.7). This status will also inform IOCS if any errors happened while the data transfer was in progress. Assuming the read request was processed without errors, IOCS will transfer the process that made the I/O request from the suspend state to the active state, where it can again compete for the CPU. This is shown as step *f* in Fig. 13.7.

13.3.2 The Input/Output Control System—IOCS

All I/O requests must go through the IOCS. This system module is responsible for checking if devices are available before beginning operations. It checks the status of peripheral devices after requests have been satisfied, and it also confirms that the requests are passed on to the correct device driver. In essence, it is a traffic controller.

Requests from a process are normally made for the following reasons:

- To transfer data (read or write)
- To check device status
- To position a device such as the movable head on a disk

I/O operations typically take a relatively long time (compared to internal CPU operations), and the calling process usually is not able to proceed until the IOCS request is complete. Therefore, we typically transfer control of the CPU to another process in the system during the I/O operation. This is done by placing the calling process in the suspend state.

The IOCS also determines which I/O device will be used for each I/O request. The sample program statement shown earlier included a reference to the device to which the I/O operation was directed, "input." In most systems this is just a logical reference that the programmer uses. It has no particular relation to the physical device that will be used. If this is the case, there must be some way to map a *logical reference* to a *physical device*. This map information is kept in a table within the operating system, often called the **logical unit table**, or **LUT**.

13.3.3 Device Drivers

The device driver is the system routine that communicates with the I/O hardware. To simplify our discussion of the device driver, we will split it into two parts: the initiation and continuation sections. The **initiator** is responsible for getting an I/O operation started. The **continuator** is responsible for responding to interrupts and for transferring data after an operation has been started.

The first function the device driver initiation section must do is to confirm that the device can do the requested operation. Most systems, for instance, will not let the user read input data from the line printer. If the driver detects a request that is not supported by the hardware to which it is attached, it will pass a parameter that tells IOCS that an error has happened. IOCS can then decide to abort the calling process or just inform that process of an I/O error and let it continue.

Before the driver initiates a data transfer, it must determine the status of the peripheral. This is usually done by loading the contents of the device's status register into the CPU and testing the various bits. After the device driver has read the peripheral's status, and determined that it is "available," the driver will send a command to the device. This is done by writing a control word into its control register.

The operation requested by the user may or may not call for the transfer of data. If the operation is a write, the driver may send a data word to the device after the write command has been sent. If the operation is a read, the device driver may wait for the data. It can do this by repeatedly loading the status word and testing whether the data is ready.

As stated in Chap. 9, waiting for a peripheral to finish an operation can be very wasteful of CPU time. A better approach is to let other processes execute while waiting for the device. Control is transferred back to the driver when an interrupt signal is received from the device. The driver then does the necessary functions, such as loading the data into the CPU from the peripheral and storing it into a buffer.

Eventually, all the data requested by the process will be read in from the peripheral or written to the peripheral. The driver will recognize the situation when the count of remaining transfers reaches zero. When this happens, the driver does a final check of the status and returns control to IOCS.

IOCS must now do something with the process that made the request. If no errors were detected during the I/O operation, IOCS will move it from the suspend state to the active state. If an error happened, IOCS has several options:

- Request that the operation be retried.
- Abort the process and send an error message to the operator.
- Send the status to the process and let it decide what to do about the problem.

Whatever the situation, this completes the handling of the I/O request by the OS.

13.4 FILES AND INFORMATION STORAGE

In this section we will discuss files, with particular emphasis on how the OS maintains and accesses them. We will keep the discussion generic so that it does not become burdened with details of particular devices or systems.

13.4.1 Conceptual Notions of Files

When studying operating systems, we frequently encounter the term *mapping* (see memory management, Sec. 13.2, for example). By this we mean that the user specifies a logical structure and the system *maps* it to a physical structure. We have such a situation with files. The user specifies a file design, and the system creates a mapping function to the physical hardware on which the file resides.

Ideally, we would like the OS to make this mapping as "transparent" as possible. That is, the user should be able to operate strictly on the "logical plane" and not worry about the physical details. ("Details" include such things as disk surface, track, sector, and offset for a particular file reference.) A user should also be able to change devices, such as disk to tape, without affecting his or her logical view of the file.

13.4.2 Levels of OS File Support

Given that the OS maps a user's logical file to some physical representation, an important question arises. At what level should this mapping be done? Should the OS provide powerful information management facilities that the user can then incorporate into logical designs? Or, should the OS provide only a very basic information storage facility and require the user to add any desired features? Phrased another way, the question is one of structure. How much structure should be furnished by the OS and how much by the user?

Existing operating systems vary greatly in their answers to this question. Some treat every file as a sequence of bytes and therefore require the user to provide most of the structure. Others have file managers that are very powerful, giving the user a variety of information structures and access techniques. Still others strive for some compromise position. Files can be either byte streams or groups of records, with access to records by one of a limited set of standard techniques.

13.4.3 Characteristics of Files

When we think of a computer file, there are several characteristics that come to mind. Is the file a sequence of bytes, or is it structured in some way such as fields within records? What is the name of the file, and is there a descriptive qualifier present? What other attributes does the file have? How will the file be accessed?

In the last section, we said that all files on a system could be just arbitrary sequences of bits or bytes. In a different approach, the file manager might allow the user to specify more structure.

A common structure is to have a file made up of logical units called records, as shown in Fig. 13.8. Designers often think of each **record** as corresponding to some entity in the "real world," such as a person, a part, or a job. These records will be divided into **fields**, each field giving the value of one descriptor of the entity. Fields are normally considered the smallest unit of access in this type of file structure.

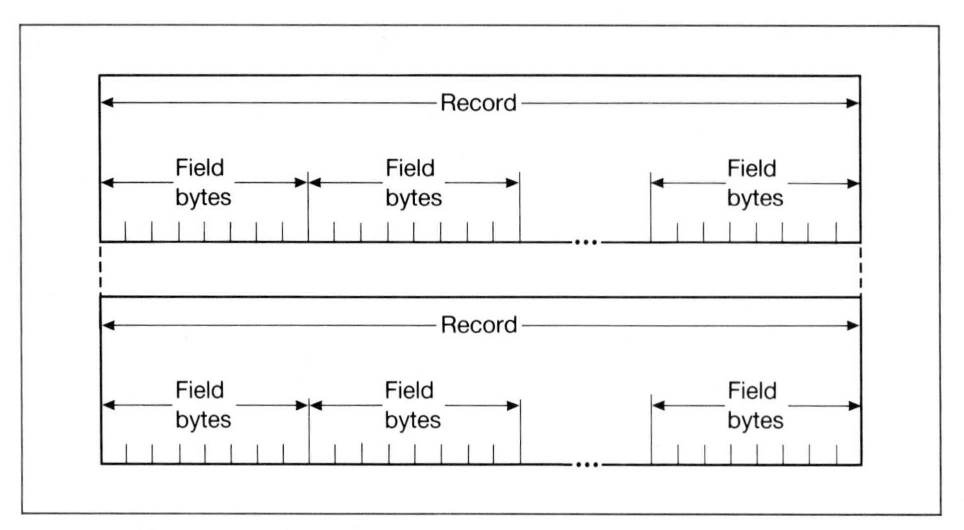

FIGURE 13.8 Common file structure.

We now examine names used to identify files. A **file name** can be more than a simple character string, such as CHAPTER13, by which we reference a file. One common extension to the name is a **qualifier**, a character string that is appended to the basic file name. This qualifier conveys some important information about the file. For example, a standard IBM DOS syntax is to have the qualifier be three characters separated from the rest of the name by a period, such as CHAP10.WDS. In this example, the .WDS could show that the file was a document created by a certain word processor.

The basic file name with its qualifier is often called the **local name**. This term implies that we can use such names only for files in the part of the file system that we are currently referencing. Files on the same device, but in a different area, might be specified by a way to get to that other area (the **path name**) and the local name, such as \BOOK\CHAP10.WDS. Files on a different device might be specified by a device designator, path name, and local name (such as A:\BOOK\CHAP10.WDS). These more comprehensive file identifiers are often called **full names** or **global names**.

Besides a name, most operating systems will also keep a variety of other information about a file. These items of information are often called **attributes** and include name of owner, creation date, last change date, etc. They are usually grouped together into a standard format descriptor or header associated with every file in the system. This descriptor might be a part of the file itself, or it might be stored as a completely separate entity.

Another important aspect of files is how they are organized. Earlier in the section we considered structure. Now we will look at organization and show you how it is an associated, but separate, notion.

Organization of a file refers to the *logical level at which the file will be accessed.* It also refers to the *logical method in which this access will be done.* We might say that in a particular application the logical access is at the record *level*—for example, we are accessing personnel "records." Two standard access *methods* which would be applied to such records are sequential access and direct (also called random) access. These were discussed in Chap. 10.

13.4.4 Finding Files

We have already discussed the idea that the OS maps a user's logical file to a physical file. Part of this mapping will involve finding the start of that physical file given the user's logical file name. The data structure used in this operation is the **file directory**, also sometimes called a **catalog** or **table of contents**.

Our first idea for elements in such a directory would probably be pairs of items:

Logical file name ↔ beginning location of physical file

How should we structure these pairs of items? A first approach might be a linear list of the name/starting-point pairs. This method is usually used only for very small directories, since search time for larger directories is prohibitive.

We could improve search times by having levels in the directory and therefore having more but smaller search spaces. For instance, suppose that our company has 10 divisions, each division has 20 teams of 5 people, and each person typically saves 200 files. With one large linear directory, that's 200,000 directory entries, and an average search would call for examining 100,000 of them. Depending on the speed of the hardware and the current load on the system, this could cause a noticeable delay in responses to user requests.

If instead we structured our directory as levels corresponding to the company, our search spaces would be:

Find division	directory of 10 entries
Find team	directory of 20 entries
Find person	directory of 5 entries
Find file	directory of 200 entries

We now have four searches but at most 200 items in any one of those searches. The average search time for this case would require around 117 accesses; this is about an 850/1 decrease from the previous technique. A diagram of this directory structure is shown in Fig. 13.9.

The fully qualified file name for this directory structure could then be something like \division\team\person\file. This type of directory scheme is often used with other names for the levels being account, group, user, and file. A sample file reference might be \COMPUTER_SCIENCE\TEACHERS\JIM\BOOK_FIGURES.

The levels need not be predefined. To expand to a fully general tree directory, we add 1 bit to our data element. This bit shows whether the pointer goes to another directory or to the file. With this structure, the user can put in as many levels as he

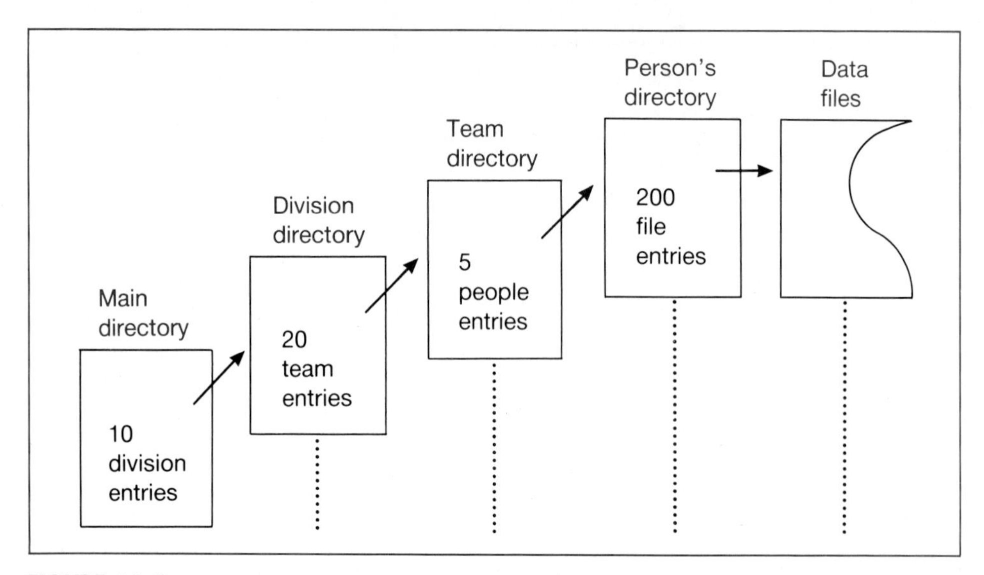

FIGURE 13.9 Example of a multilevel directory structure.

or she wants. The system simply scans directories until it finds a bit showing that the pointer in this entry goes to the target file. Trees are popular directory structures; UNIX, for example, uses a variation of the technique described above.

13.5 SHORT-TERM SCHEDULING

In this section we look at how the CPU is allocated among the processes which are ready to run, that is, are *active*. The system routine that does this task is called the **short-term scheduler**, **CPU scheduler**, or sometimes just the **scheduler**.

13.5.1 First-Come–First-Serve Scheduling

The first scheduling policy we will examine is **first-come–first-serve, FCFS**. Another name given to this policy is FIFO, for first-in–first-out. This is the basic policy chosen to provide service in many applications besides CPU scheduling.

There are several reasons one might choose FCFS as the scheduling policy for the CPU. First, it is simple; this means that a scheduler using an FCFS policy could be fast and efficient. Besides being efficient in terms of execution time, the policy needs very little information about the processes being scheduled. Also, it seems like a fair way to service requests.

We now ask the question, "Is this a good scheduling policy?" Before you answer, think about the application of this policy at the grocery store. Assume you are the newly arriving customer at the checkout register and there are n customers ahead of you. They are average customers; that is, they have two carts full of $250.00 worth of everything from apricots to zucchini. Now if all you want is a box of breakfast cereal and a diet soda, and your assignment is due, you probably will not like this policy. Customer dissatisfaction with FCFS is what led to a change of scheduling policies at grocery stores. The policy was changed to give preferential service to customers with only a few purchases. The mechanism used was the express checkout line you now find in most larger grocery stores.

If we want our CPU to provide fast service to short processes, the next step seems obvious. We install the equivalent of an express line.

13.5.2 Round-Robin Scheduling

As timesharing systems became popular, a new scheduling policy was needed to reduce the time a user had to wait for a response from his or her terminal. The policy often chosen is to force a long process to go through the active and execute queues repeatedly. The amount of time a process can spend in the execute state on each cycle is limited. If a process does not finish its work in one of those fixed amounts of time, it is forcibly removed (preempted) from the execute state. It is then placed back on the end of the active queue. The process must then work its way around to the head of that queue again before it is allowed to execute. This method is called **round-robin, RR**, scheduling.

The reasons given for using the FCFS scheduling policy were related to simplicity, efficiency, and fairness. The reasons for using RR scheduling are more subjective. The primary reason is that users are happier and more productive if they get fast responses to their simple requests. They may even pay more money to use a system if they are guaranteed fast response every time they input a line.

If you are one of the system users whose processes are mostly short, you will probably be satisfied with this policy. But, if you just submitted a process that will calculate π to 10,000 digits and then print the result, you may be very displeased with the performance of the system. What are the implications? You probably cannot make everyone happy with one scheduling policy. Because short processes get better service than long ones, RR scheduling is often described as being unfair to long processes.

In an attempt to have their cake and eat it too, designers have changed the RR scheduler so that it is fairer to long processes. They typically change the execute time interval given to long processes. One of the mechanisms that has been suggested for implementing a fairer policy for long processes is to double the time interval given a process each time it does not finish executing during its current allotted interval.

The purpose of such a change to the RR scheduler is to reduce the wait time for a long process when it is competing for service with a number of short processes. To do this, we are forced to accept longer response times for the short processes, since they must wait for the extended execute time allocated to the long process. Whether this is a good tradeoff depends on the way the system is being used.

13.6 RESOURCE MANAGEMENT

The allocation of resources, including the CPU and memory, is one of the primary functions of an operating system. Because of this, resource allocation is the subject of this section. We will first look at the different types of resources found in a computer system. Then we will consider two quite different approaches to the task of allocation. **Static allocation** implies that a process must specify and receive all the resources it will ever need before it begins execution; **dynamic allocation** allows the process to ask for and receive resources during execution.

13.6.1 Resource Classification

For our purposes, we will place resources into one of two categories: serially reusable and non-serially reusable. With one class, **serially reusable**, or sharable, the resource can be taken away from a process temporarily before that process has finished using it. With the other class, **non-serially reusable**, or nonsharable, a process must be allowed to keep the resource continuously until that process has finished using it.

Some examples will clarify the distinction between these two resource types. One of the resources on a system that must be shared by processes is the CPU. On a multiuser system we allow a process to use the CPU for some period of time, then we take the CPU away, *preempt* it, and allocate it to another process. As long as all the CPU's registers and internal status are saved and restored correctly, this causes no problems other than slowing down individual processes. Because the CPU can be shared in this serial fashion, the CPU is classified as a serially reusable resource.

Now consider one of the system's other resources, the line printer. Suppose two processes are trying to share the printer in the same way they shared the CPU. One process will send output to the printer and then the printer will be preempted. When the second process sends its output to the printer, you can probably imagine what will happen; the output from the processes will be interleaved and therefore will be useless. This means that the printer, when allocated, must stay allocated to one process until that process is finished with it. Because the printer cannot be shared between processes, it is classified as a non-serially reusable resource.

Managing serially reusable resources is generally an easier task than managing non-serially reusable resources. In the remainder of the section, we will examine the methods used to allocate non-serially reusable resources within the system.

13.6.2 Static Resource Allocation

As an analogy, let's consider one of the authors painting his toolshed. Before beginning, he must get the necessary resources. With static allocation, he would generate a single composite request for all resources needed or potentially needed to do this task, probably:

- A brush

- A can of paint
- A ladder
- A scraper

If he couldn't get any one of these resources, he would be denied permission to start painting.

The computer version of static allocation is similar. Before a process can be considered for scheduling, it must specify and get all the resources it might need during its life in the system. This must be a "worst-case" specification—if a process might need three or five tape drives, depending on the data, it must request and be allocated all five before it can proceed. The request must also be completely comprehensive; if the process plans on plotting one curve after several hours of computation, it must still request and be allocated the plotter before it can initiate the first calculation.

13.6.3 Dynamic Resource Allocation

Using static allocation in our painting analogy, all the listed resources had to be acquired before painting could begin. From the author's point of view, only a subset of these are essential to start:

- A brush
- A can of paint

The remaining resources:

- A ladder
- A scraper

are needed only when special conditions are encountered during the painting.

Given our analogy and the compare/contrast discussions in the previous section, an extensive description of dynamic allocation in operating systems is probably unnecessary. Resources are requested chronologically as they are needed. If a resource only might be needed, it is not requested until needed. If a process might need three or five tape drives, depending on the data, it requests three as they are needed and only requests the other two if they are needed. If a resource is not needed to start execution, that resource can be requested later when it is needed.

One can see immediately that resource utilization should be higher with dynamic allocation. The long idle periods from allocation to usage and usage to release are no longer necessary. Furthermore, because the available resources are more efficiently used, turnaround time should decrease. Each process in the mix can go as far as possible with the resources it currently has; it doesn't have to wait to get everything before it starts.

13.7 LONG-TERM SCHEDULING

Have you ever gone into a bakery, been given a number, and watched as the number of the customer being served moved toward your number? Or have you tried to log onto a system and been told "No more sessions being initiated at this time" or "You are number 7 on the queue"? In these situations, you have met some long-term schedulers.

If everyone had all the resources they would ever want, then there would be no need for long-term schedulers. If architects could build an infinite stadium or concert hall (in which you could still see and hear from the last row of seats), then any number of people could attend every event. If computers had an infinite number of CPUs, memory locations, ports, terminals, etc., then any number of users could log on and run any number of programs.

However, they don't, so we can't. Current stadiums and halls have a limited number of seats. Current computers have limited resources, and this implies a limited number of processes in the system at one time. Because of these limits, we need **long-term schedulers** to act as "gatekeepers" and regulate admissions.

Before continuing, we should clarify the difference between short-term and long-term schedulers. The short-term scheduler works with processes already in the system. It determines which of these active processes will be executed next. The long-term scheduler works with requests to create processes. It determines which of these requests will be serviced next and, hence, which of the corresponding processes will be created.

To incorporate the long-term scheduler into our OS schema, we establish a list of "waiting requests," each of which is intended to create a process to be run. When the system can accommodate another process, the long-term scheduler chooses a request from this "waiting" list. A process identifier is created and entered into the system, therefore creating the corresponding process.

To automate our long-term scheduler, we need to consider various criteria on which to base decisions. Two of the more obvious candidates are arrival time and predefined priorities; let's call these group 1. We have also mentioned CPU time, memory, and estimated output, which will be called group 2. There is an important difference between these two groups of scheduling criteria—the first group consists of known quantities and the second consists of estimates. As illustrations of the difference, the system can record the arrival time, but often the user can only estimate CPU time and lines of output.

A simple scheduling policy is to take the processes in the order they arrive. The long-term scheduler here uses FCFS, and the scheduling criterion is based strictly on arrival time. A problem with this can be shown with an example and a little analysis.

Suppose four process creation requests are pending. Assume (1) that the long-term scheduler took these requests one at a time, using FCFS, and (2) that the resulting processes ran as shown in Fig. 13.10*a*. Furthermore, assume that time in the system is directly proportional to CPU time used. The ratio of elapsed wall-clock time to time in the system, and hence to CPU time used, is then as shown in Fig. 13.10*b*.

(a)	(b)
No. 1 spent 2 minutes in system	No. 1: 2/2
No. 2 spent 60 minutes in system	No. 2: 62/60
No. 3 spent 2 minutes in system	No. 3: 64/2
No. 4 spent 2 minutes in system	No. 4: 66/2

FIGURE 13.10 Four processes executing in the system: (*a*) times in system; (*b*) ratios of elapsed wall-clock times to times in system.

The long process has a ratio of about 1/1, while the short processes have an average ratio of about 22/1. From these, we can see that FCFS tends to favor the long processes.

We can swing the pendulum totally in the other direction and always schedule the shortest process next. This approach obviously favors the short processes. However, it does not seem to overly penalize long processes. As can be seen in the previous scenario, the short processes would have an average ratio of $(2/2 + 4/2 + 6/2)/3 = 2/1$, while the long process will have 66/60, which is still only a little over 1/1.

The ratio would become worse if the relative lengths of the long processes to the short processes shrink. (A "long" process of 4 instead of 60 would have a ratio of 10/4.) Perhaps an even more critical problem happens when many requests for short processes are entering the system while requests for long processes are waiting. The requests for long processes keep getting pushed back and might have to wait forever.

In an attempt to be more universally "fair," we could turn to what might be called *hybrid* policies. These policies consider more than one criterion, or one criterion but with some tempering factors. However, we must stress an important point. It is absolutely impossible to be completely fair. Anything done for the benefit of some subset of the processes will adversely affect another subset to some extent.

One tempering factor used in some hybrid policies is called *aging*. As a request for a process waits, the scheduling metrics are changed by some function of the wait time. As a simple illustration, we might define

$$T' = \frac{\text{estimated CPU time}}{1 + \text{wait\_time}/k} \tag{13.1}$$

and apply a shortest process next to this new value T' instead of just the estimated CPU time. (Here k gives a relative measure of importance to the wait_time.) Using our previous numerical example and a k of 2, the equation shows that the request for the 60-min process would have the same T' after waiting 58 min as a newly arrived request for a 2-min process.

This last technique seems quite appealing and fair. It has been used with various changes in a number of operating systems. Names incorporating phrases such as **response ratio** and **penalty ratio** have been given to these long-term schedulers.

13.8 SUMMARY

The long-term scheduler is usually considered the outermost component of the operating system. You would typically interact with this outer layer via a command interpreter or shell, for example, when you type RUN MYPROG. But, you might be using an application program, such as a spreadsheet, and not interact directly with the outer layer of the OS. We consider such applications as layers on top of the OS and treat them separately in Chap. 16.

In this chapter we have looked at the functions that an OS does and the components that make up a typical OS. Our discussions of the components have been necessarily brief, but they have given you an introduction to the subject.

PROBLEMS

13.1 We said that the OS is an interface between user processes and the hardware. What are some other examples of interfaces in a computer system?

13.2 What are some noncomputer examples of interfaces?

13.3 What are some of the guardian services an OS provides?

13.4 What are the typical functions of a memory manager?

13.5 Why didn't early computer systems need a memory manager?

13.6 Develop a memory map similar to Fig. 13.2 using decimal numbering and 1000-word pages.

13.7 Develop a memory map similar to Fig. 13.2 using binary numbering and 1024-word pages.

13.8 Develop a memory map similar to Fig. 13.2 using hexadecimal numbering and 256-word pages.

13.9 Why isn't it necessary for all the pages of a process to be loaded into memory before the process starts execution?

13.10 The OS must maintain a page availability table so that it will know which pages are not being currently used. Propose a format for this table.

13.11 What is the technique of bringing pages of a process into memory only as they are needed called?

13.12 How could the page access rights problem discussed at the beginning of Sec. 13.2.5 be corrected?

13.13 Why is the programmer's knowledge about his or her program's structure important in memory mapping?

13.14 How could you determine the length requirements for each of the SMT fields shown in Fig. 13.4?

13.15 How can memory-bounds checking be done in segment mapping? How could it be done in page mapping?

13.16 Why would the OS have to keep a space availability table for segment mapping? Propose a format for this table.

13.17 Why are we not able to do segment mapping with a concatenation procedure as we did in paging?

13.18 What part of Fig. 13.6 is done by the compiler? What part is done by the OS?

13.19 Suppose two users, call them *A* and *B*, are on a time-shared system that has a paged memory allocation technique. Furthermore, suppose that in memory these two users' pages are interleaved; i.e., there is one page of *A*'s process, one of *B*'s, one of *A*'s, such as in Fig. P13.19. User *A* now stores something into program location 2100. Why won't that damage *B*'s program?

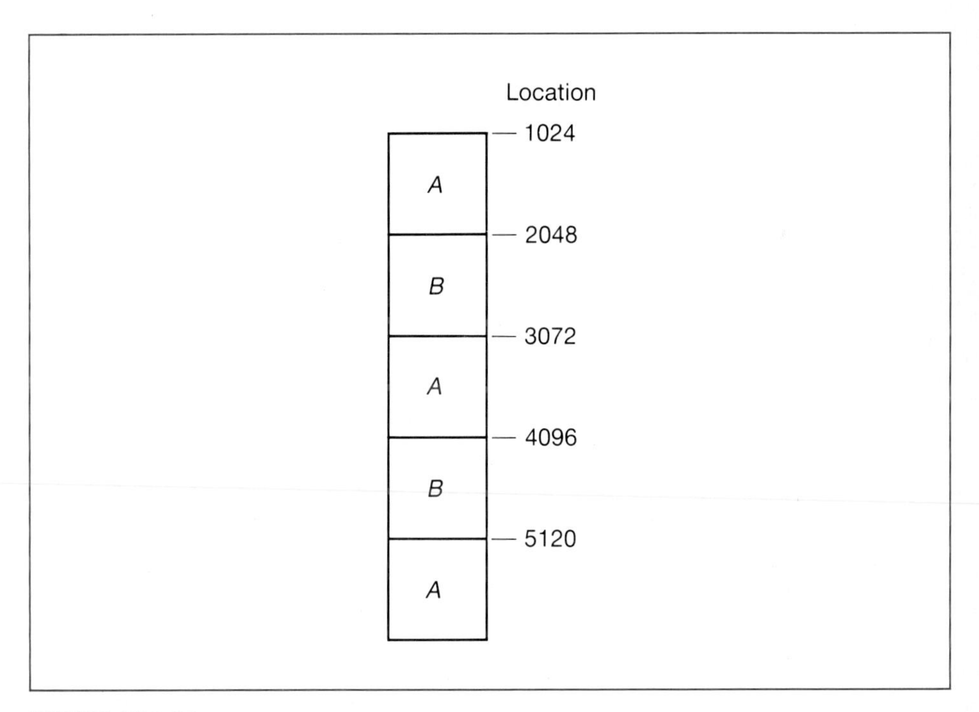

FIGURE P13.19

13.20 Compare the length of the PMT in a page-mapped system to the length of the SMT under segment mapping. How would these lengths be affected if we were to switch to paged/segment mapping?

13.21 In Fig. 13.7 one of the functions of IOCS is listed as Suspend caller. Why is this done?

13.22 What information is kept in the LUT? Describe a format for a LUT.

13.23 If your computer system has qualifiers on file names, list those you can find and describe what each means.

13.24 Design a record for a typical file application, such as a payroll or inventory system.

13.25 Describe typical applications for (*a*) a sequential-access file organization and (*b*) a direct-access file organization.

13.26 For your computer's file system, describe what local names, path names, and global names look like.

13.27 What is the smallest element within a file that can be accessed?

13.28 We said in Sec. 13.4.4 that the average search time for the division-team-person-file directory structure would be around 117 accesses. Why?

13.29 We said that round-robin scheduling was "similar" to the express line at a grocery store. How is it different?

13.30 How do banks, grocery stores, and bakeries solve the scheduling problem?

13.31 What would be some noncomputer examples of static resource allocation?

13.32 Make a list of (*a*) serially reusable and (*b*) non-serially reusable resources on your system.

13.33 What is the "best" order to schedule the processes in Fig. 13.10?

13.34 Graph the ratio of elapsed wall-clock time to time in the system for the long and short processes as the long process in Fig. 13.10*b* goes from 60 to 4 min.

13.35 Graph T' versus wait time, Eq. (13.1), for various values of estimated CPU time.

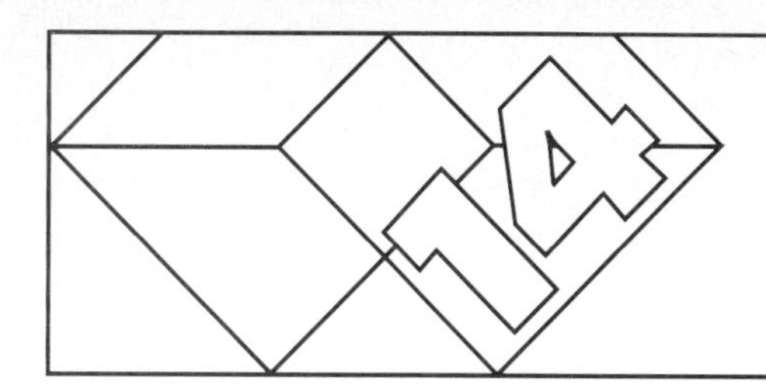

A Layered Operating System and Its Kernel

There are different ways to organize and structure an operating system. The one we prefer is to have a set of cooperating, concurrent processes that communicate according to a well-defined, layered pattern. Such a pattern is often described as analogous to a column made up of several independent blocks, as shown in Fig. 14.1 [Pinkert and Wear, 1989]. At the base of the column is the kernel. In this chapter, we will describe a typical kernel and illustrate a possible layering for an operating system.

14.1 THE KERNEL

The **kernel** is the lowest level of the operating system, the interface between the hardware and the rest of the software. The kernel is responsible for coordinating the operations of the processes which form that operating system.

Each system seems to have different functions embedded in the kernel. The UNIX kernel [Thompson, 1978; Ritchie and Johnson, 1978] contains most of the features of the OS described in the last chapter. The kernel described by Lister [1979] has only three elementary functions. We will follow Lister's approach and break the kernel into three parts:

- The dispatcher, which is responsible for placing processes into execution
- The semaphore control, which controls access to the semaphores used for process synchronization

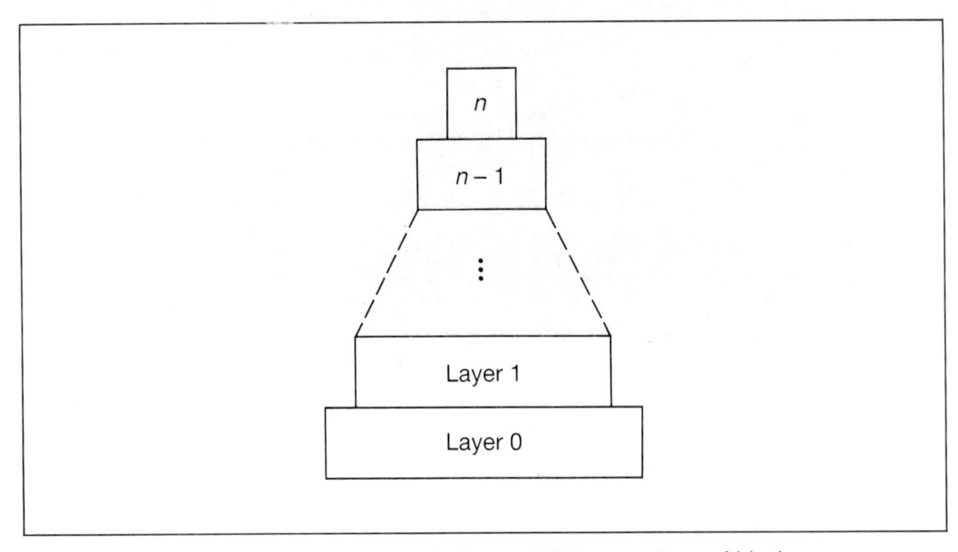

FIGURE 14.1 Pictorial representation of a layered OS as a column of blocks.

- The first-level interrupt handler, FLIH, which analyzes all interrupts and starts the correct handler

These three parts will be discussed in the following sections.

14.2 THE PROCESS DISPATCHER AND THE IDLE PROCESS

The first of our three components, the dispatcher, has a simple job. It takes a process from the active state and places that process into the execution state. To do this, it must load the CPU's environment (such as registers and status flags) with the values specified in the process descriptor. The dispatcher then branches to the first instruction of the process.

The dispatcher's first goal is to find a process to dispatch. We assume the short-term scheduler has already done any necessary ordering of the active queue. The dispatcher takes the first process on the active queue.

A pertinent question now is, "What happens if there is no process on the active queue?" A straightforward approach is to create an **idle process**, a process that is always around as a "last resort" if the system has nothing else to run. The idle process always runs at the lowest priority on the system. Its priority is never increased by any of the priority-change schema often applied to other processes. The idle process is created when the system is started up and is always available to execute. (Another name given to such processes that run when there is nothing else to do is *ghost* process.) What does this process do when it executes? To answer a question with a question, who cares?

Some systems we have seen have interesting idle processes. A common one on earlier systems was to blink a portion of the display or a light in a distinctive fashion. Besides giving the operator a hypnotic pattern to lull him or her to sleep, the rate at which the pattern changed showed how busy the CPU was. If the lights seemed to dim and stop moving completely, then the company was really getting its money's worth out of the CPU. On other systems, the idle process does some operation that might not be worth the cost of necessary CPU time but is potentially useful.

System designers have tried to come up with other work that can be done by the idle process. Two common examples are running diagnostics and archiving files to a backup device. There are many diagnostics that can be run on both the hardware and software to check for potential failures. On systems that have backup storage for disk files, files that have been changed since they were last saved are rewritten to provide current backups. Having the idle process do this provides an inexpensive way to guarantee that user files are safe in the event of a system failure.[1]

14.3 SEMAPHORES FOR CONTROLLING PROCESS SYNCHRONIZATION

In Chap. 12, we said that a system needs some form of process synchronization. For this discussion, we will assume that semaphores, as we described them earlier, are used to provide such synchronization. Our kernel will then need to include a procedure that controls access to semaphores.

This procedure accepts two commands from executing processes, Signal and Wait. It can only be entered by making a system request, which guarantees the integrity of the semaphore and prevents accidental change.

After a semaphore has been declared, a process uses it by requesting entry with a procedure call such as WAIT(buffer_ready). The compiler might convert this into call SEMAPHORE_CONTROL(wait,buffer_ready).

The SEMAPHORE_CONTROL procedure first checks to see if the requested semaphore has been declared. If it has not, the requesting process is aborted. If the semaphore exists, the status of the semaphore is checked. If it is 1, then it is changed to 0 and the calling process is allowed to continue. If the semaphore is 0, then the calling process is moved to the wait queue for that semaphore.

When a process requests a signal on a semaphore, the queue for the specified semaphore is checked. If there is a process waiting on the semaphore, the signaling process and the process that had been waiting on the semaphore are both moved to the active state. If no process is waiting on the semaphore, the value of the semaphore is changed to 1 and the signaling process is allowed to continue.

[1] To do file archiving, the idle process would have to use a special I/O routine that does not suspend the calling process. The idle process must always be active or executing; it can never be suspended.

14.4 THE FIRST-LEVEL INTERRUPT HANDLER—FLIH

The third part of our kernel deals with interrupts. The **first-level interrupt handler**, or **FLIH**, is the system routine that is executed each time a hardware or software interrupt is acknowledged. The exact operations of the FLIH are dependent in part on the way the interrupt hardware is designed.

The structure of the interrupt system may determine whether a first-level interrupt handler, separate from the interrupt service routines, is needed. If all interrupts cause control to be transferred to one routine, then a FLIH is useful. On systems that transfer control to a different routine for each interrupt, the functions of the FLIH may be embedded in the service routines.

The FLIH does two primary functions. The first is to save the status of the interrupted process and transfer control to a new process, an operation called context switching. (This was discussed in Chap. 9.) The program status variables that are saved include the contents of the program counter, the accumulator, and the condition register.

The second function of the FLIH is to determine which routine will handle the interrupt and then to call it. How does the FLIH find the correct routine? One way is to put the address of the routine into the interrupt table. Figure 14.2 shows an interrupt table where the routine address is included with the interrupt number and the current buffer address.

When an interrupt happens, the interrupt number is compared with the entries in the interrupt table to see if the interrupt was valid. If it is a valid interrupt, the interrupt processing routine (the driver for I/O interrupts) is called. For our example, the driver address is passed with the buffer information so that a driver can service multiple devices.

If the interrupt number is not found in the interrupt table, an "Illegal Interrupt" message is sent to the operator. An illegal interrupt might result from a hardware failure or an electric transient caused by turning power on to a device.

After the routine has processed the interrupt, it goes back to the FLIH if there were no changes in the active state. If the interrupt did cause a process to be added to the active state, the driver will not return to the FLIH. It will pass control to

FIGURE 14.2 An interrupt table for a system that vectors all interrupts to a FLIH.

Interrupt number	Interrupt handler	Buffer address
1	18120	2420
2	20420	0000
3	20420	4620
4	21002	4080
5	22080	0000

IOCS. If control is passed back to the FLIH, it restores the state of the CPU and passes control to the process that was interrupted.

The operations done by the kernel are quite limited: dispatching processes, controlling semaphores for synchronizing processes, and deciding what to do when an interrupt happens. From the explanation of these three functions, we saw that each is a simple and distinct operation.

In the next section, we will show how the various parts of an OS can all be connected to form a system. To do this, we will start with a description of a layered approach for the structure of the OS.

After showing how all the pieces can be put together, we follow the execution of a process from creation through termination. Lastly, we will see how the parts of the system can be loaded into memory and how system processes can be initiated.

14.5 A LAYERED ORGANIZATION FOR OPERATING SYSTEMS

Now that we understand the purpose of the components of our OS, we need to define a structure that will allow them to work together. Layering has become popular in both hardware and software. The layered structure for OS construction has been used on several systems [Dijkstra, 1968; Brinch Hansen, 1970]. The popularity and portability of UNIX has also contributed to the belief that a layered OS is superior to other design approaches [Thompson, 1978]. An excellent reference to layered operating systems can be found in Lister [1979].

The base of our layered OS structure is the kernel. As described in the previous section, it contains the first-level interrupt handler, the dispatcher, and facilities for process synchronization. This base for our column is shown in Fig. 14.3. This part of the system is very hardware-dependent and can be thought of as the first layer of software covering the system hardware.

The next layer is the memory manager. Since it must work with the memory-mapping hardware, it is also somewhat hardware-dependent. It uses some services provided by the kernel to do its operations. Memory fault interrupts, memory address violation interrupts, and memory error interrupts are detected by the kernel

FIGURE 14.3 Foundation layer of column—the kernel.

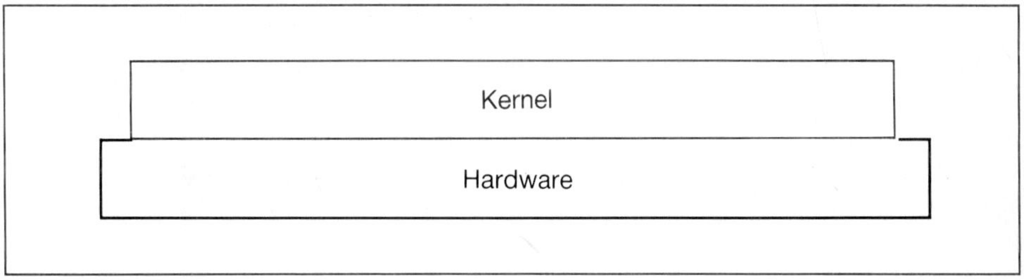

Kernel
Hardware

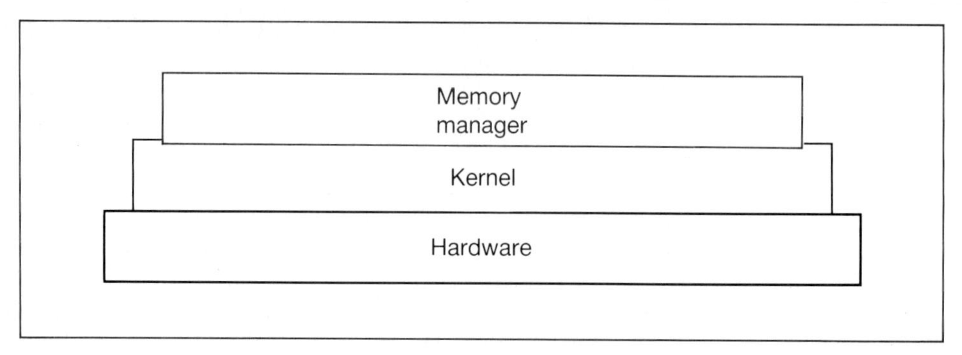

FIGURE 14.4 Adding the memory manager to the system.

software. Control is passed to routines within the memory manager for resolution. Figure 14.4 shows the structure with the memory manager software added.

The third layer of the system is the IOCS. Again, some of this software is very much hardware-dependent. I/O drivers must work directly with the hardware of the machine. The I/O layer is also dependent on the kernel to process interrupts and pass control to the correct driver routines. It may also need the services of the memory manager to get I/O buffer space. The system with the I/O layer is shown in Fig. 14.5.

Layers 1 to 3 are the parts of the OS most dependent on the underlying hardware. Some software in these layers will probably be written in assembly language or in a system's programming language because of the need to access special hardware features and to build a fast, efficient system. The layers beyond layer 3 become increasingly less dependent on the hardware.

The file manager is our next layer. Often, it can be written with little regard to the physical hardware that will be used to support it. The I/O system, memory manager, and kernel isolate it from the actual hardware. The file manager does not care how interrupts from the disk are processed or how the driver gets the status of

FIGURE 14.5 Adding the IOCS layer to the system.

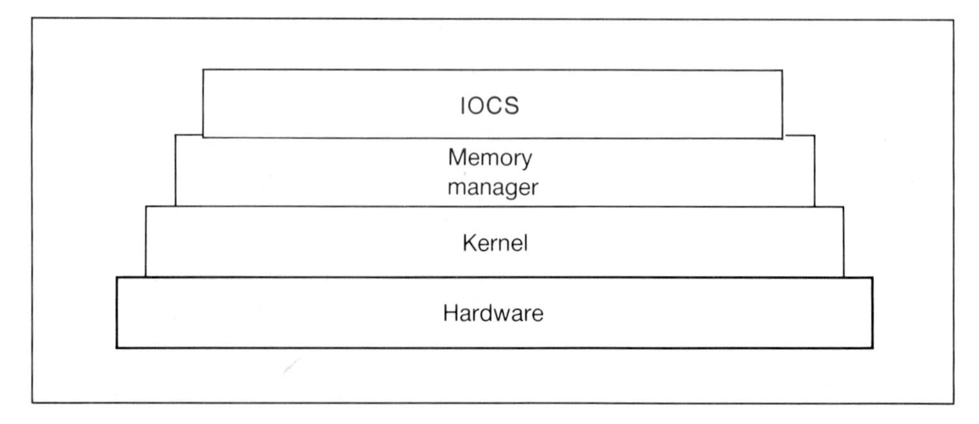

the disk. All it needs to know is how to use these services provided by the lower layers of the system.

This leads us to an important characteristic and feature of layered systems. Layered systems must be designed so that each layer has a precise set of rules that tell outer layers how to access it. This set of rules is usually called the **interface requirement**. As long as the interface requirement is strictly followed, one of the layers of the system can be changed without affecting the code in any of the other layers. For instance, suppose after working on our system for some time we decide to add a new type of disk drive to the system. As long as the driver for the new disk accepts the same requests and returns the same information, the other layers need not even know a change has happened.

As an example of such an interface, consider a call to the I/O control system, or IOCS. It might take the form

```
IOCS(command,data1,input_buffer,length,status);
```

where the parameters are as follows:

- *command* is the I/O command, such as read or write.
- *data1* is the logical unit number for the tape unit assigned to the process by the resource manager.
- *input_buffer* is the address of the data buffer.
- *length* is the number of characters to read.
- *status* is where the status of the operation is returned.

As long as this interface is maintained, IOCS could be completely rewritten without any of the other layers being affected.

Since the outer layers of the OS are not dependent on the hardware characteristics, they are often written in high-level languages. Languages currently popular for this type of software include C, Modula-2, Pascal, and systems programming languages. The layered system with the file manager added is shown in Fig. 14.6.

The next layer we will add is also not dependent on the hardware. The short-term scheduler is responsible for maintaining the active state in the system. When it is invoked, it orders the list of active processes. It then calls the dispatcher within the kernel to place the highest-priority process into execution. The short-term scheduler is shown as the fifth layer of software in Fig. 14.7.

The short-term scheduler is responsible for managing one resource, the CPU. Since our system has many other resources, we must find a place for the manager of these other resources. We will place the resource manager at layer 6. One could probably make a good argument for swapping the short-term scheduler and the resource manager. Processes may need the resource manager after the short-term scheduler has provided its service. This would be the case when dynamic resource allocation is used. Therefore, though we show the resource manager as layer 6 in Fig. 14.8, some systems might have it as layer 5.

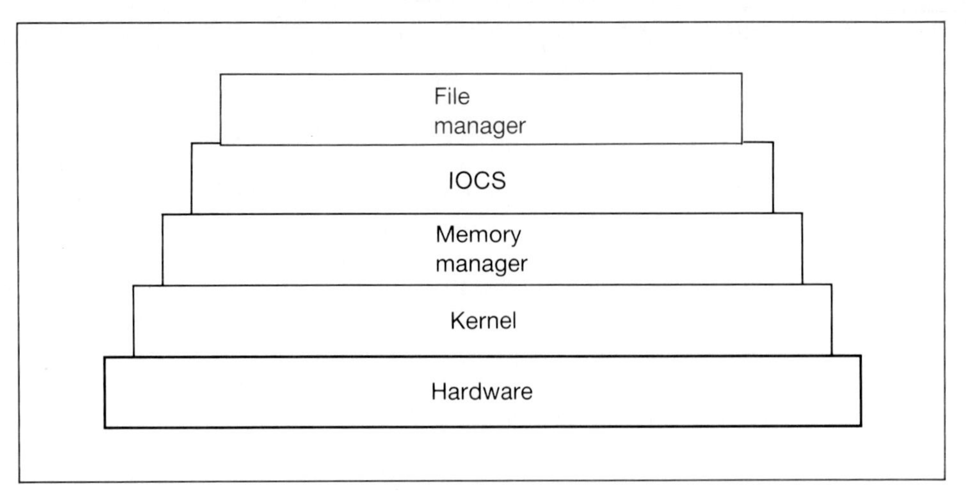

FIGURE 14.6 Adding the file manager to the system.

We have been working our way from the lower hardware-dependent layers of the system toward the outer layers. These are the ones with which the user most often interacts. Layer 7, shown in Fig. 14.9, is the long-term scheduler. This layer controls the set of processes in the system and is responsible for creating and terminating processes. It uses the capabilities of all the lower layers to complete its operations. For example, it calls:

- The file manager to find program files

FIGURE 14.7 Adding the short-term scheduler to the system.

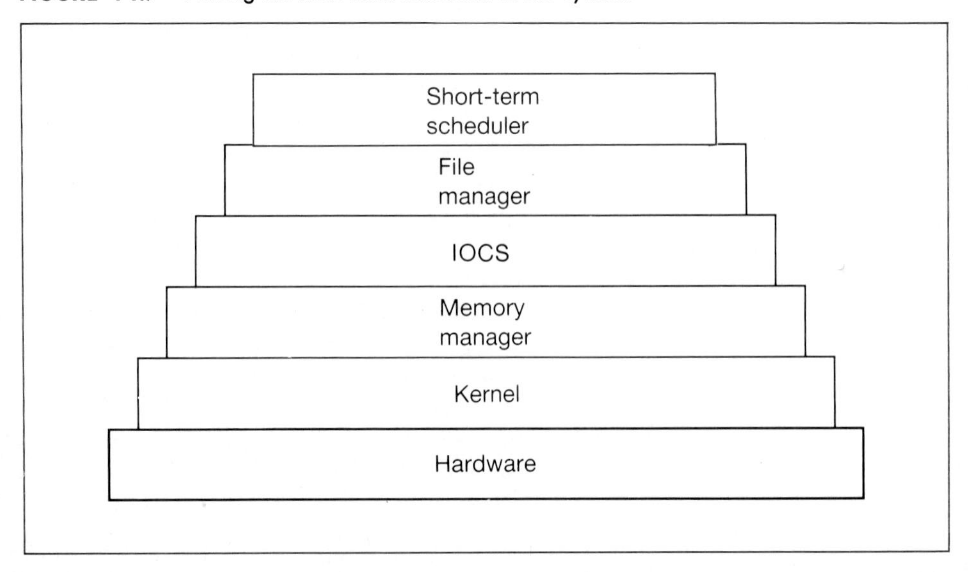

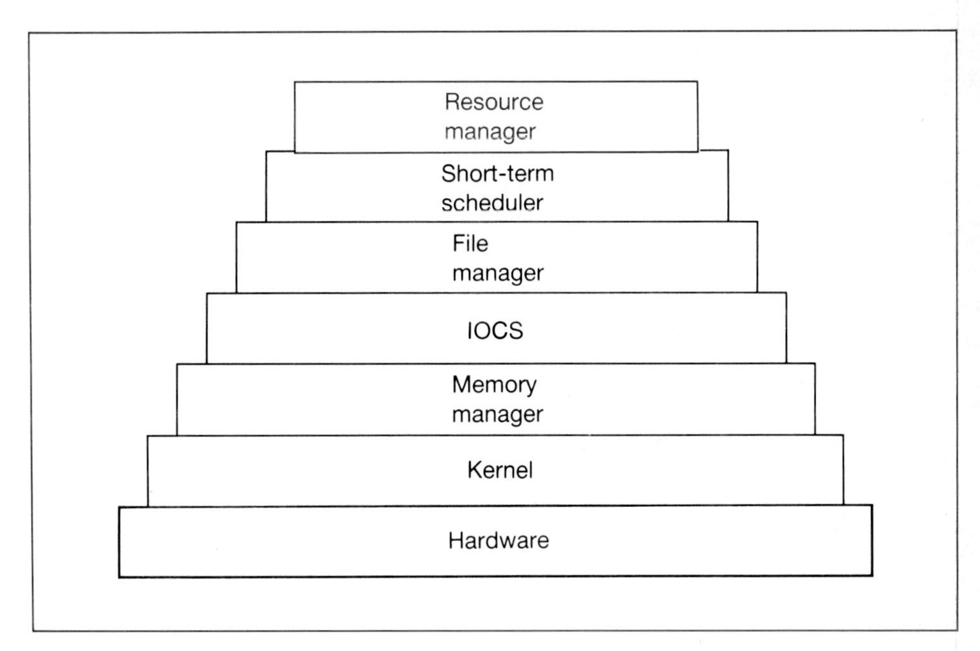

FIGURE 14.8 Adding the resource manager to the system.

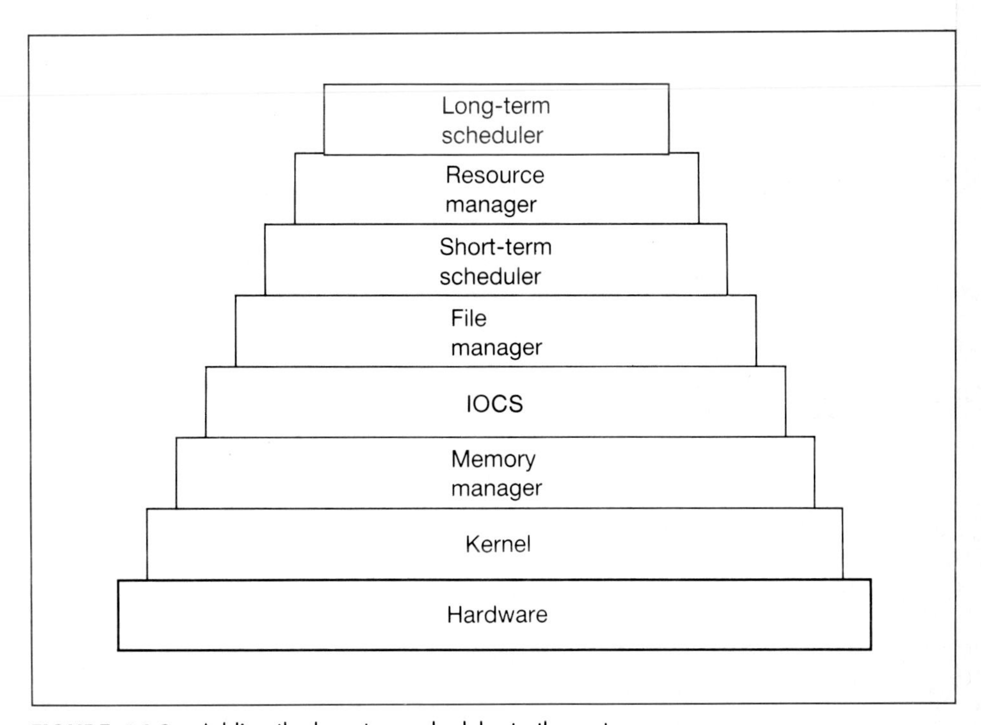

FIGURE 14.9 Adding the long-term scheduler to the system.

- The memory manager to find space for new processes
- The IOCS to read in programs

IOCS in turn uses the kernel to handle interrupts.

The last system layer on the operating system provides the system with its personality. This outer cap, or **shell**, is the part of the system with which the user directly interacts. It often determines how well users like the system. The generic name for this layer is the *command interpreter*. When a user logs onto a system, it is usually the command interpreter that sends the prompt to the screen. The command interpreter is then responsible for analyzing user requests and calling on system routines to satisfy these requests. A number of command interpreters have been designed for the UNIX system; one of the more well known is the Bourne shell [Bourne, 1978].

Because of the growing popularity of menus on microcomputer systems, several companies have developed shells that run on MS-DOS. They are designed to hide the details of DOS from the user. Instead of typing DOS commands, the user chooses the commands from a series of menus. This type of shell is popular with

FIGURE 14.10 Adding the command interpreter to the system.

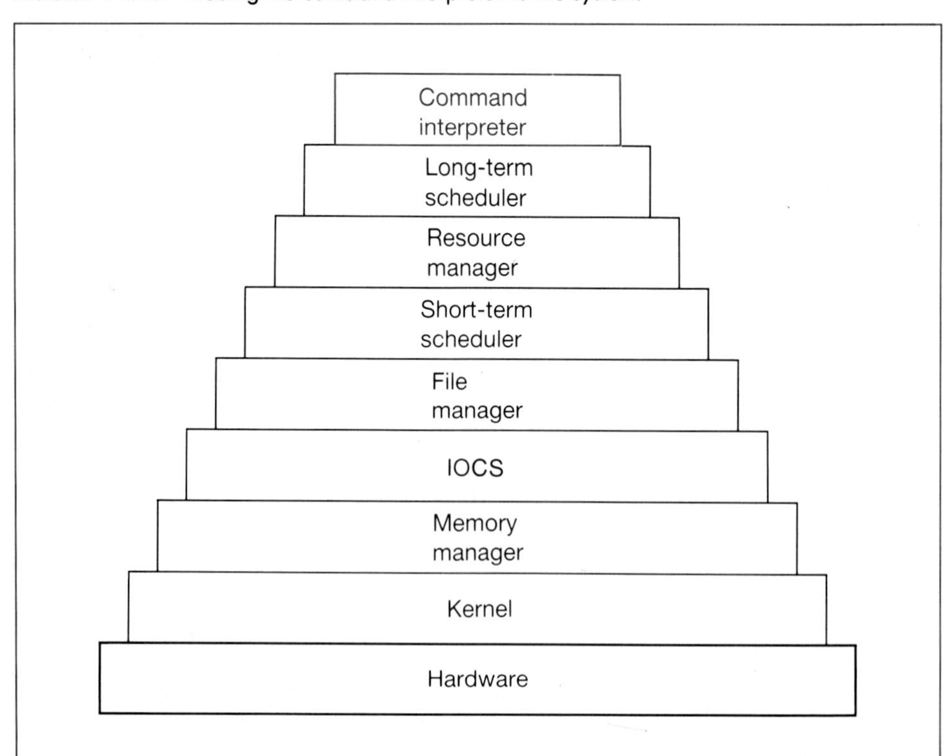

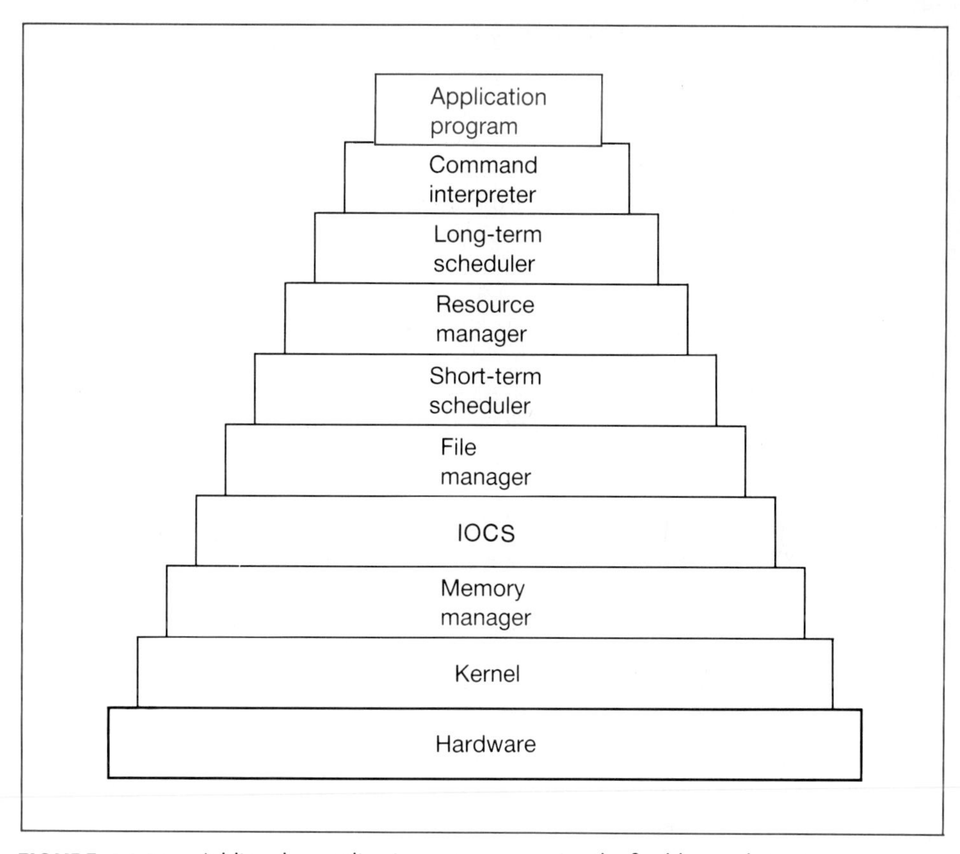

FIGURE 14.11 Adding the application program to give the final layered structure.

users who are poor typists or who do not use the system often enough to memorize DOS commands. Figure 14.10 shows the command interpreter layer.

The last layer is the user layer. This is also sometimes called the application layer, since it contains the "applications" programs for our system. Sometimes the shells described in the previous paragraph are in reality just application-level programs. They give a friendlier user interface to the basic command interpreter. A complete diagram of the layered system is shown in Fig. 14.11.

If you look at layered descriptions of other systems, don't be surprised to see systems with more (or fewer) layers than we have shown. A common layer that is added to many systems is a database management package. Other systems will add layers that just provide a remote terminal interface to some special application program. Layering is an important and useful tool in developing large complicated systems, and its use and acceptance is growing.

We have put the pieces of the system together. In the next section, we will follow the execution of a user command through the various layers of that system.

14.6 A SAMPLE PROCESS LIFE CYCLE

To illustrate the interactions of the layers of the operating system, we will follow the execution of an application program at layer 9. The relevant code for the process is shown in Fig. 14.12. The statements within the program that need OS services are labeled with numbers so that we may follow their execution.

As we follow this execution, we will describe a number of procedure calls made to various parts of the OS. For simplicity, we will not include all the parameters in these calls, just those necessary for your general understanding.

First, the executable file containing this program must be read into memory. Next, the PCB for the corresponding process must be moved to the active state. After these operations, the program will be in memory. The starting address of the program will be stored in a restart address field of the PCB. When the process has worked its way to the head of the active list, it will be dispatched and start executing.

The code for the program, which is now executing as a process, will at some time need OS services to continue. We assume the compiler has converted statement 1 into the following code:

```
RESOURCE_MGR(allocate,mag_tape,data1,status);
```

where the parameters are as follows:

- *allocate* is the request for a new device.
- *mag_tape* is the class of device requested.
- *data1* is the logical unit reference number for the device.
- *status* is where the result of the request is returned.

The resource manager will check to see if any resources of type *mag_tape* are available. If there is one available, it will be allocated to paycheck by attaching a resource control block to the process's PCB. An entry will also be made in the LUT so that the process can access the tape with future I/O requests. If there is no *mag_tape* that can be assigned, the process will be placed on the blocked queue until a tape drive becomes available.

After the application has successfully acquired the right to access a magnetic tape, statement 2 will be executed. The object code generated for this statement by the compiler will be

```
IOCS(read,data1,input_buffer,length,status);
```

where the parameters are as follows:

- *read* is the I/O command.

```
     program(paycheck,std_out);
     •
     •
     •
 1.  open(mag_tape as data1);
     •
     •
     •
 2.  read(data1,line);
     •
     •
     •
 3.  open(c:employee as data2);
     •
     •
     •
 4.  write(data2,line2);
     •
     •
     •
 5.  open(std_out);
     •
     •
     •
 6.  write(line3);
     •
     •
     •
 7.  close(data2);
     •
     •
     •
 8.  close(data1);
     •
     •
     •
 9.  close(std_out);
     •
     •
     •
10.  end;
```

FIGURE 14.12 Parts of a sample application program.

- *data1* is the logical unit number for the tape unit assigned to the process by the resource manager.
- *input_buffer* is the address of the data buffer.
- *length* is the number of characters to read.
- *status* is where the status of the operation is returned.

IOCS will examine the request to verify the logical unit number and the command and then call the device driver with

DRIVER_INITIATOR(*read*,*data1*,*input_buffer*,*length*);

where all the parameters are defined as above. When control passes back to IOCS, the PCB of the process making the request is moved from execute to I/O suspend. It will stay there until the read operation on the magnetic tape is finished.

When the data has all been transferred from tape to memory, an interrupt will be requested by the magnetic tape controller. The FLIH will be activated and will save the status of the machine. It will then go to the interrupt table to find the address of the driver that is responsible for acknowledging the interrupt. The FLIH will call that driver.

The device driver will check status and do all its necessary housekeeping. It will then call IOCS with a completion request:

IOCS(*completion*,*data1*,*status*);

where the parameters are as follows:

- *completion* tells IOCS an I/O operation has finished.
- *data1* is the device number.
- *status* returns any error conditions.

When IOCS determines that the data transfer happened without error, it will remove the process from the suspend state. It passes the process's PCB to the short-term scheduler with

SHORT_TERM_SCHED(*pcb_pointer*);

where *pcb_pointer* points to the PCB for the process that just finished its I/O request. The short-term scheduler will put the PCB passed to it onto the active list and then order the list. Once the list has been ordered, it is time to call the dispatcher with

DISPATCHER;

The dispatcher needs no parameters. It simply takes the PCB at the head of the active list and places that PCB into the execution state.

The next statement that needs system intervention is the open, statement 3. The following code will have been generated for this statement:

```
FILE_MGR(open,c:employee,data2,read,status);
```

where the parameters are as follows:

- *open* is the file operation requested.
- *c:employee* is the disk drive and name of the file.
- *data2* is the pointer to a block of information that the process will use to reference the file.
- *read* is the type of access requested.
- *status* is the status returned after the request has been processed.

Statement 4 is a request to the system to write a line of data to the disk file. The code for the statement will be

```
FILE_MGR(write,line2,length,data2,status);
```

where the parameters are as follows:

- *write* is the file manager command.
- *line2* is the pointer to the data buffer.
- *length* is the number of bytes that will be transferred.
- *data2* is the pointer to the control block opened for the file.
- *status* is the status of the request.

Since the process accesses the standard output device, *std_out*, that device must also be opened. We will assume the standard output device is the terminal or file where results are sent. The open statement might not appear in the actual program. It might be generated as a result of the program statement that referenced *std_out*. Whether it is called implicitly or explicitly by the program in a statement such as number 5, the following code will be generated:

```
FILE_MGR(open,std_out,write,status);
```

where the parameters are similar to those used in the previous file open statement.

All further writes that do not specify a device will be directed to the *std_out* device. Therefore, statement 6 will be converted to the following by the compiler:

```
IOCS(write,std_out,line3,status);
```

where again the parameters are similar to those in previous IOCS commands.

After the program has finished with the I/O devices, the close commands in statements 7, 8, and 9 are executed to close the files and return the resources to the system. The code-generated statement 7 is

```
FILE_MGR(close,keep,data2,status);
```

where the parameters are as follows:

- *close* is the file manager command.
- *keep* specifies that the file is to be made permanent instead of temporary.

and the other parameters are as defined above. The code for statement 8 is

```
RESOURCE_MGR(return,mag_tape,data1,status);
```

where *return* is the resource manager command and the other parameters are as defined above.

The final statement of the program causes the following code to be generated:

```
LONG_TERM_SCHEDULER(exit);
```

where *exit* requests termination of the executing process. This brings us full circle in our trip through a typical process life cycle.

Up to this point, we have looked at the creation, execution, and termination of a user process. We saw how the user program code was converted to calls to the OS when system services were needed. We followed these calls from the layer-9 process down through the other layers of the system. However, we have yet to examine how the system is first loaded into memory and how it starts execution.

14.7 SYSTEM START-UP—BOOTSTRAPPING THE SYSTEM

The question we now want to address is, "How do you get all those layers started?" Since no two systems start up in exactly the same way, we will again give generic answers to the question.

For any computer to start executing, it must have some instructions to execute. Most systems today have some nonvolatile memory which stores a program that can run when the system is turned on. Often, this is some form of read-only memory that cannot be accidently destroyed by the user. On older machines that did not have a ROM, when power was first turned on, the user:

1. Halted the machine.
2. Loaded the start-up program into the memory by "toggling" the appropriate switches on the front panel.
3. Loaded the starting address into the address register.
4. Then pressed RUN to initialize the system.

Most current machines use an interrupt that is activated when the power is turned on or when the RESET switch is depressed.

The start-up program, or **bootstrap**, will begin loading the system routines into memory. Creating an entire OS from a set of disk files is not a trivial task, and space in read-only memory is often limited. Because of this, starting up the system is usually a multistep operation. The ROM resident bootstrap program itself usually only reads the first few sectors from track 0 of the system disk drive. It then transfers control to this second part of the bootstrap program. This program then continues constructing the OS. This second part of the bootstrap program resides on disk and can be made as large and complex as necessary to load in the remainder of the OS.

There is another function that is done by most start-up programs. The start-up program is usually responsible for running some diagnostic checks on the system. If it detects any failures, it tries to report them to the operator and then halts the computer. Assuming there were no failures, we can get on with building the operating system.

The system loader will first bring in the kernel, memory manager, and the I/O system from disk files. It will then build some tables necessary for the system such as the interrupt table, logical unit table, and memory-mapping tables. The loader will work its way out through the layers of the system from the memory manager to the long-term scheduler.

At this point, the loader has created a system that can execute processes. The loader is now ready to create the processes that are a necessary part of the operating system. In most systems, processes must have "parents," that is, processes that created them. This being the case, one special process must be created that can be the parent of all other processes; let's call this process FIRST.

After the special first process is created, it will use the system facilities to create other necessary system processes. Usually this set includes the following:

- IDLE: the idle process
- REMOVE: the process responsible for removing terminated processes
- INITIATE: the process to select requests from the process creation request list
- LOG_ON: the process to create command interpreter processes for terminals

A scaled-down family tree of processes within the system is shown in Fig.14.13. As can be seen, all processes in the system have a parent except FIRST.

To enable users to access the system from terminals, a command process is created for each terminal by the LOG_ON process. This might sound like a waste of memory, but it isn't if the system's memory management scheme supports code sharing. Once a command interpreter has been created, a terminal user can then access the system through that process.

Later, when a user logs off the system, the LOG_ON process will be notified that one of its children, the command interpreter, has terminated. The LOG_ON process will wait for another logon request, and the cycle starts again.

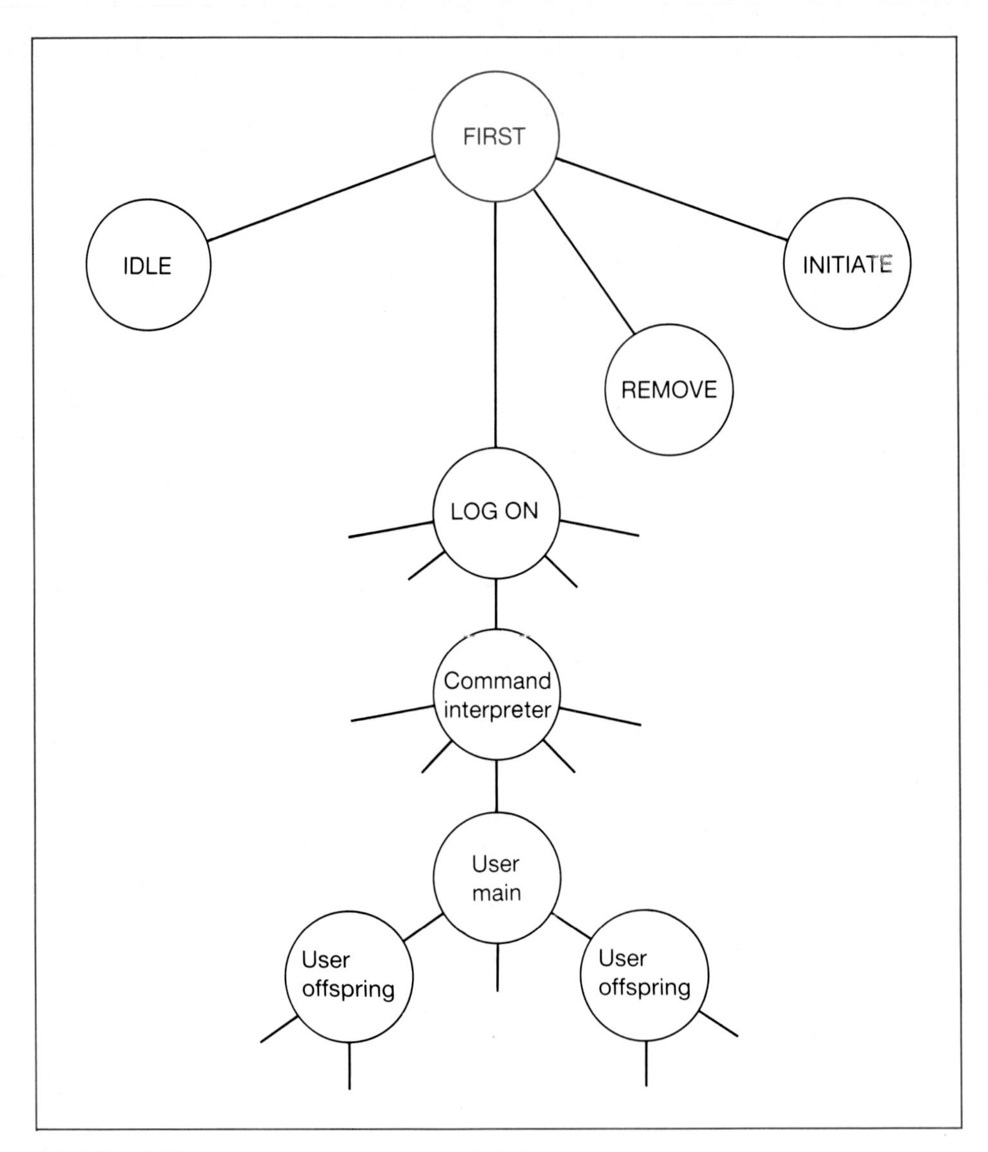

FIGURE 14.13 Family tree of processes within the system.

14.8 SUMMARY

At the beginning of this chapter, we emphasized the idea of an OS designed as a central nucleus controlling a set of concurrent processes. We have now presented a detailed organization for such an operating system. That organization is based on layering. The kernel forms the bottom layer, and successive layers provide additional capabilities: memory management, I/O, files, scheduling, resource allocation, command interpretation, and user interfacing.

Operations within the layered structure were illustrated with a sample user request. We followed the request through the system, looking at the functions performed in various layers and the interfaces between the layers.

Layering is an important concept that is used in a number of other areas of computing, and you will encounter it many more times in your studies. An example that you will often see in the literature is the seven-layer protocol defined by the International Standards Organization (ISO) for the Open Systems Interconnect (OSI) networking standard.

Lastly, we looked at how the system starts operating. The operation of the bootstrap program was described, and we saw how processes were created to service the various users of the system.

PROBLEMS

14.1 Develop a table defining the actions that the SEMAPHORE_CONTROL takes when a semaphore is requested.

14.2 Recall the OS components described in the previous chapter. Which of them might you put in the kernel?

14.3 What precautions must we take if we want an idle process to do file backups?

14.4 What happens if there are no processes in the active state and no specific idle process (such as the one in Prob. 14.3) has been defined?

14.5 What problems could you envision if one or more processes accidently waited on the wrong semaphore or for a semaphore that will never be signaled?

14.6 In Prob. 14.5, how could you tell if a process was waiting for a semaphore that will never be signaled or just waiting on one that takes a long time?

14.7 We said that if there is a process waiting on a semaphore, then the signaling process and the process that had been waiting on the semaphore are both moved to the active state. Why shouldn't the signaler just be allowed to continue executing?

14.8 What problem do you see if the following two processes are allowed to execute concurrently?

```
PROCESS 1            PROCESS 2
  (Other code)         (Other code)
WAIT(SEM_A)          WAIT(SEM_B)
  (Other code)         (Other code)
SIGNAL(SEM_B)        SIGNAL(SEM_A)
  (Other code)         (Other code)
```

14.9 If there is more than one process waiting on a semaphore, which process should be signaled?

14.10 We said that when an interrupt happens, "the interrupt number is compared with the entries in the interrupt table to see if the interrupt was valid." How do you think this is done? Would there be a "subroutine" to do it?

14.11 How could the memory manager detect an address violation? How could it signal the OS that the violation has occurred?

14.12 How does having separate device drivers, rather than incorporating their functions directly into IOCS, make the I/O layer less hardware-dependent?

14.13 We said that some characteristics of disk drives might affect the performance of the file manager. Give an example.

14.14 Why are I/O drivers hardware-dependent?

14.15 How can we ensure that adding a new peripheral device to the I/O system will not affect other layers of the OS?

14.16 Why do we have the short-term scheduler order the active queue, then call the dispatcher? Why not just have the dispatcher do the ordering itself?

14.17 How does the long-term scheduler do its job?

14.18 What makes the resource manager different from the short-term scheduler?

14.19 Why is the design of the command interpreter critical to the success of a system?

14.20 What does the application layer do?

14.21 How can the bootstrap program read in the initial parts of the OS if it doesn't have IOCS, the file manager, etc., loaded yet?

14.22 Recently, one of the authors had to remove and recondition the hard drive from his computer because track 0 failed. That same drive had other bad tracks, but the system just flagged and ignored them. Why was track 0 so important?

14.23 Describe any checks that you see done when you turn on your computer.

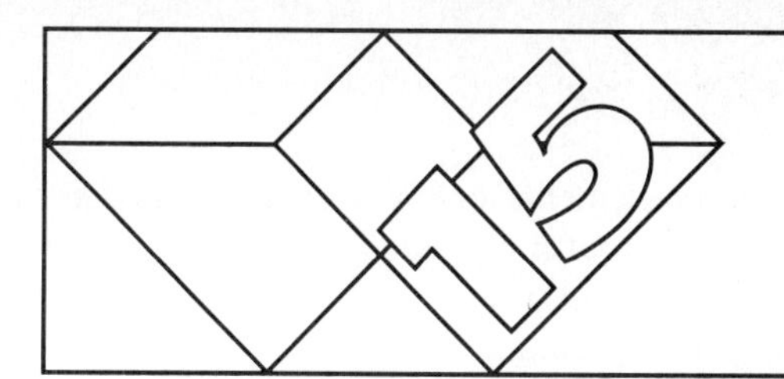

Programming Languages

In the previous three chapters, we looked at operating systems. We saw that a computer user normally communicates with an operating system via a command interpreter. Three standard formats for command interpreter interfaces are the menu form, the icon form, and the command line form.

Let's consider the command line form, in which we have monitor commands such as

```
COPY a:\accounting\master.payroll   b:\accounting\backup.payroll
```

(This particular command tells the OS to copy a file from drive a to drive b, renaming it from master to backup.) The set of all command lines for the interface of a given operating system comprises a *command language* interface on that operating system.

In this chapter, we will discuss another class of computer languages—programming languages. Our main goals will be to trace some of the major developments in programming languages and to cite some of the main differences among the languages you might be using. We will illustrate several of the languages with sample programs.

We also want to help you understand some of the many terms you will see associated with programming languages. These include translators, compilers, interpreters, and assemblers. We begin the chapter with our discussion of this terminology.

15.1 LANGUAGE TRANSLATORS

Many of you have written programs in a high-level language such as Fortran, Pascal, C, or Ada. Unfortunately, the statements in such programs *cannot* usually be executed directly by a computer. One or more elements of the computer's system software must first be invoked to convert the high-level language statements into a different language that the computer *can* execute directly.

The language a given computer understands is called the computer's **machine language**, **native language**, or **machine code**, and it varies from computer to computer. The software that converts from a higher-level language to machine language is called, not surprisingly, a **language translator**. The language that is input to a translator is called the **source language**, and the output of the translator is called the **target language**.

Translator is a generic name that applies to a number of different translation programs. We often use the more specific terms *assembler* and *compiler*. An **assembler** translates from **assembly language** (a symbolic version of the computer's machine language) to the corresponding machine language, which the computer can execute directly. This process is shown in Fig. 15.1.

A **compiler** translates from a high-level, human-oriented language such as Pascal or Fortran to a low-level, machine-oriented language. The low-level language generated by a compiler might be a given computer's machine language or it might be that computer's assembly language. In the latter case, the compiler's output would subsequently be sent to the assembler for another translation. These two alternatives are illustrated in Fig. 15.2.

A compiler might also generate a more generic **intermediate language** not corresponding directly to any particular machine. Such an approach saves compiler writers from having to redo their translation for every different machine language. In this case, the intermediate language is subsequently translated further to obtain the corresponding machine language. One possible scenario is shown in Fig. 15.3.

As an example of a compilation to assembly language, consider the following generic high-level statement, which replaces an accumulating sum with its next incremental value:

```
S ← (T/L) + S; {Increment sum by next term in the series expansion.}
```

(The sentence enclosed in braces is a comment that is ignored by the translator.) This statement would be translated into a sequence of assembly-language statements like those shown in Fig. 15.4. (In this figure we have added the comments to help you understand the program; the compiler would not normally comment its own assembly-language output.)

FIGURE 15.1 Translation by an assembler.

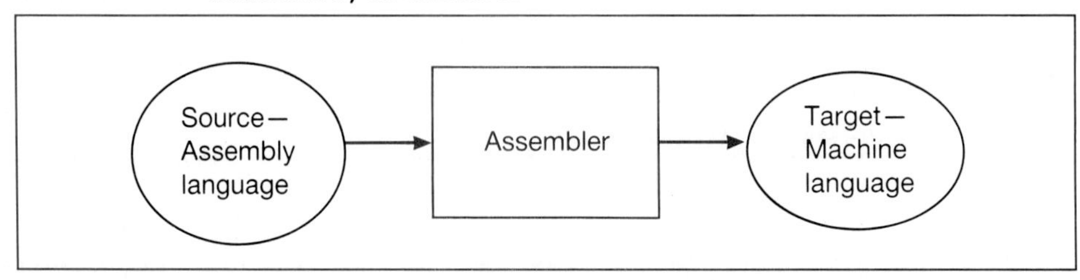

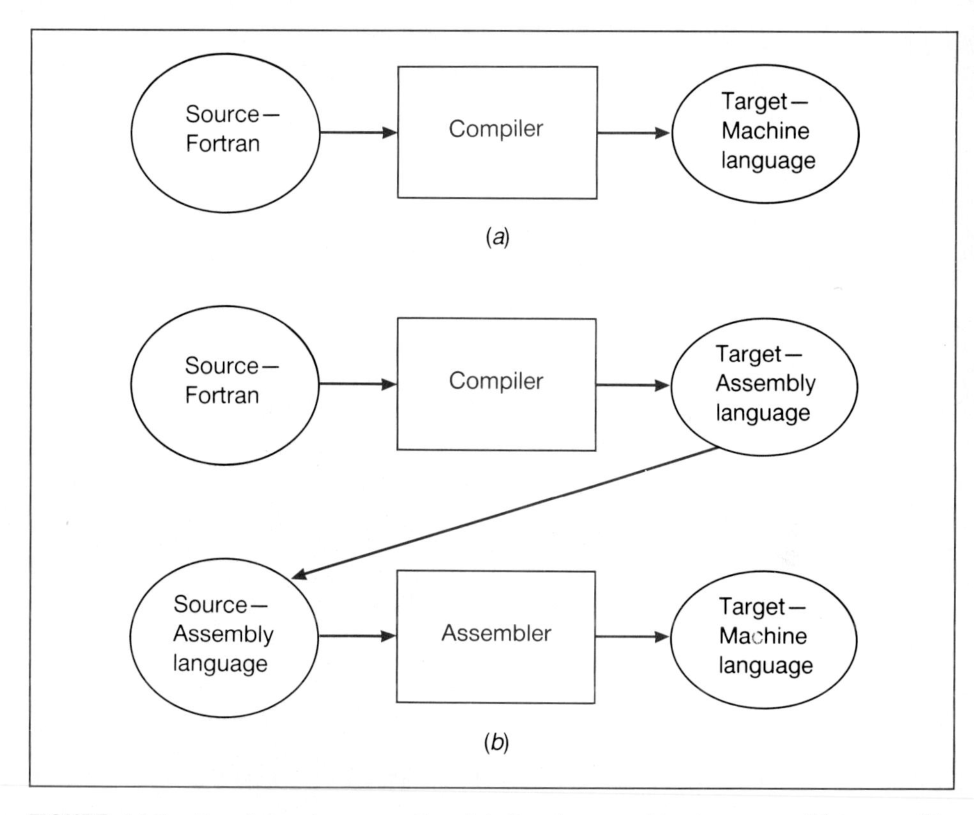

FIGURE 15.2 Translation by a compiler: (*a*) directly to machine language; (*b*) to assembly language and then to machine language.

The assembly statements in Fig. 15.4 might in turn be translated into the machine-language statements shown in Fig. 15.5. (We have included the corresponding assembly code as comments.)

These machine-language statements contain the opcode and operand address fields as described extensively in Part II of this text. These instructions can be directly executed by the hardware.

15.2 COMPILED VERSUS INTERPRETED PROGRAM EXECUTION

The aim of the language translator is to produce code that can be executed efficiently on the CPU. However, sometimes system designers are also concerned with helping the user develop his or her program, even if it means that the program will execute very slowly during development. This is especially true if the user is a novice and makes numerous errors that must be found and corrected.

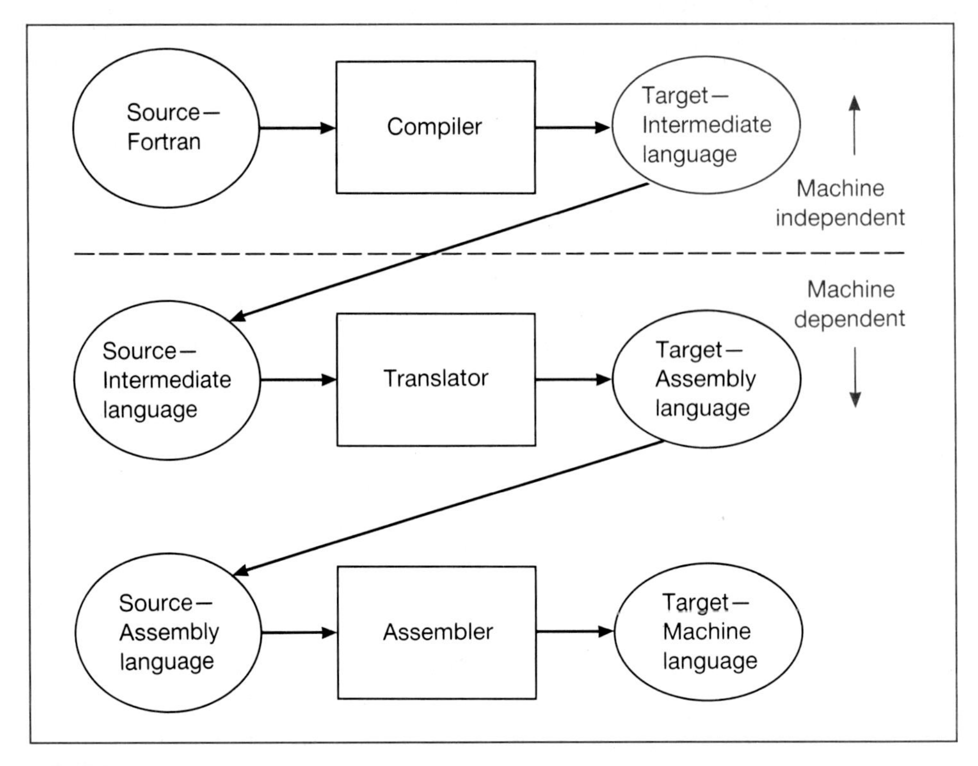

FIGURE 15.3 Translation to an intermediate language by a compiler, followed by two subsequent translations to obtain machine language.

```
MOV  AX, T ;     MOVE T INTO ACCUMULATOR AX
CWD        ;     EXTEND SIGN OF T INTO DX
IDIV L     ;     DIVIDE DX:AX BY L
ADD  AX, S ;     ADD S TO QUOTIENT IN AX
MOV  S, AX ;     STORE RESULT INTO S
```

FIGURE 15.4 Assembly version of sample high-level statement.

In such a case, a designer might create an *interpreter* instead of a compiler. An **interpreter** does not produce target code but rather takes the statements in the source language and does what they specify. That is, it *interprets* them. This operation is done under the interpreter's control, slowly, but with many error checks. For example, a programmer might have written the following statement:

```
distance := sqrt((x[i] − x[j])↑2 + (y[i] − y[j])↑2);
```

HEX ADDRESS	CONTENTS	INSTRUCTION
0000	A10002	MOV AX, T
0003	99	CWD
0004	F73E0004	IDIV L
0008	03030000	ADD AX, S
000C	A30000	MOV S, AX

FIGURE 15.5 Machine-language statements corresponding to sample high-level statement.

An interpreter could do the specified operations, while checking such things as:

- Have the variables x, y, i, and j been properly declared?
- Are i and j the right type variables for array indices?
- Are i and j within the range specified for x and y indices?
- Are x and y the right type for the function sqrt?
- Have all the variables referenced on the right of := been defined?

If any errors are found, the interpreter can inform the user and allow him or her to correct the error interactively. The corrected statement can then be reinterpreted immediately, as opposed to recompiling the entire program and then reexecuting it.

Some of you might have used an interpreted language, without knowing it. BASIC is often run with an interpreter rather than a compiler. Some other high-level languages also have interpreted versions to help with learning and debugging.

Compilers could add code in the translation process to do many of the same checks during execution that an interpreter performs during interpretation. However, this would greatly slow program execution. Therefore, system software often includes both interpreters and compilers. Compilers are good for generating efficient machine code. Interpreters are good for developing the code quickly; the tradeoff is execution speed. Compiled code executes more rapidly than interpreted code.

15.3 EARLY COMPUTER PROGRAMS

As we described in Chap. 1, early computers were controlled by external devices, such as plugboards. To "program" such a computer, you rewired that plugboard. Then came von Neumann and the idea of internal stored-program computers. Earlier in this chapter we described machine languages, and naturally an initial approach to programming a stored-program computer was to use its machine language.

We can speak from considerable experience in saying that machine-language programming is, in general, not fun. To give you some flavor of it, let's consider the

HEX ADDRESS	CONTENTS
0000	50
0001	06
0002	55
0003	B80010
0006	8EC0
0008	8BEC
000A	26A00A00
000E	256000
0011	3D6000
0014	75F4
0016	8B4608
0019	26A20000
001D	5D
001E	07
001F	58
0020	C3

FIGURE 15.6 Sample machine-language program.

machine language for the Intel 8086. A program to write a character to the screen is shown in Fig. 15.6.

We will not go into detail about how this program functions (although it should become clearer in the next section). Our main goal is to illustrate what actual machine code looks like.

Machine language is numeric, rather than symbolic, and so it is more difficult to comprehend (if you are a human, and not a computer). A programmer usually has to write many more instructions to implement an algorithm in machine language than in a high-level language. In machine language we are normally forced to use *numeric addresses* that depend on the *entire* program as a unit. The addition of new instructions will change the locations of all subsequent instructions and thus require changes to all addresses that reference these locations. For a large machine-language program, such changes can require an enormous effort.

Nowadays, very little programming is done in machine language. Probably the main reason for knowing about it is to debug low-level programs. If you have an extremely tough debugging problem, you might need to dump areas of memory and study the machine code to see what is happening at the lowest level. Even when you hear someone say a module or part of a module was done in "machine language," they probably do not mean that literally. The actual coding was probably done in a language that is the next topic of our progression—assembly language.

15.4 ASSEMBLY LANGUAGE

Assembly language might have been one of the earliest examples of a popular philosophy in computing, "We'll make the computer do more of the grunt work so that we can make life easier for the programmer." Assembly language is very close to machine language, but it is a more comprehensible, easier-to-use language in which we can employ symbolic representations rather than strict numeric coding. It also eliminates some of the problems of coding in machine languages, like the address changes described in the previous section.

Let's illustrate these ideas with the assembly language introduced at the beginning of the chapter. With this assembly language, we can use more human-oriented names for the operation codes rather than numbers. We can also use variable names rather than memory locations. An assembly program to write a character to the screen is shown in Fig. 15.7.

Once we have entered this program, we can have the assembler translate it into corresponding machine-language code. Such a translation is illustrated in Fig. 15.8.

In an earlier section, we gave you a machine-language program without explaining its operation. However, we did promise to clarify it for you later. That program is the machine language produced in Fig. 15.8.

Most assemblers have many more features than we could demonstrate with this simple example. Just to name a few, and give corresponding illustrations, most assemblers:

- Have statements to allocate blocks of memory:

```
BLOCK DEFSTORG 1000
```

- Have statements to define various constants rather than requiring the programmer to specify the corresponding hex code:

```
INTEGR1   INT    -2345
REAL1     REAL   7.89
STRN1     ASCI   "THIS IS A STRING"
```

- Allow symbolic names for registers, comparison flags, and other things that are specified numerically in machine code:

```
FLAGNN   EQU   1100B   {FLAGS FOR "NOT NEG"}
CNTR     EQU   7       {ACC 7 IS COUNTER}
```

- Allow address expressions:

```
LOAD ARRAY_STRT + 4
```

- Have symbols for quantities such as the current value of the program counter:

```
ENTER   *-1   { * IS CURRENT PC}
```

- Allow the use of parameterization so that a user can edit a few lines at the beginning of the program, assemble it, and then have a custom program

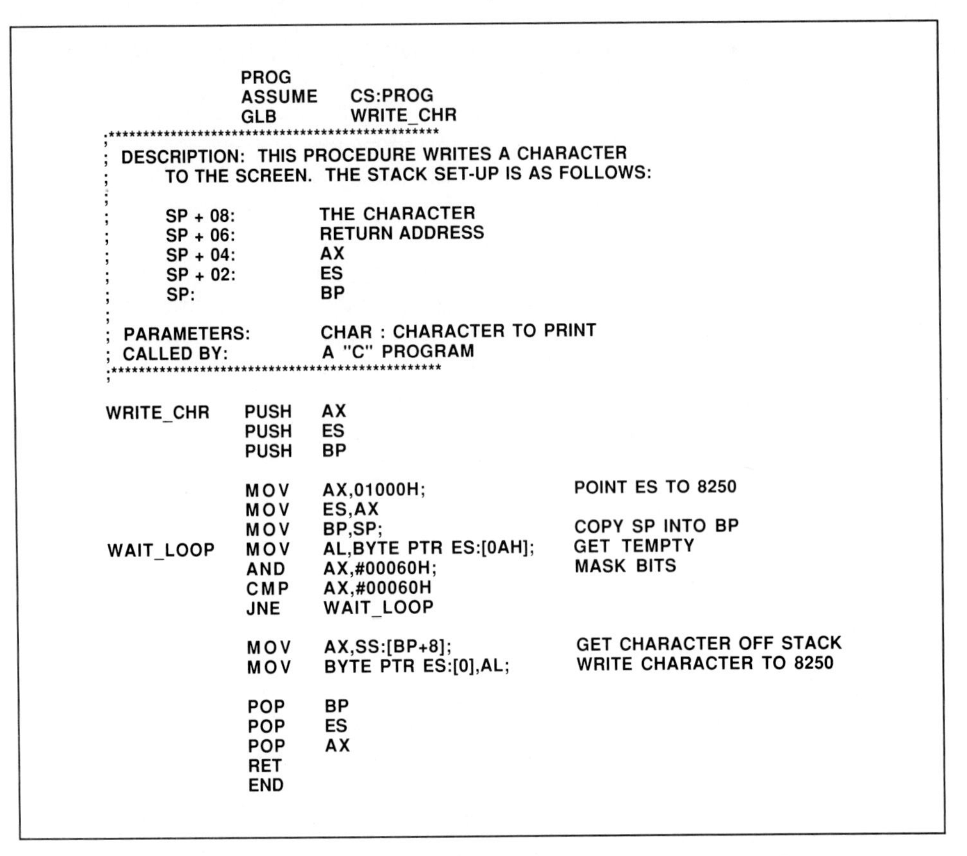

```
              PROG
              ASSUME    CS:PROG
              GLB       WRITE_CHR
;*********************************************
;
; DESCRIPTION:  THIS PROCEDURE WRITES A CHARACTER
;      TO THE SCREEN.  THE STACK SET-UP IS AS FOLLOWS:
;
;      SP + 08:          THE CHARACTER
;      SP + 06:          RETURN ADDRESS
;      SP + 04:          AX
;      SP + 02:          ES
;      SP:               BP
;
; PARAMETERS:            CHAR : CHARACTER TO PRINT
; CALLED BY:             A "C" PROGRAM
;*********************************************

WRITE_CHR     PUSH   AX
              PUSH   ES
              PUSH   BP

              MOV    AX,01000H;                  POINT ES TO 8250
              MOV    ES,AX
              MOV    BP,SP;                      COPY SP INTO BP
WAIT_LOOP     MOV    AL,BYTE PTR ES:[0AH];       GET TEMPTY
              AND    AX,#00060H;                 MASK BITS
              CMP    AX,#00060H
              JNE    WAIT_LOOP

              MOV    AX,SS:[BP+8];               GET CHARACTER OFF STACK
              MOV    BYTE PTR ES:[0],AL;         WRITE CHARACTER TO 8250

              POP    BP
              POP    ES
              POP    AX
              RET
              END
```

FIGURE 15.7 Sample assembly-language program—writing a character to the screen.

```
ADDRESS    CONTENTS

0000       50                WRITE_CHR   PUSH    AX
0001       06                            PUSH    ES
0002       55                            PUSH    BP
0003       B80010                        MOV     AX,01000H
0006       8EC0                          MOV     ES,AX
0008       8BEC                          MOV     BP,SP
000A       26A00A00          WAIT_LOOP   MOV     AL,BYTE PTR ES:[0AH]
000E       256000                        AND     AX,#00060H
0011       3D6000                        CMP     AX,#00060H
0014       75F4                          JNE     WAIT_LOOP
0016       8B4608                        MOV     AX,SS:[BP+8]
0019       26A20000                      MOV     BYTE PTR ES:[0], AL
001D       5D                            POP     BP
001E       07                            POP     ES
001F       58                            POP     AX
0020       C3                           RET
```

FIGURE 15.8 Translation of assembly program in Fig. 15.7.

tailored for his or her system:

```
TAPES   PARM  5    {NUMBER OF TAPES}
DISKS   PARM  2    {NUMBER OF DISKS}
MEMSZ   PARM  64   {MEMORY SIZE, IN KB}
```

The assembly statements described above are called **pseudoinstructions**, because they give information to the assembler but do not directly generate any machine code.

This concludes our discussion of machine and assembly languages. We now move on to languages with which you are probably more familiar, high-level programming languages.

15.5 FORTRAN

Scientific computation was one of the earliest uses of computers. Programs written for these applications often had many needs in common:

- The need to store and use numeric data in scientific notation
- The need to store and access information in arrays
- The need for a variety of complex input and output formats
- The need for efficient scientific formula expressions

You might think these features have always been available, but this was not the case prior to 1956. Scientists and engineers wanting to solve equations and display the results in a simple, readable form were forced to write hundreds or thousands of lines of complex assembly-language code.

In the summer of 1954, a project was begun at IBM to create a new environment for scientific programmers. The goal of the project "was to reduce by a large factor the task of preparing scientific problems for IBM's next large computer, the 704" [Backus et al., 1967]. The name given the project was Fortran, short for *for*mula *tran*slation. Upon completion $2\frac{1}{2}$ years later, the scientific programming community was given a language that significantly decreased the time needed to develop and maintain programs.

Fortran has undergone many changes since its first introduction, and several newer languages are now replacing it in some areas. In spite of this, it is still the language of choice for many scientists and engineers. There were numerous early versions of Fortran, such as Fortran II and Fortran IV. The major versions to note are the ANSI Standard Fortrans adopted in 1966 and 1977.

To illustrate some of the features of Fortran, we will present and discuss a small example program (Fig. 15.10). This program implements an algorithm to compute the natural logarithm of a number, as specified by the flowchart shown in Fig. 15.9.

Please note carefully that in our illustration the Fortran program is set in black type. The red numbers to the left in Fig. 15.10 are *not* part of the code; they have

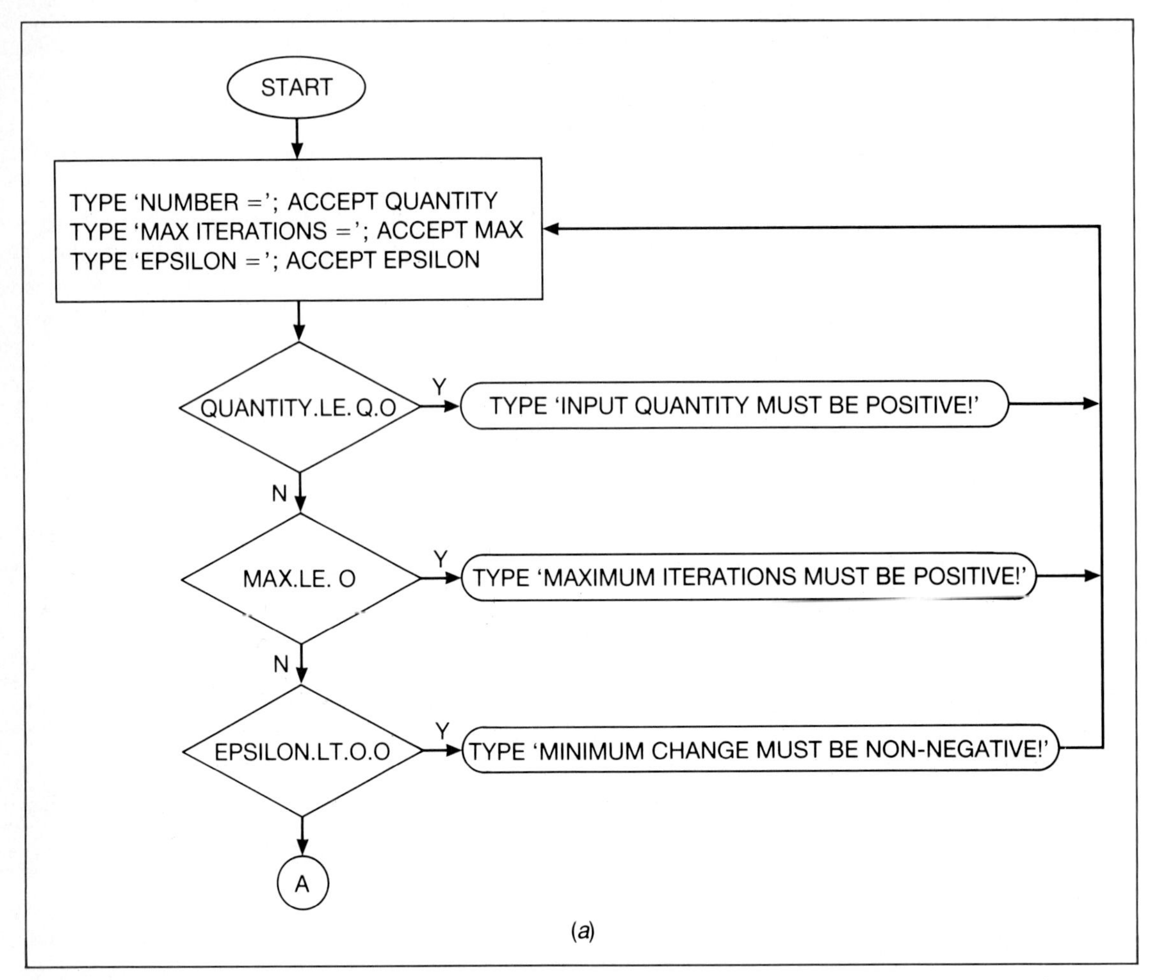

FIGURE 15.9 Flowchart for natural logarithm program: (*a*) input section; (*b*) initialization section; (*c*) loop to compute logarithm.

been added to simplify discussion of the program. This same convention will also be followed in later coding examples.

The following points refer to the red numbers in our Fortran illustration:

1. Comments are specified by putting a C in the first column of the line.

2. Variables are explicitly declared by giving a type specification and then listing variables which are of that type. We have illustrated REAL and INTEGER; there are also types such as COMPLEX and CHARACTER.

3, 5. Fortran allows easy specification of input and output. We have shown stream-directed I/O on the standard input and output units. A programmer can also specify different devices and specify the format of the data.

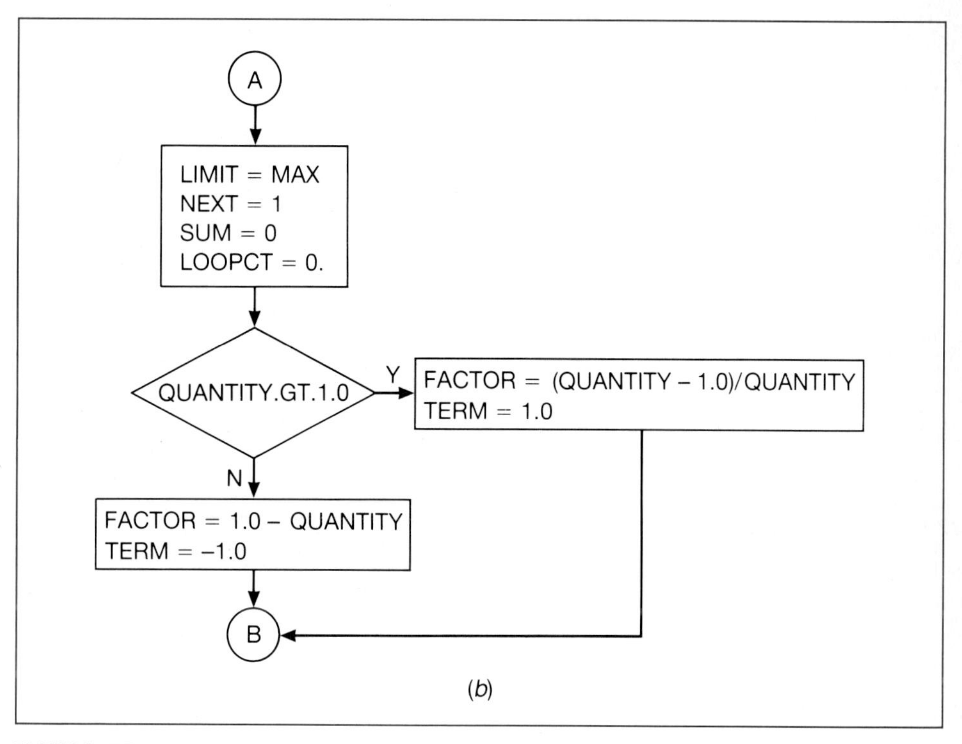

FIGURE 15.9 *(Cont.)*

4. This DO-REPEAT is one form of looping. The DO matches a later REPEAT, and the code between these two statements is repeated until an EXIT statement in the loop is executed.

6, 7. Conditional statements can be done with an IF-THEN-ELSE construct.

8. This statement is perhaps a prototype illustration of why Fortran was developed. We used this same expression in our discussion of machine and assembly language, and it took five instructions of relatively obtuse code. This statement is much more human-oriented.

There are other important features of Fortran that we do not have space to discuss. They include facilities to manipulate arrays of information rather than single variables and facilities to write independent functions and subroutines that can be called by programs.

15.6 COBOL

The other major application for computers in the 1950s was commercial data processing. One of the largest users of electronic data processing was and still is the

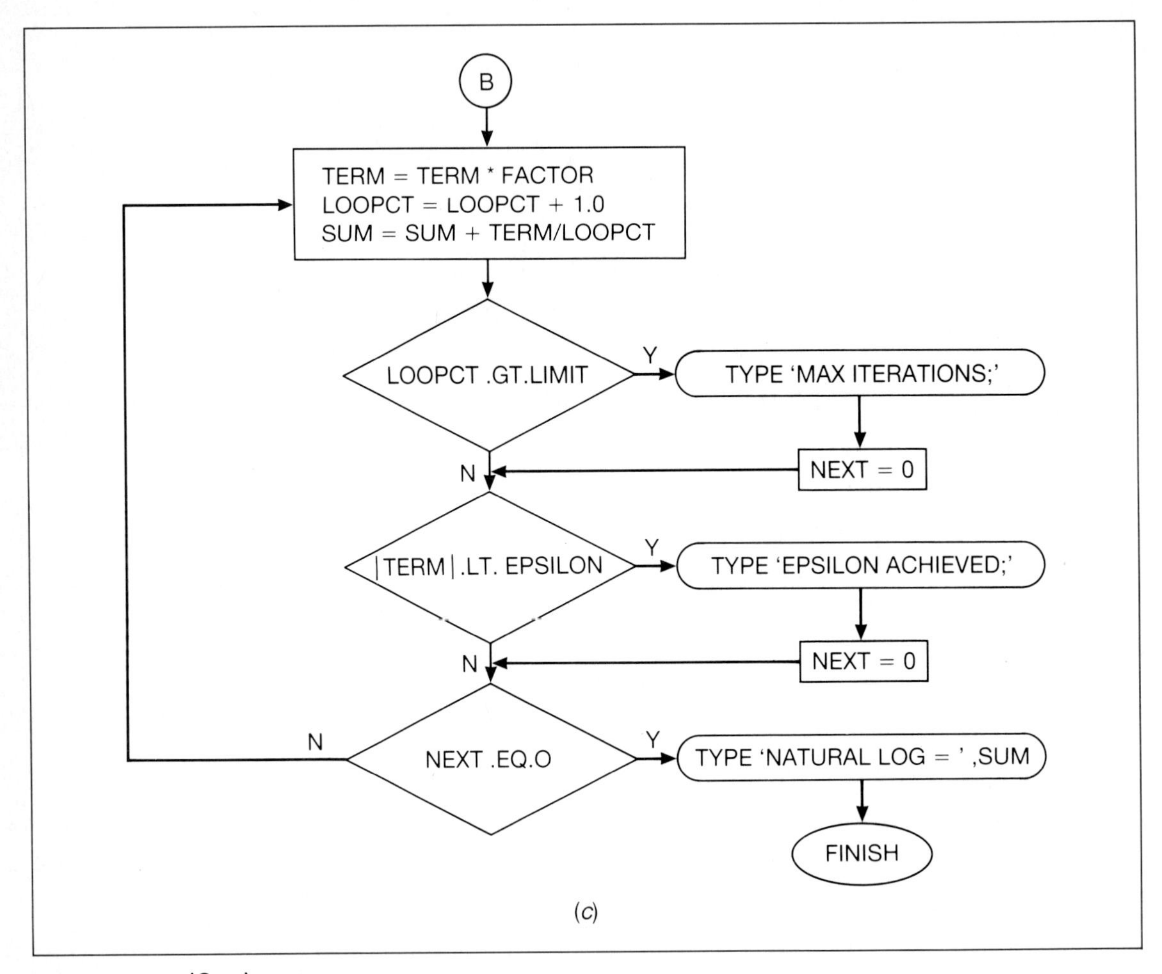

FIGURE 15.9 *(Cont.)*

U.S. government. "In May 1959, a meeting was called in the Pentagon for the purpose of considering the desirability and feasibility of establishing a common language for the adaptation of electronic computers to data processing" [Sammet, 1967]. In August 1961, the government published a document that defined Cobol, an acronym for *co*mmon, *b*usiness-*o*riented *l*anguage.

Whereas Fortran was designed to simplify the development of scientific programs, Cobol was designed to meet similar needs of the data processing community. Its strengths are in the following areas:

- Storing and manipulating financial data
- Accessing data files in secondary storage
- Manipulating nonnumeric data
- Generating reports

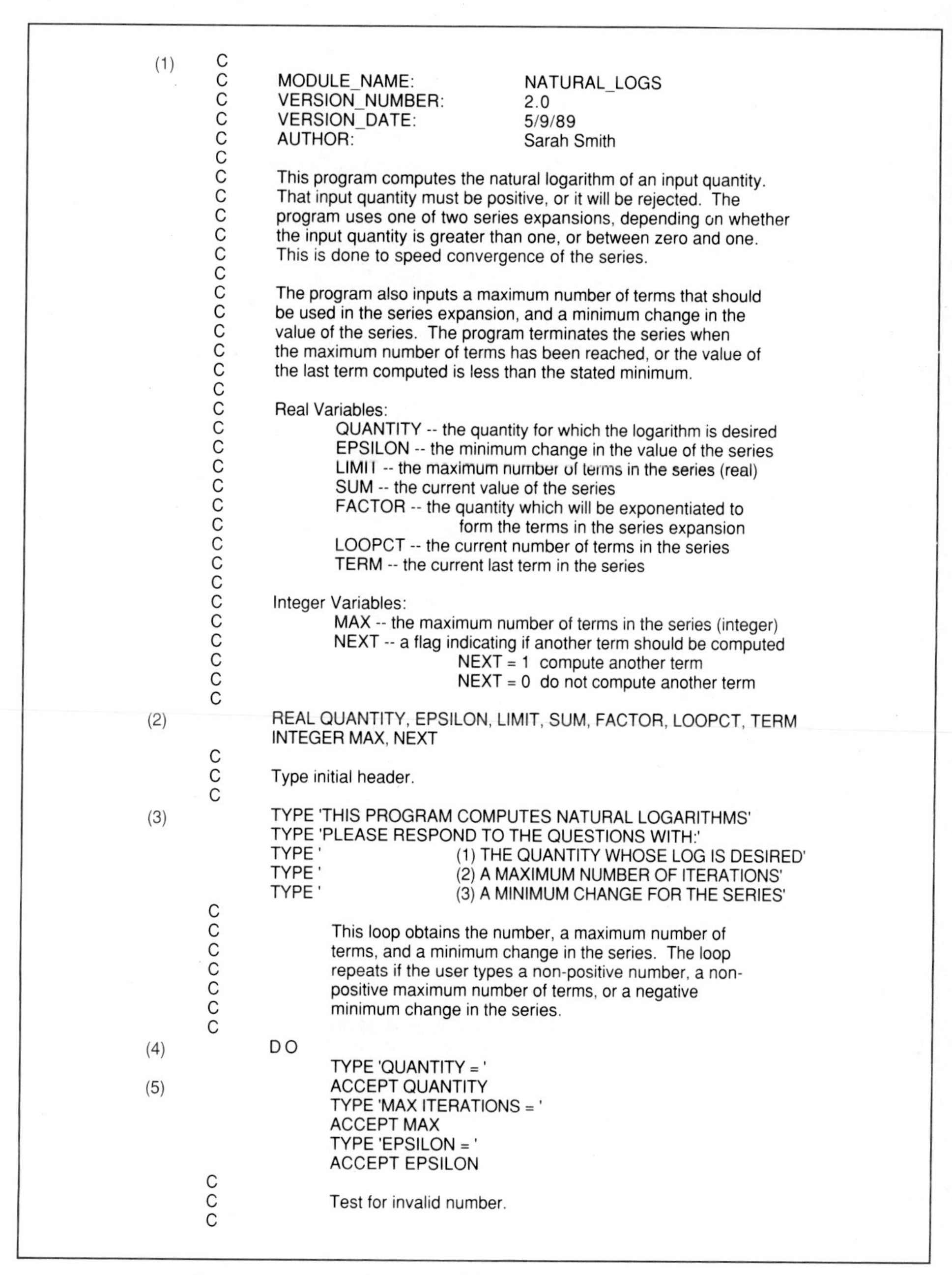

```
(1)   C
      C     MODULE_NAME:              NATURAL_LOGS
      C     VERSION_NUMBER:           2.0
      C     VERSION_DATE:             5/9/89
      C     AUTHOR:                   Sarah Smith
      C
      C     This program computes the natural logarithm of an input quantity.
      C     That input quantity must be positive, or it will be rejected. The
      C     program uses one of two series expansions, depending on whether
      C     the input quantity is greater than one, or between zero and one.
      C     This is done to speed convergence of the series.
      C
      C     The program also inputs a maximum number of terms that should
      C     be used in the series expansion, and a minimum change in the
      C     value of the series. The program terminates the series when
      C     the maximum number of terms has been reached, or the value of
      C     the last term computed is less than the stated minimum.
      C
      C     Real Variables:
      C           QUANTITY -- the quantity for which the logarithm is desired
      C           EPSILON -- the minimum change in the value of the series
      C           LIMIT -- the maximum number of terms in the series (real)
      C           SUM -- the current value of the series
      C           FACTOR -- the quantity which will be exponentiated to
      C                       form the terms in the series expansion
      C           LOOPCT -- the current number of terms in the series
      C           TERM -- the current last term in the series
      C
      C     Integer Variables:
      C           MAX -- the maximum number of terms in the series (integer)
      C           NEXT -- a flag indicating if another term should be computed
      C                       NEXT = 1  compute another term
      C                       NEXT = 0  do not compute another term
      C
(2)         REAL QUANTITY, EPSILON, LIMIT, SUM, FACTOR, LOOPCT, TERM
            INTEGER MAX, NEXT
      C
      C     Type initial header.
      C
(3)         TYPE 'THIS PROGRAM COMPUTES NATURAL LOGARITHMS'
            TYPE 'PLEASE RESPOND TO THE QUESTIONS WITH:'
            TYPE '                  (1) THE QUANTITY WHOSE LOG IS DESIRED'
            TYPE '                  (2) A MAXIMUM NUMBER OF ITERATIONS'
            TYPE '                  (3) A MINIMUM CHANGE FOR THE SERIES'
      C
      C           This loop obtains the number, a maximum number of
      C           terms, and a minimum change in the series. The loop
      C           repeats if the user types a non-positive number, a non-
      C           positive maximum number of terms, or a negative
      C           minimum change in the series.
      C
(4)         DO
                  TYPE 'QUANTITY = '
(5)               ACCEPT QUANTITY
                  TYPE 'MAX ITERATIONS = '
                  ACCEPT MAX
                  TYPE 'EPSILON = '
                  ACCEPT EPSILON
      C
      C           Test for invalid number.
      C
```

FIGURE 15.10 Fortran program for natural logarithm: (*a*) program listing; (*b*) two sample runs.

```
(6)                 IF (QUANTITY .LE. 0.0) THEN
                          TYPE 'NUMBER MUST BE POSITIVE!'
          C
          C         Test for invalid number of iterations.
          C
(7)                 ELSEIF (MAX .LE. 0) THEN
                          TYPE 'MAXIMUM ITERATIONS MUST BE POSITIVE!'
          C
          C         Test for invalid minimum change in series.
          C
                    ELSEIF (EPSILON .LT. 0.0) THEN
                          TYPE 'MINIMUM CHANGE MUST BE NON-NEGATIVE!'
                    ELSE
                          EXIT
                    ENDIF
              REPEAT
          C
          C   The maximum number of iterations is input as an integer,
          C   which is more natural for the user. Here it is converted to
          C   a real to save repeated conversions in a later IF test.
          C
              LIMIT = MAX
          C
          C   Assume that there will be another term computed.
          C
              NEXT = 1
          C
          C   Initialize the series summation and the counter of terms.
          C
              SUM = 0.
              LOOPCT = 0.

          C
          C   Decide which series to use.
          C
              IF (QUANTITY .GT. 1.0) THEN
          C
          C         Series for number greater than 1. The jth term in this
          C         series is ((Number - 1.0) j ) / ( j*(Number) j )

                    FACTOR = (QUANTITY - 1.0)/QUANTITY
                    TERM = 1.0
              ELSE
          C
          C         Series for positive number less than 1. The jth term in this
          C         series is ((-1) j * (1.0 - Number) j ) / j
          C
                    FACTOR = 1.0 - QUANTITY
                    TERM = -1.0
              ENDIF
          C
          C   Loop to compute the series. Loop terminates when the
          C   maximum number of terms have been used, or the size
          C   of the last term is less than the specified minimum.
          C
              DO
          C
          C         Compute term by using repeated multiplication of
          C         previous term, rather than raising to power each time.
          C
                    TERM = TERM * FACTOR
          C
Add one to counter, and add term to series summation.
```

FIGURE 15.10 *(Cont.)*

```
      C
                  LOOPCT = LOOPCT + 1.0
(8)               SUM = SUM + TERM/LOOPCT
      C
      C           Check for loop termination conditions. First one is
      C           maximum number of terms. Flag is changed to zero
      C           to indicate subsequent exit from loop.
      C
                  IF (LOOPCT .GT. LIMIT) THEN
                        TYPE 'MAX ITERATIONS; '
                        NEXT = 0
                  ENDIF
      C
      C           Second termination condition is less than minimum
      C           change in value of approximation.
      C
                  IF (ABS(TERM) .LT. EPSILON) THEN
                        TYPE 'EPSILON ACHIEVED; '
                        NEXT = 0
                  ENDIF
      C
      C           Check flag for possible loop exit.
      C
                  IF (NEXT .EQ. 0) EXIT
            REPEAT
      C
      C     Approximation completed. Type result.
      C
            TYPE 'NATURAL LOG = ',SUM
            STOP
            END
```

(a)

FIGURE 15.10 *(Cont.)*

```
                                   11 20
NUMBER = 20
MAX ITERATIONS = 100
EPSILON = .001
MAX ITERATIONS; NATURAL LOG =  2.990

NUMBER = 20
MAX ITERATIONS = 200
EPSILON = .001
EPSILON ACHIEVED; NATURAL LOG =  2.996
```

(b)

FIGURE 15.10 *(Cont.)*

```
IDENTIFICATION DIVISION.
PROGRAM-ID.
        UPDATE-ACCOUNTS-RECEIVABLE.
AUTHOR.
        SALLY SMITH.
INSTALLATION.
        CSU-CHICO.
DATE-WRITTEN.
        NOVEMBER 9, 1989.
DATE-COMPILED.
        DECEMBER 20, 1989.
SECURITY.
        UNCLASSIFIED.
REMARKS.
        THIS PROGRAM INPUTS A MASTER FILE CONTAINING CURRENT
        RECEIVABLES AND A TRANSACTION FILE CONTAINING PURCHASES
        THIS PERIOD. (NOTE -- EVERY MASTER FILE ENTRY MUST HAVE A
        CORRESPONDING TRANSACTION FILE ENTRY, EVEN IF PURCHASES
        ARE ZERO!) THE PROGRAM PRODUCES A NEW MASTER FILE WITH
        UPDATED RECEIVABLES BALANCES. IT ALSO DISPLAYS THE
        CUSTOMER NUMBER OF ANY CUSTOMER WHO HAS GONE OVER
        HIS/HER CREDIT CHECK POINT, AND MARKS A FLAG IN THE NEW
        CUSTOMER RECORD TO INDICATE THIS FACT.
ENVIRONMENT DIVISION.
CONFIGURATION SECTION.
SOURCE-COMPUTER.
        WXY-440-D30.
OBJECT-COMPUTER.
        WXY-440-D30.
INPUT-OUTPUT SECTION.
        SELECT TRANSACTIONS-THIS-PERIOD ASSIGN TO UT-1234-S-TRANSACT.
        SELECT OLD-MASTER-FILE ASSIGN TO UT-1234-S-OLDMASTR.
        SELECT NEW-MASTER-FILE ASSIGN TO UT-1234-S-NEWMASTR.
DATA DIVISION.
FILE SECTION.
FD      OLD-MASTER-FILE
        BLOCK CONTAINS 20 RECORDS,
        LABEL RECORDS ARE OMITTED.
01      OLD-CUSTOMER-RECORD.
        02    CUSTOMER-NUMBER              PICTURE 9(6).
        02    CUSTOMER-NAME                PICTURE X(20).
        02    CUSTOMER-ADDRESS             PICTURE X(30).
        02    PREVIOUS-BALANCE             PICTURE 99999V99.
        02    CREDIT-CHECK-POINT           PICTURE 99999V99.
        02    CREDIT-CHECK-FLAG            PICTURE X.
FD      TRANSACTIONS-THIS-PERIOD
        BLOCK CONTAINS 120 RECORDS,
        LABEL RECORDS ARE OMITTED.
01      TRANSACTION-INFORMATION.
        02    CUSTOMER-NUMBER              PICTURE 9(6).
        02    PURCHASES-THIS-PERIOD        PICTURE 9999V99.
FD      NEW-MASTER-FILE
        BLOCK CONTAINS 20 RECORDS,
        LABEL RECORDS ARE OMITTED.
```

FIGURE 15.11 Cobol program to update an accounts receivable file: (*a*) program listing; (*b*) sample run.

```
01     NEW-CUSTOMER-RECORD.
       02     CUSTOMER-NUMBER              PICTURE 9(6).
       02     CUSTOMER-NAME                PICTURE X(20).
       02     CUSTOMER-ADDRESS             PICTURE X(30).
       02     NEW-BALANCE                  PICTURE 99999V99.
       02     CREDIT-CHECK-POINT           PICTURE 99999V99.
       02     CREDIT-CHECK-FLAG            PICTURE X.
PROCEDURE DIVISION.
OPEN-FILES-AND-PRINT-HEADER.
       OPEN INPUT OLD-MASTER-FILE, TRANSACTIONS-THIS-PERIOD,
              OUTPUT NEW-MASTER-FILE.
       DISPLAY 'LIST OF CUSTOMERS OVER CREDIT CHECK POINTS'.
PROCESS-CUSTOMER-RECORDS.
       READ OLD-MASTER-FILE AT END GO TO
       END-OF-RECORD-PROCESSING.
       READ TRANSACTIONS-THIS-PERIOD.
       MOVE CORRESPONDING OLD-CUSTOMER-RECORD TO
              NEW-CUSTOMER-RECORD.
       ADD PURCHASES-THIS-PERIOD TO PREVIOUS-BALANCE,
              GIVING NEW-BALANCE.
       IF NEW-BALANCE IS GREATER THAN CREDIT-CHECK-POINT
              OF NEW-CUSTOMER-RECORD,
                     DISPLAY CUSTOMER-NUMBER OF
                            NEW-CUSTOMER-RECORD,
                     MOVE 'Y' TO CREDIT-CHECK-FLAG.
       WRITE NEW-CUSTOMER-RECORD.
       GO TO PROCESS-CUSTOMER-RECORDS.
END-OF-RECORD-PROCESSING.
       DISPLAY 'END OF CUSTOMER LIST'.
       CLOSE OLD-MASTER-FILE, TRANSACTIONS-THIS-PERIOD,
              NEW-MASTER-FILE.
       DISPLAY 'END OF TRANSACTION PROCESSING'.
       STOP RUN.
```

(*a*)

```
LIST OF CUSTOMERS OVER CREDIT CHECK POINTS
123123
234234
END OF CUSTOMER LIST
END OF TRANSACTION PROCESSING
```

(*b*)

FIGURE 15.11 (*Cont.*)

Like Fortran, Cobol has undergone many changes and improvements in the years since its introduction. Because it has evolved and its use was required on many government projects, Cobol remains popular with many data processing programmers.

One of the major goals of the development effort was to make Cobol code itself readable and easily understandable. Some people refer to this characteristic as *self-*

documenting code, with the implication that the code itself provides an important part of the documentation. For example, a Cobol statement can be

```
MULTIPLY HOURS–WORKED BY PAY–RATE GIVING GROSS–PAY
```

You can see how this statement indicates quite clearly what it is doing.

The designers of Cobol used a considerable amount of terminology from the English language. Procedural statements are called sentences, and the first word of every sentence is a verb. This verb is imperative, telling the computer what action to perform. By having the verb first in the sentence, the job of the compiler is simplified. Verbs are also examples of *reserved words*, which cannot be used as data names by the programmer. This also simplifies compilation and can eliminate some potentially confusing constructs.

Cobol formalized the notion of *declarations*. These are statements that do not indicate execution-time processing but rather give data formats and other information to the compiler.

Programs in Cobol have four separate divisions: IDENTIFICATION, ENVIRONMENT, DATA, and PROCEDURE. The separation of DATA and PROCEDURE was an especially important innovation in Cobol. Nowadays, some experts on good programming practice stress the importance of not embedding part of the data description into the procedures. For example, if a program extracts five characters as the zip code, then it has an embedded description. The change to nine-digit zip codes could cause serious problems for such a program. If the size is specified in the data division as an explicit data description, then the change requires only a few keystrokes and a subsequent recompilation.

Cobol introduced several other innovative features, including facilities to manipulate arrays of information rather than single variables, facilities to easily create highly structured and informative reports, and expanded conditional and looping statements.

A sample Cobol program is shown in Fig. 15.11. It does not implement the natural logarithm algorithm, since that is not a suitable application for Cobol. Instead, we do a simple file update. Also, this program does not have the extensive documentation of other examples. This was done to give you a taste of self-documenting code.

15.7 BASIC

Assembly languages, and the high-level languages Fortran and Cobol, were developed in the era of large mainframe computers. On these systems programmers submitted decks of cards and waited for error listings. In 1964 a new approach to developing programs was introduced.

Dartmouth professors John Kemeny and Thomas Kurtz [Kemeny and Kurtz, 1985], expanding on ideas developed at MIT and Bell Labs, built a timesharing operating system for a new computer. One of their goals was to make computing services readily available to a wide variety of students, and they felt a new language was needed. To satisfy this need they created *B*eginner's *A*ll-purpose *S*ymbolic *I*nstruction *C*ode, BASIC. In designing BASIC, they combined and simplified features from Fortran and assembly language.

BASIC was a success, not just because it was a relatively simple language, but also because of the way users interact with it. BASIC was designed as an interpreted language. As we stated earlier, the interpreter analyzes each line as the user types it. If an error is detected, a message is immediately displayed so that the user can correct it.

Early implementations of BASIC required that the user type a line number at the beginning of each source line. The line number served two purposes: (1) it specified the order of execution, and (2) lines could be edited easily or replaced by referencing the line number.

Another noticeable difference between the original BASIC and other languages is that each line explicitly included a command word. A typical replacement statement is shown below:

```
10 LET A = SIN(B)
```

Some more recent implementations of BASIC, such as TrueBASIC [TrueBASIC Inc., 1985], do not require line numbers and Let commands.

Because of its simplicity, BASIC was the natural choice of personal computer developers. It was the language supplied on all the IBM PC and Apple II computers. A whole generation of new users were first introduced to computer programming with BASIC.

We will illustrate a BASIC program with our natural logarithm computation. Since original BASIC does not have a REPEAT UNTIL or similar construct, we have modified the flowchart slightly to use a FOR statement. This modification is shown in Fig. 15.12 and the program in Fig. 15.13.

The following points refer to the red numbers in the BASIC program:

1. Comments are specified by putting a REM in the first three columns of the line.
2. BASIC allows easy specification of input and output.
3. The conditional branch, IF statement, references a line number within the program.
4. This is one form of looping. The FOR matches a later NEXT, and code between the two statements is repeated for the specified range of INDEX.
5. This statement illustrates the expression capabilities of BASIC, similar to those of Fortran.

In the past few years BASIC has lost much of its popularity. Newer, more powerful languages like Pascal, C, and Ada have replaced BASIC in many places. These languages are perceived as being more suitable for developing well-structured programs that can be more easily developed and maintained.

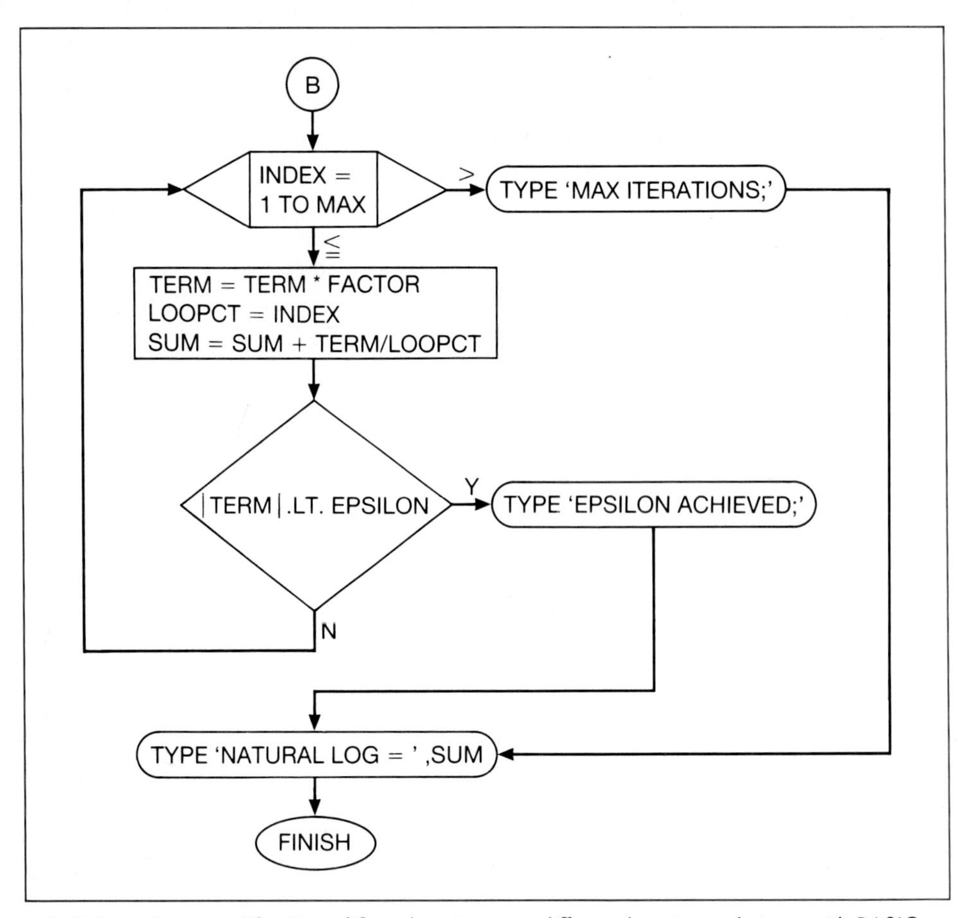

FIGURE 15.12 Modification of flowchart to use a different looping technique with BASIC.

15.8 PASCAL

Pascal is one of several languages whose roots go back to Algol 60. In some ways Algol was to the European computing community what Fortran was to U.S. computer users. Both Algol and Fortran simplified writing and maintaining scientific programs. Both gave the user the ability to express mathematical formulas in a concise manner, and both made it possible to easily split programs into subroutines. There are, however, some significant differences between the languages.

Algol introduced the concept of strong type checking. Fortran, on the other hand, lets the user mix variables of different types almost at will. This gives the programmer great flexibility but also makes it impossible for the compiler to detect some programming errors. Pascal follows in the Algol tradition and enforces strong type checking. Probably one of the main reasons for this is that Pascal was designed as a teaching language. The designers felt that forcing students to pay attention to

```
(1)  10    REM
     20    REM    MODULE_NAME:          NATURAL_LOGS
     30    REM    VERSION_NUMBER:        1.0
     40    REM    VERSION_DATE:        4/20/90
     50    REM    AUTHOR:            Jim Pinkert
     60    REM
     70    REM    This program computes the natural logarithm of an input number.
     80    REM    That input number must be positive, or it will be rejected. The
     90    REM    program uses one of two series expansions, depending on whether
    100    REM    the input number is greater than one, or between zero and one.
    110    REM    This is done to speed convergence of the series.
    120    REM
    130    REM    The program also inputs a maximum number of terms that should
    140    REM    be used in the series expansion, and a minimum change in the
    150    REM    value of the series.  The program terminates the series when
    160    REM    the maximum number of terms has been reached, or the value of
    170    REM    the last term computed is less than the stated minimum.
    180    REM
    190    REM    Real Variables:
    200    REM       NUMBER  -- the number for which the logarithm is desired
    210    REM       EPSILON  -- the minimum change in the value of the series
    220    REM       SUM  -- the current value of the series
    230    REM       FACTOR  -- the quantity which will be exponentiated to
    240    REM             form the terms in the series expansion
    250    REM       LOOPCT  -- the current number of terms in the series
    260    REM       TERM  -- the current last term in the series
    270    REM
    290    REM       MAX -- the maximum number of terms in the series
    300    REM       INDEX -- counter in loop
    310    REM
    320    REM
    330    REM    Print initial header.
    340    REM
    350       PRINT "THIS PROGRAM COMPUTES NATURAL LOGARITHMS"
    360       PRINT "PLEASE RESPOND TO THE QUESTIONS WITH:"
    370       PRINT "    (1) THE NUMBER WHOSE LOG IS DESIRED"
    380       PRINT "    (2) A MAXIMUM NUMBER OF ITERATIONS"
    390       PRINT "    (3) A MINIMUM CHANGE FOR THE SERIES"
    400    REM
    410    REM       This loop obtains the number, a maximum number of
    420    REM       terms, and a minimum change in the series.  The loop
    430    REM       repeats if the user PRINTs a non-positive number, a
    440    REM       non-positive maximum number of terms, or a negative
    450    REM       minimum change in the series.
    460    REM
    470    REM **** START OF INPUT LOOP ****
    480    REM
(2) 490       PRINT "NUMBER = "
    500       INPUT NUMBER
    510       PRINT "MAX ITERATIONS = "
    520       INPUT MAX
    530       PRINT "EPSILON = "
    540       INPUT EPSILON
    550    REM
    560    REM       Test for invalid number.
    570    REM
(3) 580       IF NUMBER > 0.0 THEN 630
    590       PRINT "NUMBER MUST BE POSITIVE"
```

FIGURE 15.13 TrueBASIC program to calculate natural logs by series expansion: (*a*) program listing, (*b*) sample run that reached the max iterations; (*c*) sample run that reached the epsilon limit.

```
      600   GOTO 490
      610  REM
      620  REM       Test for invalid number of iterations.
      630  REM
      640     IF MAX > 0 THEN 680
      650     PRINT "MAXIMUM ITERATIONS MUST BE POSITIVE"
      660     GOTO 490
      670  REM
      680  REM       Test for invalid minimum change in series.
      690  REM
      700     IF EPSILON >= 0.0 THEN 790
      710     PRINT "MINIMUM CHANGE MUST BE NON-NEGATIVE "
      720     GOTO 490
      730  REM
      740  REM    Input values are all valid.
      750  REM    Start loop to compute logarithm.
      760  REM
      770  REM    Initialize the series summation.
      780  REM
      790     LET SUM = 0.
      800  REM
      810  REM    Decide which series to use.
      820  REM
      830     IF NUMBER < 1.0 THEN 950
      840  REM
      850  REM   Series for number greater than 1.  The jth term in this
      860  REM   series is ((Number - 1.0) j ) / ( j*(Number) j )
      870  REM
      880     LET FACTOR = (NUMBER - 1.0)/NUMBER
      890     LET TERM = 1.0
      900     GOTO 1020
      910  REM
      920  REM   Series for positive number less than 1.  The jth term
      930  REM   in this series is ((-1) j * (1.0 - Number) j ) / j
      940  REM
      950     LET FACTOR = 1.0 - NUMBER
      960     LET TERM = -1.0
      970  REM
      980  REM   Loop to compute the series.  Loop terminates when the
      990  REM   maximum number of terms have been used, or the size
     1000  REM   of the last term is less than the specified minimum.
     1010  REM
 (4) 1020      FOR INDEX = 1 TO MAX
     1030  REM
     1040  REM      Compute term by using repeated multiplication of
     1050  REM      previous term, rather than raising to power each time.
     1060  REM
     1070      LET TERM  = TERM  * FACTOR
     1080  REM
     1090  REM      Set LOOPCT  (real) to INDEX (integer)  and add term to
     1100  REM      series summation.
     1110  REM
     1120      LET LOOPCT  = INDEX
 (5) 1130      LET SUM  = SUM  + TERM /LOOPCT
     1140  REM
     1150  REM      Check for loop termination condition -- TERM  less than
     1160  REM      minimum change in value of approximation.
     1170  REM
     1180      IF ABS(TERM ) >= EPSILON  THEN 1210
     1190      PRINT "EPSILON ACHIEVED; "
     1200      GOTO 1290
 (4) 1210      NEXT INDEX
     1220  REM
     1230  REM   Maximum iterations performed.
     1240  REM
     1250      PRINT "MAXIMUM ITERATIONS;"
     1260  REM
     1270  REM   Approximation completed.  Print result.
     1280  REM
     1290      PRINT "NATURAL LOG = ",SUM
     1300      END
```

(*a*)

FIGURE 15.13 (*Cont.*)

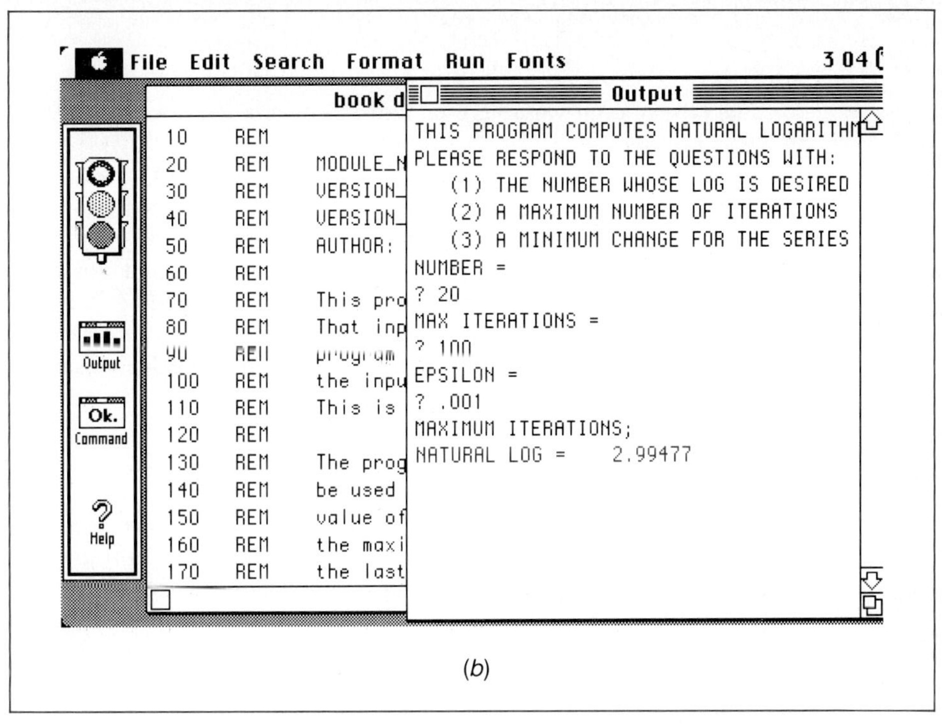

(*b*)

FIGURE 15.13 (*Cont.*)

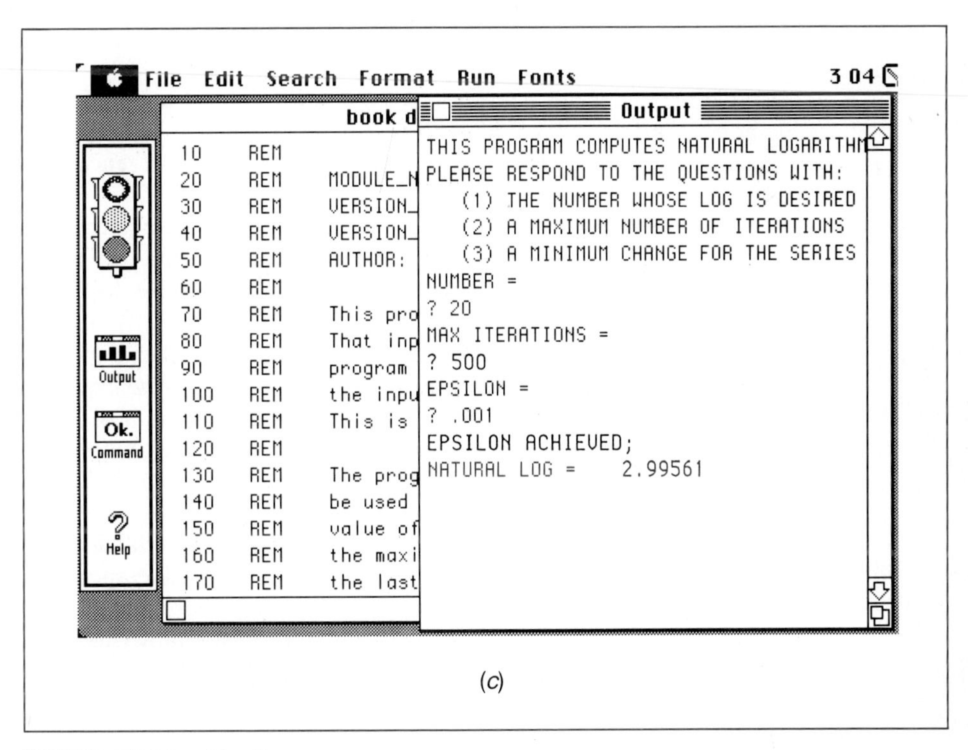

(*c*)

FIGURE 15.13 (*Cont.*)

the type of each variable used and to not mix types would result in fewer programming errors.

Many Fortran programmers had trouble adjusting to another feature of Pascal; it did not have a GOTO statement. (Some implementations included a GOTO statement, but its use was strongly discouraged.) Pascal programs were supposed to be designed and written in a "structured" fashion and not require GOTO statements. Because of its strong type checking and the emphasis on structured programming techniques, Pascal became the language of choice in many universities. In Fig. 15.14 we have rewritten in Pascal the program that calculates natural logarithms.[1]

15.9 C++

C++ is an object-oriented programming language developed at AT&T by Stroustrap [Stroustrap, 1986]. C++ is a superset of the C programming language that was developed by Dennis Ritchie to implement UNIX. The ++ operator in the C language means increment; therefore, the name C++ suggests an increment beyond C.

An **object-oriented programming language** is different from a procedural language like Pascal, C, or Fortran in several ways. In a procedural language the data and procedures that manipulate the data are grouped separately. In an object-oriented language the programmer creates objects. **Objects** contain data and methods to manipulate the object data. A **method** is a function or procedure that is used to operate on an object's data.

Three important concepts are common to object-oriented languages: encapsulation, inheritance, and polymorphism. **Encapsulation** is the grouping of data and methods into objects. Stated another way, an object encapsulates its data and methods.

Another important feature of an object-oriented language is inheritance. **Inheritance** lets you create a new object that has all the data and methods of an existing object. Suppose, for example, Jim has developed a list_handler object and Larry needs a specialized version of list_handler that scans the list from back to front. Larry can derive a new object called special_list_handler that "inherits" all the data and methods of list_handler. He then creates a new method, back_scan, and adds its associated data. Some existing methods or data of list_handler can be redefined; however, most of the code will probably not require modification. Larry will probably have to create only one new method to implement the new object, special_list_handler.

The last feature of object-oriented languages we will discuss is polymorphism. **Polymorphism** is the notion that different objects will respond to the same request appropriately. For example, suppose a programmer is using integer objects, string objects, and floating-point objects. Polymorphism allows the programmer to call a

[1] This program uses a slightly different structure than the previous Fortran and BASIC programs and has not been documented. The reason for both changes is that you will be asked to work with the program in the problems at the end of the chapter.

```
program log_series;

var
    Value : real;
    MaxIterations : integer;
    Epsilon : real;
    Result : real;

procedure ReadValues(var Value : real; var MaxIterations : integer; var Epsilon : real);

begin
repeat
    write('Number = ');
    readln(Value);
    write('Max iterations = ');
    readln(MaxIterations);
    write('Epsilon = ');
    readln(Epsilon);
   if ( Value <= 0 ) then
          writeln('Input quantity must be positive!');
   if ( MaxIterations <= 0 ) then
          writeln('Maximum iterations must be positive!');
   if ( Epsilon < 0 ) then
          writeln('Minimum change must be non-negative!');
until (Value > 0) and (MaxIterations > 0) and (Epsilon >= 0);

end;

procedure CalculateLog(Value : real; MaxIterations : integer; Epsilon : real;
                var Result : real);

var
   Next : integer;
   Counter : integer;
   Term : real;
   Power : real;
   Sum : real;

begin

Next := 1;
Sum := 0;
Counter := 0;

if (Value > 1.0) then
   begin
      Term := (Value - 1.0)/Value;
      Power := 1.0;
   end
else
   begin
      Term := 1.0 - Value;
      Power := -1.0;
   end;

repeat
   Power := Power * Term;
   Counter := Counter + 1;
```

FIGURE 15.14 Pascal program to calculate natural logs by series expansion: (*a*) program listing; (*b*) sample run that reached the max iterations; (*c*) sample run that reached the epsilon limit.

```
        Sum := Sum + Power/Counter;

        if (Counter > MaxIterations) then
          begin
             writeln('Max Iterations');
             Next := 0;
          end;
        if ( ABS(Power) < Epsilon ) then
          begin
             writeln('Epsilon achieved');
             Next := 0;
          end;
    until Next = 0;
    Result := Sum;

end;

begin
        ReadValues(Value, MaxIterations, Epsilon );
        CalculateLog(Value, MaxIterations, Epsilon, Result );
        writeln('Natural Log = ',Result:4:4);
        readln;
end.
```

(*a*)

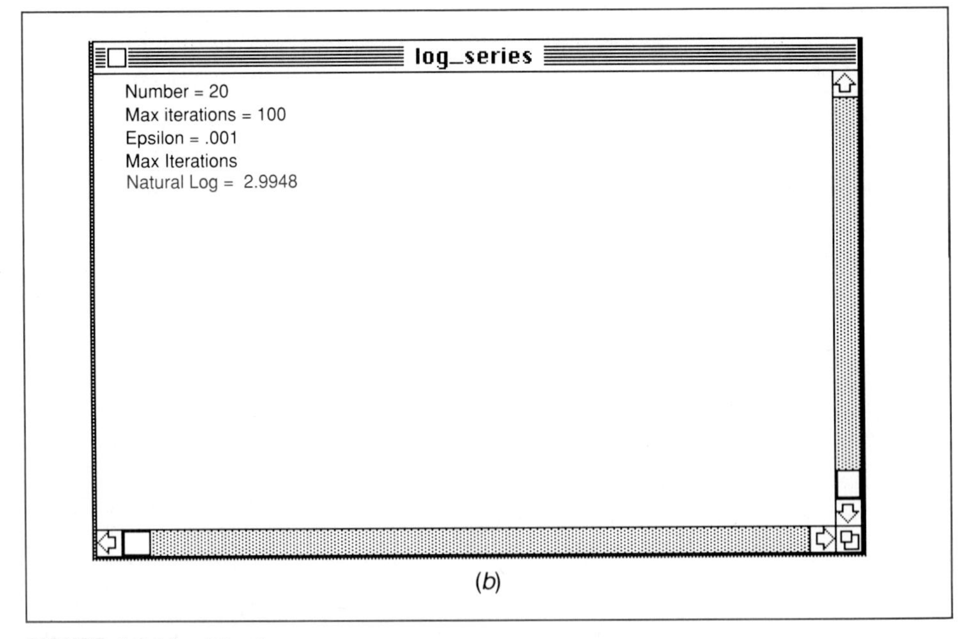

(*b*)

FIGURE 15.14 (*Cont.*)

generic "print" method for each object. The print method will be implemented differently for each type of object so that an integer object will print an integer, a string object will print a string, and a floating-point object will print a floating-point number. Polymorphism is important because it allows new types of objects to be

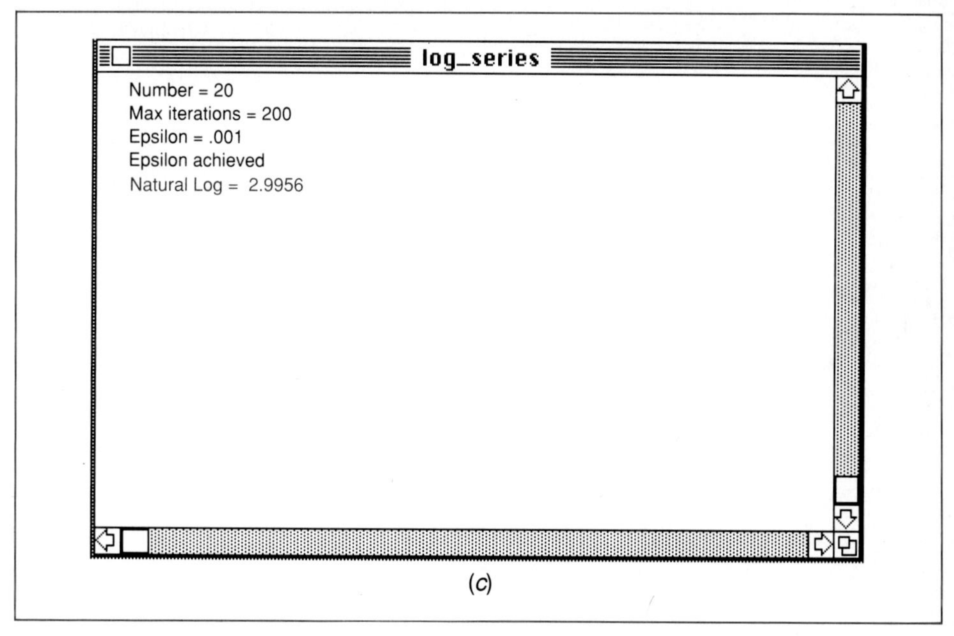

(c)

FIGURE 15.14 (*Cont.*)

added to existing systems as long as the new objects implement the correct interface methods, like print.

The following example shows encapsulation and inheritance in the C++ programming language. The keyword **class** is used to encapsulate data and methods into objects. The **typedef** statements are used to define sets containing the specified members.

The first class defined in the program in Fig. 15.15 is class Person. *Person* is a **base class** because other new classes are derived from it and inherit its attributes. Notice that Person class declares the variables eyeColor, hairColor, and sex. This data represents some attributes of a real person that are determined from the parents. If mom and dad both have blue eyes, their child is likely to have blue eyes. If mom has brown hair and dad blond, their child is likely to have brown hair because brown is a dominant hair color.

The classes *Man* and *Woman* are derived from the base class Person. This means that Man and Woman inherit the variables or attributes contained in Person. An object of type Woman can reference the variable eyeColor even though the variable was not explicitly declared in the class Woman. Notice that Man and Woman classes each declare a new method called giveChromosome. The male gives either an X or Y chromosome; the female always gives an X. The chromosome combination XX is a girl, and XY is a boy.

The example program declares three classes: Person, Man, and Woman. There is also a function called sex, which returns a new Person of either type Man or

```
#include <stream.h>
#include <libc.h>

/* - - - - - - - - - - - - - - - - - - - - - - - - - - - - - - - - - - - - - - - - - - - - - - - - - - - - - - -
|
| The types EYECOLOR, HAIRCOLOR, AND CHROMOSOME are defined below.
|
\ - - - - - - - - - - - - - - - - - - - - - - - - - - - - - - - - - - - - - - - - - - - - - - - - - - - - - - * /
typedef enum
{
   blue,
   brown
  } EYECOLOR;

typedef enum
{
   blond,
   black
  }HAIRCOLOR;

typedef enum
{
   X,
   Y
  }CHROMOSOME;

typedef enum

{
   female,
   male
   }SEX;

/ * - - - - - - - - - - - - - - - - - - - - - - - - - - - - - - - - - - - - - - - - - - - - - - - - - - - - - -
|
| CLASS Person:
|   Person contains the variables eyeColor, hairColor, and sex.  These are the attributes
|   of a real person that we are considering.  No methods or constructors are defined for Person.
|
\ - - - - - - - - - - - - - - - - - - - - - - - - - - - - - - - - - - - - - - - - - - - - - - - - - - - - - - * /
class Person
{
    public:

   EYECOLOR eyeColor;
   HAIRCOLOR hairColor;
   SEX sex;
} ;

/ * - - - - - - - - - - - - - - - - - - - - - - - - - - - - - - - - - - - - - - - - - - - - - - - - - - - - - -
|
| CLASS Woman:
|   Woman is derived from the base class Person.  Woman contains a constructor, Woman
|   which accepts the parameters tmpEyeColor and tmpHairColor.  The parameters are set to
|   default values of blue and blond respectively if no parameters are passed.
|   GiveChromosome accepts one parameter, chromo, which is always set to X.
\ - - - - - - - - - - - - - - - - - - - - - - - - - - - - - - - - - - - - - - - - - - - - - - - - - - - - - - * /
class Woman : public Person
{
    public:
```

FIGURE 15.15 A C++ program showing encapsulation and inheritance: (*a*) program listing; (*b*) sample output showing results of creating two "kids."

```
   Woman(EYECOLOR tmpEyeColor = blue, HAIRCOLOR tmpHairColor = blond)
    {
    sex = female;
    eyeColor = tmpEyeColor;
    hairColor = tmpHairColor;
    };

   void GiveChromosome(CHROMOSOME & chromo)
   {
   chromo = X;
    };
};

/*----------------------------------------------------------------
|
| CLASS Man:
|   Man is derived from the base class Person.  Man contains a constructor, Man
|   which accepts the parameters tmpEyeColor and tmpHairColor.  The parameters are set to
|   default values of brown and blond respectively if no parameters are passed.
|   GiveChromosome accepts one parameter, chromo, which is randomly set to either
|   X or Y.
\----------------------------------------------------------------*/
class Man : public Person
{
 public:
  Man(EYECOLOR tmpEyeColor = brown, HAIRCOLOR tmpHairColor = blond)
  {
  sex = male;
  eyeColor = tmpEyeColor;
  hairColor = tmpHairColor;
  };

  void GiveChromosome(CHROMOSOME & chromo)
   {
    int temp = rand();
    if ( (temp % 2) == 0) chromo = X;
    else chromo = Y;
    }
};

/*----------------------------------------------------------------
|
| FUNCTION Sex
|   The Sex function accepts two parameters, dad ( a pointer to Man object ) and mom ( a
|   ( pointer to Woman object ).  The mom and dad attributes are evaluated to determine the
|   attributes of the new child.  Also the mom and dad GiveChromosome methods are called to
|   determine the sex of the new child.  Finally, a new Man or Woman object is returned.
\----------------------------------------------------------------*/
Person* Sex(Man * dad, Woman * mom)
{
  CHROMOSOME DadsChromoType,MomsChromoType;
  EYECOLOR eyeColor;
  HAIRCOLOR hairColor;

    if (mom->eyeColor == blue && dad->eyeColor == blue )
    eyeColor = blue;
    else eyeColor = brown;

    if (mom->hairColor == blond && dad->hairColor == blond )
    hairColor = blond;
    else hairColor = black;
```

FIGURE 15.15 *(Cont.)*

```
  dad->GiveChromosome(DadsChromoType);
  mom->GiveChromosome(MomsChromoType);

  if (DadsChromoType == X && MomsChromoType == X)
   return new Woman(eyeColor, hairColor);
  else
    return new Man(eyeColor, hairColor);
}

|-----------------------------------------------------------------
|
| FUNCTION main
|   The main function declares a mom and dad of type Woman and Man respectively.  The
|   mom and dad will have attributes specified by the default values in the constructors
|   since parameters are not explicitly passed to the constructor.  The parameter child is
|   defined as a pointer to Person since it cannot be determined before a call to Sex what
|   the sex of the child will be.  After the call to Sex, the attributes of the new child are
|   displayed using the "cout" statement.
-----------------------------------------------------------------*/
void main()
{
Man dad;
Woman mom;
Person *child;

child = Sex(&dad, &mom);

cout << "Characteristics of the child\n";
cout << "Eye color is ";
if (child->eyeColor == blue) cout << "blue\n";
else cout << "brown\n";
cout << "Hair color is ";

if (child->hairColor == blond) cout << "blond\n";
else cout << "black\n";
cout << "Sex is ";
if (child->sex == female) cout << "female\n";
else cout << "male\n";
}
```

(*a*)

```
Characteristics of the child
Eye color is brown
Hair color is blond
Sex is female

Characteristics of the child
Eye color is brown
Hair color is blond
Sex is male
```

(*b*)

FIGURE 15.15 (*Cont.*)

Woman class. The new person's traits are determined by the parents. At this point we are obligated to note that C++ is not truly object-oriented, because functions and data can be declared independent of objects. In a truly object-oriented language like Smalltalk [Goldberg and Robson, 1986], functions are not declared independent of objects.

The function "main" declares a variable dad of type Man, a variable mom of type Woman, and a pointer to a child of type Person. Since the first Man and Woman do not have parents, their traits (i.e., eyeColor and hairColor) are set to the default values. These default values are specified in the constructors for each class. A **constructor** is a special method which is used to initialize each new object that is created. The constructor for the class Man is declared as Man, and the constructor for the class Woman is declared as Woman. Note that the default values specified in the constructor are blue eyes and blond hair for a Woman and brown eyes and blond hair for a Man.

The constructor for a class is called each time a new object of type class is created. Upon entering the function main, the constructors for dad and mom are called. A constructor is not called for child because child is only declared as a pointer to an object, not as an actual object. The procedure sex is called from the main loop. In sex, the new child's eye color, hair color, and sex are determined by the parents' characteristics. Once the attributes for the new child are determined, a new object of either type Man or type Woman is created using the "new" operator. When the statement new Woman (eyeColor, hairColor); is invoked, a new object of type Woman is created and the attributes are set according to the values passed into the constructor.

Sex returns a pointer to an object of type Person. The details of why a Person-type object is returned instead of a Man or Woman are beyond the scope of this chapter. However, the characteristics of the new child can be examined by looking at the new Person-type object. The "cout" statements in the main loop are used to print the attributes of the new child. A sample output is shown following the program.

15.10 PROLOG

So far, all the languages we have described individually have one thing in common—they force the programmer to write procedures that tell the computer how to solve a problem. In declarative languages, the programmer specifies a set of rules and facts, then asks the computer to check the validity of a goal based upon those rules and facts. The programmer does not specify a procedure to accomplish that check.

As a simple example, we might state a rule that names and phone numbers are related, enter pairs of names and phone numbers as a set of facts, and then ask, "Is Jim's phone number 555-5432?" The computer would then find the answer without our having to give a set of instructions like that shown in Fig. 15.16.

```
Answer := FALSE;
next := List_head;
repeat
     Get name and number from pair[next];
     If (name = "Jim") and (number = "555-5432")
          then Answer := TRUE;
     next := successor(next);
until (next = NIL) or (Answer = TRUE);
Write (Answer);
```

FIGURE 15.16 Procedural search to see if Jim's phone number is 555-5432.

Declarative languages have the ability to reach decisions based on facts and rules without being told what steps to take. Prolog, which stands for *pro*gramming in *log*ic, is probably the most well-known example of a declarative language.

Prolog has been used extensively in artificial intelligence (AI) research. It was invented by Alain Colmerauer and his colleagues at the University of Marseille and became the leading artificial intelligence language in Europe. By the late 1970s several versions of Prolog had been developed for microcomputers, and its popularity began to increase in the United States. Two events caused Prolog to catapult to the top of the list of AI languages. First, the Japanese chose to base their *fifth-generation* project (computer systems to dominate the 1990s and beyond) on Prolog. Second, Borland International created an inexpensive PC-based Prolog compiler, TurboProlog. We will now show a very simple TurboProlog program that demonstrates some of the features of the language.

One of the areas where AI researchers have made significant gains is in expert systems. An **expert system** is a program that tries to imitate the actions a human *expert* would take, given the same data. An expert system might, for example, attempt to diagnose a patient's disease based on his or her symptoms. A simple version of such a program is shown in Fig. 15.17.

TurboProlog programs are separated into four divisions: domains, predicates, goal, and clauses. **Domains** are roughly equivalent to type declarations in other languages. In our program, disease and symptom are defined as being of type "symbol." **Predicates** establish relationships between objects. As you can see, our program has one relationship: has_as_symptom. The **clauses** division contains the facts we input into the program. A typical fact is

```
has_as_symptom(flu, fever)
```

Our human interpretation of this fact is the "flu has as a symptom fever." Actually, the computer just knows there is a specific relationship between flu and fever, and it is called has_as_symptom. (Note that this is a *directed* relationship, so we cannot also infer the relationship fever has_as_symptom flu.) The **goal** division is where we ask a question of the computer. In our sample program, no goal is listed. This results in TurboProlog prompting the user for a goal when the program is executed.

```
/* Diagnostics */

/* This Prolog program contains a number of facts  */
/* (clauses) about several medical conditions.     */

/* When you run the program you can ask questions, */
/* (set goals) that the program will answer based  */
/* on the facts that have been given.              */

domains
  disease, symptom = symbol

predicates
  has_as_symptom(disease,symptom)

clauses
  has_as_symptom(cold,fever).
  has_as_symptom(cold,no_appetite).
  has_as_symptom(cold,runny_nose).
  has_as_symptom(cold,cough).
  has_as_symptom(cold,tired).

  has_as_symptom(flu,fever).
  has_as_symptom(flu,low_temperature).
  has_as_symptom(flu,muscle_ache).
  has_as_symptom(flu,tired).
  has_as_symptom(flu,dehydrated).
  has_as_symptom(flu,sweats).
  has_as_symptom(flu,cough).
  has_as_symptom(flu,high_white_count).
  has_as_symptom(flu,upset_stomach).

  has_as_symptom(allergies,normal_white_count).
  has_as_symptom(allergies,normal_temperature).
  has_as_symptom(allergies,red_eyes).
  has_as_symptom(allergies,cough).
  has_as_symptom(allergies,tired).
  has_as_symptom(allergies,runny_nose).
```

FIGURE 15.17 Prolog program to check symptoms.

Several goals, and the program's corresponding responses, are shown below. The first goal asks the program if there is a relationship between flu and fever, and the answer is True.

```
Goal: has_as_symptom(flu,fever)
True
```

The second asks if there is a relationship between fever and flu; the answer is False, since that ordering does not appear in the clauses division.

```
Goal: has_as_symptom(fever,flu)
False
```

The third goal uses a variable Symptom to ask what symptoms are related to allergies. Notice that all the symptoms related to allergies are listed, one per line.

```
Goal: has_as_symptom(allergies,Symptom)
Symptom = red_eyes
```

```
Symptom = runny_nose
Symptom = cough
Symptom = tired
Symptom = normal_white_count
Symptom = normal_temperature
6 Solutions
```

Next, we ask a more complicated question, "What disease has fever, upset_stomach, and muscle_ache as symptoms?" The answer is Disease = flu.

```
Goal: has_as_symptom(Disease,fever) AND
      has_as_symptom(Disease,upset_stomach) AND
                  has_as_symptom(Disease,muscle_ache)
Disease = flu
1 Solution
```

The final goal asks what disease has fever and bleeding. The answer is No Solution, since none of the diseases in our small list of clauses satisfy both conditions.

```
Goal: has_as_symptom(Disease,fever) AND
      has_as_symptom(Disease,bleeding)
No Solution
```

Prolog is much more powerful than our simple example can demonstrate. The authors have used it to develop programs as diverse as OS simulations and database applications. We will give you one bit of warning, though. Learning Prolog is considerably different from learning "just another procedural language." However, it can be a powerful tool when applied to the right type of problem.

15.11 SUMMARY

In this chapter we have described several computer programming languages and shown some simple examples of their uses. Because of the large number of languages that have been developed, it was impossible to discuss them all. In order to help show you how the languages relate to each other, we compiled the following figures as summaries.

Figure 15.18 is a time line that shows the major developments in procedural languages. Three languages in this group, Fortran, Cobol, and Algol, were particularly important. Not only were they used extensively in schools and industry, they were, as the figure shows, also the ancestors of many other languages.

Algol, in particular, has had a significant influence on many of today's popular languages including Ada, C, Pascal, and Modula. Many of you have used or will use these languages in school or on the job. Since the U.S. government is requiring Ada for many of the programs it is procuring, it may become as important in the future as Cobol has been in the past. We can't leave this figure without mentioning

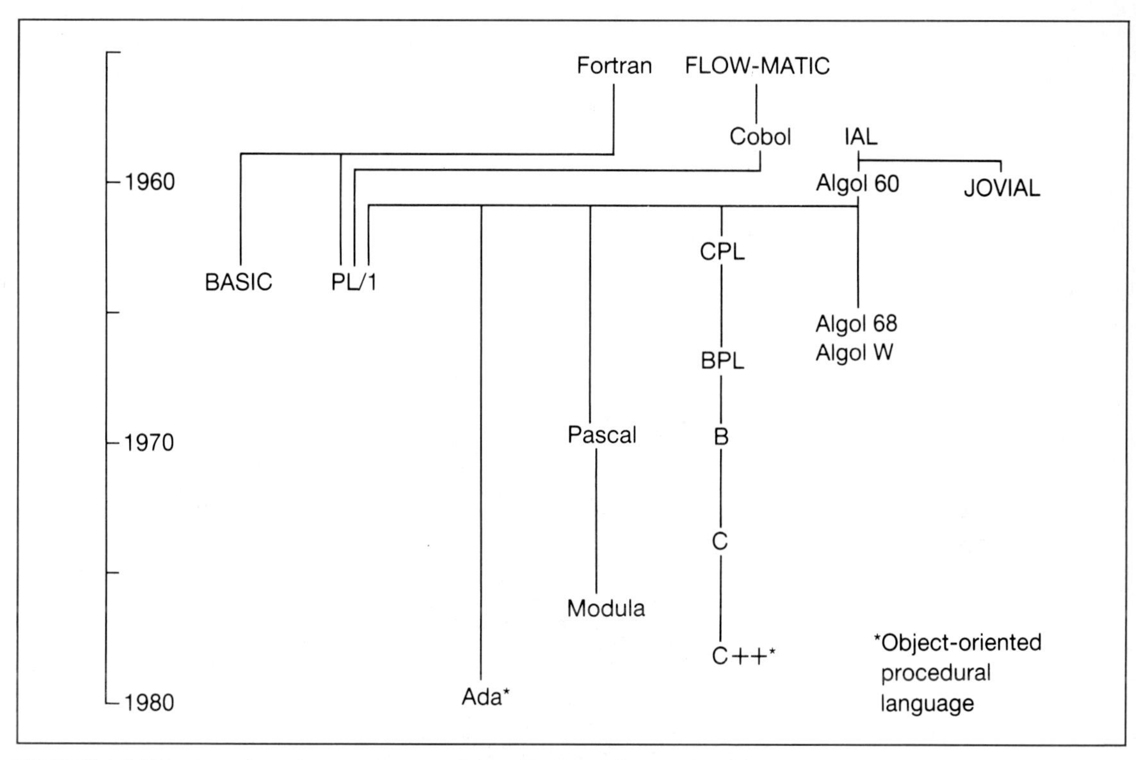

FIGURE 15.18 Time line showing history of development of procedural languages.

JOVIAL. The name comes from *J*ules' *o*wn *v*ersion of *i*nternational *a*lgorithmic *l*anguages. How could anyone omit a language with a name like that from their list of personal favorites?

As we stated earlier in the chapter, not all language translators are compilers nor are they procedural. Figure 15.19 shows a time line that describes some major developments in other areas.

The nonnumeric group includes Comit, LISP, and Snobol. These languages were developed to facilitate the manipulation of nonnumeric data—characters and strings. They have been used extensively in language research.

The interpretive time sharing languages JOSS and BASIC were developed primarily for users of time sharing systems. As we stated earlier, BASIC was designed especially for the novice or beginning programmer. The primary goal of these languages was to make programming as simple as possible.

Algebraic manipulation and equation-solving languages are another class of languages that have been used extensively in research. Mathematica [Wolfrun, 1988] is a recent addition to this group that takes advantages of the friendly user interface on current workstations and personal computers.

Object-oriented languages are in part an extension of procedural languages. As

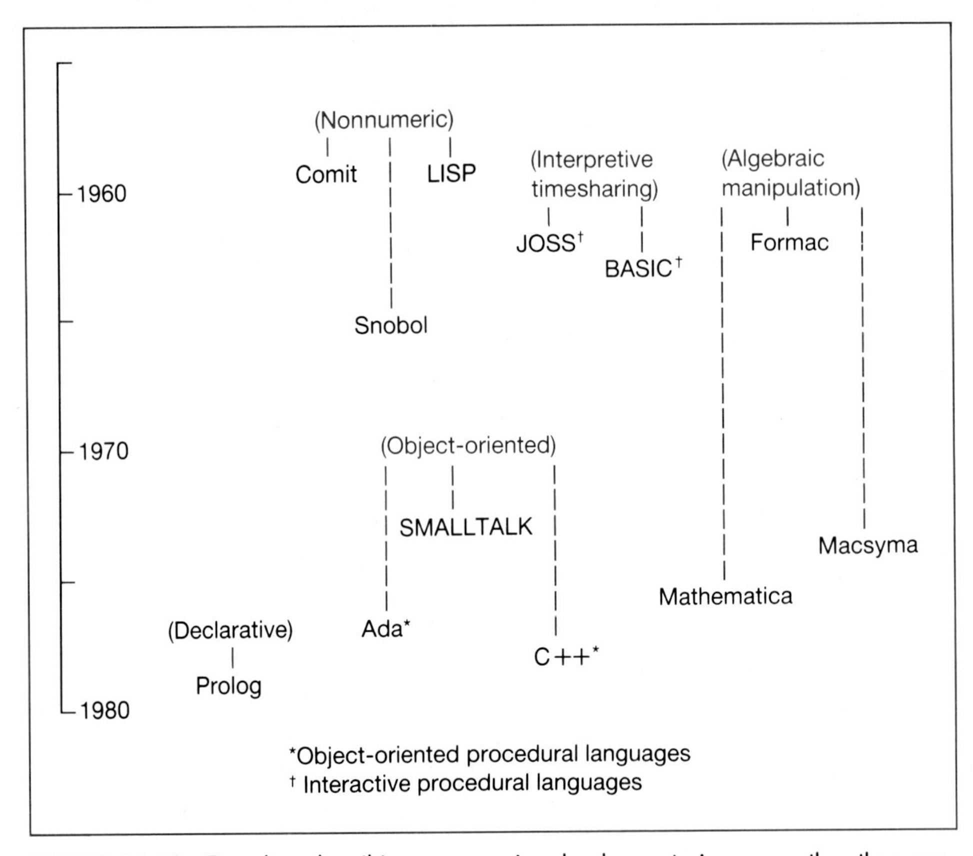

FIGURE 15.19 Time line describing some major developments in areas other than procedural languages.

we stated in Sec. 15.9, these languages emphasize encapsulation, inheritance, and polymorphism.

Figure 15.20 groups the major developments in the history of computer languages into one time line. It starts with the first, unimplemented, high-level languages in the 1940s. The first language translators that we actually used were assemblers developed in the late 1940s. This was followed by the first operational compiler in 1954 for an unnamed mathematical language on the Whirlwind computer.

The next step in the evolution of languages was procedural languages. This was followed shortly by block-structured, general-purpose procedural languages.

In recent years we have seen the development of a number of different types of languages. Some are special-purpose, and some just extensions of earlier languages. Nonprocedural and declarative languages are included in the first group, and object-oriented languages are in the second.

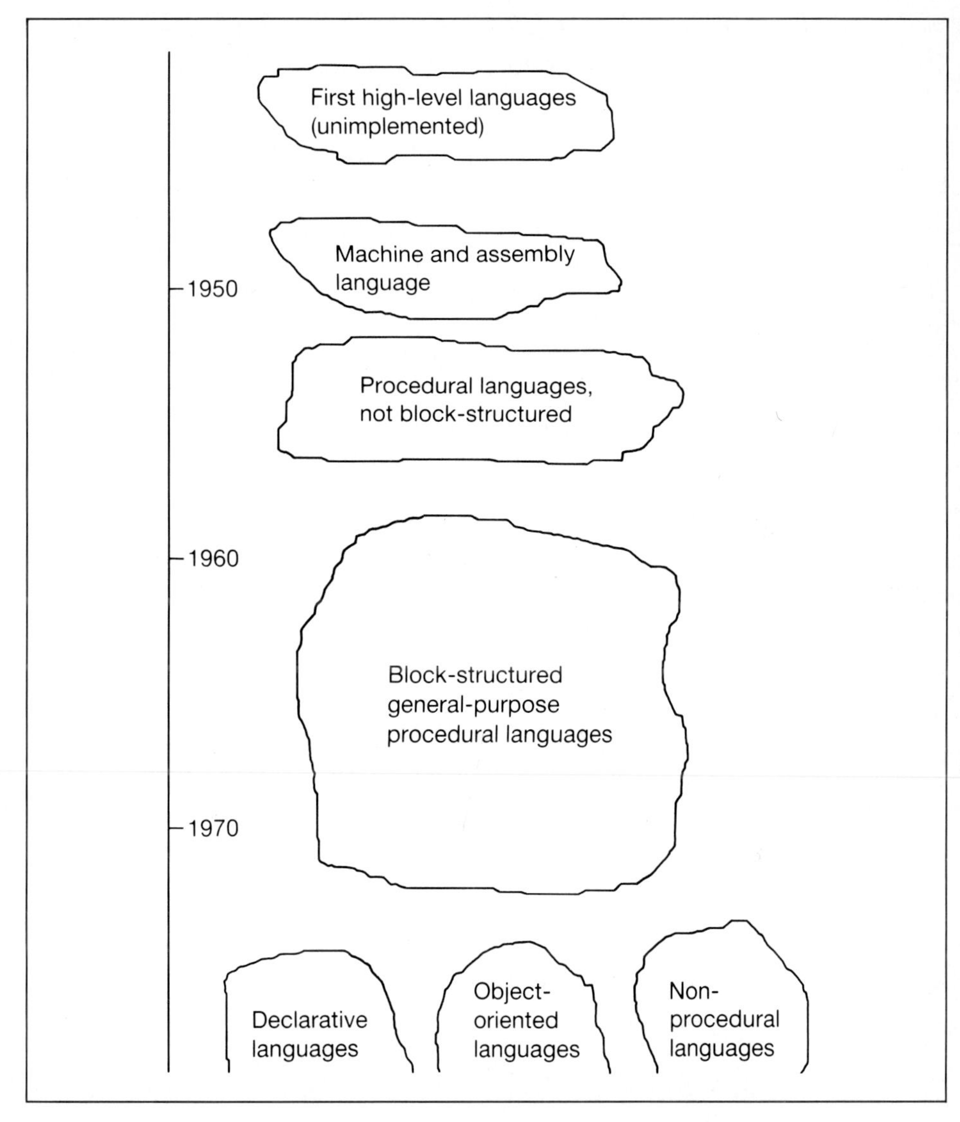

FIGURE 15.20 Major developments in the history of computer languages.

PROBLEMS

15.1 Explain the difference between machine language and assembly language.

15.2 Why do some compilers generate an intermediate language rather than a machine language?

15.3 Why do different computers have different machine languages?

15.4 Why can different computers use the same high-level language?

15.5 The process shown in Fig. 15.3 looks quite complicated compared to the one in Fig. 15.2*a*. What is the justification for having three translation steps?

15.6 Looking at the assembly- and machine-language examples in Sec. 15.4, what can you say about the architecture of the computer and its instruction set?

15.7 What characteristics of interpreters make them easier to use than compilers?

15.8 The code generated by compilers usually executes much faster than equivalent programs executed by interpreters. What are the reasons for this?

15.9 Why is it considered poor programming practice to embed data descriptions into a program?

15.10 Why were line numbers important for BASIC but not for Fortran?

15.11 What were the reasons that BASIC was so successful?

15.12 What are the disadvantages of programming in machine language?

15.13 Are there any advantages of programming in machine language?

15.14 Why do people sometimes confuse machine language and assembly language?

15.15 What do you think an *inverse assembler* does?

15.16 Why is learning a second assembly language usually relatively easy?

15.17 Referring to the program in Sec. 15.4, explain briefly the purpose of the first three instructions.

15.18 What is the purpose of the last four instructions in the program in Sec. 15.4?

15.19 Why do different computers tend to have similar assembly-language mnemonics?

15.20 What is the difference between pseudoinstructions and instructions like MOV, AND, and PUSH?

15.21 Why are standards so important in high-level languages?

15.22 Why are standards relatively unimportant in low-level languages?

15.23 How did Fortran "reduce by a large factor the task of preparing scientific programs"?

15.24 There is a general perception that programs written in assembly language are faster than equivalent programs written in a high-level language. Explain why this is true or false.

15.25 What are the divisions in a Cobol program?

15.26 What features of Cobol make it self-documenting?

15.27 What are some standard program structures found in all procedural languages?

15.28 Draw a flowchart for the Pascal version of the natural logarithm program.

15.29 What does it mean to embed the data description into the program?

15.30 Document the Pascal version of the natural logarithm program.

15.31 Why does Pascal have strong type checking?

15.32 Why does Pascal discourage the use of GOTO statements?

15.33 How do object-oriented programming languages differ from procedural programming languages?

15.34 In what ways are object-oriented programming languages similar to procedural programming languages?

15.35 What concepts do object-oriented programming languages have in common?

15.36 How does the concept of *information hiding* apply to object-oriented programming?

15.37 Why can inheritance reduce the amount of code that must be developed for a project?

15.38 Why is Prolog a popular choice for developing expert systems?

15.39 What are the divisions in a Prolog program?

15.40 What do we mean when we say that the predicates in Prolog establish directed relationships?

15.41 Create a goal for the Prolog program in Sec. 15.10 that uses logical operations other than AND. Explain what the goal does.

15.42 What is the major difference between procedure-oriented languages and non-procedure-oriented languages?

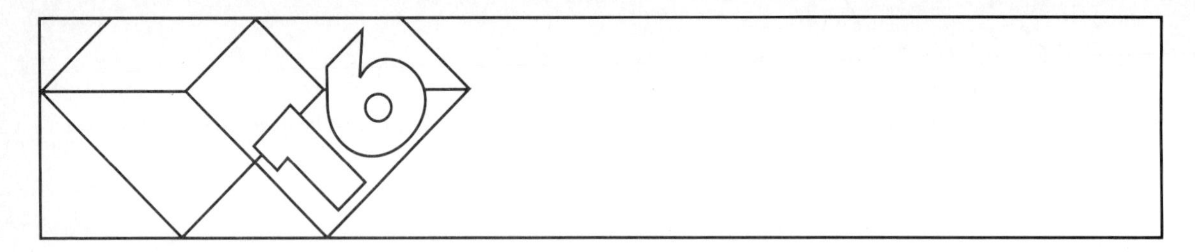

The Application Layer

In the early days of computing, computer hardware was often locked in a room that was off-limits to all but a few system programmers and the operations personnel. When a person in accounting, marketing, or engineering wanted to use the computer to analyze some data or create a report, he or she would first describe the problem to an analyst or programmer. Eventually, a programmer would then write a program to do the task. In cases like this, the analyst/programmer combination acted as an intelligent interface between the computer and the user of the system.

Some companies still operate in the manner described above, but their number is decreasing. In many cases much of the work previously done by programmers has been eliminated. Application programs, such as spreadsheets, report generators, and simulation packages, make it much easier for the user to access the facilities of the computer system. These programs are often application-dependent and are added on top of the operating system software. They make up the application layer and are the subject of this chapter.

16.1 MAKING A GENERIC COMPUTER SYSTEM USEFUL

In Part II we described the hardware components of a computer system, and in the first four chapters of Part III we described the fundamental software components of a system. The hardware and software we have discussed thus far might be described as being a general-purpose, or generic, computer system.

For many years, this was the type of system typically delivered by the computer manufacturer to the user. However, manufacturers observed that many groups of customers had needs for the same or similar program to solve particular problems. They perceived that this was a new market with great potential and started adding application software to their hardware systems.

Many people consider language compilers as part of the system software. However, the Fortran and Cobol compilers described in Chap. 15 were actually the primitive beginnings of the application layer of software. These compilers simplified the process of developing application programs on computers. Having already described compilers and language translators in Chap. 15, we will now look at other examples of application layer software.

16.2 MAKING OPERATING SYSTEM SERVICES AVAILABLE TO THE USER

On most systems, the operating system is capable of performing many services for the user, such as:

- Copying a file from one device to another
- Listing the files on a disk
- Printing the contents of a file
- Checking the time of day
- Formatting a disk

With early, primitive systems, users often had to write a program to access these services. On today's systems, there are usually a number of utility programs that provide access to these services by simply typing in a command or accessing an icon.

Those familiar with UNIX are used to accessing the operating system with terse keyboard commands such as "ls" for list. This can be very efficient and fast if you are familiar with the command structure. However, for the novice or one who does not use the system continually, it may be difficult to remember which keystrokes initiate which commands. The difficulty of accessing UNIX has been reduced somewhat by some of the command interpreter shells that have been developed for the system.

One popular set of programs for accessing system functions for the IBM PC is Norton Utilities [Peter Norton Computing, 1990]. This set of utilities performs the functions listed above in addition to a great many others. These programs also let the user initiate operations by simply pressing function keys rather than typing long sequences of commands. Pressing a function key, for instance, will display all the directories on a disk.

Making system functions readily available to the user has been expanded further on the Apple Macintosh computer [Apple Computer, 1990]. On the Macintosh, all the functions described above can be accessed by clicking the mouse on various icons and by moving the icons from place to place. Simply clicking the mouse button while the cursor is pointed to the icon for a disk will cause the disk's directory to be displayed. The evolution of the user interface is discussed in detail in Chap. 18.

Addition of new application programs has made it possible for a wide variety of users to easily access the power that computers provide. In the next section we will look at some applications that have had a significant effect on the way we use computers.

16.3 THE SPECTRUM OF APPLICATION LAYER PROGRAMS

The number and types of applications programs are growing seemingly without bounds. All you need to do to observe this is to open a user-oriented computer magazine and look at the ads. Companies are selling programs for artists, biologists, chemists, draftspersons, and probably for users in professions starting with all the other letters of the alphabet. It is not our intention here to describe all the possible applications available. We will limit our discussion in this section to a few general areas and look at the types of programs available.

16.3.1 Database Management Systems

Most application layer programs have been created as the result of a need in a specific area. A good example of this need occurred in the area of file management. Programmers saw that there were many instances where several files contained related information. The need to access this related information from multiple files resulted in the development of sophisticated database management systems.

A **database management system**, or **DBMS**, is a program or set of programs that allows the user to define and access data from a set of related files. The DBMS has become an indispensable part of most commercial data processing systems and many scientific systems. A simple example may show how these systems are used.

Consider a business that needs to keep information about all the parts that go into making its products. One possible set of files might be:

FILE	CONTENTS
Suppliers	name, address, parts supplied, price, availability, etc.
Assemblies	name, parts used, cost, quantities, etc.
Factories	name, assemblies manufactured, parts used, assemblies inventories, etc.
Warehouses	location, parts stored, assemblies stored, etc.
Parts	name, quantities, price, availability, etc.

It would be possible to store all this information in one huge file, but this has disadvantages, including unnecessary duplication of data, large file size, slow access to data, and inability to protect selective data in the file.

If we choose to use a DBMS to define and control access to the files, we might establish the relationships between files shown in Fig. 16.1. In the figure, each file is

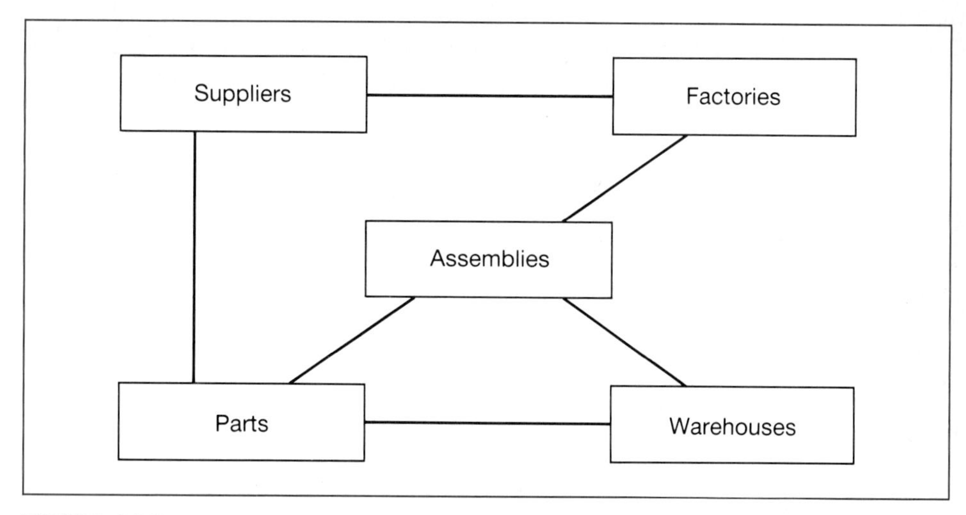

FIGURE 16.1 Files and their relations within a DBMS.

represented as a rectangle, and the bold lines show that a relationship exists between information in the files.

As Fig. 16.1 shows, the Suppliers file is related to the Parts and Factories files. Each Part entry points to one or more suppliers, and each Factories entry points to all suppliers the factory uses. If we need a particular part, for example, we simply look up the part in the Parts file. There, each part has pointers to all the suppliers who provide that part.

It would be possible to store all the supplier information with each part, but that has several disadvantages. First, there would be a lot of information duplication because the supplier data would be stored with each and every part. Duplicating supplier data obviously requires more storage, but it indirectly causes another problem. Suppose one of the suppliers moves and the supplier information were kept with each part. We would have to search through all the parts and change every occurrence of the supplier's address.

Given the example above, you might think that databases are used only in traditional data processing. This is not the case, however. Recall that in Chap. 11 we saw how hardware development tools make use of databases to keep track of the information associated with schematics, components, and printed circuit boards.

16.3.2 Spreadsheet Programs

When the first personal computers were introduced, most people thought they were just expensive toys. A few people, however, saw them as a potential business tool. Not long after the Apple II personal computer was introduced, a new application program was developed that played a significant role in changing the way people viewed the personal computer. The application program that caused this change in perception was the spreadsheet. A **spreadsheet** is, in part, a large two-dimensional array of cells, each of which can hold data, formulas, or commands.

Figure 16.2*a* shows a simple spreadsheet developed on EXCEL [Microsoft Corporation, 1990] that calculates the profit margin for a company. We assume values for projected sales increase each quarter at the rate specified by "Growth." The values for Manufacturing Costs and Marketing Costs both include a constant plus a percentage of sales. The Engineering Costs increase at a constant rate from quarter to quarter. Profit Margin is calculated by subtracting all the costs from the project-

FIGURE 16.2 EXCEL spreadsheet displaying Sales, Costs, and Profit Margin: (*a*) growth in sales = 5 percent per quarter; (*b*) growth in sales = 10 percent per quarter.

File Edit Window Select Format Options Chart Macro

B13 =B5-B7-B9-B11

ACE Projections (SS)

	A	B	C	D	E	F
1	ACE Manufacturing - Profit Analysis (1992)			Growth = 0.05		
2						
3	Quarter	1	2	3	4	Total
4						
5	Sales	$500,000	$525,000	$551,250	$578,813	**$2,155,063**
6						
7	Manufacturing Costs	$210,000	$218,000	$226,400	$235,220	**$889,620**
8						
9	Marketing Costs	$150,000	$170,000	$155,000	$140,000	**$615,000**
10						
11	Engineering Costs	$150,000	$153,000	$156,060	$159,181	**$618,241**
12						
13	Profit Margin	($10,000)	($16,000)	$13,790	$44,411	**$32,201**
14						
15						
16						

(*a*)

File Edit Window Select Format Options Chart Macro

B13 =B5-B7-B9-B11

ACE Projections (SS)

	A	B	C	D	E	F
1	ACE Manufacturing - Profit Analysis (1992)			Growth = 0.1		
2						
3	Quarter	1	2	3	4	Total
4						
5	Sales	$500,000	$550,000	$605,000	$665,500	**$2,320,500**
6						
7	Manufacturing Costs	$210,000	$226,000	$243,600	$262,960	**$942,560**
8						
9	Marketing Costs	$150,000	$170,000	$155,000	$140,000	**$615,000**
10						
11	Engineering Costs	$150,000	$153,000	$156,060	$159,181	**$618,241**
12						
13	Profit Margin	($10,000)	$1,000	$50,340	$103,359	**$144,699**
14						
15						
16						

(*b*)

ed Sales. The formula for calculating Profit Margin (cell B13) is shown on the second line of the spreadsheet:

$$B13 = B5 - B7 - B9 - B11$$

The terms B5 through B11 refer to cells within the spreadsheet; B5, for example, is a reference to the cell at the intersection of column B and row 5.

A similar formula is used to calculate the Profit Margin for the other three quarters. The column indicator, however, changes from B to C, D, or E. Spreadsheets have facilities that make it easy to duplicate formulas and automatically adjust row and column indicators. The spreadsheet is like a program in that the user enters data and instructions and results are calculated. An important difference is that the user doesn't have to learn a complex programming language to use the spreadsheet.

A second advantage of the spreadsheet is that it makes it easy to investigate different scenarios. This is sometimes referred to as the "What if?" game. If you were writing a program to make calculations similar to those in Fig. 16.2, you would probably have to recompile it or at least run it several times to look at different options. With a spreadsheet, you can see how the Profit Margin is affected by changes in sales growth by simply entering a new value into the Growth cell. When you do this, the program will recalculate all the results. Figure 16.2*b* shows how Profit Margin is affected when projected Sales grow by 10 percent rather than 5 percent per quarter.

Spreadsheets overcome another problem users have with information created by a computer program. Often programs create large amounts of data that can be difficult or impossible to interpret. Spreadsheet programs make it easy to convert columns of numbers into graphic displays. Figure 16.3 shows a graph of Profit Margin by quarters. The ability to convert easily from tabular data to graphs is a key reason so many users in a variety of fields find spreadsheets a vital tool.

It would be a mistake to think that spreadsheets are useful only for analyzing financial problems. Most spreadsheets include a wide variety of built-in functions that support scientific calculations. Because of the availability of these functions and the ease with which results can be plotted, many engineers find spreadsheets valuable for analyzing a wide range of design problems. Figure 16.4 shows the plot of a rather complicated transcendental function as an example.

In summary, spreadsheets have become a popular tool for many users because:

- They eliminate (or at least reduce) the need for programming.
- They make it easy to analyze many variations of a problem's solution.
- They can present results in readable graphic format.

16.3.3 Document Processing and Graphics

Development of the personal computer also led to the creation of other application layer programs. What the spreadsheet did for financial analysis, the word processor

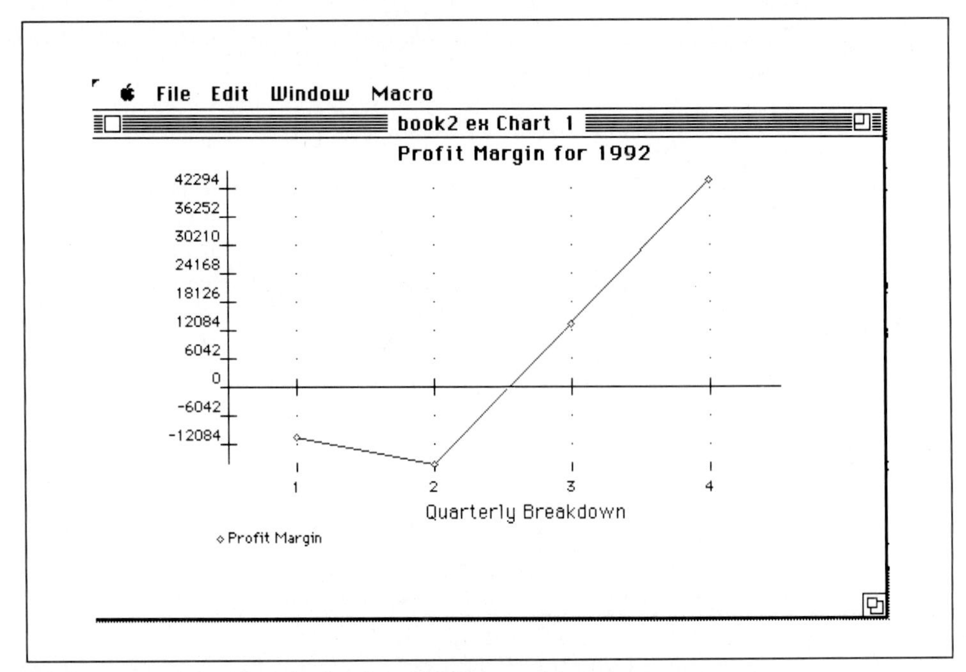

FIGURE 16.3 Graphics display from EXCEL spreadsheet program.

did for written communications. The **word processors** many of us use on our personal computers or workstations are descendents of the *line editors* that were created for mainframe and minicomputers a generation earlier. Fortunately, today's powerful word processors bear little likeness to those primitive line editors. Features often

FIGURE 16.4 EXCEL spreadsheet plot of $e^{\sin f} \cdot \sin(f^2)/5$.

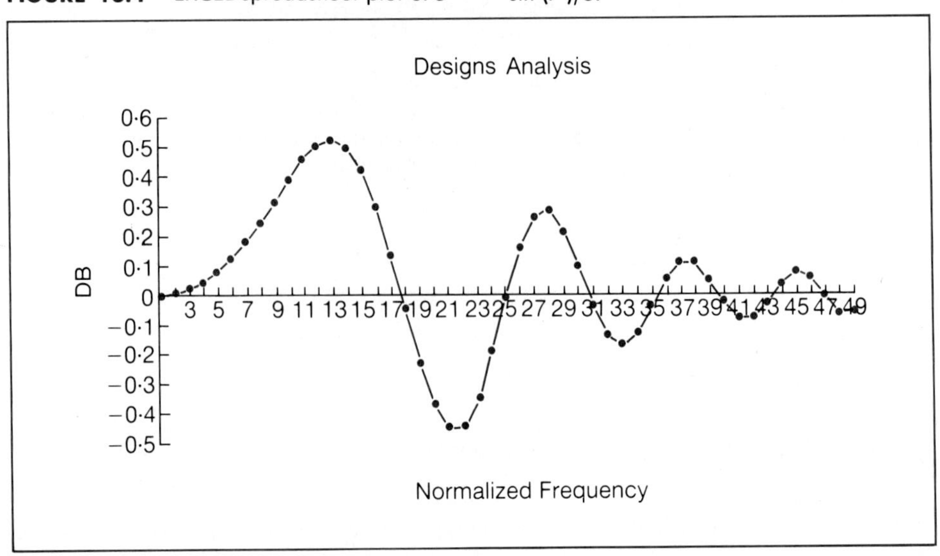

found in word processors include:

- Reformat sections or entire documents
- Move and replace blocks of text
- Use multiple character fonts
- Check spelling and grammar
- Create complex tables
- Automatically create table of contents
- Combine text and graphics

Newer word processing programs for the Macintosh, such as NISUS [Paragon Concepts, 1990], give you access to both graphics creation tools and word processing features.

Programs that help create and display graphics are another important application of computers, especially personal computers. One of the reasons that the Apple II computer became so popular was the availability of numerous recreational, educational, and tutorial programs that made extensive use of graphics.

The combination of input devices, such as mice and scanners, with new programs has had an impact on many areas of design. That combination, which let the user easily create and edit graphics, turned personal computers into design tools. Fields where these tools are used extensively include mechanical design, electronic design, video animation, and graphic arts design.

Besides the personal computer, another hardware development has had a significant effect on how we create documents with computers. The laser printer has made it possible to print documents that include both text and graphics at a quality level approaching that of professional publishers. Just as the introduction of the personal computer led to the development of a previously unheard-of application, the spreadsheet, the combination of the personal computer and the laser printer led to the development of another new application, desktop publishing.

Desktop publishing programs enable the user to design documents that combine outputs from a variety of word processors, graphic programs, and spreadsheets. These documents can be formatted and printed in a wide range of formats to meet the user's need. Common applications of desktop publishing include:

- Books and manuals
- Lecture notes
- Brochures and advertisements
- Newsletters
- Business cards
- Presentation transparencies

Figure 16.5 is a sample page created using PageMaker, a desktop publishing program from Aldus [Aldus Corporation, 1990]. The page has text imported from a word processor, a graphic imported from a graphic program, and a spreadsheet.

CHICO ELECTRONICS

1234 South Main
Suite 12A
Chico, CA 95999

ELECTRONICS FOR THE STUDENT

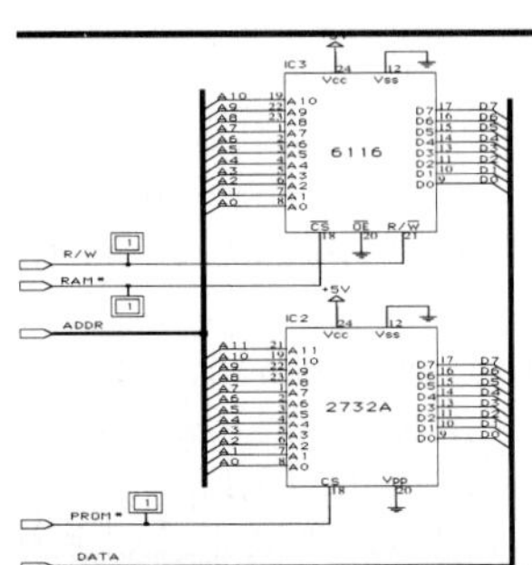

CE Amplifiers

The original "Student Special" that is the best buy in town or out. Amaze your friends with the power of electricity when it is directed to driving acoustic speakers. You will be able to blow the walls off your neighbors shower doors by switching to ***power boost max*** mode.

No other amplifier at any price can compare with this one. If you need an FHA loan to finance the purchase, we will help you apply, for a slight additional fee of course.

FEATURES

- DC to Light
- All colors
- Quick installation
- 117 vac input
- Drives any speakers
- 7 day guarantee

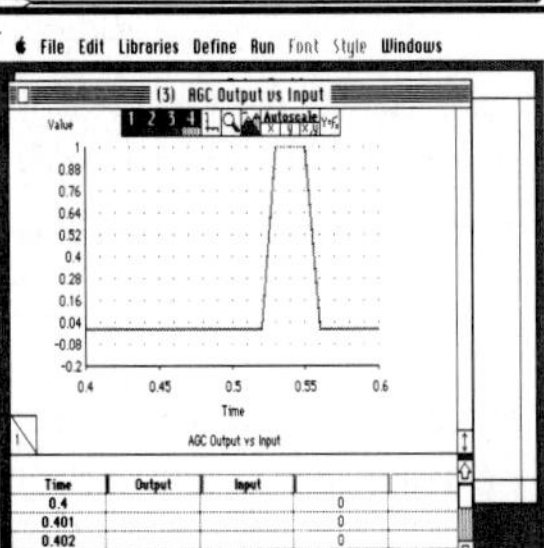

CE CD Players

The CE CD players were designed for the individual homeowner who wants to hear every last bit of noise produced by the manufacturer when they created the new high quality digitally mastered recordings.

Why wait to find out that your friends have better music systems than you? Bring in your check book and all available cash and we will fix you up with the best system in town.

This is an offer that may not be repeated again in your lifetime. It is the only chance you will have to acquire a CE CD at this low price.

FEATURES

- 0.0 volts noise
- two color plastic
- wall mount screws
- some assembly required
- simple controls
- $14.50 per CHANNEL

ACE Projections (SS)

	A	B	C	D	E	F
1	ACE Manufacturing - Profit Analysis (1992)			Growth=	0.1	
2						
3	Quarter	1	2	3	4	Total
4						
5	Sales	$500,000	$550,000	$605,000	$665,500	$2,320,500
6						
7	Manufacturing Costs	$210,000	$226,000	$243,600	$262,960	$942,560
8						
9	Marketing Costs	$150,000	$170,000	$155,000	$140,000	$615,000
10						
11	Engineering Costs	$150,000	$153,000	$156,060	$159,181	$618,241
12						
13	Profit Margin	($10,000)	$1,000	$50,340	$103,359	$144,699
14						
15						
16						

Product Specifications **Page 1**

FIGURE 16.5 Sample page created with the desktop publishing program PageMaker.

Programs are also available that let the user create animated graphic presentations. The obvious application here is the video game. Games represent a large market in themselves, but the same technology has been used for more serious applications. Animation is now used to create training and education programs for children and professionals.

16.3.4 Simulation Programs

One of the programs described in Chap. 11 was SPICE, a tool that simulates the operation of digital circuits. Simulation is a technique that can be used to study a wide variety of devices and systems. At first, simulations were written in traditional programming languages such as Fortran and assembly language. However, since many simulations have similar requirements for creating inputs and manipulating events, special languages including GPSS and SIMULA were created to simplify simulation development. These languages and others are very effective and have been used extensively.

The simulation tool we will now describe is somewhat different from the languages listed above. The program, Extend [Imagine That, 1990], could be described as an object-oriented simulation package. It allows you to use predefined "objects" and connect them together to build complex systems. The authors of Extend have created a number of different objects and included them in libraries for the user.

In Fig. 16.6 we show a simple circuit simulation that was developed using Extend. The circuit has four different types of objects: a signal source, inverters, an AND gate, and a plotter. The Sq (square wave) output from the signal generator goes to an inverter whose output is attached to one of the inputs of the AND gate.

FIGURE 16.6 An object-oriented simulation using Extend.

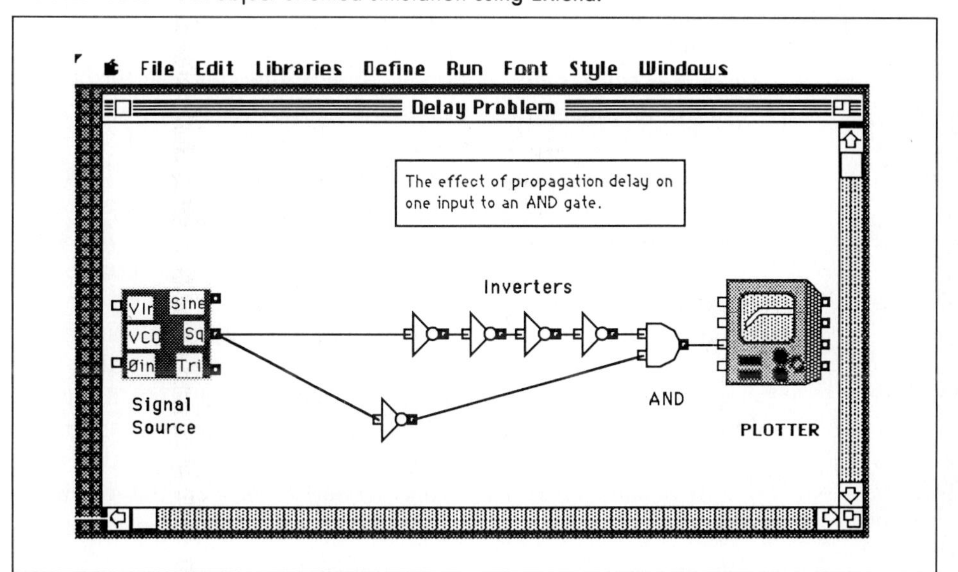

The output of the signal generator also goes to the first of four inverters. The output from the final inverter goes to the second input of the AND gate. If the inverters did nothing but invert the signal on their input, the circuit would always be zero. However, in addition to inverting the input signal, an inverter also adds a delay in the signal.

Because the inverters add delay, the output of the AND gate goes to 1 momentarily when the signal generator goes from high to low. The results of the delay can be seen in Fig. 16.7*a* and *b*. Part *a* shows the output of the AND gate for two transitions of the input signal, and part *b* shows an expanded view of one transition.

Situations like the one shown in Fig. 16.7 can lead to failures in electronic circuits. Simulation is often the easiest way to detect these and other design errors.

The simulation we created uses a few simple elements that were defined by the program's authors. Since the authors of the program could not hope to generate all the possible objects that people would want to use, the program also lets you create your own objects. You can define the symbols that are displayed when the object is selected and the list of operations the object will perform when it is activated.

Extend has several advantages over traditional simulation languages. First, it lets the user easily see how the different elements of the simulation relate to each other. Second, it provides graphical output of information rather than just tables of numbers. Extend does have one thing in common with simulation languages: complex simulations can be very costly to run in terms of computer time.

16.3.5 Productivity Tools

Not all application layer programs involve vast amounts of computations such as the spreadsheets and design tools cited above. Some have evolved just to simplify our daily work.

One new application that falls into the "noncomputational" area is electronic mail. **Electronic mail**, or **E-mail**, applications combine the storage and information processing capabilities of computers with communication networks. One of the goals of E-mail is to help create the "paperless office."

Most of us who have worked in a company or other large organization have at one time or another bemoaned the explosion of memos and other forms of paper-based communications that crossed our desks. Much of it is read and then quickly thrown away. Besides using up a lot of paper, the resulting waste must also be disposed of.

E-mail has several other advantages over the traditional methods of communication. Speed of delivery and reduced labor costs are often listed as major advantages, but there are also some potential disadvantages. First, E-mail requires significant increases in the storage capabilities of computer systems that support it. Second, communication costs can be significant for the networks that transmit the mail. This is particularly true when the networks are national or international in scope. Regardless of the potential disadvantages, most of us are quite willing to put up with them to have all the advantages that E-mail offers. (Note that we have provid-

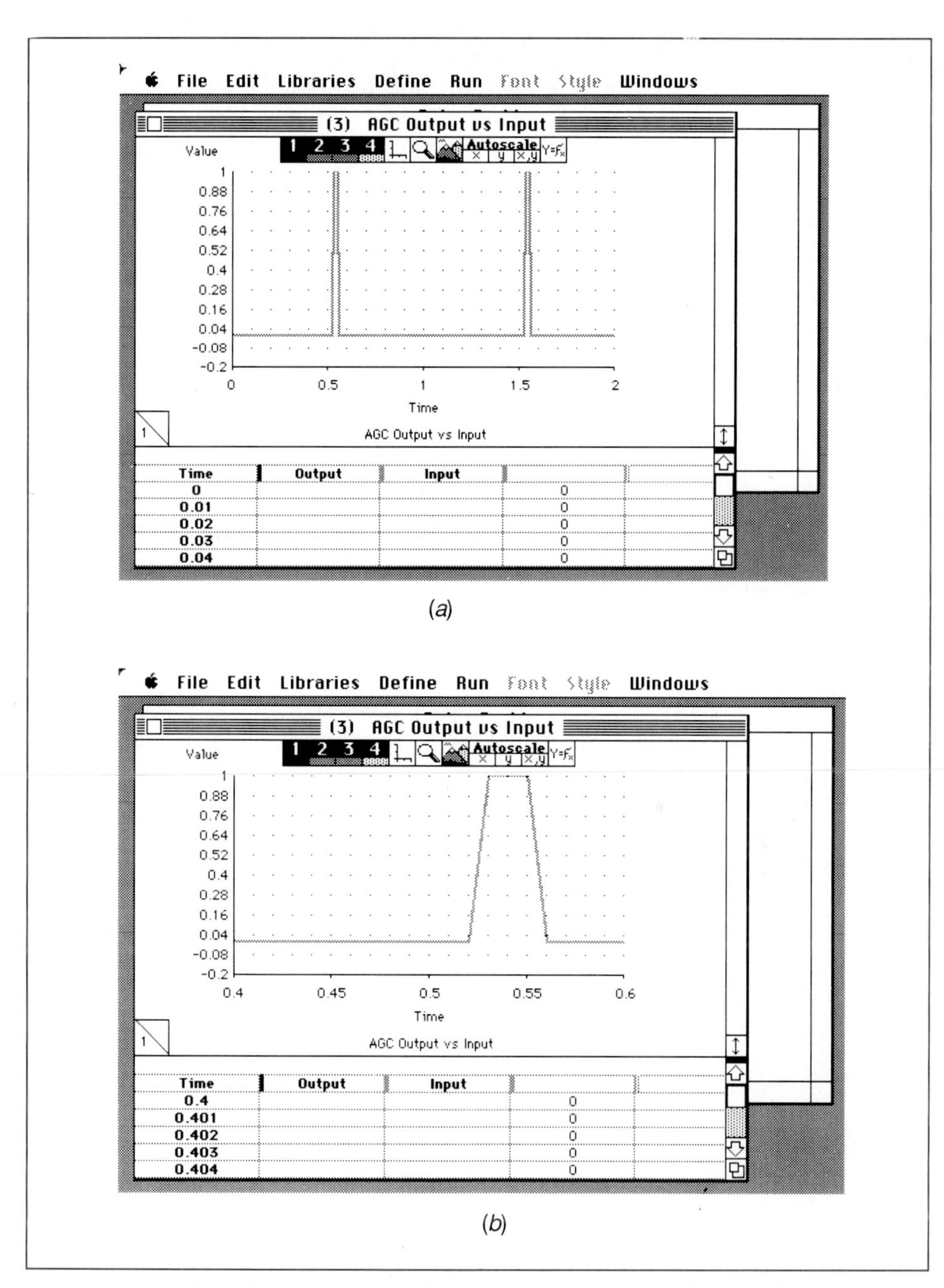

FIGURE 16.7 Extend simulation of a digital circuit: (*a*) two transitions of the input signal; (*b*) expanded view of one transition.

ed an E-mail address in this book to make it easy for readers to send us their comments and suggestions on improving this text.)

Many other productivity tools have been created to help us get our work done more efficiently. Two common tools we often find on many desktop computers are calendar and calculator programs. The calendar programs help us organize our schedules and remind us of important meetings and events. The calculator programs replace calculators that a few years ago cost as much as some computers we now have. The range of productivity tools available for personal computers is as varied as the types of users.

16.4 A NEW GENERATION OF LANGUAGES

In 1982, James Martin caused a bit of a stir in the programming world when he published the book *Application Development without Programmers* [Martin, 1982]. One of the premises of this book was that the supply of computers would far surpass the availability of programmers needed to write the programs required to make these computers productive.

In the book, Martin suggested that traditional languages such as Cobol, Fortran, and PL/1 do not permit fast application development. "Applications development did not change much for twenty years, but now a new wave is crashing in. A rich diversity of nonprocedural techniques and languages are emerging" [Martin, 1982]. At that time he felt that these languages would give end users the ability to create their own application programs without needing substantial help from a traditional programming and data processing staff.

Languages like Fortran and Cobol are often called procedure-oriented languages. In these languages, programs are created by combining blocks of instructions into procedures. The procedures are then executed in a prescribed sequence to give some desired result. Procedural languages have strict rules governing the syntax of the instructions and in some cases the way in which data elements can be manipulated. Nonprocedural languages, on the other hand, tend to be less structured and are more easily understood by humans.

Some of the languages and techniques Martin described have had a significant effect on how we access and use the data stored within the computer. In particular, SQL (pronounced "sequel" and the acronym for *s*tandard *q*uery *l*anguage) has proven to be a very useful tool for accessing data stored within a database without having to write programs in a procedural language.

Two other nonprocedural languages that have received much attention, especially for artificial intelligence and expert system applications, are LISP and Prolog. LISP was developed as a tool to analyze textual information. Prolog has been used to develop expert systems and create complex database access systems.

Martin was correct in that improved development tools are needed. However, his predictions of future trends seemed to have been based on the continuing tendency to use large mainframe computers and timesharing computers. He, like many others, failed to foresee how dramatically the personal computer and workstations would affect the way we use computers.

It is probably fair to say that as long as there are programmers, programming languages will continue to evolve. Some think the evolution will stop when our computers can listen to our spoken commands and create programs themselves based on our requests. After all, if Captain Kirk can have it, why can't we?

16.5 SUMMARY

In the last decade, the growth in the number of computer programmers has not kept pace with the growth in the number of end users. Fortunately, along with the development of inexpensive personal computers and workstations, we have seen the development of a new generation of tools. These tools have given end users the ability to develop their own applications. The combination of innovative hardware and software tools has brought the power of the computer to many new users. This, in turn, has enabled many end users to become application developers.

One advantage of having end users become developers is that there has been a rapid increase in the number and variety of applications available on the market. Unfortunately, since many of the new developers have not been schooled in programming techniques, the quality of the products is not always high. Now that we have unleashed a new generation of computer "programmers" onto the world, we need to find new and better ways to help them build higher-quality applications.

PROBLEMS

16.1 List application programs you have used.

16.2 Section 16.2 lists some services commonly provided by the OS. What are some other services provided by the OS?

16.3 What are some system services that can be accessed by clicking/dragging icons on the Apple Macintosh?

16.4 What are some system services readily available to the user on a UNIX system?

16.5 Why would a DBMS only want to protect "some" information in a file?

16.6 Using the example in Sec. 16.3.1 as a model, develop a database that would be useful for relating elements of schematics, parts lists, manufacturers lists, and printed circuit boards.

16.7 Why was development of the spreadsheet an important milestone in computer software?

16.8 What elements do you think go into your school's student records database? Draw a diagram similar to Fig. 16.1 to show the relationships.

16.9 Develop a spreadsheet to calculate your grade for this class. Include provisions for your homework assignments, quizzes, and exams.

16.10 Using a spreadsheet with which you are familiar, calculate the integral

$$\int (\sin \phi^2)^3 \, d\phi$$

between 0 and $\pi/4$ and plot the results as a function of ϕ.

16.11 Using a spreadsheet you are familiar with, plot the function

$$Y = X^4 - 1.7X^3 - 8.13X^2 + 3.545X + 10.01$$

and calculate the zeros of the function.

16.12 Show how Profit Margin would change for growth rates at −5, 0, 25, and 100 percent in the spreadsheet shown in Fig. 16.2.

16.13 Using the data from Fig. 16.2 and the results of Prob. 16.12, plot Profit Margin as a function of Growth.

16.14 What are other features of word processors that you find useful? (Refer to Sec. 16.3.3.)

16.15 What is the feature you *dislike* most on your word processor?

16.16 How could the situation described in Fig. 16.6. cause an error to occur?

16.17 Why have applications such as spreadsheets created more computer users?

16.18 Besides E-mail, what are some other productivity tools you have seen or used?

16.19 Why is creating a spreadsheet similar to programming? How does it differ from programming?

16.20 What are some additional disadvantages of E-mail compared to paper mail?

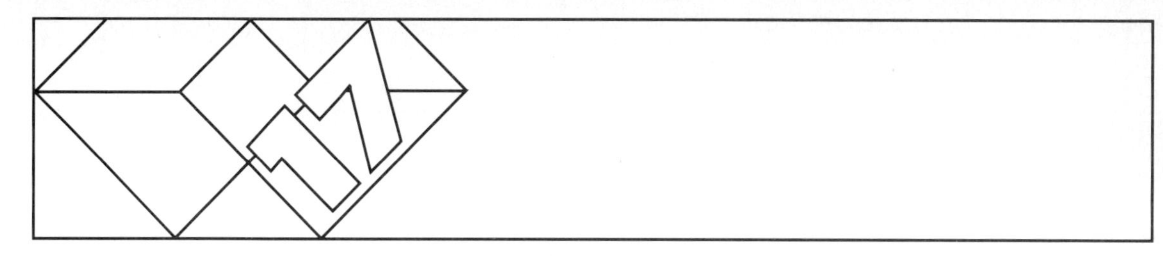

Software Tools

Software systems are steadily becoming larger, more complex, and more sophisticated. Twenty years ago, programs of 10K lines seemed exotic and enormous. Now people talk of 10M line software systems that could have major effects on the future of our earth. We learn almost daily about larger and larger software projects with more and more potential impact on society.

Because of these trends, researchers have devoted considerable effort to developing tools to help design, implement, check out, and maintain software. These tools are collectively known as **CASE**, or **computer-assisted software engineering**. In this chapter, we will look at some components of a typical CASE package.

Our presentation of CASE tools will be organized on a model for software development known as the life-cycle model. This model breaks the development process into several sequential phases. We will look at each CASE tool in the phase in which it would usually be *most* applicable. (Some tools could be used in two or more phases.)

17.1 THE LIFE-CYCLE MODEL

One version of the classic model for software development, the **life-cycle model**, is shown in Fig. 17.1. It contains five steps:

- Determine the system requirements.
- Design the system.
- Implement the design.
- Test the implementation.
- Maintain the system.

We will briefly describe each step and then look at some applicable CASE tools.

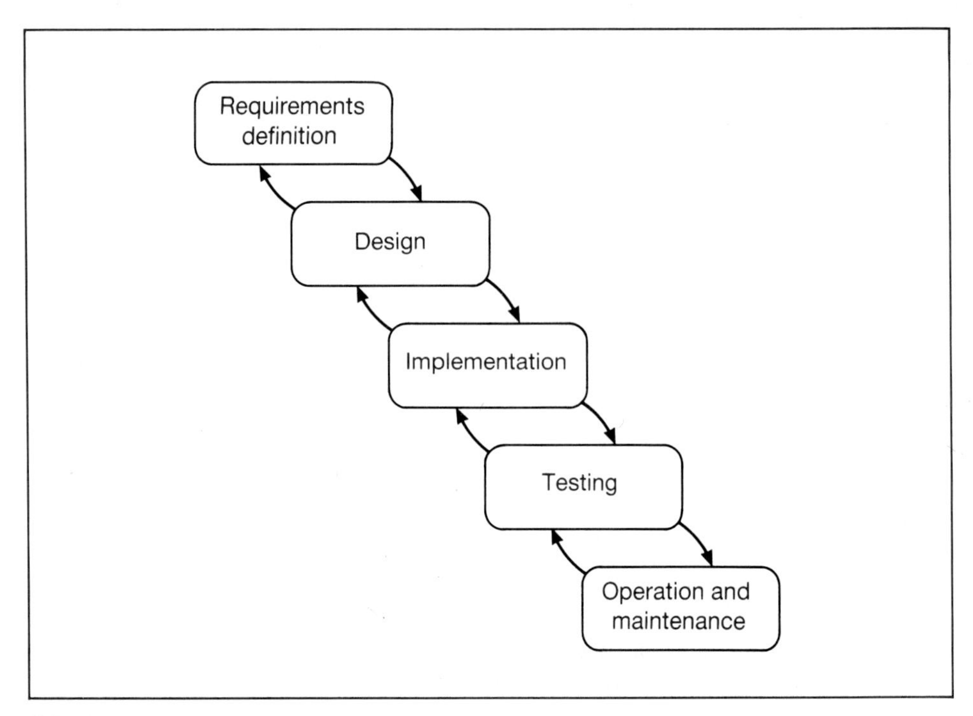

FIGURE 17.1 One version of the life-cycle model for software development.

17.2 REQUIREMENTS SPECIFICATION

Because of the proliferation of computers into every aspect of society, requirements for computer-based systems are often written by people who have little or no computer experience. The situation is similar to asking a person who is just learning to fly to specify requirements for an aircraft cockpit. These requirements are taken to an aircraft company, where an aircraft with the specified cockpit is designed, built, and delivered. The customer climbs into the plane, begins trying things out, and, not unexpectedly, finds many problems that must be corrected. All changes must be incorporated into the existing cockpit, which can lead to *kludges* (systems made up of poorly matched components) and future maintenance problems.

A better situation is to have our customer develop the specifications using an existing cockpit mock-up. This can be done as he or she is learning to fly. Ease of use and accessibility of controls could be determined, suitability of displays tried, and so on. This mock-up would obviously not fly, but it would have a number of realistic components. For example, the radar screen might show realistic displays, but they might be coming from a VCR.

The idea behind this approach to requirements definition is called **fast prototyping**. The user sees examples of what the system will do and what the interfaces will look like and gets to "try the system out." The demonstrations are normally only simulated, but they provide a valid "feel" of a proposed system. For example, a database retrieval request in a prototype might not go to an actual database. However, a sample record will be displayed in the proposed format.

To illustrate this approach, let's consider a consulting situation in which one of the authors was involved. He helped a friend with the development of a computerized system for use in her irrigation company's display room. She had never used computers before, and at that time no applicable commercial packages were available. Therefore, the project essentially was started from scratch.

The author first prototyped some sample screens that customers could access. One such screen (not the final version) is shown in Fig. 17.2. A customer could go up to the computer and use a mouse to move the cursor to a category. Clicking the mouse would display information about the available components in this category. After the customer had made a selection, the same sequence would be used to print an order form that would be processed by a clerk.

Prototyping let the owner try various screens before the system was ever designed. The internals of the system were not developed, so clicking a category did not go to a database and retrieve information. Instead, a set of dummy displays were used.

FIGURE 17.2 Prototype for a sprinkler information and purchase system.

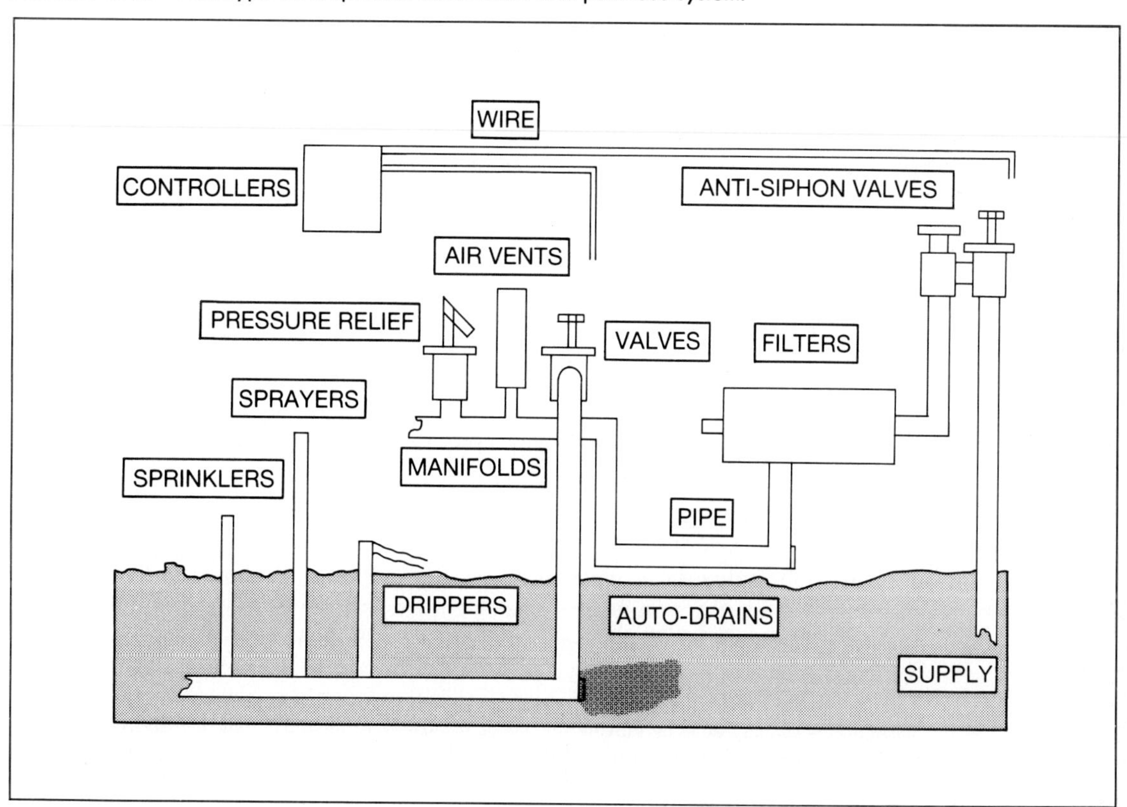

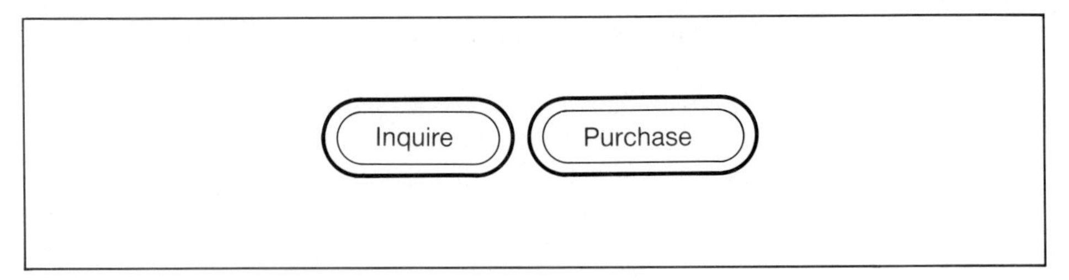

FIGURE 17.3 Initial push buttons for prototype screen.

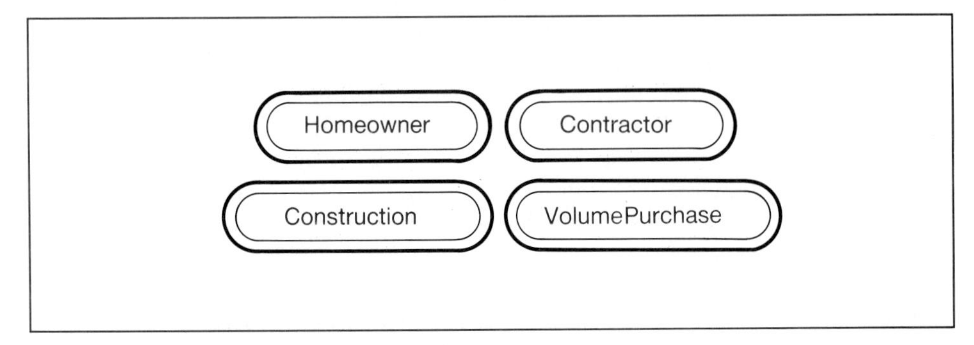

FIGURE 17.4 A second prototype of proposed push buttons.

The owner quickly realized there were some problems with the screen in Fig. 17.2. First, the Drippers category had to be divided into Emitters, Fittings, and Hose. Then she realized that some people would be using the system to buy components, while others just wanted information. Therefore, two push buttons[1] were added, as shown in Fig. 17.3.

Next, the owner realized that she had different pricing categories. Contractors received one discount, people building new homes received another, and people buying large quantities of a given component received still another.

This led to the selection categories shown in Fig. 17.4. Unfortunately, it became obvious that people could not tell ahead of time if they qualified for volume discounts. Furthermore, they might receive such discounts on some components but not on others. Therefore, the VolumePurchase selection was removed, and the volume discounts were incorporated into the pricing forms. These changes led to the configuration shown in Fig. 17.5.

Next, the focus shifted to the invoice forms that a customer filled out just before purchase. One category of these forms, as first designed, is shown in Fig. 17.6. The customer would use a mouse to make choices, then use a keyboard to type in quantity. The system would insert the price for that quantity, incorporating any volume discounts.

[1] Selection is actually done by moving the cursor inside one of the boxes and then clicking the mouse. However, people working on user interfaces call these "push buttons," as a carryover from the mechanical analogy.

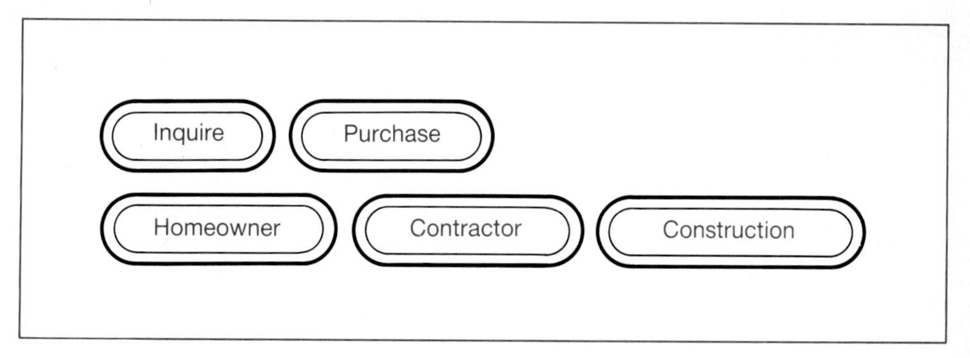

FIGURE 17.5 The final set of push buttons for the application program.

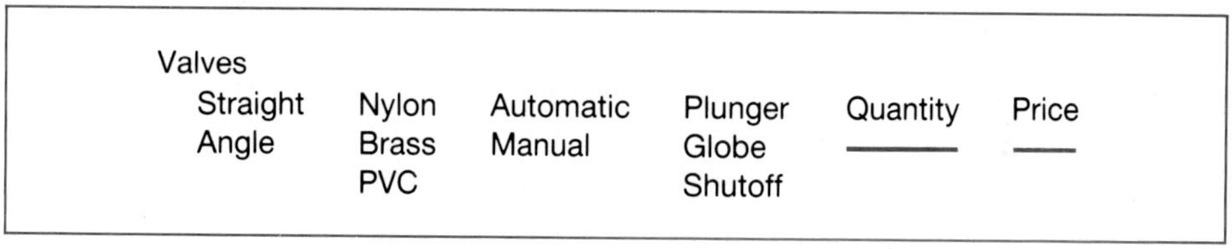

Valves

Straight	Nylon	Automatic	Plunger	Quantity	Price
Angle	Brass	Manual	Globe	———	——
	PVC		Shutoff		

FIGURE 17.6 Order specification box for our prototype.

Switching between the mouse and the keyboard did not seem convenient to the store owner. This resulted in elimination of the keyboard and incorporation of the mouse to choose parts of a number (such as the hundreds digit). The customer could adjust each digit with an up-or-down arrow. That change is shown in Fig. 17.7.

Since you should now understand how and why prototyping is done, there is no need to continue any further with this example. The key point in this approach is that before the system was ever formally designed, the user had a chance to try it out. Such trial and error resulted in many important improvements. Imagine the effort that would have been expended if the original specifications had been used to design and implement a fully working system.

Sometimes prototyping is used strictly as the basis of the requirements phase. However, some current development systems produce code as the prototypes are formulated. In such systems, prototyping spans the first three stages of the life-cycle model.

FIGURE 17.7 A mouse-driven alternative to typing in quantity desired.

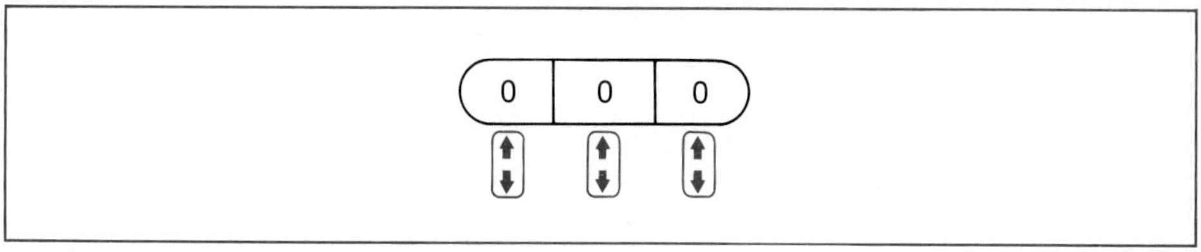

0 0 0

17.3 DESIGN

System requirements might be determined with prototyping, or they might be determined by some other means. Whatever the situation, the next step after requirements determination is system design. There are many techniques to simplify this step in the life-cycle model. One of the more popular current ones is data flow diagrams (or DFDs for short) [DeMarco, 1980]. There are several software tools available to help the designer in the generation of his or her DFDs.

Data flow diagrams (DFDs) graphically describe the flow of data among processes, files, sources, and terminators. The authors see these diagrams as an embodiment of many of the concepts in Nikolaus Wirth's insightful book entitled *Data Structures + Algorithms = Programs* [Wirth, 1976]. With DFDs we can focus on the various data items and then see what is needed to transform them (i.e., the processes).

A very simple illustration of a DFD, based on our previous sprinkler example, is shown in Fig. 17.8. In this figure, we have the following correspondences:

- Sources/terminators are shown by rectangles.
- Files are shown by two parallel lines.
- Processes are shown by ellipses (often called "bubbles").
- Data flow is shown by directed lines (lines with arrows).

The interface bubble (Get/Hndl Menu Selection) displays screens to the customer and inputs selections and specifications. If the request is for product information, this module goes to the product information database with a product identifier. The database retrieves the information, and it is displayed to the customer. If the customer is making a purchase, the module sends a sales record for each item to the module that generates invoices. When all desired purchase items have been specified, the invoice module generates the invoice and issues copies. These copies go to the customer, the cashier, and the inventory clerk (who gets the items from the stockroom). A copy of the invoice also goes to a module that updates the inventory database and the sales database.

The bubbles represent code modules, which are typically designed using structured English or decision tables. A good rule of thumb is that a module should usually be one-half to one page of code. If it is longer, then the diagram is "refined" to another level, and that module becomes a new data flow diagram. In our example, it is quite probable that the user interface would have to be further developed.

Besides the DFDs and the modules, one must maintain a **data dictionary**. This lists all the data items (labels on the directed edges, records in files) and defines them by giving their components parts. Tools for DFD development usually contain:

- A graphics editor for developing the diagrams themselves
- An editor for doing the data dictionary
- An editor for designing the modules

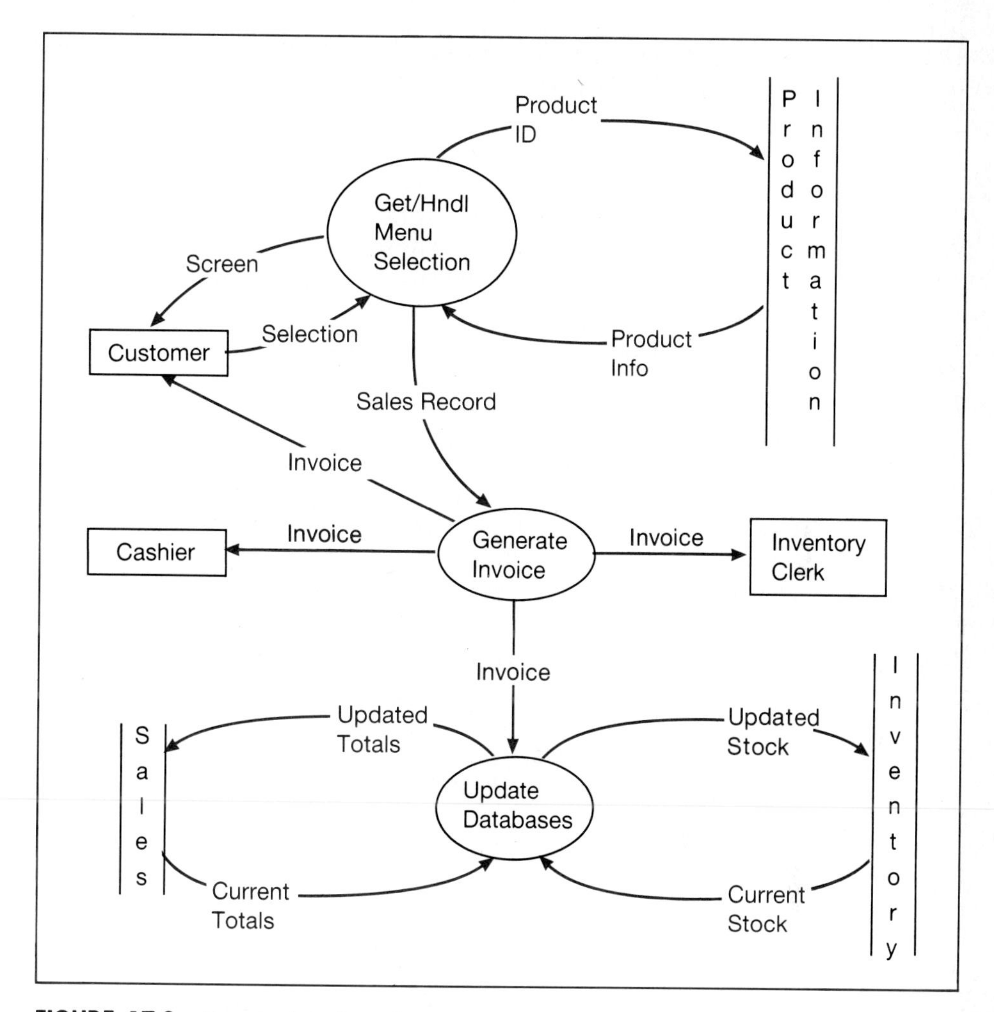

FIGURE 17.8 Sample data flow diagram.

These tools also might check for errors such as:

- A bubble that has not been defined
- Data out of a bubble containing items neither input to the bubble nor computed within the bubble
- Data items missing from the data dictionary
- Syntax errors in the structured English or data dictionary definitions
- Inconsistent references to a file

After the DFD is completed, the modules must be designed. Typically this is done using flowcharts, pseudocode [Bailey and Lundgaard, 1983], or Nassi-

Shneiderman diagrams [Pfleeger, 1987]. Like the editors for DFD development, design tools often contain facilities to help you formulate such charts or diagrams.

You have already seen a sample flowchart in Chap. 15, where we gave you one for a program to compute natural logarithms. Figure 17.9 gives a pseudocode design for that same program, with the pseudocode in the boxed regions and the other text giving documentary description. (Some people use more Pascal-like pseudocode; this particular form is known as *structured English.*) Figure 17.10 is a corresponding Nassi-Shneiderman design for the program.

It is difficult to overemphasize the importance of careful design. Novice programmers might assume that most of their errors will occur in the implementation phase, when they are coding their design. However, research has shown that nearly two-thirds of the programming errors actually occur in the design phase [Glass and Noiseux, 1981]. Furthermore, the later in the life-cycle model that the errors are discovered, the more expensive it is to fix them. A design error not discovered until the maintenance phase could easily cost 100 times as much to fix than if it were discovered in the design phase [Brooks, 1982].

17.4 IMPLEMENTATION

The next step in the life-cycle model is often the one most people think of when you mention computing system development. Having designed the system, we now implement it; that is, we *program* it.

17.4.1 Interactive Compilers and Debuggers

Probably the most common tool now used by programmers is the interactive compiler/debugger. It allows you to compile a program and then execute it using a number of interactive features that aid in debugging. Some typical features are:

- Single-statement stepping through the program
- Setting a special point in the program (called a breakpoint) and then executing up to that point
- Tracing the values of variables as the program executes
- Skipping through or around subprogram calls so that you can work on just one routine at a time
- Displaying the current status of files
- Editing the source code and then using this new edited code immediately

The pull-down "debug" menu in Fig. 17.11 shows the options available on a typical interactive compiler/debugger. This one is a Microsoft system running on a Macintosh computer [Microsoft Corporation, 1985].

Beginning of natural logarithm program

This program computes the natural logarithm of an input quantity. That input quantity must be positive, or it will be rejected. The program uses one of two series expansions, depending on whether the input quantity is greater than one, or between zero and one. This is done to speed convergence of the series.

The program also inputs a maximum number of terms that should be used in the series expansion, and a minimum change in the value of the series. The program terminates the series when the maximum number of terms has been reached, or the value of the last term computed is less than the stated minimum.

This loop obtains the quantity, a maximum number of terms, and a minimum change in the series. The loop repeats if the user types a non-positive number, a non-positive maximum number of terms, or a negative minimum change in the series.

Begin loop to obtain valid inputs

Input the quantity for which the logarithm is desired, the maximum number of terms in the series, and the minimum change in the value of the series.

Input quantity, maximum number of terms, and minimum change in series

Test for invalid number.

If the input quantity is less than or equal to zero then type an appropriate message and repeat input loop

Test for invalid number of iterations.

Otherwise, if the maximum number of terms is less than or equal to zero then type an appropriate message and repeat input loop

Test for invalid minimum change in series.

Otherwise, if the minimum change in the series is less than zero, then type an appropriate message and repeat input loop Otherwise, all input quantities are o.k. so exit the loop to obtain valid inputs End of valid input loop

The maximum number of iterations is input as an integer, which is more natural for the user. Here it is converted to a real to save repeated conversions in a later IF test.

Convert integer maximum number of terms to a real

Assume that there will be another term computed.

Set flag to indicate that another term is to be computed

Initialize the series summation and the counter of terms.

Initialize series expansion sum to zero and the number of terms in the series to zero

Decide which series to use.

FIGURE 17.9 Pseudocode for sample module design.

Series for number greater than 1.

$$\sum \frac{(\text{Quantity} - 1.0)^j}{j * (\text{Quantity})^j}$$

If the input quantity is greater than 1, then the repeated multiplier will be (quantity -1)/quantity, and all terms in the summation will be positive

Series for positive number less than 1.

$$\sum (-1)^j * \frac{(1.0 - \text{Quantity})^j}{j}$$

If the input quantity is less than 1, then the repeated multiplier will be (1 - quantity); furthermore, there will be a "-1" multiplier because the terms in the summation alternate signs

Loop to compute the series. Loop terminates when the maximum number of terms have been used, or the size of the last term is less than the specified minimum.

Begin loop to compute series expansion

Compute term by using repeated multiplication of previous term, rather than raising to power each time.

Replace current term by current term times multiplier

Add one to counter, and add term to series summation.

Add one to loop counter, and add term/loop counter to the series summation

Check for loop termination conditions. First one is maximum number of terms. Flag is changed to zero to indicate subsequent exit from loop.

If the loop counter has reached the user-specified maximum number of terms, then output an appropriate message and set flag to indicate that the loop should be terminated

Second termination condition is less than minimum change in value of approximation.

If the loop counter has reached the user-specified maximum number of terms, then output an appropriate message and set flag to indicate that the loop should be terminated

Check flag for possible loop exit.

If the flag has been set to indicate loop termination, then exit the loop

End of loop to compute series expansion

Approximation completed. Type result.

Output the series expansion

End of program

FIGURE 17.9 *(Cont.)*

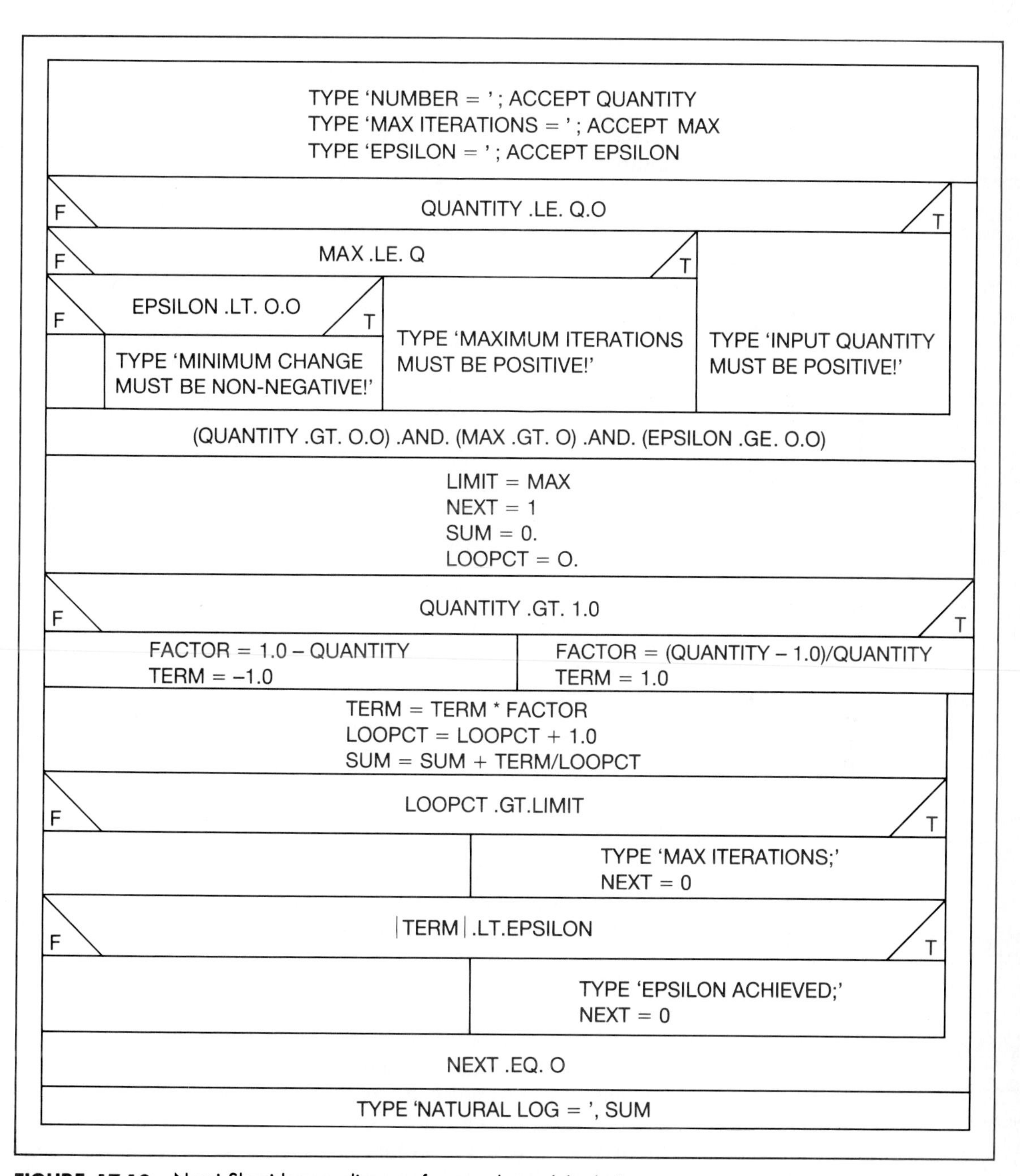

FIGURE 17.10 Nassi-Shneiderman diagram for sample module design.

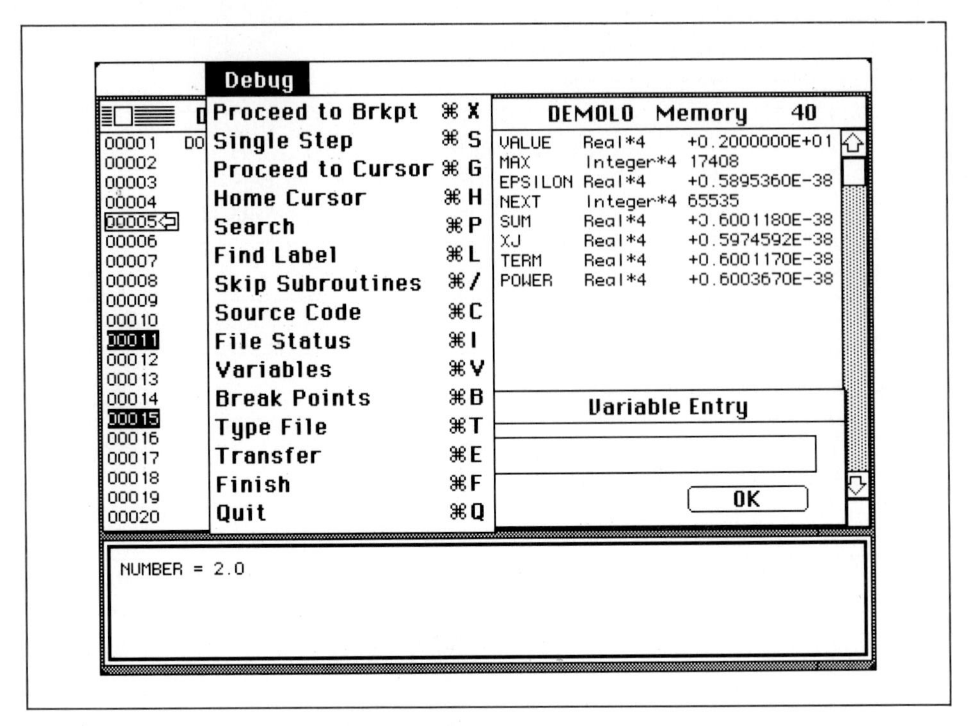

FIGURE 17.11 Sample of an interactive compiler/debugger screen during a session.

Figure 17.11 also illustrates a screen at one point in a program development session. The program has just displayed NUMBER =, and the user has responded with 2.0. This is shown in the I/O window at the bottom of the screen. At the upper left part of the screen, we see an arrow and box around 00005, which indicates that the system is about to execute statement 5. If the user selects a Single Step execution from the menu, the system would execute statement 5 and wait at statement 6. There are also two statement numbers in inverse video (00011 and 00015), showing special statements called *breakpoints*. If the user selects a Proceed to Breakpoint instead of a Single Step execution, the system would stop at the first breakpoint, statement 11.

In the right part of the screen, we see the variable trace feature. The box in the middle right allows the user to enter variables to be traced. The box in the upper right shows the type and current value of the chosen variables. Note that at this point the 2.0 has just been read, so the trace shows that VALUE now has a value of +0.2000000E + 01. The other variables are as yet undefined, so their values are indeterminate; they appear as random "junk" left from previous operations.

Interactive compilers/debuggers are available from several vendors for most popular high-level languages. There are also versions available for certain assembly languages. They can be *extremely* useful in helping programmers develop their software.

17.4.2 Interactive Input Interfaces

If requirements definition for a particular system was done via prototyping, then the user interfaces have already been designed. If not, then one of the first steps in the implementation should be the development of good user interfaces. Several tools are now available to aid a programmer in this task. These tools, called **interface builders**, aid in designing screens, windows, standard menus, pull-down and pop-up menus, help screens, error messages, and forms. Typically, the programmer:

- Enters an editor and constructs an image of a panel using special editor commands
- Defines the window which will contain that panel (including such characteristics as position, size, and title)
- Specifies the locations and types of fields in the panel

As an example of the type of window one might want, Fig. 17.12 shows an informative and well-designed help window from Microsoft WORD.

Tools for developing interfaces have been discussed here instead of in the design section because resulting interfaces are intended to be part of the *implemented* system. These tools do not just edit screen designs; they generate code to be used as

FIGURE 17.12 Help screen from Microsoft WORD.

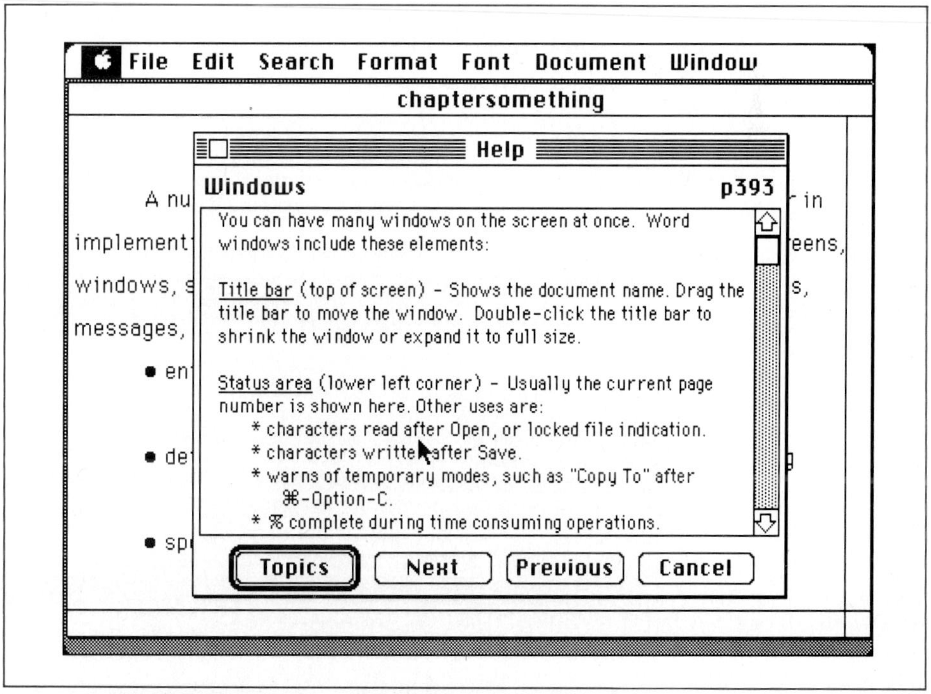

part of the delivered system. We might also want to use some tools strictly as editors to develop good interface designs during the design phase.

17.4.3 Graphics Development

Consider a scenario similar to the one described in the previous section, except in this case the programmer wants to use graphics interfaces. For example, suppose geographic regions are chosen using a mouse with a map of the United States. Corresponding sales summaries are then output as pie charts on a color graphics printer. The software involved in graphics I/O can be especially long and difficult, particularly considering the large number of different devices now available. (Recall the discussion of device drivers in Chap. 8.)

Graphics development toolkits usually contain a set of graphics functions and a set of device drivers. These can greatly ease the programmer's implementation task.

17.4.4 Database Interface Development

Many application programs access databases. There are several tools available to help you interface your program to various existing databases. Some also help you design the databases themselves, if necessary.

Figure 17.13 shows a sample design of a database for storing order forms. In this system, the designer does not have to write any code—the system does it all from the forms design. Figure 17.14 shows an inquiry of this database, using an approach known generally as *query by example.* The programmer has specified that he or she wants all order forms which have an item of type S in a quantity greater than 100.

FIGURE 17.13 Designing a database for order forms.

ORDER FORM

Order Number: 99999
Order Date: 99/99/99
Customer #: 99999-999
Name: AAAAAAAAAAAAAAA
Address: AAAAAAAAAAAAAAA
AAAAAAAAAAAAAAA
AAAAAAAAA 99999

Rcrd#	Item#	Type	Description	Quantity	Unit	Total Price
99999	99999	A	AAAAAAAAAAAA	999999	.99	999999.99

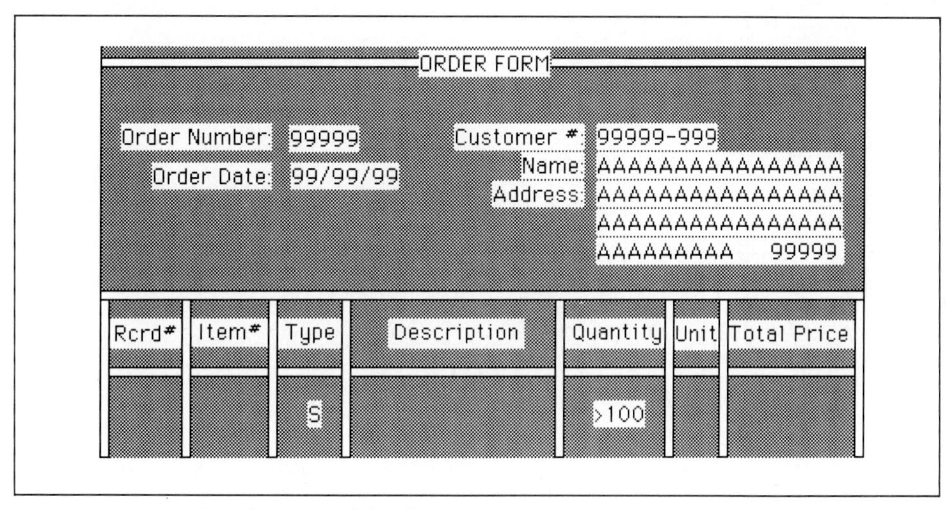

FIGURE 17.14 Sample query of database.

Again, no code is written directly by the programmer; the tool generates it all from the template.

17.4.5 Code Mapping and Navigation

An interesting approach to software development is to start with a flow or structure chart and then develop the code in a stepwise fashion. For example, suppose we are developing an interactive package that can be used as a calculator. Suppose further that a program to compute natural logs is to be developed as part of this package. The complete control map of the system will contain one module for natural logs, as shown in Fig. 17.15*a*. We expand this node into the control map, shown in part *b*. Lastly, we work on one node within that map, as shown in part *c*.

This is a simple example, and most applications of the technique would be larger. They would probably start right from the Natural Logarithm module instead of using the intermediate step shown in part *b*. However, we just want to show you the general idea.

Typically, a CASE tool for this approach would give you:

- Graphics facilities to generate the control maps
- Structure analyzers to develop control maps from existing code
- Graphics facilities to move through the control map
- Expansion capabilities so that you can focus on one node and see or develop a code map for it
- Ability to transfer to a program language editor to write code when you have expanded to enough depth
- Capabilities to look at file designs used in your program

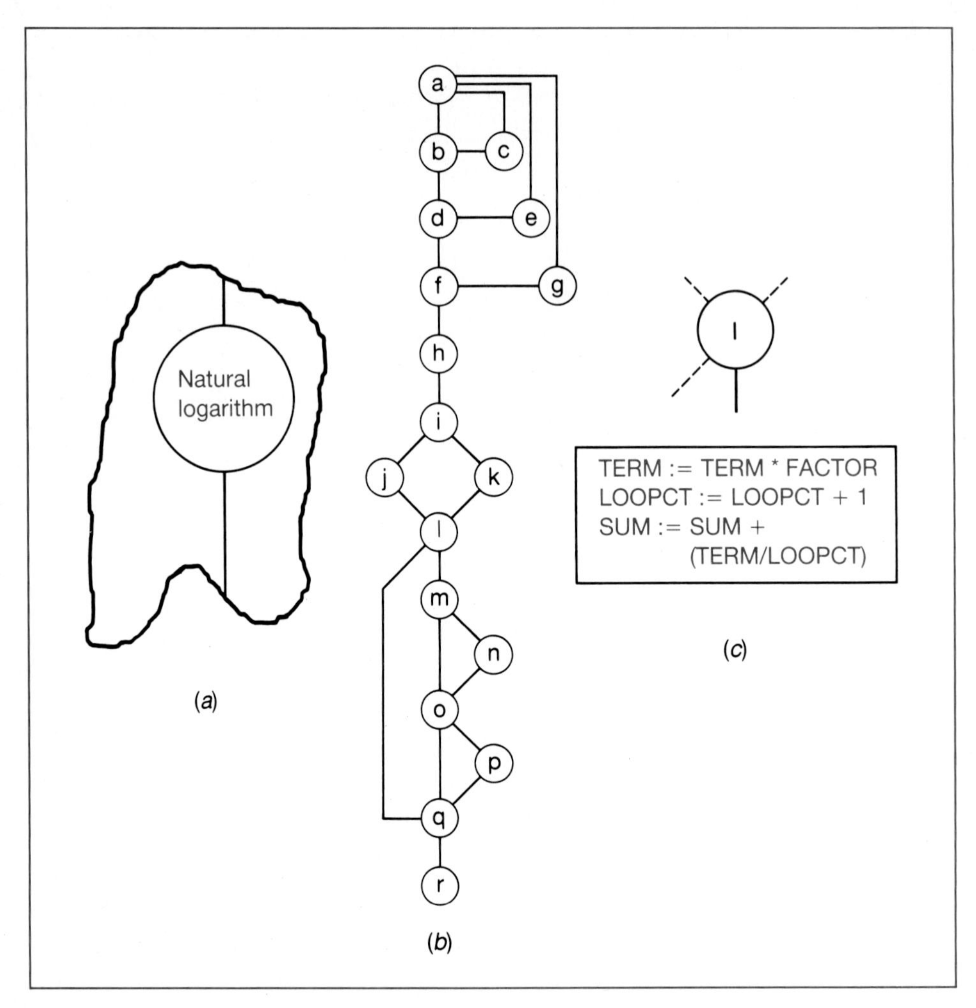

FIGURE 17.15 Development by refining structure charts: (*a*) module; (*b*) structure chart for module; (*c*) one loop within module.

17.5 TESTING

Some testing will obviously be done during implementation. However, once the system has been implemented, it should be subjected to a formal and rigorous test regime. In this section, we will look at tools to help with testing.

17.5.1 Test Coverage Analyzer

Large programs typically do not follow a linear execution path; they have many IFs and other control statements that permit alternate execution paths. In debugging such a program, it would be advantageous to know that all the possible paths have been followed, or "exercised." A test coverage analyzer can help in determining this.

Let's assume we are working on a large program called BIGJOB. Typically, a test coverage analyzer would be implemented as a **preprocessor**, a program that reads in our program BIGJOB and augments BIGJOB before passing it on to the appropriate compiler. The augmentation would usually consist of:

- Determining the possible execution paths in BIGJOB
- Generating some easy-to-follow representation of these paths
- Storing this representation into a database
- Adding counting code to BIGJOB to determine how often each of the paths is executed

After a test run of BIGJOB, the values of the counters would be incorporated into the path representation in the database. We could then study the flow patterns of the program. This process is diagramed in Fig. 17.16.

As an example, suppose we go back to the logarithm program that we introduced earlier. Let's modify the design slightly in the interest of a complete but more concise illustration. First, we'll add a loop so that we can repeat the entire program as often as desired and thus test various paths. Second, we'll simplify the input validation and test all the inputs in one step. This is done purely to cut down the number of paths for the example; the original version is more user friendly. These two changes are shown in Fig. 17.17.

FIGURE 17.16 Operation of test coverage analyzer.

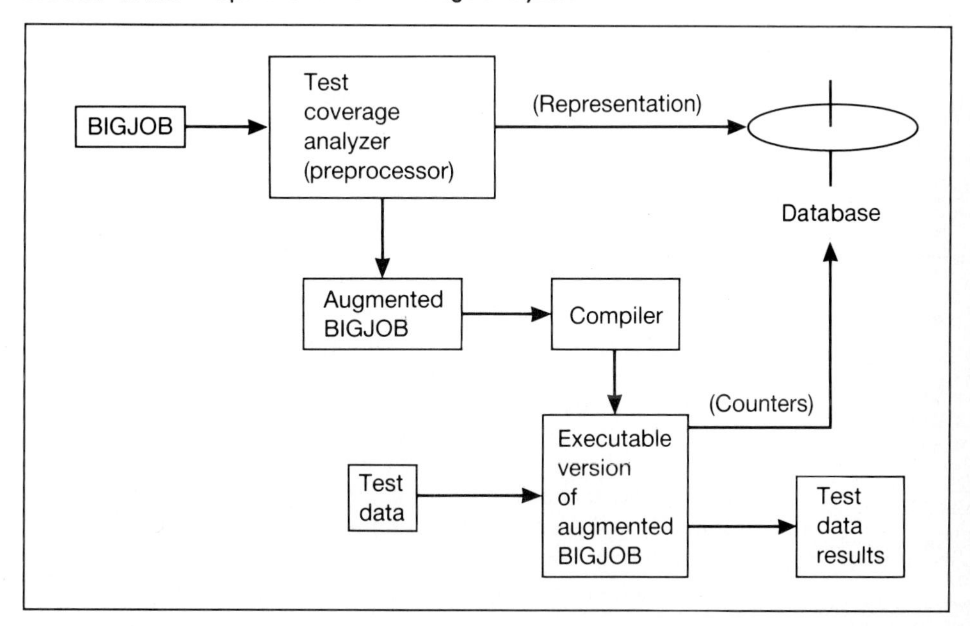

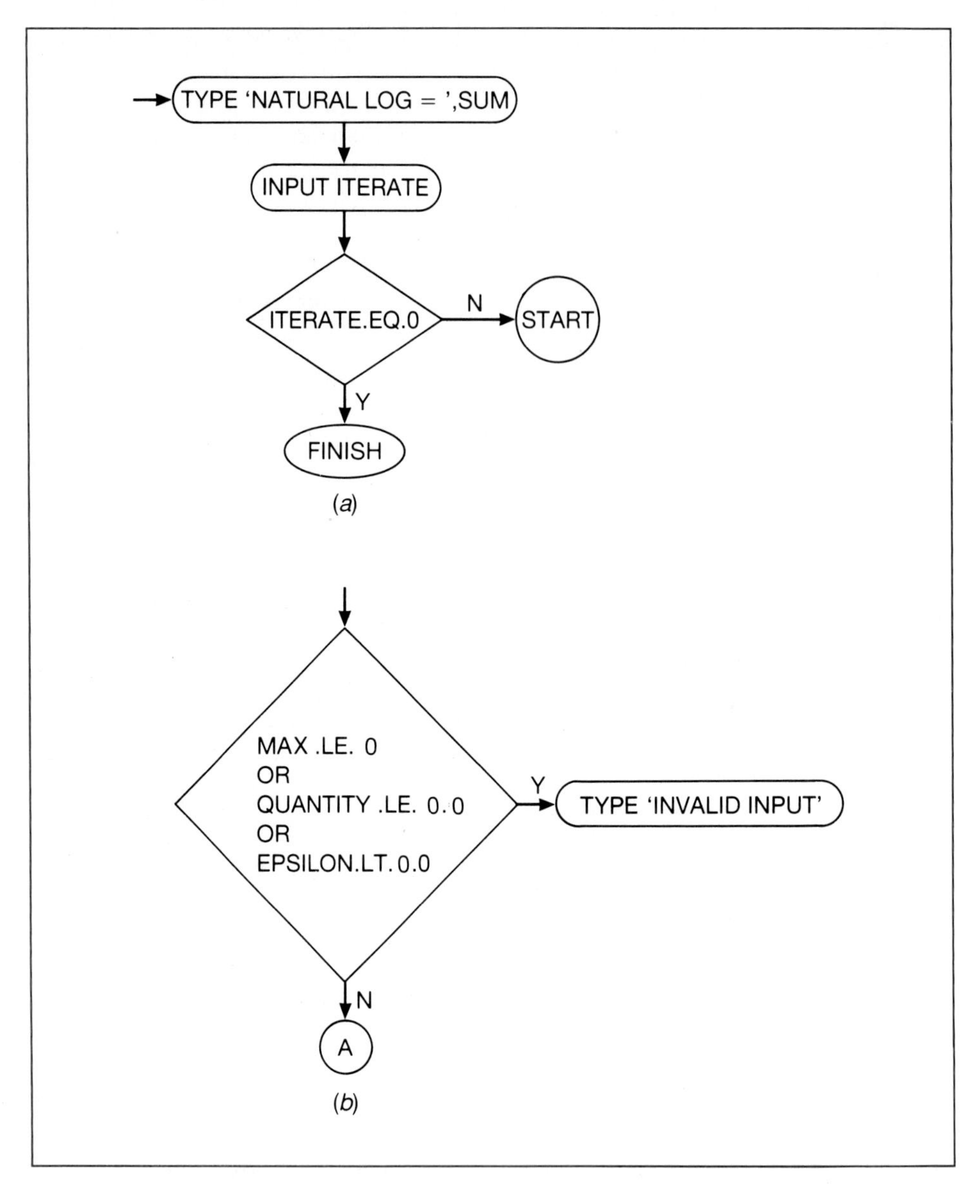

FIGURE 17.17 Changes to the flowchart for our logarithm program: (*a*) added loop for entire program; (*b*) revised input validation portion.

If we write a program corresponding to this flowchart and run it through a test coverage analyzer, we might get the structure chart shown in Fig. 17.18. (The flowchart is also included in this figure for your reference.) Such a structure chart graphically isolates the important alternative paths which our tests should cover.

The analyzer inserts code into the program to count the number of times each of these paths is executed during a given test run. A partial dump of a screen produced

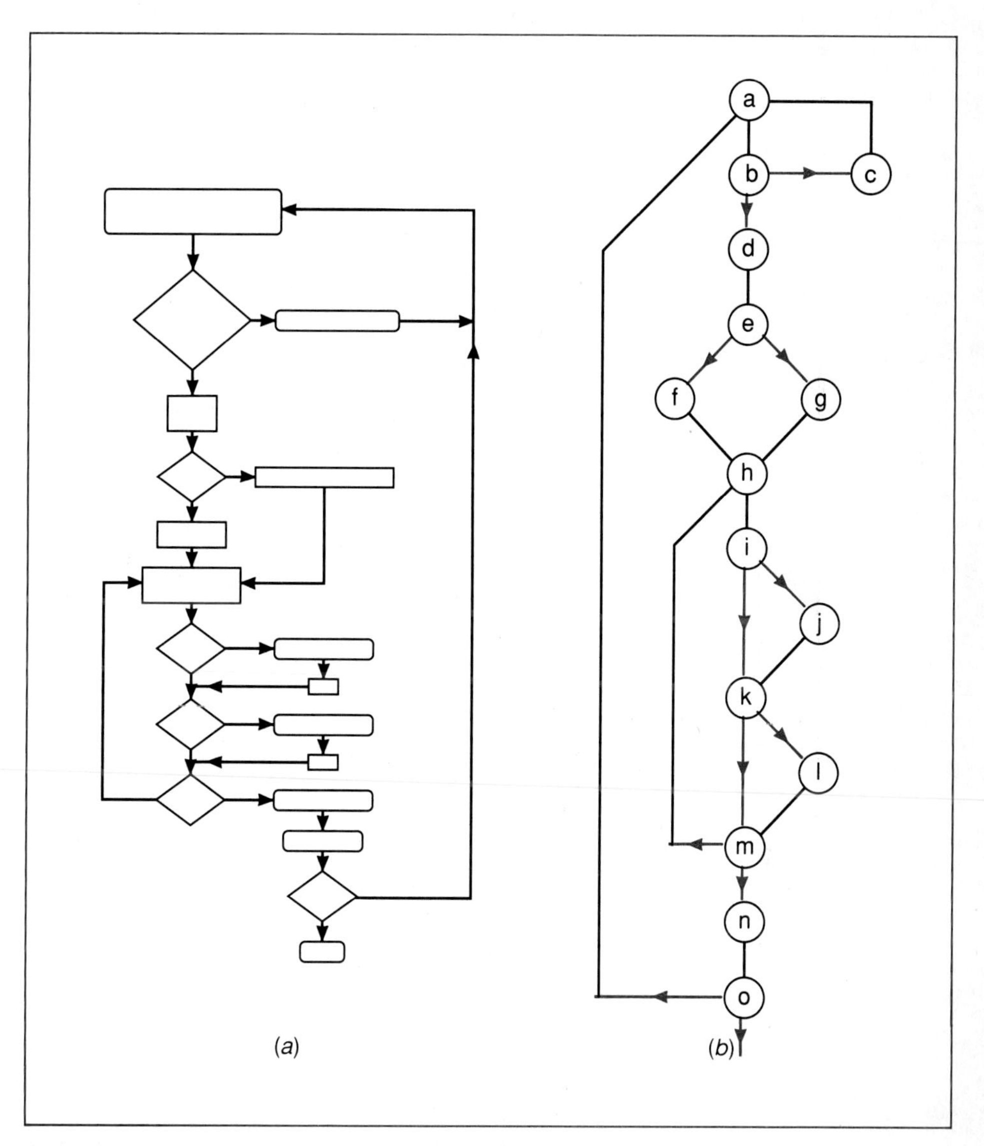

FIGURE 17.18 Sample structure chart for the log program: (*a*) flowchart from which program was written; (*b*) structure chart itself.

during a run of the preprocessed sample program is shown in Fig. 17.19. In this case, the execution path counters were written to a database and also echoed to the screen for our perusal.

The structure chart, with the number of times each path was executed during our sample run, is illustrated in Fig. 17.20. Note that we did exercise each path in this program at least once during these tests.

```
NUMBER = .1
MAX ITERATIONS = 10
EPSILON = .00001
MAX ITERATIONS; NATURAL LOG =  -2.147     TYPE 0 TO TERMINATE
NUMBER = .1
MAX ITERATIONS = 100
EPSILON = .001
EPSILON ACHIEVED; NATURAL LOG =  -2.302   TYPE 0 TO TERMINATE
NUMBER = 0.0
MAX ITERATIONS = 10
EPSILON = .01
NUMBER = 2.7182832
MAX ITERATIONS = 100
EPSILON = .0001
EPSILON ACHIEVED; NATURAL LOG =  1.000    TYPE 0 TO TERMINATE
    COUNTER #  1   REPETITIONS =  1
    COUNTER #  2   REPETITIONS =  3
    COUNTER #  3   REPETITIONS =  1
    COUNTER #  4   REPETITIONS =  2
    COUNTER #  5   REPETITIONS =  1
    COUNTER #  6   REPETITIONS =  97
    COUNTER #  7   REPETITIONS =  2
    COUNTER #  8   REPETITIONS =  96
    COUNTER #  9   REPETITIONS =  95
    COUNTER # 10   REPETITIONS =  3
    COUNTER # 11   REPETITIONS =  2
    COUNTER # 12   REPETITIONS =  1
```

FIGURE 17.19 Test run of log program.

There can be a subtle problem with test coverage analyzers. Although we might exercise each possible path at least once, we do not know that we exercised each possible *combination* of paths. Suppose a program has just executed two IF statements, call them A and B. The program might work properly if we executed:

- The THEN clause in A followed by the THEN clause in B
- The THEN clause in A followed by the ELSE clause in B
- The ELSE clause in A followed by the THEN clause in B

It might not work properly if we execute the ELSE clause in A followed by the ELSE clause in B. Our test coverage analyzer could tell us we exercised both clauses in both IF statements, but the program still might not be debugged.

The principles incorporated in test coverage analyzers can be used in another context. Suppose that the program we are developing is a real-time control routine. It must respond to critical inputs within a specified maximum amount of time. The program might appear to be debugged and working properly, but it does not meet those response-time constraints. The structure chart with counters could be used to tell us which sections of code are being executed the most. These could be replaced by assembly routines or recoded with emphasis on efficiency. This feature has proven so useful that some systems now incorporate a utility specifically for produc-

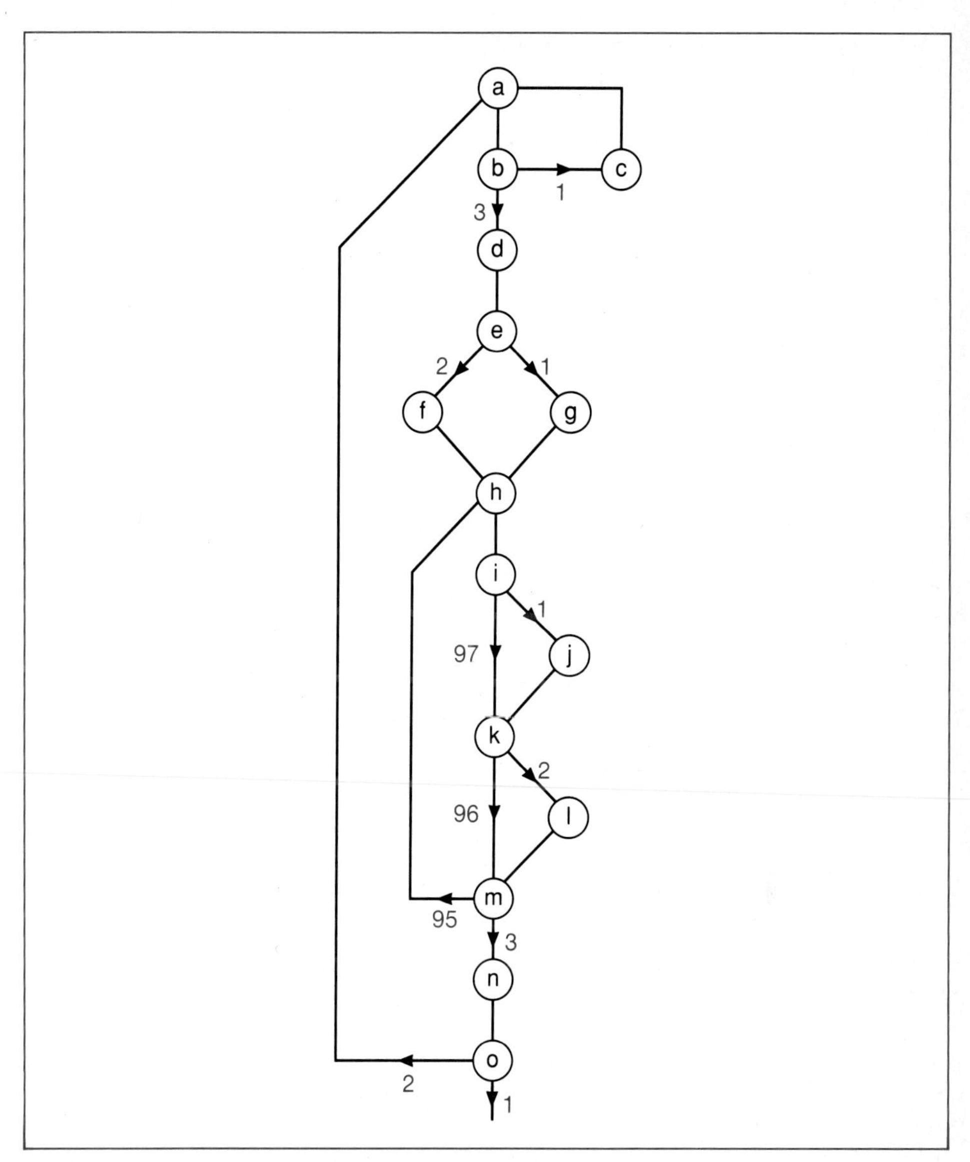

FIGURE 17.20 Structure chart of illustrative program, with counters shown on appropriate paths.

ing such an execution analysis. It is often called an **activity profiler** or some similar name.

17.5.2 Assertion Checker

Have you ever been analyzing a program error and made a statement like, "I assumed the value of this term would never be 0."? Frequently, the proper execution of a program depends on certain presumptions about relation or value. The programmer might "assume" such constraints instead of explicitly verifying them. An assertion checker gives us a formal tool to incorporate such explicit verifications.

During the design phase, the programmer records assumptions that he or she is making. During the implementation phase, these are incorporated into the software assertions. An assertion checker is then used to verify that the assertions do hold or to inform the programmer if they are ever violated.

Frequently the assertion checker is used in a manner similar to a test coverage analyzer. The user's program (with the assertions inserted into it) is run through a preprocessor. The preprocessor converts the assertions into code and adds this code to the program. The augmented code is compiled and executed, and any assertion violations are noted. A diagram of this process is shown in Fig. 17.21. (Note the similarity to Fig. 17.16.)

Some standard assertions we might use are:

- A variable is in a specified range.
- A variable is *not* in a specified range.
- A variable has one of a certain set of values.
- A variable *does not* have one of a set of prohibited values.
- A variable has subscripts within a specified range.

FIGURE 17.21 Operation of an assertion checker.

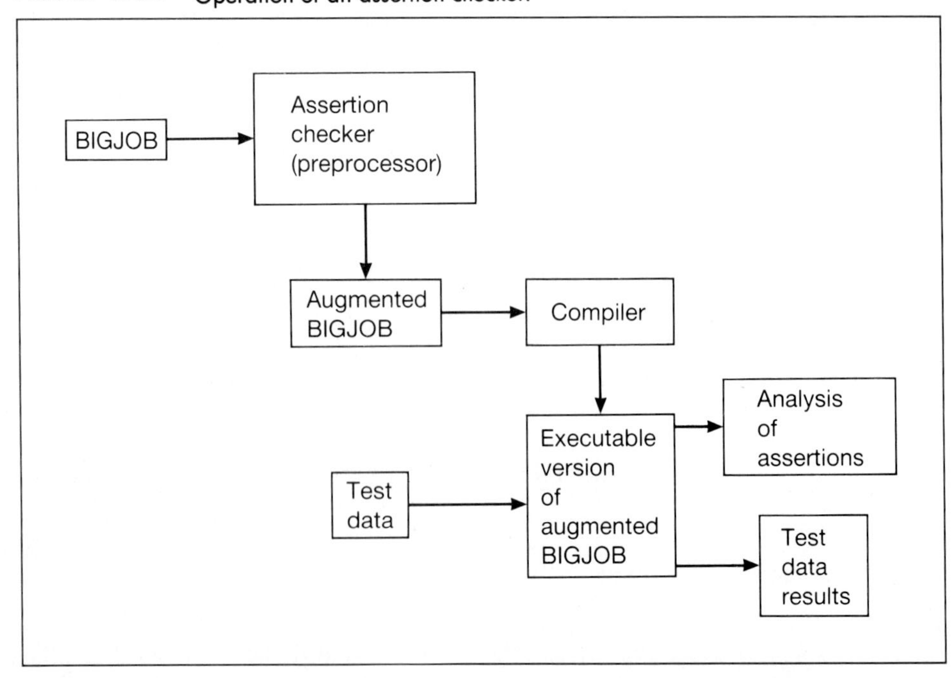

- A variable must not be changed.
- A set of variables satisfies a certain logical expression.

These might be specified in a format such as the following:

- $X \in [0.0 \ldots 100.0]$
- $Y \notin [-1.0 \ldots +1.0]$
- $J \in \{1,2,3,4,5\}$
- $M \notin \{-1, 0, +1\}$
- $R(,,,) \in [(1,1,0,-10) \ldots (25,50,100,0)]$
- LOCK L
- $(X < Y)$ & $(Y < Z)$

A violation during program execution could take the form

```
******ASSERTION VIOLATION   M = 0 AT LINE 27******
```

An interesting side effect of assertions is that they are excellent added documentation for a program. Some designers have made the format of their assertions conform to a special type of comment statement in the program's source language. During debugging, the program is run through the assertion checker, and these assertions are recognized and preprocessed. Later, the assertions are left in the program as documentation. Since they are comment statements as far as the compiler is concerned, they will be ignored during normal compilation and execution.

17.5.3 Test Data Generators

So far, our discussion has implicitly assumed that a *person* was generating data to test the programs. It is also possible for one program to generate test data for another program, using some previous concepts. For instance, in our logarithm computation program of Fig. 17.17, such a system would check the various branches of the first two IF statements. It would determine that they could be exercised by three ranges for the variable VALUE:

- $\text{VALUE} \leq 0.0$
- $0.0 < \text{VALUE} \leq 1.0$
- $\text{VALUE} > 1.0$

Randomly choosing three values in these ranges would give the desired test pattern.

You should observe that this task is conceptually much more difficult than just a structural analysis. The system must be able to deduce the relationships of variables necessary to force the program under test to take specific paths. Considerable research is going on as to how to best do this in a completely automated fashion.

Some current systems are interactive so that the programmer can help in determining certain of the relationships.

Some early test data generators used a "shotgun" approach to the problem. They did not try to "predetermine" what data needed to be generated. They simply generated massive amounts of random data and then checked if the paths were exercised.

17.5.4 Standard Test Data

A special case in choosing test data happens when there are standardized data sets available. For example, if you are writing a Cobol or Ada compiler, there are standardized *compiler validation sets* that comprehensively test those compilers.

This philosophy is a little different from the one described in the previous section. We are not trying to structurally analyze our particular program and exercise its various paths. Instead, we are looking at that program as a "box" and seeing if it generates the proper output when given a certain input. What happens internally in the box is not important.

Another form of standardized test data is a **benchmark**, a set of test data designed to compare various systems using some specified criteria. There are many standard benchmarks available. For instance, the SPEC benchmark [Weiss, 1989] compares execution times for 10 different programs to the execution times on a VAX computer. Alternatively, a company ready to buy a large computer system will often generate its own custom benchmarks to compare candidate systems. In this form of testing, the focus is normally on the system as a box, with comparisons based on correctness and speed of the output produced. The goal is not to exercise specific paths of a particular system.

17.5.5 Test Drivers

With the coming of various modular programming techniques, a programmer might need to test a component of a system independently. The parts of the system that would use this module, and hence generate test data for it, are not yet available. Some other technique must be found to generate that test data.

Often, a program called a **test driver** is written for this purpose. It might be written in a different language, might not follow all the tenets of good programming style, and might not be well-documented. It is often "throwaway" code, since its sole purpose is to generate test data until other modules are finished.

As a simple example, suppose that you are part of a team writing a compiler in C. You are responsible for the symbol table routine and have your module ready for testing. However, the modules that call the symbol table routine are not yet ready. You therefore write a program in a string-processing language such as Snobol or LISP to generate suitable test data.

17.5.6 McCabe's Complexity Metric

Earlier, we showed how structural analysis of our sample program can be used to study test coverage and to seek out possible efficiency bottlenecks. Structural analysis can also be used in connection with graphs to develop a complexity metric for a module. This complexity metric in turn points out potentially faulty software even before the testing process is begun. It shows us where we should focus our main testing efforts.

Figure 17.22 shows our logarithm flowchart and a representation of its control

FIGURE 17.22 Converting our sample logarithm program to a flow graph.

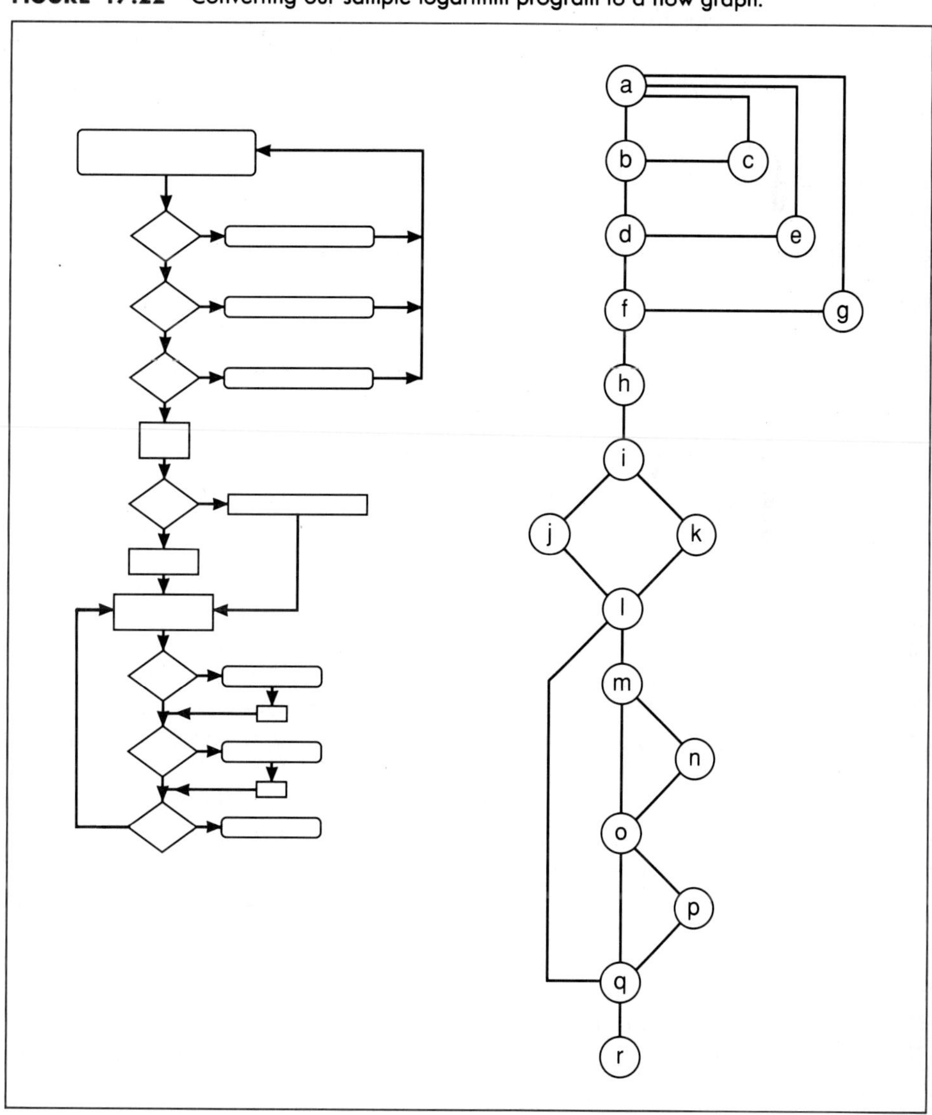

flow. The main idea here is that blocks of code are represented as single circles to reduce the number of nodes. The final step in the process is to make certain that the graph is strongly connected. This means that every node can be reached from every other node. To make our graph strongly connected, we must add one edge from node *r* to node *a*.

The McCabe complexity metric is defined as the number of edges minus the number of nodes plus 1. In the graph of Fig. 17.22, there are 25 edges (counting the added one from *r* to *a*), and 18 nodes. Therefore, the complexity metric is 8.

Research has shown a strong correlation between high defect densities and complexity metrics above 10. On this basis, we could project that our sample module should be relatively error-free, easy to test, and easy to maintain. These projections could be made even before we began the actual testing.

If our module had a complexity metric well above 10, we might consider redesigning it. If that were not feasible, we should probably at least subject the module to especially rigorous testing.

17.6 MAINTENANCE AND MISCELLANEOUS TOOLS

Some tools are used in several phases of the life cycle; editors are a good example. Others are important but just didn't seem to fit in any previous section. Two tools that fall into these categories are included here.

17.6.1 Sophisticated Make Utilities

Suppose that during integration testing of module *P* you discover a bug in module *Q*. You thought *Q* had been thoroughly debugged and was working properly. You go in and fix the bug but forget to save the newly compiled object code for *Q*. The object code module for *Q* therefore remains as before, with the bug. You now go back to working on the integration of *P*. You spend hours in frustration before you realize that during linking you were still using the old version of *Q*. This problem can be especially difficult to handle after a product has been released to manufacturing.

A Make utility can do such things as checking when modules were changed and ensuring that corresponding object code files were later replaced. It can go through your modules to see which other modules they need, then go through those modules to see which ones they need, and so on, to ensure that all necessary include files are specified. The Make file with which this utility works can be used to store compiler switches and options, file dependencies, directory paths, and other housekeeping details. Therefore, they don't have to be respecified repeatedly at execution time.

Some of these utilities also incorporate conditional constructs. This would make it possible, for example, to have a custom package run under different operating systems simply by setting a compiler parameter.

17.6.2 Rapid Tool Development Systems

Sometimes you do not really want to go through the formal software development process. You just need a "quick and dirty" routine to do something such as generating a cross-reference table. It would be desirable to do this task rapidly and with minimal errors. You might be able to use your editor directly or to write an editor macro for the task. If these are inadequate, there might be an operating system utility available to do it.

If none of these are available, you might have to write a program to do the task. Another approach might be to try to go through with the editor and do the operation manually. Both could be time-consuming, and the second option is especially prone to error.

Another approach is to use a special ultrahigh-level programming language for rapid development of such tools. These languages often have powerful string- and table-handling facilities and a simple basic syntax such as

$$\text{pattern}_i \ \{\text{action}_i\}$$

for some set of patterns and the desired corresponding actions.

As a simple example, suppose you want a list of all the places certain variables appear in your source code. You might set up a file with the list of desired variable names. Then you might specify a tool to do the actual operation, as shown in Fig. 17.23.

The first line reads each variable name (record) from the file of variable names, sets up a table entry for that name, and initializes the table entry to NIL. Line 2 specifies that an action is to be done for each record in the source code file, as specified by lines 3 through 8. Line 4 says that an action is to be done for each name in the table, as specified by lines 5 through 7. This action is to see if the name is in the source code record, and if so, add the record number to the table containing references to that variable name. Lastly, line 9 prints the results.

Actual development time for this particular tool was approximately 7 min. Consider how long it might take you to write and debug a Pascal program to do the same operation.

FIGURE 17.23 Building a cross-reference table for a set of variables.

```
01   getline("variable_list") {table[*] := ""}
02   getline("source_code")
03        {
04        name ∈ table
05             {
06             name ⊆ * {table[name] := table[name] " " #}
07             }
08        }
09    name ∈ table {printline name," ",table[name]}
```

17.7 A COMMERCIAL CASE PRODUCT

In the previous sections of this chapter, we have given generic descriptions of a number of software tools. This was done to provide comprehensive general information without necessitating the details of specific products. However, we also want you to see actual commercial software tools. Therefore, in this section we will describe MacAnalyst and MacDesigner, a powerful pair of CASE tools from Excel Software [Excel Software, 1990].

MacAnalyst is a tool that automates standard analysis techniques using data flow diagrams (DFDs). It provides a palette of specialized tools to create and revise DFDs, and it facilitates creation of a data dictionary. This data dictionary not only contains information about data and data relationships but also ties together all analysis documents to provide complete project documentation.

Data flow diagrams can be **leveled**, that is, organized into a structure in which modules on higher-level diagrams are expanded into lower-level diagrams. This is continued until transformations within a primitive module can be fully described in a short textual frame. A text editor for creating such module specifications is included in the system.

MacAnalyst provides error checking for diagraming errors, information-balancing errors between diagrams, and errors in the data dictionary. Error reports can be generated whenever desired.

A screen from MacAnalyst is shown in Fig. 17.24. The user has partially com-

FIGURE 17.24 A data flow screen from MacAnalyst.

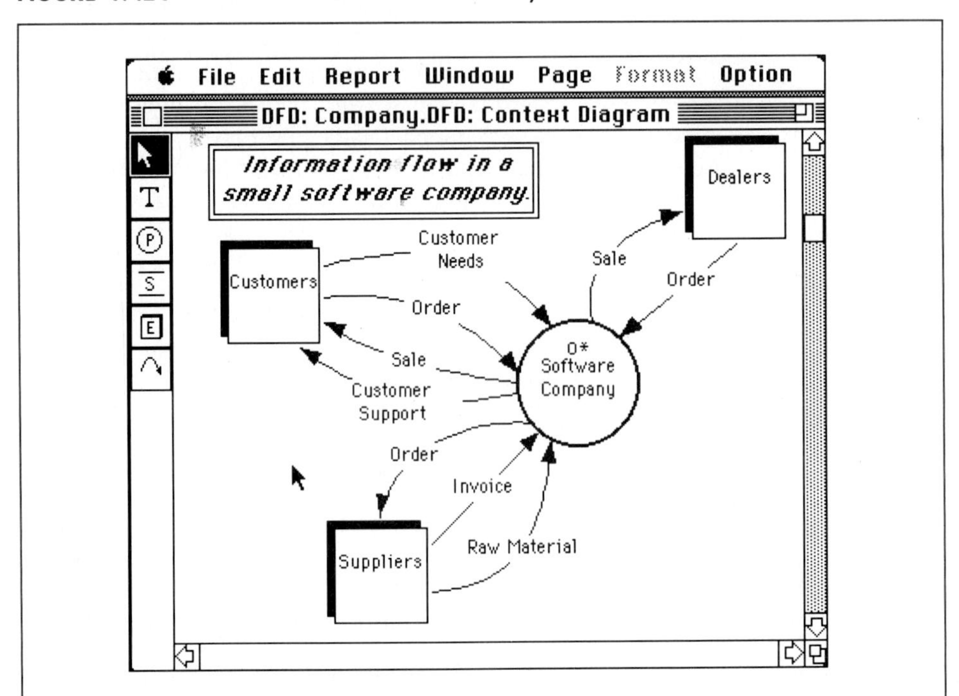

pleted the DFD for a small software company. On the left of the screen you can see the palette of tools for creating bubbles, files, external sources/sinks, and arrows. There are also tools for inserting text and for manipulating the diagram (such as moving items around). A wide variety of other facilities are provided by the pull-down menus on the top of the screen; we do not have space to describe them.

The completed DFD for the company is shown in Fig. 17.25. Note that we have extracted the DFD from the screen because of space constraints for the figures. This has been necessary in several instances throughout the section.

The bubble in the screens of Figs. 17.24 and 17.25 is labeled 0*. An asterisk by a bubble number indicates that the bubble is further described at lower "levels."

This idea of leveling is illustrated by Fig. 17.26, where bubble 0 has been expanded to its next level. Continuing the leveling, bubble 1 was then expanded again. This brought us to the level of primitive modules, at which we could move to module descriptions. Figure 17.27 shows the description of the primitive module represented by bubble 1.1.

To illustrate the data dictionary, we selected one of the data items from the completed DFD in Fig. 17.25. Its entry is shown in Fig. 17.28. MacAnalyst also allows us to view the structure of a composite data item as a tree, and the tree diagram for this data item is shown in Fig. 17.29.

This completes our brief description of MacAnalyst, a tool that automates standard analysis techniques using data flow diagrams. We have looked at DFDs, leveling, module descriptions, and data dictionaries in MacAnalyst.

The other tool in this set, MacDesigner, is used for design. We turn now to a discussion of MacDesigner and some examples from it.

FIGURE 17.25 Top-level analysis completed.

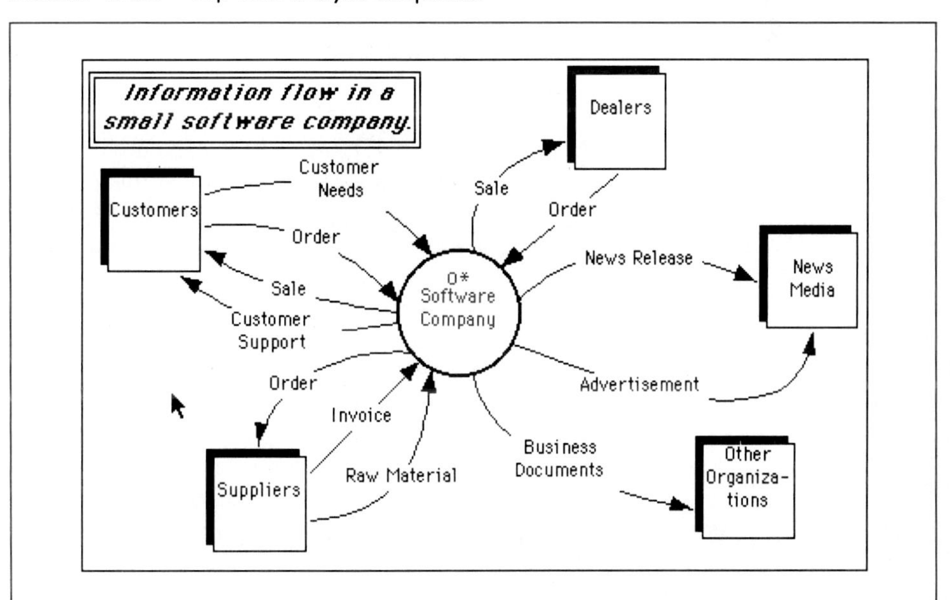

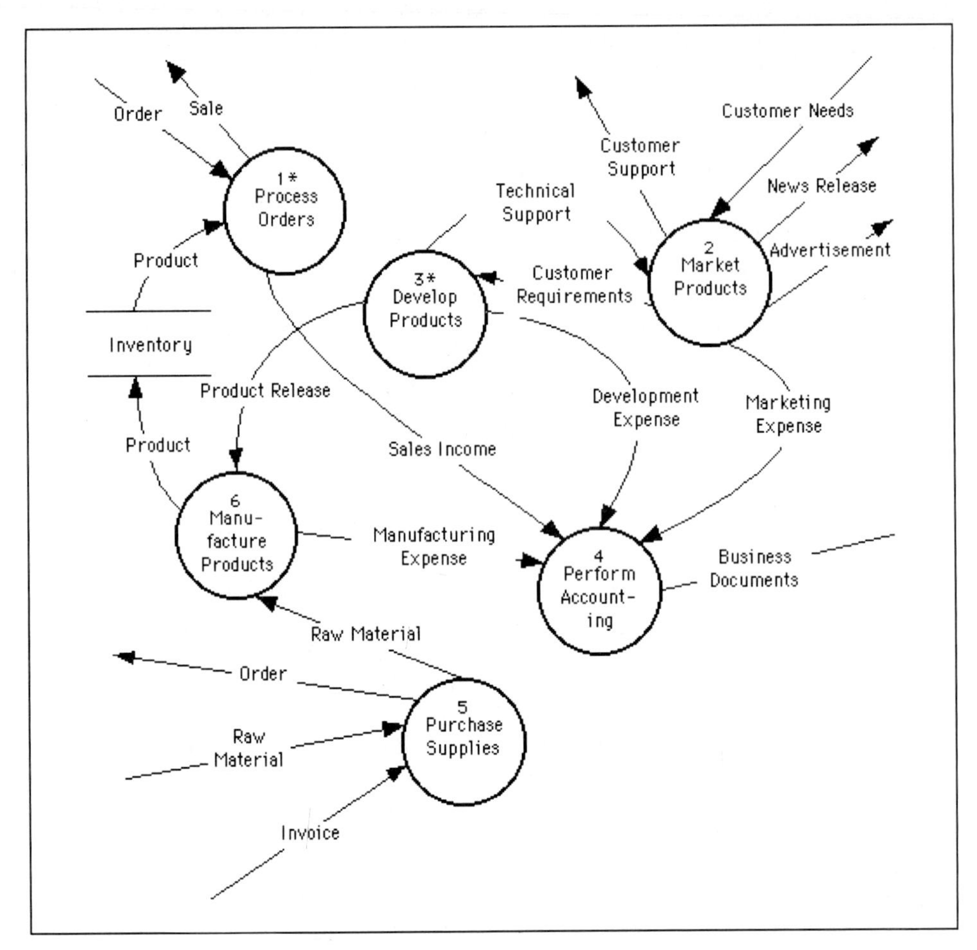

FIGURE 17.26 Expansion of bubble 0.

MacDesigner is a tool that automates the structured design process for new systems. It can also be used to document existing software systems. MacDesigner facilitates development of structure charts, which graphically specify the structure of a software system. It also facilitates development of tree diagrams, module descriptions, and a complete data dictionary. MacDesigner supports cross-reference lists to show which modules a given module *calls* and which modules a given module is *called by*. Data verification can be performed on data in the diagrams and in the data dictionary to check for complete and consistent definitions.

A screen from MacDesigner is shown in Fig. 17.30. The user has partially completed the structure chart for a small payroll system. On the left of the screen you can see the palette of tools for creating components of the structure chart. There are also tools for inserting text and for manipulating the diagram (such as moving items around). A wide variety of other facilities are provided by the pull-down menus on the top of the screen; we do not have space to describe them.

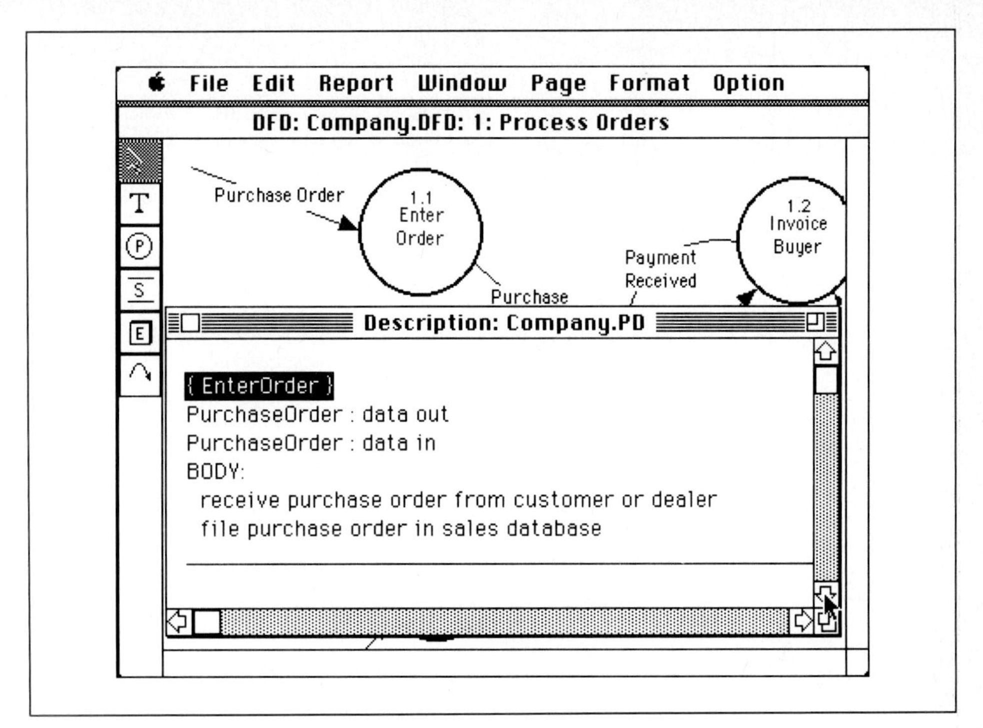

FIGURE 17.27 Description of bubble 1.1.

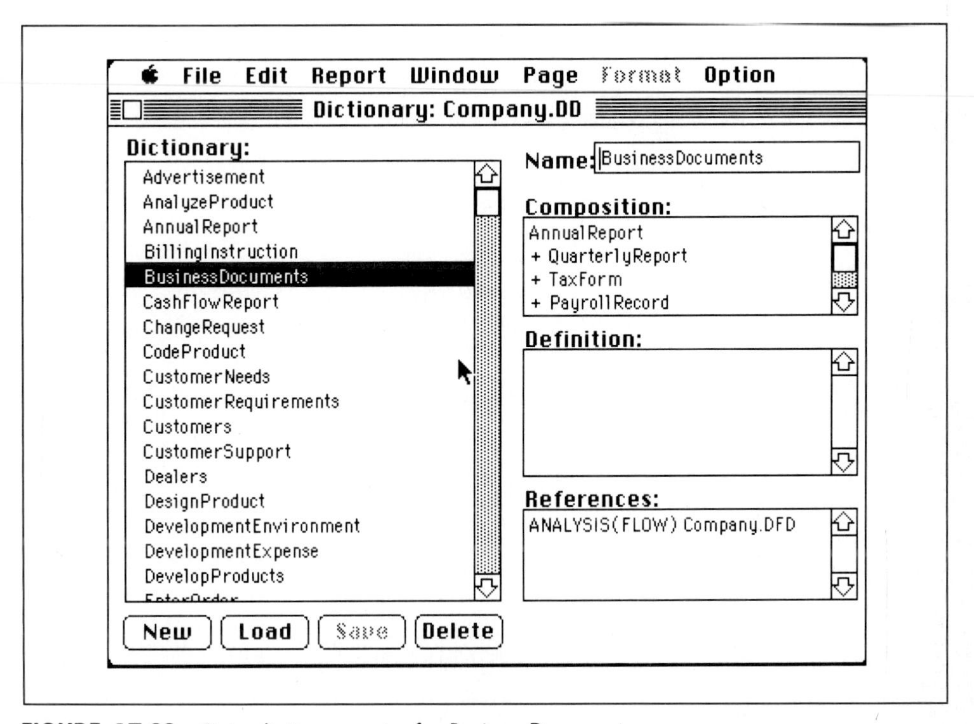

FIGURE 17.28 Data dictionary entry for BusinessDocuments.

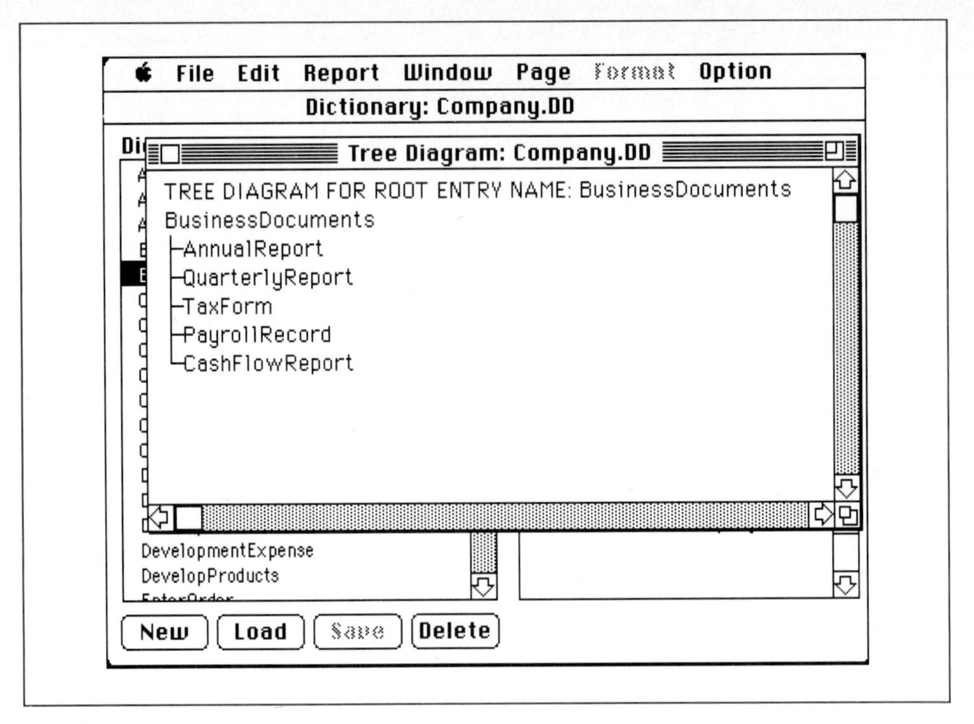

FIGURE 17.29 Tree diagram of data dictionary entry for BusinessDocuments.

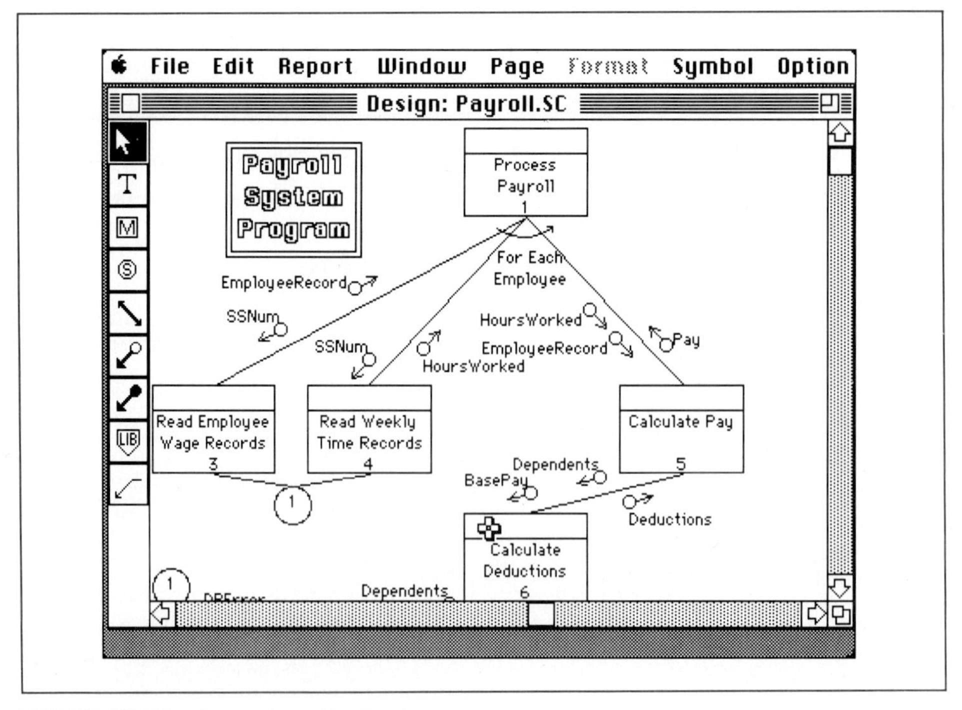

FIGURE 17.30 Screen from MacDesigner.

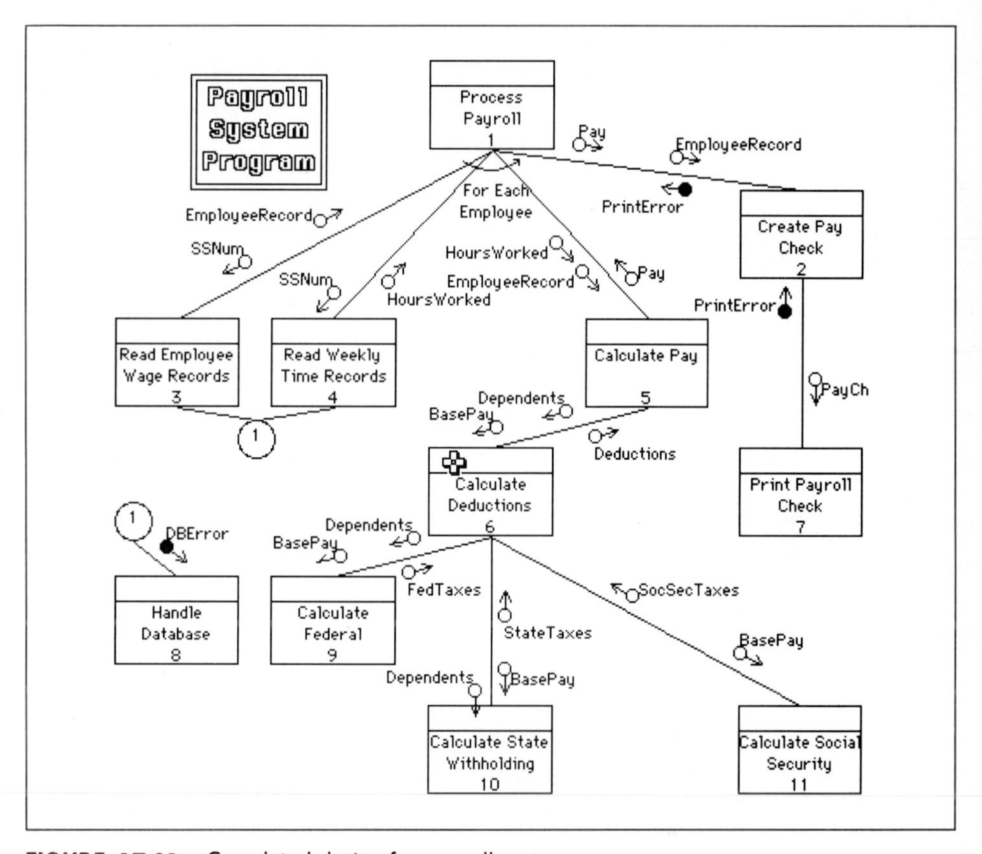

FIGURE 17.31 Completed design for payroll system.

The completed structure chart for our system is shown in Fig. 17.31. Again, we have extracted it from the screen because of space constraints for the figures.

Modules are described with textual frames, similar to the frames illustrated earlier in MacAnalyst. Figure 17.32 shows a description for one of the modules. It should be noted that such descriptions are easily obtained by simply double-clicking on an item in the structure chart.

Our sample system, like many software systems nowadays, accesses databases. The dictionary entry for such a database and its corresponding tree diagram are shown in Figs. 17.33 and 17.34.

To close this discussion, we demonstrate one feature we think is extremely useful and convenient. Often, in reviewing the description of a module, we see a data item whose structure must be verified. This could involve a fairly lengthy sequence of operations, opening and closing documents and looking up the item. With MacDesigner, a user can simply double-click on the item in the description and automatically look it up in the dictionary. This is shown in Figs. 17.35 and 17.36.

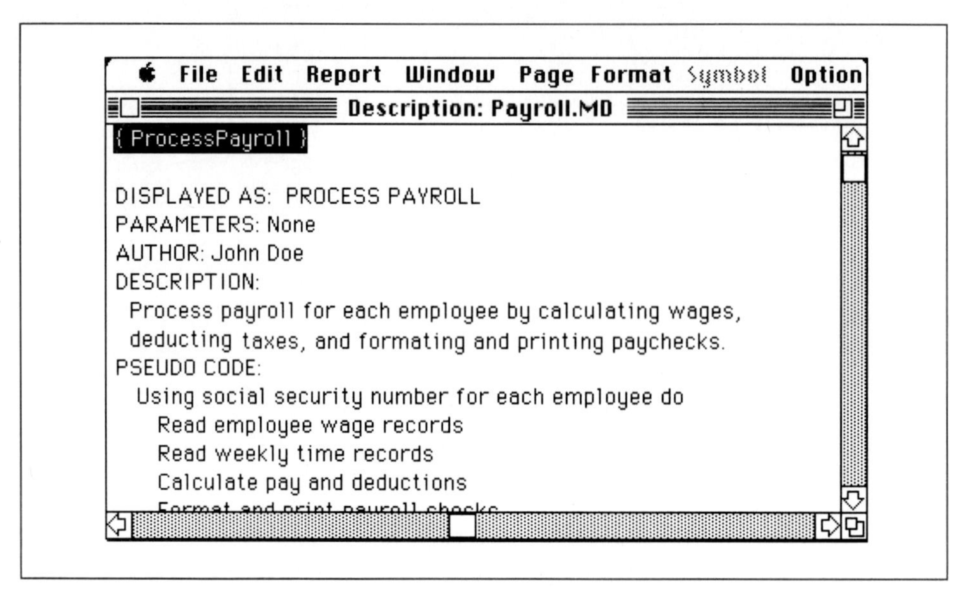

FIGURE 17.32 A module description.

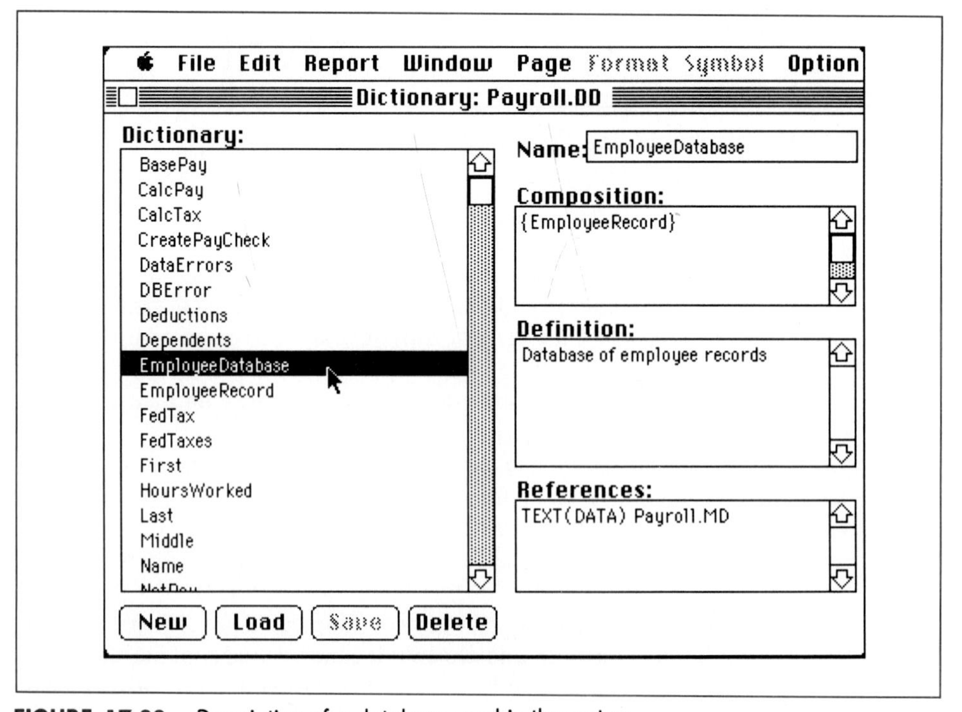

FIGURE 17.33 Description of a database used in the system.

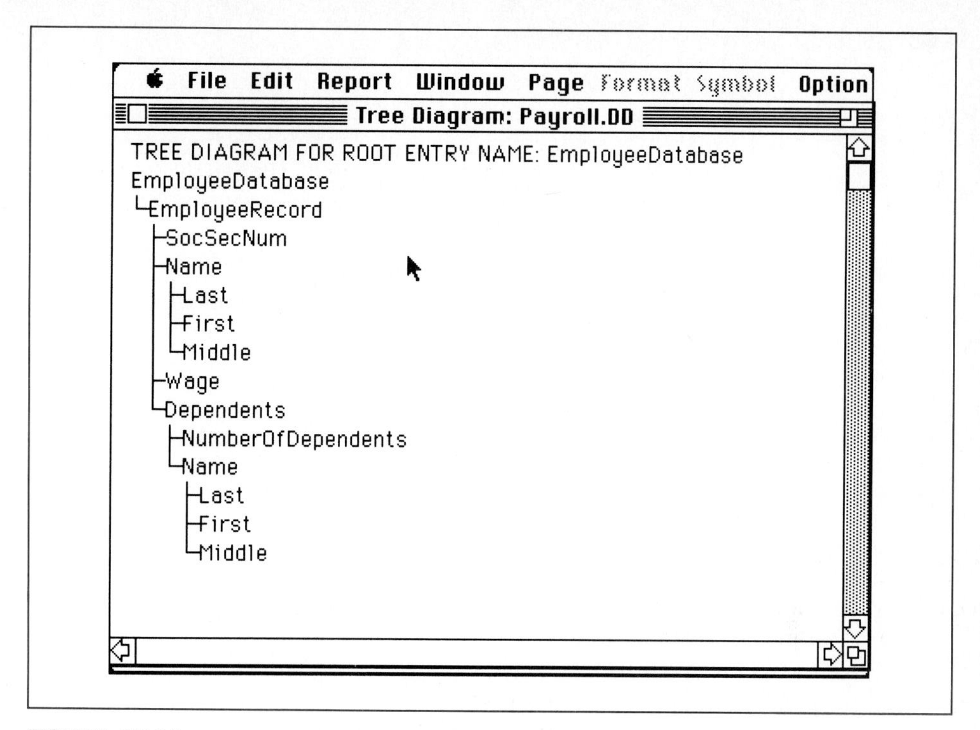

FIGURE 17.34 Tree diagram of the database shown in Fig. 17.33.

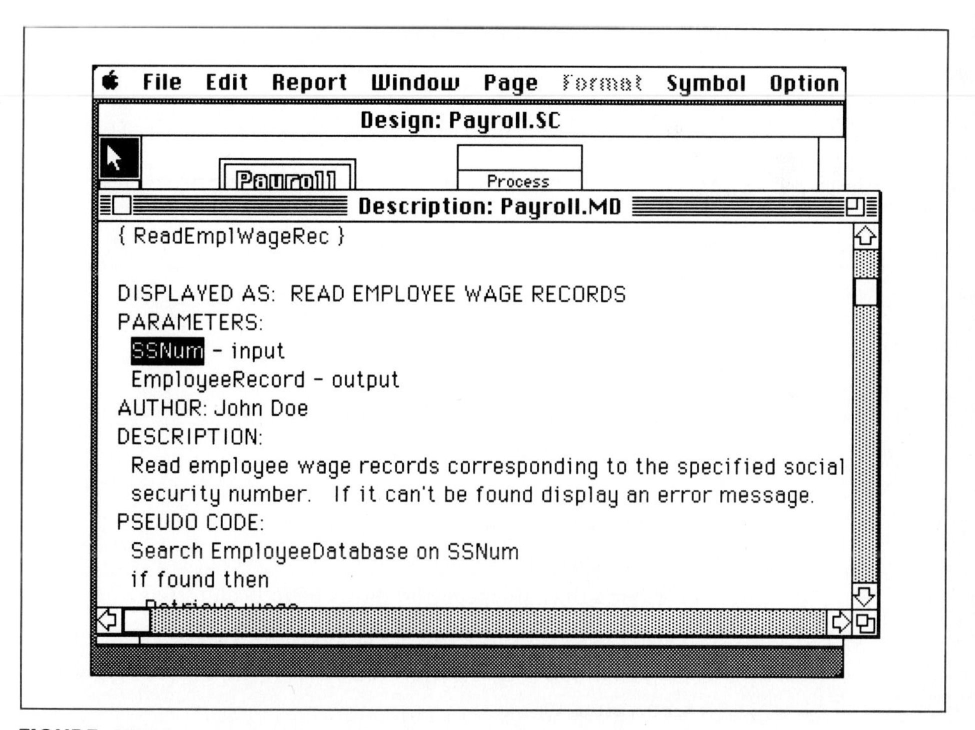

FIGURE 17.35 Highlighting a data item in one of the module descriptions.

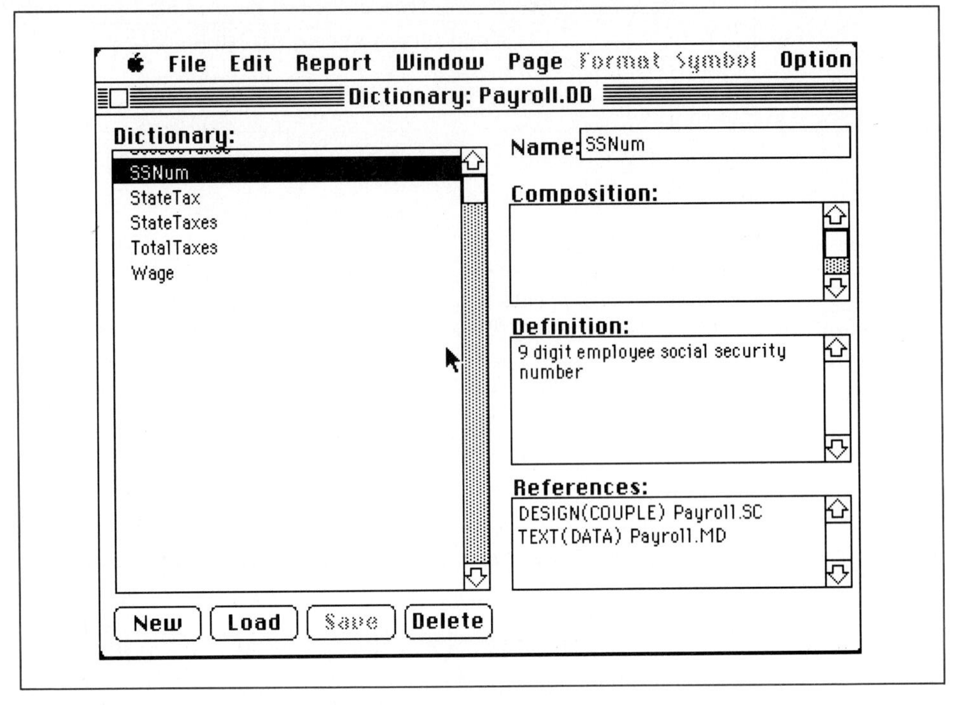

FIGURE 17.36 Getting the description of the item selected in Fig. 17.35.

17.8 SUMMARY

We have looked at general descriptions and illustrations of many software development tools, such as tools for:

- Prototyping
- Developing DFDs
- Interactive compiling/debugging
- Developing good user interfaces
- Developing graphics
- Designing databases
- Interfacing to existing databases
- Verifying that all execution paths have been tested
- Verifying that programmer assumptions hold
- Monitoring the significance of mathematical computations
- Generating test data
- Computing complexity metrics for modules

You probably will never use all of them. However, having been introduced to them, you can at least be aware of their existence if a need arises.

We have also looked at two specific examples of commercial software tools, MacAnalyst and MacDesigner. This should have helped you see how such tools are implemented and how they can be used.

PROBLEMS

17.1 Do research on the approximate sizes of some large software systems.

17.2 Check if your school has any of the CASE tools described in this chapter. If so, see if your instructor can demonstrate them or if you can try them out.

17.3 Why do you suppose we have used hardware prototypes for so long but have only fairly recently begun using software prototypes?

17.4 Have your instructor "simulate" a client who wants to automate an existing manual system, and work with him or her to develop some prototype screens for your proposed system. If you have a screen generator or good graphics package available, use it; otherwise just hand-sketch the screens.

17.5 Do a data flow diagram of the existing manual system from Prob. 17.4.

17.6 Do a data flow diagram of the proposed computerized system from Prob. 17.4.

17.7 Do a data dictionary for the existing manual system from Prob. 17.4.

17.8 Do a data dictionary for the proposed computerized system from Prob. 17.4.

17.9 If you have a database system like the one described in Sec. 17.4.4 available, work with your instructor to develop a simple record format. Then enter and retrieve some sample records.

17.10 List the steps in the life-cycle model.

17.11 Briefly describe what is done in the requirements definition step of the life-cycle model.

17.12 Briefly describe what is done in the design step of the life-cycle model.

17.13 Briefly describe what is done in the implementation step of the life-cycle model.

17.14 Briefly describe what is done in the testing step of the life-cycle model.

17.15 Briefly describe what is done in the operation and maintenance step of the life-cycle model.

17.16 Contrast prototyping and the classic approach to requirements definition.

17.17 How can you get a prototype to look like the final system when that final system has not yet been implemented?

17.18 Describe the major components of a DFD.

17.19 Briefly describe your interpretation of the system described by the DFD in Fig. P17.19.

17.20 Describe what we mean by *refining* a DFD bubble to another level (a process also called *leveling*).

17.21 List five typical features of an interactive compiler/debugger.

17.22 List five things that a tool for developing interactive input interfaces might help the programmer do.

17.23 List the steps that a programmer would typically do to use a tool for designing interactive input interfaces.

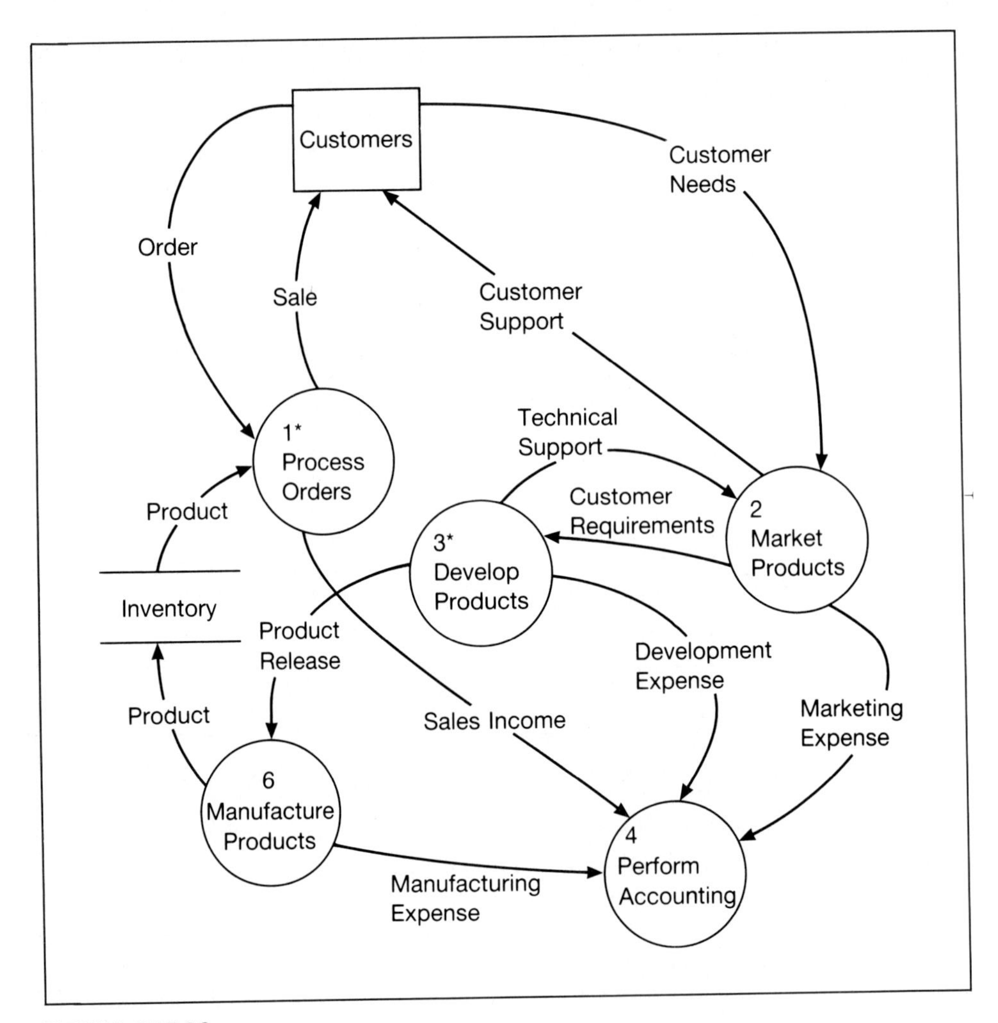

FIGURE P17.19

17.24 Why have tools for developing interactive input interfaces speeded up interactive systems development so much?

17.25 What are control maps?

17.26 How could a programmer use control maps to develop a system?

17.27 Why would a programmer want to use graphics development toolkits?

17.28 What is a preprocessor?

17.29 List some software tools that might be implemented as preprocessors.

17.30 Describe the four steps of a typical test coverage analyzer.

17.31 Describe a problem that a test coverage analyzer might miss even though all possible paths were executed at least once.

17.32 What is an activity profiler?

17.33 What is an assertion checker?

17.34 List five things an assertion checker might be used to look for.

17.35 How does a test data generator differ from a test coverage analyzer?

17.36 What is a benchmark?

17.37 How is the use of a benchmark different from the use of a test data generator?

17.38 Explain why you think McCabe's complexity metric is a good indicator of potentially problematic modules.

17.39 What is a rapid tool development system?

17.40 What do you think people mean by throwaway software?

17.41 What do you think people mean by reusable software?

17.42 What is MacAnalyst used for?

17.43 What is MacDesigner used for?

17.44 Your instructor will specify a simple programming assignment. Develop a flowchart for that program *before you code it*. (Do not include checks for data validity; that will be part of a later question.)

17.45 Develop a pseudocode design for the program in Prob. 17.44.

17.46 Develop a Nassi-Shneiderman diagram for the program in Prob. 17.44.

17.47 (*a*) If you have an interactive compiler/debugger available, use it to write the program you design in Prob. 17.44. (*b*) Otherwise, use a standard compiler; explain the errors you encountered and how you found them. How would an interactive compiler/debugger have helped?

17.48 Modify the code for your program from Prob. 17.44 to include checks for input data validity. First modify your *design*, then incorporate the changes into the code.

17.49 Do a structure chart (like Fig. 17.18) for your program from Prob. 17.48.

17.50 Modify your program from Prob. 17.48 to count the number of times each path (as determined in the previous structure chart) is executed.

17.51 Add some assertion-checking statements to your program from Prob. 17.48.

17.52 Generate a formal, complete set of test data for your program from Prob. 17.48. Explain how you developed it and how it checks all paths in your program.

17.53 Compute the complexity metric for your program from Prob. 17.48.

SYSTEM CONSIDERATIONS

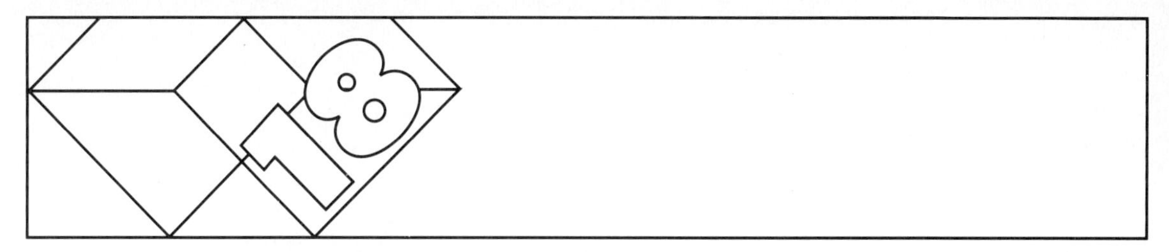

The User Interface

In previous chapters, we discussed interfaces between different layers of software and hardware. In this chapter, we will learn about the most prominent and observable interface of a computer system, the interface between human and machine. This interface, known as the **user interface**, or **UI**, defines the way in which users communicate and interact with computers.

The user interface is becoming increasingly important because more people without computing experience are using computers. A well-designed UI can make a system easy to use and therefore popular. The Apple Macintosh has gained wide acceptance in part because its designers spent considerable effort on the UI.

Designing a good UI is not an easy task, because we as engineers are trained in how to design "things," not in how people will best interact with these things. Traditionally, engineers have designed all parts of the computer, but engineers are not always well-schooled in the psychology of human/machine interaction. Also, the UI is not understood as well as components like a CPU or I/O device.

In this chapter we will discuss how UIs have evolved to their current state, the components of a UI, current design philosophies and methodologies, and examples of successful UIs.

18.1 A BRIEF HISTORY OF USER INTERFACE DEVELOPMENT

The earliest computers operated in a batch-processing mode. Later, computers were designed with an interactive user interface that improved as technology advanced. In this section, we will elaborate on batch versus interactive processing and discuss how the user interface has evolved.

Designers of the first computers concentrated on the internal structure of the machine, and they made use of existing devices for input and output. By doing so, the type of user interface they could create was limited. Early commercial computers used card readers, card punches, and teletypewriters for input and output.

Hard-copy output was commonly prepared on devices separate from the system, so a two-step process was required to obtain listings. First, the system would transfer the output to a communication medium such as magnetic tape or punched cards. An offline printer would then print the data contained on the tape or cards. Most users did not interact directly with the computer.

Those people who did interact directly with the early computers found interfaces that were more cumbersome than those used in today's systems. Such an interface typically consisted of two parts:

- A comprehensive control panel that could be used for both program debugging and system maintenance
- An electric typewriter or teletype unit that served as the keyboard and low-speed printer for the system operator

The control panel contained toggle switches for entering data or addresses, push buttons for initiating actions, and banks of lights to show the contents or status of selected registers and flip-flops. Programs could be executed in single-step mode—one instruction at a time—and the contents of the individual registers could be noted at each step. This gave the user unlimited control of the computer, but it was difficult for novices to use because:

- A thorough understanding of the computer hardware was required to debug software.
- The procedure was only applicable to assembly-level programs.

Since most of the early computers had to serve many users, computer centers operated in a batch-mode environment. The computer operator maintained a "batch" of programs in the card reader, as shown in Fig. 18.1, to ensure that time was not wasted between jobs. Job control cards were included for loading and running the program and for identifying the beginning and end of each job. Direct interaction with the computer was not only discouraged, it was usually not allowed.

The typical programmer followed the development procedure diagramed in Fig. 18.2. Programs and data were handlisted (in block letters) on coding sheets similar to those shown in Figs. 18.3 and 18.4. The coding sheets were then submitted to a keypunch group who prepared punched cards (one card per coding sheet line) by keying the program or data on the coding sheet into a keypunch unit.

The punched-card "decks" and the coding sheets were sometimes given to another group of people to check for keypunch errors. The information on the coding sheet was rekeyed into a verifier unit that simultaneously compared the keyed characters to the existing holes and identified cards on which any keyed character did not match the punched character.

The user interface for the programmer consisted of coding sheets, punched-card decks, and program listings from the line printer. These are conceptually the same

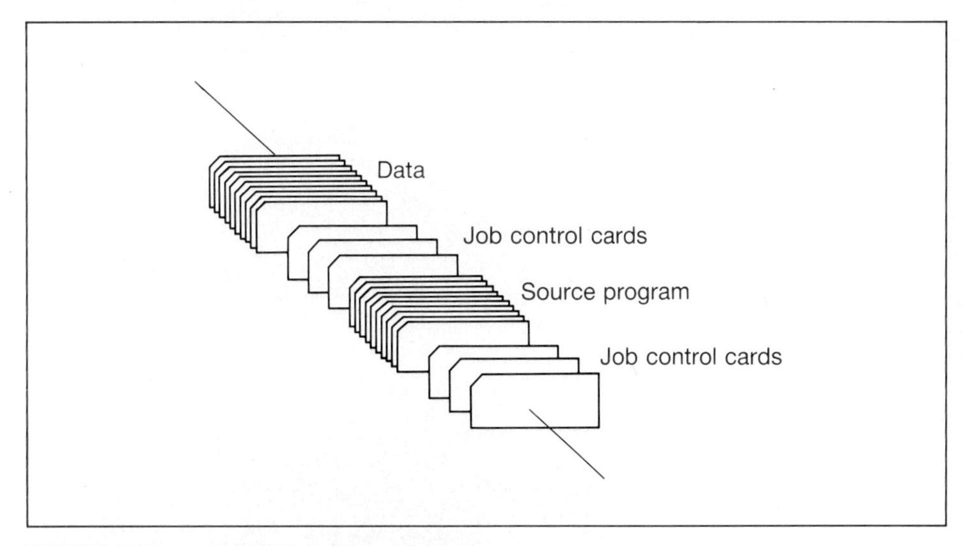

FIGURE 18.1 A batch-mode card sequence.

as with today's computers, except the media have changed. The major difference is the time frame. **Turnaround time**, the time between submitting a job and getting results back, used to be measured in hours or days. When errors were identified and corrected, new cards were punched and inserted into the program deck. The corrected program was resubmitted to the input queue at the computer center, and the programmer waited through another turnaround cycle.

The problem with this approach is the lengthy turnaround time. Good UIs must provide users with prompt feedback of errors and results. Fortunately, the batch-computing environment described above changed with the development of interactive online systems and the direct connection of teletypewriters to the system. These newer systems dramatically reduced the turnaround time by eliminating cards, job queues, and operators.

18.1.1 The Teletypewriter and Line Editors as the User Interface

The emergence of teletypewriters as the user interface device to an online system represented a major advance in the state of the art in UIs. It changed the way people interacted with computers for years to come. Programmers no longer had to wait long periods to determine whether their programs contained any errors because each user had his or her own I/O device. Punched cards were eliminated because programs were entered directly from the keyboard. Errors were detected and messages printed as they were encountered.

The UI was also improved by special programs written for interactive systems. **Line editors** provided users with the ability to edit their programs from the keyboard. Users no longer had to wait for the results to be returned by an I/O clerk in the computer center.

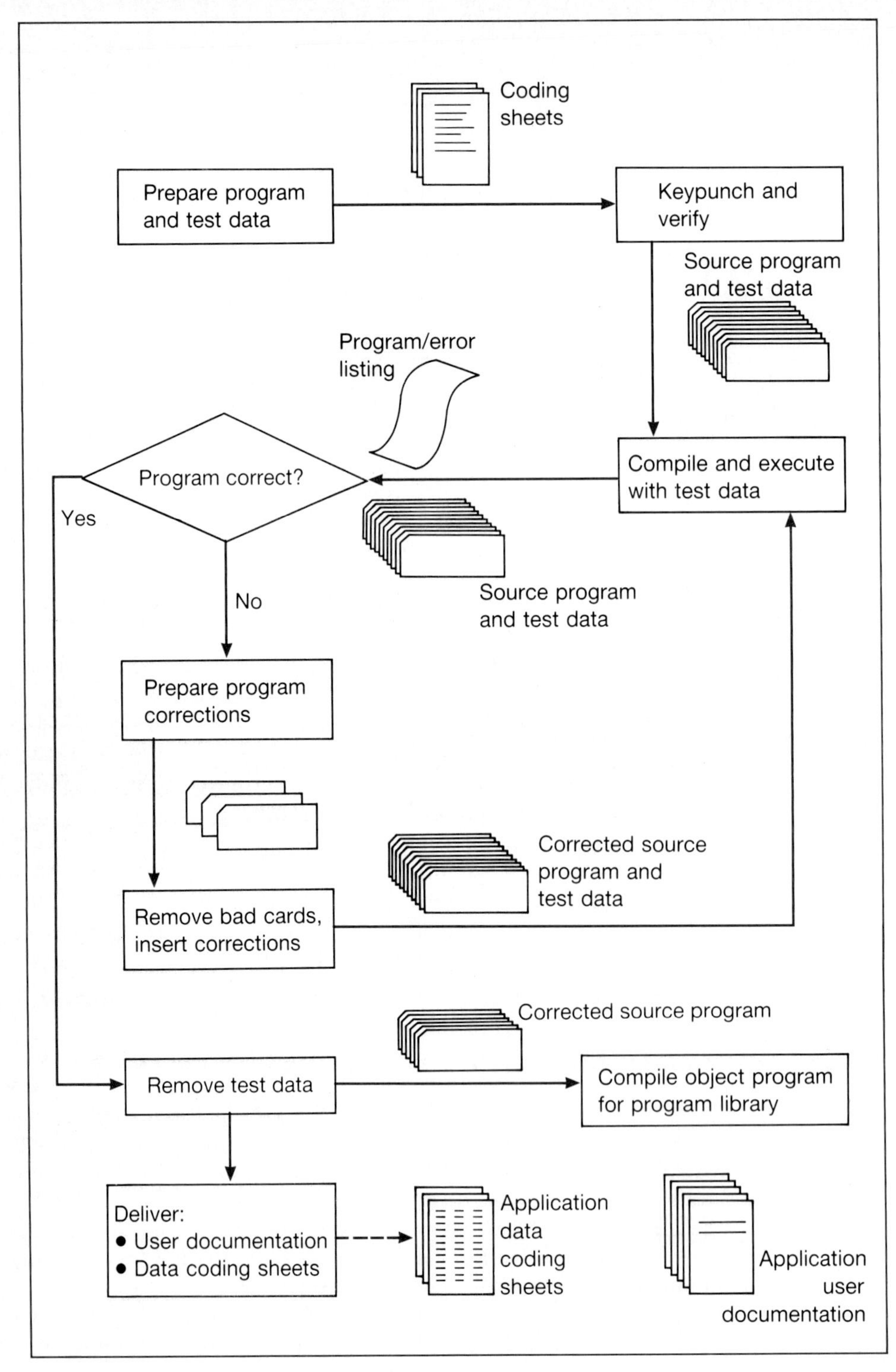

FIGURE 18.2 The punched-card program development cycle.

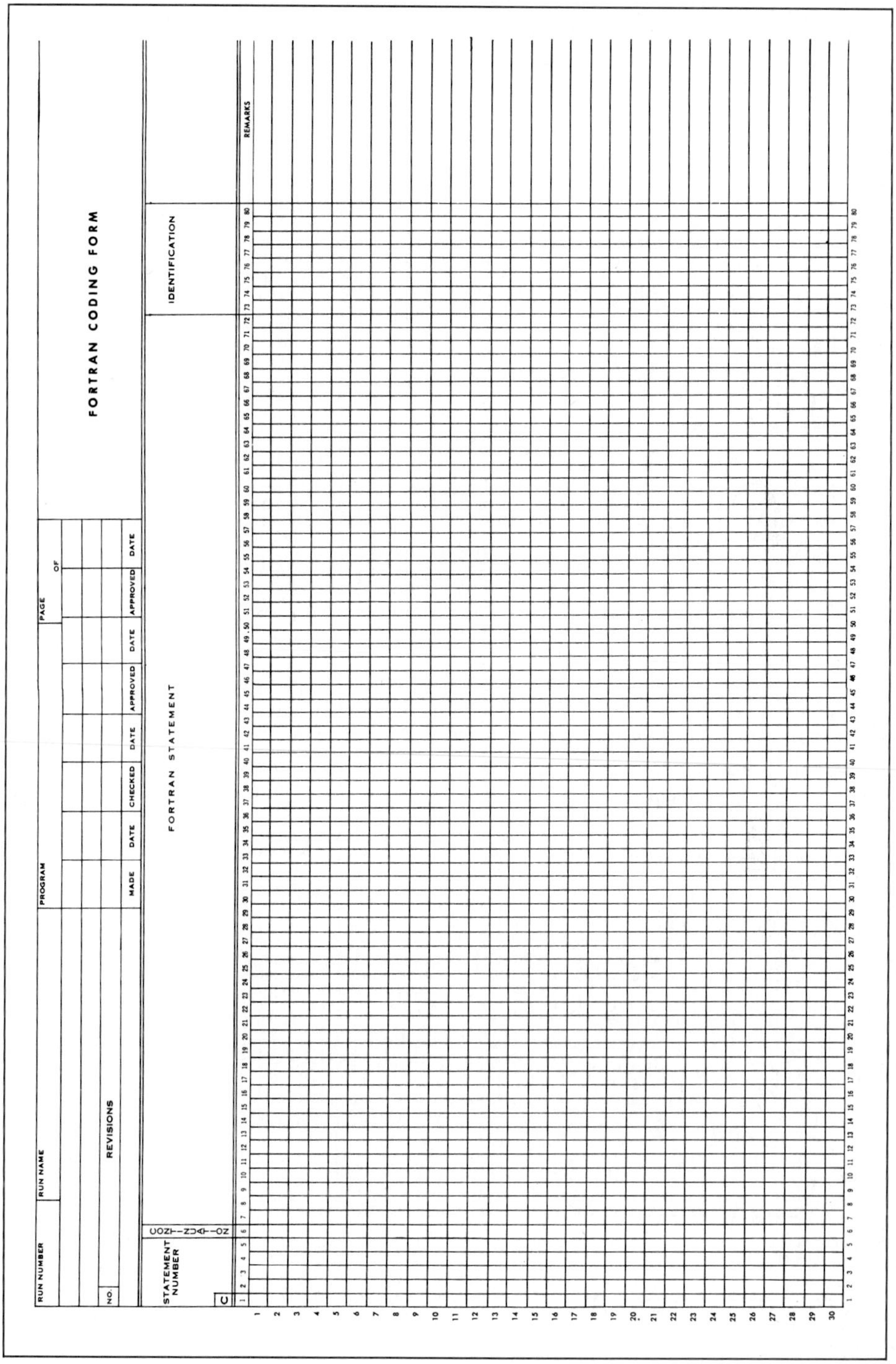

FORTRAN CODING FORM

RUN NUMBER	RUN NAME	PROGRAM						PAGE	OF
NO.	REVISIONS								
		MADE	DATE	CHECKED	DATE	APPROVED	DATE	APPROVED	DATE

C	STATEMENT NUMBER	CONTINUATION	FORTRAN STATEMENT	IDENTIFICATION	REMARKS
1	2 3 4 5	6	7 8 9 10 11 12 13 14 15 16 17 18 19 20 21 22 23 24 25 26 27 28 29 30 31 32 33 34 35 36 37 38 39 40 41 42 43 44 45 46 47 48 49 50 51 52 53 54 55 56 57 58 59 60 61 62 63 64 65 66 67 68 69 70 71 72	73 74 75 76 77 78 79 80	

Rows: 1 2 3 4 5 6 7 8 9 10 11 12 13 14 15 16 17 18 19 20 21 22 23 24 25 26 27 28 29 30

1 2 3 4 5 6 7 8 9 10 11 12 13 14 15 16 17 18 19 20 21 22 23 24 25 26 27 28 29 30 31 32 33 34 35 36 37 38 39 40 41 42 43 44 45 46 47 48 49 50 51 52 53 54 55 56 57 58 59 60 61 62 63 64 65 66 67 68 69 70 71 72 73 74 75 76 77 78 79 80

FIGURE 18.3 Fortran coding sheet.

JOB [ ] (1 2 3)

ZONE [ ] (4 5)

LAST ZONE	TYPE	DESIGN CONDITIONS: 100 CFM OA	100 CFM INFIL	TEMP RM	R.H. RM	TEMP DIFF CHILL WAT	SYST OPER TIME: BEG	END	ALLOWANCE SAFETY FACTORS %: BY PASS	SEN	LAT	FAN	SUP DUCT GAIN	SUP DUCT LEAK	RTN DUCT GAIN	RTN DUCT LEAK	PMP	PIPE LOSS	MAX. ZONE DIM.: LENGTH	WIDTH	HEIGHT	PEOPLE BTU/PERSON: SENS	LAT	NO. SIM. ZONES
	1																							

	LAST WALL	TYPE	ORIENT	GLASS 1: AREA SQ. FT.	FACT	U FACT	OVERHANG A	B	C	GLASS 2: AREA SQ. FT.	FACT	U FACT	OVERHANG A	B	C	WALL TYPE 1: AREA SQ. FT.	U FACT	DENS	COL	WALL TYPE 2: AREA SQ. FT.	U FACT	DENS	COL	WALL TYPE 3: AREA SQ. FT.	U FACT	DENS	COL	WALL TYPE 4: AREA SQ. FT.	U FACT	DENS	COL
NORTH		2	1																												
NORTHEAST		2	2																												
EAST		2	3																												
SOUTHEAST		2	4																												
SOUTH		2	5																												
SOUTHWEST		2	6																												
WEST		2	7																												
NORTHWEST		2	8																												
HORIZONTAL		2	9																												

TYPE	CEILING ROOF 1: AREA SQ. FT.	U FACT	DENS	COL	COND	CEILING ROOF 2: AREA SQ. FT.	U FACT	DENS	COL	COND	FLOOR 1: AREA SQ. FT.	U FACT	TEMP	FLOOR 2: AREA SQ. FT.	U FACT	TEMP	PART 1: AREA SQ. FT.	U FACT	TEMP	PART 2: AREA SQ. FT.	U FACT	TEMP	PART 3: AREA SQ. FT.	U FACT	TEMP
3																									

NUMBER OF PEOPLE IN ZONE EACH HOUR

TYPE	1	2	3	4	5	6	7	8	9	10	11	12	13	14	15	16	17	18	19	20	21	22	23	24
4																								

POWER LOAD IN KW EACH HOUR

TYPE	1	2	3	4	5	6	7	8	9	10	11	12	13	14	15	16	17	18	19	20	21	22	23	24
5																								

LIGHTING LOAD IN KW EACH HOUR

TYPE	1	2	3	4	5	6	7	8	9	10	11	12	13	14	15	16	17	18	19	20	21	22	23	24
6																								

LAST APPL	TYPE	APPLIANCE FACTOR: BTU / UNIT SENSIBLE	BTU / UNIT LATENT	NUMBER OF APPLIANCES EACH HOUR: 1	2	3	4	5	6	7	8	9	10	11	12	13	14	15	16	17	18	19	20	21	22	23	24
	7 1																										
	7 2																										
	7 3																										
	7 4																										
	7 5																										
	7 6																										
	7 7																										
	7 8																										
	7 9																										

1 2 3 4 5 6 7 8 9 10 11 12 13 14 15 16 17 18 19 20 21 22 23 24 25 26 27 28 29 30 31 32 33 34 35 36 37 38 39 40 41 42 43 44 45 46 47 48 49 50 51 52 53 54 55 56 57 58 59 60 61 62 63 64 65 66 67 68 69 70 71 72 73 74 75 76 77 78 79 80

FIGURE 18.4 A typical application data coding sheet.

Although the teletypewriter and line editor represented breakthroughs, their use still had several disadvantages:

- All output was still in hard-copy form (printed on paper).
- Long listings were difficult to handle.
- The process was slow.

The solution was the development of the keyboard/CRT display terminal.

18.1.2 Keyboard/CRT Display Terminals and Screen Editors as the User Interface

The cathode-ray tube, or CRT, display terminal led to the next step in the development of the user interface. It supplied essentially the same functionality as the teletypewriter, except the output was now in soft-copy form on the CRT display. It provided faster output, and the listing could be scrolled up or down. The CRT display further reduced latency between user action and computer response.

Besides reducing latency, CRT display terminals also made it possible to develop editors that worked on an entire screen, not just a single line. A **screen editor** allows the user to view and edit an entire page of the file as opposed to a line editor that allows editing of only one line at a time. Early full-screen editors allowed the user to scroll around the page using keys on the keyboard. The keyboards often included four arrow keys that pointed up, down, right, and left. Pressing any one of the arrow keys would make a pointer move in the direction of the arrow. This pointer, called a **cursor**, was a marker that represented the current place in the file.

Screen editors made text editing and word processing much easier for the user. For example, let's look at how one would add some text to a file with a line editor and a screen editor. With the line editor, the user would first have to list the file on the screen to determine the line, designated by a line number, where the text was to be added. In the simplest case, the user might just insert a few characters without increasing the line length beyond the limit, 80 characters. If it were necessary to add more characters than this, it is possible that all the lines following the insertion would have to be modified. This could be very time-consuming.

Adding text with a screen editor is much easier. All you did was position the cursor at the point of insertion and start typing. The editor program took care of adjusting all the remaining text. This greatly simplified text editing.

A disadvantage of both screen editors and line editors was that they allowed the user to enter and edit only standard text. If the user needed bold, italics, or other formatting, he or she would have to insert special commands into the text. Table 18.1 shows some typical nroff commands that are used to format text in UNIX environment.

If a user wanted to specify different fonts within a document, the .I, .B, and .R commands would be used to bracket the selected text. Likewise, if the user wanted the title line centered, the .CE command would be entered before the title line. The

TABLE 18.1 TEXT-FORMATTING COMMANDS FROM NROFF

Command	Function
.I	Begin italic font
.B	Begin bold font
.R	Begin normal font
.CE	Center the following line
.BP	Begin new page

.BP was inserted where a new page was needed. At best, these text-formatting commands were inconvenient, because the user had no way of seeing exactly how the document would turn out until it was printed.

The CRT display terminal and interactive computer system were the technologies that replaced the punched card in computer systems. However, they did not end the influence of punched cards. To maintain compatibility with existing program libraries, CRT displays were, and some still are, designed with line lengths that match the number of columns on a punched card. Older compilers and many other application programs still use the same column formats that were developed for punched cards.

18.1.3 Smart Terminals and Screen Editors as the User Interface

In the 1970s, Hewlett-Packard introduced the 2640 terminal, which had a CRT display, a keyboard, and some *processing* capabilities. The 2640 was a "smart" terminal because it had a microprocessor, memory, and local storage. The significance of the microprocessor was that the terminal was essentially a stand-alone computer. It could execute programs designed to run on its microprocessor, and it could save programs and data on a tape using its built-in cassette tape recorder.

The smart terminal has processing capability; let's look at why this is important for the user interface. When you are working on the smart terminal and running a full-screen editor, you are the only user. No other user processes will interrupt and use the terminal's CPU. Therefore, the editor will always give fast responses for your commands. The host computer to which the smart terminal is connected does not have to process "editor" requests from the terminal, as it does for the teletypewriter or CRT display terminal. If the host system for the teletypewriter or CRT display terminal is heavily loaded, response time to your edit commands might be poor and inconsistent.

Since a smart terminal has some processing power, it is an early example of distributed processing. **Distributed processing** means that the system work load is *distributed* over multiple processors rather than concentrated on one host processor. Figure 18.5 shows how interaction with the computer changes when smart terminals are used.

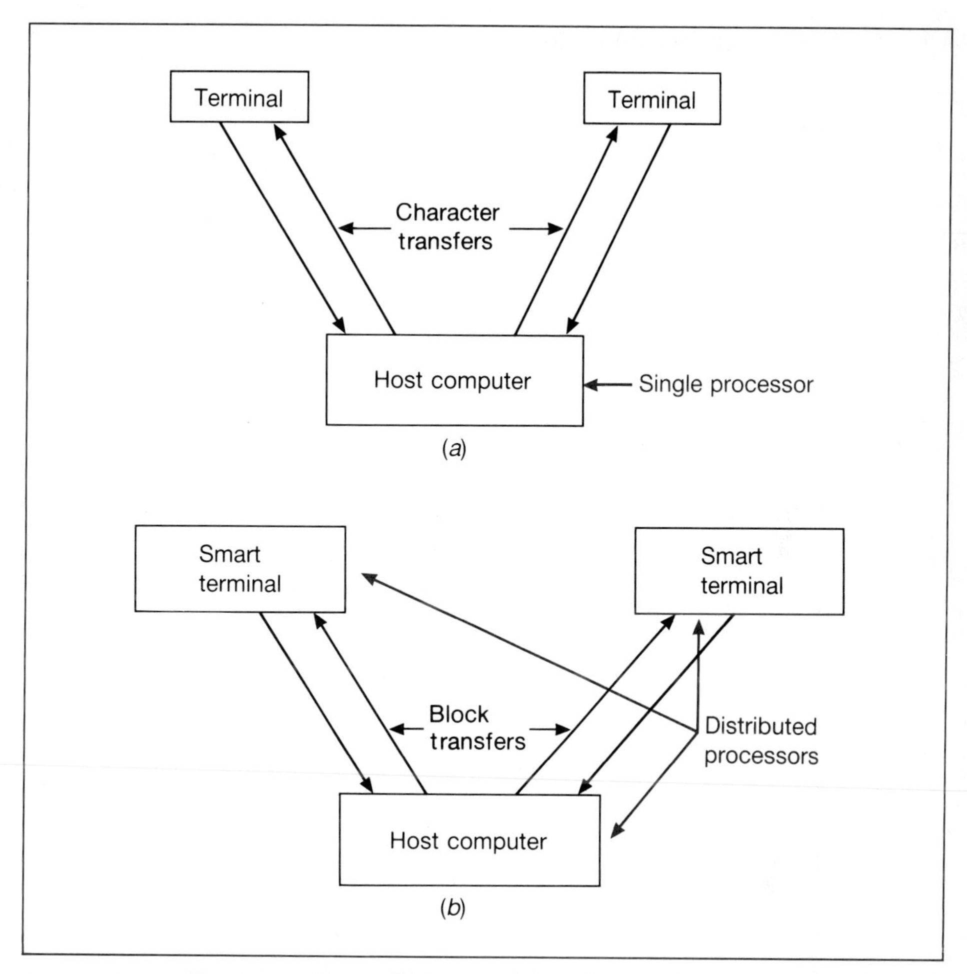

FIGURE 18.5 Two processing configurations: (*a*) single processor system; (*b*) distributed-processor system employing smart terminals.

Smart terminals improved the UI because they made full-screen editing with fast response times possible. They also helped to distribute processing, which made response times more consistent. Despite these improvements in the UI, computers were still relatively difficult for the novice to use.

To this point we have discussed how new *devices* improved the response time involved in human and computer interaction. We saw how the teletypewriter was an improvement over batch-mode processing. We also saw how the CRT and smart terminals improved response times compared to teletypewriters.

In the next two sections we will discuss two very different *styles* of interaction used to communicate with computers. The first interaction style evolved from job control languages used with batch systems. The second was developed because command languages were still difficult for the average person to use.

18.1.4 Command Language Interfaces

In the batch systems described earlier, commands were sent to the system on job control cards. The language used to write these commands was called the **job control language**, or **JCL**. With the advent of terminals and interactive access to the system, JCL cards were replaced by commands entered from the terminal. These commands, which are written in the command language of the system, provide an interface to the system for the user.

Command languages are interpreted by an operating system program called the command interpreter. The command interpreter accepts input from a keyboard and echoes the keystrokes to the monitor. After a command and any necessary parameters are entered, the command interpreter performs the specified operation. In this section we will discuss the UNIX C Shell as a specific example of a command language interface to the computer.

Table 18.2 is a very limited subset of the commands available under the UNIX C Shell, but it should give you a feel for how the interaction happens.

Suppose that a user wants to edit a file named phone_numbers that is found in the directory /usr/staff/larry/, but the user is currently in the directory /etc. Two ways that the user could access and edit the file are explained below.

First, the user could execute a Change Directory command and then execute the Edit File commands:

```
larry:/etc⟩ cd /usr/staff/larry ⟨CR⟩

larry:/usr/staff/larry⟩ gnuemacs phone_numbers ⟨CR⟩
```

The ⟨CR⟩ means that the carriage return key is pressed to enter the command. The larry:/etc⟩ is a prompt displayed by the command interpreter to signal the user that it is ready to receive a command. Notice that this prompt displays the current working directory. The **current working directory** is analogous to a cursor in a word processing program. It represents the user's *position* in the file system, like a cursor represents the user's position in a text file.

Next, the user executes the gnuemacs command with the phone_numbers file as a parameter. The gnuemacs editor will find the file phone_numbers because it is in the current working directory. If the user had tried to execute the command gnuemacs

TABLE 18.2 UNIX C SHELL COMMANDS, PARAMETERS, AND FUNCTIONS

Command	Parameters	Function
cp	⟨file1⟩ ⟨file2⟩	Copy file1 to file2
ls	⟨directory⟩	List files in the directory
lpr	⟨filename⟩	Print file ⟨filename⟩
more	⟨filename⟩	Display contents of file
mkdir	⟨dir⟩	Make a new directory ⟨dir⟩
rm	⟨filename⟩	Remove file ⟨filename⟩
gnuemacs	⟨filename⟩	Edit a file

phone_numbers from the /etc directory, the editor would not have found the desired file.

The user had to execute two commands to edit the file in the above example. The same result can be accomplished by executing a single command:

```
larry:/etc> gnuemacs /usr/staff/larry/phone_numbers <CR>
```

Because the user knows exactly which directory contains the file to be edited, the directory and file can be specified together. The gnuemacs editor will load the file phone_numbers under the directory /usr/staff/larry for editing. This type of interaction is convenient because the user can tell the computer precisely what to do with a single command.

Command interpreters are powerful tools. With them, the user can make the computer perform a variety of operations with minimal keystrokes. But command interpreters do have disadvantages. Command languages are often terse and cryptic, which makes them difficult to learn and remember. They also require the user to recall exact syntax for each command. These difficulties with using conventional command interpreters led researchers to look into other approaches to the user interface.

18.1.5 Graphical User Interfaces

One of the most significant advances in the development of user interfaces was the graphical user interface. The **graphical user interface**, or **GUI**, introduced by Xerox, reduced the problem of remembering commands by placing them in menus and by using icons. On the Xerox Star Workstation, the user does not type commands into a command interpreter [Smith et al., 1982]. All commands executed on the Star are chosen from menus with a mouse or the keyboard. Application programs like editors are started by clicking on icons or by selecting the name of the editor program from a menu.

The revolutionary GUI was pioneered at the Xerox Palo Alto Research Center (PARC) during the mid- to late-1970s. The GUI makes the computer easier to use for many reasons, some of which we will now discuss.

The July 1989 edition of *Byte Magazine* had an article entitled "A Guide To GUIs" that gave an excellent description of graphical user interfaces. It states:

> ... most GUIs consist of three major components: a *windowing system*, an *imaging model*, and an *application program interface* (API). The **windowing system** is a set of programming tools and commands for building the windows, menus, and dialog boxes that appear on the screen. It controls how windows are created, sized, and moved on the screen, and how the user moves from one window to another, among other functions. [italics and boldface supplied] [Hayes and Baran, 1989]

We will discuss only the windowing system component because it is what users interact with directly.

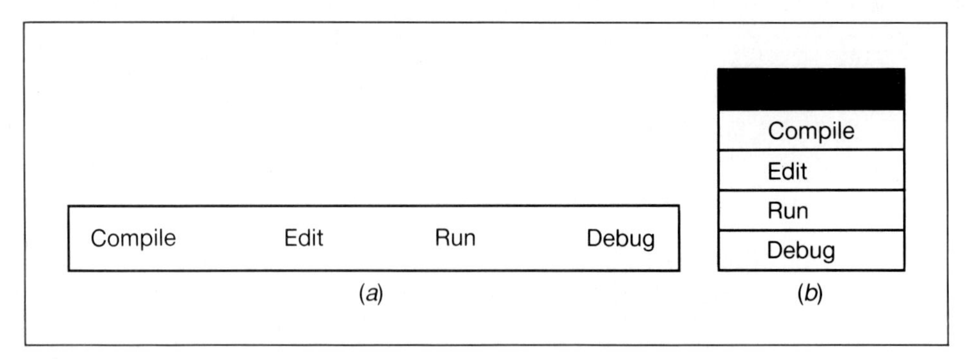

FIGURE 18.6 Two menus of commands: bar menu and pull-down menu.

As stated above, a windowing system contains tools to develop menus. A **command menu** displays a set of commands that is available on the system. Such a menu of commands is analogous to a command language with the commands displayed on the screen from which the user can choose. Figure 18.6 shows two types of menus.

Several techniques are commonly used to execute commands in a menu. First, the command could be executed if the mouse were used to click on the command in the menu. Another technique is to have the command executed when a particular key is pressed. For example, the Compile command could be selected from the menu shown in Fig. 18.6, if c or C were typed at the keyboard. Menus work well when the command needs no parameters, but how are parameters specified?

One solution is the **dialog box**, such as the one in Fig. 18.7. A dialog box allows the user to specify various parameters for a given command. Parameters that are required and mutually exclusive can be specified with *radio buttons.* Those that are fixed or optional can be specified with *check boxes.*

Radio buttons are used like buttons on a car radio. Only one station can be selected at a time. There are three sets of radio buttons shown in Fig. 18.7, one set for Pages, one set for Paper Feed, and one set for Quality. Currently, Quality is set

FIGURE 18.7 Dialog box with radio buttons, check boxes, string edit boxes, and push buttons.

Print
Printer: ImageWriter
OK
Pages: All Selection From: To:
Cancel
Copies: 1 Paper Feed: Automatic Manual
Print Hidden Text Tall Adjusted
Quality: Best Faster Draft

to Best; it could be set to either Faster or Draft, but only one option can be specified at a time.

Check boxes are useful when the command has optional parameters that are fixed. The user simply "checks," with the mouse, the options to be performed. There are two check boxes in the dialog box, one for Print Hidden Text and another for Tall Adjusted. The Print Hidden Text option is checked and the Tall Adjusted option is not, although both options could be selected simultaneously.

Not all parameters lend themselves to check boxes or radio buttons. Often, the user needs to specify file names, printer names, or other computer names as dialogs. Some systems will determine that a file name is the desired parameter and present a list of all possible file names the user can choose from. This scheme works well provided the file is available, but what if it does not exist? Say, for example, the user wants to create a new file and name it myfile. A **fill-in box** is provided for this type of interaction. Three fill-in boxes are shown in Fig. 18.7. The boxes labeled From: and To: accept text typed from the keyboard. The From: and To: boxes are used to specify which pages to print; for example, print From 2 To 4. The Copies box allows the user to specify the number of copies he or she wants printed.

The last items of interest in the dialog box shown in Fig. 18.7 are the **push buttons**. The push buttons are used to signal the computer to execute the command with the given parameters or to quit the command. The two push buttons OK and Cancel operate similar to a hardware push button; they are "pressed" by clicking on them with the mouse. Clicking the OK button makes the printer print with the parameters specified in the dialog box. Clicking Cancel causes the entire dialog box to disappear without executing the print command.

Normally, a user will set the check boxes, radio buttons, and fill-in boxes to his or her liking. He or she then clicks a push button to make the computer execute the command with the specified parameters.

Let's now turn to a discussion of windows. It is often useful to group information in separate areas on the screen. For example, the files in different directories can be grouped together. On multiprocessing computers, it is convenient to use different areas of the screen to perform different tasks. A software device used to divide the screen into logical areas is called a **window**. It is useful to have files in different directories grouped in their own window, as shown in Fig. 18.8. Windows are also useful to show several applications running concurrently, as shown in Fig. 18.9.

Normally, the windowing system allows many operations for windows, such as moving, resizing, and scrolling. Moving a window is a matter of placing it in a different position on the screen. Resizing a window requires that the window's length and width be adjusted to the desired size. Scrolling is a technique that allows hidden information to be moved up or down or right or left in the window so that it can be seen.

18.1.6 WYSIWYG

Earlier, during our discussion of full-screen editors, we listed a table of commands necessary to format text (Table 18.1). Today's word processors and desktop publi-

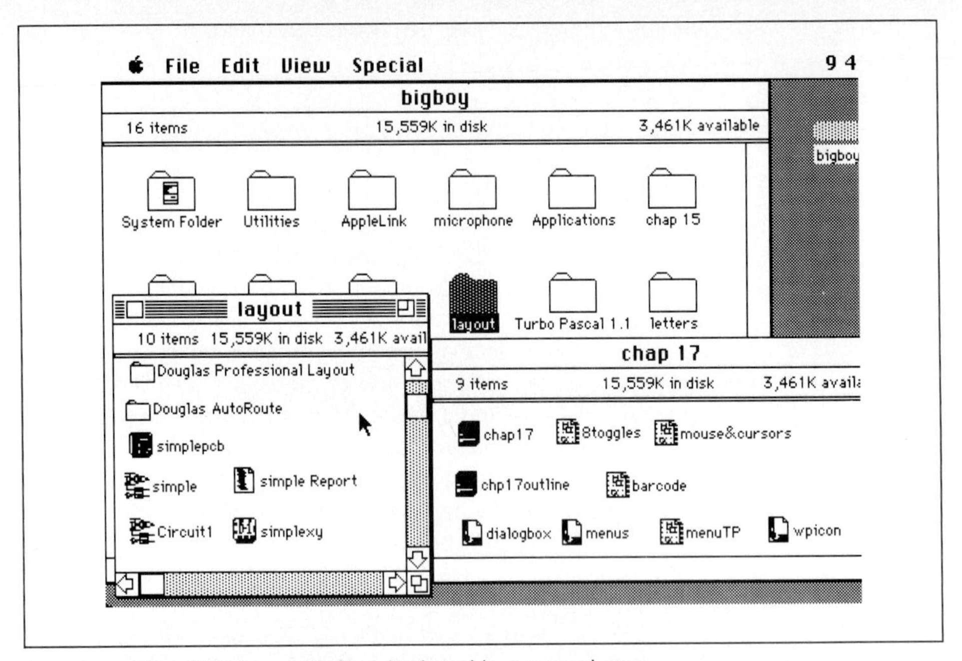

FIGURE 18.8 Windows with files displayed by icon and name.

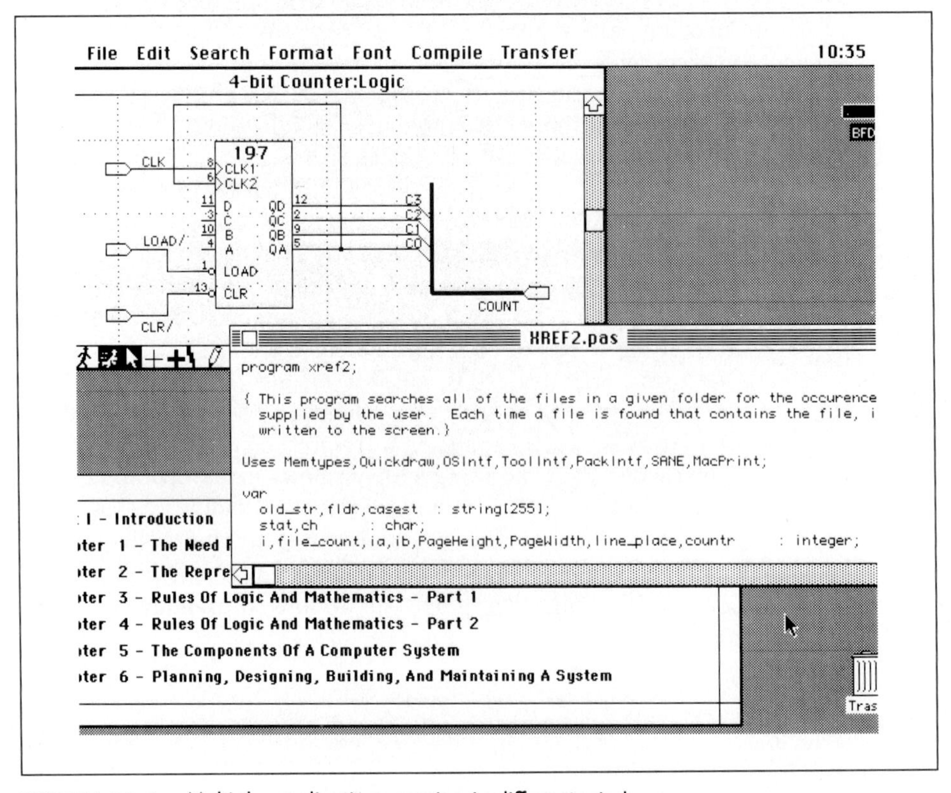

FIGURE 18.9 Multiple applications running in different windows.

shing programs are based on another concept developed at Xerox PARC—"What You See Is What You Get," or WYSIWYG, pronounced "whiz-see-wig" [Johnson et al., 1989]. These programs allow the user to choose character fonts, indent paragraphs, center text, single- and double-space lines in the text, check spelling, and do other operations like the command interpreters. The big difference is that the user immediately sees the results of this formatting on the screen—bold appears as bold, centered lines are centered, and so on. The user does not have to wait until the document is printed. *What you see* on the display is *what you get* at the printer.

Except for the GUI, all the UI concepts and devices we have discussed were first implemented for mainframe computers. The GUI was first implemented on a microprocessor-based workstation (the Xerox Star).

This section has given you some perspective on the history of the user interface. We now turn our attention to some current UI design techniques and philosophies.

18.2 ELEMENTS OF A GOOD USER INTERFACE

We implied in the introduction that it is not obvious how to design a good user interface. Because the UI determines how humans and machines communicate and interact, UI designers should know something about the psychology of human interaction. Psychologists and engineers have been working to provide some guidelines and philosophies to aid the UI designer.

A well-designed interface makes a computer easy to use. Conversely, a poorly designed interface can make an otherwise useful system difficult or impossible to use. The term **user friendly** is generally applied to computers that people find easy to learn, easy to use, and easy to understand. Obviously, user friendliness is a subjective quality that is difficult to measure.

A good interface should increase the usability of a computer by making its use intuitive and consistent. Some qualities we feel are important to a successful user interface, and the methods to determine them, are:

- It satisfies the user needs (task analysis).
- It is convenient for intended users (usability goals).
- It conforms to the user's view of the problem (user conceptual model).
- It has easily accessible help for the user (help and tutorials).
- It has uniform commands (consistency).
- It can be modified to fit individual needs (user configurability).

We will now discuss these qualities and explain how they are related to the design techniques and philosophies presented in this section.

18.2.1 Task Analysis

The first goal in designing a UI is to research the user group for whom the computer is intended. Three questions must be answered:

- Who are the users?

- What knowledge and capabilities do they have?
- What are the tasks that these users will perform?

Some questions can be asked to determine who the users are:

- What is the average age of the user group?
- What is their knowledge of computer systems?
- Will they use the system daily, weekly, monthly, or less?
- What are the typing skills of the users?
- What is the educational level of the users?

There are many other relevant questions that can be asked, but this sample should give a general understanding of the intended users.

Many mistakes can be avoided by determining who will be using the computer. For example, what do the buttons presented in Fig. 18.10 mean to you? To a person unfamiliar with the periodic table of elements, these buttons might be confusing.

As another example, a designer would not want to use a command language interface with commands like Find, Save, Transfer, Configure, and Browse to teach very young children math. The vocabulary used in command languages and menu-driven interfaces must be suitable for the educational level of the user group.

The third reason for researching the user group is to determine what tasks they will perform. Information needed about each task includes:

- How is the task performed?
- What information is needed to perform it?
- How is the task learned?
- How often is the task performed?

We will illustrate task analysis with an example where the intended users are teaching assistants (TAs) in a college English department. Purely for the sake of this discussion, let's assume an English TA will perform two primary tasks, word processing and calculating grades. The first step of task analysis is complete because we have identified the two primary tasks.

Word processing is done by typing text at the keyboard and formatting the text, preferably in a WYSIWYG format. The user would need to learn how to set

FIGURE 18.10 Buttons to request the price of silver or gold.

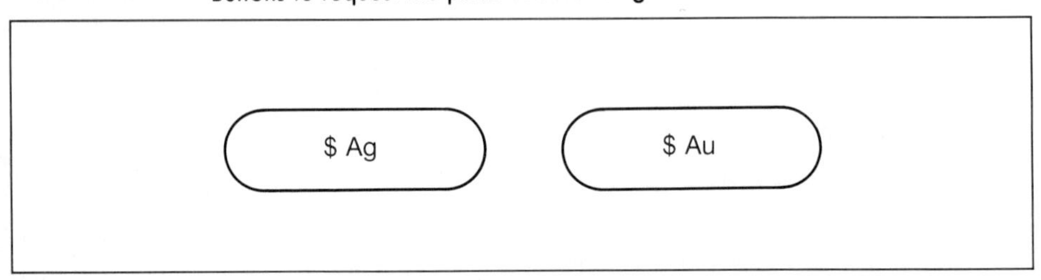

margins, how to set spacing to double or single, how to choose a text font, and how to print documents. Like many tasks, word processing is learned by doing, so an interactive tutorial in which the user gets to perform many functions provided by the application would be useful. The English TA will probably use the word processing on a daily basis.

The frequency of use will help the designer choose an interaction style. When a user does a task every day, he or she can be expected to memorize commands and functions after a short period of time. However, if the user does the task only on a monthly basis, he or she will probably need assistance in remembering commands. For a word processing program used on a daily basis, the designer might decide to associate important word processing functions to specific keys on the keyboard.

Associating functions with keys is sometimes called **key binding**. For example, the keystroke CONTROL F may make the cursor move forward one space, and CONTROL B may make the cursor move back a space. Because the function "move forward a space" is done when the CONTROL and F keys are pressed simultaneously, we say that the function is *bound* to that keystroke. Key binding is useful because the user can quickly perform many tasks. However, if the application will only be used infrequently, the designer might choose to put these functions in a menu, as shown in Fig. 18.11.

Task analysis is important because the UI designer needs to know what type of applications to provide for the user. Equally important, the designer must determine the interaction style for each task. We leave the task analysis of a spreadsheet program as a problem for the student.

18.2.2 Usability Goals

UI designs will probably never be perfect. Therefore, usability goals are defined to help designers decide when the UI is good enough, or "usable." These goals are both objective and subjective in nature.

FIGURE 18.11 A pull-down menu showing cursor movement functions.

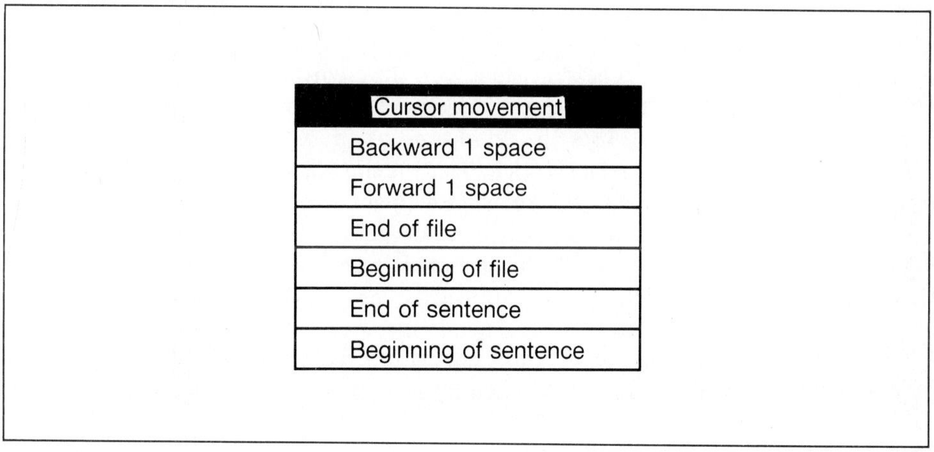

- Cursor movement
- Backward 1 space
- Forward 1 space
- End of file
- Beginning of file
- End of sentence
- Beginning of sentence

Objective usability goals can be measured by experimenting with potential users. For example, suppose we have the following objective usability goal:

> The user will be able to perform a specific task in less than 1 hour and will commit fewer than 20 errors.

After the UI has been implemented, several potential users can be asked to participate in an experiment to determine if the goal has been met. If the users perform the task in less than an hour (average) and they commit fewer than 20 errors (average), then the UI meets the goal. Designers will often specify many objective usability goals, and if users perform adequately on a certain percentage of the goals, then the UI is deemed usable.

The other class of usability goals, **subjective usability goals**, depends on the user's opinion of the UI. A subjective usability goal may be that users will rate the UI as at least a 7 on a scale of 1 to 10, where 10 is very user friendly and 1 is definitely not user friendly. User friendliness is a subjective quality of any computer because anybody who uses it will have different experiences and impressions. Some users might find that it is obvious how to perform a task and the computer is helpful in this performance, while others will find the computer frustrating and unpredictable or inconsistent.

Subjective usability goals can be measured by interviewing an experimental group of users after they have done a set of tasks designed to measure some objective usability goals. During the interview, users are asked to give their opinions of the UI. If 80 percent or more of the users (where 80 is a percentage specified before the experiment) rate the computer as user friendly and easy to use, then it meets the subjective usability goals.

Usability goals enable the designer to establish criteria for success. He or she cannot continually strive for a perfect interface and never deliver a product.

18.2.3 User Conceptual Model

The user conceptual model is how the user views the system. The model should be familiar to users so that they feel comfortable using the computer. The model should also be consistent with the environment being modeled.

Figure 18.12 shows a full-screen view from the Apple Macintosh computer. The Macintosh uses the "desktop" as the user conceptual model for the system. Note that the screen shows icons for items that might be found on a real desk. Folders are used to organize papers. Documents can be stored in a folder. Also, notice the trash can in the lower right-hand corner of the screen. There is a game called MacFootball on the desk and a picture titled Screen 3. A clock resides in the upper right corner. The user should get the feeling of sitting at a desk when viewing the Macintosh screen.

The model is consistent with a real desk. A folder, or document in a folder, can be dragged and dropped into the trash can when the user is finished. This is exactly what the user might do with a real document after he or she is finished with it. If at

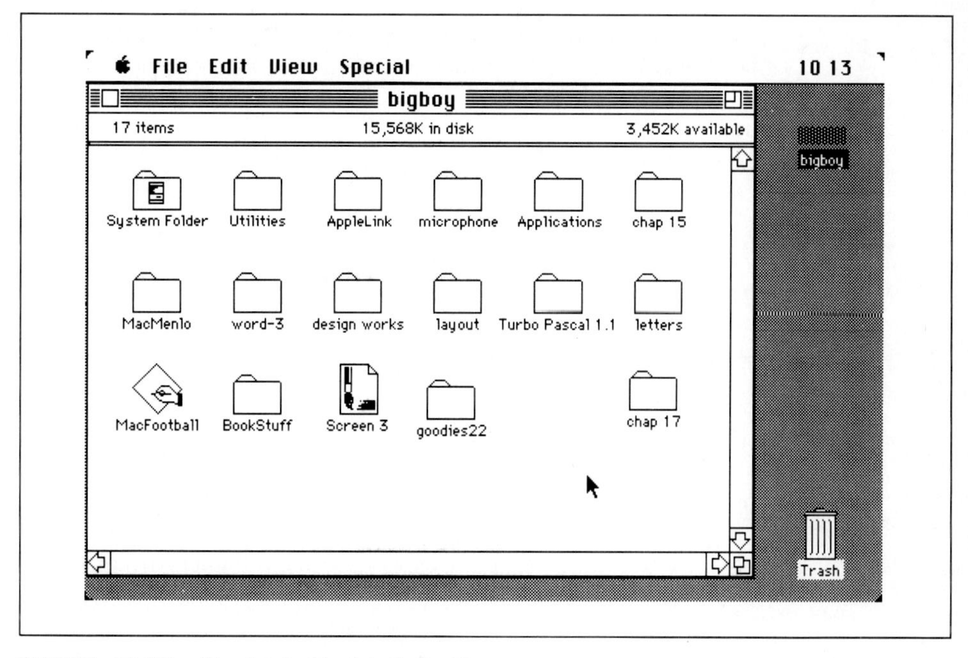

FIGURE 18.12 The Apple Macintosh Desktop.

some later time the user decides he or she really wanted the document previously thrown in the trash, it can be retrieved—if the trash hasn't been emptied.

The desktop is a natural choice for a computer intended for an office environment. A computer used to train pilots should use a model that is consistent with the cockpit of a plane. A computer used to organize the kitchen in a restaurant might be modeled after the kitchen itself. Whatever the choice, the model should be consistent with the application environment and should be familiar to the user.

18.2.4 Help and Tutorials

It is not completely intuitive how to use most programs and systems, so the user often needs some form of help. A beginning user may need help learning the commands available to him or her. Advanced users often need help recalling the syntax of rarely used commands or the function provided by a command.

The UNIX operating system provides help via the man command, which lets the user view the manual pages on the screen. In UNIX, the user can type man cp, and the computer will display the manual pages for the command cp. The manual pages for commands give a description of the necessary and optional parameters, the function of the command, and often examples of how to use the command. This type of help is good for users who have knowledge of the command set but who occasionally forget syntax or parameters. A beginning user would not benefit greatly because he or she would not know the names of the commands in the first place. A beginning user might benefit from a tutorial, however.

A **tutorial** is a tool that guides the user through one or more sessions on how to use the system or an application. Tutorials are often interactive, and they let users experiment with different commands to see their effects. They also correct errors the user may make and teach the user how to learn more about the system.

Help is a useful feature of a UI because it allows users to solve problems on their own. Help may be provided as a tutorial for beginners, manual pages for experts, or some combination of the two.

18.2.5 Consistency

A good user interface is consistent. If all programs on the computer work in a similar fashion, the user does not have to learn different commands for each new application. For instance, if all applications provide help to the user after they press the CONTROL SHIFT H keys, the user will only have to remember one key combination to get help for any application or tool. This example may seem simple and obvious, but there are many cases in which designers are inconsistent with their use of buttons, keys, or windows.

There are many aspects of consistency that apply. We mentioned earlier that the users' conceptual model should be consistent with the environment being modeled. Menus, windows, dialog boxes, buttons, color, and function keys should all be used in the same way by different applications. Inconsistency in the UI is frustrating and confusing for users and should be avoided.

Although it seems obvious that being consistent is a good feature of a UI, you might be surprised how many interfaces are inconsistent. For example, one particular minicomputer used the commands Exit, Bye, Stop, End, and Halt to do essentially the same function. Only one command is needed to specify "stop doing a function."

Another example is a menu-driven UI in which function keys are used to choose menu items. Suppose the F16 function key is normally used to exit the menu. It would be inappropriate to have some applications where the F16 function key provides help instead of exiting.

One way to achieve consistency is to outline how each function should be done and create a set of guidelines for application designers. Apple Computer strongly recommends that designers follow a set of style guidelines for developing applications for the Macintosh [Apple Computer, 1985]. These guidelines help to ensure that all applications use the same type of windows, buttons, dialog boxes, and menus.[1]

18.2.6 User Configurability

During the prototyping stage many features that the users request can be added or removed from the UI. However, no user interface will be ideal for every user, so

[1] Unfortunately, Apple has not supplied similar guidelines for using color.

users should be able to configure the system to their liking. One obvious feature that users can be given control of is color. Any user should be able to change the colors to suit his or her tastes or viewing pleasure. Another feature that users may want to change is the key bindings for various functions. Also, users may want to bind commands presented in menus to keys on the keyboard so that they don't have to take their hands off the keyboard to use a mouse. Users may want certain commands to always appear in a menu. There are many things users may want to change about the system.

For example, a user might want to change the key bindings so that different keystrokes perform different functions. It may seem obvious to one user that CONTROL R is the replace command in a text editor, while another user thinks CONTROL R should be "reverse search." Earlier, when we were discussing word processing, we mentioned that CONTROL F and CONTROL B might be used to move the cursor forward and backward one space. A user may want the forward and backward spacing to be done when he or she hits ESC⟩ and ESC⟨ instead.

Finally, we mentioned that users might want to change menu commands so that they can be done with keystrokes instead of using a mouse. Suppose that the commands to save and print files normally reside in a menu. To execute these commands, assume the user would have to choose the Save or Print commands from the file menu with the mouse. Some users prefer to type a command like CONTROL S for save and CONTROL P for print. It is simply a matter of preference, but if users cannot operate the application in a way that is comfortable to them, they may not like using it, or worse they may choose not to use the application at all.

The shell programs that run on top of the UNIX operating system allow users to **alias** commands to different names. A popular command to use under UNIX is ls -l, which says to list the long version of the current directory. Since this command is typed often, many users choose to alias ls -l to ll. A user can customize his or her shell so that when he or she types ll, the shell will pass the ls -l command to the operating system. Aliasing is useful, because extremely long commands that are used often can be reduced to a few letters.

Designers should allow users to configure the system. No UI will be perfect for every user. Users will be more productive and happy with an environment that is customized to their liking.

18.3 BUILDING A GOOD USER INTERFACE

A well-designed UI should make using the computer simple and straightforward. This is important because computers are supposed to make tasks easier to perform, and tasks are generally easier to accomplish if communication is simple.

We have mentioned several times that designing a good UI is not a trivial matter. One way to test the UI before it is implemented is to build a prototype, as described in Chap. 17. The prototype does not need to have all the functionality of the final product. The purpose of doing a prototype is to gather usability information. By building a prototype UI early in the design phase, designers can test the prototype with potential users and then analyze the interactions with the UI.

For example, suppose that the UI will be a menu-driven system. The menus and the screen layout can be implemented easily. A prototype may be as simple as a screen with menus. When a menu item is chosen by a user, a message can be displayed saying the particular task is being done.

Although this prototype does not provide any functionality, much useful information can be gathered. The command names presented in the menus may or may not be meaningful to the user. The menus may be organized in an inconsistent manner, and users can point this out. Users may find certain colors displeasing, or they may think the screen looks cluttered.

A useful tool for gathering information about each experimental user's interaction with a prototype is the videotape recorder. Each user's session can be taped and played back to designers so that they can see firsthand where the user had problems. Users can be asked to express their likes and dislikes about the UI verbally so that designers will get audio and visual feedback about the prototype.

Prototyping is an important step in the design process. Building several prototypes will help to ensure that the final product meets its usability goals; it is improbable a UI design will meet the usability goals in only one iteration. Currently, not enough is known about how users interact with machines to expect that a design will be correct the first time.

The essential message we can convey to future UI designers is that designers of the system are not usually the end users. A designer may think the interface is consistent, has a good conceptual model, and is usable. However, that belief may simply be the result of the designer knowing so much about the UI. It is important to research the potential users group to learn about them. This eliminates obvious mistakes from invalid assumptions about the group. It is also important to do prototyping to have valid potential users give feedback about the UI. Having potential users participate in the design process will help designers build better interfaces.

18.4 THE FUTURE FOR USER INTERFACES

The future for UIs is bright. We have progressed from using card punches and a batch-processing style of interaction to the GUIs supplied for many computers today. However, there is surely no reason to be satisfied with current UIs. Let's think about how the interaction between a human and a computer might ideally happen.

Future computers would ideally understand natural languages like spoken English. They would also understand gestures and the many other forms of nonverbal communication we humans use. Imagine sitting in a room with a large display device where the computer is translating your spoken English into a computer program. Or better yet, imagine describing a task to the computer and having it write the program to do the task. The dialog might go something like this:

Human: LEP, please write me a program to balance my checkbook each month.

LEP: OK, should I ask you what checks you wrote each day, or once a week?

Human: Once a week is fine.

LEP: How would you like to have the results presented?

Human: Print me a statement after each month.

LEP: Should the final balance be displayed in red if it is a negative number and in black if it is a positive number?

Human: Sure, LEP, that's fine.

LEP: The program is complete. I will ask what transactions you have done each Monday. How about a game of chess?

Human: OK LEP, pawn to king 3.

18.4.1 Voice Recognition

Research on voice recognition is currently being done at numerous places including Carnegie-Mellon University. There, researchers have developed hardware and software that is capable of understanding almost 1000 spoken words. One system they have developed, SPHINX, can achieve as high as 96.2 percent correct recognition for a vocabulary size of 997 words. This can be done independent of the speaker, but it does require some amount of structure in the sentences that are spoken [Lee, 1988]. If the ordering of the words is totally random, the error rate increases.

Eventually, CMU researchers hope to provide a speaker-independent system capable of recognizing a 1000-word vocabulary that does not depend on sentence structure. Think about how many useful commands could be performed with 1000 words. We could certainly implement a speech recognition shell, similar to the C Shell on UNIX. Instead of typing commands, users could simply say them.

18.4.2 The Data Glove

The UI will be better when computers can understand spoken words, but human beings use more than just words to communicate. We use our hands to create gestures, and we use facial expressions. Imagine listening to Eddie Murphy or Bill Cosby on the radio; they would still be funny, but not as funny as they are in person.

A device called the *Data Glove* has been designed and built by a company called VPL Research. The Data Glove is an input device that reports the position and orientation of a user's hand and the relative positions of the user's fingers. It employs directional antennas to determine the position and orientation, and it has fiber optics along the fingers. These fibers leak light when bent, and sensors measure the leakage to determine the finger position. The Data Glove has a standard RS-232 interface, so it can be used with any computer that has the software to handle it [Foley, 1987].

The Data Glove could be used to train the computer to understand how humans communicate with their hands. It could, for instance, learn to provide help with a certain component in an assembly when the user pointed to that component.

Researchers are currently using the Data Glove to have computers understand sign language. The computer can interpret different hand positions to mean the corresponding letter or word. A blind and deaf person could issue instructions to the computer using sign language, and the computer could respond with a Braille output device. Such a system is being perfected at the time of this writing.

18.4.3 Virtual Reality

Currently, researchers are working on more elaborate UI schemes, including Data Clothes and special helmets. The Data Clothes will be able to interpret the movements of the human body analogous to how the Data Glove interprets movements of the hand. The special helmets are equipped with small television screens in front of each eye.

A person wearing the computerized gloves, clothes, and helmet could interact with a completely computer-simulated environment. The computer could project images of objects, and the user could touch and feel the objects. The computer would have to simulate the weight of the objects by forcing the clothes to react to the user picking up the object. For example, as the user picked up an orange, the computer would have to provide the appropriate resistance by causing the clothes to push the hand. If this all sounds difficult, it is. But it will come.

Just imagine the "virtual or artificial realities" that could be created. The computer could simulate the structure of an atom for a chemistry student. The student could touch and feel life-size electrons. The student could force reactions to happen by aligning molecules and pushing them together [Pollack, 1989]. Obviously, there can be a huge amount of new applications and games developed when the gloves, clothes, and helmets become cost-effective.

In our lifetimes, computers will learn to speak our language instead of us having to learn to speak the computer's language. Although much useful work can be done now, current interfaces are less than ideal. Use your imagination to consider what would be best, and build it some day so that the rest of us can benefit. Today's science fiction is tomorrow's reality.

18.5 SUMMARY

User interfaces have come a long way since 1950, but in 10 or 20 years our current user interfaces should seem as distant and outdated as toggle switches and lights seem to us now. There is much that is not known about human/computer interaction. It may be useful for those interested in designing UIs to take some psychology courses. Professionals other than engineers are getting involved in designing UIs. For example, graphical designers are assisting with screen design, and psychologists are studying how to communicate effective error messages and how to present other information to users. Many people are participating in UI design to help build better machines.

An excellent source of information about user interfaces and related material is *Readings in Human-Computer Interaction*, by Bakcker and Buxton [1987]. The book is a collection of important articles written by many different UI researchers and developers.

PROBLEMS

18.1 List the problems of the card reader type of user interface.

18.2 What are some advantages of the control panel interface to the computer?

18.3 What type of commands would one expect to find on JCL cards?

18.4 In Sec. 18.1 we said that cards were often verified after they were punched. Why don't we see that operation being done on today's interactive systems?

18.5 How did interactive use of computers change the secondary storage requirements of a system?

18.6 In Sec. 18.1.1 we described some of the disadvantages of a paper-oriented user interface. List some advantages of this type of interface.

18.7 What are the advantages of the keyboard/CRT terminals over teletypewriters?

18.8 Are microprocessor-based terminals required for full-screen editors? Justify your answer.

18.9 What are the disadvantages of separate editors and text formatters compared with current word processors? What advantages do they have?

18.10 We stated in Sec. 18.1.3 that editing on a smart terminal was an example of distributed processing. What are some other examples?

18.11 MS-DOS for IBM PC compatible computers does not have exactly the same commands as UNIX. How can equivalent MS-DOS commands be created easily with a UNIX shell?

18.12 What is the equivalent of the current working directory on the GUI of the Macintosh computer?

18.13 It was stated that command interpreters require exact syntax for commands. What can be done to make it easier to correct errors in commands?

18.14 What additional features could be added to the dialog box in Fig. 18.7 to make it more complete or easier to use?

18.15 What can be done to make it easy for the user to find the format of commands?

18.16 Design a dialog box like the one in Fig. 18.7 that might be used to initiate a phone connection to a remote computer.

18.17 Explain the differences between a dialog box, a fill-in box, and a push button.

18.18 Show a possible organization of tasks (windows) on the screen for program development environment.

18.19 What are icons in a GUI? Why are icons used in the GUI?

18.20 What are some disadvantages of using icons to identify objects?

18.21 How do WYSIWYG word processors differ from programs like nroff?

18.22 In Sec. 18.2 we listed qualities we feel are important in a user interface. Rank the qualities in order of importance and give a brief justification for your choices.

18.23 What are some characteristics other than the one listed in Sec. 18.2.1 that you might want to know about users before designing the user interface?

18.24 Why do you think some people prefer the UNIX shell type of interface over the Macintosh GUI type of interface?

18.25 List five applications where you think the UNIX type of UI would be a good choice.

18.26 List five applications where you think the Macintosh type of GUI would be a good choice.

18.27 Assume you are developing a user interface for a system that doctors would use to enter and retrieve information about patients' medical histories. What questions would you want to ask doctors before designing the interface?

18.28 Describe key binding and why it is useful.

18.29 How are extra keys on keyboards used to enhance the user interface?

18.30 If you were designing a user interface for a spreadsheet program, what information would you want from potential users?

18.31 Describe four objective usability goals other than error rate.

18.32 Describe four subjective usability goals other than user friendliness.

18.33 The desktop conceptual model is what the Macintosh UI is based on. Describe the conceptual model for a different type of application.

18.34 List examples of ways you have seen programs or systems provide "help" information to users.

18.35 Some applications and systems have levels of help available for users. Pick an application or system with which you are familiar and show how the levels of help are implemented.

18.36 Give some examples where you feel the applications or systems you use do not have consistent interfaces.

18.37 What methods have you used to configure applications or systems you use?

18.38 How are programs written so that individuals can configure them easily?

18.39 What are the primary benefits of building and testing prototype user interfaces?

18.40 What features have you configured or would you like to configure on your word processor?

18.41 Other than voice and data gloves, what do you think future user interfaces might use as input devices?

18.42 Science fiction writers have already described some "virtual reality" systems. What are examples you have seen in science fiction?

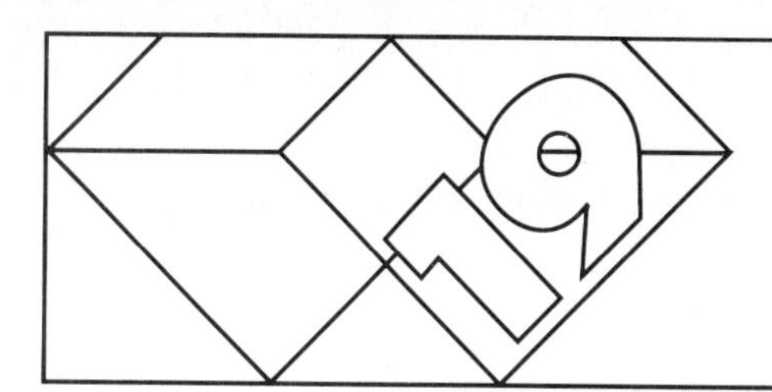

Reliable and Economical System Design

In the earlier chapters of this text we discussed the various components that make up a computer system. We also saw cases where there were several possible ways to build a particular component. For instance, the control unit could be designed as a small, fast hardwired device or as a more flexible but slower microprogrammable device. In this chapter we will suggest some criteria that can be used when making decisions about which design approach should be used.

We begin with a discussion of terms such as availability, reliability, and maintainability. We will then describe some systems that call for highly reliable hardware and software. The discussion next turns to methods that can be used to achieve high reliability in computer systems. It is not enough, however, to design only highly reliable systems; we must also create products that can be economically manufactured, tested, and maintained. These considerations lead us to the topic of life-cycle costs of a product.

The final section of the chapter describes the topic that probably presents the greatest challenge to the designer of any system—design tradeoffs. Given a set of requirements, many designers can create a product that meets those requirements. What often separates average designs from an exceptional design are the tradeoffs the designers choose to make among speed, cost, and quality.

19.1 DEFINITIONS AND CONCEPTS

In this chapter we will make extensive use of several *-ability* terms. Most of these terms have one thing in common; they refer to our *ability* to do some operation, such as manufacture a product. When we talk about the manufacturability of a product, for instance, we are describing the ease (or difficulty) with which we can build the product.

The following terms define concepts that should be important to all designers:

- **Affordability**: the measure of a system's ability to provide its services to a customer at a reasonable cost
- **Availability**: the measure of a system's ability to provide its services to a customer at a given instant in time
- **Maintainability**: the measure of how easy (or difficult) it is to keep a system operating after it has been installed
- **Manufacturability**: the measure of how easy (or difficult) it is to manufacture a product
- **Reliability**: the likelihood (probability) that a product will remain operating correctly in an interval $[t, t + \Delta t]$ assuming it is operating correctly at time t
- **Testability**: the measure of how easy it is to test a system to determine that it is operating correctly or to determine that it has failed
- **Usability**: the measure of how easy it is for the user to learn and work with the system (Sometimes this is known as user friendliness.)

Two more terms are becoming important to designers of computer-based systems:

- **Fault tolerance**: the ability of a system to continue supplying reliable service after one or more parts within the system have failed
- **Redundancy**: adding elements to a system beyond the minimum needed to meet its performance specifications; redundant components are often added to increase a system's fault tolerance.

In recent years many Americans have been made painfully aware of the need to produce reliable, affordable products. In the 1950s and 1960s, American-built automobiles were in great demand worldwide because of their high quality and economical prices. Then, in the late 1950s, German automakers started exporting a very low priced, high-quality car to the United States—the Volkswagen. U.S. automakers for the most part chose to ignore this upstart. They continued producing the same type of cars that had made them successful for decades.

In the 1960s the Japanese followed the German lead and also started exporting large numbers of low-cost, highly reliable cars to the United States. Still, the automobile companies were slow to change their methods. Because of this unwillingness or inability to change, the market share of U.S.-built passenger cars dropped from 94 percent in 1965 [U.S. Department of Commerce, 1980] to 72 percent in 1986 [U.S. Department of Commerce, 1988].

Unfortunately, the automotive industry is not the only one to experience declining market shares because of foreign competition. Television and audio component manufacturing has nearly disappeared from this country. Several factors have contributed to the decline of these industries. Two are nearly always listed as reasons for the success of foreign products over U.S.-built products: reliability (or quality) and cost.

The U.S. computer industry has not yet suffered as much as some industries mentioned above. However, we as designers and builders of computer systems must heed the warning signs and place a much higher emphasis on reliability and quality. If we do not, it is only a matter of time before the United States will lose a significant share of the computer systems market to foreign competition.

Many of today's computer users are no longer willing to accept systems that fail during normal operation. Our society has become too dependent on computer systems to tolerate their failure. In the next section, we will describe some computer applications that demand highly reliable designs.

19.2 RELIABLE AND FAULT-TOLERANT SYSTEMS

It is probably correct to say there has always been a demand for reliable, fault-tolerant computer systems. In the past, however, many users were willing to sacrifice reliability and availability for greater speed and reduced cost. Since the first computers were installed, most users have lived with the fact that their computers will fail occasionally.

Sometimes the failure of a computer is little more than an inconvenience. In other cases it can be very expensive and sometimes life-threatening. We will next examine some applications where computer system failure cannot be tolerated. We will then look briefly at the design of two commercially available fault-tolerant computer systems.

19.2.1 The Evolution of Reliable and Fault-Tolerant Systems

Highly reliable computer systems have been in use for quite some time. However, until the last decade high costs limited their use mainly to military and space applications.

One of the earliest examples of a military computer system built to provide a high degree of reliability was the SAGE system [Enslow, 1974]; it contained two separate computers. When one malfunctioned, a switch was thrown that replaced the unit that had failed with the spare. The spare (redundant) unit was always powered up and ready to be placed into service; this mode of operation is called **hot standby**. Although the system operated successfully for many years, its design was not copied for use in commercial systems. The redundancy needed to meet the reliability requirements made the cost of the system prohibitively high for commercial systems.

Fault-tolerant computers did not make significant inroads into commercial applications until 1976. At that time Tandem Computers introduced a system that reduced the redundancy needed to achieve fault tolerance. Since that time several other companies including Stratus and Sequoia have produced fault-tolerant systems for the commercial market.

One of the largest current applications for fault-tolerant systems, dollarwise, is online transaction processing, or OLTP. Two of the more visible examples of this application are automated teller machines, or ATMs, and credit card verification systems.

Automated tellers are data entry devices that send financial information from remote locations to a central computer system. People typically use ATMs when they want to withdraw money from their account. When they input a withdrawal request, a message is sent to a central computer to check if there is enough money in the account. If the person has enough money in his or her account, a message is sent back to the ATM, telling it to dispense the money. Both banks and their customers demand reliable systems when financial transactions are involved.

Credit card systems call for highly reliable systems for the same reason ATMs do. They are used to extend credit, and this is essentially the same as giving someone cash. It is obvious why banks are concerned about the reliability of ATM and credit card systems.

Customers are often willing to pay more for a computer system if a manufacturer can show that its system will not fail simply because one component within the system fails. A question at the end of the chapter asks you to calculate the cost to a bank of not having access to its deposits for a weekend.

Certain types of control systems represent another area where high reliability is important. Typical examples of these systems are flight control and process control systems. One of the simpler applications of process control is a traffic light controller. At the other end of the complexity scale we might find a reactor control system. In these systems safety is the primary reason for wanting fault tolerance, not loss of money.

Since neither designers nor the components they specify are perfect, even highly reliable systems may fail. In applications where human safety is involved, designers usually try to prevent catastrophic consequences from happening when the computer systems fail. One technique that is used in many of these systems is a fail-safe design. A **fail-safe design** is one that will lead to a safe condition even if the control computer system fails. In the reactor control system, for instance, the safe condition would be to shut down the reactor if any anomalous condition occurred. A question at the end of the chapter asks you to consider what a fail-safe design for a traffic light controller might do.

Fault-tolerant computers are common where the loss of the products or systems is significantly more than the cost of the computers controlling them. The nuclear reactor cited above is an example of such a system. Other systems that fall into this category are:

- Medical monitoring systems
- Aircraft and rapid transit systems
- Air traffic control systems

- Refineries and chemical plants
- Power distribution systems

How systems detect a failure so that they can enter their fail-safe mode will now be considered.

19.2.2 Techniques for Detecting Errors

Error detection is the first step in creating systems that can continue to run correctly after a component has failed. We have already described a common method used to detect error in storage devices—parity. When we add parity to a byte or word, we attach an extra, redundant bit to the word, and that bit makes the count of 1s in the word add up to an even (even parity) or odd (odd parity) number. Use of parity in memory and on buses was discussed in Chaps. 7 and 9.

Before a word is stored into memory, written to a peripheral, or sent out on a communication line, a parity bit is added to it when reliable operation is required. Then, when the word is read, the parity of the word is checked to see if it has changed. If the parity is correct, we assume no error happened while writing, storing or reading, or sending the word. Parity is an inexpensive way to check for errors, but it is not perfect. You are asked in one of the problems at the end of the chapter to describe the type of errors a single parity bit can detect.

Because a single parity bit will not detect all types of errors, other techniques are also needed. In Chap. 10 we mentioned that digital magnetic tape recorders attach an extra character at the end of a record that checks the parity of the data in each track of the tape. This character is sometimes called a checksum character because each bit in the character can be calculated by adding all the 1s in that track together and taking the result modulo 2.

Both magnetic tapes and disks also use another technique to try to guarantee that the data read is the same as the data that was written; they add a **cyclic redundancy check character,** or **CRC character**, to the data in the record. The CRC character is generated by sending all the data bits in a record through a feedback shift register. A **feedback shift register** shifts the data bits through one at a time and XORs the contents of some of the bits with the current data bit. The character in the shift register is added to the data record, and when the record is read, the process is repeated and the CRC characters are compared. If they are the same, the record is assumed to be correct.

The disk record with CRC that was described in Chap. 10 is reproduced in Fig. 19.1*a*. Figure 19.1*b* shows a magnetic tape record that has all three types of checks: parity on each byte, a CRC character, and a checksum character.

The techniques described above are very useful for detecting errors in memory words and blocks of data. As we saw earlier, parity can also be used on registers and data buses within a computer. Suppose, however, that you want to find out if an entire CPU has failed. How can we check for this type of failure?

One method that has been used successfully on many systems is to duplicate the CPU and compare the output of the two CPUs to see if they agree. The second

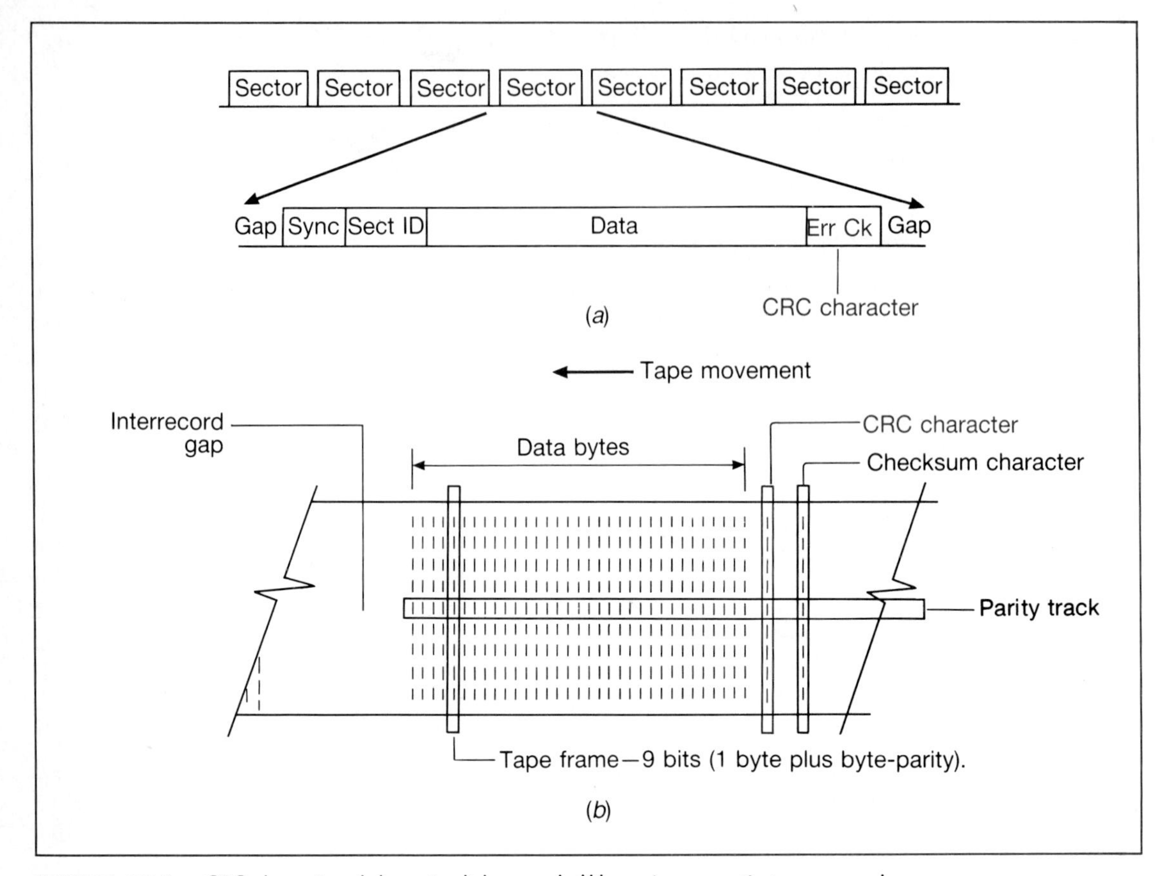

FIGURE 19.1 CRC character: (*a*) use in disk record; (*b*) use in magnetic tape record.

computer is another example of a redundant module. If they do not agree, one of the CPUs has failed. Unfortunately, just comparing the results does not show which one has failed. Detecting the failure can help prevent the system from continuing with calculations when it is not working correctly.

19.2.3 Fault-Tolerant Designs

Fault-tolerant designs not only detect failures, they continue to operate correctly in the presence of a failure. To go beyond just error detection to fault tolerance calls for more redundancy. The computers in the space shuttle [Sklaroff, 1976] serve as a good example of this technique. The shuttle has five computers, and during critical times of the mission, four of the computers calculate control information. Here there are three redundant CPUs. The results of the calculations are compared by a voter. A **voter** can be a circuit or program that compares several inputs and gives an output that is the same as the one specified by the majority.

In the space shuttle system if one of the four computers disagrees with the majority, it is removed electronically from the system, and the system then operates with three computers voting on results. If a second CPU disagrees with the other two, it is also removed from the system. When the system is operating with only two CPUs, voting no longer makes sense. When the system is down to two CPUs and there is a disagreement, the fifth computer takes over control. This computer executes a different version of the software. This reduces the chance that the fifth computer will fail because of a design problem with the software.

The concept of separate versions of programs, ***n*-version programming**, has been examined by Avizienis and Chen among others [Avizienis and Chen, 1977]. They have published papers that describe the advantages of running several versions of a program in parallel and voting on the results.

Using multiple CPUs or other functional units and voting on the results is usually called ***n*-modular redundancy**. In the simplest case where three units operate in parallel and a voter chooses the correct answer, the system uses **triple modular redundancy (TMR)**. The circuit gives the correct result as long as two or three modules are working properly *and* the voter is working correctly. A diagram of a TMR system is shown in Fig. 19.2.

There are many other fault-tolerant techniques used in the different parts of a system. The most common one is error-correcting codes for memory. The method, based on work by Hamming, adds additional parity bits to the data word so that errors can be corrected when they occur [Hamming, 1950].

When your goal is to design and produce a computer system that will not fail, one of the first things you do is try to make each part as reliable as possible. In the next section we will examine techniques used to design and build reliable computer systems.

FIGURE 19.2 Device using triple modular redundancy.

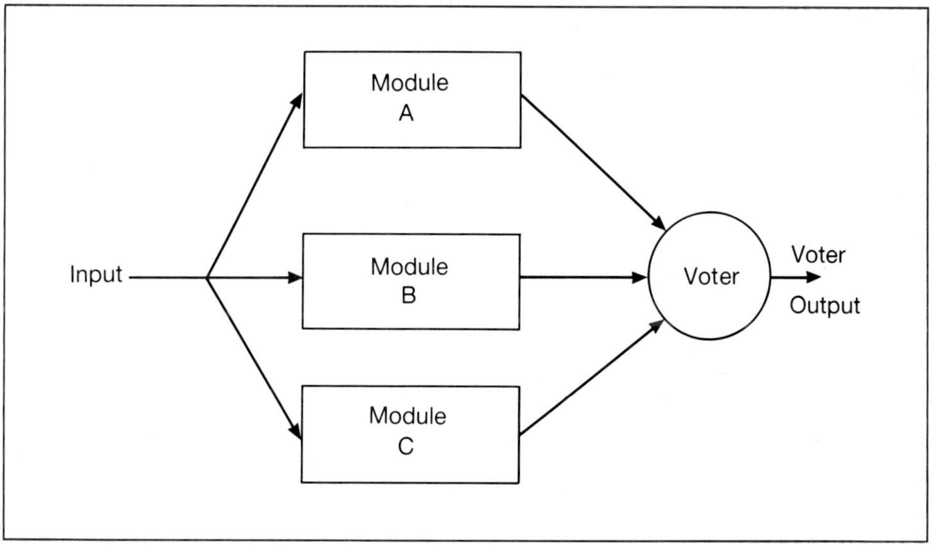

19.3 DESIGNING FOR RELIABILITY

The subject of this section could easily fill an entire book; many books have been written on it [Anderson and Randell, 1979; Glass, 1979]. Since this is only an introduction, we will limit our discussion to three topics:

- Controlling the environment
- Choosing and evaluating components
- Creating conservative designs

One important term that comes up when discussing reliability is failure rate. **Failure rate** is a measure of failures that a component or system will experience in a fixed period. For example, we might observe by testing many lots of components that, on average, a typical integrated circuit fails after 10^4 h. This gives a failure rate of

$$\text{IC failure rate} = \frac{1}{10^4\ \text{h}}$$
$$= 10^{-4}\ \text{failures/h}$$

A related measure is the **mean time to failure**, or **MTTF**, the mean (average) time until a component fails. It is calculated as 1/(failure rate).

$$\text{IC MTTF} = \frac{1}{10^{-4}/\text{h}}$$
$$= 10^4\ \text{h}$$

Most component and system manufacturers will provide reliability information about their components and systems in the form of failure rates and MTTFs. The rates and MTTFs are often specified as a function of some environmental parameter such as temperature.

19.3.1 Controlling the Environment

When we mention environment, one of the first things that probably comes to mind is the physical environment: heat, moisture, mechanical shock, and so on. These tend to put stress on the system's hardware components. Although it is not as obvious, the software in a system can also be "stressed" by the environment. Software stress typically happens when programs and data structures reach their design limits or there are too many users on the system. A good design goal is to reduce all types of stress in a system and thereby improve its reliability.

Failure rates for semiconductor components have been shown to be a function of the operating temperature of the device. Figure 19.3 shows that the failure rate of a chip drops dramatically as the operating temperature is lowered.

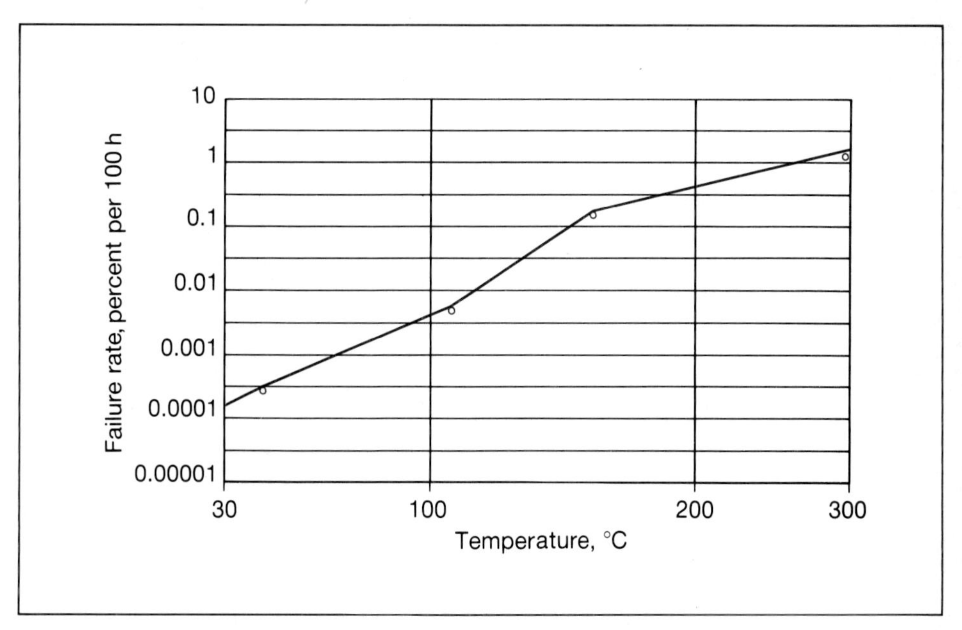

FIGURE 19.3 Typical component failure rate as a function of temperature.

If we are interested in creating a reliable design, one of our design goals should be to reduce the operating temperature. There are several ways to do this. A common one is to place components that generate significant amounts of heat on heat sinks. A **heat sink** is a device, usually metallic, that has a large surface area from which its heat may be dissipated to the surrounding environment. Figure 19.4 shows an integrated circuit with a built-in heat sink.

Assuming the heat sink does its job, heat is removed from the device and transferred to the air that flows around the sink. This means that the surrounding air

FIGURE 19.4 An integrated circuit with heat sink.

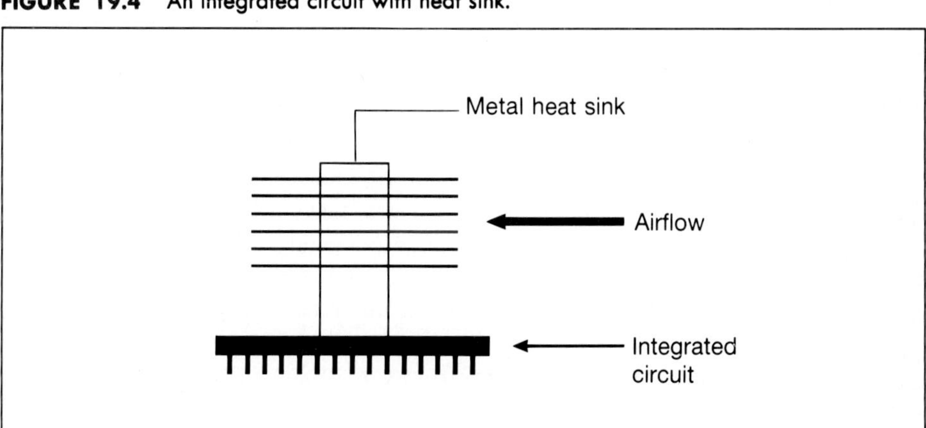

will be heated, which in turn means that the temperature of the entire system will increase. The next step should be obvious; get rid of the heated air.

Getting rid of some heated air may be done simply by making holes in the case containing our device, but this is often not enough. The usual approach is to put fans in the enclosure to force the heated air out and replace it with cooler air. (This assumes the air outside the case is cooler than the air inside the case.) On some systems, heat sinks and fans are not enough to keep the operating temperature low. Large computers sometimes pump liquid coolant through the system to reduce the temperature.

The cost of controlling the environment increases as we go from heat sinks to fans to refrigerators. As designers, we must choose the tradeoff between how reliable the system should be versus how much the customer can afford or is willing to pay for this reliability.

Unwanted electric signals put another form of stress on a system. You might have seen the results of this type of stress if you plugged your personal computer into the same power circuit as your TV. This can cause strange lines on the TV every time the disk was accessed. Here the stress is in the form of electric interference.

Whenever electric current is sent from one point to another, an electromagnetic field is created. If this field is not contained, it acts just like the field from a radio transmitter. When the field reaches another circuit, it can induce a signal into that circuit. For a radio transmitter, this is a desirable effect; for computer circuits, it is usually *very undesirable*. The field adds unwanted signals to the circuit, creating another form of stress. As was the case with heat stress, good design techniques will help reduce the electric interference stress on a system.

To eliminate the propagation of the unwanted fields, shielding of cables carrying electric signals and proper packaging of components are critical. Figure 19.5 shows how shielding can be added around the wires in a cable to reduce emissions. The shield tends to direct the unwanted field to ground and therefore prevent it from propagating to other circuits.

Just as shielding can prevent signals from escaping from cables, good packaging of circuit boards can keep signals within the enclosure that holds the boards. If you compare the enclosures of new personal computers with those of 5 years ago, you can see some improvements that have been made.

- Holes in the enclosures have been reduced or eliminated.
- Nonconducting plastic cases have been replaced with metal ones.
- Cables are grounded.

You will have a better understanding of why these techniques are effective if you study electric fields.

Anyone who has had to work in the sun on a hot summer day or who has dropped a camera into the pool can easily understand why stress reduces the reliability of components. It is not as easy to imagine how software can be stressed.

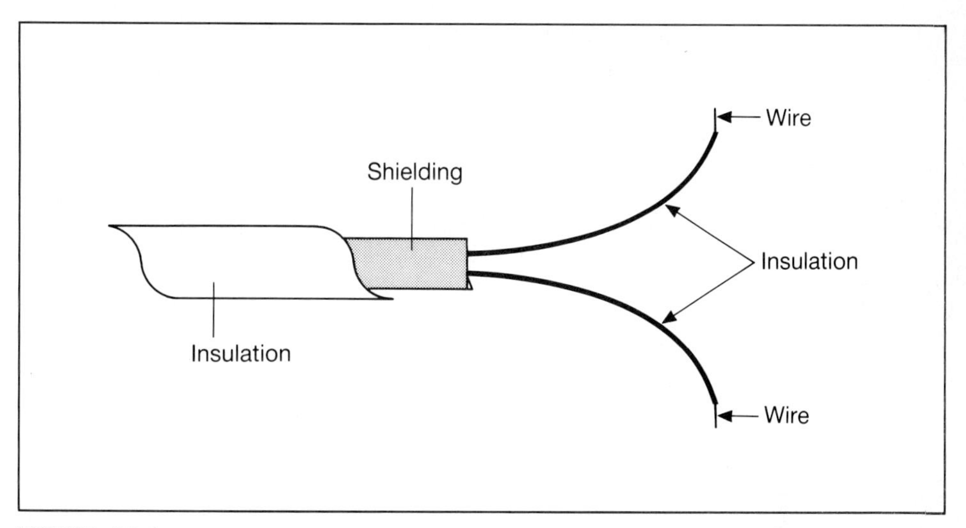

FIGURE 19.5 Cable with shielding to reduce electromagnetic emissions.

Often, software "breaks," or does not operate properly, when there is a heavy work load on the system or when there is a very high utilization of resources such as memory or the CPU. High resource utilization is often stressful to the software.

Studies have shown that the software tends to fail under these conditions because one or more boundary conditions occur. A **boundary condition** happens when a limit on some resource is reached. For example, if a disk is used for secondary storage and we reach the point where all the space is in use and we try to allocate more space to a file, we have reached a boundary condition. The reason that boundary conditions often cause problems is that it is difficult, if not impossible, to account for all possible boundary conditions when we are designing and testing the system. Software often fails when boundary conditions that have not been tested are reached.

Testing is one method used to try to eliminate the effects of boundary conditions. Unfortunately, testing all possible boundary conditions can be difficult for the following reasons:

- We must be able to define where all the boundaries occur.
- We need to know what to do when the conditions occur.
- We should take care of the boundary conditions without reducing performance of the processor significantly.

In this section we have focused on heat and boundary conditions as causes of failures. There are also other environmental factors such as moisture and mechanical shock that must be considered when designing a product. Component selection is an important technique that is used to reduce a system's sensitivity to environmental effects and therefore improve its reliability.

19.3.2 Selecting and Evaluating Components

After a designer specifies the requirements for a product, he or she must pick the hardware and software components that will be put together to make the product. Usually, there will be a wide variety of components that are "functionally equivalent." By this we mean that under normal circumstances they produce the same results given the same inputs.

For example, suppose your design calls for a gate that generates the function

output = NOT(inputA · inputB)

One possible choice is a 7400 integrated circuit [Texas Instruments, 1981] that is manufactured by many different companies. Besides the 7400, it is possible to find dozens of other integrated circuits that will generate this function. They differ in such characteristics as:

- Operating temperature
- Power consumption
- Operating speed
- Physical packaging
- Output current capacity

The designer has the opportunity to decide which component will meet her or his requirements most closely at the best price.

There are also component decisions that must be made throughout the software design process. First, we may have to decide which operating system should be incorporated into the system. Even then, there may be more choices. Having chosen UNIX, for instance, we may find there are several versions supplied by different manufacturers.

After deciding on an operating system, we may be required to decide which application software will be included in the system. There may be several different C or Pascal compilers from which to choose. There will also be a large variety of databases and applications programs available. The designer has to review and compare the specifications for each of the products to determine which product most closely meets the requirements at the best price.

Component selection is only one step in the design and manufacturing process. Only the most naive or inexperienced designers will believe that all components behave exactly as the manufacturer claims. Most companies test some, if not all, of the components they build to be sure that components meet their specifications.

Testing devices such as integrated circuits requires that they be inserted into a test device and their parameters measured over a variety of operating conditions. For software, we must test to see that the program operates as specified in the user's manual. In both cases, the testing process can be long and expensive. This is another price we must pay to build a reliable product.

19.3.3 Creating Conservative Designs

Creating conservative designs is a concept that we can take from the older civil and mechanical engineering disciplines. Most civil or structural engineers will not subject a beam in a bridge to the maximum load for which it is rated. A safety factor is added to all designs, and the larger the safety factor, the more conservative the design.

We can apply the same idea to computer systems. Just because a manufacturer claims that a component will operate at 50°C and at 25 MHz does not mean that we can reliably run it at those limits for a long time. We will provide cooling so that the temperature stays below 30°C and design the circuit so that it runs at only 18 MHz if we are conservative. There are many other areas where we will have the freedom to make choices that are conservative and hopefully more reliable.

How can this technique be applied to software? If the manufacturer claims that a program can accept and process 1000 elements in 10 s, we should be able to test the program and verify the claim. Once that has been done, we would like to be able to count on the program always accepting and processing 1000 elements within 10 s. However, the fact that the program processed one set of 1000 elements does not guarantee that it will successfully process a different set of 1000 elements. A conservative design might only send 750 elements to the program and give it 15 s to complete the processing.

A conservative functional design does not by itself guarantee a reliable, high-quality product. If a product is going to be successful in the marketplace, we must be able to manufacture it, test it, and maintain it. Some techniques used to design and build these systems are discussed in the next section.

19.4 PRODUCING AFFORDABLE, HIGH-QUALITY PRODUCTS

Creating a design that meets a set of functional requirements is the first step in building a highly reliable product. If we want to produce the product economically and still maintain high reliability, the product must be designed so that quality can be maintained throughout the manufacturing process.

In most design classes you take in school, the emphasis will be placed on creating a design that meets a set of specifications passed out by the instructor. The time available in class is seldom sufficient to allow you to take the next step, creating a design that can yield a manufacturable product. In industry, however, you will find that this is one of your most important responsibilities.

19.4.1 The Effects of Component Tolerance on Reliability

As part of the design of a device or program, we will normally build a prototype, which is usually a *hand-built model* of the product. We must be very cautious when making assumptions about the manufacturability of a product based on how the prototype works. There is an old saying among engineers, "You can make one of anything work." One reason for this is that when building a prototype, we tend to handpick components so that performance is optimized.

Unfortunately, on the production line there is no time to handpick and optimize components. This means there is a finite probability that every component that goes into a product can be at the limit of the range of acceptability. If the design works correctly only when all the components have "typical" values, there is a very good chance that any unit manufactured with components that are not typical will fail. The lesson we should learn from this is that *reliable designs should be based on component parameters that can take on worst-case values.*

Tolerances, variations from the desired value, apply to most physical components within a product. Electronic components all have tolerances. Integrated circuits, for instance, have propagation delay times and ranges of output voltages. Mechanical components also have tolerances on dimensions and operating temperatures. Even software has tolerances; programs, for instance, may only be able to accept inputs within a given range or may have execution times that can vary within a given range. Before any product is placed into production, designers should verify that it works correctly when components are at the limit of their tolerances.

19.4.2 Testing and Product Reliability

One of the important steps in producing a reliable product is testing. After the product rolls off the assembly line, it should go to a final test. The purpose of the final test is to verify that the product meets the specifications to which it was designed. Cars are checked to assure that the engines start and run smoothly, that the wheels go around, and that the brakes work. TV sets are tested to confirm that they can receive all channels clearly and that all controls function properly. With computer-based products, it may be much more difficult to verify that all the specifications are met.

There are several reasons why computer-based products are difficult to test. First, for any moderately complicated computer-based product, there are many execution paths through the circuits and programs. To determine whether one of these products always generates the correct output, it is necessary to exercise every possible combination of paths. Usually, it requires far too much time to test all possible execution paths. One of the questions at the end of the chapter asks you to calculate the time it would take to completely test a memory module to verify that all possible combinations of data can be stored and retrieved without error.

One way the designer can help reduce this problem is to add testing routines to the software that is shipped with the product. If you have watched the display screen while the computer is starting, you may have seen an example of these test routines. Usually some messages stating that memory is being tested or that the system has passed a self-test will be displayed. These tests, which are not required to meet the functional specifications for the product, are one way the software designer can help to create a testable product.

Hardware designers also need to consider testability when creating their design. A problem that can happen with hardware designs is that it may only be possible to attach test probes to a limited number of input and output pins. If so, it may not be possible to detect an internal failure in a product. Hardware designers can help alleviate this problem by adding extra circuits used only for testing. Although this again adds nothing to the functionality of the product, it can contribute to the testability of the product.

19.4.3 Product Maintenance

So far we have limited our discussions to how we can create a more reliable and manufacturable product. No matter how good our designs and how carefully the products are manufactured, systems will eventually fail. Because of this, there is usually another important characteristic: the maintainability of the product. Although in many areas we seem to be moving toward a "throwaway" society, there are still many areas where the customer is willing to pay for products that are not only reliable but are also maintainable. The cost of maintaining a system can be greater than the cost of acquisition if it is not designed for low maintenance.

In the previous section we discussed how software test routines and hardware test circuits can be used to verify that a device is working. These tests, called **diagnostics** or **self-tests**, can also be used to determine which component has failed. Once we find the component that has failed, we should be able to repair or replace the component quickly. Our ability to do this is another measure of the quality of our design.

After testing has determined which component has failed, the next step is to replace that component. How we design a system can also affect the ease with which parts can be replaced. To reduce costs, some products are built on a single, large printed circuit board that is then bolted to the chassis. Although this can reduce manufacturing costs, it usually leads to very high maintenance costs. The alternative approach is to design each part of the system on a small, replaceable printed circuit that can be plugged into a connector. This method can simplify maintenance procedures but adds to manufacturing costs.

There are other methods that could be used to improve the maintainability of a product. **Remote testing** is an increasingly important technique that may help improve maintainability. With this technique, we can have a maintenance computer call the computer that has failed and diagnose the problem. A set of test signals is sent to the "sick" computer that will hopefully determine what part of the computer has failed and which components must be repaired or replaced.

As we have stated, the design for a product should not be considered complete just because it meets a set of functional requirements. To be successful, most products must also be manufacturable, testable, and maintainable. Products that are hurried through the design phase may have lower initial costs because some design expenses have been eliminated. These products also tend to be expensive to manufacture and maintain. If more design time is spent making the product easier to manufacture and maintain, the initial price may have to be increased, but reduced maintenance costs should more than compensate for this. The customer prefers a product that has a low initial price and low maintenance costs, but given a choice, they will usually pay a little more to get a system that is reliable and will cost less ultimately.

19.5 THE TOTAL COST OF OWNERSHIP

Anyone who has bought a complex product such as a car or television set is probably aware that the initial cost of the product represents only a fraction of the total cost associated with owning and using the product. If we buy a car, we will also need gas, oil, and insurance, and unless we are very fortunate, we will have to pay for repairs occasionally. Even our highly reliable TV may end up costing more than we first estimated. First, we must provide power, and what good is the TV without a cable service so that we can watch movie channels, sports channels, and so forth? Most products have costs beyond the initial purchase price associated with their use.

19.5.1 Initial Price versus Total Cost—The User's Dilemma

When you go out to buy a product, there are many factors that normally influence your decision to buy. One obvious consideration is the initial cost; others include the cost of operation and the cost of maintenance. Manufacturers know this, and to persuade us to buy their products, they will often try to convince us that buying their products will lead to the lowest purchase price, operating cost, and maintenance.

We see examples of this all the time in the automotive industry. If sales are slow, a company will offer a rebate to lower the initial cost. Some dealers will also offer to give 500 gallons of gasoline free with the purchase of any new car to reduce operating costs. Then to show that maintenance costs are low, they offer a 50,000-mile "limited warranty." The purpose of these sales gimmicks is to try to convince buyers that their product has the lowest overall cost of ownership.

To be competitive in the marketplace, most manufacturers try to produce products that have the lowest possible cost of ownership. A natural question to ask is, "How can one reduce the cost of ownership?" We will take a simplistic look at this problem and ignore some important considerations such as marketing and overhead costs.

19.5.2 Reducing Design Costs

One of the first costs incurred in creating a product is the cost of designing the product. To stay in business a company must recover the cost of design when it sells the product. Reducing design costs should enable a company to make a given profit with a lower selling price or by selling fewer units.

A way to reduce the cost of design is to spend less time designing the product. If we do this, however, we may have to cut corners and therefore create a less reliable design. There may be an alternative, however. If designers are given better tools, such as those described in Chaps. 11 and 17, they may be able to design a better product in less time. If we buy versatile design tools, they can be used on many designs, and we can spread the cost of the tools over many products.

Another possible way to reduce design costs is to hire an expert instead of hiring and training a less-experienced person. Some unethical companies may even try to hire an employee from a competing company and effectively steal their competitor's design by acquiring the knowledge of the designer. Other companies pay individuals to steal complete designs, including drawings and programs. Ethical people should not allow themselves to be involved in such practices.

19.5.3 Reducing Manufacturing Costs

Manufacturing costs are an obvious candidate when it is necessary to reduce the cost of a product. Three major elements of manufacturing costs are:

- Cost of the manufacturing facility
- Cost of components
- Cost of labor

As designers we may not have much control over the cost of the facility. However, we do have control over the second two elements.

Component selection can be critical to the overall success of a product. In an earlier section of this chapter, we pointed out there are often a wide variety of components that may satisfy our functional requirements. A good designer can pick the component that meets all the requirements and does it at the best possible price.

Designers should establish what, if anything, they are sacrificing when they specify a less expensive component. Specifically, they need to be sure that when choosing less expensive components to reduce the manufacturing cost, reliability is not reduced and future maintenance costs are not increased.

The designer can also have a direct effect on the labor costs associated with producing a product. If the design yields a product that can be assembled and tested entirely with robots, labor costs will be reduced. But, if the complexity of the design calls for extensive human intervention, labor costs will increase. One technique that can simplify the manufacturing process is to use components that are "standard" size and that do not require special fixtures for mounting or handling. Another common practice is to use preferred parts, those that have been proven to be reliable and economical.

19.5.4 Reducing Operating Costs

After buying and installing a product, there are two other expenses that the user will incur: operating and maintenance costs. If your only experience with computers has been with personal computers, you might think that operating costs will be insignificant. This is not the case, however, for larger computer systems. The three components of operating cost that we will examine are power, cooling, and operator costs.

The cost of power for a computer system will obviously vary with the size of the system. Power consumption for large systems can be significant. It can often be reduced by choosing components that consume less power.

Power consumption can be doubly expensive. First, building larger-capacity power supplies is more expensive than building smaller ones. Second, once power has been consumed, the resulting heat must be removed from the system. This requires ventilation, fans, and possibly special air-conditioning. In the 1970s large IBM computers required that liquid coolant be pumped through the system to remove the heat produced by the circuits. One of the significant advantages that Amdahl Computer Corporation claimed for its IBM-compatible mainframe computers was that they needed no special coolant. Amdahl designed heat sinks for their integrated circuits that eliminated the need for special coolant. Their reduced power consumption and cooling requirements enabled Amdahl to offer customers significantly lower operating costs.

The cost of labor has increased significantly in recent years. Because of this, it is highly desirable to have a system that needs little or no operator intervention in normal operation. Good hardware and software designs can reduce the need for operator personnel. Hardware designers can try to eliminate components that require button pushing and media replacement. Software designers can create software that is user friendly and does not call for highly skilled operators or frequent operator intervention.

If we are to produce products that are inexpensive to operate, the design goals must reflect these needs. It is not enough to produce a product that meets functional goals and then later try to change it to meet operational requirements.

19.5.5 Reducing Maintenance Costs

One of the major problems that computer manufacturers have faced in recent years is the need for larger and better-trained field service organizations. This has tended to drive up the cost of maintenance for the customer.

We have already discussed some ways to reduce maintenance costs. One way is to make the product testable, which can reduce maintenance in two ways. First, it can reduce the time needed to isolate a failure. Second, it may make it possible to have a technician with minimal experience actually replace the faulty part. This latter benefit will only accrue if the product has also been designed with maintainability in mind. Some companies are trying to simplify maintenance and repair procedures to the point that the unsophisticated user can maintain the equipment without the help of a field service person.

One technique to reduce maintenance costs, especially on large computer systems, is to make extensive use of remote diagnostics, which were described earlier. This method of testing makes it possible to have a few highly trained individuals test machines at remote locations without having to travel to the site. This can reduce travel time for maintenance personnel and therefore save the customer money. It can also reduce the total number of highly trained service personnel a company needs to have available to meet its service requirements.

Running diagnostics on a computer that has failed seems to present a major problem. How can you test a computer that has failed? To solve this problem, some companies have added a diagnostic computer to the main computer, and its sole purpose is to test the main computer and determine what part of the main computer has failed. This is another example of adding redundant components to a system. The diagnostic computer does not increase the power of the system but does increase its availability and maintainability.

What we see here again is a tradeoff. We add extra hardware and software to a system which increases its initial cost so that we may reduce maintenance costs. Analyzing tradeoffs and choosing the best option is probably the most important talents a designer can develop.

19.6 TRADEOFFS AND THE DESIGN PROCESS

Being faced with tradeoffs is one of the most universal problems that confronts designers and engineers in all fields. In this section we will look at a few of the tradeoffs that the computer system designer will face.

19.6.1 Space versus Time

The space/time tradeoff is one of the first that programmers observe. It manifests itself in several ways. One place we see it is when we have the choice of writing a program as a long "in-line" sequence of instructions or a shorter version that makes extensive use of procedure calls (or subroutines). A program written using in-line code is longer but usually executes more rapidly. On the other hand, one written using procedure calls is usually shorter but requires more execution time.

Sorting a large file of information stored on disk provides another example of the space/time tradeoff. If there is enough main memory to hold the entire file, the sort can be executed more quickly than if the data must be sorted by reading and writing records to disk.

Space/time tradeoffs also occur in hardware. If we add optional hardware, such as an arithmetic coprocessor, we increase the hardware "space." However, the additional hardware enables the software to run programs faster.

Many other instances of the space/time tradeoff happen in both software and hardware.

19.6.2 Speed versus Cost

The speed/cost tradeoff comes up in many places other that computer product design. Have you noticed the price difference between a Porsche and a Volkswagen?

In computer hardware we see many examples of this tradeoff; memory chips are one. If you look up the price of memory chips, you often see that the same capacity chip is available with several different access times. The price increases with decreasing access time. Besides costing more, the higher-speed chips usually need more power. Other components, such as processors, also exhibit this characteristic.

19.6.3 Cost versus Quality

The initial cost of higher-quality components is usually greater than the cost of components whose quality is lower or is unknown. A simple example of this can be found in the price of resistors. Companies price resistors (and many other electric/electronic components) based on how close their values are to the value listed on the data sheet; typical tolerances are 20, 10, 5, and 1 percent. The price increases with decreasing tolerance-range decreases. One reason for this is that the manufacturer must test the components more rigorously to guarantee that they meet the specified tolerance.

19.6.4 Quality versus Speed

Some types of printers serve as good examples of the quality versus speed tradeoff. Many dot matrix printers have two or three modes of printing. The draft mode forms characters with a relatively small number of dots for each character. Since only a few dots are needed for each character, characters can be printed quickly. The letter-quality mode, on the other hand, may use twice as many dots when creating each character. This takes more time but produces a more readable character font. Thus, we see there is a tradeoff between speed and quality.

A similar tradeoff happens when sending information over phone lines. Modems, devices that convert digital information to tones, may send data at rates between 10 and 960 characters per second. Experience shows that the error rate when transmitting at the lower speed is lower than the error rate when transmitting at the higher speed.

19.6.5 The Decision Triangle

All the tradeoffs described above show that we cannot have our cake and eat it too. If we want more speed, we will have to pay more for it. If we want higher quality, we may have to accept lower speeds. The speed/cost/quality tradeoffs can be shown nicely in a decision triangle. The idea behind the decision triangle is that for a given product it is relatively easy to create a design that meets two of the three requirements. It is much more difficult to create a design that meets all three of the requirements. Figure 19.6 illustrates this concept.

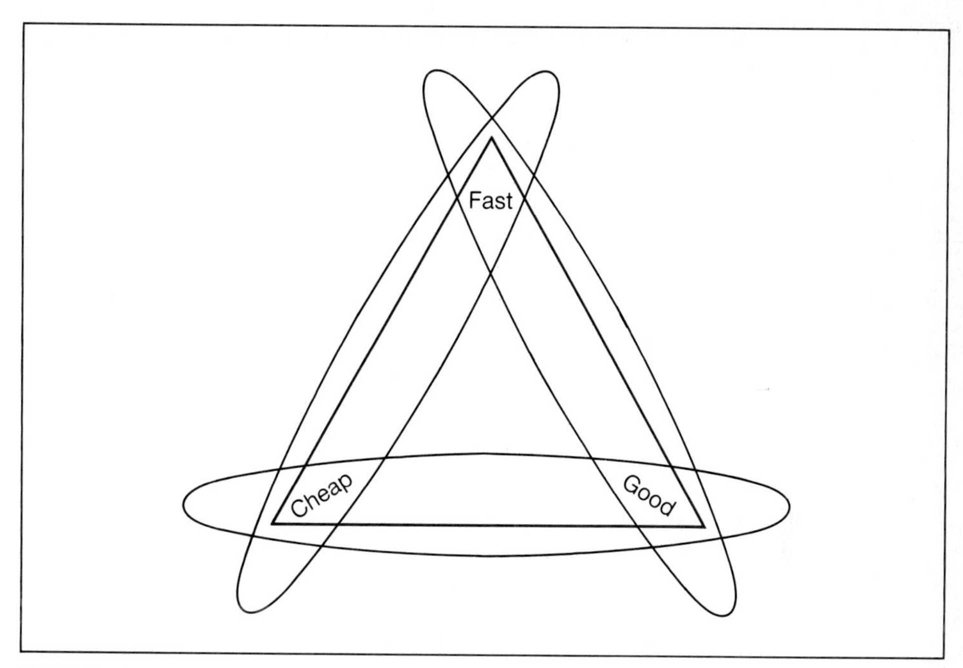

FIGURE 19.6 The decision triangle for speed, cost, and quality.

19.7 SUMMARY—THE BOTTOM LINE

A phrase that has become popular lately is the *bottom line.* This term usually relates to the profit (or loss) associated with some commodity. When we talk about manufacturing a product, the bottom line refers to how much we make selling the product after we subtract all the costs associated with designing, manufacturing, marketing, and selling it.

Companies are not in business to lose money; therefore, management must be convinced that a product will contribute profit to the company's bottom line before they will decide to produce it.

When a company commits to building a product, it also commits to investing in engineering time, facilities, and tools. The company expects a positive **return on investment**, or **ROI**. Many companies calculate a projected ROI for a product before deciding to proceed with the design and manufacturing of a product.

The projected ROI is calculated as follows:

$$\text{ROI} = \frac{\text{sales revenues} - \text{production costs} - \text{sales and marketing costs}}{\text{required investment}}$$

First, estimate total sales revenue for the product. Then, subtract all expenses associated with producing the product. Next, estimate the total amount that will have to be invested to create the product. If the ratio of income to investment is not greater than the company-established goal, the project is usually canceled.

This has been a very simplistic discussion of ROI, but it helps make an important point. Companies design and build products because they expect to make money from the product. Designers can significantly improve the chance of making money on a project by creating an economical and reliable design.

PROBLEMS

19.1 Select two of the *-ability* terms in Sec. 19.1 and show why they can place conflicting demands on the designs.

19.2 Select two of the *-ability* terms in Sec. 19.1 and show how satisfying one can also help satisfy the other.

19.3 Describe two additional terms that should be added to the *-ability* list.

19.4 Briefly describe several process control systems where fault tolerance is important.

19.5 Describe a noncomputer example that uses a hot standby.

19.6 What are some industries where U.S. products compete well in international markets? Why do they compete successfully?

19.7 Even fault-tolerant systems can fail when certain combinations of individual components fail. If you were designing a fault-tolerant traffic light control system, what would you try to have the system do if the control computers halted?

19.8 Besides traffic control and reactor control systems, describe another system that has a fail-safe property.

19.9 What types of errors can parity detect? (*Hint*: Try changing 1, 2, 3, ... bits in a correct word.)

19.10 Using the magnetic tape record in Fig. 19.1*b*, show how the combination of tape frame parity and checksum character could be used to find *and* correct a single bit failure in the record.

19.11 Will the TMR system in Fig. 19.2 *always* continue to operate correctly if one of the modules fails? Justify your answer.

19.12 What are the boundary conditions associated with a program that maintains a linked list of records?

19.13 The file manager portion of the OS manages space on disk drives. What boundary conditions would you expect it to encounter?

19.14 Computer systems usually fail more frequently when there are many users accessing system resources. Other than Murphy's law, why might this happen?

19.15 With MTTFs in the range of 10^8 h, it takes a long time for an average component to fail. What is an alternate way to measure failure rates other than observing one unit?

19.16 Give an example of how the environment can be controlled to reduce software stress.

19.17 A common type of electronic system failure called "Infant mortality" occurs within the first few weeks of operation. How can this be prevented?

19.18 If you were evaluating two compilers (for the same language), what characteristics would you compare?

19.19 Besides taking time and costing money, what are some other disadvantages of component testing for physical devices?

19.20 How do you verify the quality of programs you write for classes? How would your procedures have to change if your programs were going to be sold as products?

19.21 List two examples of software tolerances.

19.22 A typical semiconductor memory chip can hold about 4 million bits of information. If it takes 50 ns to read/write 1 bit of data from/to the chip, calculate the time needed to test all possible combinations of data that the chip can hold.

19.23 Assume there are 10 possible boundary conditions on the resources used by the system software. Estimate the number of individual tests that would be needed to guarantee that the system operated correctly under these conditions.

19.24 Give two other examples of where conservative design principles can be applied to the design of computer hardware.

19.25 Give two other examples of where conservative design principles can be applied to the design of computer software.

19.26 Give two other examples that show the space/time tradeoff.

19.27 Give two other examples that show the speed/cost tradeoff.

19.28 Give two other examples that show the cost/quality tradeoff.

19.29 Give two other examples that show the quality/speed tradeoff.

19.30 How much would it cost a bank if, because of a computer failure, it could not transfer $20 million in deposits to the reserve bank (where it earns interest) over a weekend?

19.31 In Sec. 19.4 we stated that one computer could remotely test another one. What are some problems with this approach to diagnosis?

19.32 Go to an electronics parts store or find an electronics parts catalog and price several different components that are available with different tolerances (resistors and capacitors, for instance). Plot the cost as a function of tolerance for the selected devices. Comment on your results. Is the relationship the same for different components?

19.33 What are some ways to reduce design costs other than having good tools?

19.34 Repeat Prob. 19.32 but plot cost as a function of speed for some devices such as memory chips, logic circuits, and processors.

19.35 List methods that reduce the cost of manufacturing (*a*) hardware and (*b*) software.

19.36 Explain why sorting a file can be done much faster when there is enough main memory to contain the file.

19.37 The circuit in Fig. P19.37 is a simple voltage divider. The equation for the output voltage is

$$V_o = \frac{R_2 V_i}{R_1 + R_2}$$

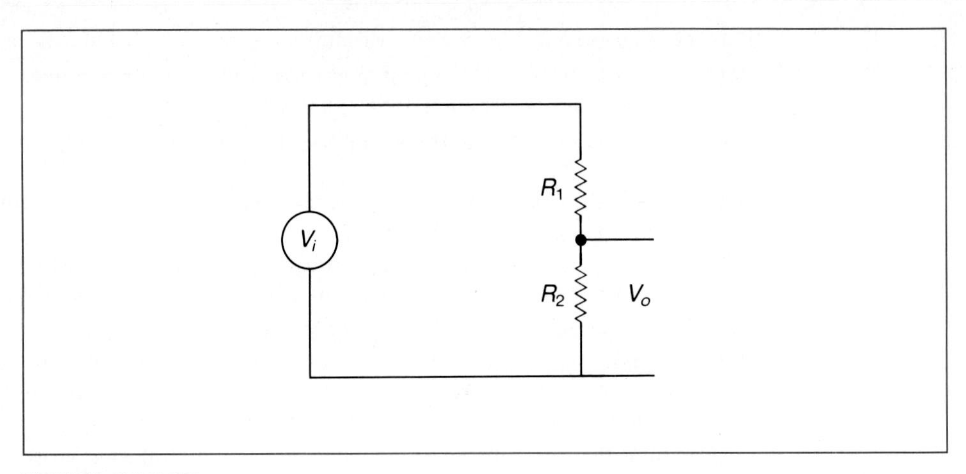

FIGURE P.19.37

Compute the range of values V_o, assuming R_1 and R_2 have the same nominal values and that the tolerances for R_1 and R_2 are 1, 5, 10, and 20 percent.

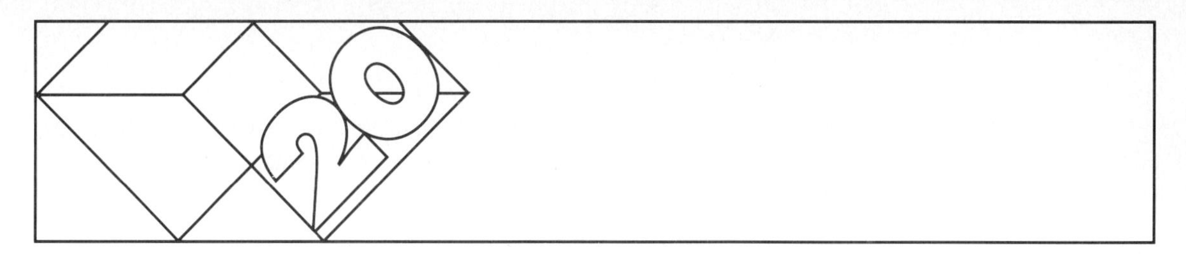

The Design Process

Our goal with this book has been to provide you with an integrated introduction to the concepts upon which good computer design decisions are based. In keeping with this goal, we began with the mathematics that form the theoretical bases for computing. We then proceeded to the hardware and software components that make up computer systems. Lastly, we described tools used in hardware and software design, and important considerations involved in making design decisions.

Having presented these many related (but diverse) concepts, we now want to show you how it is all "brought together." Therefore, we will end our presentation with an introduction to the design process. We will briefly describe the way computing systems projects are initiated and implemented in a commercial environment.

We will spend much of this chapter working with the life-cycle model, both the classic version and a revised version. Before entering that discussion, though, we want to stress three important activities that should be considered throughout the system life cycle: documentation, test planning, and concurrent engineering.

20.1 DOCUMENTATION

There are two general types of documentation associated with a system development project. Technical documentation defines the hardware, software, maintenance, support, and use of the system. Management documentation describes the project purpose, schedule, personnel requirements, budget, etc. Both types are important in various phases of the life cycle.

In earlier chapters of this book, you have been introduced to several different forms of technical documentation. You have probably also seen technical documentation in your programming courses and in the manuals for one or more computer systems. Think about the times when you needed some information from a manual. Was it easy to find, and if you found it, was it informative and easy to understand?

Consider the very real possibility that at some point in your career you might have to fix or change a complex system. What if this system has inadequate, inaccurate, or completely outdated documentation? You might have to spend inordinate amounts of time just to figure out how the current system works. You might even have to scrap the existing system and start over.

Software seems particularly prone to poor documentation. This might be because many programmers think their code is totally logical and needs no explanation. Furthermore, the programmer is often the designer, the developer, and the tester. You frequently hear the excuse that there is no time or that the documentation requirements are too rigid or irrelevant.

Many documentation styles have been proposed for computer systems. Glass and Noiseux [1981][1] suggest that good documentation must contain both historic and current components. The historic documentation should consist of:

- *Design notes.* Beyond just the formal design representation, the original design notes are themselves very useful. They help readers understand facets such as why things were done the way they were. They should be stored chronologically and cross-referenced by subject.
- *Problem reports.* As with the design notes, problem reports should be stored chronologically and cross-referenced. They should be kept in a working file until the problem is resolved, and then placed into a historical file.
- *Improvement suggestions.* These suggestions should be gathered into a central folder as they arise, then periodically reviewed and considered for incorporation.
- *Version descriptions.* Each version should have associated documentation containing complete descriptions of the changes that were incorporated into the product. There should be references to resolved problem reports and implemented change requests. There should also be a discussion of the changes' effects on the system and its users. (You might hear the term *engineering change control* used in this regard.)

These suggestions apply to both hardware and software, and we strongly endorse their use.

20.1.1 Hardware Documentation

Hardware documentation has historically been more complete than software documentation. There are probably two main reasons.

- Hardware documentation is necessary because hardware is typically designed by engineers, prototyped by technicians, and then manufactured and tested by factory personnel.

[1] This book is thorough, informative, and well-written. It concentrates on software, but many ideas are equally applicable to hardware. We strongly recommend it.

TABLE 20.1 ITEMS TYPICALLY CONTAINED IN CURRENT HARDWARE DOCUMENTATION

Level	Documentation
System level	System and component block diagrams
	Physical layout diagrams showing location of modules
	Cabling diagrams
	Power and grounding requirements
	Allowable temperature ranges
	Cooling loads
Logic level	Schematics showing component interconnections
	Functional specifications of the various chips (logic levels, pin assignments, timing requirements, etc.)
	Board layouts
	Bus configurations and protocols
	Communications interfaces and protocols
	Test procedures for the system and its individual modules
Circuit level	Wiring lists
	Circuit-board masks (for making the boards)
	Connector lists—types and pin assignments
	Pin assignments
	Voltage levels and current requirements
	Critical signal and waveform patterns
	Schematics for discrete-element circuits with circuit elements (resistor, capacitor, inductor, diode, transistor, etc.) and the descriptions of these elements (type, size, power rating, and tolerances)
	Material lists

- Hardware documentation often includes fully defined descriptions of commercially available logic chips and microprocessors, and agreed-upon industry interface and protocol standards.

In the previous section, we described *historic* documentation applicable to both hardware and software. Table 20.1 summarizes items typically contained in each of the three levels of *current* hardware documentation.

Drawings, such as schematics and board layouts, are traditionally given separate drawing and revision numbers. These numbers are used to identify and keep track of the separate components and assemblies. Space is also included on the drawings for change notices, and some drawings, such as assembly drawings, may also include a bill of materials.

20.1.2 Software Documentation

Like current hardware documentation, current software documentation also begins at the system level. Typically, the top-level software system would be described by block diagrams and system/application user manuals. However, the primary software documentation is at the program level.

Current software documentation at the program level embraces three major concepts. The first is stylistic and is used throughout the program: code should be self-documenting. Such self-documenting code is usually characterized by the following list:

- The design documentation is inserted as commentary wherever possible.
- Names emphasize function but also include structure for complex systems. This can be done by including module and data block names as part of each name used in the module.
- Code is structured and uses indentation to emphasize the structure.

The other two facets of current software documentation are more traditional: top-level and detail-level software definition.

Some items that Glass and Noiseux [1981] suggest for top-level software definition are:

- *An overall structure summary.* This is a textual and graphical description of the system. Functional block diagrams with execution order and data flow can be most useful.
- *An overall database summary.* This is a textual and graphical description of databases, record structures, indexing, storage method, etc.
- *An underlying design philosophy.* A description including what the system does, what design approach was used, and why it was done the way it was.
- *Pointers to detail-level information.* These are "indexes" to help the reader find specific information that he or she is seeking.

They suggest the following for the detail-level documentation:

- *At the start of each program part*, explain what the code does and how it does it. For complex sections, explain why the particular approach was used.
- *At the beginning of each major module*, specify module name, version, date, developer, inputs and outputs, limitations, restrictions, assumptions, error messages and actions, and general operation. Include a maintenance history, listing for each change the date, maintainer's name, purpose of change, and scope.
- *At the start of each subfunction*, give an explanation of that subfunction.
- *For each interface*, define that interface and give a reference to the other side of the interface.
- *For each group of related declarations*, describe the role and constitution of the group.
- *For each declaration*, explain the role of each item and the meaning of any special values.

These lists are comprehensive, and maybe not all aspects are appropriate for all projects. We included them as suggested guidelines, not as rigid rules. Different companies, applications, and people will have their own specific approaches. *The important feature is that documentation is there, that it includes relevant information, and that it is kept up to date.*

In Chaps. 15 and 17, we presented several versions of an algorithm to compute natural logarithms. In Fig. 20.1, we have given you an undocumented Fortran program to implement that algorithm. Please scan the undocumented code in this figure, then look back at the documented versions in those earlier chapters. We trust this will impress upon you the value of good documentation.

FIGURE 20.1 An example of an undocumented program to compute natural logarithms.

```
DO
TYPE 'NUMBER = '
ACCEPT C
TYPE 'MAX ITERATIONS = '
ACCEPT I
TYPE 'EPSILON = '
ACCEPT D
IF (C .LE. 0.0) THEN
TYPE 'NUMBER MUST BE POSITIVE!'
ELSEIF (I .LE. 0) THEN
TYPE 'MAXIMUM ITERATIONS MUST BE POSITIVE!'
ELSEIF (D .LT. 0.0) THEN
TYPE 'MINIMUM CHANGE MUST BE NON-NEGATIVE!'
ELSE
EXIT
ENDIF
REPEAT
A = I
J = 1
E = 0.
B = 0.
IF (C .GT. 1.0) THEN
F = (C - 1.0)/C
G = 1.0
ELSE
F = 1.0 - C
G = -1.0
ENDIF
DO
G = G * F
B = B + 1.0
E = E + G/B
IF (B .GT. A) THEN
TYPE 'MAX ITERATIONS; '
J = 0
ENDIF
IF (ABS(G) .LT. D) THEN
TYPE 'EPSILON ACHIEVED; '
J = 0
ENDIF
IF (J .EQ. 0) EXIT
REPEAT
TYPE 'NATURAL LOG = ',E
END
```

20.2 TEST PLANNING

Unfortunately, too many designers postpone all thought of testing until after the system has been designed and built or programmed. The consequence of the postponement can be very costly.

All too frequently, we hear of computer projects being behind schedule and over-budget. We hear of computer systems that "do not do what they are supposed to." A common reason is that neither the customers nor the developers fully understood what was needed, and neither recognized this fact. Developing a test concept at the beginning of the project helps to define the requirements and specify how you will be sure these requirements have been met.

Another reason for developing a test plan early is that we must have the means to measure system performance and verify correct system operation. This often calls for test facilities at key points in both hardware and software, and these test facilities must be included at design time. Not doing so can lead to time-consuming changes in the system and delays to the project schedule.

The specifics of how testing is done varies from project to project, and it is unnecessary to detail them here. Some will be illustrated in our discussion of the life-cycle model. The reason for introducing this topic here is to raise your awareness of the importance of considering testing throughout product development.

20.3 CONCURRENT ENGINEERING

One striking characteristic that you will notice in most models of computer system development is an implicit *sequentiality*. The classic life-cycle diagram shown in Fig. 20.2 implies that one group does something, then the next group takes over, then the next, until the product is in the customer's hands.

For example, it might seem that an analyst figures out the requirements and hands the requirements to a designer. The designer designs the system and hands the designs to a developer. The developer implements the design and then passes the system to a tester. The tester checks out the system, and it is cleared for release. Marketing advertises the system, sales sells the system, and maintenance maintains it.

This approach is known by various names and has been used on many systems. A descriptive term we have seen used for this process is *over the wall*. That is, each group completes its part of the project and tosses the results "over the wall" to the next group. There is little or no interaction between the various groups.

Perhaps you have seen or heard of noncomputer examples of this approach and its shortcomings. For many years the name Edsel (an automobile developed by the Ford Motor Company) was synonymous with a good product that was inappropriate for existing market trends. Other automobiles became famous (infamous) because simple maintenance operations, such as changing a spark plug or replacing a switch, required an inordinately large amount of time. On a recent four-wheel-

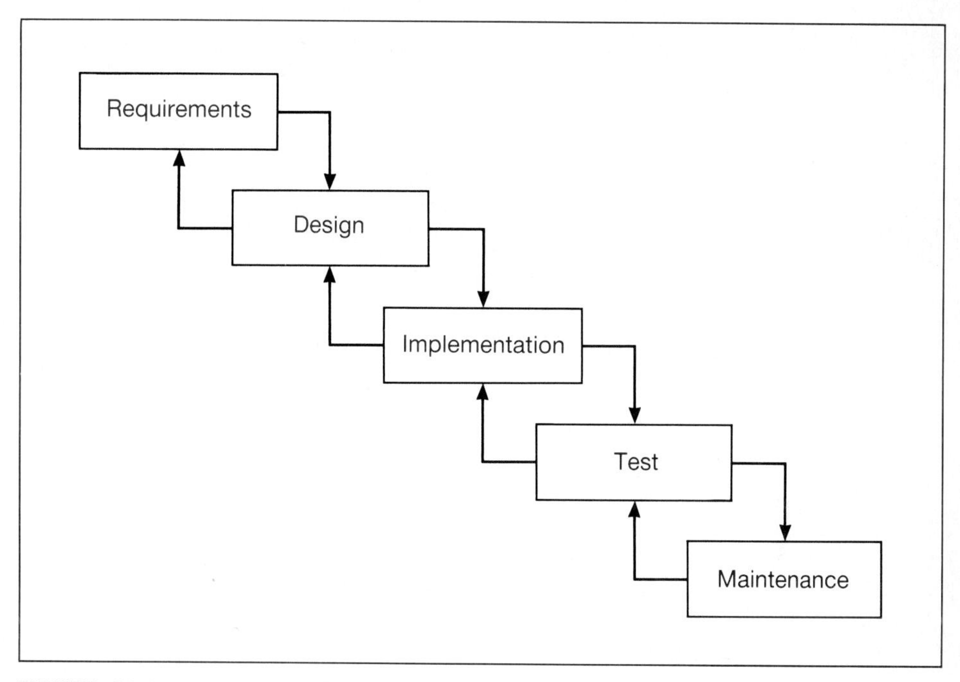

FIGURE 20.2 The classic version of the life-cycle model for system development.

drive vehicle, the rear hatch pops up so high when opened that many people cannot reach the handle to close it. These and a host of other examples show what happens when the various groups involved in system development do not interact throughout the design process.

The authors have been involved in numerous computer projects where this sequence was used, exactly as described. In the past this was often the accepted approach to system design. Today, however, it is perceived as antiquated, cumbersome, inefficient, and wasteful.

A more modern approach to system development is called concurrent engineering. **Concurrent engineering** emphasizes the philosophy that all departments involved with a project should participate in decisions throughout the entire project. To design a product that can be manufactured, tested, marketed, and maintained requires input from many different groups within the company. Not only are their inputs needed, they are needed early in the project life cycle. All the groups must work on their own parts of the project concurrently.

For instance, maintenance's narrow task is to maintain a product after it has been manufactured and delivered. However, maintenance (and all other groups involved in a project) should participate in all phases of the design process. In the examples cited above, if someone from marketing or maintenance had reviewed the designs, they might have pointed out the problems before the designs were released to manufacturing.

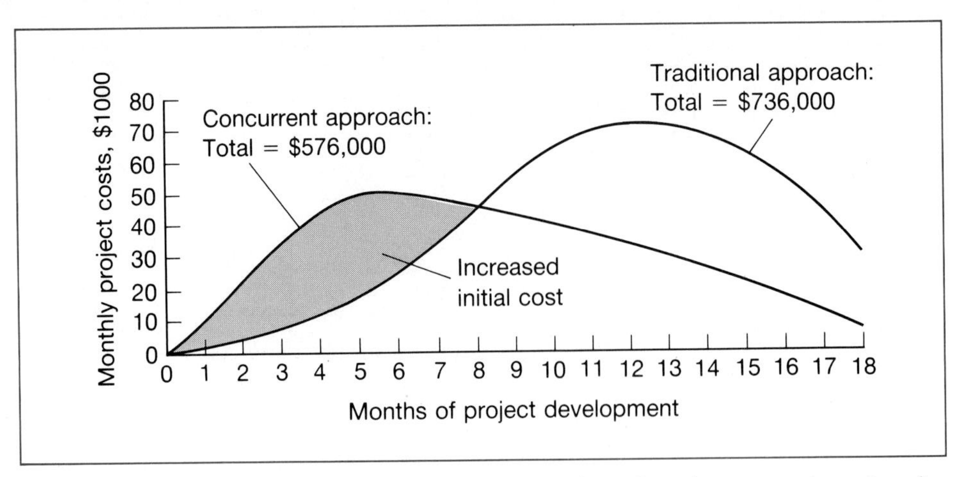

FIGURE 20.3 Distribution of project costs for traditional and concurrent engineering approaches to the design process.

Therefore, keep in mind as we discuss models of the product life cycle that there is considerable concurrent activity in a good systems development environment. Marketing can be directly involved in requirements definition via prototyping (compare Chaps. 17 and 18). Hardware and software documentation can be started as the system is being designed and implemented. Test mechanisms can be specifically designed into the system, rather than "tacked on" later. Manufacturing can make any modifications to their processes necessary to produce the product efficiently. Maintenance can point out potential service problems while it is still easy and inexpensive to alter the design. There are many benefits to concurrent engineering; only a few have been listed here. The key is to remember concurrency rather than sequentiality in the design process.

There is an apparent drawback to the concurrent engineering approach. It requires higher initial expenditures, in both time and money, than the traditional approach. This is because more people are involved in early phases of the project.

Figure 20.3 shows this higher initial cost. However, the total project costs *are reduced.* In the figure, the total cost of the traditional approach is $736,000, and the total cost of the concurrent engineering approach is $576,000.

This closes our discussion of documentation, test planning, and concurrent engineering. We turn now to our models of system development.

20.4 THE LIFE-CYCLE MODEL

The classic paradigm for computer system development is the life-cycle model introduced in Chap. 17 and repeated earlier in this chapter as Fig. 20.2. It contains five steps:

- Determine the system requirements.

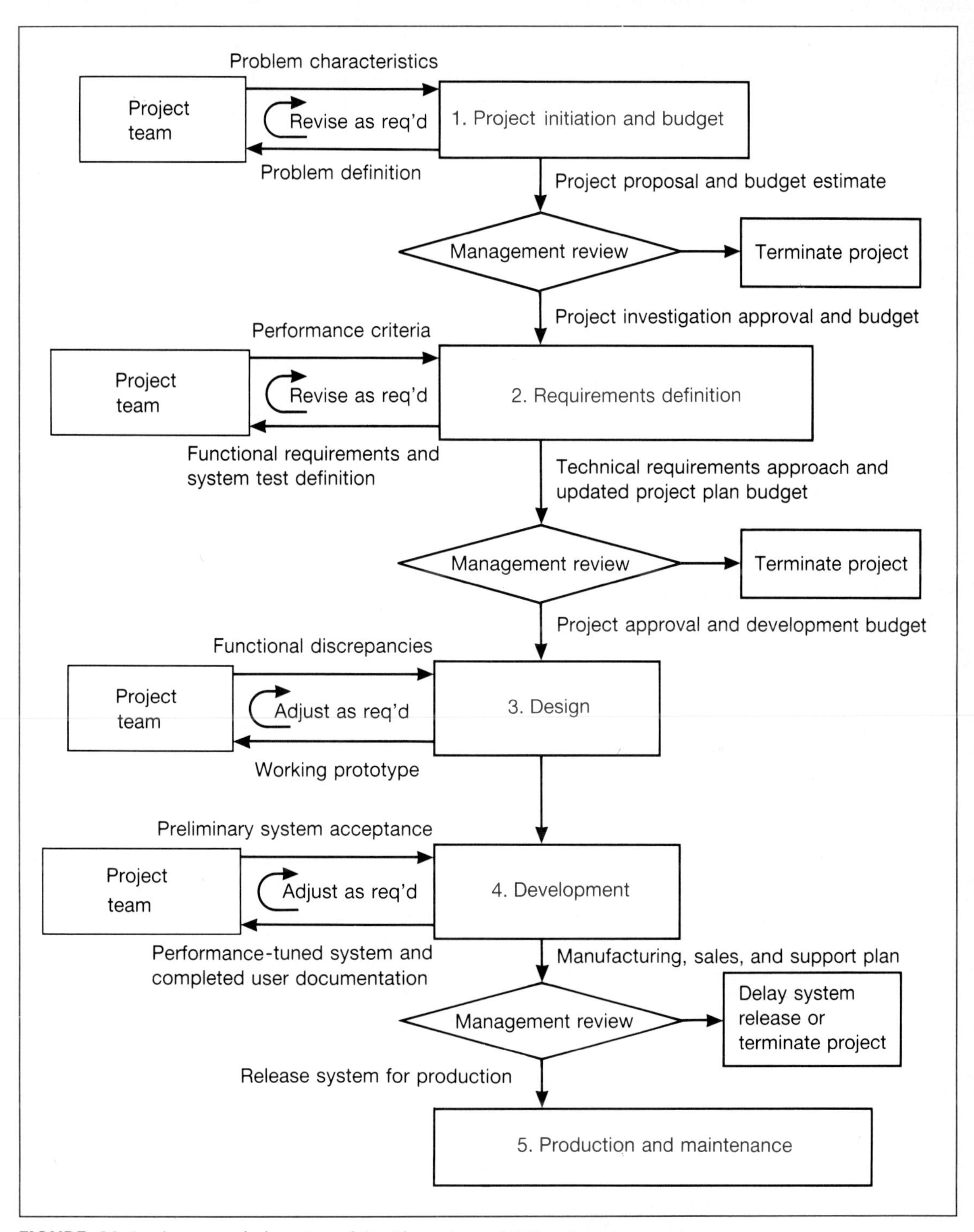

FIGURE 20.4 An expanded version of the life-cycle model showing phase activities.

- Design the system.
- Implement the design.
- Test the implementation.
- Maintain the system after it is installed.

This looks and sounds straightforward, but there are four major shortcomings to this model:

- It does not describe how a project gets started.
- It does not detail what is to be done in each of the steps.
- It implies that testing need not be considered until after the system has been developed.
- It reinforces separation and isolation of different elements of the project team.

Therefore, we will use an alternate, expanded version of the life-cycle model that was designed to answer these concerns. It is shown in Fig. 20.4.

20.5 THE LIFE-CYCLE PROJECT ACTIVITIES

The life-cycle model shown in Fig. 20.2 can be used for projects that are hardware-only, software-only, or that include both hardware and software development. The model as shown applies to new product development, but it can also be used in the development of a product upgrade. With these considerations in mind, we are ready to elaborate the activities done in each of the phases.

20.5.1 Phase 1: Project Initiation

The project initiation phase is sometimes considered to be a prephase to an actual project. A problem has been identified, but no development commitment has been made. No one yet knows how much it will cost or how long it will take. They don't know what effect it will have on other projects or on the company's position in the marketplace. In short, there is not yet enough information to make an informed decision whether to go ahead with the development. The project initiation phase provides a structure for the development of the information necessary to support that decision.

The activities for this phase are shown in Fig. 20.5. They begin with the identification of a problem or need and end with a management decision whether to go ahead with the project. Typically, management wants to know:

- Who will buy it.
- How much it will cost to design, produce, and maintain.
- How long it will take to design.
- What the return on investment will be.

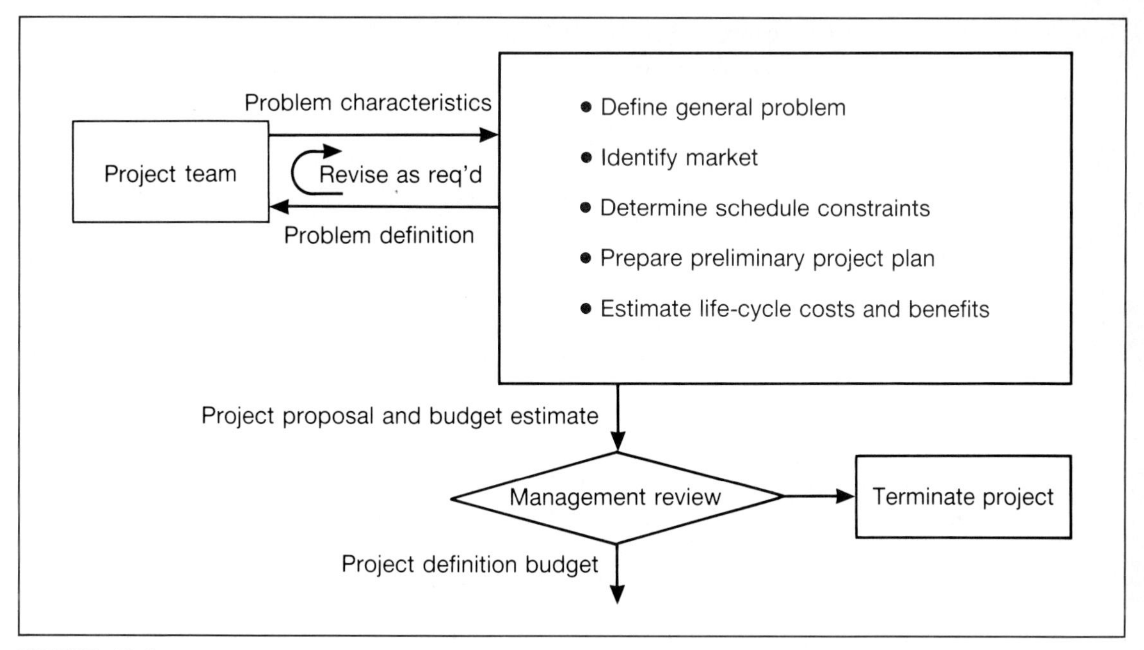

FIGURE 20.5 Project initiation phase activities.

The questions about time and cost imply the need for a **project schedule** and a **project budget**. Pfleeger [1987][1] says that a project schedule:

- Describes the development cycle for a project
- Enumerates the phases of the project
- Breaks each phase into discrete tasks
- Shows the interactions among these tasks
- Estimates the time that each task will take

Once we have a schedule, we can continue forward to determine personnel needs and then estimate costs.

Now you might be concerned about an apparent paradox. We have not yet entered the requirements and design stages of the life-cycle model. How in the world can we try to establish a schedule or a budget? Pfleeger has an excellent discussion of this, based on the notions of deliverables, milestones, schedule phases, work breakdown structures, and activity graphs.

She defines **deliverables** as anything that the client expects to see produced or demonstrated as part of the system development. They could include specific soft-

[1] Like Glass and Noiseux [1981], this book is thorough and informative. It is exceptionally well-written, interesting, and fun to read. Also, like Glass and Noiseux, this book concentrates on software, but many ideas are equally applicable to hardware. We strongly recommend it.

ware and hardware subsystems, demonstrations of functionality and reliability, and documentation. All deliverables should be carefully listed and defined.

Once we have defined the deliverables, we can analyze development of the system and specify certain critical points, or **milestones**. These are clearly defined points that show a measured progress. Often, they are a completion of one of the deliverables on our list. If we were building a workstation, then one of the *deliverables* could be the keyboard design. Completion of that design would be tagged as a *milestone*.

By using such an approach, we can separate the total system development into a succession of stages. Furthermore, we can separate each stage into a succession of steps and each step into a succession of tasks. Completion of specific steps or tasks can then be milestones. The result of this process is known as a **work breakdown structure**.

We can then analyze the work breakdown structure to determine interdependencies of tasks and to determine which tasks can be done in parallel. A good way to display interdependencies and concurrency is with a graph, suitably called an **activity graph**. The nodes in the graph are the milestones of the project. The lines or edges between the nodes are the activities that must take place to go from one node to the other.

Next, we attach a time estimate to each edge in the graph. Since nodes are milestones, an estimate on the edge between node A and node B would be the time needed to reach milestone B *after* reaching milestone A. That is, the edge represents the time to complete the task corresponding to milestone B and to produce the corresponding deliverable. This activity graph with attached time estimates (or a corresponding bar chart) gives us our schedule for both individual stages and the overall project. Once we have the schedule, we can continue forward to determine personnel needs and then estimate costs.

As a simple example of these concepts, consider one author's approach to lawn maintenance. Table 20.2 gives the work breakdown structure and Table 20.3 the milestones for this project. Figure 20.6 shows the activity graph with typical time estimates. Figure 20.7 is a bar chart taken directly from the activity graph in Fig. 20.6.

Note that, as stated earlier, these graphs and charts illustrate concurrency allowable within any imposed constraints. The bar chart in Fig. 20.7 shows activities that can be done concurrently. As the chart shows, Trim back can be done at the same time as Mow front. In this example, taking advantage of concurrency could save approximately $2\frac{1}{2}$ h in project completion time. However, it would necessitate hiring an assistant. Tradeoffs of completion time, personnel, equipment, facilities, and other aspects are a crucial part of planning and budgeting.

Having completed the graphs, charts, and other tasks in phase 1, a preliminary budget is developed. Then a management review is held to discuss the technical considerations of the project and the potential costs and benefits to the company. Assuming we get the go-ahead, the next phase is requirements definition and specification.

TABLE 20.2 WORK BREAKDOWN STRUCTURE

Stage 1: Preparation
- Step 1.1: Prepare mower
  - Task 1.1.1: Add gas if needed
  - Task 1.1.2: Add oil if needed
  - Task 1.1.3: Inspect blade; sharpen or replace if needed
- Step 1.2: Prepare string trimmer
  - Task 1.2.1: Add gas/oil mixture if needed
  - Task 1.2.2: Inspect shaft; lube if needed
  - Task 1.2.3: Add string if needed

Stage 2: Mowing and trimming
- Step 2.1: Trimming
  - Task 2.1.1: Trim front
  - Task 2.1.2: Trim back
- Step 2.2: Mowing
  - Task 2.2.1: Mow front
  - Task 2.2.2: Mow back
- Step 2.3: Clean equipment
  - Task 2.3.1: Clean trimmer
  - Task 2.3.2: Clean mower

Stage 3: Termination duties
- Step 3.1: Raking
  - Task 3.1.1: Rake front
  - Task 3.1.2: Rake back
- Step 3.2: Put tools away
- Step 3.3: Haul clippings to compost pile

TABLE 20.3 MILESTONES

1.1.1: Gas added to mower if needed
1.1.2: Oil added to mower if needed
1.1.3: Mower blade inspected; repaired if needed
1.2.1: Gas/oil mixture added to trimmer if needed
1.2.2: Shaft of trimmer inspected; lubed if needed
1.2.3: String added to trimmer if needed

2.1.1: Front trimmed
2.1.2: Back trimmed
2.2.1: Front mowed
2.2.2: Back mowed
2.3: Equipment cleaned

3.1: Raking completed
3.2: Tools put away
3.3: Clippings hauled to compost pile

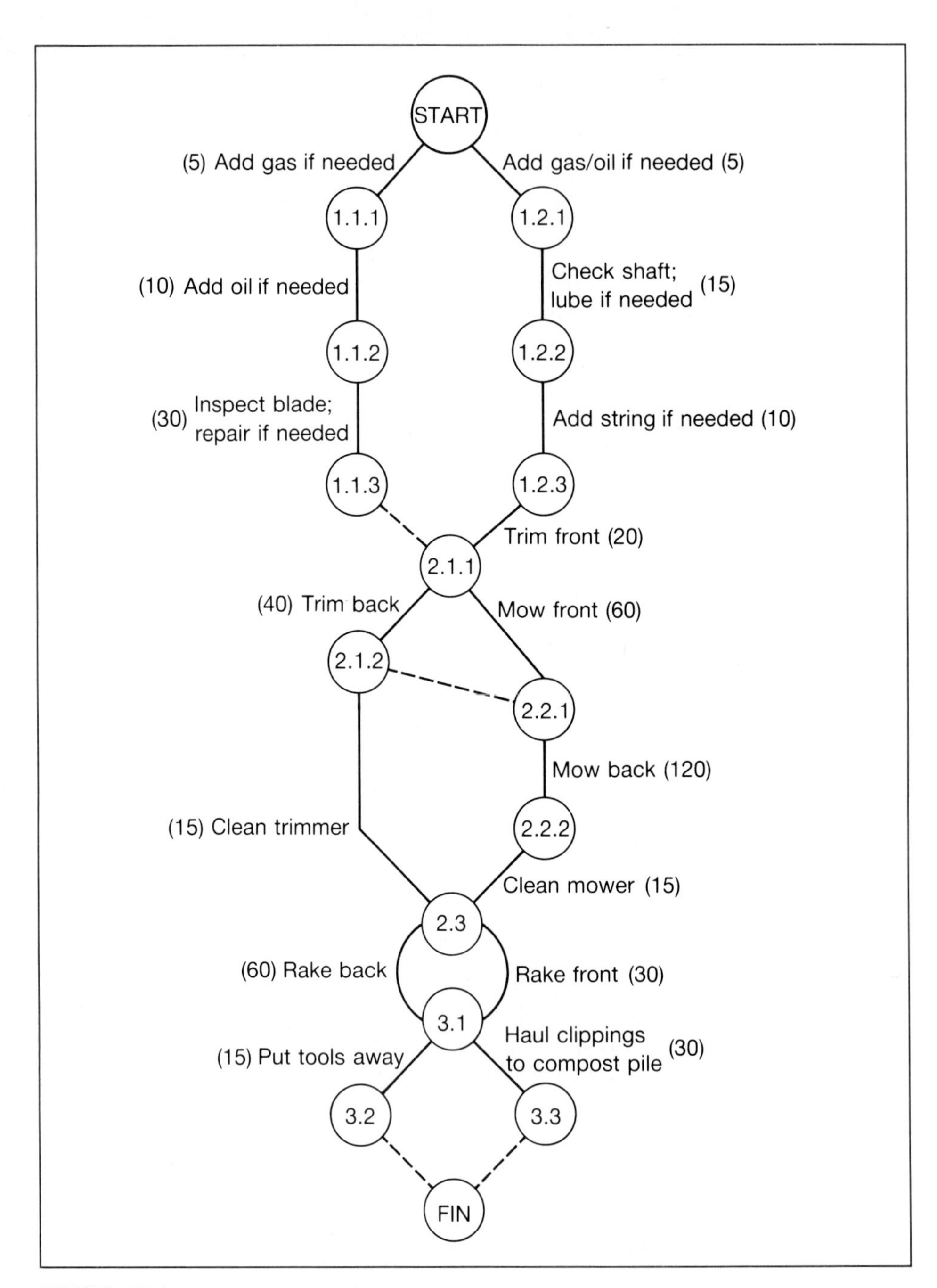

FIGURE 20.6 Activity graph with times.

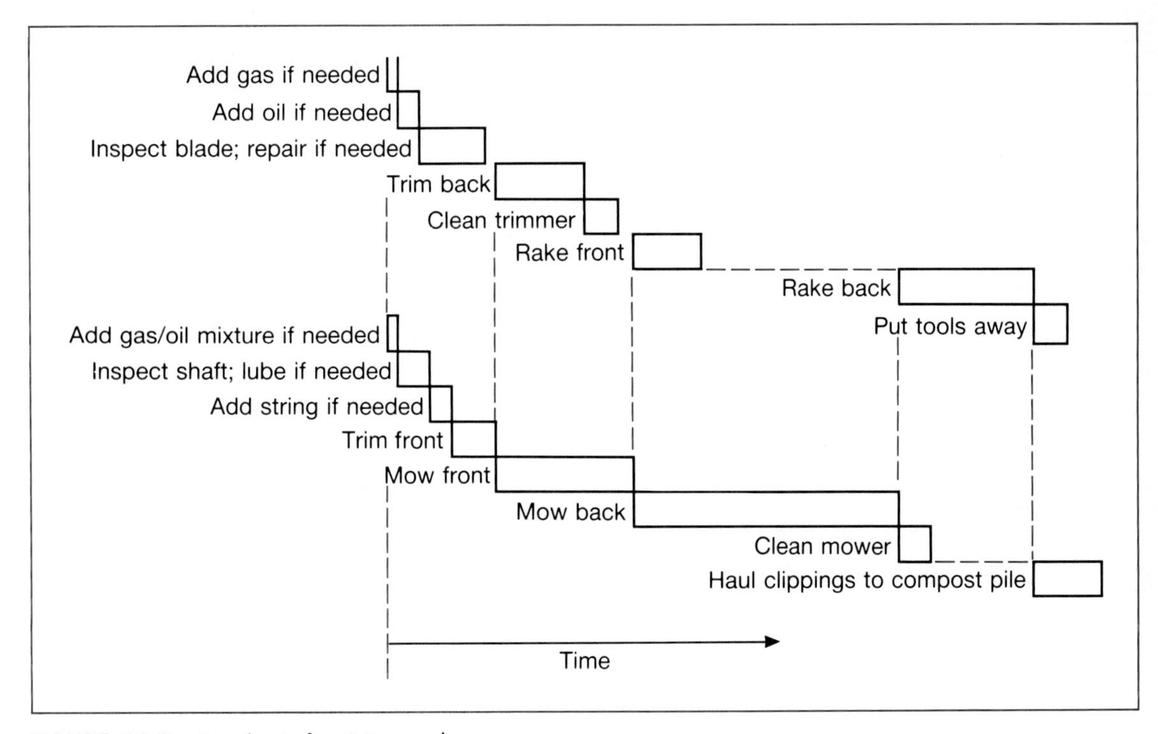

FIGURE 20.7 Bar chart of activity graph.

20.5.2 Phase 2: Requirements Definition and Specification

This phase (shown in Fig. 20.8) is similar to the first step in the classic life-cycle model. We have to determine precisely *what the system is to do*, but not how it is to do it. That can be a major stumbling block. Many projects have failed because not enough attention was paid to determining what is needed. If the designers do not get the requirements firmly established, then they leave themselves open to criticism later in the project. Their design might be rejected because the customer says, "That is not what I wanted."

Requirements definition is usually an iterative process where multiple meetings are held with the project team and possibly the customer to determine what is needed. At the same time that the requirements are being determined, we need to reach an important agreement. How will we know the product produced was the one that was intended? The system test plan will answer this question. If we are designing a large system, such a plan might not be easy to define. However, it is usually faster to do it while each of the requirements is in mind than wait until later. Also, development of this plan often identifies requirements that were overlooked.

In Chaps. 17 and 18, we suggested that prototypes can be used to aid in the requirements definition process. Prototypes give a feel of how the system will work

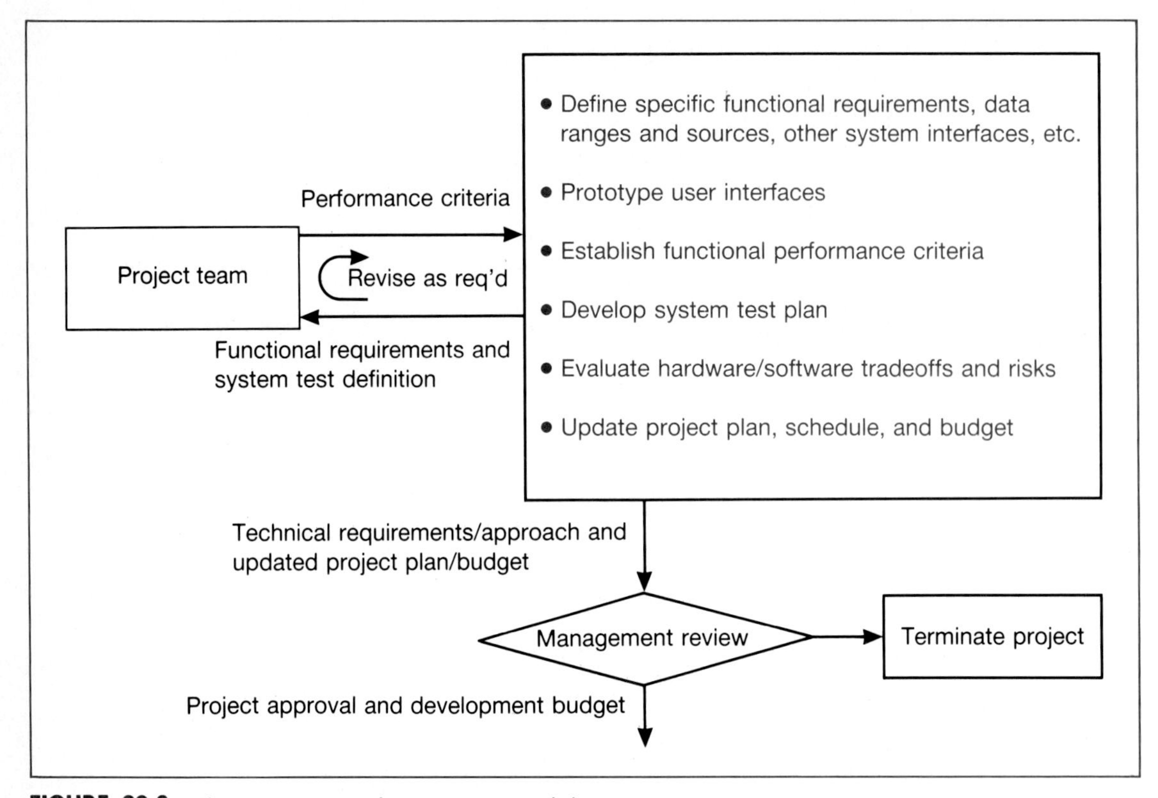

FIGURE 20.8 The requirements phase activities and decision points.

and help identify deficiencies in the design. This whole process might seem overly long and drawn out, particularly for small projects, but our experience has shown that ultimately it saves time and costs.

Each task should be fully defined. The technical decisions and test criteria that result from this phase become the basis for the design. The updated work breakdown structure, schedule, and budget define the fiscal and staffing requirements to get it done.

At the end of this phase there is another management review. Again, the project might be terminated or moved into phase 3.

20.5.3 Phase 3: Design and Prototyping

The requirements phase defined what each of the modules is to do. The design phase is where we decide *how it is to be done*. Figure 20.9 shows the activities done in the design phase.

The feasibility study at the end of the requirements phase evaluated the choices and defined the approach that will be taken in designing a product. We next need to translate this approach into an actual design.

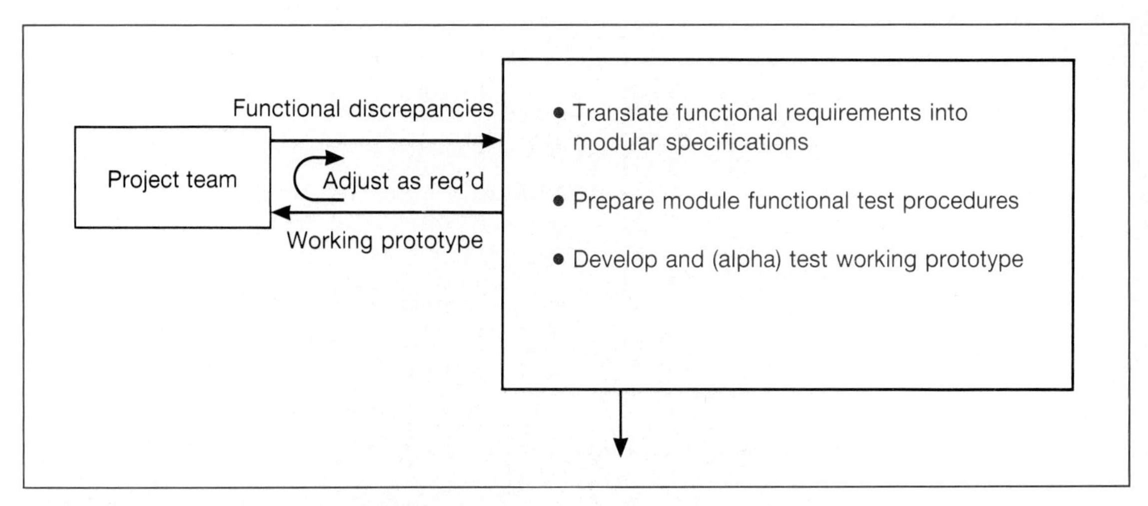

FIGURE 20.9 The design phase activities.

The first hardware task is to identify the major modules that make up the system. We need to translate the functional requirements into module specifications. Once that is done, the actual design of the individual hardware modules can proceed in parallel.

The situation is similar for software design. The first task is to define the main module and the module interfaces (calling procedures, data structures, etc.). Then the functional requirements can be specified, and the design can proceed on each of the software modules.

The actual design processes will typically make use of the hardware and software design tools that were discussed in Chaps. 11 and 17. The importance of meticulous design is difficult if not impossible to overemphasize. Novice programmers might assume that most of their errors will occur in the development phase, when they are coding their design, but research has shown that nearly two-thirds of the programming errors occur before coding begins [Glass and Noiseux, 1981].

Furthermore, design errors can be difficult to find. The later in the life cycle they are found, the more costly they are. Software errors found after the product is released can cost 100 times more to fix than if they had been found in the design phase [Brooks, 1982]. A similar situation exists with the hardware.

After all the individual components and modules have been tested, the system can be integrated and testing can begin. The first level of system testing is often called an **alpha test**. This test gives the opportunity to exercise the new system before any production units are built. Preliminary user documentation is also developed as a parallel activity during this phase.

20.5.4 Phase 4: Development

This phase is where the hardware and software prototypes are turned into a manufacturable system. It is also where the documentation is finished and where sales

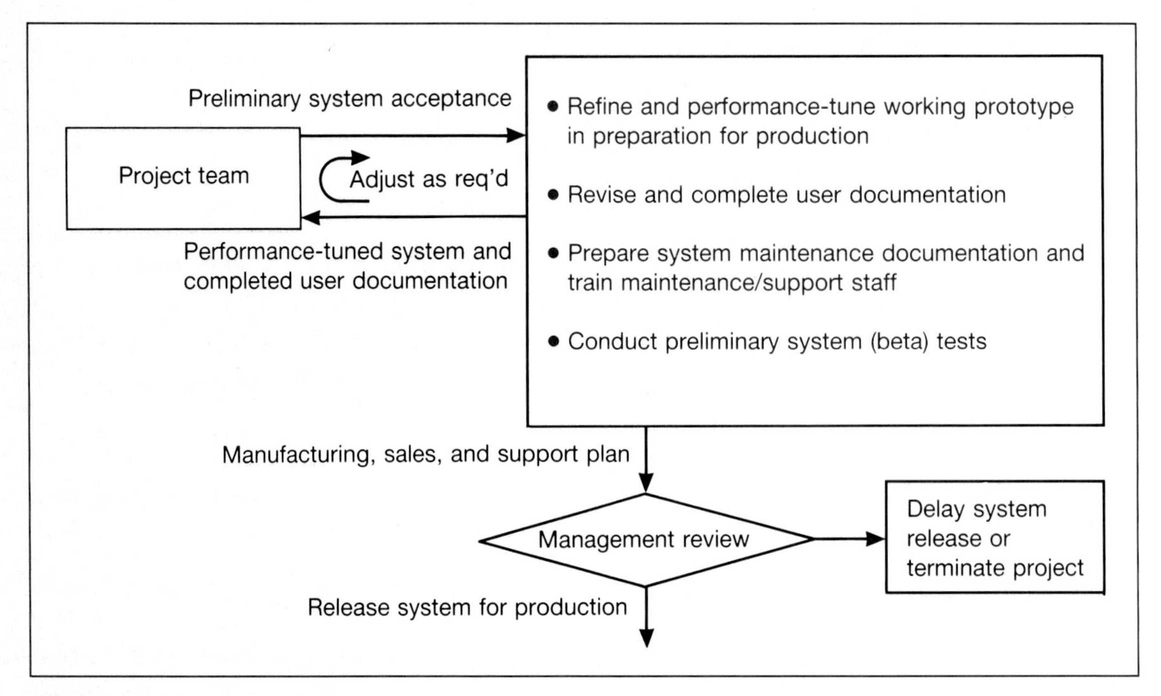

FIGURE 20.10 Development phase activities.

and maintenance personnel receive their final training on how to sell and support the system. As shown in Fig. 20.10, it is the final phase before releasing the system for production.

You might wonder why this phase is necessary. We already have a working system, so why can't we just build it and sell it just as it is? Unfortunately, that happens more often than we like. There is always a temptation to release the system immediately after it is working, but that is usually not a good idea.

You have probably seen news reports announcing recalls for various cars to correct some deficiency. Recalls usually don't happen in the computer industry. Instead, software corrections are incorporated into new "release levels," and hardware changes are incorporated during system maintenance or as hardware upgrades. If such errors happen too often, however, they can adversely affect the company's reputation and future sales.

The first units that have been built and tested may have special circuitry and programs that will be of no use to customer or maintenance personnel. Furthermore, the prototypes might have been patched and modified in the testing process. These conditions require that, before any production begins, we take the prototypes and do things like:

- Incorporate corrections.
- Remove unneeded hardware and software.

- Verify that performance objectives have been met.
- Verify that manufacturability and serviceability goals have been met.

After all these steps, preliminary versions of the manufactured product should be tested in user environments. These tests are often called **beta tests**. With these tests we can get customer reaction and feedback before release to manufacturing and authorization to go into full production.

The development phase ends with a management review of the product and the market. This review studies the recommendations of the beta tests to reassess the competition and to verify that:

- The product is ready for release.
- Manufacturing can produce the quantities necessary to support expected sales.
- Field support will be able to maintain the systems once these systems have been delivered.

This review also provides management a final opportunity to examine the costs and proposed pricing structure. It gives them a chance to time the product announcement and release for sale. The result of the review is a decision to proceed with production, delay production, or scrap the project completely. (A delayed release could give time to incorporate some changes that were identified earlier or suggested after beta tests.)

20.5.5 Phase 5: Production and Maintenance

The last phase of the life cycle is shown in Fig. 20.11. This is where the product is put into production and is installed at customer sites. The production and maintenance phase is usually also the longest phase of the life cycle. Systems typically have to be maintained long after production and sales have stopped. Manufacturers usually guarantee to support and maintain a product for a certain number of years after sale.

As shown in Fig. 20.11, systems must be tested after they are produced. This test period can vary from minutes to weeks. It depends on the complexity of the system,

FIGURE 20.11 Implementation/production phase activities.

- Production
- System testing
- Sales
- Customer training
- Maintenance and support

the history of previous system failures, and the testing philosophy of the organization.

If we are producing a large hardware system, it might have to be disassembled and then reassembled after delivery. Or, our system might have to be integrated into some existing system. In any case, the system should be subjected to a formal and rigorous system test regime. This regime should be specifically designed to test all the required features of the system before it leaves the factory. It should then be retested when it is installed at the customer site.

System testing typically begins at the bottom and proceeds in a sequence of steps:

1. Individual component or module testing
2. Integration testing as the modules are combined into subsystems
3. Total system testing

This is a simple concept that is applicable to a wide range of system sizes. Unfortunately, people often do not follow this concept when they are up against a deadline.

For example, we frequently see students in upper-division programming classes and in hardware labs skipping the lower-level testing procedures. They go directly to the system level because the assignment is due. Then they are bogged down trying to find errors that probably could have been found more easily at the component or subsystem levels.

After the system has been produced and installed, it must be maintained. This means taking a product that has already been delivered and is operational, and keeping it functional. Glass and Noiseux [1981] define maintenance as:

- Ensuring continued system reliability
- Fixing bugs found after delivery
- Changing the system to meet new requirements
- Converting to a new language, operating system, or chip
- Improving system maintainability
- Enhancing the system, such as increasing efficiency
- Ensuring that documentation is kept accurate and up to date

Most of these should be clear, except for maybe "improving system maintainability." This means that maintenance changes are sometimes made not to correct errors or alter system operation but to make the system easier to maintain. For example, you might rewrite a poorly structured module, generalize a special-purpose module, or change names in a non-self-documenting module. Also, it may be necessary to redesign and replace a hardware module that has a history of frequent failure.

The maintenance function can be considered as having three subcomponents:

- Preventive maintenance, to increase the system's useful life
- Adaptive maintenance, to adapt the system to a changing environment (such as a new operating system)
- Corrective maintenance, to fix errors

Many people used to think that once a product has been released for production, the engineering and design work is finished. This is definitely not true! On some computer-based systems, it has been estimated that corrective maintenance and product enhancements can require from 20 to 70 percent of the total project costs.

Glass and Noiseux [1981] have an interesting diagram of the maintenance process. Part of that diagram is adapted in Fig. 20.12. The figure shows that there is a dialog between users and the maintenance organization that helps define needed improvements and bug fixes. The circle labeled Change control represents the process for controlling and monitoring all changes to the system.

On large systems it is especially important to carefully monitor all changes that are proposed and implemented. Suppose different programmers are working on modifications to the same program for separate purposes. If there is no coordination, the two changes may create conflicts. In addition, Change control must also make sure that any modifications to the system are documented in the appropriate user and maintenance manuals.

FIGURE 20.12 An overview of maintenance.

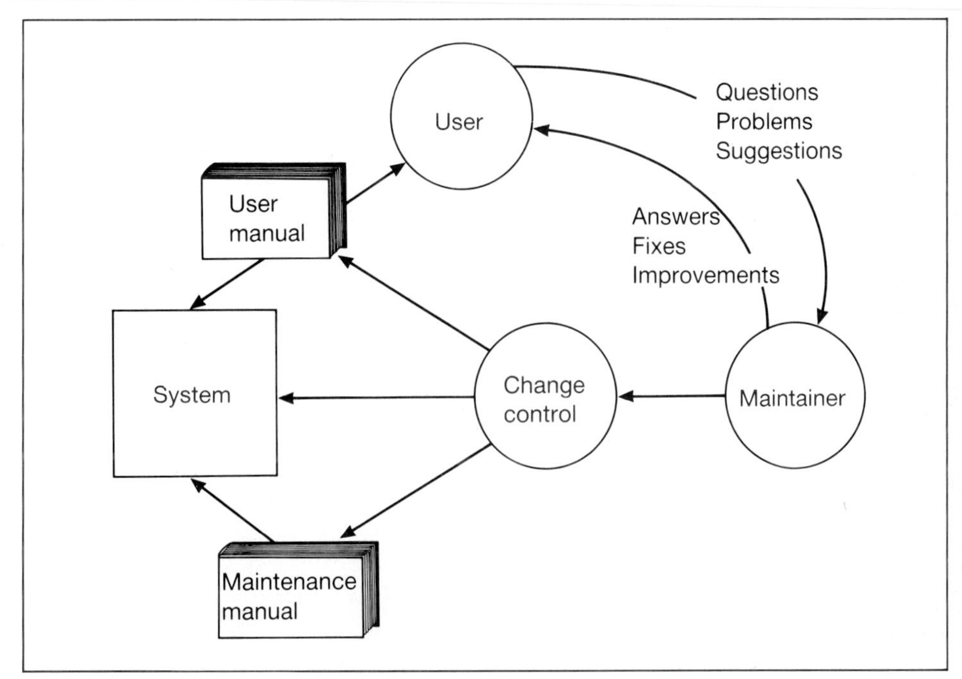

20.6 SUMMARY

This chapter has concentrated on a definition of the project life-cycle model and an explanation of how that model could be used. We are not trying to "convert" you to any specific techniques or paradigms; instead, we hope that you understand the importance of using some solid, well-structured approaches to the design process.

Throughout the chapter we have emphasized the benefits of concurrent engineering in the design process. Although this approach requires higher up-front costs, the benefits seem to far outweigh this disadvantage.

To compete in the international marketplace in the future, U.S. companies must improve their design processes. This will require adopting new methods and techniques to increase productivity and quality.

Bibliography

Aldus Corporation: *Aldus PageMaker Users Manual*, Seattle, 1990.

Anderson, T., and B. Randell (eds.): *Computing System Reliability*, Cambridge University Press, Cambridge, 1979.

Apple Computer: *Inside Macintosh Volumes I, II, and III*, Addison-Wesley, Reading, Mass., 1985.

———: *Macintosh IIci*, Cupertino, Calif., 1990.

Avizienis, A., and L. Chen: "On the Implementation of N-Version Programming for Software Fault-Tolerance during Program Execution," *Proceedings COMPSAC 77*, pp. 149–155, 1977.

Backus, J. W., R. J. Beeber, S. Best, R. Goldberg, L. M. Haibt, H. L. Herrick, R. A. Nelson, D. Sayre, P. B. Sheridan, H. Stern, L. Ziller, R. A. Hughes, and R. Nutt: "The FORTRAN Automatic Coding System," in S. Rosen (ed.), *Programming Systems and Languages*, McGraw-Hill, New York, 1967.

Bailey, T. E., and K. Lundgaard: *Program Design with Pseudocode*, Brooks/Cole, Monterey, Calif., 1983.

Bakcker, R. M., and W. A. S. Buxton: *Readings in Human-Computer Interaction*, Morgan Kaufmann, San Mateo, Calif., 1987.

Bell, C. G., and A. Newell: *Computer Structure: Readings and Examples*, McGraw-Hill, New York, 1971.

Bourne, S. R.: "The UNIX Shell," *Bell Sys. Tech. J.*, vol. 57, no. 6, pt. 2, pp. 1971–2020, July–August 1978.

Brinch Hansen, P.: "The Nucleus of a Multiprogramming System," *Comm. ACM*, vol. 13, no. 4, April 1970.

Brooks, F. P.: *The Mythical Man-Month*, Addison-Wesley, Reading, Mass., 1982.

Capilano Computing: *DesignWorks*, Vancouver, Canada, 1990.

DeMarco, T.: *Structured Analysis and System Specification*, Prentice-Hall, Englewood Cliffs, N.J., 1980.

Deutsch Research: *MacSPICE Professional*, Palo Alto, Calif., 1990.

Dijkstra, E. W.: "Cooperating Sequential Processes," EDW123, Mathematics Department, Technological University, Eindhoven, Netherlands, September 1965. Reprinted in F. Genuys (ed.), *Programming Languages*, Academic, London, 1968.

———: "The Structure of 'THE' Multiprogramming System," *Comm. ACM*, vol. 18, no. 8, pp. 341–346, May 1968.

Douglas Electronics: *Professional System, Electronic Design and Manufacturing System*, San Leandro, Calif., 1990.

Encyclopedia Britannica: 15th ed., Chicago, 1985.

Enslow, P. H., Jr. (ed.): *Multiprocessors and Parallel Processing*, Wiley, New York, 1974.

Excel Software: *Introducing MacAnalyst 2.0 and MacDesigner 3.0*, Marshalltown, Iowa, 1990.

Foley, J. D.: "Interfaces for Advanced Computing," *Scientific American*, October 1987.

Gibson, J. C.: "The Gibson Mix," Rep. TR00.2043, IBM Systems Development Div., Poughkeepsie, N.Y., 1970.

Glass, R. L.: *Software Reliability Guidebook*, Prentice-Hall, Englewood Cliffs, N.J., 1979.

——— and R. A. Noiseux: *Software Maintenance Guidebook*, Prentice-Hall, Englewood Cliffs, N.J., 1981.

Goldberg, A., and D. Robson: *Smalltalk-80: The Language and Its Implementation*, Addison-Wesley, Reading, Mass., 1986.

Goldstine, H. H.: *The Computer from Pascal to von Neumann*, Princeton University Press, Princeton, N.J., 1972.

Grun, Bernard: The Timetables of History (based on Werner Stein, *Kulturfahrplan*), Simon & Schuster, New York, 1975.

Hamming, R. W.: "Error Detecting and Error Correcting Codes," *Bell Sys. Tech. J.*, vol. 26, no. 2, pp. 147–160, April 1950.

Hayes, F., and N. Baran: "A Guide to GUIs," *Byte Magazine*, July 1989.

Hennessy, J. L.: "VLSI Processor Architecture," *IEEE Trans. Comput.*, vol. C-33, pp. 1221–1246, December 1984.

Ifrah, Georges: *From One to Zero—A Universal History of Numbers* (English translation), Viking Penguin, New York, 1985.

Imagine That: *Extend*, San Jose, Calif., 1990.

Johnson, J., T. Roberts, W. Verplank, D. Smith, C. Irby, M. Beard, and K. Mackey: "The Xerox Star: A Retrospective," *IEEE Computer*, September 1989.

Kemeny, J. G., and T. E. Kurtz: *Back to Basic: The History, Corruption, and Future of the Language*, Addison-Wesley, Reading, Mass., 1985.

Kidder, T.: *The Soul of a New Machine*, Avon, New York, 1985.

Lee, K.: "Large-Vocabulary Speaker-Independent Continuous Recognition: The SPHINX System," Carnegie-Mellon University, Pittsburgh, April 1988.

Lister, A. M.: *Fundamentals of Operating Systems*, 2d ed., Macmillan, London, 1979.

Mano, M. M.: *Computer Engineering Hardware Design*, Prentice-Hall, Englewood Cliffs, N.J., 1988.

Martin, J.: *Application Development without Programmers*, Prentice-Hall, Englewood Cliffs, N.J., 1982.

Microsoft Corporation: *Microsoft FORTRAN Compiler*, Redmond, Wash., 1985.

———: *Microsoft Excel*, Redmond, Wash., 1990.

Paragon Concepts: *NISUS*, Del Mar, Calif., 1990.

Patterson, D. A.: "Reduced Instruction Set Computers," *Comm. ACM*, vol. 28, no. 4, April 1985.

Peter Norton Computing: *Norton Utilities*, Santa Monica, Calif., 1990.

Pfleeger, S. L.: *Software Engineering, The Production of Quality Software*, Macmillan Publishing, New York, 1987.

Pinkert, J. R., and L. L. Wear: *Operating Systems: Concepts, Policies, and Mechanisms*, Prentice-Hall, Englewood Cliffs, N.J., 1989.

Pollack, A.: "What Is Artificial Reality? Wear a Computer and See," *The New York Times*, vol. CXXXVIII, no. 47836, Apr. 10, 1989.

Radin, G.: "The 801 Microcomputer," Proc. Symposium Architectural Support for Programming Languages and Operating Systems, Palo Alto, Calif., March 1983.

Ritchie, D. M., and K. Thompson: "The UNIX Time-Sharing System," *Bell Sys. Tech. J.*, vol. 57, no. 6, pt. 2, pp. 1905–1929, July–August 1978.

Sammet, J. E.: "Basic Elements of COBOL 61," in S. Rosen (ed.), *Programming Systems and Languages*, McGraw-Hill, New York, 1967.

Sklaroff, J. R.: "Redundancy Management Techniques for Space Shuttle Computers," *IBM Journal of Research and Development*, 1976.

Smith, D. C., C. Irby, R. Kimball, B. Verplank, and E. Harslem: "Designing the Star User Interface," *Byte Magazine*, April 1982.

Stroustrap, B.: *The C++ Programming Language*, Addison-Wesley, Reading, Mass., 1986.

Texas Instruments: *The TTL Data Book for Designer Engineers*, Dallas, 1981.

Thompson, K.: "UNIX Implementation," *Bell Sys. Tech. J.*, vol. 57, no. 6, pt. 2, pp. 1931–1946, July–August 1978.

TrueBASIC: *TrueBASIC Reference Manual*, Addison-Wesley, Reading, Mass., 1985.

Tuinenga, P. W.: *SPICE A Guide to Circuit Simulation and Analysis Using PSPICE*, Prentice-Hall, Englewood Cliffs, N.J., 1988.

U.S. Department of Commerce: *Statistical Abstract of the United States*, 1980.

———: *Statistical Abstract of the United States*, 1988.

Weiss, R.: "SPEC Benchmarks Arrive," *Electronic Engineering Times*, Oct. 9, 1989.

Weste, N. H. E.: *Principles of CMOS VLSI Design—A Systems Perspective*, Addison-Wesley, Reading, Mass., 1985.

Wilkes, M. V., and J. B. Stringer: "Microprogramming and the Design of the Control Circuits in an Electronic Digital Computer," *Proc. Cambridge Philosophical Society*, vol. 49, pp. 230–238, 1953.

Wirth, N. K.: *Algorithms + Data Structures = Programs*, Prentice-Hall, Englewood Cliffs, N.J., 1976.

Wolfrun, S.: *Mathematica: A System for Doing Mathematics by Computer*, Addison-Wesley, Reading, Mass., 1988.

Answers to Selected Problems

CHAPTER 1

1.1 There must be (*a*) an identified problem or need and (*b*) an available technology or a technology that is easy to develop.

1.3 It was easier to assign existing letter forms to language sounds than to create and agree upon entirely new forms.

1.5 The development of an alphabet.

1.7 To clarify grammar.

1.9 The shape or form of the letters is not really critical so long as the letter symbols are unique. Of course, handwriting or printing could be made more difficult by making shapes complex.

1.11 Punctuation separates and identifies the words, sentences, and paragraphs of the text and provides information about breathing pauses and word accents. Without it, our written language would be much more difficult to read and understand.

1.13 It contained all the critical factors—(1) separate symbols for each digit, including zero, (2) positional number system, and (3) representation of fractions—needed to support computation, and it allowed the representation of both large and fractional numbers.

1.15 The symbol zero is required in a place-value system to allow the representation of a null value in a digit position.

1.17 Their number systems would not support complex calculations.

1.19 (*a*) 1162; (*c*) 18.

1.21 Subtraction table for Roman numbers

	I	II	III	IV	V
I	*	I	II	III	IV
II	†	*	I	II	III
III	†	†	*	I	II
IV	†	†	†	*	I
V	†	†	†	†	*

* Not representable because the number system did not include the value zero.

† Not representable because the number system did not provide for negative numbers.

1.23 It seemed to be the logical thing to do since software and hardware design were perceived as being separate functions.

1.25 Not everyone needs to understand both design functions, and there is a larger pool of people who understand one but not both.

1.27 Each person needs a path to every other person, which would give $n(n-1)$ paths. However, since path direction is unimportant, half the paths can be removed. Therefore the number is

$$\frac{n(n-1)}{2}$$

1.29 Law and medicine.

1.31 A person who knows both hardware and software design, because a VCR is a small, specialized project and assigning two people to the task might be overkill.

CHAPTER 2

2.1 (family name) = {mother's name, father's name, child one's name, child two's name, ...}

2.3 The number of people in your immediate family.

2.5

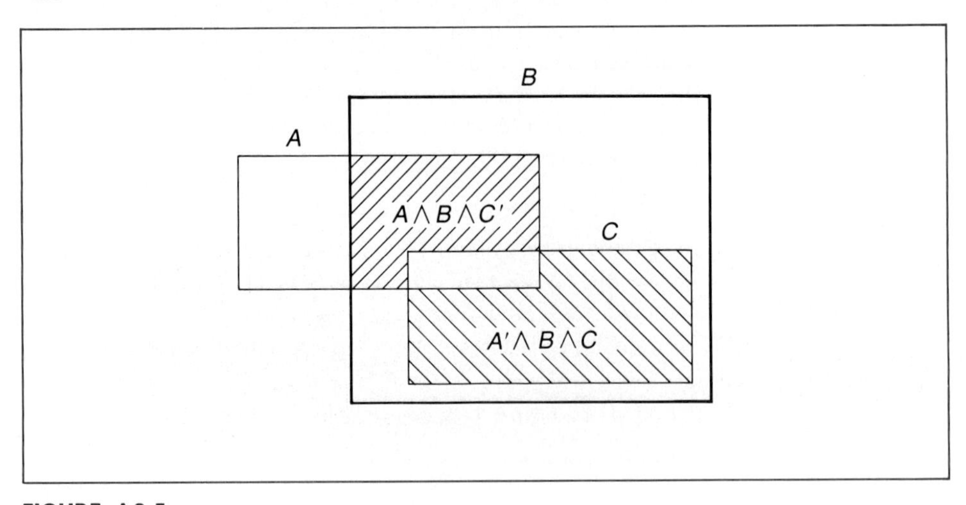

FIGURE A2.5

2.7 (*a*) The areas 1, 2, 3, 4, 5, 6, 7
(*c*) The areas 5, 6, 7
(*e*) The areas 2, 3, 4, 5, 6, 7
(*g*) The areas 0, 1, 2, 3, 5, 6, 7

2.9 (*a*) $B + AC$; the areas 2, 3, 5, 6, 7
(*c*) $A(B + C) + BC'$; the areas 2, 5, 6, 7

2.11 (*a*) $A'BC + A'BC' + ABC' + AB'C + ABC$
(*c*) $ABC + ABC' + A'BC' + AB'C$

2.12 (*a*) $\sum(1, 3, 4, 6, 7) = X'Y'Z + X'YZ + XY'Z' + XYZ' + XYZ$
(*c*) $\sum(0, 2, 4, 6) = X'Y'Z' + X'YZ' + XY'Z' + XYZ'$
(*e*) $\prod(0, 1, 3, 4) = (X + Y + Z) \cdot (X + Y' + Z') \cdot (X' + Y + Z) \cdot (X' + Y' + Z)$
2.13 (*a*) $\sum(1, 3, 4, 6, 7) = X'Z + YZ + XZ'$

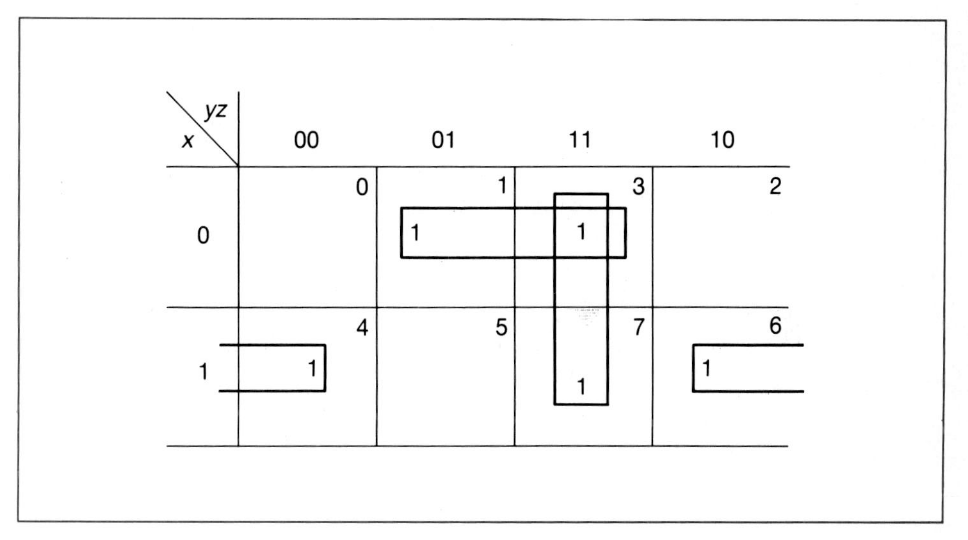

FIGURE A2.13(1)

(*c*) $\sum(0, 2, 4, 6) = Z'$

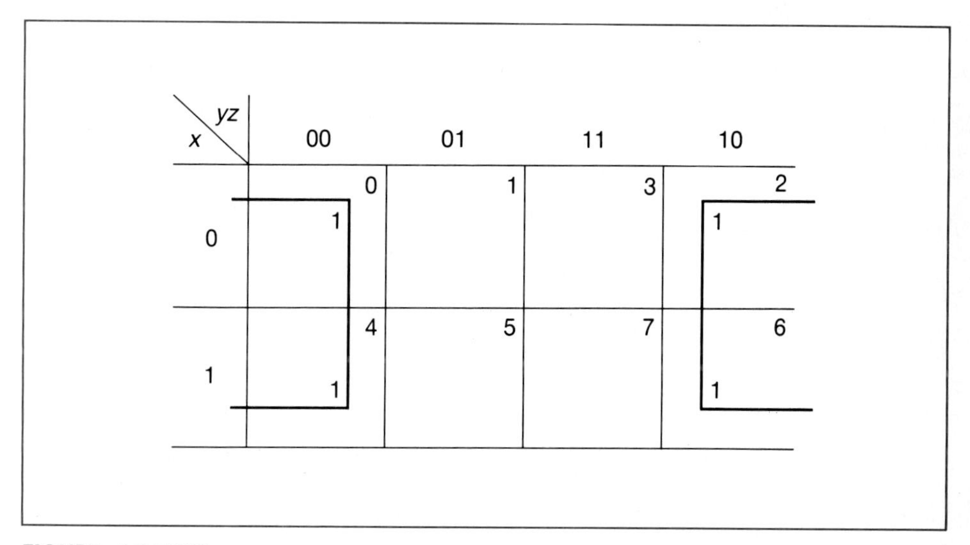

FIGURE A2.13(2)

(*e*) $\prod(0, 1, 3, 4) = XZ + YZ' = (Y + Z)(X + Z')$

x \ yz	00	01	11	10
0	0 (cell 0)	0 (cell 1)	0 (cell 3)	1 (cell 2)
1	0 (cell 4)	1 (cell 5)	1 (cell 7)	1 (cell 6)

FIGURE A2.13(3)

2.15 It isolates the output of the flip-flop from the input and synchronizes the change (of the flip-flop) to the clock.

CHAPTER 3

3.1

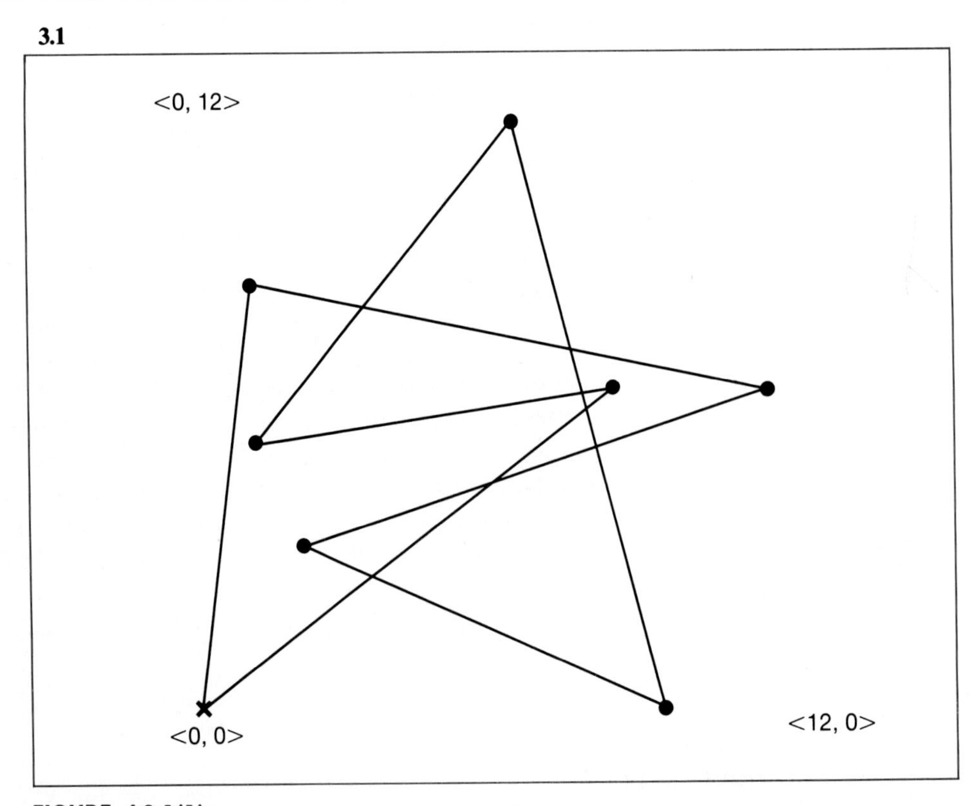

FIGURE A3.1(1)

Better ordering:

$$\langle\langle 2, 3\rangle, \langle 1, 5\rangle, \langle 1, 8\rangle, \langle 11, 6\rangle, \langle 8, 6\rangle, \langle 11, 6\rangle, \langle 0, 9\rangle\rangle$$

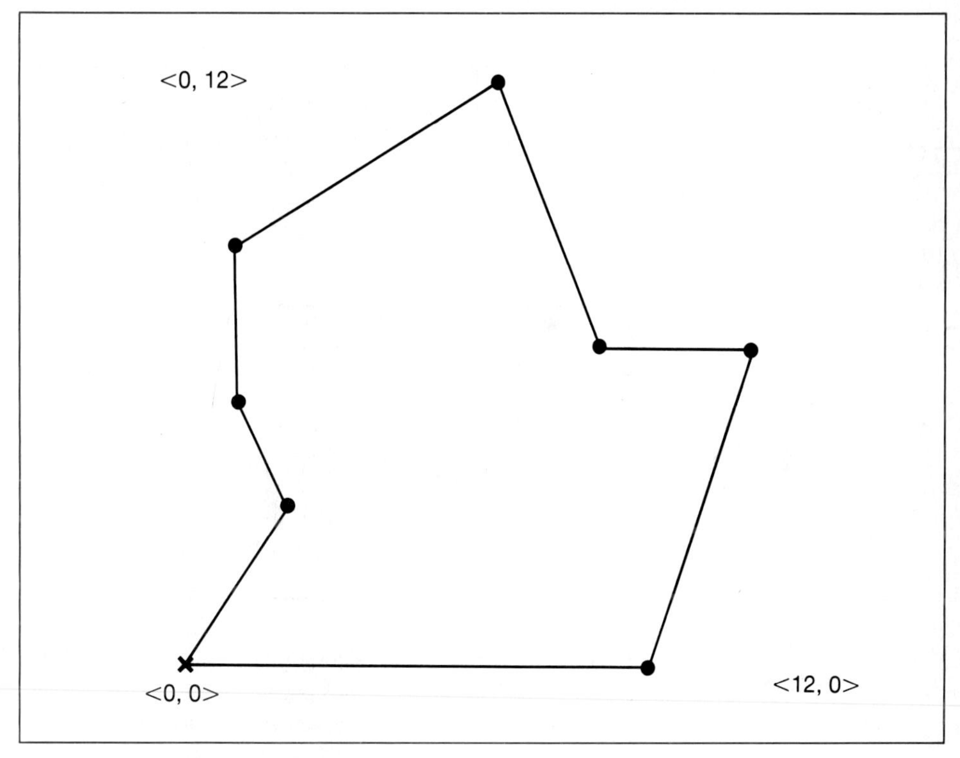

FIGURE A3.1(2)

3.3 An additional pause at the end of a letter signifies the end of a word.

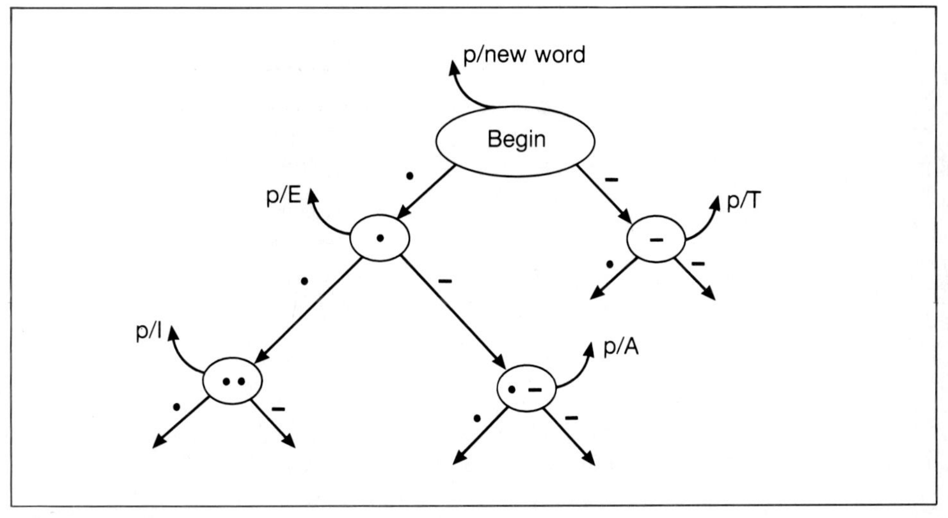

FIGURE A3.3

3.5 No. It requires too much time to send characters, since it requires multiple key presses for each character rather than one keystroke per character.

3.7 One of the codes was SHIFT, which was a qualifier for the next character. The SHIFT allowed the reuse of the same codes for lowercase as well as uppercase. This is similar to using the SHIFT key to specify the case of a letter on today's keyboards.

3.9 The byte has become the standard size for memory storage. Many devices are organized around the 8-bit byte. Peripheral devices transfer bytes, and many memory modules are 1 byte wide. Also, a word needs to be an integer number of bytes long to prevent the ASCII or EBCDIC characters from being split and stored in different words.

3.10 (*a*) $19A_{16} = 1\ 1001\ 1010_2 = 410_{10}$
(*c*) $398_{16} = 11\ 1001\ 1000_2 = 920_{10}$

3.11 (*a*) $28_{10} = 11100_2 = 1C_{16}$
(*c*) $255_{10} = 1111\ 1111_2 = FF_{16}$

3.12 (*a*) $-29_{10} = 1111\ 1111\ 1110\ 0011_2$
(*c*) $-64_{10} = 1111\ 1111\ 1100\ 0000_2$

3.13 (*a*) -4959; (*c*) -3590.

3.15 (*a*) 0.48299112E67; (*c*) 0.256E41.

3.17 All three have the same maximum error, which is 0.5.

3.19 On the average, it is more accurate than the other methods of rounding, since it doesn't always round up when the fraction is 0.5.

3.20 (*a*) $3612_{10} = 0011\ 0110\ 0001\ 0010_2$; (*c*) $5_{10} = 0000\ 0101_2$.

3.21 As an additional digit (either leftmost or rightmost) in the representation.

CHAPTER 4

4.1 Adding numbers with an abacus is done the same way that numbers are added manually—one digit at a time.

$$\begin{array}{r} 813 \\ 8 \\ \hline 821 \\ \\ 20 \\ \hline 841 \end{array}$$

The difference is in how we represent and manipulate the individual values in each place position. Let's consider the *suan pan* first. Zero out any current value and load the first number. This is a partial sum.

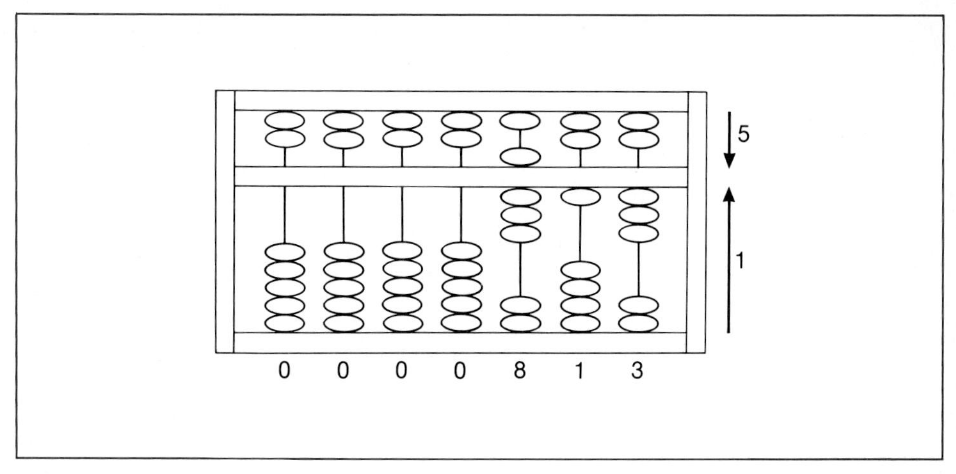

FIGURE A4.1(1)

Add the units digit of the second number, 8. This is done by adding a 3 and then a 5, but we have a problem. Three of the available 1s have already been used for the current value of the partial sum, and there are not enough 1s left to add 3 directly to the units digit. But we can add 3 as $5 - 2$ and then add a second 5 to make 8. This creates another problem. We now have an 11 ($10 + 1$) in the units digit.

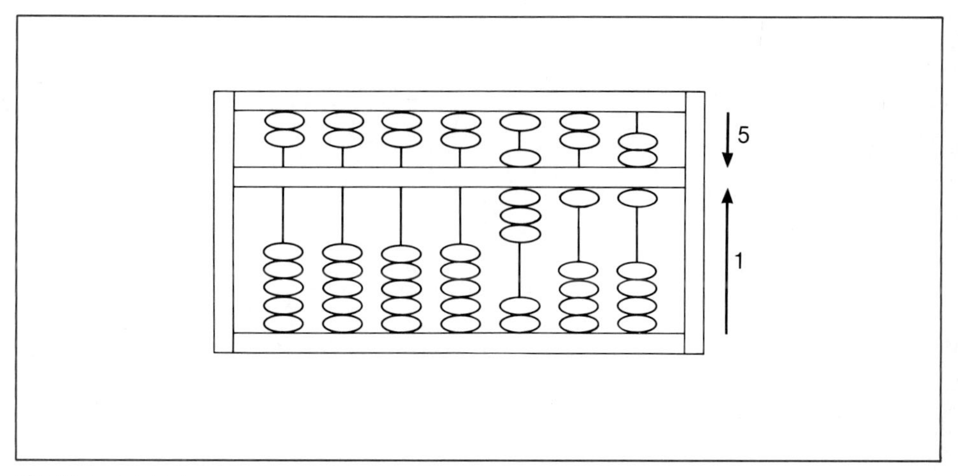

FIGURE A4.1(2)

The 10 is a carry, and it is removed by subtracting the 10 from the units digit and adding 1 to the tens digit.

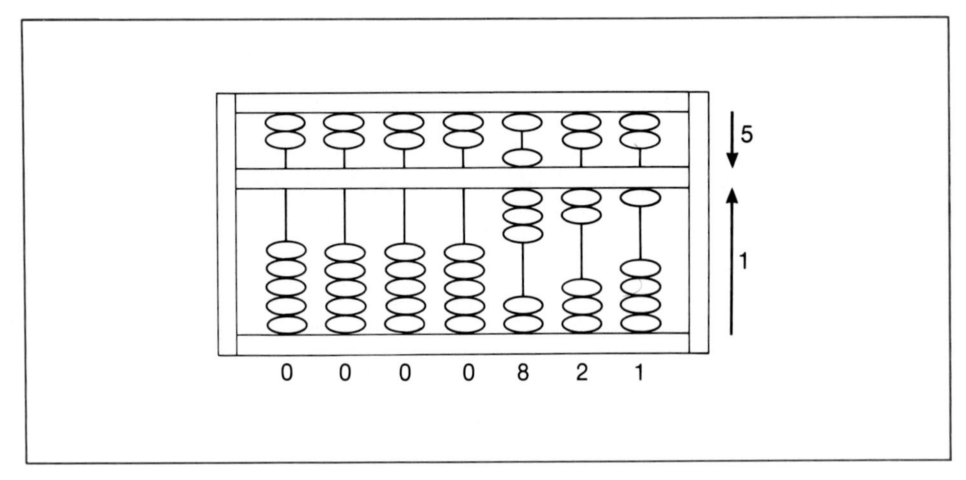

FIGURE A4.1(3)

Add 2 to the tens digit and the operation is complete.

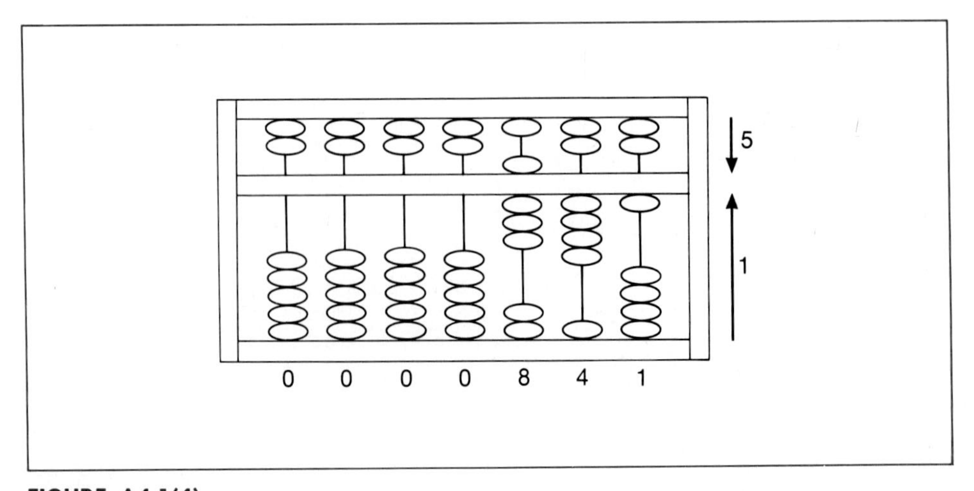

FIGURE A4.1(4)

Now consider the *soroban*. Zero the current value and load the first number.

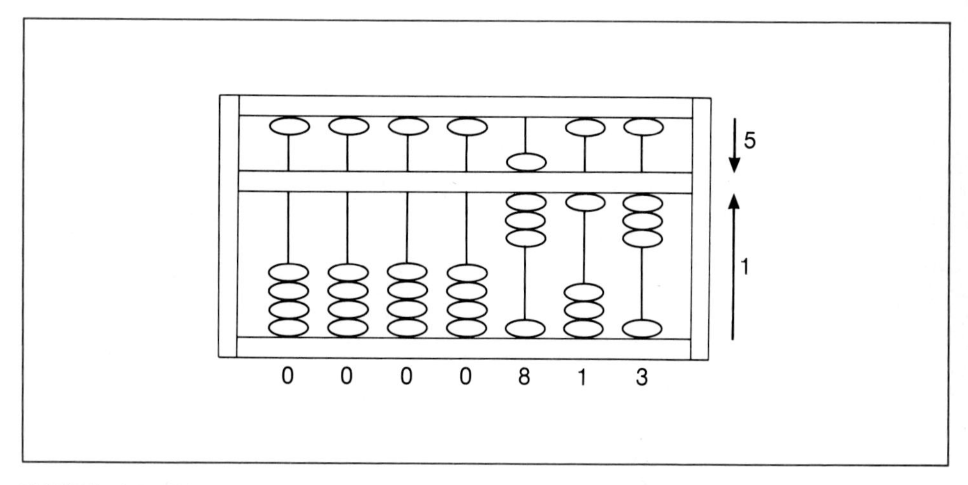

FIGURE A4.1(5)

Add the units digit of the second number, 8. We use essentially the same procedure as with the *suan pan* for the 3, (5 − 2), but the additional 5 is handled differently. We don't have a second 5 in the units digit, so we add 1 to the tens digit and subtract 5 from the units digit (5 + 10 − 5).

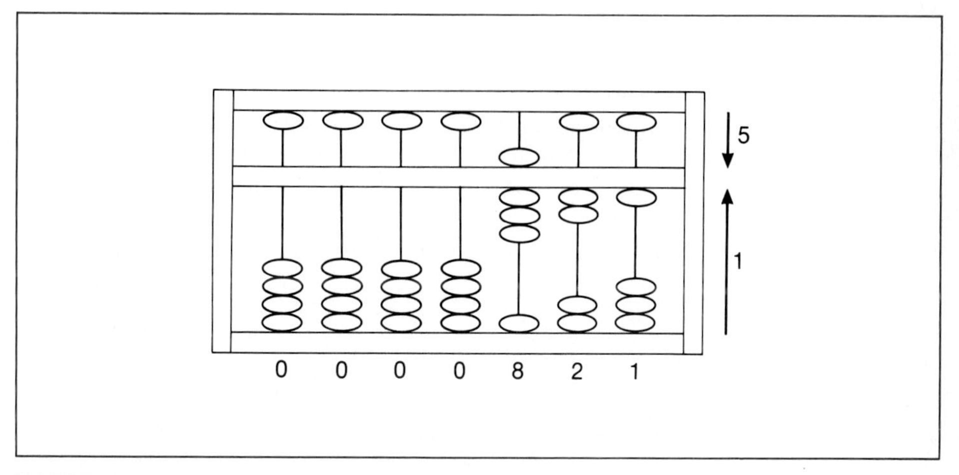

FIGURE A4.1(6)

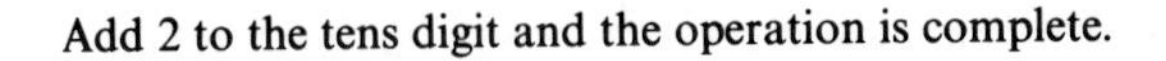
Add 2 to the tens digit and the operation is complete.

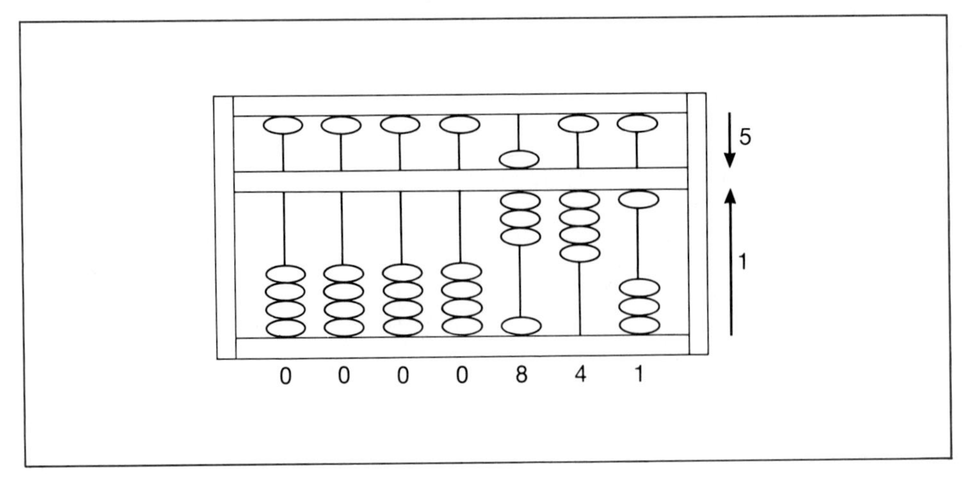

FIGURE A4.1(7)

Many people consider the *soroban* to be the better of the two devices, because carries are handled more directly and the number of operations is somewhat smaller. In actual practice, however, it is more a matter of preference and what you have learned to use.

4.3 Multiplying two numbers on an abacus is just a series of adds, as shown below.

	21 as multiplier	34 as multiplier	
21 as multiplier	34	21	34 as multiplier
	340	21	
	340	21	
	714	21	
		210	
		210	
		210	
		714	

Use 21 as the multiplier, because the number of operations is dependent on the sum of the digits in the multiplier.

4.5 Using existing methods and technology to process census data would have required more than the 10 years between census dates. Hollerith extended the ideas developed for sorting cards into additional ways for processing data other than just sorting and counting.

4.7 The program for the abacus is stored in the memory of the user. The abacus has no processing ability by itself.

4.9 In the early days of computing, CPU usually referred to the control unit, main memory, the arithmetic unit, and the I/O unit. Nowadays, CPU usually refers only to the control and arithmetic units. Memory and I/O are considered separate parts of the computer.

4.11 When the computer is turned on or reset, an address is loaded into the program counter. This address points to a location in nonvolatile memory that contains the first instruction of the bootstrap program.

4.13 A computer instruction contains a command, called the opcode, and zero or more objects, called the operands.

4.15 Primary memory usually contains the parts of programs that are currently executing along with their data. It also contains programs and data used by the operating system.

Secondary memory contains the parts of currently executing programs (and data) that have not been recently accessed. It also contains files that users have created recently or files that users expect to work on or use in the near future.

Archival memory contains backup copies of programs and data. It may also have programs that are only run infrequently, such as at tax time.

4.17 Primary memory must be faster, since it has to supply instructions and data at the rate at which the CPU needs them.

4.19 Integer add, subtract, multiply, and divide are included in most arithmetic units. The ALU may have to support variations of these operations based on the type of data the ALU will be expected to process.

4.21 The I/O unit is needed so that the ALU and memory can communicate with secondary memory devices and with users.

4.23 The operating system manages resources, provides an interface between the user and the hardware, coordinates the work load on the computer, protects system resources from illegal access, controls who can use the system, optimizes use of resources, keeps track of who is using the system's resources, and provides services to users' programs.

4.25 A compiler translates from a human-oriented, high-level language to a machine-oriented, low-level language. It also checks for errors to help the user find mistakes in programs.

4.27 Pascal, Fortran, Cobol, C, Prolog, LISP, C++, BASIC, and assembly language are examples of source languages.

4.29 If users never made mistakes, the Undelete command would probably be unnecessary. However, since people do make errors, Undelete often saves us from the drudgery of reentering data and programs after we have erroneously deleted a file.

4.31 In addition to requiring a password when a person logs onto a system, some file systems require users to enter a password for a file before they can access the file. This is important on systems that have multiple users.

4.33 They all use resources provided by the operating system. They all give the user an additional level of functionality above that provided by the system software and hardware. Each one is customized to provide a specific set of functions to help solve one problem.

CHAPTER 5

5.1 Fetch, decode, and execute.

5.3 LOAD R1, COUNT

This instruction tells the computer (implied subject) to load (verb) the contents of COUNT (direct object) into register number 1 (indirect object).

5.5 With a straightforward interpretation, the instruction has 5 bits for the opcode and 11 bits for the operand. Therefore the computer could have $2^5 = 32$ instructions and $2^{11} = 2048$ memory locations. We qualified the answer with "straightforward interpretation," because we could do things like have one opcode specify "I/O operation" and then use the operand field for further specifications. We could use the operand field as a displacement from a segment register. There are many variations of an apparently straightforward instruction format.

5.7 Data movement, arithmetic, logical, shifting, branching, testing, procedure calls/returns, and I/O.

5.9

```
                LOAD    START       Load the accumulator with the desired
                                      starting value
                CMP     TARGET      Compare the accumulator to the
                                      target value, in case the value
                                      is already greater or equal
                JGT     ERROR       If the value is greater, then skip this
                                      loop; something is wrong
                JEQ     ENDLOOP     If the value is equal, then skip this
                                      loop; we're already done
LOOP            ADD     =1          Increment the accumulator by a
                                      location whose contents are 1
                CMP     TARGET      Compare the accumulator to the
                                      target value
                JLT     LOOP        If the value is less, then repeat this
                                      loop; otherwise, we're done
ENDLOOP         ...
  :
ERROR           ...
```

5.11

```
SWAP M     Exchange the contents of the accumulator and the
             contents of memory location M
LDIM  V    Load the contents of the address field into the
             accumulator, treating it as a signed integer
```

5.13

```
SGN  M     Determine the arithmetic sign of the value at location M
JXX  L     Jump to location L if the sign determination indicated:
             JPS   Jump positive
             JNG   Jump negative
             etc.
```

5.15 Instruction execution sequence

Location	Contents	Stack contents (after instruction)
	⋮	

5.15 Instruction execution sequence (*Cont.*)

Location	Contents	Stack contents (after instruction)
ADD1	JSB L1	ADD1 + 1
L1	⋮	
ADD2	JSB L2	ADD2 + 1 ADD1 + 1
L2	⋮	
	RTN	ADD1 + 1
ADD2 + 1	⋮	
ADD3	JSB L2	ADD3 + 1 ADD1 + 1
L2	⋮	
ADD4	JSB L3	ADD4 + 1 ADD3 + 1 ADD1 + 1
L3	⋮	

5.15 Instruction execution sequence (*Cont.*)

Location	Contents	Stack contents (after instruction)
	RTN	ADD3 + 1 ADD1 + 1 (empty)
ADD4 + 1	⋮	
	RTN	ADD1 + 1 (empty) (empty)
ADD3 + 1	⋮	
	RTN	(empty) (empty) (empty)
ADD1 + 1	⋮	

5.17 Call forwarding on your telephone. A person dials your phone number, and if you have set the forwarding, then the system takes this initial number as an "indirect address." It dials the other number, and the caller contacts you at the other location.

5.19

		Loop to add together (COUNT) numbers, contained in memory locations ARRAY through ARRAY + (COUNT) − 1; if (COUNT) = 0, it is assumed that the sum is zero
LOAD	= 0	Load accumulator with zero
STORE	SUM	Initialize sum
LOAD	COUNT	Load the accumulator with desired number of values to add
CMP	= 0	Compare accumulator to a location with zero, in case the count is incorrect or zero
JLT	ERROR	If the value is less, then skip this loop; something is wrong
JEQ	ENDLOOP	If the value is equal, then skip this loop; we're already done

```
            ADD      = ARRAY     Add the address of the start of the
                                   array to the counter; the result
                                   is one location beyond the end
                                   of the desired block to be
                                   summed
            SUB      = 1         Subtract 1, to point to last location
                                   to be summed
LOOP        STORE    POINTER     Store this address in a location to be
                                   used for indirection
            LOAD     SUM         Load the current value of sum
            ADD      I,POINTER   Increment the accumulator by the
                                   contents of the next location
                                   to be summed
            STORE    SUM         Store the current value of sum
            LOAD     POINTER     Load the current value of the
                                   pointer
            SUB      = 1         Decrement by 1
            CMP      = ARRAY     Compare the pointer to the address
                                   of the last location to be
                                   summed
            JGE      LOOP        If the value is greater or equal, then we
                                   still have more values to add, so
                                   repeat this loop; otherwise,
                                   we're done
ENDLOOP     ...
    :
ERROR       ...
```

5.21

a	*b*	*t*	Output
0	0	0	0
0	1	0	0
1	0	0	0
1	1	0	1
0	0	1	z
0	1	1	z
1	0	1	z
1	1	1	z

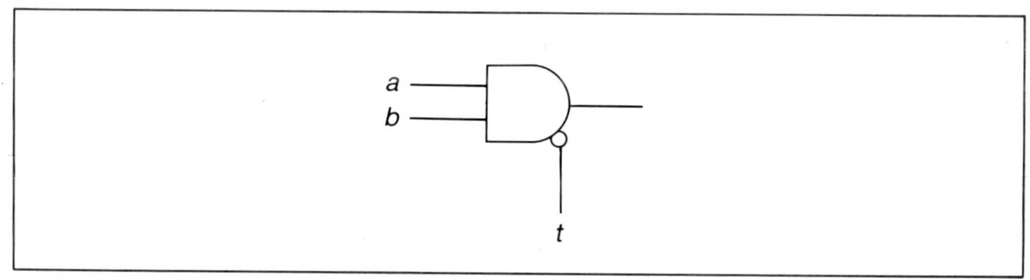

FIGURE A5.21

5.23 The control unit.

5.25 The control unit extracts the return address from the top of a stack, from a special register, or from wherever else the system has stored the appropriate return. It then puts this value into the program counter so that that location will be the next one fetched.

5.27 Count up such things as the number of gates that must be controlled, the number of ALU functions, and the number of memory banks—anything that a bit in the microinstruction must control.

5.29 Because the control unit retrieves a sequence of values from a memory, which is generally slower than generating control signals with logic circuits in a hardwired system.

5.31 These programs are exactly analogous to the ADD instruction in Fig. 5.16, except we would have a 1 under the AND, OR, or exclusive OR rather than under the + in location 135.

5.33 This is an idea known as vertical microinstruction coding. Suppose we have 16 functions for a certain ALU. Instead of allocating 16 bits, one for each function, we allocate 4 bits. We then use a logic circuit to decide which of the $2^4 = 16$ functions we want to execute for each of the 16 possible 4-bit patterns. The advantage is, as the problem suggests, a smaller microinstruction. The disadvantage is the time it takes for the logic circuit to "translate" the 4 bits.

5.35 The disadvantage is the time the decoder requires to decide which function is being specified by the given 4 bits.

5.37 The logic that maps instruction opcodes to microprograms.

5.39 Direct address/Immediate data
Shift logical or arithmetic

5.41 One of the operands could specify the number of positions to shift. We might even have an operand specify the direction and type of shift.

5.43 An instruction whose operation depends on a preceding instruction—we can't execute this instruction until the preceding one is *completely* finished, thus contradicting the idea of pipelining. An instruction that modifies program control, such as a Skip or a Jump—we don't know which path will be taken, so we don't know which instructions to start feeding into the pipeline.

5.45 So that one clock can regulate all the cycles and they can be done in parallel.

5.47 With a variable-length instruction, you don't know how much to decode until you've determined what type of instruction it is. That is, you must partially decode it before going on to the rest of the decoding.

5.49 To say that what is currently in the pipeline is useless and must be discarded. For example, the program might have done a jump and the pipeline was decoding the instructions immediately following that jump.

CHAPTER 6

6.1 DRAM: 1 megabit costs $13, or $0.0013/bit
SRAM: 32768 × 8 costs $10, or $0.0038/bit
Register: 8-bit shift register costs 60 cents, or 7.5 cents/bit

6.3 Probably floating-point numbers. Character strings. Program constructs such as records or structures.

6.5 It has a different function and usually a different structure and must be faster.

6.7 It is similar to the notion of an array in a program. With a one-dimensional array, you reference a particular item in a linear sequence of items, such as LINEAR[7]. With a one-dimensional memory, the decoder picks a particular location in a linear sequence of locations. With a two-dimensional array, you specify two subscripts; you reference a particular item in a "plane" of items by giving two coordinates of the plane, such as PLANE[7, 13]. With a two-dimensional memory, the decoder picks a particular location in a "plane" of locations by determining a row and a column address.

6.9 Mainly because it reduces the number of address pins required on the chip and eliminates an extra chip.

6.11 With static memory, the cells maintain their current state until changed. With dynamic memory, the cells must be refreshed periodically or they will lose their current state.

6.13 Normally, a manufacturer would increase both the row and the column size. If the row size were doubled and the column size were doubled, then the entire chip capacity would increase by a factor of 4. You might wonder why we assumed doubling, rather than increasing by 50 percent or something. If you increase the size, you must increase the addressing space. Adding 1 bit to the address space implies doubling its size.

6.15 Have the system begin a main memory access at the same time it is accessing cache. If a cache hit occurs, the main memory access could be canceled. If a cache miss occurs, the main memory access will already be in progress, and the system can just wait for it to complete.

6.17 If the GOTOs went to relatively distant code segments, then they would be constantly changing localities. This would in turn result in many cache misses and would slow down execution of the program.

6.19 Suppose the system brings in blocks of data from disk to memory and/or from main memory to cache. If the compiler stores arrays by row, then the program segment that accesses by row order would have all its accesses within a given block and then move on to the next block, etc. The program segment that accesses by column order, however, would have its accesses repeatedly going to different blocks, requiring many swaps.

6.21

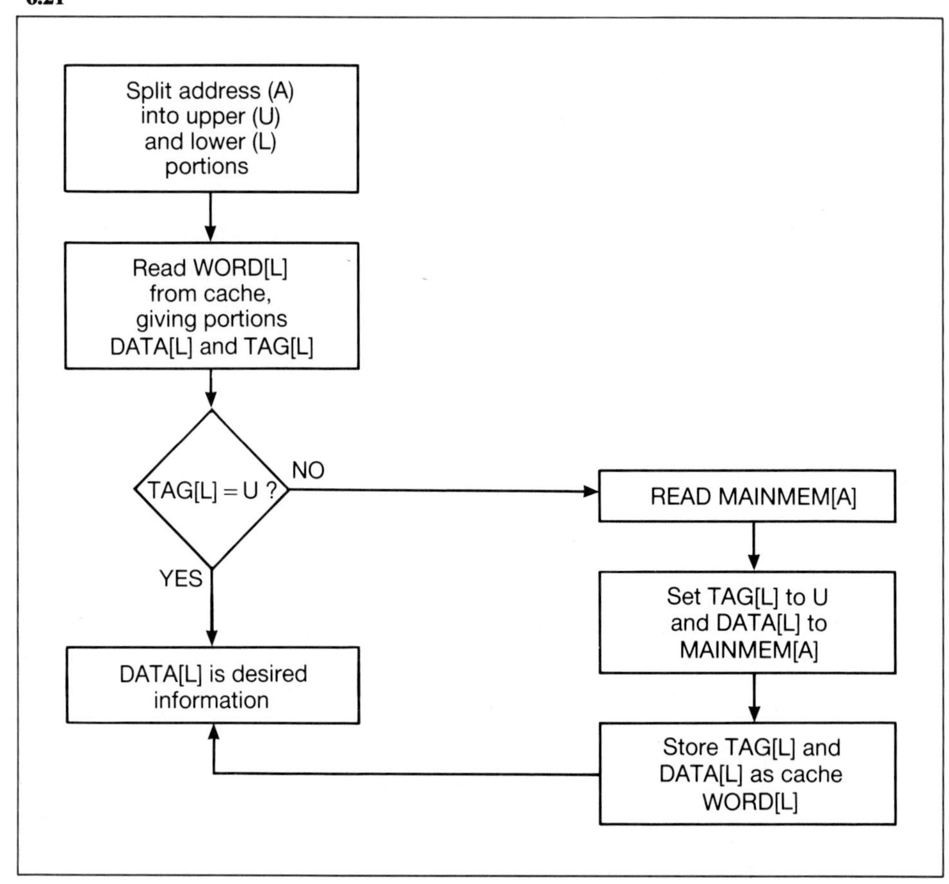

FIGURE A6.21

6.23 Besides being cheaper, disk memory is permanent. With a removable disk, it is easy to save information, make backups of information, and send information to other people.

6.25 The major slowdowns in disk access are the time to move the head to the proper track and the time for the appropriate sector on that track to move under the head. With a head-per-track construction, you have eliminated one of those slowdowns.

6.27 Because magnetic fluxes stay in their particular orientation until they are changed by a sufficiently strong magnetic field. Hence, assuming you keep the devices away from such a magnetic field, they are inherently stable.

6.29 To minimize cost while providing the required speed and capacity.

CHAPTER 7

7.1 They are generated in the control unit by the instruction decoding process. Depending on whether the control is microprogrammed or hardwired, they come from either the microprogram instruction register or the decoding matrix.

7.3 Because going *into* the registers we simply have to load what's on the bus or not load it. Our problem, and the need for tristate gates, is coming *out* of the registers. Without tristate gates, the output of many registers would be contributing simultaneous signals to the bus.

7.5 Mainly simplicity and resulting cost, since you have essentially half the number of bus lines.

7.7 36 lines, 4 in control + 16 in data + 16 in address.

7.9 Parity bits, one for each byte of data and one for the address.

7.11 The sender's phone or the receiver's phone could be busy. Circuitry between the phones could be busy (e.g., when you try calling home on Mother's Day). The sender must know the receiver's phone number, not just name or address. The receiver might be in the shower or otherwise not able (or not willing) to answer.

7.13 Use one or more multiple-input OR gates, depending on how many units there are and how many inputs your gates can have.

7.15 If you have any deterministic approach, it is probably not going to be considered fair by everyone. Perhaps the only fair way is a random selection, such as the following:

1. Start with the sequence $S = \langle 1, 2, 3, \ldots, n \rangle$.
2. Set SIZE to n.
3. Pick a random number r between 1 and SIZE; s_r is the unit whose request line is checked.
4. After s_r has finished, or if s_r does not want the bus, remove s_r from S.
5. Decrement SIZE by 1.
6. If size > 0, return to step 3; otherwise, return to step 1.

7.17 If a given unit's address appeared on the address line and that unit wanted the bus, then the Bus Busy line from that unit would go to 1.

7.19 The phone system, where many phone calls are transmitted on the same physical cable. FM stereo radio, where channels "left + right" are sent as one signal. (Channels "left − right" are also sent as one signal on a different frequency.) Roads, airports, train tracks, etc., could also be considered as multiplexed resources, since different vehicles use them.

7.21 Lower hardware cost and reduce space cost.

7.23 They would not affect the transfer on the bus per se, since they would also be sent in parallel. However, it would take time to generate these parity bits prior to putting the entire set of signals on the bus. It would also take time after the signals have been received to check the parity. Therefore, the effective transfer rate is decreased.

7.25 If our transmission rate was n bits/s, then with the parity bit it takes $9/n$ s to send 8 bits. Thus our effective rate is $8/(9/n) = \frac{8}{9}n$ bits/s. Similarly, for 16 bits, it would be $\frac{16}{17}n$ bits/s.

7.27 Usually the receiver notifies the sender that the message must be retransmitted.

CHAPTER 8

8.1 The ALU is the part of the computer that executes arithmetic and logic operations. Probably at a minimum we want (1) arithmetic operations for integers and floating-point numbers, (2) standard logic operations, (3) shifts, (4) compares and jumps, (5) data moves, and (6) I/O.

8.3 A big advantage is that you can have one set of jumps for a number of different compares (e.g., string, logical, and arithmetic). Also, the jump need not be executed right with the compare. However, this arrangement does require two instructions to compare and jump. Also, the status indicators must be saved and restored at various times.

8.5 Mod and exponentiation would be dyadic.

8.7 One is a circuit and one is a poisonous snake.

8.9

X_i	Y_i	Borrow_{i-1}	Difference_i	Borrow_i
0	0	0	0	0
0	0	1	1	1
0	1	0	1	1
0	1	1	0	1
1	0	0	1	0
1	0	1	0	0
1	1	0	0	0
1	1	1	1	1

8.11 The half-adder has three gate levels, so it would take 30 ns to produce the sum and carry bits. The next stage must wait for this carry, then pass it through another half-adder (30 ns) and an OR gate (10 ns) for a total of 40 ns to produce its carry. Each of the next two stages will similarly require 40 ns, giving a total of 150 ns.

8.13 Looking at the truth table for the half-adder, we see that it is identical to an exclusive OR. The output is a 1 if either of the input bits is a 1, but not if both input bits are 1 (or both input bits are 0).

X_i	Y_i	Sum_i
0	0	0
0	1	1
1	0	1
1	1	0

X_i	Y_i	$\oplus_i$
0	0	0
0	1	1
1	0	1
1	1	0

8.15 First define $A = B$ as $(AB + A'B')$. The boolean expression is then just $(A_1 = B_1)(A_2 = B_2)(A_3 = B_3)(A_4 = B_4)$.

8.17 This would just be the exclusive OR function.

8.19 To provide high-speed temporary storage facilities and sometimes to provide certain machine operations such as shifting.

8.21 The only difference is that the accumulator clear might change some status bits in the ALU.

8.23 The idea would be to toggle (change state) when the input and the current state don't agree. However, this is just our standard exclusive OR, so we use that exclusive OR as the flip-flop toggle input (comparing the data bit to the current flip-flop state).

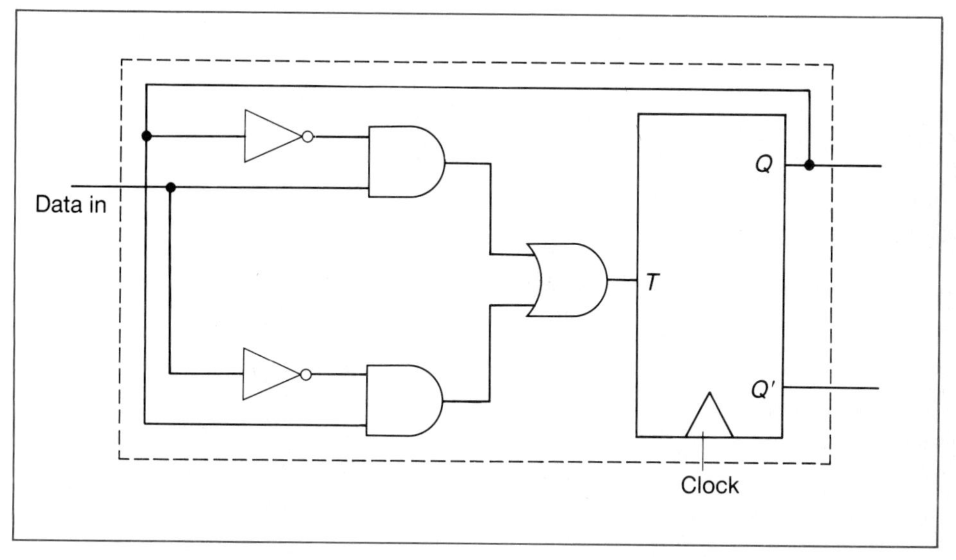

FIGURE A8.23

8.25 They would be generated in the control unit of the CPU. This control unit would decide, based on the current opcode, which signal to generate. Depending on whether the control is microprogrammed or hardwired, the actual signals would come from the control store register or the decoder matrix.

8.27 On a left shift, it becomes the effective "leftmost" bit. With a normal left shift, it would be discarded. On other shifts, it would move to the rightmost position or to the concatenated register, depending on the type of left shift. For right shifts, it has no special significance.

8.29 Because only one of the logic functions can be selected at a time.

8.31

Acc	0	0	0	0	0	0	0	0	0	0	
+	0	0	0	0	0	0	0	0	0	0	(01011)
	0	0	0	0	0	0	0	0	0	0	
Shift	0	0	0	0	0	0	0	0	0	0	
+						1	1	1	0	1	(01011)
	0	0	0	0	0	1	1	1	0	1	
Shift	0	0	0	0	1	1	1	0	1	0	
+	0	0	0	0	0	0	0	0	0	0	(01011)
	0	0	0	0	1	1	1	0	1	0	

8.31 *(Cont.)*

Shift	0	0	0	1	1	1	0	1	0	0	
+	0	0	0	0	0	1	1	1	0	1	(01011)
	0	0	1	0	0	1	0	0	0	1	
Shift	0	1	0	0	1	0	0	0	1	0	
+	0	0	0	0	0	1	1	1	0	1	(01011)
	0	1	0	0	1	1	1	1	1	1	

8.33 Let's define the number of bits in an integer M as length $(M) = \lceil \log_2 M \rceil$. Then the time to add two numbers, M and N, would be some circuit constant times max (length (M), length (N)), since we have to propagate the carry down the longest of the two numbers. To multiply M by N as shown in this chapter, we would start with a register whose length is length (M) + length (N). We would repeatedly add M or 0 to this register, with each such add taking a circuit constant times (length (M) + length (N)). The number of such adds would be length (N), giving a total of a circuit constant times length (N) times (length (M) + length (N)). To see a little more simple expression, let's assume L = length (M) = length (N). Then the time to add is proportional to L, while the time to multiply is proportional to L^2.

8.35 Because you have to do the equivalent of the integer operation on the fractional part of the number, *plus* doing the shifting, alignment, normalization, etc.

8.37 The functional unit is a combinational circuit, not a sequential circuit. It cannot store values for the next clock cycle. This circuit shows only one bus, so we would have to try and gate the first number into the functional unit during one cycle, then gate the second number in during the second cycle and do the operation. However, because the functional unit could not store the first number, this first number would be lost before the second number was available.

8.39 We could transfer the sign directly. The fraction would have to be rounded to fit into 23 bits and the exponent adjusted accordingly (if necessary). Finally, the exponent would have to be reduced to fit into 8 bits. Note that the exponent could not be rounded—if it didn't fit, an error would have to be generated.

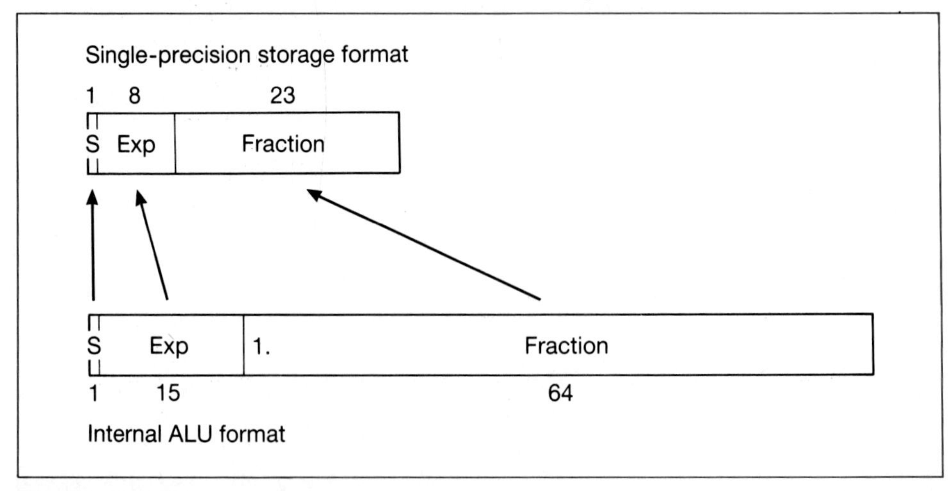

FIGURE A8.39

8.41 Before anything else is done. Since a special result will occur, there is no sense doing all the operations first and then finding out they were unnecessary.

8.43 Standards for threads, standards for bolt strengths, standards for various classes and diameters of pipe, measurement standards, time standards, weight standards.

CHAPTER 9

9.1 Mouse, joystick, plotter, light pen, modem, touch screen, track ball, tape drive, graphics tablet, optical scanner.

9.3 Characteristics of the hardware, such as the voltage necessary to cause a circuit element to change state.

9.5 We could double the rate by (*a*) halving the clock cycle, which is probably not possible, or (*b*) doubling the data bus to 16 bits.

9.7 Address—how many peripheral controllers we want to address on the I/O bus. Data—how fast the devices attached to the controllers (and subsequently to the bus) can transfer data, and how much money we want to spend.

9.9 Typically, they have to communicate with peripheral controllers that are slower and farther away than main memory. Also, the buses to these controllers may have lower bandwidths.

9.11 Waiting for I/O on a device that is turned off, nonexistent, or broken.

9.13 Looking at Fig. 9.7, we could temporarily replace the STP READER command by an unconditional jump to WAIT1. This gives us 10 instructions to read in 2 bytes of data, or 5 instructions per byte. Assuming all instructions operate in one clock cycle, we would have 500 ns/byte, or 1/(500 ns) = 2,000,000 bytes/s.

9.15 Simpler, cheaper, and they assumed that users of these systems would not be multitasking I/O operations anyway.

9.17 Assuming it is a long operation, we could ignore the initiation and termination overhead on the CPU. Thus we just have to move the data when it is available. For each interrupt we first save the PC and ACC. Assuming 1 byte at a time (which is generally not the case, but serves to illustrate), we would move the byte off the bus and into a register and then store the register into memory. After the transfer we would restore the PC and ACC. Thus we would use 6 instructions per byte, or 3600 instructions per second, for a 600 bytes/s rate. Our machine is doing 1/(100 ns), or 10,000,000 instructions per second. We are therefore saving (10,000,000 – 3600)/10,000,000 = 99.964 percent of the time required by wait loop I/O.

9.19 On many computers, the program counter, comparison indicators, and other status information are kept in a special register. Assuming this is the case, it would take (1 + number of accumulators) · (100 ns) to save the current context and the same amount of time to load a new context. For 16 accumulators, this is 3400 ns.

9.21 They could be stored in a special register or in the address portion of the return instruction. Neither of these is the most desirable, because they cause problems with recursion, code modification, and nested calls. Probably the best way is to keep them in a stack area of memory.

9.23 Because the interrupt servicing (and thus the information in the vector) will vary depending on what device the interrupt came from. The controller is the one who knows about this and therefore the one who must generate the vector.

9.25 It must do a limited set of operations we normally associate with the CPU, such as handling interrupts, moving data, and manipulating and testing counters.

9.27 It might be working on an instruction that currently does not need the system bus. If the CPU does need the bus, it must wait, locked out by the bus control logic.

9.29 Change the priority so that the CPU would be given priority over the DMA every other cycle.

9.31 In Prob. 9.17, we used 6 instructions per byte as the load on the CPU. For 128 characters, this would be 768 instructions. Assuming the CPU was not delayed by being locked out of the system bus, we would save $768 \cdot 100$ ns $= 76.8\ \mu$s.

9.33 It must increment/decrement counters, perhaps assemble bytes into words, and do other simple calculations such as addressing.

9.35 Yes. They can eliminate the CPU having to fetch instructions for each I/O data transfer.

9.37 The CPU can treat all devices like memory, hardware costs are lowered (fewer buses), the system can easily incorporate different-speed memories, the instruction set size is reduced (no I/O), and different devices on the bus can communicate directly.

9.39 The individual components are so fast, and the organization allows so many things to be done in parallel, that a single bus would be a real bottleneck.

9.41 Probably not. In earlier answers we ignored the overhead of initiating and terminating I/O operations because we assumed long data transfers. If most such transfers are short, we get to the point where this overhead is comparable to the time for the CPU to do the operations directly itself. (Refer to Fig. 9.13 for DMA programming.)

CHAPTER 10

10.1 No change would be needed unless the controller needed to convert EBCDIC to ASCII.

10.3
1,200 bps: $t = 833.33\ \mu$s
4,800 bps: $t = 208.33\ \mu$s
9,600 bps: $t = 104.17\ \mu$s
19,200 bps: $t = 52.08\ \mu$s

10.5 Convert among various voltage levels, compensate for timing differences, provide device selection and control, provide temporary data storage, convert among various data formats, and provide for interconnections with differing cable connectors.

10.7 To shift more of the burden away from the CPU. The more "grunt work" the controller can do (such as conversions between data formats), the less has to be done by the CPU. This is usually a good tradeoff because the microprocessor in the peripheral controller is usually cheaper and less powerful than the main CPU.

10.9 Audio input device, scanner (converts pictures to an input format such as a bit map), bar code reader (such as the ones at supermarket checkouts), modem (for phone communication), and video cameras.

10.11 They send two signals to the controller. These signals are either voltage levels proportional to the X and Y position or pulses whose numbers are proportional to the X and Y distances moved.

10.13 Probably convenience; it seems easier and more stable for us to move the mouse on the flat surface than to direct the light pen at the screen.

10.15 Aircraft controls used them.

10.17 $640 \cdot 400 \cdot 8 = 2{,}048{,}000$ bits (256 colors require 8 bits, since $2^8 = 256$).

$$\frac{2{,}048{,}000}{9{,}600} = 213 \text{ s, or approximately } 3\tfrac{1}{2} \text{ min}$$

10.19 The print hammer must pass over the ribbon three times, once for each of the three primary colors.

10.21 Because the characters must be formed from very tiny line segments, which is slow, cumbersome, and inaccurate on plotters.

10.23 Mainly because of weight and size. On a tape, the seven or nine tracks are perpendicular to the movement of the tape, all tracks are the same length, and the head doesn't move. The disk heads must be moved quickly and weight reduces the speed.

10.25 The cushion of air separating the head from the rapidly moving disk is no longer there, so the head contacts the surface of the disk. Usually, this scrapes away magnetic material and damages the disk.

10.27 $31416 \cdot 165 \cdot 165 - 31416 \cdot 75 \cdot 75 = 8.553 - 1.767 = 6.786$

$$\frac{150{,}000{,}000}{6.786} = 22{,}104{,}332 \text{ bytes/in}^2$$

These calculations assume that the outer tracks have more data than inner ones, since they have more area, and that there are no gaps between sectors or tracks.

10.29 Music and showing movies at home.

10.31 Semiconductor: 5–100 ns
Hard disk: 4–20 ms
Floppy disk: 100–200 ms

10.33 $655{,}360/6250 = 10.5$ inches per record + 0.25-in gap = 10.75 inches per record, or $6{,}553{,}600/1075 = 6096$ bits/in effective density = 6096 bytes/in for the eight tracks plus parity. $150{,}000{,}000/6096 = 24{,}606$ in, and $24{,}606/25 = 984$ s, or almost $16\tfrac{1}{2}$ min.

10.35 (*a*) Typically something using graphics, such as computer-aided architectural design. (*b*) Typically programming or word processing, where displays are textual.

10.37 Advantages—sharing data and having a large, sophisticated file system available to all users for a much lower cost per computer. Disadvantages—problems of shared access and time to transfer data on a relatively long cable.

10.39 Distributed intelligence spreads the processing over multiple devices rather than forcing one device to perform all computations.

10.41 Keyboard—fastest possible human typing
Printer—the speed of the interface to the computer

CHAPTER 11

11.1 A logical component is a symbol representing a logical operation, such as AND or OR. A physical component is the electronic device that actually implements the operation.

11.3 You can enter the design language program using a simple keyboard and a system with no graphics capabilities.

11.5 The first is the logic symbol for an invertor. The second is the transistor-level representation of a CMOS implementation of an invertor.

11.7

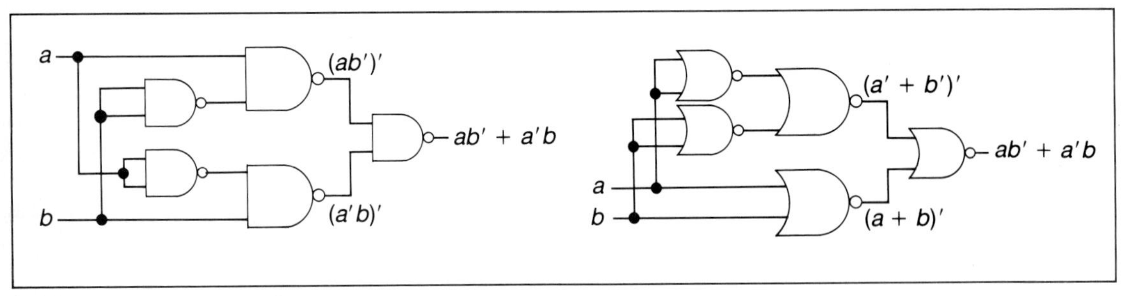

FIGURE A11.7

11.9 It lists components that can be put onto a schematic capture page. It also gives information about the component, such as the graphical symbol and the terminal assignments.

11.11 The graphical equivalent of a pin on an integrated circuit component or a connection on a discrete component such as a resistor.

11.13 The same pins are used for two purposes: to transfer and address or to transfer a data item.

11.15 These terms refer to the number of gates on the chip:

SSI: several gates
MSI: 10–100 gates
LSI: 100–1000 gates
VLSI: thousands or tens of thousands of gates
ULSI: tens or hundreds of thousands of gates

11.17 Parts list = list of components used in a schematic capture design
Net list = list of connections between the components

11.19 It helps the designer place the nets or traces on a board. This involves aspects such as minimizing route length while avoiding congested areas and minimizing the number of crossover points.

11.21 The physical placement tool works with the footprint of the component, that is, its physical embodiment rather than its logical symbol. The placement tool helps the designer physically place the component on the board and assign gates.

11.23 Because most components will have connections to both power and ground, and because the use of planes helps with the electrical noise problem.

11.25 Pins from components go through holes in the board. After the components have been placed onto the board, the board is passed over a molten wave of solder so that the protruding pins and traces on the board just touch the solder. The molten solder fills in the area between the walls of the hole and the pin, holding the component in place and establishing electric contact.

11.27 The footprint of the component is its physical outline rather than its logical symbol. It is used to determine where the component will be physically placed on the board.

11.29 Defining the characteristics of the circuit board itself, such as its dimensions, number of layers, and type of mounting.

11.31 Often, a physical device will contain more than one gate. For example, a device might contain four AND gates, four OR gates, or six INVERTORS. Gate assignment is deciding which gate in the device will be assigned to which corresponding logical element on the schematic diagram.

11.33 A Manhattan route uses only horizontal and vertical traces (not diagonals). The optimal route is the shortest distance between two points using only horizontal and vertical traces. (Of course, this route must also take into account other components that lie between the two points to be connected.)

11.35 There are a footprint of a component and nets on the bottom of the board (solid lines). There are also vias leading to the other side of the board and nets on that side (dashed lines).

11.37 Ultimately, we would like to minimize the total length of the traces. However, we must take into consideration other factors, such as avoiding congested areas and separating some pairs of nets.

11.39 The delay in a signal going from one point to another point. This delay is caused by the time it takes the electric signal to travel along the trace and the time it takes for intervening electronic components to change their state.

11.41 The physical length of the trace, the number of components connected to the trace, and the characteristics of the device that is putting the signal onto the trace.

11.43 The oscilloscope is a tool that displays current and voltage waveforms on a video screen. It is useful for hardware debugging, doing things such as showing propagation delays and signal distortions.

11.45 A distorted signal might not reach the minimum value necessary to cause a component to change states. A delayed signal might reach a component after the clock signal has shut off and thus not result in the correct logical operation.

11.47 At 25 MHz, the time between rising edges would be 40 ns. If the memory takes 20 ns to fetch the data, then we have 20 ns for the request to reach memory and the data to be sent back to the CPU. At a propagation of 0.5 ft/ns, the maximum length would be 10 ft.

CHAPTER 12

12.1 A state field in a process control block indicates which state (active, execute, suspend, blocked) the corresponding process is in.

12.3 A process is the unique execution of a program with a specific set of data.

12.5 It is waiting for a prespecified time to resume execution.

12.7 It has used up all its allocated time, or a higher-priority process needs to execute, or an interrupt occurs.

12.9 *Simultaneously* means that two things are occurring at precisely the same time. This is also the meaning of *true concurrency*. However, *apparent concurrency* means that two things appear to be happening simultaneously, but actually they happen in a very small time interval.

12.11 More than one processor.

12.13

```
{producer}                            {consumer}
produce object                        wait(access_to_pool_of_objects)
wait(access_to_pool_of_objects)       if pool of objects is not empty then
add item to pool of objects             remove object from pool of objects
signal(access_to_pool_of_objects)     signal(access_to_pool_of_objects)
```

12.15 We could think of traffic signal lights as granting access to a shared resource, the intersection. Hence they are analogous to a semaphore.

CHAPTER 13

13.1 The interface between the user and the system. The interface between external devices and the CPU. The interface between a personal computer and a network.

13.3 Guarding access to files. Guarding access to resources. Guarding access to memory. Protecting one user from another user. Protecting itself from the users. Guarding accounting information.

13.5 There was only one user using the system, and he or she had access to everything.

13.7

Memory page number	Actual memory locations
0000	0000000000000 – 0000111111111
0001	0001000000000 – 0001111111111
0010	0010000000000 – 0010111111111
0011	0011000000000 – 0011111111111
⋮	⋮
1111	1111000000000 – 1111111111111

13.9 The process can be executing only one instruction when it starts, which means it can require only one page. This is an extreme case, since most systems initially give a process a set of several pages. However, this extreme case does show us that the process does not need all its pages when it starts.

13.11 Demand paging.

13.13 Because the programmer will know the best way to split up his or her program and what pieces of the program need to be in memory at the same time. For example, the system might split a loop right in the middle, and potentially these two pages would be constantly swapped in and out as the loop executes.

13.15 Compare the offset to the segment length. Compare page field to maximum pages allocated to program.

13.17 Since segment sizes are user-defined, we cannot assign fixed numbers of bits for segment starting address and displacement within the segment, as we could for paging. (As an example, if a page is 1024 bytes and memory is 4,194,304 bytes, then we use 12 bits for every starting address and 10 bits for every displacement.) Without being able to do this, we cannot have two standard-size components for our concatenation.

13.19 Location 2100 is relative to *A*'s program; it is not absolute address 2100, which is in *B*'s area. The system will determine that *A*'s address 2100 is a displacement of 52 from the start of *A*'s page 3. Let's say *A*'s page 3 starts at 5120. Then *A*'s address 2100 will map to absolute address 5120 + 52 = 5172.

13.21 So that the process will not waste valuable CPU time waiting for the I/O operation to be completed.

13.23 Some examples are:

.PAS, .COB, .FOR, .ASM for files containing programs in that language
.BAK for a backup copy of a file
.BIN for the object code output of a compiler
.DAT for a data file

13.25 A standard application for a sequential-access file is one in which the data is input and processed in a sequential manner. For example, we might have time cards input in order of employee number and process the payroll file sequentially. A standard application for a direct-access file is one in which data is input and processed in random order. For example, we might have orders for parts coming in via customers' phone calls and update inventory using direct access based on part number.

13.27 A field.

13.29 In an express line in a grocery store, the customer is completely checked out before the checker goes to the next customer. In round-robin scheduling, a process might require more time than is allocated for the standard CPU quantum. In this case, the system moves on to the next process and comes back to the preempted process later.

13.31 Renting a carpet cleaning machine, where you are given a number of attachments and solvents without determining if you need them all. Renting a car, where you are given a jack, spare tire, and emergency phone number with the hope that you won't need them.

13.33 We can't say, because we don't know what the definition of "best" is for this application.

13.35

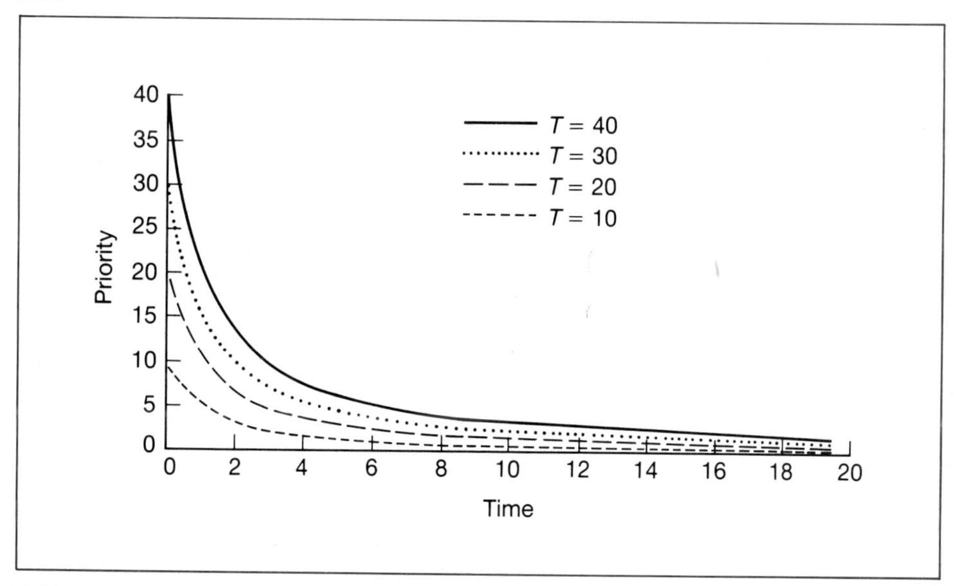

FIGURE A13.35

CHAPTER 14

14.1 WAIT:

See if the semaphore requested has been declared.
If it has not, the requesting process is aborted.
If the semaphore exists, the status of the semaphore is checked.
If it is 1, then it is changed to 0 and the calling process is allowed to continue.
If the semaphore is 0, then the calling process is moved to the wait queue for that semaphore.

SIGNAL:

See if the semaphore requested has been declared.
If it has not, the requesting process is aborted.
Check the queue for that semaphore.
If a process is waiting on the queue, then both the signaling process and the process that had been waiting on the semaphore are moved to the active state.
If no process is waiting on the semaphore, the value of the semaphore is changed to 1 and the signaling process is allowed to continue.

14.3 We must make certain that the idle process stays in execute while the I/O is being done (which is called I/O without wait). Otherwise, if the idle process is suspended, we might encounter a situation where there is no process available to execute.

14.5 The system would start building up a list of processes that could never finish. This uses up process ID numbers and other resources.

14.7 Because the process that was signaled might have a higher priority. Also, while the signal was serviced, other processes might have become available for execution. One of them might also have higher priority.

14.9 There is no "pat" answer to this question. The system has to implement some policy to make that decision. As with long-term scheduling, there is no "best" approach in every case.

14.11 Nowadays, checks on memory bounds are often done in hardware. For example, assuming a segmented system, the segment length could be kept in a register connected to a comparator. Before a memory reference is initiated, the displacement of the desired location from the start of the segment is gated to that comparator. If the displacement of the desired location is greater than the segment length, then the operating system is notified via an interrupt.

14.13 The seek time and the latency time will determine how fast the file manager can access a given sector on the disk. The transfer rate of the disk will affect how fast the file manager can transfer information to/from that sector.

14.15 Make certain that IOCS has a standard set of calling sequences for all common types of peripherals. Then, when the new device is added, we simply tell the other layers, "Use this calling sequence with these parameters."

14.17 It has an equation that determines the priority of each process creation request waiting on its queue. When a new process can be created, the long-term scheduler applies this equation to its queue and selects the highest-priority request. It creates this process and puts the resulting PCB into the active state.

14.19 Because it defines the way the system interfaces with users.

14.21 Because it has its own small versions of the necessary modules built in. For example, the bootstrap would have a driver for the disk, since it needs to read from the system disk. However, since the bootstrap is the only process running, its version of IOCS could be much simpler than the multiuser version normally used by the OS.

CHAPTER 15

15.1 Both define operations that can be directly executed by the hardware. Machine language uses a numeric notation that can be directly loaded into memory. Assembly language uses mnemonics and symbolic notations that are easier for the programmer to understand. Assembly-language programs must be run through a translator before they can be executed.

15.3 Because they incorporate different hardware, and the machine language directly reflects the hardware. As a sidelight, it is interesting to note that some new machines have been designed so that they could execute a popular assembly language. In this case we have role reversal, where the hardware is designed for the assembly language rather than vice versa.

15.5 Two reasons. Going directly from a high- or intermediate-level language to machine language makes such a translator more complex than it has to be. An assembler already exists, and it is easier to produce assembly code. This rationalizes the middle step in Fig. 15.3. However, in general most machines have different assembly languages. Therefore, even if we used the procedure in Fig. 15.2*b*, we would have a significant job for each new machine and especially for each new language. If we use the procedure in Fig. 15.3, one group writes one universal translator for the given source language. Each manufacturer then has a significantly easier task of writing translators from a machine-like, but machine-independent, intermediate language to the assembly language (or languages) for its machines.

15.7 They usually do many more error checks than the corresponding compiler. Furthermore, they allow an error to be corrected as it is encountered and the corrected version immediately reexecuted. With compilers, we would have to make the correction, recompile the entire program, and then reexecute it.

15.9 Because it might be necessary at a later time to change the data format, such as making zip codes nine digits. If the data description is embedded, then all code incorporating this embedded description must be modified when the data format is changed.

15.11 It is a relatively simple language, and it is normally interpreted. This makes it easier for novices to learn and use than the other high-level languages.

15.13 If the corresponding assembler is powerful and general enough, then we can't think of much cause to program directly in machine language. Perhaps a person who could program directly in machine language would be able to patch a program directly, without having to go through an assembler. This would be significant if some major real-time system crashed and needed to be fixed without recompiling the entire system.

15.15 It takes machine code and produces a corresponding assembly version. This makes it easier for a human to understand what a certain code segment is doing.

15.17 PROG says this is the start of a new segment.
ASSUME says the following will be a code segment named PROG.
GLB says label WRITE_CHR should be known by entire system.

15.19 Most computers are able to execute essentially the same fundamental set of operations (cf Chap. 5). Also, new assembly users of the system can easily relate to the assembly language. It is similar to learning a new dialect of a human language you already know, rather than learning a completely new language.

15.21 So that programs and programmers can move easily from machine to machine. For example, assume a company writes all its software in high-level language XYZ and at some later time needs to buy a new machine. If the corresponding language on the new machine has significant differences, then considerable resources will be "wasted" in converting software and retraining programmers.

15.23 By incorporating a human-oriented, mathematics-type syntax rather than assembly language.

15.25 Identification, environment, data, and procedure.

15.27 Looping structures (like Do While), decision structures (like If-Then-Else), structures to branch to other modules and then return (like Call), and array structures.

15.29 Having code that directly depends on the data being in a certain format. For example, we might have code in the program which says to move address[zip_start] through address[zip_start + 4], with our goal being to extract the zip code. When the zip code goes to nine digits, this code has to be changed.

15.31 Because it was intended to be a good teaching language. A common programming error is to perform an operation with the wrong data type, and Pascal wants to catch such errors.

15.33 In an object-oriented language, the programmer creates objects. These objects contain ("encapsulate") data and methods to manipulate the data. Also, new objects can inherit data and methods of an existing object. In object-oriented languages, we also have polymorphism, which allows different objects to respond appropriately to the same request.

15.35 Encapsulation, inheritance, and polymorphism.

15.37 Because new objects can inherit the data and methods of existing objects.

15.39 Domains, predicates, goals, and clauses.

15.41 Find all diseases that do not have a fever or a cough as a symptom:

NOT (has_as_symptom(Disease,fever) OR has_as_symptom(Disease,cough))

CHAPTER 16

16.1 Spreadsheets, word processors, compilers, etc.

16.3 Setting the alarm clock. Using calculator facilities. Finding a file. Deleting a file. Getting information about a file. Formatting a disk. Copying, cutting, and pasting text. Formatting text for printout. Checking spelling. Sorting.

16.5 It might be necessary for a given user to access part of the database records, such as name and address, but not others, such as salary and medical records.

16.7 They changed many peoples' impressions about computers from expensive toys to really useful business tools. They also introduced the idea that users could do complex calculations without mastering a high-level programming language such as Cobol.

16.9 Approximate locations are: −2.5, −1.5, 0.7, and 3.05.

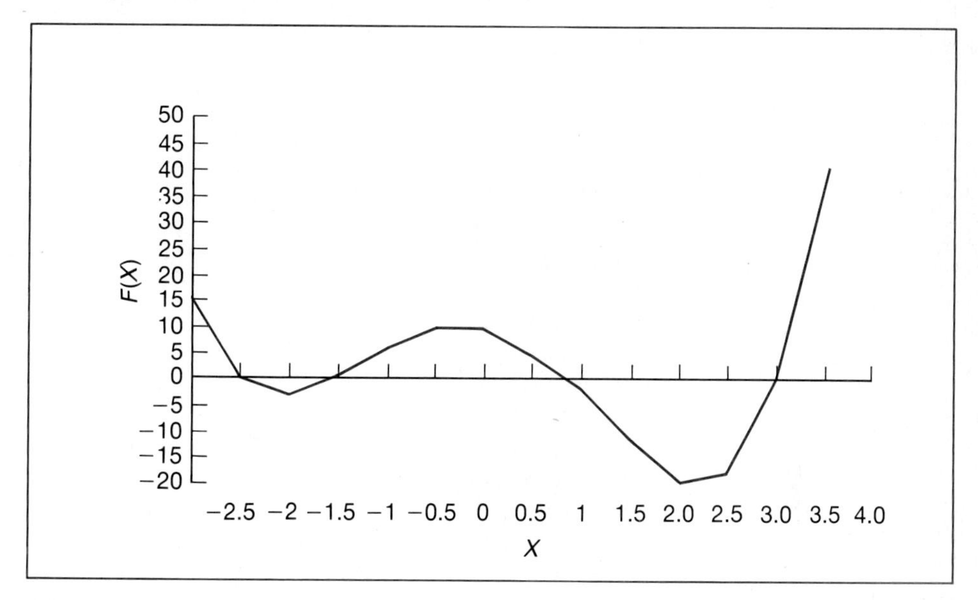

FIGURE A16.9

16.11

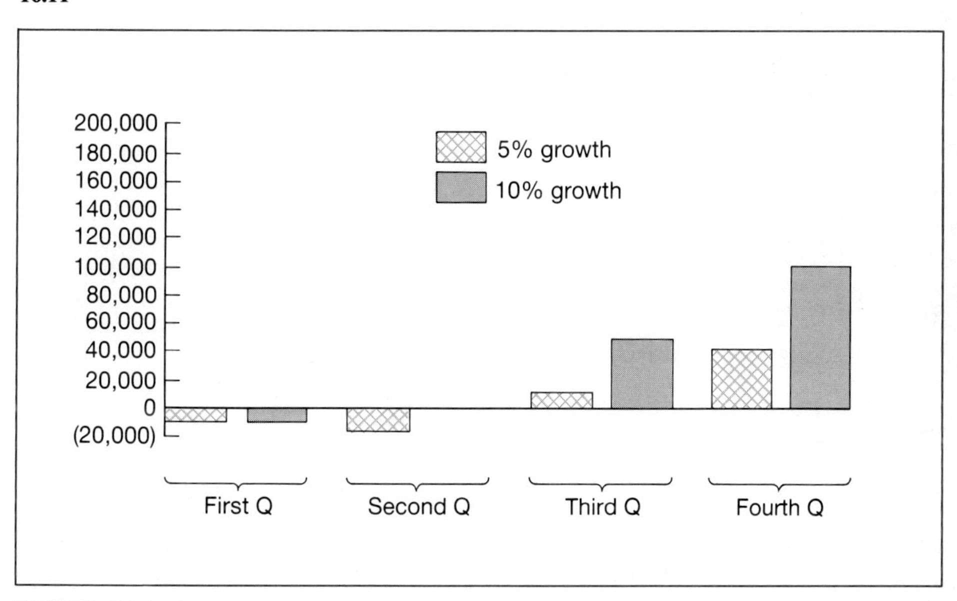

FIGURE A16.11

16.13 On ours, there were two concerns. First, you could not cut pieces of text vertically; for example, one might want the first 10 columns of every line on a page. Second, you could not wrap text around a figure; once the figure was inserted, the word processor took the height of the figure as the height of the line.

16-7 Setting the alarm clock. Using calculator facilities. Finding a file. Deleting a file. Getting information about a file. Formatting a disk. Copying, cutting, and pasting text. Formatting text for printout. Checking spelling. Sorting.

16-9 It might be necessary for a given user to access part of the database records, such as name and address, but not others, such as salary and medical records.

16-11 They changed many peoples' impressions about computers, from expensive toys to really useful business tools. They also introduced the idea that users could do complex calculations without mastering a high-level programming language such as COBOL.

FIGURE A16.13(1)

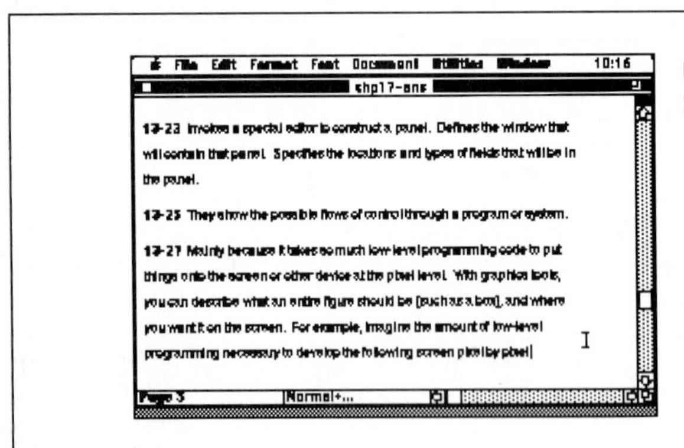

Mainly because it takes so much low-level programming code to put things onto the screen or other device at the pixel level. With graphics tools, you can describe what an entire figure should be (such as a box), and where you want it on the screen. For example, imagine the amount of low-level programming necessary to develop the adjacent screen dot to dot.

FIGURE A16.13(2)

16.15 They can be used for sophisticated applications without learning computer programming languages.

16.17 You enter commands and data, analogous to the commands of a program and the data that is entered when the program is executed. However, spreadsheets can be used for sophisticated applications without learning computer programming languages.

CHAPTER 17

17.1 The SDI system is estimated to be 10 million lines of code.

17.3 In engineering, prototypes have been used for many years. Thus, the people developing the hardware had backgrounds in prototyping. Many early software developers came from other disciplines not accustomed to prototyping, such as business or mathematics. Furthermore, until recently, it seemed that to build a software prototype one needed to essentially program the system anyway. Prototyping tools have changed this.

17.11 We determine exactly what it is that the client needs for his or her system. Frequently, this is now done with the aid of prototyping.

17.13 Implementation is where we write the programs and build the hardware.

17.15 Operation is where we run the released system. Maintenance involves such things as fixing bugs, modifying the system to meet new conditions, and improving existing hardware or software subsystems.

17.17 Usually this is done with special tools that allow you to display the menus, icons, etc., without having the software that implements the operations. For example, you could display a database record even though the database itself does not yet exist.

17.19 Customers order products from the sales division of this company. They also communicate special needs to the marketing division of the company. Marketing provides customer support for existing products and makes suggestions to R&D based on customer communications. R&D provides technical support to marketing and designs new products to be manufactured. Manufacturing builds the products and places them into inventory. Sales sells them to customers and removes them from inventory. Accounting processes expenses from marketing, R&D, and manufacturing, and income from sales.

17.21 Single-stepping through a program. Displaying and modifying values of variables. Executing up to a specified point (breakpoint). Modifying the code and recompiling. Skipping the branches to external subprograms.

17.23 Invokes a special editor to construct a panel. Defines the window that will contain that panel. Specifies the locations and types of fields that will be in the panel.

17.25 They show the possible flows of control through a program or system.

17.27 Mainly because it takes so much low-level programming code to put things onto the screen or other device at the dot level. With graphics tools, you can describe what an entire figure should be (such as a box) and where you want it on the screen. For example, imagine the amount of low-level programming necessary to develop the following screeen dot by dot.

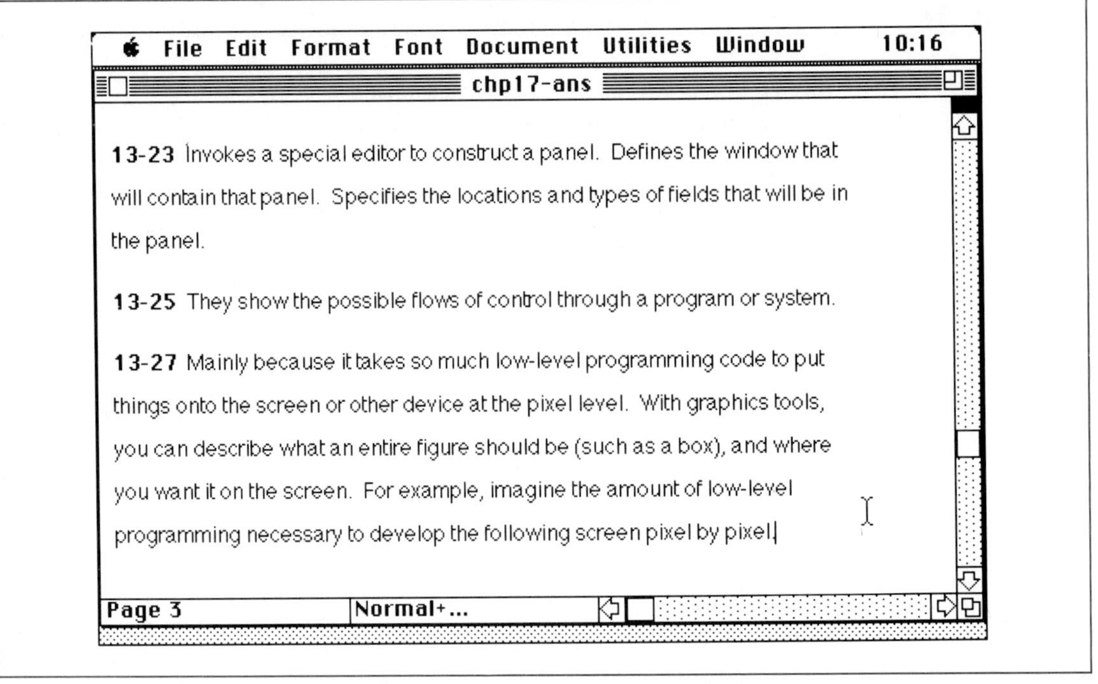

FIGURE A17.27

17.29 Test coverage analyzer, assertion checker, complexity metric evaluator.

17.31 Suppose you have two IF tests someplace, call them A and B. Furthermore, suppose the program works in all cases except when both A and B take the FALSE branches. You might test A TRUE and B TRUE, A TRUE and B FALSE, and A FALSE and B TRUE. According to the test coverage analyzer, all paths have been executed at least once.

17.33 You give it certain assertions that you made when writing the program, such as, "The tolerance can never be zero or negative." It then verifies that these assertions hold during execution.

17.35 A test coverage analyzer just tells you that all paths have been checked; you must furnish the data to accomplish this. A test coverage analyzer actually generates the data to make sure that all paths are checked.

17.37 Usually, a benchmark is a standardized set of test data that is given to many different programs, such as a standard Cobol program to test Cobol compilers. A test coverage analyzer generates data to make sure that all paths are checked in a specific program, such as your particular Cobol compiler.

17.39 It is a very high level language that is used to rapidly develop tools that you need for a specific application. Typically, it is used when the tool you are currently using, such as an editor, does not have a certain feature you need.

17.41 You save software that you have developed in libraries. Later, instead of always developing things from scratch, you check the library to see if something similar exists and can be used as is or with small modifications.

17.43 It automates the structured design process for system development.

CHAPTER 18

18.1 Inability to do things interactively. Difficulty in correcting errors. Slow turnaround. Physical problems of handling and storing large decks.

18.3 Maximum CPU time, maximum number of lines, maximum memory usage, password, priority, definition of I/O devices, file definition and manipulation commands.

18.5 The system needed larger and faster secondary storage devices. The user now stored data and programs on these devices rather than on cards and needed to access them at speeds commensurate with the interactive system.

18.7 Speed, ability to do full-screen editing and graphics.

18.9 One needs to switch constantly to different programs to do the operations; furthermore, sometimes the files are not compatible between programs. However, sometimes one editor does not provide all the features a user might want; with separate programs, a user can pick each one specifically for a given requirement.

18.11 The MS-DOS commands are input to the shell, and the shell converts each one to a corresponding UNIX command or set of commands.

18.13 A monitor command can put the previous (incorrect) command back up on the screen. A *mini-line-editing* facility can then be used to correct the command, and it can be retried. *Help* systems could be available to show the user correct syntax.

18.15 A help system could be available to show the user the syntax of a desired command. Alternatively, if the user did not know the appropriate command name, the help system could display commands by type (such as file manipulation). This display would include a description of what each command does.

18.17 A dialog box allows a user to specify various parameters for a given command. Such a dialog box can contain fill-in boxes and push buttons. A fill-in box allows a user to enter a character string, such as a file name or a page number. A push button allows a user to make a selection by moving the cursor to the desired button, then clicking the mouse.

18.19 They are symbols representing components of a system, such as a file or an application program. They make it easier for the user to see such things as function (for example, the trash can icon representing a place to dispose of files). They can also show what application created a given file.

18.21 They show things directly on the screen, like **BOLD** when the word is boldface. Other types might use .BBOLD.B to indicate that the word between the .B's is boldface.

18.23 Are they using the system to learn about computers, or are they using the system for a job? What portion of their job will involve the computer? How crucial is response time, such as a library catalog system versus a medical information system in an emergency room? Will the users be writing programs or using applications?

18.25 Advanced computer classes in a university. Applications development organizations. Maintenance programming. System development organizations. Research labs. Computer companies. Applications where the users are skilled computer personnel rather than end users (such as a database development team as opposed to the people using the database after it has been developed).

18.27 What method of retrieval would they prefer for a person, such as name or hospital ID number? What type of input would they prefer for standard questions; for example, should sex be entered as M or F or via a radio button type of interface? Should standard required information be entered via a sequence of questions, or via a "page" shown on the screen, with fill-in boxes for answers? Should the system know about standard tests; for example, if the doctor says the patient had an XYZ test, should it ask for the sequence of results involved in such a test? What types of information retrieval are desired; will the doctor want to ask questions such as, "List any times at which the patient had symptom A and symptom B but not symptom C"?

18.29 Function keys can invoke certain operations. Arrow keys can move the cursor around. Command key + a letter can be used rather than typing a command or selecting it from a menu. The option key can select alternate uses for standard keys, such as Greek letters or mathematical symbols. The escape keys can get the user out of a command or system. Editing keys, such as delete and insert, can be used rather than a mouse or menu.

18.31 Transactions processed per time unit. Average time a customer has to wait for the computer user to provide desired information. Physical movement by the user to process a given transaction; for example, how many times back and forth between mouse and keyboard? Number of suggestions or complaints from the user. Number of operations necessary to perform each type of transaction.

18.33 One of the graphics packages used to do the original draft of this book uses a workshop type of conceptual model. For example, there is a , , , , and .

18.35 In Microsoft WORD on the Macintosh, you can invoke Help from the desktop accessories, giving

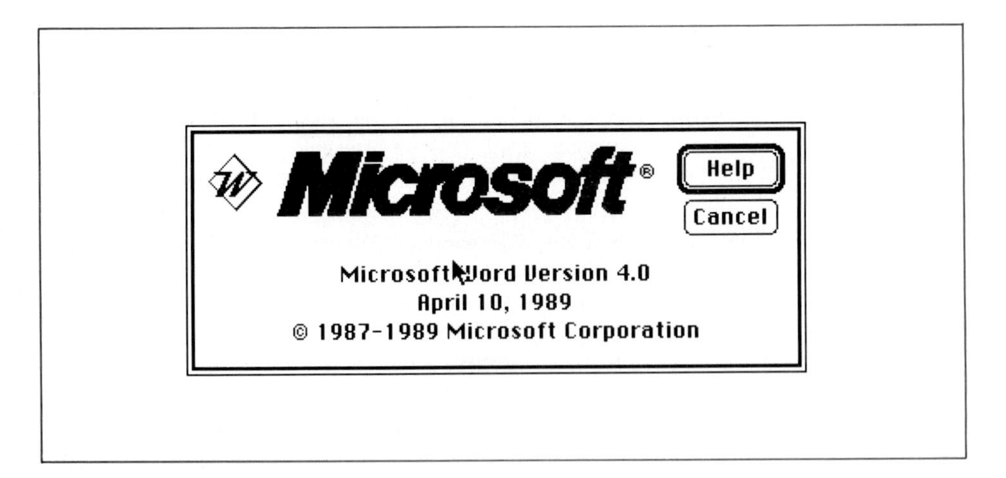

FIGURE A18.35(1)

By selecting Help, you can get the list of topics. The first option is information about the Help system itself.

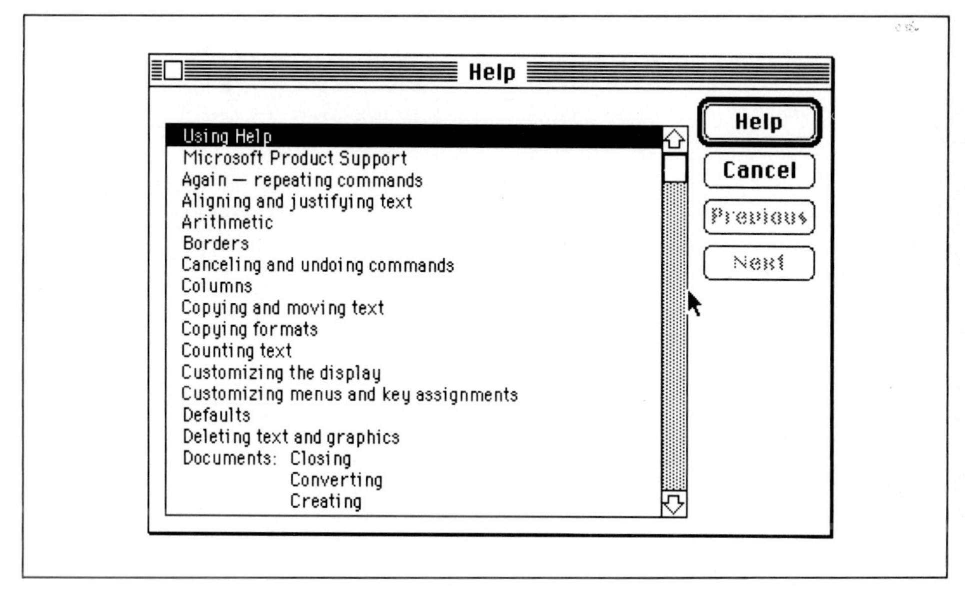

FIGURE A18.35(2)

Selecting this gives you

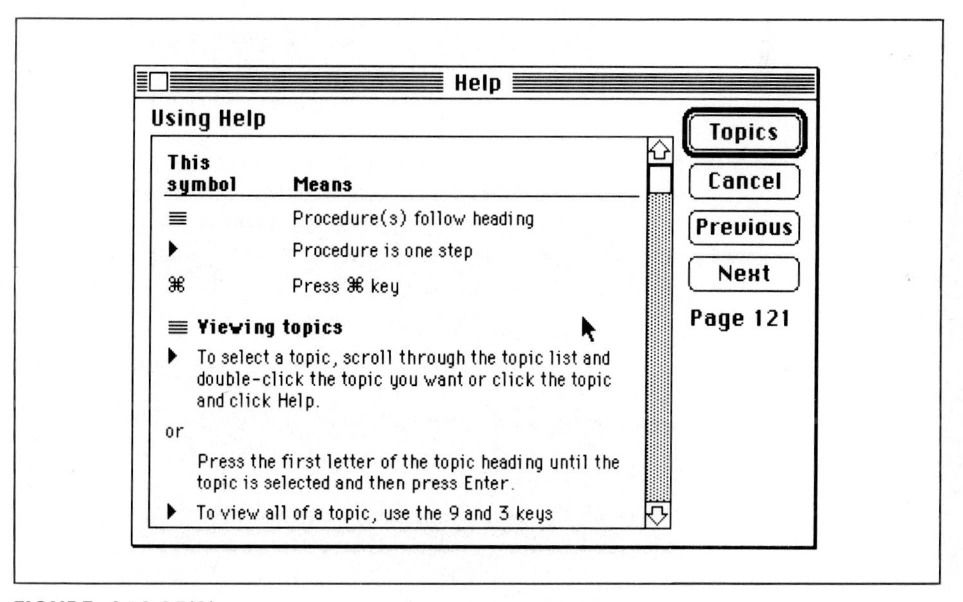

FIGURE A18.35(3)

You can also scroll through the topics and pick something such as Defaults, which is shown below.

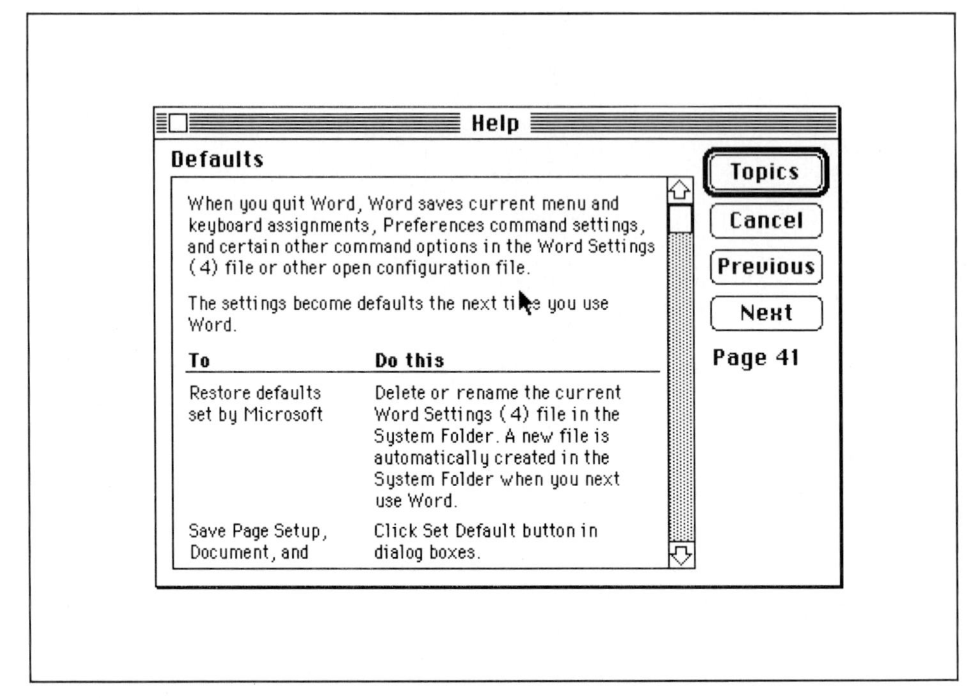

FIGURE A18.35(4)

18.37 Some systems ask you for values when you boot them. Others have a default file, which you set up as desired. Others have parameters in the program, which you set via an editor before you compile the application. Others have special commands, which you can invoke when running the application.

18.39 Mainly, you get to see if you like the interface before you actually program it. Also, you might find things that have been forgotten in the specifications. You can get good ideas about testing the system when it has been developed.

18.41 People are now talking about eye movement to direct the cursor. Special devices for physically impaired individuals are looking at signals directly from the nerve cells.

CHAPTER 19

19.1 If we stress reliability, using redundancy, for example, then we will reduce the affordability of the product. If we stress maintainability and testability, we might make the product more difficult to manufacture, thus reducing manufacturability.

19.3 "Portability" might be important in some instances. For high-security installations, "non-snoopability" is vital; this implies such things as added shielding to prevent leaks of electromagnetic radiation. Perhaps low power consumption or low heat dissipation is important, although we couldn't think of an *-ability* term for these.

19.5 Commercial aircraft with a copilot—this copilot is able to fly the plane should something happen to that pilot. (Although he or she also assists the pilot with some duties, this assistance could be provided by a less-qualified person were it not for the importance of the fully qualified "hot standby.")

19.7 Shift to a blinking red light on all streets, thus requiring traffic to take turns using the intersection.

19.9 An odd number of errors. If there are an even number of errors, such as two or four, then the parity will be the same as the correct word.

19.11 No. The voter could fail on its own, and thus the system would not function properly.

19.13 The disk being full. This would be especially critical if some files were open and users were adding data to these files.

19.15 Test many units, and use statistical projection to estimate MTTF. For example, we might know from experience that a certain type of component fails according to the curve shown in figure *a* below. We select a set of components and test them for the portion of the curve shown in figure *b* and expanded in figure *c*. From this, we estimate the overall curve and the MTTF.

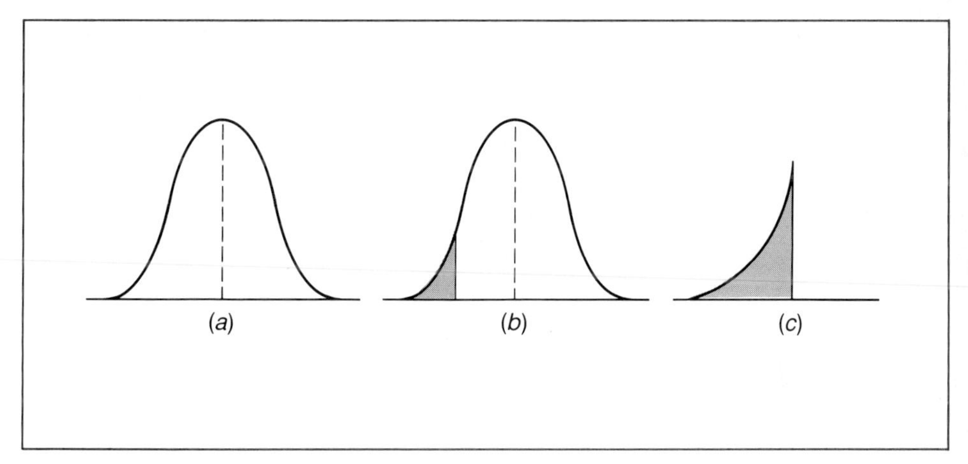

FIGURE A19.15

19.17 One common way is called *burn-in*. Before being released, the components are operated for the length of time during which most infant mortality occurs.

19.19 The devices are then, in some sense, "used." As an analogy, consumers can understand a new car having a certain number of miles on it, but if there are too many, then the car usually must be sold as a demonstrator.

19.21 A certain system will respond within a specified time range. A certain computation will be done with a specified accuracy range. Memory requirements for a typical application will be in a certain range.

19.23 Let's assume we classify a resource operation as not in use, normal, close to boundary, and at boundary. The number of configurations of every resource in every configuration would then be $4^{10} = 1{,}048{,}576$.

19.25 An operating system might claim that it handles n users; we might design our application for $0.75n$ users. A networking system might claim to transfer n packets per second; we might design our message processor to assume $0.75n$ per second.

19.27 Printers and modems frequently exhibit definite speed/cost tradeoffs.

19.29 Graphics packages usually manipulate higher-resolution images much more slowly. CAD packages might take much longer to make a higher-tolerance cut on a piece of metal.

19.31 Cost, speed, and error rates of the connecting lines. Also, if the problems in the computer to be tested are related to any of the hardware and/or software controlling the remote connections, it might not be possible to execute the remote tests.

19.33 Have certain standardized approaches to specified design tasks. Reuse existing designs (with necessary modifications) whenever possible. Eliminate some features of the system.

19.35 Picking components that meet all the requirements at the best possible price. Reducing labor costs. Using "standard" components, that do not require specialized mounting, handling, etc. Use parts that have proven to be reliable and economical.

19.37 Let $R1 = xR$ and $R2 = yR$. Then our equation for voltage becomes $[y/(x + y)](V_i)$. For the various tolerances, we have:

$$1\%: \frac{0.99}{1.01 + 0.99} = 0.495 \qquad \frac{1.01}{0.99 + 1.01} = 0.505$$

$$5\%: \frac{0.95}{1.05 + 0.95} = 0.475 \qquad \frac{1.05}{0.95 + 1.05} = 0.525$$

$$10\%: \frac{0.90}{1.1 + 0.90} = 0.45 \qquad \frac{1.1}{0.90 + 1.1} = 0.55$$

$$20\%: \frac{0.80}{1.20 + 0.80} = 0.40 \qquad \frac{1.20}{0.80 + 1.20} = 0.60$$

Index